CW01263349

THE RF TRANSMISSION SYSTEMS
HANDBOOK

ELECTRONICS HANDBOOK SERIES

Series Editor:
Jerry C. Whitaker
Technical Press
Morgan Hill, California

PUBLISHED TITLES

AC POWER SYSTEMS HANDBOOK, SECOND EDITION
Jerry C. Whitaker

THE COMMUNICATIONS FACILITY DESIGN HANDBOOK
Jerry C. Whitaker

THE ELECTRONIC PACKAGING HANDBOOK
Glenn R. Blackwell

POWER VACUUM TUBES HANDBOOK, SECOND EDITION
Jerry C. Whitaker

THERMAL DESIGN OF ELECTRONIC EQUIPMENT
Ralph Remsburg

THE RESOURCE HANDBOOK OF ELECTRONICS
Jerry C. Whitaker

MICROELECTRONICS
Jerry C. Whitaker

SEMICONDUCTOR DEVICES AND CIRCUITS
Jerry C. Whitaker

SIGNAL MEASUREMENT, ANALYSIS, AND TESTING
Jerry C. Whitaker

ELECTRONIC SYSTEMS MAINTENANCE HANDBOOK, SECOND EDITION
Jerry C. Whitaker

DESIGN FOR RELIABILITY
Dana Crowe and Alec Feinberg

THE RF TRANSMISSION SYSTEMS HANDBOOK
Jerry C. Whitaker

THE RF TRANSMISSION SYSTEMS HANDBOOK

Edited by
Jerry C. Whitaker

CRC PRESS

Boca Raton   London   New York   Washington, D.C.

Library of Congress Cataloging-in-Publication Data

The RF transmission systems handbook / edited by Jerry C. Whitaker.
    p. cm.
  Includes bibliographical references and index.
  ISBN 0-8493-0973-5 (alk. paper)
  1. Radio—Transmitters and transmission. I. Whitaker, Jerry C.

TK6561 .R52 2002
621.384'11—dc21   2002017434

This book contains information obtained from authentic and highly regarded sources. Reprinted material is quoted with permission, and sources are indicated. A wide variety of references are listed. Reasonable efforts have been made to publish reliable data and information, but the authors and the publisher cannot assume responsibility for the validity of all materials or for the consequences of their use.

Neither this book nor any part may be reproduced or transmitted in any form or by any means, electronic or mechanical, including photocopying, microfilming, and recording, or by any information storage or retrieval system, without prior permission in writing from the publisher.

All rights reserved. Authorization to photocopy items for internal or personal use, or the personal or internal use of specific clients, may be granted by CRC Press LLC, provided that $1.50 per page photocopied is paid directly to Copyright Clearance Center, 222 Rosewood Drive, Danvers, MA 01923 USA The fee code for users of the Transactional Reporting Service is ISBN 0-8493-0973-5/02/$0.00+$1.50. The fee is subject to change without notice. For organizations that have been granted a photocopy license by the CCC, a separate system of payment has been arranged.

The consent of CRC Press LLC does not extend to copying for general distribution, for promotion, for creating new works, or for resale. Specific permission must be obtained in writing from CRC Press LLC for such copying.

Direct all inquiries to CRC Press LLC, 2000 N.W. Corporate Blvd., Boca Raton, Florida 33431.

Trademark Notice: Product or corporate names may be trademarks or registered trademarks, and are used only for identification and explanation, without intent to infringe.

Visit the CRC Press Web site at www.crcpress.com

© 2002 by CRC Press LLC

No claim to original U.S. Government works
International Standard Book Number 0-8493-0973-5
Library of Congress Card Number 2002017434
Printed in the United States of America 1 2 3 4 5 6 7 8 9 0
Printed on acid-free paper

Preface

Radio frequency (RF) transmission is one of the oldest forms of electronics. From the days of Hertz and Marconi, RF transmission has pioneered the art and science of electronics engineering. It has also served as the basis for a myriad of related applications, not the least of which includes audio amplification and processing, video pickup and reproduction, and radar. RF technology has reshaped our national defense efforts, radically changed the way we communicate, provided new products and services, and brought nations together to celebrate good times and mourn bad times.

RF is an invisible technology. It is a discipline that often takes a back seat to other subjects taught in colleges and universities. Yet RF transmission equipment has reshaped the way we live.

This book is intended to serve the information needs of persons who specify, install, and maintain RF equipment. The wide variety of hardware currently in use requires that personnel involved in RF work be familiar with a multitude of concepts and applications. This book examines a wide range of technologies and power devices, focusing on devices and systems that produce in excess of 1 kilowatt (kW).

Extensive theoretical dissertations and mathematical explanations have been included to the extent that they are essential for an understanding of the basic concepts. Excellent reference books are available from this publisher that examine individual RF devices and the underlying design criteria. This book puts the individual elements together and shows how they interrelate.

The areas covered by *The RF Transmission Systems Handbook* range from broadcasting to electronic counter-measures. The basic concepts and circuit types of all major RF applications are covered. The generous use of illustrations makes difficult or complex concepts easier to comprehend. Practical examples are provided wherever possible. Special emphasis is given to radio and television hardware because these applications provide examples that can readily be translated to other uses.

The RF Transmission Systems Handbook is divided into the following major subject areas:

- **Applying RF Technology**. Common uses of radio frequency energy are examined and examples given. This treatment includes an overview of RF bands, modulation methods, and amplifier operating classes.
- **Solid-State Power Devices**. The operating parameters of semiconductor-based power devices are discussed, and examples of typical circuits are given. Included is an outline of the basic principles of bipolar and FET semiconductors, including potential failure modes.
- **Power Vacuum Tube Devices**. The basic principles and applications of gridded vacuum tubes are outlined, and example circuits provided.
- **Microwave Power Tubes**. The operating principles of classic microwave devices and new high-efficiency tubes are given. This treatment reviews the basic concepts of klystrons, traveling-wave tubes, and other microwave power devices.

- **RF Components and Transmission Line.** The operation of hardware used to combine and conduct RF power are explained, including coaxial transmission line, waveguide, hot-switches, and circulators.
- **Antenna Systems.** An overview of antenna theory and common designs is given and basic operating parameters are described. Examples are provided of antennas used in radio and TV broadcasting, satellite service, and radar.

RF power technology is a complicated but exciting science. It is a science that advances each year. The frontiers of higher power and higher frequency continue to fall as new applications drive new developments by manufacturers. Radio frequency technology is not an aging science. It is a discipline that is, in fact, just reaching its stride.

Jerry C. Whitaker
Editor

Editor

Jerry Whitaker is technical director of the Advanced Television Systems Committee, Washington D.C. He previously operated the consulting firm Technical Press. Whitaker has been involved in various aspects of the communications industry for more than 25 years. He is a fellow of the Society of Broadcast Engineers (SBE) and an SBE-certified Professional Broadcast Engineer. He is also a member and fellow of the Society of Motion Picture and Television Engineers, and a member of the Institute of Electrical and Electronics Engineers. Whitaker has written and lectured extensively on the topic of electronic systems installation and maintenance.

Whitaker is the former editorial director and associate publisher of *Broadcast Engineering* and *Video Systems* magazines. He is also a former radio station chief engineer and TV news producer.

Whitaker is the author of several books, including:

- *The Resource Handbook of Electronics*, CRC Press, 2000
- *The Communications Facility Design Handbook*, CRC Press, 2000
- *Power Vacuum Tubes Handbook*, 2nd edition, CRC Press, 1999
- *AC Power Systems*, 2nd edition, CRC Press, 1998
- *DTV Handbook*, 3rd edition, McGraw-Hill, 2000
- Editor-in-Chief, *NAB Engineering Handbook*, 9th edition, National Association of Broadcasters, 1999
- Editor-in-Chief, *The Electronics Handbook*, CRC Press, 1996
- Co-author, *Communications Receivers: Principles and Design*, 3rd edition, McGraw-Hill, 2000
- *Electronic Display Engineering*, McGraw-Hill, 2000
- Co-editor, *Standard Handbook of Video and Television Engineering*, 3rd edition, McGraw-Hill, 2000
- Co-editor, *Information Age Dictionary*, Intertec/Bellcore, 1992
- *Radio Frequency Transmission Systems: Design and Operation*, McGraw-Hill, 1990

Whitaker has twice received a Jesse H. Neal Award Certificate of Merit from the Association of Business Publishers for editorial excellence. He also has been recognized as Educator of the Year by the Society of Broadcast Engineers. He resides in Morgan Hill, California.

Contributors

John R. Brews
Electrical and Computer Engineering
 Department
University of Arizona
Tucson, Arizona

Anthony J. Ferraro
Department of Electrical Engineering
Pennsylvania State University
University Park, Pennsylvania

Timothy P. Hulick
Electrical Engineering Consultant
Schwenksville, Pennsylvania

Robert Kubichek
Department of Electrical Engineering
University of Wyoming, Laramie
Laramie, Wyoming

Ken Seymour
Sprynet
Santa Rosa, California

Sidney Soclof
Department of Electrical Engineering
California State University, Los Angeles
Los Angeles, California

Gerhard J. Straub
Hammett & Edison, Inc.
San Francisco, California

Douglas H. Werner
Pennsylvania State University
University Park, Pennsylvania

Pingjuan L. Werner
Department of Electrical Engineering
Pennsylvania State University
Benezett, Pennsylvania

Jerry C. Whitaker
Technical Press
Morgan Hill, California

Rodger E. Ziemer
University of Colorado
Colorado Springs, Colorado

Contents

1 Applications of RF Technology *Jerry C. Whitaker* **1**-1
    1.1   Introduction **1**-1
    1.2   Broadcast Applications of RF Technology **1**-8
    1.3   Nonbroadcast Applications **1**-23

2 Electromagnetic Spectrum *Jerry C. Whitaker* **2**-1
    2.1   Introduction **2**-1

3 Amplitude Modulation *Robert Kubichek* **3**-1
    3.1   Amplitude Modulation **3**-1

4 Frequency Modulation *Ken Seymour* **4**-1
    4.1   Introduction **4**-1
    4.2   The Modulated FM Carrier **4**-1
    4.3   Frequency Deviation **4**-3
    4.4   Percent of Modulation in FM **4**-4
    4.5   Modulation Index **4**-4
    4.6   Bandwidth and Sidebands Produced by FM **4**-5
    4.7   Narrow-Band vs. Wide-Band FM **4**-7
    4.8   Phase Modulation **4**-7
    4.9   FM Transmission Principles.. **4**-7
    4.10 FM Reception Principles **4**-11

5 Pulse Modulation *Rodger E. Ziemer* **5**-1
    5.1   Introduction **5**-1
    5.2   The Sampling Theorem **5**-1
    5.3   Analog-to-Digital Conversion **5**-4
    5.4   Baseband Digital Pulse Modulation **5**-5
    5.5   Detection of Pulse Modulation Formats.. **5**-10
    5.6   Analog Pulse Modulation **5**-10

6 Digital Modulation *Rodger E. Ziemer* **6**-1
    6.1   Introduction **6**-1
    6.2   Detection of Binary Signals in Additive White Gaussian Noise **6**-1
    6.3   Detection of *M*-ary Signals in Additive White Gaussian Noise............ **6**-8

xi

6.4	Comparison of Modulation Schemes	6-15
6.5	Higher Order Modulation Schemes	6-20

7 High-Power Vacuum Devices *Jerry C. Whitaker* 7-1
7.1	Introduction	7-1
7.2	Electron Optics	7-2
7.3	Diode	7-5
7.4	Triode	7-6
7.5	Tetrode	7-9
7.6	Pentode	7-12
7.7	High-Frequency Operating Limits	7-13
7.8	Device Cooling	7-15

8 Microwave Vacuum Devices *Jerry C. Whitaker* 8-1
8.1	Introduction	8-1
8.2	Grid Vacuum Tubes	8-5
8.3	Klystron	8-9
8.4	Traveling Wave Tube	8-14
8.5	Crossed-Field Tubes	8-20

9 Bipolar Junction and Junction Field-Effect Transistors *Sidney Soclof* 9-1
9.1	Bipolar Junction Transistors	9-1
9.2	Amplifier Configurations	9-3
9.3	Junction Field-Effect Transistors	9-5

10 Metal-Oxide-Semiconductor Field-Effect Transistor *John R. Brews* 10-1
10.1	Introduction	10-1
10.2	Current-Voltage Characteristics	10-3
10.3	Important Device Parameters	10-4
10.4	Limitations on Miniaturization	10-11

11 Solid-State Amplifiers *Timothy P. Hulick* 11-1
11.1	Linear Amplifiers and Characterizing Distortion	11-2
11.2	Nonlinear Amplifiers and Characterizing Distortion	11-8
11.3	Linear Amplifier Classes of Operation	11-10
11.4	Nonlinear Amplifier Classes of Operation	11-17

12 Coaxial Transmission Lines *Jerry C. Whitaker* 12-1
12.1	Introduction	12-1
12.2	Coaxial Transmission Line	12-2
12.3	Electrical Considerations	12-6
12.4	Coaxial Cable Ratings	12-7

13 Waveguide *Jerry C. Whitaker* 13-1
13.1 Introduction 13-1
13.2 Ridged Waveguide 13-3
13.3 Circular Waveguide 13-4
13.4 Doubly Truncated Waveguide 13-4
13.5 Impedance Matching 13-6

14 RF Combiner and Diplexer Systems *Jerry C. Whitaker* 14-1
14.1 Introduction 14-1
14.2 Passive Filters 14-2
14.3 Four-Port Hybrid Combiner 14-4
14.4 Non-Constant-Impedance Diplexer 14-6
14.5 Constant-Impedance Diplexer 14-8
14.6 Microwave Combiners 14-12
14.7 Hot-Switching Combiners 14-13
14.8 High-Power Isolators 14-17

15 Radio Wave Propagation *Gerhard J. Straub* 15-1
15.1 Introduction 15-1
15.2 Radio Wave Basics 15-1
15.3 Free Space Path Loss 15-3
15.4 Reflection, Refraction, and Diffraction 15-4
15.5 Very Low Frequency (VLF), Low Frequency (LF), and Medium Frequency (MF) Propagation 15-7
15.6 HF Propagation 15-9
15.7 VHF and UHF Propagation 15-12
15.8 Microwave Propagation 15-14

16 Antenna Principles *Pingjuan L. Werner, Anthony J. Ferraro, and Douglas H. Werner* ... 16-1
16.1 Antenna Types 16-1
16.2 Antenna Bandwidth 16-2
16.3 Antenna Parameters 16-3
16.4 Antenna Characteristics 16-13
16.5 Apertures 16-26
16.6 Wide-Band Antennas 16-27

17 Practical Antenna Systems *Jerry C. Whitkaer* 17-1
17.1 Introduction 17-1
17.2 Antenna Types 17-6
17.3 Antenna Applications 17-11
17.4 Phased-Array Antenna Systems 17-23

xiii

18 Preventing RF System Failures *Jerry C. Whitaker* 18-1
- 18.1 Introduction 18-1
- 18.2 Routine Maintenance 18-2
- 18.3 Klystron Devices 18-8
- 18.4 Power Grid Tubes 18-9
- 18.5 Preventing RF System Failures 18-18
- 18.6 Transmission Line/Antenna Problems 18-21
- 18.7 High-Voltage Power Supply Problems 18-24
- 18.8 Temperature Control 18-28

19 Troubleshooting RF Equipment *Jerry C. Whitaker* 19-1
- 19.1 Introduction 19-1
- 19.2 Plate Overload Fault 19-2
- 19.3 RF System Faults 19-5
- 19.4 Power Control Faults 19-11

20 RF Voltage and Power Measurement *Jerry C. Whitaker* 20-1
- 20.1 Introduction 20-1
- 20.2 RF Power Measurement 20-5

21 Spectrum Analysis *Jerry C. Whitaker* 21-1
- 21.1 Introduction 21-1

22 Testing Coaxial Transmission Line *Jerry C. Whitaker* 22-1
- 22.1 Introduction 22-1
- 22.2 Testing Coaxial Lines 22-3

23 Safety Issues for RF Systems *Jerry C. Whitaker* 23-1
- 23.1 Introduction 23-1
- 23.2 Electric Shock 23-4
- 23.3 Polychlorinated Biphenyls 23-11
- 23.4 OSHA Safety Requirements 23-16
- 23.5 Beryllium Oxide Ceramics 23-19
- 23.6 Corrosive and Poisonous Compounds 23-19
- 23.7 Nonionizing Radiation 23-20
- 23.8 X-Ray Radiation Hazard 23-22
- 23.9 Hot Coolant and Surfaces 23-22
- 23.10 Management Responsibility 23-22

Index I-1

1
Applications of RF Technology

Jerry C. Whitaker
Editor

1.1 Introduction 1-1
   Modulation Systems • Spread-Spectrum Systems • RF Power Amplifiers • Frequency Sources • Operating Class • Operating Efficiency • Broadband Amplifier Design • Amplifier Compensation • Stagger Tuning • Matching Circuits • Power Combining • Output Devices

1.2 Broadcast Applications of RF Technology 1-8
   AM Radio Broadcasting • Shortwave Broadcasting • FM Radio Broadcasting • Television Broadcasting

1.3 Nonbroadcast Applications 1-23
   Satellite Transmission • Radar • Electronic Navigation • Microwave Radio • Induction Heating

1.1 Introduction

Radio frequency (RF) power amplifiers are used in countless applications at tens of thousands of facilities around the world. The wide variety of applications, however, stem from a few basic concepts of conveying energy and information by means of a radio frequency signal. Furthermore, the devices used to produce RF energy have many similarities, regardless of the final application. Although radio and television broadcasting represent the most obvious use of high-power RF generators, numerous other common applications exist, including:

- Induction heating and process control systems
- Radio communications (two-way mobile radio base stations and cellular base stations)
- Amateur radio
- Radar (ground, air, and shipboard)
- Satellite communications
- Atomic science research
- Medical research, diagnosis, and treatment

Figure 1.1 illustrates the electromagnetic spectrum and major applications.

Modulation Systems

The primary purpose of most communications systems is to transfer information from one location to another. The message signals used in communication and control systems usually must be limited in frequency to provide for efficiency transfer. This frequency may range from a few hertz for control systems to a few megahertz for video signals. To facilitate efficient and controlled distribution of these signals,

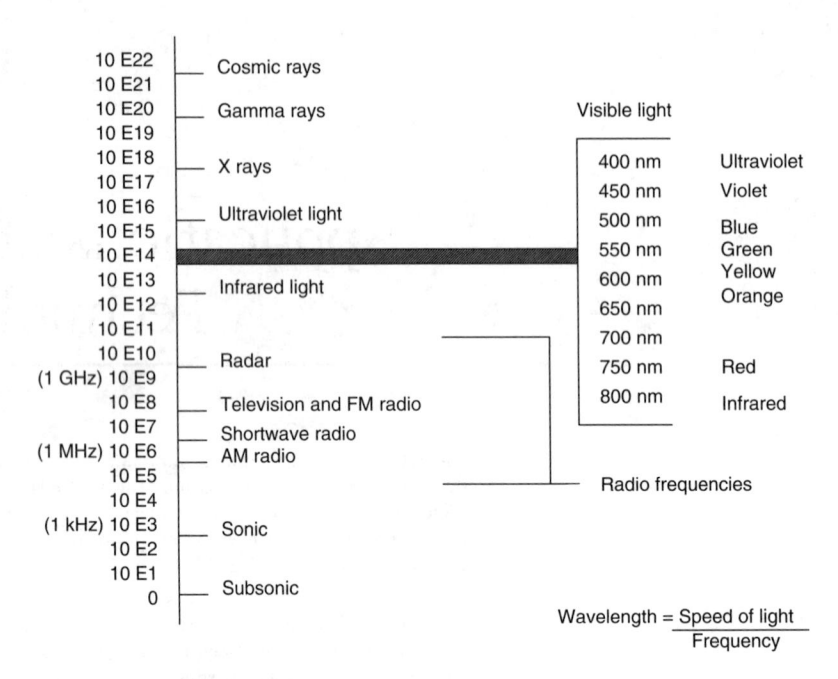

```
                10 E22  ─ Cosmic rays
                10 E21
                10 E20  ─ Gamma rays            Visible light
                10 E19
                10 E18  ─ X rays                ┌─────────┬──────────┐
                10 E17                          │ 400 nm  │ Ultraviolet │
                10 E16  ─ Ultraviolet light     │ 450 nm  │ Violet      │
                10 E15                          │ 500 nm  │ Blue        │
                10 E14  ━━━━━━━━━━━━━━━━━━━━━━━ │ 550 nm  │ Green       │
                10 E13  ─ Infrared light        │ 600 nm  │ Yellow      │
                10 E12                          │ 650 nm  │ Orange      │
                10 E11                          │ 700 nm  │             │
                10 E10  ─ Radar                 │ 750 nm  │ Red         │
       (1 GHz) 10 E9                            │ 800 nm  │ Infrared    │
                10 E8   ─ Television and FM radio
                10 E7   ─ Shortwave radio
       (1 MHz) 10 E6   ─ AM radio
                10 E5                           ── Radio frequencies
                10 E4
       (1 kHz) 10 E3   ─ Sonic
                10 E2
                10 E1   ─ Subsonic
                     0
                                                Wavelength = Speed of light
                                                             ─────────────
                                                              Frequency
```

FIGURE 1.1 The electromagnetic spectrum.

an *encoder* is generally required between the source and the transmission channel. The encoder acts to *modulate* the signal, producing at its output the *modulated waveform*. Modulation is a process whereby the characteristics of a wave (the *carrier*) are varied in accordance with a message signal, the modulating waveform. Frequency translation is usually a by-product of this process. Modulation can be continuous, where the modulated wave is always present, or pulsed, where no signal is present between pulses. There are a number of reasons for producing modulated waves, including:

- *Frequency translation.* The modulation process provides a vehicle to perform the necessary frequency translation required for distribution of information. An input signal can be translated to its assigned frequency band for transmission or radiation.
- *Signal processing.* It is often easier to amplify or process a signal in one frequency range as opposed to another.
- *Antenna efficiency.* Generally speaking, for an antenna to be efficient, it must be large compared with the signal wavelength. Frequency translation provided by modulation allows antenna gain and beamwidth to become part of the system design considerations. Use of higher frequencies permits antenna structures of reasonable size and cost.
- *Bandwidth modification.* The modulation process permits the bandwidth of the input signal to be increased or decreased as required by the application. Bandwidth reduction can permit more efficient use of the spectrum, at the cost of signal fidelity. Increased bandwidth, on the other hand, permits increased immunity to transmission channel disturbances.
- *Signal multiplexing.* In a given transmission system, it may be necessary or desirable to combine several different signals into one baseband waveform for distribution. Modulation provides the vehicle for such *multiplexing*. Various modulation schemes allow separate signals to be combined at the transmission end, and separated (*demultiplexed*) at the receiving end. Multiplexing can be accomplished using *frequency-domain multiplexing* (FDM) or *time-domain multiplexing* (TDM).
- *Modulation of a signal does not come without undesirable characteristics.* Bandwidth restriction or the addition of noise or other disturbances are the two primary problems faced by the transmission system designer.

Spread-Spectrum Systems

The specialized requirements of the military led to the development of *spread-spectrum* communications systems. As the name implies, such systems require a frequency range substantially greater than the basic information-bearing signal. Spread-spectrum systems have some or all of the following properties:

- Low interference to other communications systems
- Ability to reject high levels of external interference
- Immunity to jamming by hostile forces
- Provides for secure communications paths
- Operates over multiple RF paths

Spread-spectrum systems operate with an entirely different set of requirements than the transmission systems discussed previously. Conventional modulation methods are designed to provide for the easiest possible reception and demodulation of the transmitted intelligence. The goals of spread-spectrum systems, on the other hand, are secure and reliable communications that cannot be intercepted by unauthorized persons. The most common modulation and encoding techniques in spread-spectrum communications include:

- *Frequency hopping*, where a random or *pseudorandom number* (PN) sequence is used to change the carrier frequency of the transmitter. This approach has two basic variations: *slow frequency hopping*, where the hopping rate is smaller than the data rate; and *fast frequency hopping*, where the hopping rate is larger than the data rate. In a fast frequency hopping system, the transmission of a single piece of data occupies more than one frequency. Frequency hopping systems permit multiple-access capability to a given band of frequencies because each transmitted signal occupies only a fraction of the total transmitted bandwidth.
- *Time hopping*, where a PN sequence is used to switch the position of a message-carrying pulse within a series of frames.
- *Message corruption*, where a PN sequence is added to the message before modulation.
- *Chirp spread spectrum*, where linear frequency modulation of the main carrier is used to spread the transmitted spectrum. This technique is commonly used in radar and has also been applied to communications systems.

In a spread-spectrum system, the signal power is divided over a large bandwidth. The signal, therefore, has a small average power in any single narrowband slot. This means that a spread-spectrum system can share a given frequency band with one or more narrowband systems.

RF Power Amplifiers

The process of generating high-power RF signals has been refined over the years to an exact science. Advancements in devices and circuit design continue to be made each year, pushing ahead the barriers of efficiency and maximum operating frequency. Although different applications place unique demands on the RF design engineer, the fundamental concepts of RF amplification are applicable to virtually any system.

Frequency Sources

Every RF amplifier requires a stable frequency reference. At the heart of most systems is a quartz crystal. Quartz acts as a stable high Q mechanical resonator. Crystal resonators are available for operation at frequencies ranging from 1 kHz to 300 MHz and beyond.

The operating characteristics of a crystal are determined by the *cut* of the device from a bulk "mother" crystal. The behavior of the device strongly depends on the size and shape of the crystal and the angle of the cut. To provide for operation at a wide range of frequencies, different cuts, vibrating in one or more selected modes, are used.

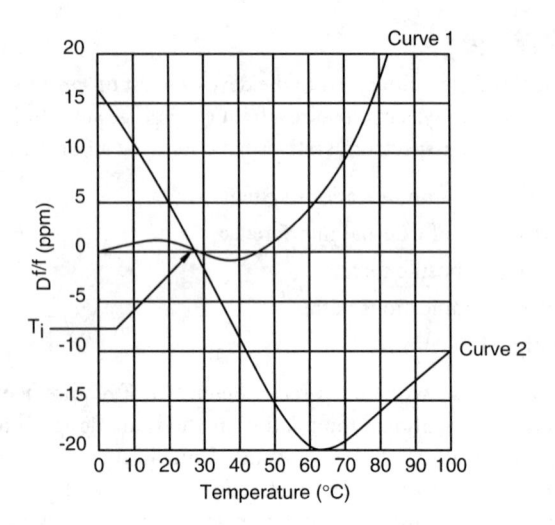

FIGURE 1.2 The effects of temperature on two types of AT-cut crystals.

Crystals are temperature sensitive, as shown in Fig. 1.2. The extent to which a device is affected by changes in temperature is determined by its cut and packaging. Crystals also exhibit changes in frequency with time. Such *aging* is caused by one or both of the following:

- Mass transfer to or from the resonator surface
- Stress relief within the device itself

Crystal aging is most pronounced when the device is new. As stress within the internal structure is relieved, the aging process slows.

The stability of a quartz crystal is inadequate for most commercial and industrial applications. Two common methods are used to provide the required long-term frequency stability:

- *Oven-controlled crystal oscillator:* a technique in which the crystal is installed in a temperature-controlled box. Because the temperature is constant in the box, controlled by a thermostat, the crystal remains on-frequency. The temperature of the enclosure is usually set to the *turnover temperature* of the crystal. (The turnover point is illustrated in Fig. 1.2.)
- *Temperature-compensated crystal oscillator* (TCXO): a technique where the frequency-vs.-temperature changes of the crystal are compensated by varying a load capacitor. A thermistor network is typically used to generate a correction voltage that feeds a varactor to re-tune the crystal to the desired on-frequency value.

Operating Class

Power amplifier (PA) stage operating efficiency is a key element in the design and application of an RF system. As the power level of an RF generator increases, the overall efficiency of the system becomes more important. Increased efficiency translates into lower operating costs and usually improved reliability of the system. The operating mode of the final stage, or stages, is the primary determining element in the maximum possible efficiency of the system.

All electron amplifying devices are classified by their individual *class of operation*. Four primary class divisions apply to RF generators:

- *Class A:* a mode wherein the power amplifying device is operated over its linear transfer characteristic. This mode provides the lowest waveform distortion, but also the lowest efficiency. The basic operating efficiency of a class A stage is 50%. Class A amplifiers exhibit low intermodulation distortion, making them well suited for linear RF amplifier applications.

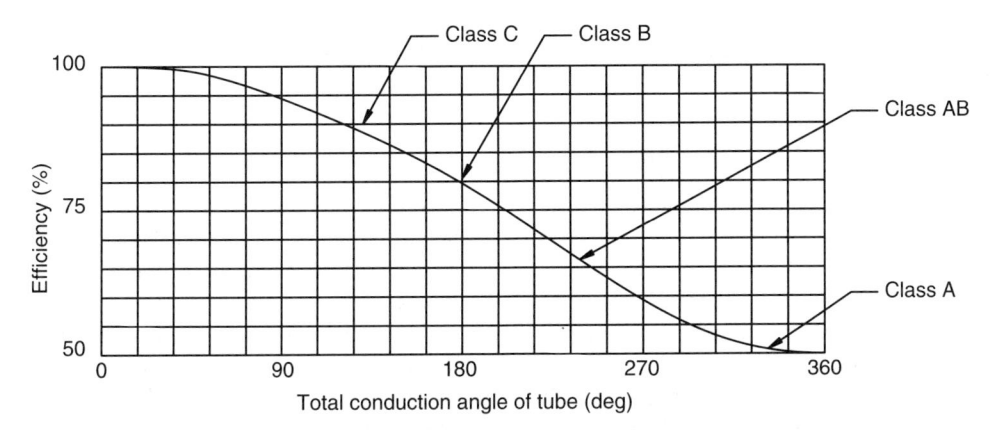

FIGURE 1.3 Plate efficiency as a function of conduction angle for an amplifier with a tuned load.

- *Class B:* a mode wherein the power amplifying device is operated just outside its linear transfer characteristic. This mode provides improved efficiency at the expense of some waveform distortion. Class AB is a variation on class B operation. The transfer characteristic for an amplifying device operating in this mode is, predictably, between class A and class B.
- *Class C:* a mode wherein the power amplifying device is operated significantly outside its linear transfer characteristic, resulting is a pulsed output waveform. High efficiency (up to 90%) can be realized with class C operation; however, significant distortion of the waveform will occur. Class C is used extensively as an efficient RF power generator.
- *Class D:* a mode that essentially results in a switched device state. The power amplifying device is either *on* or *off*. This is the most efficient mode of operation. It is also the mode that produces the greatest waveform distortion.

The angle of current flow determines the class of operation for a power amplifying device. Typically, the conduction angle for class A is 360°; class AB is between 180° and 360°; class B is 180°; and class C is less than 180°. Subscripts can also be used to denote grid current flow in the case of a power vacuum tube device. The subscript "1" means that no grid current flows in the stage; the subscript "2" denotes grid current flow. Figure 1.3 charts operating efficiency as a function of the conduction angle of an RF amplifier.

The class of operation is not directly related to the type of amplifying circuit. Vacuum tube stages may be grid- or cathode-driven without regard to the operating class. Similarly, solid-state amplifiers may be configured for grounded emitter, grounded base, or grounded collector operation without regard to the class of operation.

Operating Efficiency

The design goal of all RF amplifiers is to convert input power into an RF signal at the greatest possible efficiency. DC input power that is not converted to a useful output signal is, for the most part, converted to heat. This heat represents wasted energy, which must be removed from the amplifying device. Removal of heat is a problem common to all high-power RF amplifiers. Cooling methods include:

- Natural convection
- Radiation
- Forced convection
- Liquid
- Conduction
- Evaporation

The type of cooling method chosen is dictated in large part by the type of active device used and the power level involved. For example, liquid cooling is used almost exclusively for high-power (100 kW) vacuum tubes; conduction is used most often for low-power (20 W) transistors.

Broadband Amplifier Design

RF design engineers face a continuing challenge to provide adequate bandwidth for the signals to be transmitted, while preserving as much efficiency as possible from the overall system. These two parameters, while not mutually exclusive, often involve trade-offs for both designers and operators.

An ideal RF amplifier will operate over a wide band of frequencies with minimum variations in output power, phase, distortion, and efficiency. The bandwidth of the amplifier depends to a great extent on the type of active device used, the frequency range required, and the operating power. As a general rule, bandwidth of 20% or greater at frequencies above 100 MHz can be considered *broadband*. Below 100 MHz, broadband amplifiers typically have a bandwidth of one octave or more.

Most development in new broadband designs focuses on semiconductor technology. Transistor and MOSFET (metal oxide semiconductor field effect transistor) devices have ushered in the era of *distributed amplification*, where multiple devices are used to achieve the required RF output power. Semiconductor-based designs offer benefits beyond active device redundancy. Bandwidth at frequencies above 100 MHz can often be improved because of the smaller physical size of semiconductor devices, which translates into reduced lead and component inductance and capacitance.

Amplifier Compensation

A variety of methods can be used to extend the operating bandwidth of a transistorized amplifier stage. Two of the most common methods are series- and shunt-compensation circuits, shown in Fig. 1.4. These two basic techniques can be combined, as shown. Other circuit configurations can be used for specific requirements, such as phase compensation.

Stagger Tuning

Several stages with narrowband response (relative to the desired system bandwidth) can be cascaded and, through the use of *stagger tuning*, made broadband. While there is an efficiency penalty for this approach, it has been used for years in all types of equipment. The concept is simple: offset the center operating frequencies (and, therefore, peak amplitude response) of the cascaded amplifiers so the resulting passband is flat and broad.

For example, the first stage in a three-stage amplifier is adjusted for peak response at the center operating frequency of the system. The second stage is adjusted above the center frequency, and the third stage is adjusted below center. The resulting composite response curve yields a broadband trace. The efficiency penalty for this scheme varies, depending on the power level of each stage, the amount of stagger tuning required to achieve the desired bandwidth, and the number of individual stages.

Matching Circuits

The individual stages of an RF generator must be coupled together. Rarely do the output impedance and power level of one stage precisely match the input impedance and signal-handing level of the next stage. There is a requirement, therefore, for broadband matching circuits. Matching at RF frequencies can be accomplished with several different techniques, including:

- *Quarter-wave transformer:* a matching technique using simply a length of transmission line 1/4-wave long, with a characteristic impedance of:

$$Z_{line} = \sqrt{Z_{in} \times Z_{out}} \tag{1.1}$$

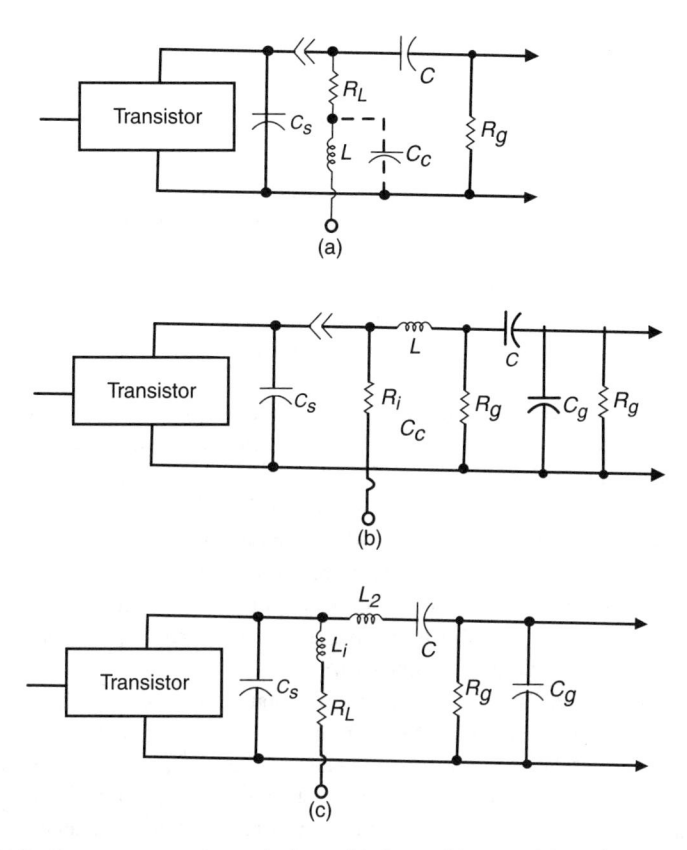

FIGURE 1.4 High-frequency compensation techniques: (a) shunt, (b) series, (c) combination of shunt and series. (After Fink, D. and Christiansen, D., Eds., *Electronics Engineer's Handbook*, 3rd ed., McGraw-Hill, New York, 1989.)

where Z_{in} and Z_{out} are the terminating impedances. Quarter-wave transformers can be cascaded to achieve more favorable matching characteristics. Cascaded transformers permit small matching ratios for each individual section.

- *Balun transformer:* a transmission-line transformer in which the turns are physically arranged to include the interwinding capacitance as a component of the characteristic impedance of the transmission line. This technique permits wide bandwidths to be achieved without unwanted resonances. Balun transformers are usually made of twisted wire pairs or twisted coaxial lines. Ferrite toroids can be used as the core material.
- Other types of lumped reactances.
- Short sections of transmission line.

Power Combining

The two most common methods of extending the operating power of semiconductor devices are *direct paralleling* of components and *hybrid splitting/combining*. Direct paralleling has been used for both tube and semiconductor designs; however, application of this simple approach is limited by variations in device operating parameters. Two identical devices in parallel do not necessarily draw the same amount of current (supply the same amount of power to the load). Paralleling at UHF frequencies and above can be difficult because of the restrictions of operating wavelength.

The preferred approach involves the use of identical stages driven in parallel from a *hybrid coupler*. The coupler provides a constant-source impedance and directs any reflected energy from the driven stages to a *reject port* for dissipation. A hybrid coupler offers a voltage standing wave ratio or VSWR-canceling effect that improves system performance. Hybrids also provide a high degree of isolation between active devices in a system.

Output Devices

Significant changes have occurred within the past 10 years or so with regard to power amplifying devices. Vacuum tubes were the mainstay of RF transmission equipment until advanced semiconductor components became available at competitive prices. Many high-power applications that demanded vacuum tubes can now be met with solid-state devices arranged in a distributed amplification system. Metal oxide silicon field effect transistor (MOSFET) and bipolar components have been used successfully in radio and television broadcast transmitters, shortwave transmitters, sonar transmitters, induction heaters, and countless other applications.

Most solid-state designs used today are not simply silicon versions of classic vacuum tube circuits. They are designed to maximize efficiency through class D switching and maximize reliability through distributed amplification and redundancy.

The principal drawback to a solid-state system over a vacuum tube design of comparable power is the circuit complexity that goes with most semiconductor-based hardware. Preventive maintenance is reduced significantly, and — in theory — repair is simpler as well in a solid-state system. The parts count in almost all semiconductor-based hardware, however, is significantly greater than in a comparable tube system. Increased parts translate (usually) into a higher initial purchase price for the equipment and increased vulnerability to device failure of some sort.

Efficiency comparisons between vacuum tube and solid-state systems do not always yield the dramatic contrasts expected. While most semiconductor amplifiers incorporate switching technology that is far superior to class B or C operation (not to mention class A), power losses are experienced in the signal splitting and combining networks necessary to make distributed amplification work.

It is evident, then, that vacuum tubes and semiconductors each have their benefits and drawbacks. Both technologies will remain viable for many years to come. Vacuum tubes will not go away, but are moving to higher power levels and higher operating frequencies.

1.2 Broadcast Applications of RF Technology

Broadcasting has been around for a long time. Amplitude modulation (AM) was the first modulation system that permitted voice communications to take place. This simple modulation system was predominant throughout the 1920s and 1930s. Frequency modulation (FM) came into regular broadcast service during the 1940s. Television broadcasting, which uses amplitude modulation for the visual portion of the signal and frequency modulation for the aural portion of the signal, became available to the public in the mid-1940s. More recently, digital television (DTV) service has been launched in the United States and elsewhere using the conventional television frequency bands and 6-MHz bandwidth of the analog system, but with digital modulation.

AM Radio Broadcasting

AM radio stations operate on 10-kHz channels spaced evenly from 540 to 1600 kHz. Various classes of stations have been established by the Federal Communications Commission (FCC) and agencies in other countries to allocate the available spectrum to given regions and communities. In the United States, the basic classes are *clear*, *regional*, and *local*. Current practice uses the CCIR (international) designations as class A, B, and C, respectively. Operating power levels range from 50 kW for a clear channel station to as little as 250 W for a local station.

High-Level AM Modulation

High-level anode modulation is the oldest and simplest way of generating a high power AM signal. In this system, the modulating signal is amplified and combined with the dc supply source to the anode of

Applications of RF Technology

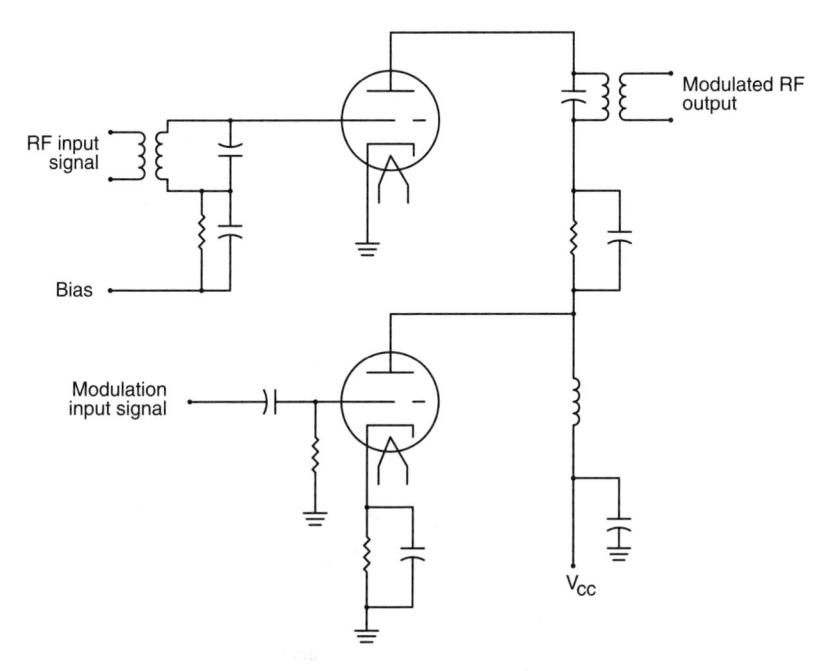

FIGURE 1.5 Simplified diagram of a high-level, amplitude-modulated amplifier.

the final RF amplifier stage. The RF amplifier is normally operated class C. The final stage of the modulator usually consists of a pair of tubes operating class B in a push–pull configuration. A basic high-level modulator is shown in Fig. 1.5.

The RF signal is normally generated in a low-level transistorized oscillator. It is then amplified by one or more solid-state or vacuum tube stages to provide final RF drive at the appropriate frequency to the grid of the final class C amplifier. The audio input is applied to an intermediate power amplifier (usually solid state) and used to drive two class B (or class AB) push–pull output devices. The final amplifiers provide the necessary modulating power to drive the final RF stage. For 100% modulation, this modulating power is equal to 50% of the actual carrier power.

The modulation transformer shown in Fig. 1.5 does not usually carry the dc supply current for the final RF amplifier. The modulation reactor and capacitor shown provide a means to combine the audio signal voltage from the modulator with the dc supply to the final RF amplifier. This arrangement eliminates the necessity of having dc current flow through the secondary of the modulation transformer, which would result in magnetic losses and saturation effects. In some newer transmitter designs, the modulation reactor has been eliminated from the system, thanks to improvements in transformer technology.

The RF amplifier normally operates class C with grid current drawn during positive peaks of the cycle. Typical stage efficiency is 75 to 83%. An RF tank following the amplifier resonates the output signal at the operating frequency and, with the assistance of a low-pass filter, eliminates harmonics of the amplifier caused by class C operation.

This type of system was popular in AM broadcasting for many years, primarily because of its simplicity. The primary drawback is low overall system efficiency. The class B modulator tubes cannot operate with greater than 50% efficiency. Still, with inexpensive electricity, this was not considered to be a significant problem. As energy costs increased, however, more efficient methods of generating high-power AM signals were developed. Increased efficiency normally came at the expense of added technical complexity.

Pulse-Width Modulation

Pulse-width modulation (PWM), also known as pulse-duration modulation (PDM), is one of the most popular systems developed for modern vacuum tube AM transmitters. Figure 1.6 shows the basic PDM scheme. The PDM system works by utilizing a square-wave switching system, illustrated in Fig. 1.7.

The PDM process begins with a signal generator (see Fig. 1.8). A 75-kHz sine wave is produced by an oscillator and used to drive a square-wave generator, resulting in a simple 75-kHz square wave. The square wave is then integrated, resulting in a triangular waveform that is mixed with the input audio in a summing circuit. The resulting signal is a triangular waveform that rides on the incoming audio. This

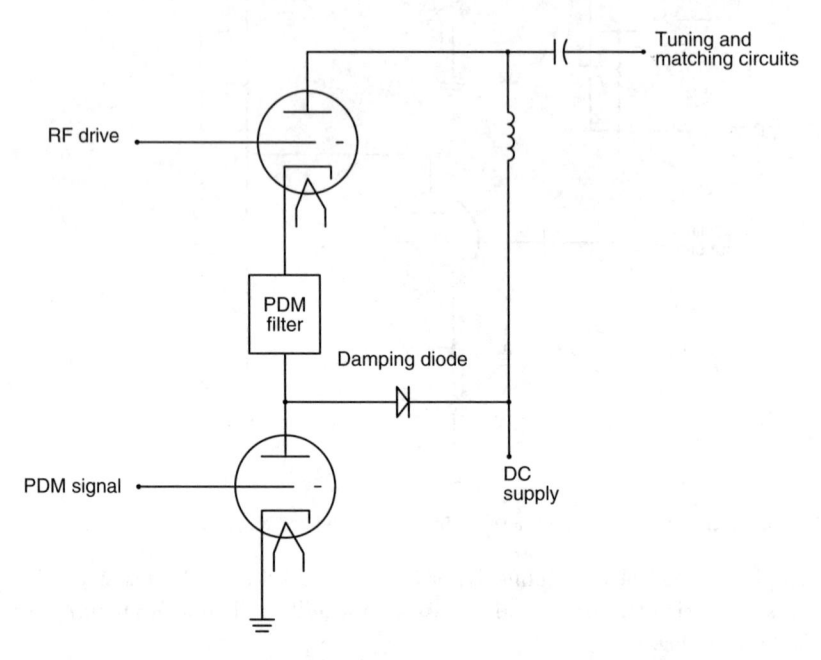

FIGURE 1.6 The pulse-duration modulation (PDM) method of pulse-width modulation.

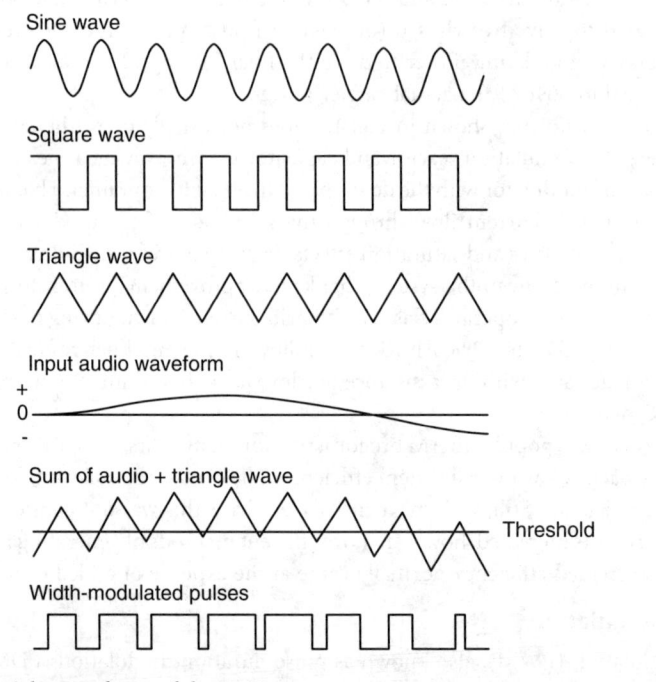

FIGURE 1.7 The principles waveforms of the PDM system.

Applications of RF Technology

composite signal is then applied to a threshold amplifier, which functions as a switch that is turned on whenever the value of the input signal exceeds a certain limit. The result is a string of pulses in which the width of the pulse is proportional to the period of time the triangular waveform exceeds the threshold. The pulse output is applied to an amplifier to obtain the necessary power to drive subsequent stages. A filter eliminates whatever transients may exist after the switching process is complete.

The PDM scheme is, in effect, a digital modulation system with the audio information being sampled at a 75-kHz rate. The width of the pulses contains all the audio information. The pulse-width-modulated signal is applied to a *switch* or *modulator tube*. The tube is simply turned *on*, to a fully saturated state, or *off* in accordance with the instantaneous value of the pulse. When the pulse goes positive, the modulator tube is turned on and the voltage across the tube drops to a minimum. When the pulse returns to its minimum value, the modulator tube turns off.

This PDM signal becomes the power supply to the final RF amplifier tube. When the modulator is switched on, the final amplifier will experience current flow and RF will be generated. When the switch or modulator tube goes off, the final amplifier current will cease. This system causes the final amplifier to operate in a highly efficient class D switching mode. A dc offset voltage to the summing amplifier is used to set the carrier (no modulation) level of the transmitter.

A high degree of third-harmonic energy will exist at the output of the final amplifier because of the switching-mode operation. This energy is eliminated by a third-harmonic trap. The result is a stable amplifier that normally operates in excess of 90% efficiency. The power consumed by the modulator and its driver is usually a fraction of a full class B amplifier stage.

The damping diode shown in the previous figure is included to prevent potentially damaging transient overvoltages during the switching process. When the switching tube turns off the supply current during a period when the final amplifier is conducting, the high current through the inductors contained in the PDM filters could cause a large transient voltage to be generated. The energy in the PDM filter is returned to the power supply by the damping diode. If no alternative route is established, the energy will return by arcing through the modulator tube itself.

The PWM system makes it possible to completely eliminate audio frequency transformers in the transmitter. The result is wide frequency response and low distortion. It should be noted that variations on this amplifier and modulation scheme have been used by other manufacturers for both standard broadcast and shortwave service.

Digital Modulation

Current transmitter design work for AM broadcsting has focused almost exclusively on solid-state technology. High-power MOSFET devices and digital modulation techniques have made possible a new generation of energy-efficient systems, with audio performance that easily surpasses vacuum tube designs.

Most solid-state AM systems operate in a highly efficient class D switching mode. Multiple MOSFET driver boards are combined through one of several methods to achieve the required carrier power.

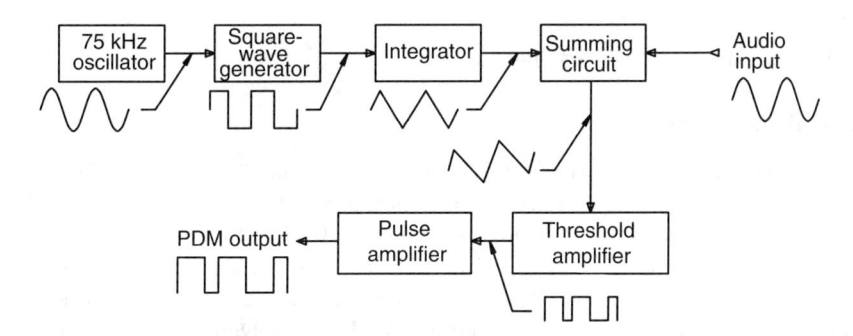

FIGURE 1.8 Block diagram of a PDM waveform generator.

Shortwave Broadcasting

The technologies used in commercial and government-sponsored shortwave broadcasting are closely allied with those used in AM radio. However, shortwave stations usually operate at significantly higher powers than AM stations.

International broadcast stations use frequencies ranging from 5.95 to 26.1 MHz. The transmissions are intended for reception by the general public in foreign countries. Table 1.1 shows the frequencies assigned by the Federal Communications Commission (FCC) for international broadcast shortwave service in the United States. The minimum output power is 50 kW. Assignments are made for specific hours of operation at specific frequencies.

TABLE 1.1 Operating Frequency Bands for Shortwave Broadcasting

Band	Frequency (kHz)	Meter Band (m)
A	5,950–6,200	49
B	9,500–9,775	32
C	11,700–11,975	25
D	15,100–15,450	19
E	17,700–17,900	16
F	21,450–21,750	14
G	25,600–26,100	11

Very high-power shortwave transmitters have been installed to serve large geographical areas and to overcome jamming efforts by foreign governments. Systems rated for power outputs of 500 kW and more are not uncommon. RF circuits designed specifically for high power operation are utilized.

Most shortwave transmitters have the unique requirement for automatic tuning to one of several preset operating frequencies. A variety of schemes exist to accomplish this task, including multiple exciters (each set to the desired operating frequency) and motor-controlled variable inductors and capacitors. Tune-up at each frequency is performed by the transmitter manufacturer. The settings of all tuning controls are stored in memory. Automatic retuning of a high-power shortwave transmitter can be accomplished in less than 30 seconds in most cases.

Power Amplifier Types

Shortwave technology has advanced significantly within the last 5 years, thanks to improved semiconductor devices. High-power MOSFETs and other components have made solid-state shortwave transmitters operating at 500 kW and more practical. The majority of shortwave systems now in use, however, use vacuum tubes as the power-generating element. The efficiency of a power amplifier/modulator for shortwave applications is of critical importance. Because of the power levels involved, low efficiency translates into higher operating costs.

Older, traditional tube-type shortwave transmitters typically utilize one of the following modulation systems:

- Doherty amplifier
- Chireix outphasing modulated amplifier
- Dome modulated amplifier
- Terman-Woodyard modulated amplifier

FM Radio Broadcasting

FM radio stations operate on 200-kHz channels spaced evenly from 88.1 to 107.9 MHz. In the United States, channels below 92.1 MHz are reserved for noncommercial, educational stations. The FCC has established three classifications for FM stations operating east of the Mississippi River and four classifications for stations west of the Mississippi. Power levels range from a high of 100 kW *effective radiated power* (ERP) to 3 kW or less for lower classifications. The ERP of a station is a function of transmitter power output (TPO) and antenna gain. ERP is determined by multiplying these two quantities together and allowing for line loss.

A transmitting antenna is said to have "gain" if, by design, it concentrates useful energy at low radiation angles, rather than allowing a substantial amount of energy to be radiated above the horizon (and be lost in space). FM and TV transmitting antennas are designed to provide gain by stacking individual radiating elements vertically.

At first examination, it might seem reasonable and economical to achieve licensed ERP using the lowest transmitter power output possible and highest antenna gain. Other factors, however, come into play that make the most obvious solution not always the best solution. Factors that limit the use of high-gain antennas include:

- Effects of high-gain designs on coverage area and signal penetration
- Limitations on antenna size because of tower restrictions, such as available vertical space, weight, and windloading
- Cost of the antenna

Stereo broadcasting is used almost universally in FM radio today. Introduced in the mid-1960s, stereo has contributed in large part to the success of FM radio. The left and right sum (monophonic) information is transmitted as a standard frequency-modulated signal. Filters restrict this *main channel* signal to a maximum of about 17 kHz. A pilot signal is transmitted at low amplitude at 19 kHz to enable decoding at the receiver. The left and right difference signal is transmitted as an amplitude-modulated subcarrier that frequency-modulates the main FM carrier. The center frequency of the subcarrier is 38 kHz. Decoder circuits in the FM receiver matrix the sum and difference signals to reproduce the left and right audio channels. Figure 1.9 illustrates the baseband signal of a stereo FM station.

Modulation Circuits

Early FM transmitters used *reactance modulators* that operated at low frequency. The output of the modulator was then multiplied to reach the desired output frequency. This approach was acceptable for monaural FM transmission but not for modern stereo systems or other applications that utilize subcarriers on the FM broadcast signal. Modern FM systems all utilize what is referred to as *direct modulation*. That is, the frequency modulation occurs in a modulated oscillator that operates on a center frequency equal to the desired transmitter output frequency. In stereo broadcast systems, a composite FM signal is applied to the FM modulator.

Various techniques have been developed to generate the direct-FM signal. One of the most popular uses a variable-capacity diode as the reactive element in the oscillator. The modulating signal is applied to the diode, which causes the capacitance of the device to vary as a function of the magnitude of the modulating signal. Variations in the capacitance cause the frequency of the oscillator to vary. Again, the magnitude of the frequency shift is proportional to the amplitude of the modulating signal, and the rate of frequency shift is equal to the frequency of the modulating signal.

The direct-FM modulator is one element of an FM transmitter exciter, which generates the composite FM waveform. A block diagram of a complete FM exciter is shown in Fig. 1.10. Audio inputs of various types (stereo left and right signals, plus subcarrier programming, if used) are buffered, filtered, and preemphasized before being summed to feed the modulated oscillator. It should be noted that the oscillator is not normally coupled directly to a crystal, but a free-running oscillator adjusted as closely as possible to the carrier frequency of the transmitter. The final operating frequency is carefully maintained by an automatic frequency control system employing a *phase-locked loop* (PLL) tied to a reference crystal oscillator or frequency synthesizer.

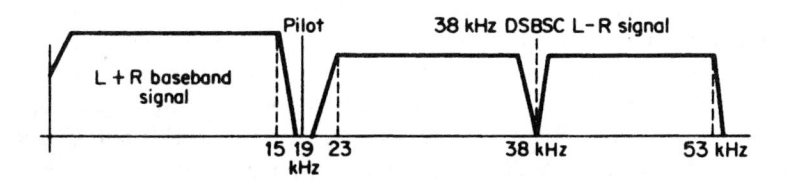

FIGURE 1.9 Composite baseband stereo FM signal.

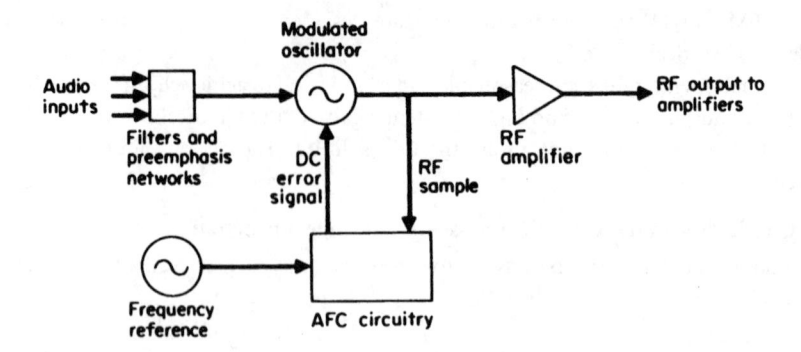

FIGURE 1.10 Block diagram of an FM exciter.

A solid-state class C amplifier follows the modulated oscillator and raises the operating power of the FM signal to 20 to 30 W. One or more subsequent amplifiers in the transmitter raise the signal power to several hundred watts for application to the final power amplifier stage. Nearly all current high-power FM transmitters utilize solid-state amplifiers up to the final RF stage, which is generally a vacuum tube for operating powers of 5 kW and above. All stages operate in the class C mode. In contrast to AM systems, each stage in an FM power amplifier can operate class C because no information is lost from the frequency-modulated signal due to amplitude changes. As mentioned previously, FM is a constant-power system.

Auxiliary Services

Modern FM broadcast stations are capable of not only broadcasting stereo programming, but one or more subsidiary channels as well. These signals, referred to by the FCC as *Subsidiary Communications Authorization* (SCA) services, are used for the transmission of stock market data, background music, control signals, and other information not normally part of the station's main programming. These services do not provide the same range of coverage or audio fidelity as the main stereo program; however, they perform a public service and can represent a valuable source of income for the broadcaster.

SCA systems provide efficient use of the available spectrum. The most common subcarrier frequency is 67 kHz, although higher subcarrier frequencies may be utilized. Stations that operate subcarrier systems are permitted by the FCC to exceed (by a small amount) the maximum 75-kHz deviation limit under certain conditions. The subcarriers utilize low modulation levels, and the energy produced is maintained essentially within the 200-kHz bandwidth limitation of FM channel radiation.

FM Power Amplifiers

Most high-power FM transmitters manufactured today employ cavity designs. The 1/4-wavelength cavity is the most common. The design is simple and straightforward. A number of variations can be found in different transmitters but the underlying theory of operation is the same. The goal of any cavity amplifier is to simulate a resonant tank circuit at the operating frequency and provide a means to couple the energy in the cavity to the transmission line. Because of the operating frequencies involved (88 to 108 MHz), the elements of the "tank" take on unfamiliar forms.

A typical 1/4-wave cavity is shown in Fig. 1.11. The plate of the tube connects directly to the inner section (tube) of the plate-blocking capacitor. The blocking capacitor can be formed in one of several ways. In at least one design, it is made by wrapping the outside surface of the inner tube conductor with multiple layers of insulating film. The exhaust chimney/inner conductor forms the other element of the blocking capacitor. The cavity walls form the outer conductor of the 1/4-wave transmission line circuit. The dc plate voltage is applied to the PA (power amplifier) tube by a cable routed inside the exhaust chimney and inner tube conductor. In this design, the screen-contact fingerstock ring mounts on a metal plate that is insulated from the grounded-cavity deck by a blocking capacitor. This hardware makes up

Applications of RF Technology   1-15

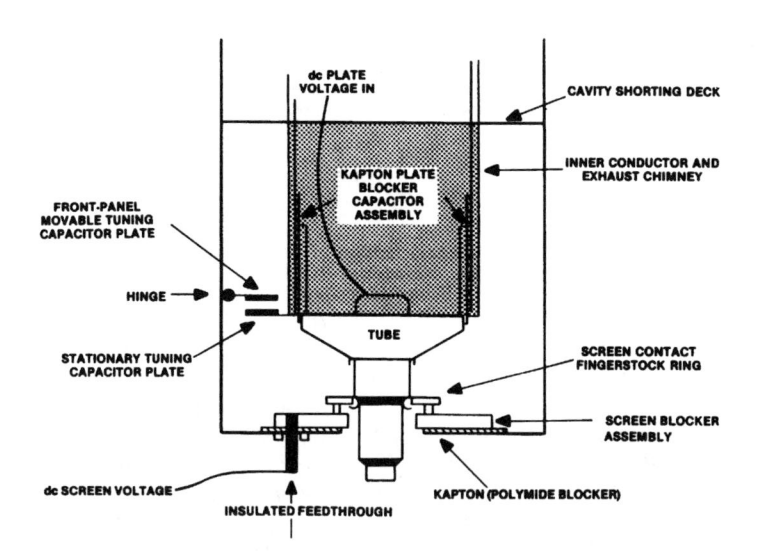

FIGURE 1.11 Physical layout of a common type of 1/4-wave PA cavity for FM broadcast service.

the screen-blocker assembly. The dc screen voltage feeds to the fingerstock ring from underneath the cavity deck through an insulated feedthrough.

Some transmitters that employ the 1/4-wave cavity design use a grounded-screen configuration in which the screen contact fingerstock ring is connected directly to the grounded cavity deck. The PA cathode then operates at below ground potential (i.e., at a negative voltage), establishing the required screen voltage for the tube.

Coarse tuning of the cavity is accomplished by adjusting the cavity length. The top of the cavity (the cavity shorting deck) is fastened by screws or clamps and can be raised or lowered to set the length of the assembly for the particular operating frequency. Fine-tuning is accomplished by a variable-capacity plate-tuning control built into the cavity. In the example, one plate of this capacitor, the stationary plate, is fastened to the inner conductor just above the plate-blocking capacitor. The movable tuning plate is fastened to the cavity box, the outer conductor, and is mechanically linked to the front-panel tuning control. This capacity shunts the inner conductor to the outer conductor and varies the electrical length and resonant frequency of the cavity.

Television Broadcasting

Television transmitters in the United States operate in three frequency bands:

- Low-band VHF: channels 2 through 6 (54–72 MHz and 76–88 MHz)
- High-band VHF: channels 7 through 13 (174–216 MHz)
- UHF: channels 14 through 69 (470–806 MHz). UHF channels 70 through 83 (806–890 MHz) have been assigned to land mobile radio services. Certain TV translators may continue to operate on these frequencies on a secondary basis.

Because of the wide variety of operating parameters for television stations outside the United States, this section focuses primarily on TV transmission as it relates to the United States (Table 1.2 shows the frequencies used by TV broadcasting). Maximum power output limits are specified by the FCC for each type of service. The maximum effective radiated power for low-band VHF is 100 kW; for high-band VHF, it is 316 kW; and for UHF, it is 5 MW.

The second major factor that affects the coverage area of a TV station is antenna height, known in the broadcast industry as *height above average terrain* (HAAT). HAAT takes into consideration the effects of the geography in the vicinity of the transmitting tower. The maximum HAAT permitted by the FCC for

TABLE 1.2 Channel Designations for VHF and UHF Television Stations in the U.S.

Channel Designation	Frequency Band (MHz)	Channel Designation	Frequency Band (MHz)	Channel Designation	Frequency Band (MHz)
2	54–60	30	566–572	57	728–734
3	60–66	31	572–578	58	734–740
4	66–72	32	578–584	59	740–746
5	76–82	33	584–590	60	746–752
6	82–88	34	590–596	61	752–758
7	174–180	35	596–602	62	758–764
8	180–186	36	602–608	63	764–770
9	186–192	37	608–614	64	770–776
10	192–198	38	614–620	65	776–782
11	198–204	39	620–626	66	782–788
12	204–210	40	626–632	67	788–794
13	210–216	41	632–638	68	794–800
14	470–476	42	638–644	69	800–806
15	476–482	43	644–650	70	806–812
16	482–488	44	650–656	71	812–818
17	488–494	45	656–662	72	818–824
18	494–500	46	662–668	73	824–830
19	500–506	47	668–674	74	830–836
20	506–512	48	674–680	75	836–842
21	512–518	49	680–686	76	842–848
22	518–524	50	686–692	77	848–854
23	524–530	51	692–698	78	854–860
24	530–536	52	698–704	79	860–866
25	536–542	53	704–710	80	866–872
26	542–548	54	710–716	81	872–878
27	548–554	55	716–722	82	878–884
28	554–560	56	722–728	83	884–890
29	560–566				

a low- or high-band VHF station is 1000 ft (305 m) east of the Mississippi River, and 2000 ft (610 m) west of the Mississippi. UHF stations are permitted to operate with a maximum HAAT of 2000 ft (610 m) anywhere in the United States (including Alaska and Hawaii).

The ratio of visual output power to aural power can vary from one installation to another; however, the aural is typically operated at between 10 and 20% of the visual power. This difference is the result of the reception characteristics of the two signals. Much greater signal strength is required at the consumer's receiver to recover the visual portion of the transmission than the aural portion. The aural power output is intended to be sufficient for good reception at the fringe of the station's coverage area, but not beyond. It is of no use for a consumer to be able to receive a TV station's audio signal, but not the video.

In addition to the full-power stations discussed previously, two classifications of low-power TV stations have been established by the FCC to meet certain community needs. They are:

- *Translators:* low-power systems that rebroadcast the signal of another station on a different channel. Translators are designed to provide "fill-in" coverage for a station that cannot reach a particular community because of the local terrain. Translators operating in the VHF band are limited to 100 W power output (ERP), and UHF translators are limited to 1 kW.
- *Low-power television* (LPTV): a service established by the FCC to meet the special needs of particular communities. LPTV stations operating on VHF frequencies are limited to 100 W ERP and UHF stations are limited to 1 kW. LPTV stations originate their own programming and can be assigned by the FCC to any channel, as long as full protection against interference to a full-power station is afforded.

Applications of RF Technology

Television Transmission Standards

Analog television signals transmitted throughout the world have the following similarities.

- All systems use two fields interlaced to create a complete frame.
- All contain luminance, chrominance, syncronization, and sound components.
- All use amplitude modulation to put picture information onto the visual carrier.
- Modulation polarity, in most cases, is negative (greatest power output from the transmitter occurs during the sync interval; least power output occurs during peak white).
- The sound is transmitted on an aural carrier that is offset on a higher frequency than the visual carrier, using frequency modulation in most cases.
- All systems use a vestigial lower sideband approach.
- All systems derive a luminance and two-color difference signals from red, green, and blue components.

There the similarities stop and the differences begin. There are three primary color transmission standards in use.

- NTSC (National Television Systems Committee): used in the United States, Canada, Central America, most of South America, and Japan. In addition, NTSC has been accepted for use in various countries or possessions heavily influenced by the United States. The major components of the NTSC signal are shown in Fig. 1.12.
- PAL (Phase Alternation each Line): used in England, most countries and possessions influenced by the British Commonwealth, many western European countries, and China. Variation exists in PAL systems.
- SECAM (SEquential Color with [Avec] Memory): used in France, countries and possessions influenced by France, the U.S.S.R. (generally the Soviet Bloc nations, including East Germany), and other areas influenced by Russia.

The three standards are incompatible for the following reasons.

- Channel assignments are made in different frequency spectra in many parts of the world. Some countries have VHF only; some have UHF only; others have both. Assignments with VHF and UHF do not necessarily coincide between countries.
- Channel bandwidths are different. NTSC uses a 6-MHz channel width. Versions of PAL exist with 6-MHz, 7-MHz, and 8-MHz bandwidths. SECAM channels are 8-MHz wide.
- Vision bands are different. NTSC uses 4.2 MHz. PAL uses 4.2 MHz, 5 MHz, and 5.5 MHz, while SECAM has 6-MHz video bandwidth.
- The line structure of the signals varies. NTSC uses 525 lines per frame, 30 frames (60 fields) per second. PAL and SECAM use 625 lines per frame, 25 frames (50 fields) per second. As a result, the scanning frequencies also vary.
- The color subcarrier signals are incompatible. NTSC uses 3.579545 MHz, PAL uses 4.43361875 MHz, while SECAM utilizes two subcarriers, 4.40625 MHz and 4.250 MHz. The color subcarrier values are derived from the horizontal frequencies in order to interleave color information into the luminance signal without causing undue interference.
- The color encoding system of all three standards differ.
- The offset between visual and aural carriers varies. In NTSC, is it 4.5 MHz; in PAL, the separation is 5.5 or 6 MHz, depending on the PAL type; and SECAM uses 6.5-MHz separation.
- One form of SECAM uses positive polarity visual modulation (peak white produces greatest power output of transmitter) with amplitude modulation for sound.
- Channels transmitted on UHF frequencies may differ from those on VHF in some forms of PAL and SECAM. Differences include channel bandwidth and video bandwidth.

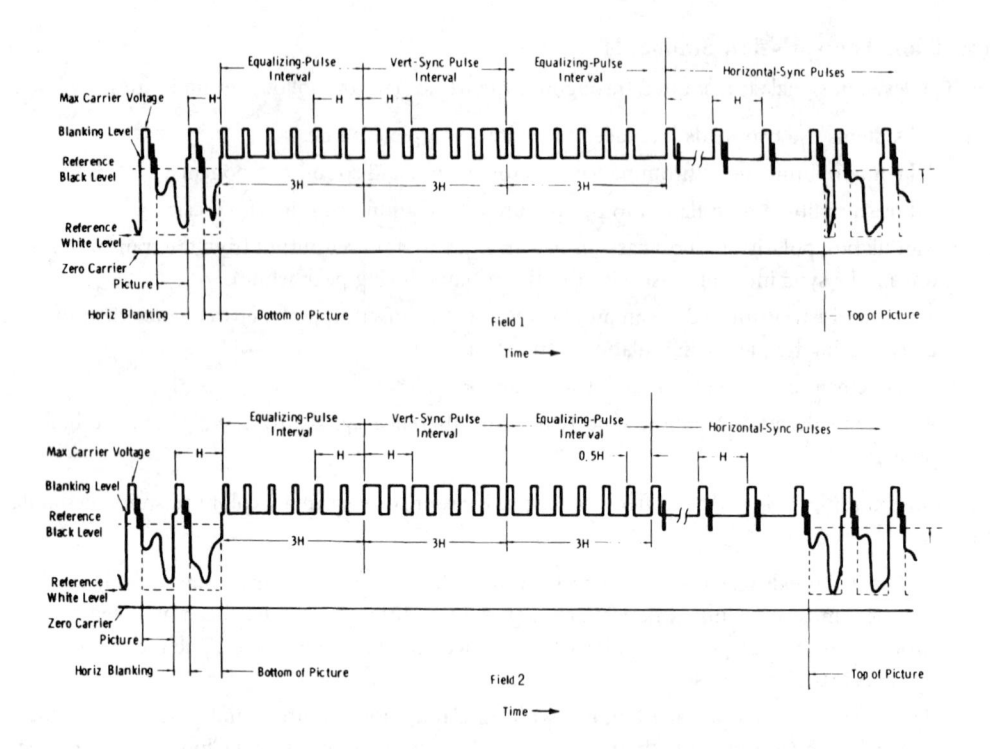

FIGURE 1.12 The major components of the NTSC television signal. H = time from start of line to the start of the next line. V = time from the start of one field to the start of the next field.

It is possible to convert from one television standard to another electronically. The most difficult part of the conversion process results from the differing number of scan lines. In general, the signal must be disassembled in the input section of the standards converter, and then placed in a large dynamic memory. Complex computer algorithms compare information on pairs of lines to determine how to create the new lines required (for conversion to PAL or SECAM) or how to remove lines (for conversion to NTSC). Non-moving objects in the picture present no great difficulties, but motion in the picture can produce objectionable artifacts as the result of the sampling system.

Transmitter Design Considerations

An analog television transmitter is divided into two basic subsystems: (1) the *visual* section, which accepts the video input, amplitude-modulates an RF carrier, and amplifies the signal to feed the antenna system; and (2) the *aural* section, which accepts the audio input, frequency-modulates a separate RF carrier, and amplifies the signal to feed the antenna system. The visual and aural signals are usually combined to feed a single radiating antenna. Different transmitter manufacturers have different philosophies with regard to the design and construction of a transmitter. Some generalizations can, however, be made with respect to basic system design. Transmitters can be divided into categories based on the following criteria:

- Output power
- Final stage design
- Modulation system

Output Power

When the power output of a TV transmitter is discussed, the visual section is the primary consideration. Output power refers to the peak power of the visual stage of the transmitter (*peak of sync*). The

FCC-licensed ERP is equal to the transmitter power output times feedline efficiency times the power gain of the antenna.

A low-band VHF station can achieve its maximum 100-kW power output through a wide range of transmitter and antenna combinations. A 35-kW transmitter coupled with a gain-of-4 antenna would do the trick, as would a 10-kW transmitter feeding an antenna with a gain of 12. Reasonable parings for a high-band VHF station would range from a transmitter with a power output of 50 kW feeding an antenna with a gain of 8, to a 30-kW transmitter connected to a gain-of-12 antenna. These combinations assume reasonable feedline losses. To reach the exact power level, minor adjustments are made to the power output of the transmitter, usually by a front panel power control.

UHF stations that want to achieve their maximum licensed power output are faced with installing a very high power (and very expensive) transmitter. Typical pairings include a transmitter rated for 220 kW and an antenna with a gain of 25, or a 110-kW transmitter and a gain-of-50 antenna. In the latter case, the antenna could pose a significant problem. UHF antennas with gains in the region of 50 are possible but not advisable for most installations because of the coverage problems that can result. High-gain antennas have a narrow vertical radiation pattern that can reduce a station's coverage in areas near the transmitter site. Whatever way is chosen, getting 5-MW ERP is an expensive proposition. Most UHF stations therefore operate considerably below the maximum permitted ERP.

Final Stage Design

The amount of output power required of a transmitter will have a fundamental effect on system design. Power levels usually dictate whether the unit will be of solid-state or vacuum tube design; whether air, water, or vapor cooling must be used; the type of power supply required; the sophistication of the high-voltage control and supervisory circuitry; and whether *common amplification* of the visual and aural signals (rather than separate visual and aural amplifiers) is practical.

Tetrodes are generally used for VHF transmitters above 5 kW and for low-power UHF transmitters (below 5 kW). As solid-state technology advances, the power levels possible in a reasonable transmitter design steadily increase. As of this writing, all-solid-state VHF transmitters of 60 kW have been produced.

In the realm of UHF transmitters, the klystron and related devices reign supreme. Klystrons use an electron bunching technique to generate high power (55 kW from a single tube is not uncommon) at UHF frequencies. They are currently the first choice for high-power service. Klystrons, however, are not particularly efficient. A stock klystron with no special circuitry might be only 40% efficient. Various schemes have been devised to improve klystron efficiency, one of the oldest being *beam pulsing*. Two types of pulsing are in common use:

- *Mod-anode pulsing*, a technique designed to reduce power consumption of the device during the color burst and video portion of the signal (and thereby improve overall system efficiency).
- *Annular control electrode* (ACE) *pulsing*, which accomplishes basically the same thing by incorporating the pulsing signal into a low-voltage stage of the transmitter, rather than a high-voltage stage (as with mod-anode pulsing).

Variations of the basic klystron intended to improve UHF transmitter efficiency include the following:

- The *inductive output tube* (IOT): a device that essentially combines the cathode/grid structure of the tetrode with the drift tube/collector structure of the klystron.
- The *multi-stage depressed collector* (MSDC) *klystron:* a device that achieves greater efficiency through a redesign of the collector assembly. A multi-stage collector is used to recover energy from the electron stream inside the klystron and return it to the beam power supply.
- *Modulation system:* a number of approaches may be taken to amplitude modulation of the visual carrier. Most systems utilize low-level, intermediate frequency (IF) modulation. This approach allows superior distortion correction, more accurate vestigial sideband shaping, and significant economic advantages to the transmitter manufacturer.

Elements of the Transmitter

An analog television transmitter can be divided into four major subsystems:

- The exciter
- Intermediate power amplifier (IPA)
- Power amplifier
- High-voltage power supply

Figure 1.13 shows the audio, video, and RF paths for a typical design. The exciter includes of the following circuits:

- Video input buffer
- Exciter-modulator
- RF processor

Depending on the design of the transmitter, these sections may be separate units or simply incorporated into the exciter itself. A power supply section supplies operating voltages to the various subassemblies of the transmitter.

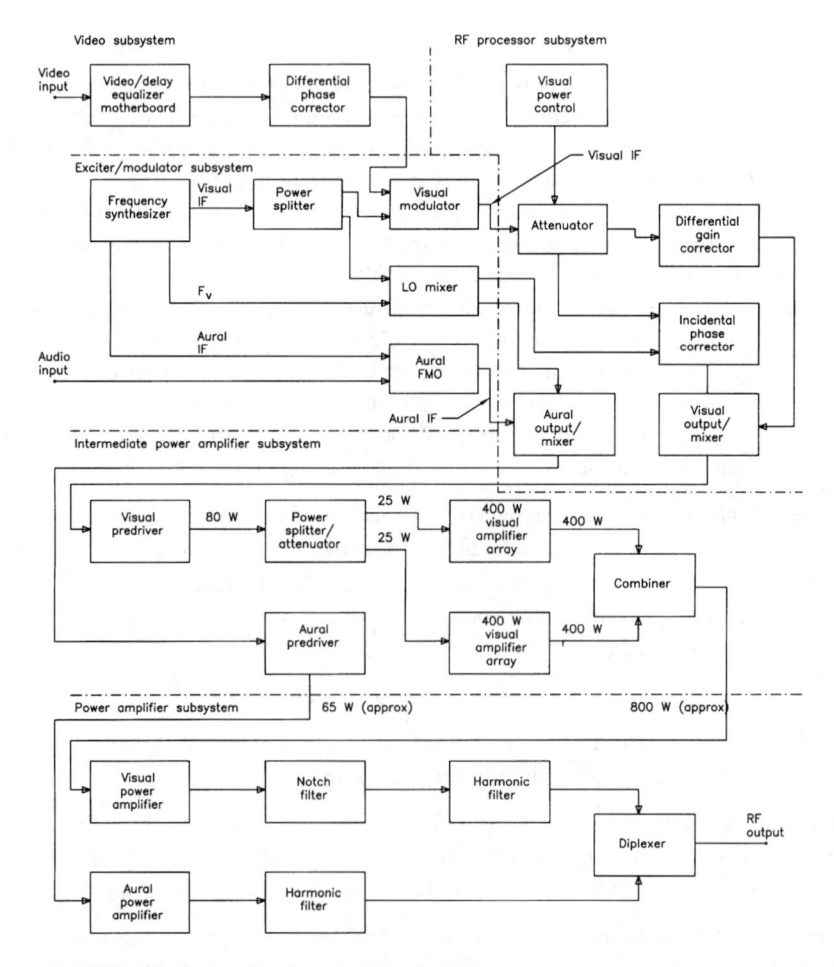

FIGURE 1.13 Simplified block diagram of a television transmitter.

Intermediate Power Amplifier

The function of the IPA is to develop the power output necessary to drive the power amplifier stages for the aural and visual systems. A low-band, 16- to 20-kW transmitter typically requires about 800 W RF drive, and a high-band, 35- to 50-kW transmitter needs about 1600 W. A UHF transmitter utilizing a high-gain klystron final tube requires about 20 W drive, while a UHF transmitter utilizing a klystrode tube needs about 80 W. Because the aural portion of a television transmitter operates at only 10 to 20% of the visual power output, the RF drive requirements are proportionately lower.

Virtually all transmitters manufactured today utilize solid-state devices in the IPA. Transistors are preferred because of their inherent stability, reliability, and ability to cover a broad band of frequencies without retuning. Present solid-state technology, however, cannot provide the power levels needed by most transmitters in a single device. To achieve the needed RF energy, devices are combined using a variety of schemes.

A typical "building block" for a solid-state IPA provides a maximum power output of approximately 200 W. To meet the requirements of a 20-kW low-band VHF transmitter, a minimum of four such units would have to be combined. In actual practice, some amount of *headroom* is always designed into the system to compensate for component aging, imperfect tuning in the PA stage, and device failure.

Most solid-state IPA circuits are configured so that in the event of a failure in one module, the remaining modules will continue to operate. If sufficient headroom has been provided in the design, the transmitter will continue to operate without change. The defective subassembly can then be repaired and returned to service at a convenient time.

Because the output of the RF up-converter is about 10 W, an intermediate amplifier is generally used to produce the required drive for the parallel amplifiers. The individual power blocks are fed by a splitter that feeds equal RF drive to each unit. The output of each RF power block is fed to a hybrid combiner that provides isolation between the individual units. The combiner feeds a bandpass filter that allows only the modulated carrier and its sidebands to pass.

The inherent design of a solid-state RF amplifier permits operation over a wide range of frequencies. Most drivers are broadband and require no tuning. Certain frequency-determined components are added at the factory (depending on the design); however, from the end-user standpoint, solid-state drivers require virtually no attention. IPA systems are available that cover the entire low- or high-band VHF channels without tuning.

Advances continue to be made in solid-state RF devices. New developments promise to substantially extend the reach of semiconductors into medium-power RF operation. Coupled with better devices are better circuit designs, including parallel devices and new push–pull configurations. Another significant factor in achieving high power from a solid-state device is efficient removal of heat from the component itself.

Power Amplifier

The power amplifier (PA) raises the output energy of the transmitter to the required RF operating level. As noted previously, solid-state devices are increasingly being used through parallel configurations in high-power transmitters. Still, however, the majority of television transmitters in use today utilize vacuum tubes. The workhorse of VHF television is the tetrode, which provides high output power, good efficiency, and good reliability. In UHF service, the klystron is the standard output device for transmitters above 20 kW.

Tetrodes in television service are operated in the class B mode to obtain reasonable efficiency while maintaining a linear transfer characteristic. Class B amplifiers, when operated in tuned circuits, provide linear performance because of the *fly-wheel effect* of the resonance circuit. This allows a single tube to be used instead of two in push–pull fashion. The bias point of the linear amplifier must be chosen so that the transfer characteristic at low modulation levels matches that at higher modulation levels. Even so, some nonlinearity is generated in the final stage, requiring differential gain correction. The plate (anode) circuit of a tetrode PA is usually built around a coaxial resonant cavity, which provides a stable and reliable tank.

UHF transmitters using a klystron in the final output stage must operate class A, the most linear but also most inefficient operating mode for a vacuum tube. The basic efficiency of a non-pulsed klystron is approximately 40%. Pulsing, which provides full available beam current only when it is needed (during peak of sync), can improve device efficiency by as much as 25%, depending on the type of pulsing used.

Two types of klystrons are presently in service: integral cavity and external cavity devices. The basic theory of operation is identical for each tube; however, the mechanical approach is radically different. In the integral cavity klystron, the cavities are built into the klystron to form a single unit. In the external cavity klystron, the cavities are outside the vacuum envelope and bolted around the tube when the klystron is installed in the transmitter.

A number of factors come into play in a discussion of the relative merits of integral-vs.-external cavity designs. Primary considerations include operating efficiency, purchase price, and life expectancy.

The PA stage includes a number of sensors that provide input to supervisory and control circuits. Because of the power levels present in the PA stage, sophisticated fault-detection circuits are required to prevent damage to components in the event of a problem either external to or inside the transmitter. An RF sample, obtained from a directional coupler installed at the output of the transmitter, is used to provide automatic power-level control.

The transmitter system discussed thus far assumes separate visual and aural PA stages. This configuration is normally used for high-power transmitters. Low-power designs often use a combined mode in which the aural and visual signals are added prior to the PA. This approach offers a simplified system, but at the cost of additional pre-correction of the input video signal.

PA stages are often configured so that the circuitries of the visual and aural amplifiers are identical. While this represents a good deal of "overkill" insofar as the aural PA is concerned, it provides backup protection in the event of a visual PA failure. The aural PA can then be reconfigured to amplify both the aural and visual signals at reduced power.

The aural output stage of a television transmitter is similar in basic design to an FM broadcast transmitter. Tetrode output devices generally operate class C, providing good efficiency. Klystron-based aural PAs are used in UHF transmitters.

Coupling/Filtering System

The output of the aural and visual power amplifiers must be combined and filtered to provide a signal that is electrically ready to be applied to the antenna system. The primary elements of the coupling and filtering system of a TV transmitter are:

- Color notch filter
- Aural and visual harmonic filters
- Diplexer

In a low-power transmitter (below 5 kW), this hardware may be included within the transmitter cabinet itself. Normally, however, it is located external to the transmitter.

Color Notch Filter

The color notch filter is used to attenuate the color subcarrier lower sideband to the –42 dB requirements of the FCC. The color notch filter is placed across the transmitter output feedline. The filter consists of a coax or waveguide stub tuned to 3.58 MHz below the picture carrier. The Q of the filter is high enough so that energy in the vestigial sideband is not materially affected, while still providing high attenuation at 3.58 MHz.

Harmonic Filters

Harmonic filters are used to attenuate out-of-band radiation of the aural and visual signals to ensure compliance with FCC requirements. Filter designs vary, depending on the manufacturer; however, most are of coaxial construction utilizing components housed within a prepackaged assembly. Stub filters are also used, typically adjusted to provide maximum attenuation at the second harmonic of the operating frequency of the visual carrier and the aural carrier.

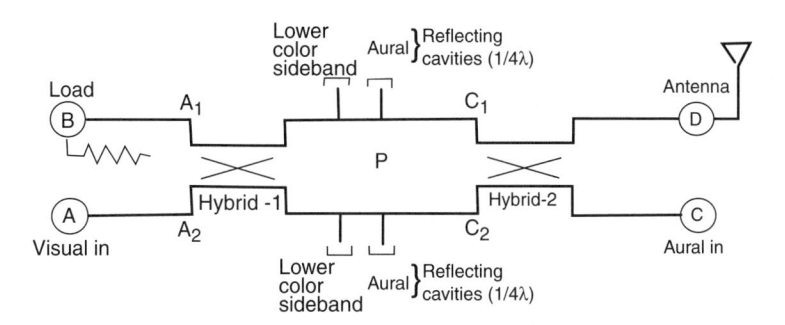

FIGURE 1.14 Functional diagram of a notch diplexer, used to combine the aural and visual outputs of a television transmitter for application to the antenna.

Diplexer/Combiner

The filtered visual and aural outputs are fed to a diplexer where the two signals are combined to feed the antenna (see Fig. 1.14). For installations that require dual-antenna feedlines, a hybrid combiner with quadrature-phased outputs is used. Depending on the design and operating power, the color notch filter, aural and visual harmonic filters, and diplexer can be combined into a single mechanical unit.

A hybrid combiner serves as the building block of the notch diplexer, which combines the aural and visual RF signals to feed a single-line antenna system and provide a constant impedance load to each section of the transmitter.

The notch diplexer consists of two hybrid combiners and two sets of reject cavities. The system is configured so that all of the energy from the visual transmitter passes to the antenna (port D), and all of the energy from the aural transmitter passes to the antenna. The phase relationships are arranged so that the input signals cancel at the resistive load (port B). Because of the paths taken by the aural and visual signals through the notch diplexer, the amplitude and phase characteristics of each input do not change from the input ports (port A for the visual and port C for the aural) and the antenna (port D), thus preserving signal purity.

1.3 Nonbroadcast Applications

Radio and television broadcasting are the most obvious applications of RF technology. In total numbers of installations, however, nonbroadcast uses for RF far outdistance radio and TV stations. Applications range from microwave communications to induction heating. Power levels range from a few tens of watts to a million watts or more. The areas of nonbroadcast RF technology covered in this section include:

- Satellite transmission
- Radar
- Electronic navigation
- Induction heating

Satellite Transmission

Commercial satellite communication began on July 10, 1962, when television pictures were first beamed across the Atlantic Ocean through the Telstar 1 satellite. Three years later, the INTELSAT system of *geostationary* relay satellites saw its initial craft, Early Bird 1, launched into a rapidly growing communications industry. In the same year, the U.S.S.R. inaugurated the Molnya series of satellites traveling in an elliptical orbit to better meet the needs of that nation. The Molnya satellites were placed in an orbit inclined about 64° relative to the equator, with an orbital period half that of the Earth.

All commercial satellites in use today operate in a geostationary orbit. A geostationary satellite is one that maintains a fixed position in space relative to Earth because of its altitude, roughly 22,300 miles above the Earth. Two primary frequency bands are used: the *C-band* (4–6 GHz) and the *Ku-band* (11–14 GHz). Any satellite relay system involves three basic sections:

- An *uplink* transmitting station, which beams signals toward the satellite in its equatorial geostationary orbit
- The satellite (the space segment of the system), which receives, amplifies, and retransmits the signals back to Earth
- The *downlink* receiving station, which completes the relay path

Because of the frequencies involved, satellite communication is designated as a microwave radio service. As such, certain requirements are placed on the system. Like terrestrial microwave, the path between transmitter and receiver must be line-of-sight. Meteorological conditions, such as rain and fog, result in detrimental attenuation of the signal. Arrangements must be made to shield satellite receive antennas from terrestrial interference. Because received signal strength is based on the inverse square law, highly directional transmit and receive parabolic antennas are used, which in turn requires a high degree of aiming accuracy. To counteract the effects of galactic and thermal noise sources on low-level signals, amplifiers are designed for exceptionally low noise characteristics. Figure 1.15 shows the primary elements of a satellite relay system.

Satellite Communications

The first satellites launched for INTELSAT and other users contained only one or two radio relay units (*transponders*). Pressure for increased satellite link services has driven engineers to develop more economical systems with multiple transponder designs. Generally, C-band satellites placed in orbit now typically have 24 transponders, each with 36-MHz bandwidths. Ku-band systems often use fewer transponders with wider bandwidths.

Users of satellite communication links are assigned to transponders generally on a lease basis, although it may be possible to purchase a transponder. Assignments usually leave one or more spare transponders aboard each craft, allowing for standby capabilities in the event a transponder should fail.

By assigning transponders to users, the satellite operator simplifies the design of uplink and downlink facilities. The Earth station controller can be programmed according to the transponder of interest. For example, a corporate video facility may need to access four or five different transponders from one satellite. To do so, the operator needs only to enter the transponder number (or carrier frequency) of interest. The receiver handles retuning and automatic switching of signals from a dual-polarity feedhorn on the antenna.

Each transponder has a fixed center frequency and a specific signal polarization. For example, according to one frequency plan, all odd-numbered transponders use horizontal polarization while the even-numbered ones use vertical polarization. Excessive deviation from the center carrier frequency by one signal does not cause interference between two transponders and signals because of the isolation provided by cross-polarization. This concept is extended to satellites in adjacent parking spaces in *geosynchronous* orbit. Center frequencies for transponders on adjacent satellites are offset in frequency from those on the first craft. In addition, an angular offset of polarization is employed. The even and odd transponder assignments are still offset by 90° from one another. As spacing is decreased between satellites, the polarization offset must be increased to reduce the potential for interference.

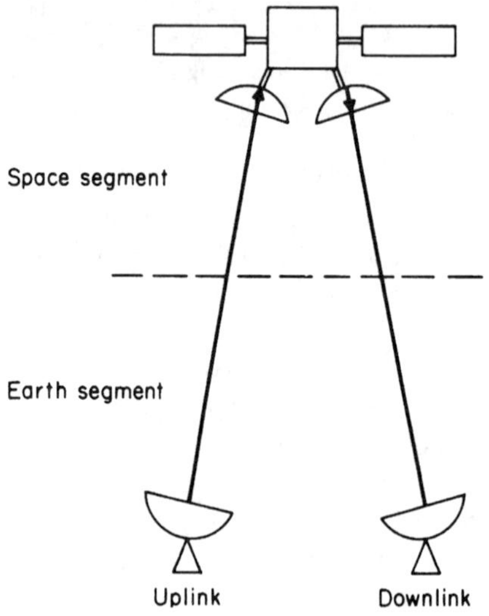

FIGURE 1.15 A satellite communications link consists of an uplink, the satellite (as the space segment), and a downlink.

Satellite Uplink

The ground-based transmitting equipment of the satellite system consists of three sections: baseband, intermediate frequency (IF), and radio frequency (RF).

The baseband section interfaces various incoming signals with the transmission format of the satellite being used. Signals provided to the baseband section may already be in a modulated form with modulation characteristics (digital, analog, or some other format) determined by the terrestrial media that brings the signals to the uplink site. Depending on the nature of the incoming signal (voice, data, or video), some degree of processing will be applied. In many cases, multiple incoming signals will be combined into a single composite uplink signal through multiplexing.

When the incoming signals are in the correct format for transmission to the satellite, they are applied to an FM modulator, which converts the composite signal upward to a 70-MHz intermediate frequency. The use of an IF section has several advantages:

- A direct conversion between baseband and the output frequency presents difficulties in maintaining frequency stability of the output signal.
- Any mixing or modulation step has the potential of introducing unwanted by-products. Filtering at the IF may be used to remove spurious signals resulting from the mixing process.
- Many terrestrial microwave systems include a 70 MHz IF section. If a signal is brought into the uplink site by terrestrial microwave, it becomes a simple matter to connect the signal directly into the IF section of the uplink system.

From the 70-MHz IF, the signal is converted upward again, this time to the output frequency of 6 GHz (for C-band) or 14 GHz (for Ku-band) before application to a high-power amplifier (HPA). Conventional Earth station transmitters operate over a wide power range, from a few tens of watts to 12 kW or more. Transmitters designed for deep space research can operate at up to 400 kW.

Several amplifying devices are used in HPA designs, depending on the power output and frequency requirements. For the highest power level in C- or Ku-band, klystrons are employed. Devices are available with pulsed outputs ranging from 500 W to 5 kW, and a bandwidth capability of 40-MHz. This means that a separate klystron is required for each 40 MHz wide signal to be beamed upward to a transponder.

The *traveling wave tube* (TWT) is another type of vacuum power device used for HPA transmitters. While similar in some areas of operation to klystrons, the TWT is capable of amplifying a band of signals at least ten times wider than the klystron. Thus, one TWT system can be used to amplify the signals sent to several transponders on the satellite. With output powers from 100 W to 2.5 kW, the bandwidth capability of the TWT offsets its much higher price than the klystron in some applications.

Solid-state amplifiers based on MOSFET technology can be used for both C- and Ku-band uplink HPA systems. The power capabilities of solid-state units are limited, 5 to 50 W or so for C-band and 1 to 6 W for Ku-band. Such systems, however, offer wideband performance and good reliability.

Uplink Antennas

The output of the HPA, when applied to a parabolic reflector antenna, experiences a high degree of gain when referenced to an ideal isotropic antenna (dBi). For example, large reflector antennas approximately 10 m in diameter offer gains as high as 55 dB, increasing the output of a 3-kW klystron or TWT amplifier to an effective radiated power of 57 to 86 dBW. Smaller reflector sizes (6 to 8 m) can also be used, with the observation of certain restrictions in regard to interference with other satellites and other services. Not surprisingly, smaller antennas provide lower gain. For a 30-m reflector, such as those used for international satellite communications, approximately 58 dB gain can be achieved. Several variations of parabolic antenna designs are used for satellite communications services, including the following (see Fig. 1.16):

- *Prime focus, single parabolic reflector:* places the source of the signal to be transmitted in front of the reflector precisely at the focal point of the parabola. Large antennas of this type commonly employ a feedhorn supported with a tripod of struts. Because the struts, the waveguide to the feedhorn, and the horn assembly itself are located directly within the transmitted beam, every effort is made to design these components with as little bulk as possible, yet physically strong enough to withstand adverse weather conditions.

- *Offset reflector:* removes the feedhorn and its support from the radiated beam. Although the reflector maintains the shape of a section of a parabola, the closed end of the curve is not included. The feedhorn, while still located at the focal point of the curve, points at an angle from the vertex of the parabola shape.
- *Double reflector:* the primary reflector is parabolic in shape while the subreflector surface, mounted in front of the focal point of the parabola, is hyperbolic in shape. One focus of the hyperbolic reflector is located at the parabolic focal point, while the second focal point of the subreflector defines the position for the feedhorn signal source. Signals reflected from the hyperbolic subreflector are spread across the parabolic prime reflector, which then directs them as a parallel beam toward the satellite. This two-reflector antenna provides several advantages over a single-reflector type: (1) the overall front-to-back dimension of the two-reflector system is shorter, which simplifies mounting and decreases wind-loading; (2) placement of the subreflector closer to the main reflector generates less spillover signal because energy is not directed as closely to the edge of the main reflector; and (3) the accuracy of the reflector surfaces is not as stringent as with a single-reflector type of structure.

The antenna used for signal transmission to the satellite can also be used to receive signals from the satellite. The major change needed to provide this capability is the addition of directional switching or coupling to prevent energy from the transmitter HPA from entering the receiver system. Switching devices or *circulators* use waveguide characteristics to create a signal path linking the transmitter signal to the antenna feedhorn, while simultaneously providing a received signal path from the feedhorn to the receiver input.

Signal Formats

The signal transmitted from the uplink site (or from the satellite, for that matter) is in the form of frequency modulation. Limitations are placed on uplinked signals to avoid interference problems resulting from excessive bandwidth. For example, a satellite relay channel for television use typically contains

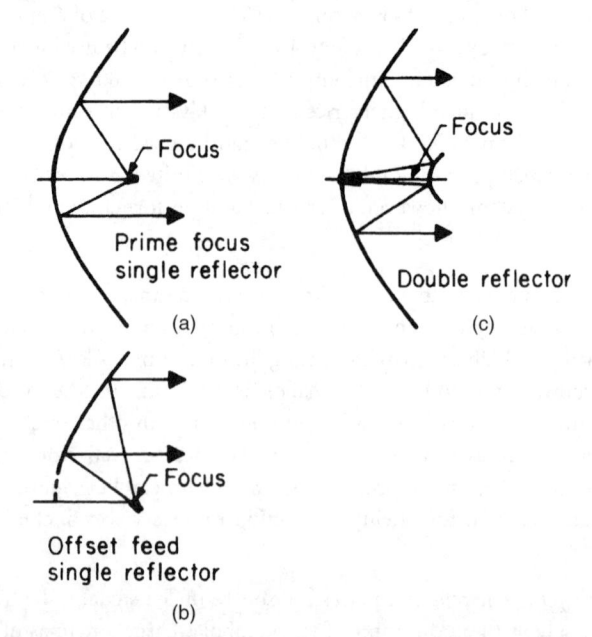

FIGURE 1.16 Satellite transmitting/receiving antennas: (a) prime focus, single reflector; (b) offset feed, single reflector; (c) double reflector.

only a single video signal and its associated audio. Audio is carried on one or more subcarriers that are stacked onto the video signal. To develop the composite signal, each audio channel is first modulated onto its subcarrier frequency. Then, each of the subcarriers and the main channel of video are applied as modulation to the uplink carrier. The maximum level of each component is controlled to avoid overmodulation.

In the case of telephone relay circuits, the same subcarrier concept is used. A number of individual voice circuits are combined into groups, which are then multiplexed to subcarriers through various digital means. The result is that thousands of telephone conversations can occur simultaneously through a single satellite.

Satellite Link

Like other relay stations, the communications spacecraft contains antennas for receiving and retransmission. From the antenna, signals pass through a low-noise amplifier before frequency conversion to the transmit band. A high-power amplifier feeds the received signal to a directional antenna, which beams the information to a predetermined area of the Earth to be served by the satellite (see Fig. 1.17).

Power to operate the electronics hardware is generated by solar cells. Inside the satellite, storage batteries, kept recharged by the solar cell arrays, carry the electronic load, particularly when the satellite is eclipsed by the Earth. Figure 1.18 shows the two most common solar cell configurations.

Power to the electronics on the craft requires protective regulation to maintain consistent signal levels. Most of the equipment operates at low voltages, but the final stage of each transponder chain ends in a high-power amplifier. The HPA of C-band satellite channels may include a traveling wave tube or a solid-state power amplifier (SSPA). Ku-band systems typically rely on TWT devices. Klystrons and TWTs require multiple voltages levels. The filaments operate at low voltages but beam focus and electron collection electrodes require voltages in the hundreds and thousands of volts. To develop such a range of voltages, the satellite power supply includes voltage converters.

From these potentials, the klystron or TWT produces output powers in the range of 8.5 to 20 W. Most systems are operated at the lower end of the range to increase reliability and life expectancy. In general, the lifetime of the spacecraft is assumed to be 7 years.

A guidance system is included to stabilize the attitude of the craft as it rotates around the earth. Small rocket engines are provided for maintaining an exact position in the assigned geostationary arc (see Fig. 1.19). This work is known as *station-keeping*.

Satellite Antennas

The antenna system for a communications satellite is really several antennas combined into a single assembly. One is for receiving signals from Earth. Another, obviously, is for transmitting those signals

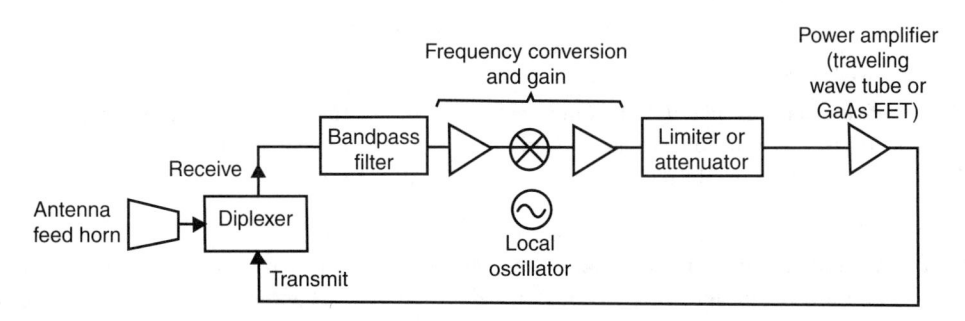

FIGURE 1.17 Block diagram of a satellite transponder channel.

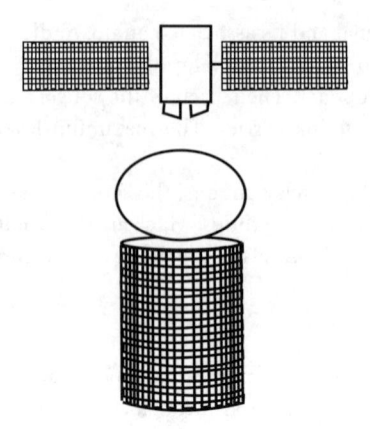

FIGURE 1.18 The two most common types of solar cell arrays used for communications satellites.

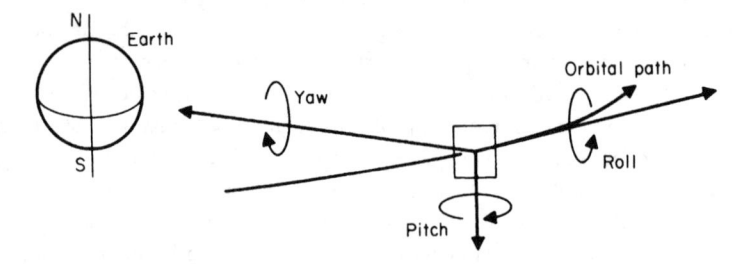

FIGURE 1.19 Attitude of the spacecraft is determined by pitch, roll, and yaw rotations around three reference axes.

back to Earth. The transmitting antenna can be made of more than one section to handle the needs of multiple signal beams. Finally, a receive-transmit beacon antenna provides communication with the ground-based satellite control station.

At the receiving end of the transponder, signals coming from the antenna are split into separate bands through a channelizing network, allowing each input signal to be directed to its own receiver, processing amplifier, and HPA. At the output, a combiner brings all channels together again into one signal to be fed to the transmitting antenna.

The approach to designing the complex antenna system for a relay satellite depends a good deal on horizontal and vertical polarization of the signals as a means to keep incoming and outgoing information separated. Multilayer, dichroic reflectors that are sensitive to the polarizations can be used for such purposes. Also, multiple feedhorns may be needed to develop one or more beams back to Earth. Antennas for different requirements may combine several antenna designs, but nearly all are based on the parabolic reflector. The parabolic design offers a number of unique properties. First, rays received by such a structure that are parallel to the feed axis are reflected and converged at the focus. Second, rays emitted from the focal point are reflected and emerge parallel to the feed axis. Special cases may involve some use of spherical and elliptical reflector shapes, but the parabolic is of most importance.

Satellite Downlink

Satellite receiving stations, like uplink equipment, perform the function of interfacing ground-based equipment to satellite transponders. Earth stations consist of a receiving antenna, *low noise amplifier* (LNA), 4-GHz (C-band) or 11-GHz (Ku-band) tuner, 70-MHz IF section, and baseband output stage.

Downlink Antennas

Antenna type and size for any application are determined by the mode of transmission, band of operation, location of the receiving station, typical weather in the receiving station locale, and the

Applications of RF Technology

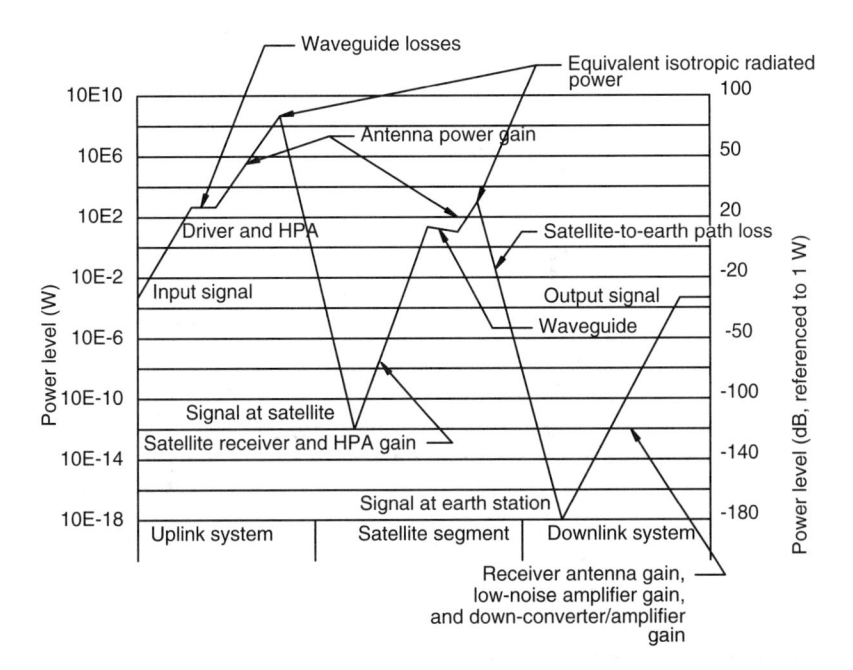

FIGURE 1.20 The power levels in transmission of an analog TV signal via satellite.

required quality of the output signal. Digital transmissions allow a smaller main reflector to be used because the decoding equipment is usually designed to provide error correction. The data stream periodically includes information to check the accuracy of the data and, if errors are found, to make corrections. Sophisticated error concealment techniques make it possible to correct errors to a certain extent. Greater emphasis is placed on error correction for applications involving financial transactions or life-critical data, such as might be involved with a manned space flight. For entertainment programming, such as TV broadcasts and telephone audio, absolute correction is less critical and gives way primarily to concealment techniques.

Receiving antennas for commercial applications, such as radio/TV networks, cable TV networks, and special services or teleconferencing centers, generally fall into the 7- to 10-m range for C-band operation. Ku-band units can be smaller. Antennas for consumer and business use may be even more compact, depending on the type of signal being received and the quality of the signal to be provided by the downlink. The nature of the application also helps determine if the antenna will be strictly parabolic, or if one of the spherical types, generally designed for consumer use, will be sufficient.

In general, the gain and directivity of a large reflector are greater than for a small reflector. The size of the reflector required depends on the level of signal that can be reliably received at a specific location under the worst-possible conditions. Gain must be adequate to bring the RF signal from the satellite to a level that is acceptable to the electronics equipment. The output signal must maintain a signal-to-noise ratio sufficiently high that the receiver electronics can recover the desired signal without significant degradation from noise.

It is instructive to consider the power budget of the downlink, that is, a calculation of positive and negative factors determining signal level. Figure 1.20 shows an analysis of both the uplink and downlink functions, as well as typical values of gain or loss. From this figure the need for receiving equipment with exceptional low noise performance becomes more obvious. One of the most critical parts of the receiver is the low noise amplifier (LNA) or *low noise conversion* unit (LNC), which is the first component following the antenna to process the signal. Such devices are rated by their *noise temperature*, usually a number around 211 K. The cost of an LNA or LNC increases significantly as the temperature figure goes down.

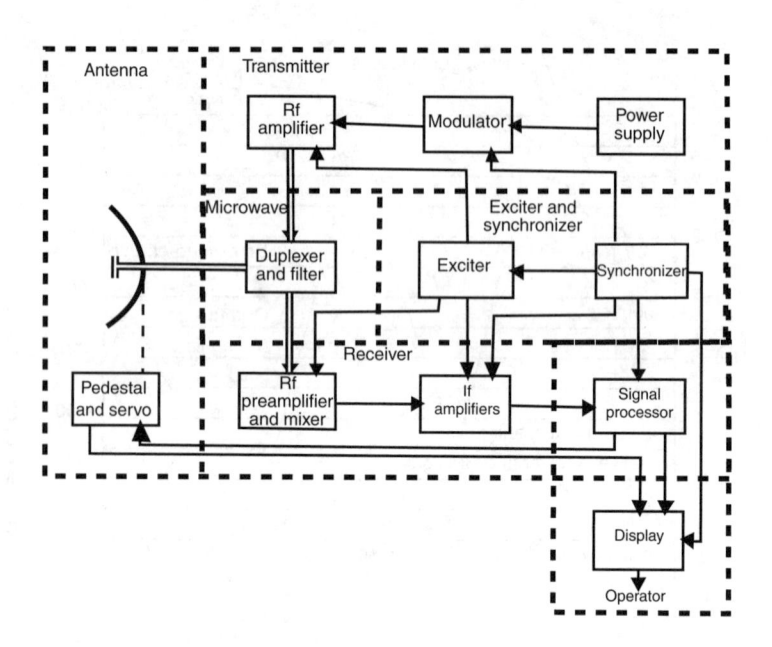

FIGURE 1.21 Simplified block diagram of a pulsed radar system.

Radar

The word "radar" is an acronym for *radio detection and ranging*. The name accurately spells out the basic function of a radar system. The measurement of target angles is an additional function of most radar equipment. Doppler velocity can also be measured as an important parameter. A block diagram of a typical pulsed radar system is shown in Fig. 1.21. Any system can be divided into six basic subsections:

- *Exciter and synchronizer:* controls the sequence of transmission and reception
- *Transmitter:* generates a high-power RF pulse of specified frequency and shape
- *Microwave network:* couples the transmitter and receiver sections to the antenna
- *Antenna system:* consists of a radiating/receiving structure mounted on a mechanically steered, servo-driven pedestal. A *stationary array*, which uses electrical steering of the antenna system, can be used in place of the mechanical system shown in Fig. 1.21.
- *Receiver:* selects and amplifies the return pulse picked up by the antenna
- *Signal processor and display:* integrates the detected echo pulse, synchronizer data, and antenna pointing data for presentation to an operator

Radar technology is used for countless applications. Table 1.3 lists some of the more common uses.

Radar Parameters

Because radar systems have many diverse applications, the parameters of frequency, power, and transmission format also vary widely. There are no fundamental bounds on the operating frequencies of radar. In fact, any system that locates objects by detecting echoes scattered from a target that has been illuminated with electromagnetic energy can be considered radar. While the principles of operation are similar regardless of the frequency, the functions and circuit parameters of most radar systems can be divided into specific operating bands. Table 1.4 shows the primary bands in use today. As shown in this table, letter designations have been developed for most of the operating bands.

Radar frequencies have been selected to minimize atmospheric attenuation by rain and snow, clouds, and fog, and (at some frequencies) electrons in the air. The frequency bands must also support wide bandwidth radiation and high antenna gain.

Applications of RF Technology

TABLE 1.3 Typical Radar Applications

Air surveillance	Long-range early warning, ground-controlled intercept, acquisition for weapon system, height finding and three-dimensional radar, airport and air-route surveillance
Space and missile surveillance	Ballistic missile warning, missile acquisition, satellite surveillance
Surface-search and battlefield surveillance	Sea search and navigation, ground mapping, mortar and artillery location, airport taxiway control
Weather radar	Observation and prediction, weather avoidance (aircraft), cloud-visibility indicators
Tracking and guidance	Antiaircraft fire control, surface fire control, missile guidance, range instrumentation, satellite instrumentation, precision approach and landing
Astronomy and geodesy	Planetary observation, earth survey, ionospheric sounding

Source: Fink, D. and Christiansen, Eds., *Electronics Engineers' Handbook*, 3rd ed., McGraw-Hill, New York, 1989, Table 302. IEEE standard 521–1976.

TABLE 1.4 Radar Frequency Bands

Name	Frequency Range	Radiolocation Bands based on ITU Assignments in Region II
VHF	30–300 MHz	137–144 MHz
UHF	300–1,000 MHz	216–225 MHz
P-band[b]	230–1,000 MHz	420–450 MHz
		890–940[a] MHz
L-band	1,000–2,000 MHz	1,215–1,400 MHz
S-band	2,000–4,000 MHz	2,300–2,550 MHz
		2,700–3,700 MHz
C-band	4,000–8,000 MHz	5,255–5,925 MHz
X-band	8,000–12,500 MHz	8,500–10,700 MHz
Ku-band	12.5–18 GHz	13.4–14.4 GHz
		15.7–17.7 GHz
K-band	18–26.5 GHz	23–24.25 MHz
Ka-band	26.5–40 GHz	33.4–36 MHz
Millimeter	>40 GHz	

[a] Sometimes included in L-band.
[b] Seldom used nomenclature.

Source: Fink, D. and Christiansen, Eds., *Electronics Engineers' Handbook*, 3rd ed., McGraw-Hill, New York, 1989, Table 302. IEEE standard 521–1976.

Transmission Equipment

The operating parameters of a radar transmitter are entirely different from the other transmitters discussed thus far. Broadcast and satellite systems are characterized by medium-power, continuous-duty applications. Radar, on the other hand, is characterized by high-power, pulsed transmissions of relatively low duty cycle. The unique requirements of radar have led to the development of technology that is foreign to most communications systems.

Improvements in semiconductor design and fabrication have made solid-state radar sets practical. Systems producing kilowatts of output power at frequencies of 2 GHz and above have been installed. Higher operating powers are achieved using parallel amplification.

A typical radar system consists of the following stages:

- *Exciter:* generates the necessary RF and local-oscillator frequencies for the system
- *Power supply:* provides the needed operating voltages for the system
- *Modulator:* triggers the power output device into operation; pulse-shaping of the transmitted signal is performed in the modulator stage.
- *RF amplifier:* converts the dc input from the power supply and the trigger signals from the modulator into a high-energy, short-duration pulse

Antenna Systems

Because the applications for radar vary widely, so do antenna designs. Sizes range from less than one foot to hundreds of feet in diameter. An antenna intended for radar applications must direct radiated power from the transmitter to the azimuth and elevation coordinates of the target. It must also serve as a receiver antenna for the echo.

There are three basic antenna designs for radar:

- *Search antenna:* available in a wide variety of sizes, depending on the application. Most conventional search antennas use mechanically scanned hornfeed reflectors. The horn radiates a spherical wavefront that illuminates the antenna reflector, the shape of which is designed to focus the radiated energy at infinity. The radiated beam is usually narrow in azimuth and wide in elevation (fan shaped).
- *Tracking antenna:* intended primarily to make accurate range and angle measurements of the position of a particular target. Such antennas use circular apertures to form a pencil beam of about 1° in the X and Y coordinates. Operating frequencies in the S, C, and X bands are preferred because they allow a smaller aperture for the same transmitted beamwidth. The tracking antenna is physically smaller than most other types of comparable gain and directivity. This permits more accurate pointing at a given target.
- *Multifunction array:* an electrically steered antenna used for both airborne and ground-based applications. An array antenna consists of individual radiating elements that are driven together to produce a plane wavefront in front of the antenna aperture. Most arrays are flat, with the radiating elements spaced about 0.6 wavelength apart. Steering is accomplished by changing the phase relationships of groups of radiating elements with respect to the array.

Electronic Navigation

Navigation systems based on radio transmissions are used every day by commercial airlines, general aviation aircraft, ships, and the military. Electronic position-fixing systems are also used in surveying work. While the known speed of propagation of radio waves allows good accuracies to be obtained in free space, multipath effects along the surface of the Earth are the primary enemies of practical airborne and shipborne systems. A number of different navigation tools, therefore, have evolved to obtain the needed accuracy and coverage area.

Electronic navigation systems can be divided into three primary categories:

- *Long-range systems:* useful for distances of greater than 200 mi, are primarily used for transoceanic navigation.
- *Medium-range systems:* useful for distances of 20 to 200 mi, are mainly employed in coastal areas and above populated land masses.
- *Short-range systems:* useful for distances of less than 20 mi, are used for approach, docking, or landing applications.

Electronic navigation systems can be further divided into *cooperative* or *self-contained*. Cooperative systems depend on transmission, one- or two-way, between one or more ground stations and the vehicle. Such systems are capable of providing the vehicle with a location fix, independent of its previous position. Self-contained systems are entirely contained in the vehicle and may be radiating or nonradiating. In general, they measure the distance traveled and have errors that increase with time or distance. The type of system chosen for a particular application depends on a number of considerations, including how often the location of the vehicle must be determined and the accuracy required.

Because aircraft and ships may travel to any part of the world, many electronic navigation systems have received standardization on an international scale.

Virtually all radio frequencies have been used in navigation at one point or another. Systems operating at low frequencies typically use high-power transmitters with massive antenna systems. With few

Applications of RF Technology

exceptions, frequencies and technologies have been chosen to avoid dependence on ionospheric reflection. Such reflections can be valuable in communications systems but are usually unpredictable.

Direction Finding

Direction finding (DF) is the oldest and most widely used navigation aid. The position of a transmitter can be determined by comparing the arrival coordinates of the radiated energy at two or more known points. Conversely, the position of a receiving point can be determined by comparing the direction coordinates from two or more known transmitters.

The weakness of this system is its susceptibility to site errors. The chief weapon against error is the use of a large DF antenna aperture. In many cases, a multiplicity of antennas, suitably combined, can be made to favor the direct path and discriminate against indirect paths (see Fig. 1.22).

Ship navigation is a common application of DF. Coastal beacons operate in the 285- to 325-kHz band specifically for ship navigation. This low frequency provides ground-wave coverage over seawater to about 1000 mi. Operating powers vary from 100 W to 10 kW. A well-designed shipboard DF system can provide accuracies of about ±2° under typical conditions.

Two-Way Distance Ranging

By placing a transponder on a given target, automatic distance measuring can be accomplished, as illustrated in Fig. 1.23. The system receives an interrogator pulse and replies to it with another pulse, usually on a different frequency. Various codes can be employed to limit responses to a single target or class of target.

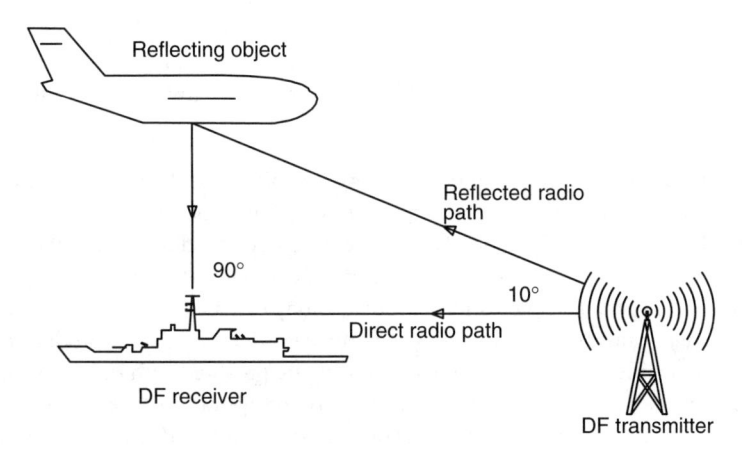

FIGURE 1.22 Direction finding error resulting from beacon reflections.

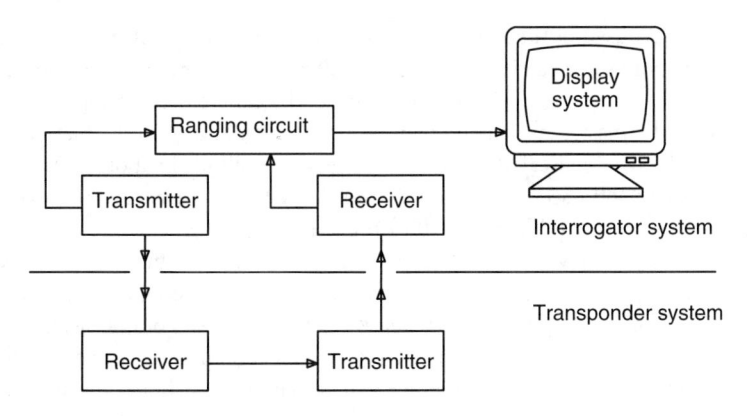

FIGURE 1.23 The concept of two-way distance ranging.

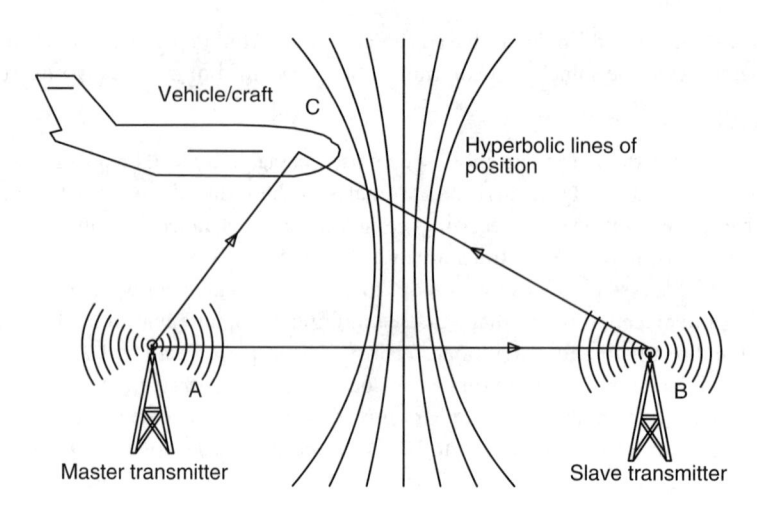

FIGURE 1.24 The concept of differential distance ranging (hyperbolic).

Distance-measuring equipment (DME) systems are one application of two-way distance ranging. An airborne interrogator transmits 1-kW pulses at a 30-Hz rate on one of 126 channels spaced 1 MHz apart. (The operating band is 1.025 to 1.150 GHz). A ground transponder responds with similar pulses on another channel 63 MHz above or below the interrogating channel.

In the airborne set, the received signal is compared with the transmitted signal, their time difference derived, and a direct digital reading of miles is displayed with a typical accuracy of ±0.2 mi.

Ground transponders are arranged to handle interrogation from up to 100 aircraft simultaneously.

Differential Distance Ranging

Two-way ranging requires a transmitter at both ends of the link. The differential distance ranging system avoids carrying a transmitter on the vehicle by placing two on the ground. One is a master and the other a slave repeating the master (see Fig. 1.24). The receiver measures the difference in the arrival of the two signals. For each time difference, there is a *hyperbolic line of position* that defines the target location. (Such systems are known as *hyperbolic* systems.) The transmissions may be either pulsed or continuous-wave using different carrier frequencies. At least two pairs of stations are needed to produce a fix.

If both stations in a differential distance ranging system are provided with stable, synchronized clocks, distance measurements can be accomplished through one-way transmissions whose elapsed time is measured with reference to the clocks. This mode of operation is referred to as *one-way distance ranging*. The concept is illustrated in Fig. 1.25.

Loran C

Hyperbolic positioning is used in the Loran C navigation system. Chains of transmitters, located along coastal waters, radiate pulses at a carrier frequency of 100 kHz. Because all stations operate on the same frequency, discrimination between chains is accomplished by different pulse-repetition frequencies. A typical chain consists of a master station and two slaves, about 600 mi from the master. Each antenna is 1300 ft high and is fed 5-MW pulses, which build up to peak amplitude in about 50 μsec and then decay to zero in approximately 100 μsec. The slow rise and decay times are necessary to keep the radiated spectrum within the assigned band limits of 90 to 100 kHz.

To obtain greater average power at the receiver without resorting to higher peak power, the master station transmits groups of nine pulses, 1 msec apart. These groups are repeated at rates ranging from 10 to 25 per second. Within each pulse, the RF phase can be varied for communications purposes.

Coverage of Loran C extends to all U.S. coastal areas, plus certain areas of the North Pacific, North Atlantic, and Mediterranean. There are currently 17 chains employing about 50 transmitters.

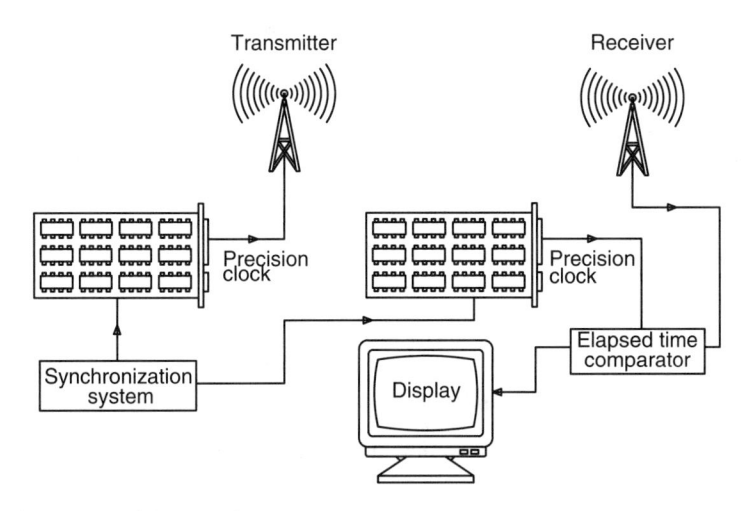

FIGURE 1.25 The concept of one-way distance ranging.

Omega

Omega is another navigation system based on the hyperbolic concept. The system is designed to provide worldwide coverage from just eight stations. Omega operates on the VLF band, from 10 to 13 kHz. At this low frequency, skywave propagation is relatively stable. Overall accuracy is on the order of 1 mi, even at ranges of 5000 mi.

There are no masters or slaves; each station transmits according to its own standard. Each station has its own operating code and transmits on one frequency at a time for a minimum of about 1 sec. The cycle is repeated every 10 sec. These slow rates are necessary because of the high Qs of the transmitting antennas. A simple Omega receiver monitors for signals at 10.2 kHz and compares emissions from one station against those of another by using an internal oscillator. The phase difference data are transferred to a map with hyperbolic coordinates.

Most Omega receivers are also able to use VLF communications stations for navigation. There are about ten such facilities operating between 16 and 24 kHz. Output powers range from 50 kW to 1 MW. Frequency stability is maintained to 1 part in 10^{12}. This allows one-way DME to be accomplished with a high degree of accuracy.

Microwave Radio

Microwave radio relay systems carry considerable long-haul telecommunications in the United States and other countries. The major common-carrier bands and their applications are shown in Table 1.5. The goal of microwave relay technology has been to increase channel capacity and lower costs. Solid-state

TABLE 1.5 Common-Carrier Microwave Frequencies Used in the U.S.

Band (GHz)	Allotted Frequencies (MHz)	Bandwidth (MHz)	Application
2	2,110–2,130	20	Limited
	2,160–2,180		
4	3,700–4,200	20	Major long-haul microwave relay band
6	5,925–6,425	500	Long and short haul
11	10,700–11,700	500	Short haul
18	17,700–19,700	1,000	Short haul, limited use
30	27,500–29,500	2,000	Short haul, experimental

devices have provided the means to accomplish this goal. Current efforts focus on the use of fiber-optic landlines for terrestrial long-haul communications systems. Satellite circuits have also been used extensively for long-haul, common-carrier applications.

Single-sideband amplitude modulation is used for microwave systems because of its spectrum efficiency. Single-sideband systems, however, require a high degree of linearity in amplifying circuits. Several techniques have been used to provide the needed channel linearity. The most popular is amplitude predistortion to cancel the inherent nonlinearity of the power amplifier.

Induction Heating

Induction heating is achieved by placing a coil carrying alternating current adjacent to a metal workpiece so that the magnetic flux produced induces a voltage in the workpiece. This causes current flow and heats the workpiece. Power sources for induction heating include:

- Motor-generator sets, which operate at low frequencies and provide outputs from 1 kW to more than 1 MW.
- Vacuum-tube oscillators, which operate at 3 kHz to several hundred MHz at power levels of 1 kW to several hundred kilowatts. Figure 1.26 shows a 20-kW induction heater using a vacuum tube as the power generating device.
- Inverters, which operate at 10 kHz or more at power levels of as much as several megawatts. Inverters utilizing thyristors (silicon controlled rectifiers) are replacing motor-generator sets in high-power applications.

Dielectric Heating

Dielectric heating is a related application for RF technology. Instead of heating a conductor, as in induction heating, dielectric heating relies on the capacitor principle to heat an insulating material. The material to be heated forms the dielectric of a capacitor, to which power is applied. The heat generated is proportional to the *loss factor* (the product of the dielectric constant and the power factor) of the material. Because the power factor of most dielectrics is low at low frequencies, the range of frequencies employed for dielectric heating is higher than for induction heating. Frequencies of a few megahertz to several gigahertz are common.

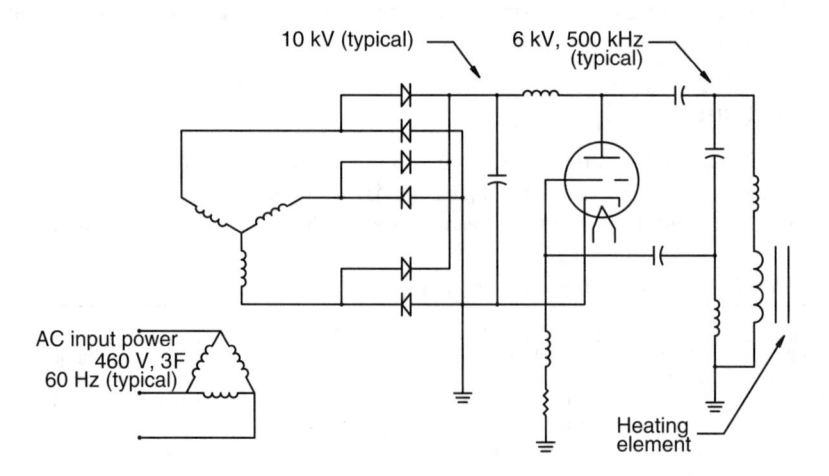

FIGURE 1.26 A 20-kW induction heater circuit.

References

1. Fink, D. and D. Christiansen, Eds., *Electronics Engineers' Handbook*, 3rd ed., McGraw-Hill, New York, 1989.

Bibliography

Jordan, Edward C., Ed., *Reference Data for Engineers: Radio, Electronics, Computer and Communications*, 7th ed., Howard W. Sams, Indianapolis, IN, 1985.

Hulick, Timothy P., Using tetrodes for high power UHF, *Proceedings of the Society of Broadcast Engineers National Convention*, Vol. 4, 52–57, 1989.

The Laboratory Staff, *The Care and Feeding of Power Grid Tubes*, Varian Eimac, San Carlos, CA, 1982.

Benson, K. B., Ed., *Television Engineering Handbook*, McGraw-Hill, New York, 1986.

Benson, K. B. and J. C. Whitaker, *Television and Audio Handbook for Technicians and Engineers*, McGraw-Hill, New York, 1989.

2
Electromagnetic Spectrum

Jerry C. Whitaker
Editor

2.1 Introduction 2-1
Operating Frequency Bands

2.1 Introduction

The usable spectrum of electromagnetic-radiation frequencies extends over a range from below 100 Hz for power distribution to 1020 Hz for the shortest x-rays. The lower frequencies are used primarily for terrestrial broadcasting and communications. The higher frequencies include visible and near-visible infrared and ultraviolet light, and x-rays.

Operating Frequency Bands

The standard frequency band designations are listed in Tables 2.1 and 2.2. Alternate and more detailed subdivision of the VHF, UHF, SHF, and EHF bands are given in Tables 2.3 and 2.4.

Low-End Spectrum Frequencies (1 to 1000 Hz)

Electric power is transmitted by wire but not by radiation at 50 and 60 Hz, and in some limited areas, at 25 Hz. Aircraft use 400-Hz power to reduce the weight of iron in generators and transformers. The restricted bandwidth that would be available for communication channels is generally inadequate for voice or data transmission, although some use has been made of communication over power distribution circuits using modulated carrier frequencies.

Low-End Radio Frequencies (1000 Hz to 100 kHz)

These low frequencies are used for very long-distance radio-telegraphic communication where extreme reliability is required and where high-power and long antennas can be erected. The primary bands of interest for radio communications are given in Table 2.5.

Medium-Frequency Radio (20 kHz to 2 MHz)

The low-frequency portion of the band is used for around-the-clock communication services over moderately long distances and where adequate power is available to overcome the high level of atmospheric noise. The upper portion is used for AM radio, although the strong and quite variable sky wave occurring during the night results in substandard quality and severe fading at times. The greatest use is for AM broadcasting, in addition to fixed and mobile service, LORAN ship and aircraft navigation, and amateur radio communication.

TABLE 2.1 Standardized Frequency Bands

Extremely low-frequency (ELF) band:	30–300 Hz	(10–1 Mm)
Voice-frequency (VF) band:	300 Hz–3 kHz	(1 Mm–100 km)
Very low-frequency (VLF) band:	3–30 kHz	(100–10 km)
Low-frequency (LF) band:	30–300 kHz	(10–1 km)
Medium-frequency (MF) band:	300 kHz–3 MHz	(1 km–100 m)
High-frequency (HF) band:	3–30 MHz	(100–10 m)
Very high-frequency (VHF) band:	30–300 MHz	(10–1 m)
Ultra high-frequency (UHF) band:	300 MHz–3 GHz	(1 m–10 cm)
Super high-frequency (SHF) band:	3–30 GHz	(1–1 cm)
Extremely high-frequency (EHF) band:	30–300 GHz	(1 cm–1 mm)

Source: Whitaker, Jerry C., Ed., *The Electronics Handbook*, CRC Press, Boca Raton, FL, 1996. Used with permission.

TABLE 2.2 Standardized Frequency Bands at 1 GHz and Above

L-band:	1–2 GHz	(30–15 cm)
S-band:	2–4 GHz	(15–7.5 cm)
C-band:	4–8 GHz	(7.5–3.75)
X-band:	8–12 GHz	(3.75–2.5 cm)
Ku-band:	12–18 GHz	(2.5–1.67 cm)
K-band:	18–26.5 GHz	(1.67–1.13 cm)
Ka-band:	26.5–40 GHz	(1.13–7.5 mm)
Q-band:	32–50 GHz	(9.38–6 mm)
U-band:	40–60 GHz	(7.5–5 mm)
V-band:	50–75 GHz	(6–4 mm)
W-band:	75–100 GHz	(4–3.33 mm)

Source: Whitaker, Jerry C., Ed., *The Electronics Handbook*, CRC Press, Boca Raton, FL, 1996. Used with permission.

TABLE 2.3 Detailed Subdivision of the UHF, SHF, and EHF Bands

L-band:	1.12–1.7 GHz	(26.8–17.6 cm)
LS-band:	1.7–2.6 GHz	(17.6–11.5 cm)
S-band:	2.6–3.95 GHz	(11.5–7.59 cm)
C(G)-band:	3.95–5.85 GHz	(7.59–5.13 cm)
XN(J, XC)-band:	5.85–8.2 GHz	(5.13–3.66 cm)
XB(H, BL)-band:	7.05–10 GHz	(4.26–3 cm)
X-band:	8.2–12.4 GHz	(3.66–2.42 cm)
Ku(P)-band:	12.4–18 GHz	(2.42–1.67 cm)
K-band:	18–26.5 GHz	(1.67–1.13 cm)
V(R, Ka)-band:	26.5–40 GHz	(1.13 cm–7.5 mm)
Q(V)-band:	33–50 GHz	(9.09–6 mm)
M(W)-band:	50–75 GHz	(6–4 mm)
E(Y)-band:	60–90 GHz	(5–3.33 mm)
F(N)-band:	90–140 GHz	(3.33–2.14 mm)
G(A)-band:	140–220 GHz	(2.14–1.36 mm)
R-band:	220–325 GHz	(1.36–0.923 mm)

Source: Whitaker, Jerry C., Ed., *The Electronics Handbook*, CRC Press, Boca Raton, FL, 1996. Used with permission.

TABLE 2.4 Subdivision of the VHF, UHF, SHF Lower Part of the EHF Band

A-band:	100–250 MHz	(3–1.2 m)
B-band:	250–500 MHz	(1.2–60 cm)
C-band:	500 MHz–1 GHz	(60–30 cm)
D-band:	1–2 GHz	(30–15 cm)
E-band:	2–3 GHz	(15–10 cm)
F-band:	3–4 GHz	(10–7.5 cm)
G-band:	4–6 GHz	(7.5–5 cm)
H-band:	6–8 GHz	(5–3.75 cm)
I-band:	8–10 GHz	(3.75–3 cm)
J-band:	10–20 GHz	(3–1.5 cm)
K-band:	20–40 GHz	(1.5 cm–7.5 mm)
L-band:	40–60 GHz	(7.5–5 mm)
M-band:	60–100 GHz	(5–3 mm)

Source: Whitaker, Jerry C., Ed., *The Electronics Handbook*, CRC Press, Boca Raton, FL, 1996. Used with permission.

TABLE 2.5 Radio Frequency Bands

Longwave broadcasting band:	150–290 kHz
AM broadcasting band:	550–1640 kHz (1.640 MHz) (107 channels, 10-kHz separation)
International broadcasting band:	3–30 MHz
Shortwave broadcasting band:	5.95–26.1 MHz (8 bands)
VHF television (channels 2–4):	54–72 MHz
VHF television (channels 5–6):	76–88 MHz
FM broadcasting band:	88–108 MHz
VHF television (channels 7–13):	174–216 MHz
UHF television (channels 14–83):	470–890 MHz

Source: Whitaker, Jerry C., Ed., *The Electronics Handbook*, CRC Press, Boca Raton, FL, 1996. Used with permission.

High-Frequency Radio (2 to 30 MHz)

This band provides reliable medium-range coverage during daylight and, when the transmission path is in total darkness, worldwide long-distance service, although the reliability and signal quality of the latter is dependent to a large degree upon ionospheric conditions and related long-term variations in sun-spot activity affecting sky-wave propagation. The primary applications include broadcasting, fixed and mobile services, telemetering, and amateur transmissions.

Very High and Ultrahigh Frequencies (30 MHz to 3 GHz)

VHF and UHF bands, because of the greater channel bandwidth possible, can provide transmission of a large amount of information, either as television detail or data communication. Furthermore, the shorter wavelengths permit the use of highly directional parabolic or multielement antennas. Reliable long-distance communication is provided using high-power tropospheric scatter techniques. The multitude of uses include, in addition to television, fixed and mobile communication services, amateur radio, radio astronomy, satellite communication, telemetering, and radar.

Microwaves (3 to 300 GHz)

At these frequencies, many transmission characteristics are similar to those used for shorter optical waves, which limit the distances covered to line-of-sight. Typical uses include television relay, satellite, radar, and wide-band information services. (See Tables 2.6 and 2.7.)

Infrared, Visible, and Ultraviolet Light

The portion of the spectrum visible to the eye covers the gamut of transmitted colors ranging from red, through yellow, green, cyan, and blue. It is bracketed by infrared on the low-frequency side and ultraviolet (UV) on the high-frequency side. Infrared signals are used in a variety of consumer and industrial equipment for remote controls and sensor circuits in security systems. The most common use of UV waves is for excitation of phosphors to produce visible illumination.

X-Rays

Medical and biological examination techniques and industrial and security inspection systems are the best-known applications of x-rays. X-rays in the higher-frequency range are classified as hard x-rays or gamma-rays. Exposure to x-rays for long periods can result in serious irreversible damage to living cells or organisms.

TABLE 2.6 Applications in the Microwave Bands

Aeronavigation:	0.96–1.215 GHz
Global positioning system (GPS) downlink:	1.2276 GHz
Military communications (COM)/radar:	1.35–1.40 GHz
Miscellaneous COM/radar:	1.40–1.71 GHz
L-band telemetry:	1.435–1.535 GHz
GPS downlink:	1.57 GHz
Military COM (troposcatter/telemetry):	1.71–1.85 GHz
Commercial COM and private line of sight (LOS):	1.85–2.20 GHz
Microwave ovens:	2.45 GHz
Commercial COM/radar:	2.45–2.69 GHz
Instructional television:	2.50–2.69 GHz
Military radar (airport surveillance);	2.70–2.90 GHz
Maritime navigation radar:	2.90–3.10 GHz
Miscellaneous radars:	2.90–3.70 GHz
Commercial C-band satellite (SAT) COM downlink:	3.70–4.20 GHz
Radar altimeter:	4.20–4.40 GHz
Military COM (troposcatter):	4.40–4.99 GHz
Commercial microwave landing system:	5.00–5.25 GHz
Miscellaneous radars:	5.25–5.925 GHz
C-band weather radar:	5.35–5.47 GHz
Commercial C-band SAT COM uplink:	5.925–6.425 GHz
Commercial COM:	6.425–7.125 GHz
Mobile television links:	6.875–7.125 GHz
Military LOS COM:	7.125–7.25 GHz
Military SAT COM downlink:	7.25–7.75 GHz
Military LOS COM:	7.75–7.9 GHz
Military SAT COM uplink:	7.90–8.40 GHz
Miscellaneous radars:	8.50–10.55 GHz
Precision approach radar:	9.00–9.20 GHz
X-band weather radar (and maritime navigation radar):	9.30–9.50 GHz
Police radar:	10.525 GHz
Commercial mobile COM [LOS and electronic news gathering (ENG)]:	10.55–10.68 GHz
Common-carrier LOS COM:	10.70–11.70 GHz
Commercial COM:	10.70–13.25 GHz
Commercial Ku-band SAT COM downlink:	11.70–12.20 GHz
Direct broadcast satellite (DBS) downlink and private LOS COM:	12.20–12.70 GHz
ENG and LOS COM:	12.75–13.25 GHz
Miscellaneous radars and SAT COM:	13.25–14.00 GHz
Commercial Ku-band SAT COM uplink:	14.00–14.50 GHz
Military COM (LOS, mobile, and tactical):	14.50–15.35 GHz
Aeronavigation:	15.40–15.70 GHz
Miscellaneous radars:	15.70–17.70 GHz
DBS uplink:	17.30–17.80 GHz
Common-carrier LOS COM:	17.70–19.70 GHz
Commercial COM (SAT COM and LOS):	17.70–20.20 GHz

TABLE 2.6 (CONTINUED) Applications in the Microwave Bands

Private LOS COM:	18.36–19.04 GHz
Military SAT COM:	20.20–21.20 GHz
Miscellaneous COM:	21.20–24.00 GHz
Police radar:	24.15 GHz
Navigation radar:	24.25–25.25 GHz
Military COM:	25.25–27.50 GHz
Commercial COM:	27.50–30.00 GHz
Military SAT COM:	30.00–31.00 GHz
Commercial COM:	31.00–31.20 GHz
Navigation radar:	31.80–33.40 GHz
Miscellaneous radars:	33.40–36.00 GHz
Military COM:	36.00–38.60 GHz
Commercial COM:	38.60–40.00 GHz

Source: Whitaker, Jerry C., Ed., *The Electronics Handbook*, CRC Press, Boca Raton, FL, 1996. Used with permission.

TABLE 2.7 Satellite Frequency Allocations

Band	Uplink	Downlink	Satellite Service
VHF		0.137–0.138	Mobile
VHF	0.3120–0.315	0.387–0.390	Mobile
L-band		1.492–1.525	Mobile
	1.610–1.6138		Mobile, radio astronomy
	1.613.8–1.6265	1.6138–1.6265	Mobile LEO
	1.6265–1.6605	1.525-1.545	Mobile
		1.575	Global Positioning System (GPS)
		1.227	GPS
S-band	1.980–2.010	2.170–2.200	MSS. Available in U.S. in 2005
	(1.980–1.990)		
	2.110–2.120	2.290–2.300	Deep-space research
		2.4835–2.500	Mobile
C-band	5.85–7.075	3.4–4.2	Fixed (FSS)
	7.250–7.300	4.5–4.8	FSS
X-band	7.9–8.4	7.25–7.75	FSS
Ku-band	12.75–13.25	10.7–12.2	FSS
	14.0–14.8	12.2–12.7	Direct Broadcast (BSS) (U.S.)
Ka-band		17.3–17.7	FSS (BSS in U.S.)
	22.55–23.55		Intersatellite
	24.45–24.75		Intersatellite
	25.25–27.5		Intersatellite
	27–31	17–21	FSS
Q-band	42.5–43.5, 47.2–50.2	37.5–40.5	FSS, MSS
	50.4–51.4		Fixed
		40.5–42.5	Broadcast Satellite
V-band	54.24–58.2		Intersatellite
	59–64		Intersatellite

Note: Allocations are not always global and may differ from region to region in all or subsets of the allocated bands.

Sources: Final Acts of the World Administrative Radio Conference (WARC-92), Malaga–Torremolinos, 1992; 1995 World Radiocommunication Conference (WRC-95). Also, see Gagliardi, R.M., *Satellite Communications*, van Nostrand Reinhold, New York, 1991.

References

Whitaker, Jerry C., Ed., *The Electronics Handbook*, CRC Press, Boca Raton, FL, 1996.

3
Amplitude Modulation

Robert Kubichek
University of Wyoming

3.1 Amplitude Modulation **3**-1
 Frequency Domain Concepts • Linear Systems • Concept of
 Amplitude Modulation • Double Sideband-Suppressed Carrier
 (DSB-SC) • Noise Effects • Superheterodyne Receivers

3.1 Amplitude Modulation

There are two basic types of communication systems, *baseband* systems and *passband* systems. In baseband systems, the signal is transmitted without modifying the frequency content. A simple intercom is an example of this approach. Here, a microphone senses the input or **message signal**, and injects the resulting signal $m(t)$ into a cable or *channel*. At the receiver, the signal is filtered to remove noise, amplified and reproduced into sound using a speaker. In passband communication systems, the message signal is *modulated* by translating its spectrum to a new frequency location called the *carrier frequency*, f_c. **Modulation** offers numerous advantages over baseband communication including:

- Maximized efficiency: Signals can be modulated into regions of the spectrum where there is lower noise and interference, or better propagation characteristics.
- Frequency-domain multiplexing (FDM): Multiple signals can be modulated into nonoverlapping frequency bands and transmitted simultaneously over the same channel. In commercial broadcasting, for example, this allows AM and FM signals to be multiplexed onto the same radio frequency channel.
- Physical considerations: The physical size of antennas and other electronic components tend to decrease with increasing frequency. This makes it feasible to build smaller radio receivers, transmitters, and antennas when higher frequencies are used.

Frequency Domain Concepts

To understand the basic concepts of modulation, we first review time- and frequency-domain representations of signals. The information present in $m(t)$ can be completely specified by a complex function of speech amplitude vs. frequency, $M(f)$, obtained using the *Fourier transform*. Since $M(f)$ contains all of the information in $m(t)$, it is possible to go back and forth between $m(t)$ and $M(f)$ using the forward and inverse Fourier transforms.

Forward transform:

$$M(f) = \int_{-\infty}^{\infty} m(t) e^{-j2\pi ft} dt \qquad (3.1)$$

Inverse transform:

$$m(t) = \int_{-\infty}^{\infty} M(f)e^{j2\pi ft} df \qquad (3.2)$$

Although we usually think of positive-valued frequencies, Eq. (3.1) shows that negative frequency values are also mathematically valid. We can get a better feel for this idea by considering the signal $\sin 2\pi f_c t$, which is a pure tone with frequency f_c Hz. The relationship $\sin 2\pi f_c t = -\sin 2\pi (-f_c)t$ shows that a pure tone at $-f_c$ Hz has the same magnitude as the positive frequency pure tone, but is different in phase by 180°.

Modulation Theorem

The *modulation theorem* states that when any signal is multiplied by a sine-wave signal of frequency f_c, the resulting signal has a spectrum similar to the original, but translated out to frequency $\pm f_c$. That is, consider a signal $x(t)$ produced by multiplying $m(t)$ against a tone signal,

$$x(t) = m(t)\cos(2\pi f_c t) \qquad (3.3)$$

The frequency-domain result is

$$X(f) = 0.5\{M(f-f_c) + M(f+f_c)\} \qquad (3.4)$$

where $M(f-f_c)$ and $M(f+f_c)$ represent the entire message spectrum translated to two new locations, the carrier frequency f_c and the negative carrier frequency $-f_c$, respectively. This multiplication process is called *mixing* or *heterodyning*. Figure 3.1 shows an example of the modulation theorem where $M(f)$ represents the baseband spectrum and $X(f)$ is the modulated spectrum. A second example in Fig. 3.2 shows the passband signal $x(t)$ from Fig. 3.1 being modulated to produce components at $2f_c$, 0, and $-2f_c$ Hz. This approach can be used in a receiver to *demodulate* the passband signal and reproduce the original baseband signal. This is accomplished simply by removing the spectral components at $\pm 2f_c$ using a filter.

The spectrum $M(f)$ is a complex function that can be represented either by real and imaginary or by magnitude and phase components $|M(f)|$ and $\phi(f)$, respectively. When $m(t)$ is real valued, the magnitude spectrum is an even function ($|M(f)| = |M(-f)|$) and phase is an odd function ($\phi(f) = -\phi(-f)$) of frequency. Most basic communication concepts can be illustrated using only the magnitude spectrum $|M(f)|$; this will be the approach taken in this chapter.

Bandwidth: Baseband vs. Passband

Bandwidth is defined as the range of positive frequencies occupied by a signal. Thus, for the baseband signal shown in Fig. 3.1(b), the bandwidth equals the highest frequency present, B Hz. By comparison, the modulated signal shown in Fig. 3.1(c) has a bandwidth of $2B$ Hz, which occupies double the bandwidth of the baseband signal and, therefore, represents poor spectral efficiency.

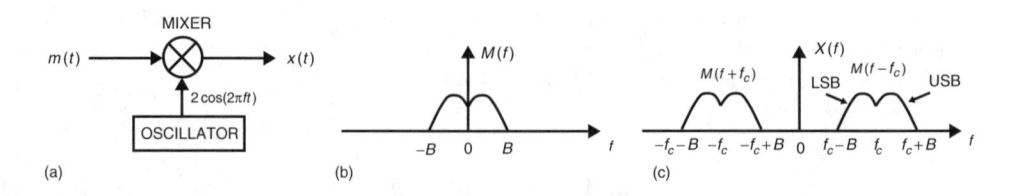

FIGURE 3.1 Example of the modulation theorem: (a) the modulator, (b) the input message spectrum, and (c) the modulated spectrum composed of upper and lower sidebands (USB and LSB).

Amplitude Modulation 3-3

Linear Systems

Low-pass and *bandpass* filters (denoted LPF and BPF, respectively) are fundamental components in amplitude modulation systems. These filters, along with the communication channel itself, are often modeled as *linear systems*, which implies that system output amplitude is strictly proportional to the system input amplitude. Figure 3.3 shows a linear system with the input and output amplitude spectra given by $X(f)$ and $Y(f)$, respectively. The proportionality between input and output at any given frequency is described by the system *transfer function* $H(f)$: $Y(f) = X(f) H(f)$. Figure 3.3 also shows transfer functions for both low-pass and bandpass filters.

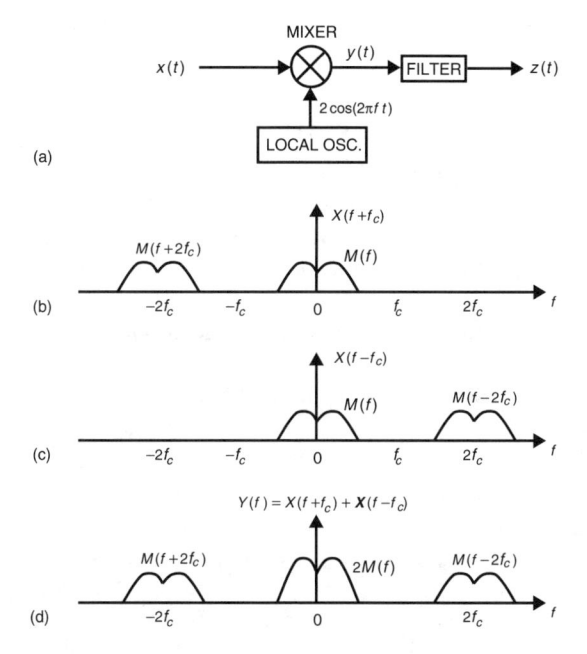

FIGURE 3.2 Example of modulation theorem used in demodulation: (a) the demodulator, (b) and (c) the two components of the output signal, and (d) the combined demodulator output spectrum before filtering.

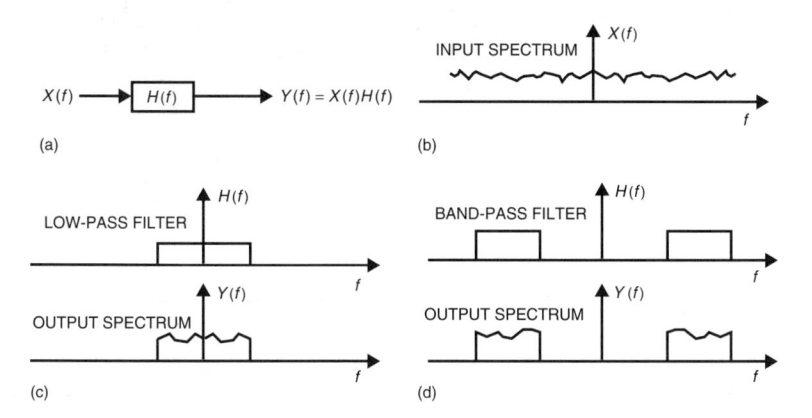

FIGURE 3.3 Example of low-pass and bandpass linear filters: (a) the linear system, (b) the input signal spectrum, (c) low-pass filter and output spectrum, and (d) a bandpass filter and output spectrum.

Concept of Amplitude Modulation

We can write Eq. (3.1) in more general terms,

$$w(t) = A(t) \cos[2\pi f_c t + \theta(t)] \quad (3.5)$$

where both the amplitude and phase are allowed to vary as functions of time. If the message signal $m(t)$ affects only the amplitude $A(t)$, the resulting signal is termed **amplitude modulation**. This generic category includes the well-known commercial broadcast amplitude modulation (AM) system, as well as **double sideband (DSB), single sideband (SSB),** and **vestigial sideband (VSB)** transmission. When the message signal is transmitted by modulating the phase component $\theta(t)$, the result is called *angle modulation*. This includes *frequency modulation* (FM) and *phase modulation* (PM) techniques.

Double Sideband-Suppressed Carrier (DSB-SC)

A transmission system directly implementing Eq. (3.3) is shown in Fig. 3.1(a) and is called double-sideband (DSB) modulation. The two lobes observed in the magnitude spectrum in Fig. 3.1(c) are called the *upper sideband* (USB) and *lower sideband* (LSB). The USB and LSB are mirror images of each other (but with opposite phases) and each contains sufficient information to reconstruct the message signal $m(t)$.

A transmitter requires an *oscillator* to generate the **carrier** signal $c(t) = \cos(2\pi f_c t)$ and a *mixer* to multiply $m(t)$ and $c(t)$.

An alternative modulation system is shown in Fig. 3.4. This technique uses an electronic switch rather than a carrier and a mixer. To see why this works, consider that switching $m(t)$ on and off is the same as multiplying $m(t)$ with a square wave $s(t)$ having frequency f_c; $x(t) = m(t)s(t)$. Using Fourier series, $s(t)$ can be written as a trigonometric expansion:

$$s(t) = 0.5 + a_1 \cos 2\pi f_0 t + a_3 \cos 2\pi 3 f_0 t + a_5 \cos 2\pi 5 f_0 t + \cdots$$

where a_i represent the ith fourier series coefficient. The switching output is then given by

$$x(t) = 0.5 m(t) + a_1 m(t) \cos 2\pi f_0 t + a_3 m(t) \cos 2\pi 3 f_0 t + a_5 m(t) \cos 2\pi 5 f_0 t + \cdots$$

Applying the modulation theorem to each term, we see that the expansion represents an infinite sum of DSB signals with passbands centered at 0, $\pm f_0$, $\pm 3f_0$, $\pm 5f_0$, etc. A bandpass filter selects the desired DSB signal at $\pm f_c$ and attenuates the remaining undesired harmonics as shown in Fig. 3.5.

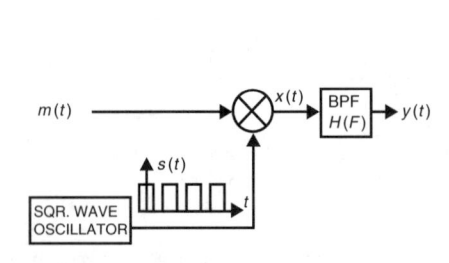

FIGURE 3.4 DSB-SC modulation using a switching circuit.

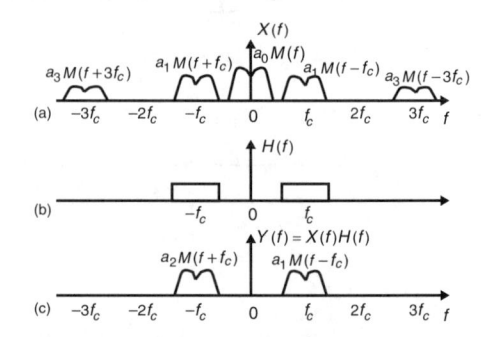

FIGURE 3.5 The switched signal spectrum, $X(f)$, can be filtered to create a DSB signal: (a) the switched signal spectrum, (b) the bandpass filter response, and (c) the DSB output spectrum.

Amplitude Modulation 3-5

Figure 3.2 shows a simple design for a receiver (or **detector** or *demodulator*). The receiver requires a local oscillator to generate 2 cos($2\pi f_c t$), a mixer, and a low-pass filter at the mixer output to remove the components at $\pm 2f_c$. Receiver operation is described in Table 3.1 in both time- and frequency-domain equations.

Ideally, the receiver oscillator, or *local* oscillator, produces a carrier signal in phase with the transmitter oscillator, that is, $C_{trans}(t) = C_{rcvr}(t) = \cos(2\pi f_c t)$; this is called *coherent detection* or *product detection*. Keeping both oscillators perfectly in phase, however, can be quite difficult. To find out what happens if the transmitter carrier $C_{trans}(t)$ and receiver carrier $C_{rcvr}(t)$ have different phases, we can write $C_{trans}(t) = \cos(2\pi f_c t)$ and $C_{rcvr}(t) = \cos(2\pi f_c t + \varphi)$, where φ represents phase error. An analysis similar to Table 3.1 gives the result $z(t) = m(t) \cos(\varphi)$. This shows that any nonzero phase errors will decrease the receiver output amplitude by an amount $\cos(\varphi)$, the worst case occurring when $\varphi = \pi/2$, which produces zero output.

To minimize phase error effects, DSB systems often employ *phase-locked loops* in the receivers to lock the local oscillator into phase synchony with the received signal. In another approach, a pilot signal is transmitted simultaneously with the DSB signal but on a different frequency. The pilot is typically a sine wave in phase with $C_{trans}(t)$ and chosen to be harmonically related to f_c. It is used at the receiver to generate an in-phase local oscillator signal.

Double Sideband Plus Carrier (DSB+C). Figure 3.6 shows a DSB system similar to that in Fig. 3.1 except for the constant DC level added to $m(t)$ prior to the mixer stage. The mixer output signal is, thus,

$$x(t) = [A + m(t)] \cos(2\pi f_c t) = A \cos(2\pi f_c t) + m(t) \cos(2\pi f_c t) \qquad (3.6)$$

This consists of a simple DSB signal (the second term) plus a carrier of amplitude A (the first term). The carrier concentrates all of its energy at frequency f_c and shows up in the magnitude spectrum as a spike. This is the classic AM transmission scheme and is also known as DSB plus carrier (DSB+C). The system described in the last section is often called DSB-suppressed carrier or DSB-SC to differentiate it from DSB+C.

Figure 3.7 illustrates the DSB-SC and DSB+C waveforms. In DSB+C, the message signal is contained in the positive (or negative) *envelope* and the carrier is confined within the envelope boundary. No simple relationship between $m(t)$ and the envelope is evident in DSB-SC waveform.

TABLE 3.1 Demodulation Equations

Description	Time Domain	Frequency Domain
Received signal	$x(t) = m(t) \cos(2\pi f_c t)$	$X(f) = 0.5\{M(f+f_c) + M(f-f_c)\}$
Mixer output	$y(t) = 2x(t) \cos(2\pi f_c t)$	$Y(f) = 2*0.5\{X(f+fc) + X(f-f_c)\}$
	$= 2m(t) \cos(2\pi f_c t) \cos(2\pi f_c t)$	$= 0.5M(f+2f_c) + 0.5M(f+f_c-f_c)$
		$+0.5M(f-f_c+f_c) + 0.5M(f-2f_c)$
Some algebra: (use the trig identity: $\cos^2 A = 0.5 + 0.5 \cos(2A)$	$y(t) = m(t) + m(t) \cos(4\pi f_c t)$	$Y(f) = 0.5M(f+2f_c) + M(f) + 0.5M(f-2f_c)$
Low-pass filter output	$z(t) = m(t)$	$Z(f) = M(f)$

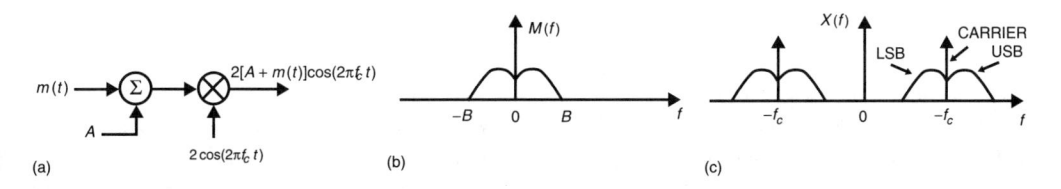

FIGURE 3.6 Insertion of carrier producing DSB+C signal showing (a) DSB+C modulator, (b) the message spectrum, and (c) the DSB+C spectrum.

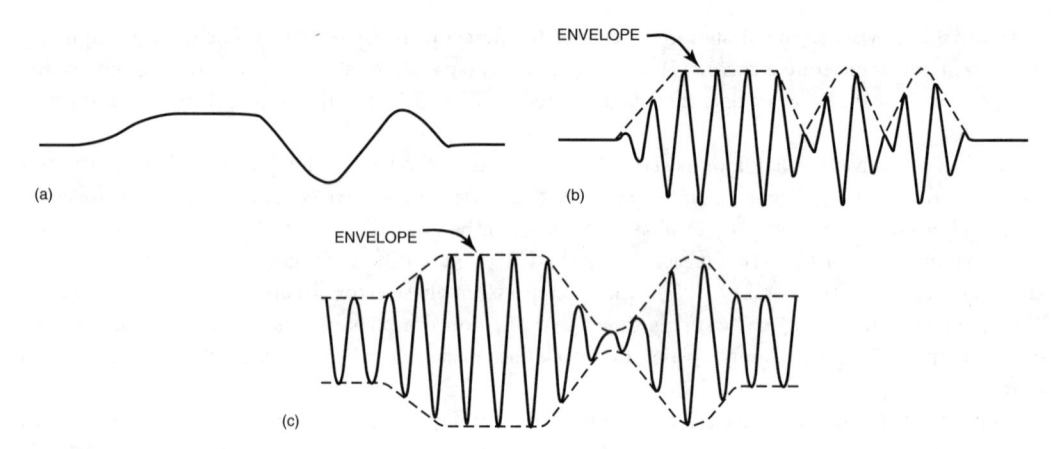

FIGURE 3.7 The envelope of the DSB-SC signal (b) cannot be used to reconstruct the message signal (a). In contrast, the message is preserved in the DSB+C envelope (c).

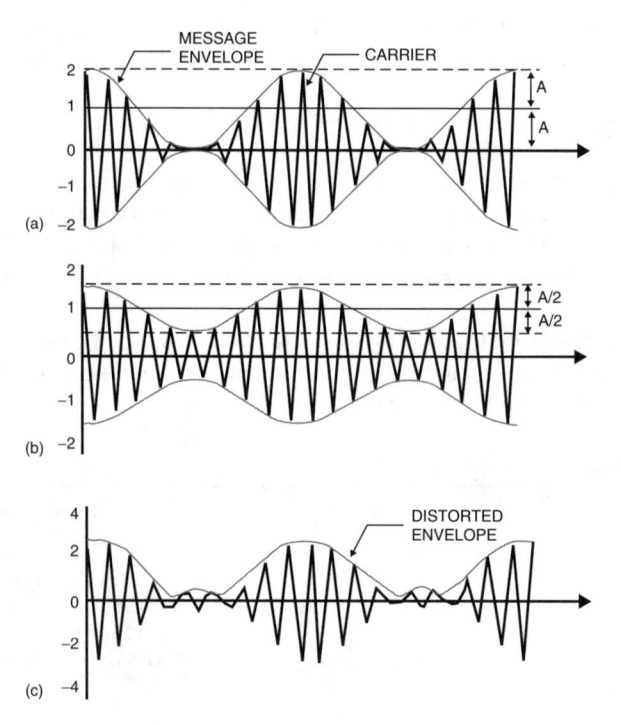

FIGURE 3.8 DSB+C signal with (a) 100%, (b) 50%, and (c) 120% modulation index.

The amount of signal impressed on the carrier is measured by a *modulation index*, defined as

$$m = -\min[m(t)]/A$$

which is sometimes stated as a percentage. Figure 3.8, for example, shows AM signals with 100, 50, and 120% modulation indices. The 120% modulation case shows that the envelope containing $m(t)$ is distorted, making it impossible to recover $m(t)$ during demodulation. This is called *overmodulation*.

The modulation index is also indicative of the power efficiency of the AM transmission. From Eq. (3.6) it is clear that the term due to the carrier conveys no useful information, whereas the sideband term contains all of the message signal content. Efficiency can be defined as the ratio of energy containing

Amplitude Modulation

useful information to the total transmitted energy, and can be written in terms of the modulation index:

$$E = m^2/(2 + m^2) \times 100\%$$

The highest efficiency is at 100% ($m = 1$) modulation and results in only 33% efficiency. Lower modulation values result in even lower efficiencies. By contrast, suppressed-carrier DSB systems are 100% efficient since they waste no energy in a carrier; this is a key disadvantage of AM systems.

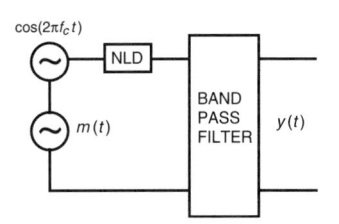

FIGURE 3.9 DSB+C modulator using a nonlinear device (NLD).

The DSB+C signal can be generated without a mixing circuit by using a nonlinear device (NLD) such as a diode, as in Fig. 3.9. For example, suppose the input x and output y of an NLD can be modeled as:

$$y = a_1 x + a_2 x^2$$

Noting that the NLD input is $[\cos 2\pi f_c t + m(t)]$, the output is given by

$$y(t) = a_1[\cos(2\pi f_c t) + m(t)] + a_2[\cos(2\pi f_c t) + m(t)]^2$$

which can be simplified to give

$$
\begin{aligned}
y(t) &= a_1 m(t) + a_2 m^2(t) + a_2/2 & \text{baseband terms} \\
&+ a_1 \cos(2\pi f_c t) + 2a_2 m(t) \cos(2\pi f_c t) & \text{passband terms} \\
&+ a_2/2 \cos(4\pi f_c t) & \text{double frequency term}
\end{aligned}
$$

A bandpass filter centered at f_c will reject all but the passband term at f_c giving the outputs $z(t) = a_1[1 + Cm(t)] \cos(2\pi f_c t)$, with $C = 2a_2/a_1$, which is the desired DSB+C signal. The ease of this method often makes it the preferred choice, especially for high-power applications such as broadcast AM transmitters.

The simplicity of generating DSB+C signals makes it attractive for use in DSB-SC generation. Figure 3.10 shows a *balanced modulator* consisting of two AM modulators whose inputs are opposite in sign and whose outputs are subtracted from each other. By inspection, the balanced modulator output is $y(t) = 2m(t) \cos(2\pi f_c t)$; in other words, the carriers of the two DSB+C modulators have been suppressed to yield a DSB-SC signal. The balanced modulator can be thought of as a multiplier or mixer circuit operating on $m(t)$ and $\cos(2\pi f_c t)$.

The DSB+C signal can be demodulated coherently, as described for DSB-SC signals. The real benefit of using DSB+C systems, however, is the ability to receive using *envelope detection* methods.

These receivers do not require coherent local oscillators and are much cheaper and easier to build than coherent receivers. Envelope detection is illustrated by the simple systems shown in Fig. 3.11. In Fig. 3.11(a), a diode clips the incoming signal to isolate the positive envelope of the signal. A resistor-capacitor (RC) stage is used to low-pass filter the diode output to remove most of the carrier component. The resulting signal is capacitively coupled to remove the unwanted DC component and then amplified for output through earphones or a speaker. Figure 3.11(b) shows a more general design called *rectifier detection*. A potential drawback with these techniques occurs if $m(t)$ contains significant low-frequency information: this energy will be attenuated by the capacitive coupling.

Single-Sideband

Double-sideband methods transmit redundant information since the baseland signal is duplicated in both the upper and lower sidebands. Consequently, the required bandwidth is twice the needed amount. In contrast, a single-sideband signal contains only the USB or LSB, thus reducing bandwidth by one-half. This is the primary advantage of SSB communication.

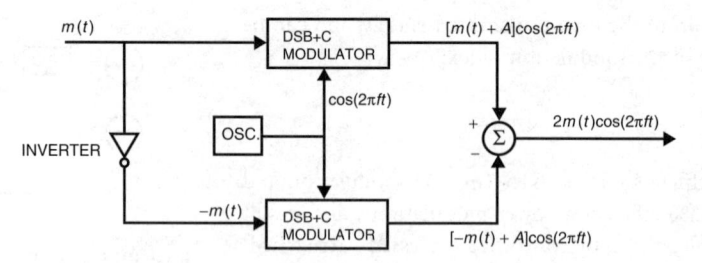

FIGURE 3.10 Balanced modulator to generate DSB-SC signals.

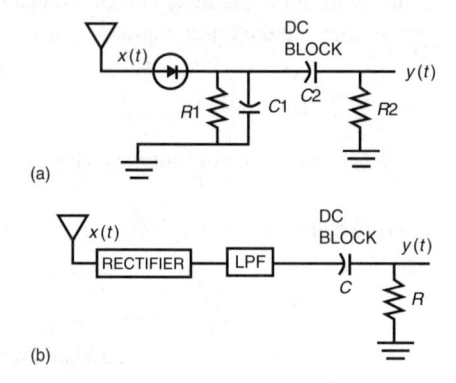

FIGURE 3.11 DSB+C detection using (a) envelope detection, and (b) rectifier detection.

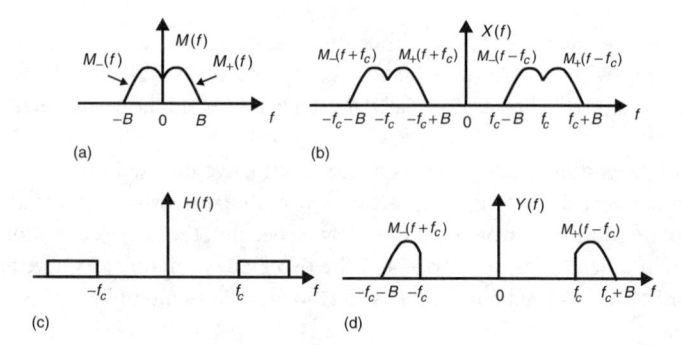

FIGURE 3.12 USB single sideband generation using a sharp BPF to remove the LSB: (a) shows the message spectrum, (b) the DSB-SC spectrum, (c) the sideband filter response, and (d) the resulting USB-SSB spectrum.

The most common technique for generating SSB signals is by applying a sharp bandpass filter to the DSB-SC signal, as shown in Fig. 3.12. This sideband filter passes only the desired sideband and suppresses the other sideband before the signal is amplified and transmitted.

To help understand SSB operation, we note that the baseband signal can be expressed as the sum of a negative frequency part $M_-(f)$ and a positive frequency part $M_+(f)$: $M(f) = M_-(f) + M_+(f)$, as shown in Fig. 3.12. Similarly, the passband DSB-SC signal can be written as

$$X(f) = 0.5\{M_-(f+f_c) + M_+(f+f_c) + M_-(f-f_c) + M_+(f-f_c)\}$$

The USB-SSB signal shown in Fig. 3.12(d) is created by suppressing the lower sidebands, $M_+(f+f_c)$ and $M_-(f-f_c)$, giving

$$Y_{USB}(f) = 0.5\{M_-(f+f_c) + M_+(f-f_c)\}$$

Amplitude Modulation

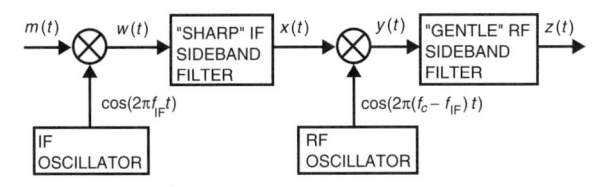

FIGURE 3.13 SSB generation using an intermediate frequency stage followed by a second modulator stage.

A significant disadvantage of SSB transmission results from the need for sideband filters with extremely sharp rolloff characteristics. Such "brick wall" response is difficult to achieve in the real world, especially at high frequencies used in many communications systems. A typical bandpass filter (BPF) response has a much more gentle rolloff characteristic that could result in attenuating the low-frequency part of the message signal, or in passing part of the unwanted sideband, thus causing signal distortion. Consequently, single-sideband transmission typically suffers from poor low-frequency response.

This problem can be partially addressed by creating a DSB-SC signal at a low intermediate frequency f_{IF}, where $f_{IF} \ll f_c$, as shown in Fig. 3.13. A sharp sideband filter (which is much easier to implement at low frequency) is used to create an SSB-SC signal at frequency f_{IF}. This signal is modulated up to the desired carrier frequency where there are again two sidebands present. Since these are separated by a distance of $2f_{IF}$, the undesired sideband can be removed using an easily implemented BPF with gentle rolloff response as in Fig. 3.14.

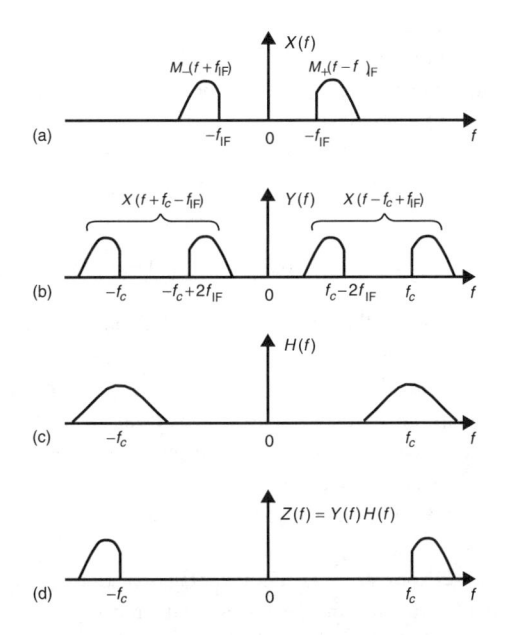

FIGURE 3.14 SSB modulation: (a) intermediate frequency SSB signal $X(f)$, (b) after modulation to passband, (c) gentle rolloff sideband filter response, and (d) removal of unwanted sideband to produce the final SSB signal $Z(f)$.

An alternative method for SSB generation not requiring sideband filters is called the *phase-shifting method*. Mathematical analysis shows that the time-domain representations of the USB and LSB signals can be written as:

$$y_{USB}(t) = m(t)\cos(2\pi f_c t) - m_H(t)\sin(2\pi f_c t)$$

and

$$y_{LSB}(t) = m(t)\cos(2\pi f_c t) - m_H(t)\sin(2\pi f_c t)$$

where $m_H(t)$ is the *Hilbert transform* of $m(t)$ computed as a $-90°$ phase shift of all frequency components. In both cases the SSB signal is made up of a DSB signal (the left-hand term) combined with a second term that cancels out the undesired sideband. The right-hand term is equal in magnitude to the DSB term, but has one sideband exactly 180° out of phase. Thus, one of the sidebands is canceled when this term is added to or subtracted from the DSB component. Figure 3.15 shows a phase-shifting USB modulator implementing these equations. The primary disadvantage of this approach is the need for precision wideband phase-shift networks that can be difficult to implement.

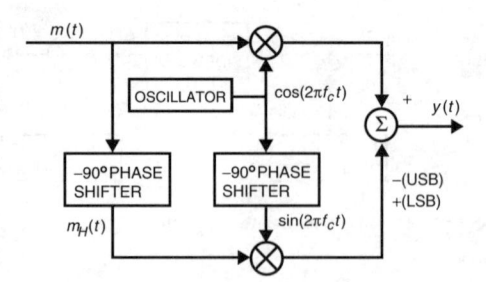

FIGURE 3.15 SSB modulator using the phase-shifting method.

Although SSB generation is rather complex, detection can be accomplished using a coherent demodulation system identical to that used for DSB-SC signals (Fig. 3.2(a)). For example, when a USB-SSB signal $Y_{USB}(f)$ is mixed with the local oscillator signal $2\cos(2\pi f_c t)$, the modulation theorem gives $Z(f) = 2*0.5\{Y_{USB}(f+f_c) + Y_{USB}(f-f_c)\}$, or

$$Z(f) = 0.5\{M_-(f+f_c+f_c) + M_+(f-f_c+f_c) + M_-(f+f_c-f_c) + M_+(f-f_c-f_c)\}$$
$$= 0.5M_-(f+2f_c) + 0.5M_+(f-2f_c) + M_+(f) + M_-(f)$$
$$= 0.5M_-(f+2f_c) + 0.5M_+(f-2f_c) + M(f)$$

The result is the desired baseband signal plus unwanted harmonics at $\pm 2f_c$, which are removed by the receiver LPF.

As with the DSB receiver, achieving coherent detection requires a local oscillator in-phase with the transmitter oscillator. This can be done by transmitting a pilot signal, using a phase-locked loop (PLL) in the receiver circuit, or by simply tuning the local oscillator for the best output signal.

Alternatively, a carrier can be inserted into the SSB signal, which can then be demodulated using simple envelope detection. This SSB+C approach has the advantages of both low bandwidth and low-complexity receiver design, but is inefficient in power use due to the added carrier and usually performs poorly in the presence of noise.

Vestigial Sideband (VSB)

The chief disadvantage of SSB is poor low-frequency response resulting from the difficulty of realizing perfect sideband filters or, for the phase-shifting method, perfect wideband 90° phase shifters. On the other hand, double sideband modulation methods have good low-frequency response, but suffer from excessive bandwidth. **Vestigial sideband (VSB)** modulation is a compromise between these two techniques, offering excellent low frequency response along with reasonably low bandwidth.

A VSB signal can be generated using a sideband filter similar to that used in SSB transmission. For VSB, however, a controlled portion of the rejected sideband is retained. Figure 3.16 shows a comparison of SSB and

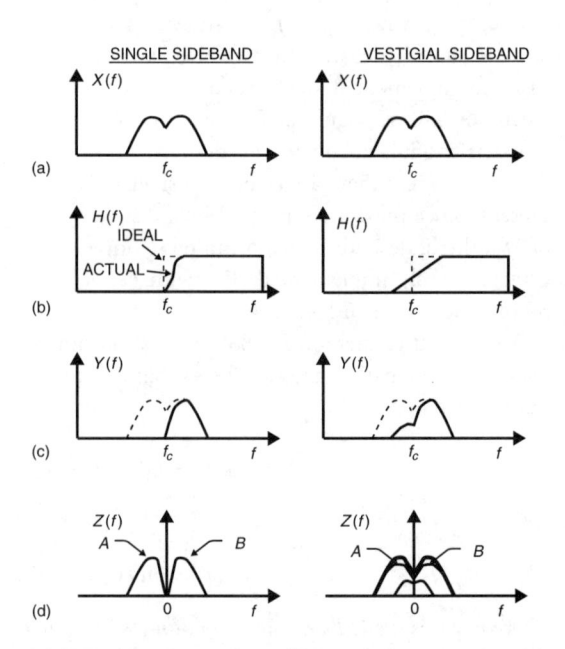

FIGURE 3.16 Comparison of SSB and VSB generation: (a) DSB-SC spectrum, (b) SSB and VSB sideband filter response, (c) spectra of filtered signal, and (d) reconstruction of baseband signal.

Amplitude Modulation

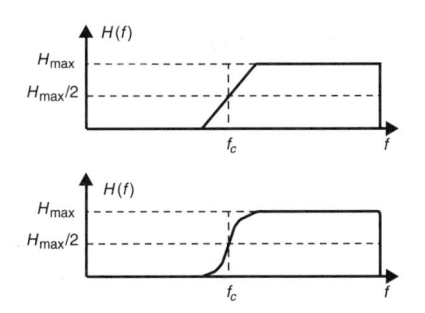

FIGURE 3.17 Two examples of sideband filter responses having vestigial symmetry.

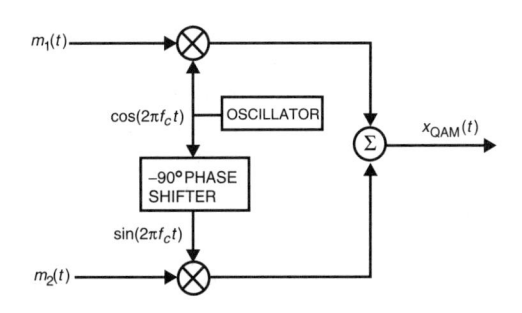

FIGURE 3.18 QAM modulator.

VSB sideband filters and the corresponding output signals. Note that low-frequency energy in the demodulated baseband signal is significantly attenuated by the SSB sideband filter. In contrast, no distortion is apparent in the demodulated VSB signal. As shown in the figure, this is because the negative and positive frequency components (labeled A and B) add constructively at baseband to perfectly reconstruct the low-frequency portion of the band. Achieving this result requires that the pass-band filter has *vestigial symmetry*.

This simply means that the filter's frequency response must have odd symmetry about frequency f_c and amplitude $H_{max}/2$, where H_{max} is the maximum filter response. Two examples are shown in Fig. 3.17.

As with DSB-SC and SSB-SC, a carrier can be added to the transmitted VSB signal allowing inexpensive envelope detection to be used in the receiver. This scheme is used in commercial broadcast television and makes it possible to mass produce relatively inexpensive high-quality receivers.

Quadrature Amplitude Modulation (QAM)

Single-sideband transmission makes very efficient use of the spectrum; for example, two SSB signals can be transmitted within the bandwidth normally required for a single DSB signal. However, DSB signals can achieve the same efficiency by means of *quadrature amplitude modulation* (QAM), which permits two DSB signals to be transmitted and received simultaneously using the same carrier frequency. Suppose we want to transmit two message signals, $m_1(t)$ and $m_2(t)$. The QAM-DSB modulator shown schematically in Fig. 3.18 can be represented mathematically as

$$x(t) = m_1(t) \cos(2\pi f_c t) + m_2(t) \sin(2\pi f_c t)$$

The detection circuit shown in Fig. 3.19 can be used to receive the QAM signal. To show that the two message signals are fully recovered, we see that the two outputs are

$$\begin{aligned} y_1(t) &= \text{LPF}\{2x_{QAM}(t)\cos(2\pi f_c t)\} \\ &= \text{LPF}\{2m_1(t)\cos^2(2\pi f_c t) + 2m_2(t)\sin(2\pi f_c t)\cos(2\pi f_c t)\} \\ &= m_1(t) \end{aligned}$$

and, similarly,

$$\begin{aligned} y_2(t) &= \text{LPF}\{2x_{QAM}(t)\sin(2\pi f_c t)\} \\ &= \text{LPF}\{2m_1(t)\cos(2\pi f_c t)\sin(2\pi f_c t) + 2m_2(t)\sin^2(2\pi f_c t)\} \\ &= m_2(t) \end{aligned}$$

where LPF{·} represents the low-pass filtering operation, which eliminates the double frequency terms at $2f_c$. Thus, the two DSB signals coexist separately within the same bandwidth by virtue of the 90° phase shift between them. The signals are said to be in **quadrature**. Demodulation uses two local oscillator signals that are also in quadrature (a sine and cosine signals). The chief disadvantage of QAM is the need for a coherent local oscillator at the receiver exactly in-phase with the transmitter oscillator signal. Slight errors in phase or frequency can cause both loss of signal and interference between the two signals (called *cochannel interference* or *crosstalk*).

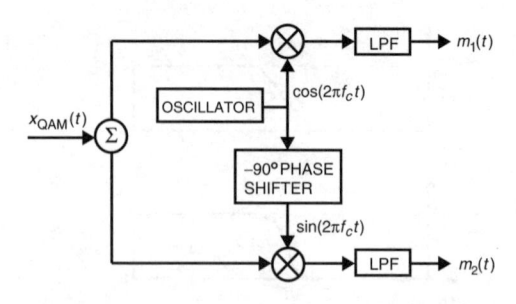

FIGURE 3.19 QAM demodulator.

Noise Effects

Real-world communication channels adversely affect the received signal due to noise and attenuation. A communication receiver designed to operate in the presence of noise is similar to the design shown, but with an additional bandpass filter inserted immediately before the receiver input (see Fig. 3.20). The filter removes noise and interference energy outside the passband.

The *input signal-to-noise ratio* (SNR_i) is defined as the ratio of received signal power to noise power at the band-pass filter output: $SNR_i = S_i/N\,B$, where S_i is the received signal power, N is the spectral noise power in watts per hertz, and B is the bandwidth of the baseband signal $m(t)$. An *output signal-to-noise ratio* (SNR_o) is also defined and is measured at the receiver output. This latter measure is indicative of the receiver's output quality since it measures the noise content of the output signal. Another useful measure of performance is the ratio of output to input SNR values, i.e., $P = SNR_o/SNR_i$. Values of P equal to 1.0 or greater indicate good performance, whereas smaller values of P imply poorer system performance. Table 3.2 shows P values for each amplitude modulation method. A few points are worth discussing in greater detail. First, these results show that for a fixed level of transmitted power, the suppressed-carrier systems all produce about the same level of noise and signal power at the receiver output. In particular, we note that SSB-SC requires the same transmitter power as DSB-SC, even though DSB-SC uses both sidebands whereas SSB uses only one. The reason for this is that none of the DSB-SC energy is wasted: the receiver utilizes both sidebands to reproduce the output signal. The remaining systems, DSB+C, SSB+C, and VSB+C all require much more transmitter power to achieve the same output SNR as the suppressed carrier methods, assuming envelope detection is being used. Another fact not shown in the table is that envelope detection systems exhibit a *threshold effect*. That is, when the input signal SNR_i drops below a certain value, the output SNR begins to decline at a much more rapid rate. For DSB+C signals, this threshold occurs when received SNR is below approximately 10 dB, which is a much lower level than the 30 dB or so needed for good reception. For most practical applications, the threshold effect has a minimal effect on amplitude modulation systems.

Superheterodyne Receivers

Early receiver designs used a tunable local oscillator (LO) to select the desired station and a bandpass filter to reject stations at nearby frequencies, out-of-band noise, and interference. The filter was *ganged*

TABLE 3.2 Performance of Modulation Techniques in Noise: Performance Ratio P

Modulation Technique	Suppressed Carrier	Carrier Present
DSB	1	$P_m^2/[A^2 + P_m^2]$ using envelope detection P_m is average power in $m(t)$
SSB	1	Much less than DSB+C
VSB	≈1	Much less than DSB+C
QAM	1	Not applicable

Amplitude Modulation

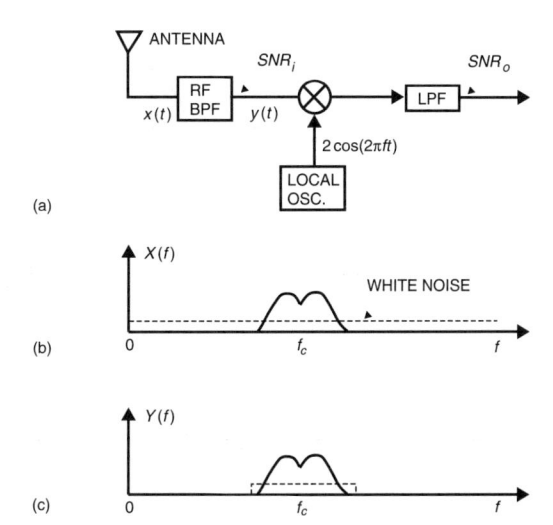

FIGURE 3.20 Receiver noise considerations: (a) demodulator using RF bandpass filter showing where SNR_i and SNR_o are measured, (b) received signal spectrum, and (c) bandpass output spectrum.

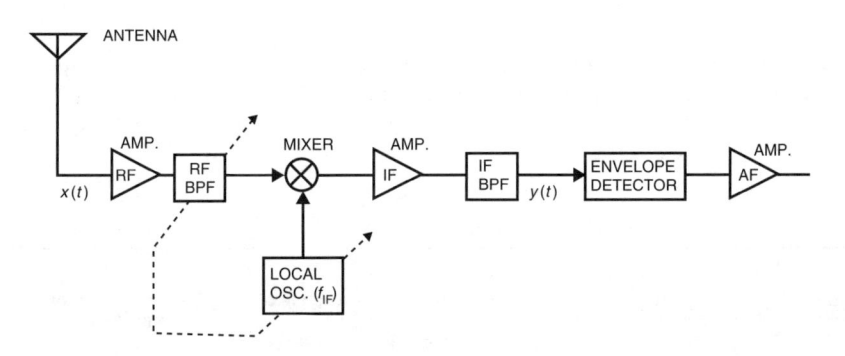

FIGURE 3.21 Superheterodyne AM (DSB+C) receiver.

with the local oscillator so that its center frequency tracked the oscillator frequency. However, designing the tunable bandpass filter with sharp cut-off response is difficult. A cheaper and much more effective approach called the **superheterodyne** receiver is shown in Fig. 3.22. The front-end *RF section* of this receiver uses an inexpensive ganged tunable filter with a gentle rolloff response to eliminate most noise and interference. It is standard practice for the LO to be tuned to 455 kHz above the desired station (hence, the name super heterodyning) for voice reception. (Automobile receiver designs often use an IF of 262.5 kHz.) This translates the desired station to a fixed *intermediate frequency* (IF) of 455 kHz. The IF section contains a sharp-rolloff bandpass filter; this is practical because of the fixed center frequency f_{IF} and because f_{IF} is relatively low. The IF-BPF filter provides the needed *selectivity* to reject adjacent stations that were not removed by the RF bandpass filter. An envelope detector follows the IF stage to produce the audio output signal.

Although an RF bandpass filter with sharp rolloff is not required in this technique, there is a limit on how broad its response can be. Consider a station located at a frequency $2f_i = 910$ kHz above the desired frequency. If this station is not removed by the RF section, it will be removed to the IF frequency on top of the desired station and cause unacceptable interference, as shown in Fig. 3.21. The unwanted signal is called an *image* and must be removed at the receiver's front end for acceptable receiver performance.

Superheterodyning is also used in TV, FM, and other receiver designs where the use of an intermediate frequency permits effective IF filtering strategies. The relative merits of the modulation schemes discussed in this chapter are summarized in Table 3.3.

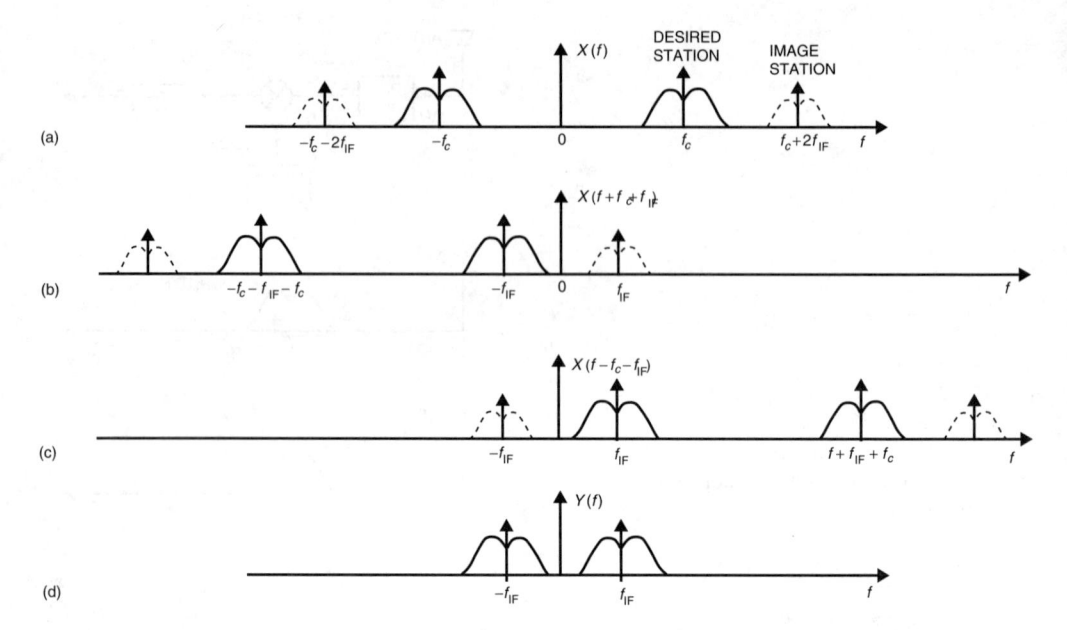

FIGURE 3.22 Superheterodyne receiver operation: (a) the desired signal and undesired image spectra (shows as a dotted line), (b) and (c) the two components of the mixer output, and (d) the IF bandpass output given by $Y(f) =$ BPF $(X(f + f_c + f_{IF}) + X(f - f_c - f_{IF}))$. Notice that the image signal must be removed by the RF stage or it will interfere with the desired signal in the IF stage.

TABLE 3.3 Comparison of Amplitude Modulation Techniques

Modulation Scheme	Advantages	Disadvantages	Comments
DSB-SC	Good power efficiency. Good low-frequency response.	More difficult to generate than DSB+C. Detection requires coherent local oscillator, pilot, or phase-locked loop (PLL). Poor spectrum efficiency.	
DSB+C (AM)	Easier to generate than DSB-SC, especially at high-power levels. Inexpensive receivers using envelope detection.	Poor power efficiency. Poor spectrum efficiency. Poor low-frequency response. Exhibits threshold effect in noise.	Used in commercial AM.
SSB-SC	Excellent spectrum efficiency.	Complex transmitter design. Complex receiver design (same as DSB-SC). Poor low-frequency response.	Used in military communication systems, and to multiplex multiple phone calls onto long-haul microwave links.
SSB+SC	Good spectrum efficiency. Low receiver complexity.	Poor power efficiency. Complex transmitters. Poor low-frequency response. Poor noise performance.	
VSB-SC	Good spectrum efficiency. Excellent low-frequency response. Transmitter easier to build than for SSB.	Complex receivers (same as DSB-SC).	
VSB+C	Good spectrum efficiency. Good low-frequency response. Inexpensive receivers using envelope detection.	Poor power efficiency. Poor performance in noise.	Used in commercial TV.
QAM	Good low-frequency response. Good spectrum efficiency.	Complex receivers. Sensitive to frequency and phase errors.	Two SSB signals may be preferable.

Defining Terms

Amplitude modulation: Linear modulation schemes that imprint the message signal onto the amplitude component of a carrier. In contrast, *angle modulation* methods operate by modulating the phase of the carrier. Amplitude modulation schemes include DSB-SC, DSB+C, SSB-SC, SSB+C, VSB-C, VSB+C, and QAM, as well as numerous digital communications methods.

Carrier: A sinusoidal signal of frequency f_c (the *carrier frequency*). It is mixed with a *message* signal to produce a modulated transmitter output signal centered at f_c. In *suppressed carrier* systems, no carrier component is present in the spectrum, whereas in added-carrier systems, such as broadcast AM, the carrier contains a significant amount of power.

Detection: The process of converting the transmitted passband signal back into a baseband message signal. *Envelope detection* operates by extracting the envelope of the received signal to reconstruct the message signal, whereas *product* or *coherent detection* uses a local oscillator in phase with the transmitter oscillator along with a mixing circuit. Also known as *demodulation*.

Double sideband (DSB): An amplitude modulation transmission scheme where both sidebands are transmitted.

Message signal: The input audio signal or program that is desired to be transmitted. This can be speech, music, or a digital signal.

Modulation: The process of multiplying, or *mixing,* a message signal with a carrier signal. This causes the message spectrum to be translated out to the positive and negative carrier frequencies. In other words, it transforms a *baseband* signal (the message) into a *passband* signal (the transmitter output).

Quadrature: The condition when two signals having the same carrier frequency exhibit a 90° phase difference.

Sideband: A component of an amplitude modulated waveform. A modulated signal has a spectrum that is symmetric about frequency f_c, such that the upper-half is a mirror image of the lower-half. These are termed the *upper sideband* and *lower sideband*; each sideband contains an entire copy of the program information.

Single sideband (SSB): A transmission scheme using only one sideband to convey information.

Superheterodyne: A receiver design that mixes the input signal with a carrier tuned to f_{IF} Hz above the desired station. This translates it to an *intermediate frequency* f_{IF}, where efficient fixed frequency filters can be utilized. *Image stations* located at $f_c + 2f_{IF}$ must be removed by the RF section to prevent interference.

Vestigial sideband (VSB): A transmission scheme using one sideband plus a carefully controlled portion of the other sideband. This system provides excellent fidelity at all frequencies and is still spectrally efficient.

References

Carlson, A.B. 1986. *Communication Systems — An Introduction to Signals and Noise in Electrical Engineering,* 3rd ed. McGraw-Hill, New York.

Couch, L.W. II. 1995. *Modern Communication Systems — Principles and Applications.* Prentice-Hall, Englewood Cliffs, NJ.

Haykin, S. 1994. *Communication Systems,* 3rd ed. Wiley, New York.

Haykin, S. 1989. *An Introduction to Analog and Digital Communications.* Wiley, New York.

Lathi, B.P. 1989. *Modern Digital and Analog Communication Systems,* 2nd ed. Holt, Rinehart, and Wieston, Orlando, FL.

Proakis, J.G. and Salehi, M. 1994. *Communication Systems Engineering.* Prentice-Hall, Englewood Cliffs, NJ.

Schwartz, M. 1990. *Information Transmission, Modulation, and Noise,* 4th ed. McGraw-Hill, New York.

Shanmugam, K.S. 1979. *Digital and Analog Communication Systems.* Wiley, New York.

Ziemer, R.E. and Tranter, W.H. 1995. *Principles of Communications — Systems, Modulation, and Noise,* 4th ed. John Wiley & Sons, New York.

4

Frequency Modulation

4.1	Introduction	4-1
4.2	The Modulated FM Carrier	4-1
4.3	Frequency Deviation	4-3
	Example 4.1	
4.4	Percent of Modulation in FM	4-4
4.5	Modulation Index	4-4
	Example 4.2	
4.6	Bandwidth and Sidebands Produced by FM	4-5
	Example 4.3	
4.7	Narrow-Band vs. Wide-Band FM	4-7
4.8	Phase Modulation	4-7
4.9	FM Transmission Principles	4-7
	Direct FM Modulators • Voltage Controlled Oscillator (VCO) Direct-FM Modulators • Indirect-FM Modulators	
4.10	FM Reception Principles	4-11
	Limiters • FM Detectors • Discriminators and Ratio Detectors	

Ken Seymour
Sprynet

4.1 Introduction

Perhaps no other form of modulation has had more impact on our culture in the past 30 years than **frequency modulation** (FM). It is virtually one of the most widely used modes of modulation. From applications in commercial broadcasting, television audio, cordless phones, to cellular and mobile communications, FM is indeed both a reliable and important form of modulation. The brainchild of Edwin H. Armstrong, FM was first demonstrated in December 1933 [Lewis 1991, p. 256] as a solution to eliminate the static and noise problems that plagued AM communications.

In amplitude modulation (AM), interference, such as static, lightning, and manmade noise, cause the amplitude of an RF signal to vary widely. This is because these noises are predominately amplitude modulated signals in composition. The noise is added and superimposed on the transmitted AM signal carrying the desired intelligence. This increases the overall amplitude of the signal as shown in Fig. 4.1. These added variations are then demodulated at the receiver, and the noise is passed onto the audio section, where they are reproduced as clicks, pops, and various other objectionable noises. The problems associated with amplitude modulation are overcome in frequency modulation. FM receivers are designed to reduce the amplitude variations of an incoming signal. This is done without affecting the frequency modulated waveform that contains the desired intelligence.

4.2 The Modulated FM Carrier

The composition of the unmodulated AM carrier and FM carrier are identical. That is, the RF carrier is a sinusoidal waveform operating at a specific period or frequency. As the frequency is increased, the

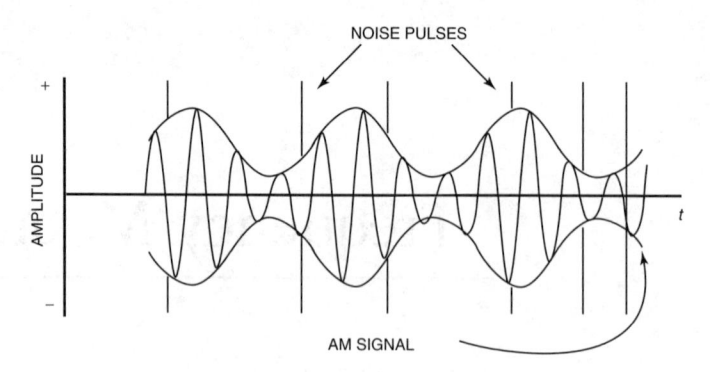

FIGURE 4.1 Typical AM signal with induced noise.

period of the waveform decreases as more cycles are completed per second. When an AM carrier is modulated, the amplitude of the carrier is affected. When the FM carrier is modulated, the frequency of the carrier varies by an amount that is proportional to amplitude of the modulating waveform; this occurs at a rate that is determined by the modulation frequency.

To better understand how the modulating signal affects a carrier that is being frequency modulated, it is best to use illustrative examples. Figure 4.2 graphically illustrates what happens when a carrier is frequency modulated. Figure 4.2(a) shows one period of an audio signal that will be used to frequency modulate a carrier. The AC signal is positive for 180° and swings negative for the remaining 180° to complete one 360° cycle.

Figure 4.2(b) illustrates the effect of how the audio signal (a) effects the carrier. At time $t = <0$, we see the RF carrier operating at a specific frequency. This is sometimes referred to as the *center* or *resting* frequency. At time $t = 0$, the modulating signal (a) is applied to the RF carrier. As the amplitude of signal (a) swings positive, the frequency of the RF carrier also begins to change. At time $t = 1$, the frequency of the modulated RF carrier has increased proportionally, resulting in a greater number of cycles occurring in a given interval of time. At time $t = 2$, the amplitude of the modulating signal (a) reaches its maximum. At the same time, the RF carrier has increased to its maximum frequency. At time $t = 3$, the modulating waveform (a) begins to decrease in amplitude and the carrier frequency (b) also begins to decrease. At time $t = 4$, the modulating frequency returns to zero and the carrier frequency returns to the resting frequency.

For the last 180° portion of the modulation cycle, the amplitude of the modulating signal goes negative. At time $t = 6$, the modulating signal (a) has decreased to its maximum negative value and the frequency of the modulated carrier (b) also reaches its minimum. In fact, it has decreased to a frequency below that of the unmodulated carrier. At time $t = 7$, the modulation amplitude begins its return journey back to zero. Then, at $t = 8$, the carrier frequency returns to the resting frequency. Figure 4.2(c) shows waveforms (a) and (b) superimposed on each other to better illustrate the waveform relationships.

When an FM signal is received, it is the amount of frequency shift that is produced in the modulated waveform that determines the audio intensity or volume that is heard on the speaker of the receiver. To summarize, the frequency modulated carrier observes these following characteristics:

- The higher the modulating amplitude, the greater is the amount of frequency shift away from the resting frequency. This form of FM is also referred to as **direct FM**.
- As the amplitude of the modulating source increases, the frequency of the carrier increases.
- As the amplitude of the modulating source decreases, the frequency of the carrier decreases.
- The amplitude of the FM modulated carrier remains constant as the amplitude of the modulating source varies.

Table 4.1 illustrates how a modulating signal affects an FM and AM modulated waveform. For example, as the amplitude of the modulating signal increases, the overall frequency swing on FM increases. With AM, the amplitude of the carrier increases.

Frequency Modulation 4-3

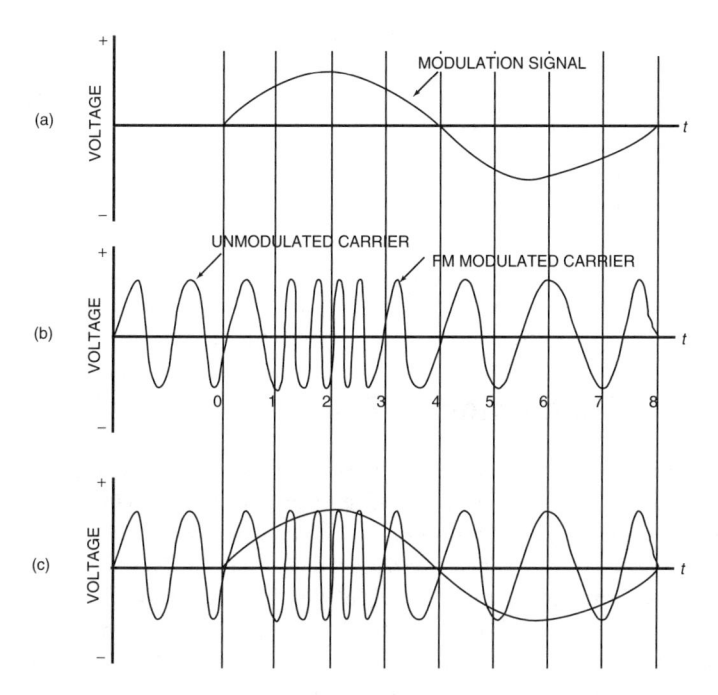

FIGURE 4.2 Frequency modulating an RF carrier.

TABLE 4.1 Summary of Modulation Effects on AM vs. FM

	Modulating Amplitude	
	Increases	Decreases
FM	Frequency swing increases	Frequency swing decreases
AM	Carrier level increases	Carrier level decreases

4.3 Frequency Deviation

As we saw in the previous section, the amplitude of the modulating signal plays an important part in the overall characteristic of the carrier frequency. The peak difference between the modulated carrier and the frequency of the carrier is known as the **frequency deviation** [Code 1993]. The peak difference between the minimum and maximum frequency values is known as the *frequency (or carrier) swing* [Code 1993]. This can be defined as:

$$\Delta f = fpc - fc$$

where:

Δf = frequency deviation
fpc = peak frequency of the modulated carrier (minimum or maximum)
fc = frequency of the carrier (unmodulated)

Example 4.1

A commercial FM broadcast station operates on a frequency of 97.1 MHz. On a modulation peak, the frequency increases to 97.13 MHz. Determine the (1) frequency deviation and (2) the frequency swing. The solution is as follows:

1. Applying the deviation equation: $fpc = 97.13$ and $fc = 97.10$. Substituting terms: $\Delta f = 97.13 - 97.10$. This results in a frequency deviation of +30 kHz.
2. The frequency swing would be the peak difference of the minimum and maximum frequency deviation or $(2)(\Delta f) = (2)(30 kHz) = 60$ kHz.

It is important to remember, that there are two cases of frequency deviation. In the preceding example, the frequency was increased as a result of the modulating signal. Likewise in FM, when a carrier is modulated, the carrier frequency also goes negative. Therefore, the swing is defined as $(2)(\Delta f)$. The FCC places limits on the amount of deviation that a frequency can swing. With commercial FM stations, the frequency deviation is +/–75 kHz or 150 kHz total frequency swing. Depending on the FM communication service, the FCC imposes different frequency deviation requirements.

4.4 Percent of Modulation in FM

In amplitude modulation, 100% modulation is defined as the point where the amplitude of the RF carrier rises to twice the normal amplitude at its maximum, and drops to zero at its minimum. Anything greater than 100% modulation, in AM, causes distortion to the modulated wave.

With FM, it is the amount of frequency deviation that determines the degree of modulation. The frequency deviation that corresponds to 100% modulation is an arbitrary value defined by the FCC or other appropriate licensing authority as related to the FM service in use. For commercial FM broadcasting, 100% modulation is reached when the carrier frequency deviation reaches 75 kHz. In television, 100% modulation is set at a frequency deviation of +/–25 kHz by the FCC.

If the carrier is modulated above 100%, distortion and spurious sidebands are not produced as in AM. To avoid interference between adjacent stations, the FCC has set the channel spacing for commercial FM at 200 kHz. This gives ample guard band for stations operating at 100% modulation, which accounts for a +/–75 kHz frequency deviation or a total frequency swing of 150 kHz.

4.5 Modulation Index

In FM, the **modulation index** is often used more frequently than **percentage** of **modulation**. Modulation index is defined as the ratio of the frequency deviation to the frequency of the modulating signal [National 1985, p. 3.3-65]. The term has no units and is expressed as a decimal. For a constant frequency deviation, the modulation index drops as the frequency of the modulating signal increases. During the transmission of an FM signal, the modulation index varies as the modulation frequency varies. As we shall see later, this relationship is important for determining the bandwidth requirements of an FM signal,

$$mf = \Delta f / fs$$

where:
mf = modulation index
Δf = frequency deviation
fs = frequency of the modulating signal

Example 4.2

A 97.1-MHz carrier frequency is modulated by a 10-kHz audio signal source. This produces a frequency deviation of +/–40 kHz. Determine the modulation index. The solution is

$$mf = 40/10$$
$$= 4$$

Frequency Modulation 4-5

TABLE 4.2 Bessel Functions: Values of Carrier and Sideband Amplitudes

Modul. Index, m_f	Carrier, A f_c	f_1	f_2	f_3	f_4	f_5	f_6	f_7	f_8	f_9	f_{10}	f_{11}	f_{12}
0.00	1.00	—	—										
0.25	0.98	0.12	0.01										
0.50	0.94	0.24	0.03	—									
1.00	0.77	0.44	0.11	0.02	—								
1.50	0.51	0.56	0.23	0.06	0.01	—							
2.00	0.22	0.58	0.35	0.13	0.03	0.01							
2.50	−0.05	0.50	0.45	0.22	0.07	0.02	—						
3.00	−0.26	0.34	0.49	0.31	0.13	0.04	0.01	—					
4.00	−0.40	−0.07	0.36	0.43	0.28	0.13	0.05	0.02	—	—			
5.00	−0.18	−0.33	0.05	0.36	0.39	0.26	0.13	0.05	0.02	0.01	—		
6.00	0.15	−0.28	−0.24	0.11	0.36	0.36	0.25	0.13	0.06	0.02	0.01	—	
7.00	0.30	0.00	−0.30	−0.17	0.16	0.35	0.34	0.23	0.13	0.06	0.02	0.01	—
8.00	0.17	0.23	−0.11	−0.29	−0.10	0.19	0.34	0.32	0.22	0.13	0.06	0.03	0.01

4.6 Bandwidth and Sidebands Produced by FM

Frequency modulation differs from amplitude modulation in that the modulated wave consists of the carrier frequency and numerous sideband components that are generated for each modulating frequency. Recall that AM consists of a carrier and an upper and lower sideband. The bandwidth of an AM signal is determined by the highest frequency of the modulating signal. For example, a carrier is modulated by an audio signal which contains frequencies up to 4000 Hz. The AM bandwidth would therefore be: $(2)(4000) = 8000$ Hz.

In FM, the amplitude of modulating signal is the primary factor in determining the amount of bandwidth. This was illustrated in Fig. 4.2 where the difference in the amplitude of the audio modulating signal produced a difference frequency change, or deviation. This deviation could shift the frequency of the carrier by 75 kHz or more, depending on the amplitude of the modulating signal.

The sidebands generated in FM are spaced on both sides of the carrier at frequency intervals equal to the modulating frequency and its multiples. To better understand all of the components of FM sidebands, we will analyze an FM modulated carrier using *Bessel functions* and *spectral diagrams*. Figure 4.3 illustrates the mathematical relationship of the Bessel function, the FM waveform theoretically contains an infinite number of side frequencies on both sides of the carrier. The side frequencies are spaced at intervals that correspond to the modulating frequency, which can be represented as:

$$f_c +/- f_s +/- 2f_s +/- 3f_s +/- 4f_s \cdots \infty$$

Notice that the Bessel functions illustrated in Fig. 4.3 resemble damped sine waves. The Bessel curves enable us to understand the component of the FM sidebands. The number of sidebands depends totally on the modulation index (m_f). Table 4.2 is a tabular representation of the Bessel curve illustrated in Fig. 4.3. It shows the amplitudes of the sidebands for audio modulating harmonics where they decrease to a value close to the amplitude of the unmodulated carrier. The following example illustrates how to determine FM sidebands using the Bessel tables.

Example 4.3

Analyze and draw a spectral diagram of: (1) an AM carrier modulated 100% by a 12.5-kHz audio source and (2) an FM carrier modulated by an 12.5-kHz audio source. Assume a frequency deviation of +/−75 kHz. The solution is as follows:

1. The spectral diagram is easy to illustrate as shown in Fig. 4.4. Notice that there are two significant sidebands, upper and lower. The total bandwidth is 25 kHz.
2. Next, we need to determine the modulation index. This is defined as

$$mf = \Delta f/fs$$

Substituting terms, $mf = 75{,}000/12{,}500 = 6$.

Referring to Table 4.2, with a modulation index of 6, we see that there are 10 harmonics plus the carrier. The amplitudes of the sideband harmonics are taken from the table and are illustrated in the spectral diagram shown in Fig. 4.5.

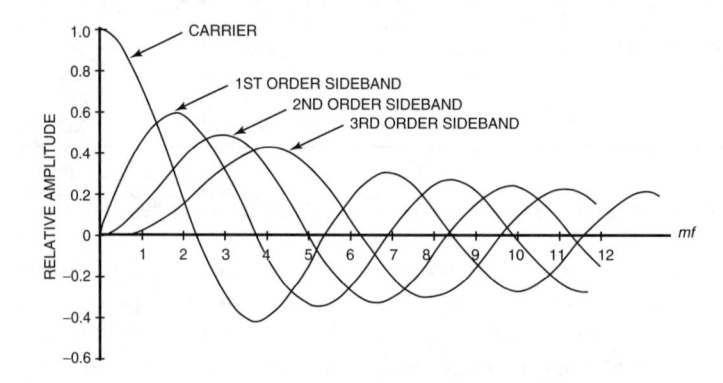

FIGURE 4.3 Relationship of carrier and sideband amplitudes to modulation index mf (Bessel functions).

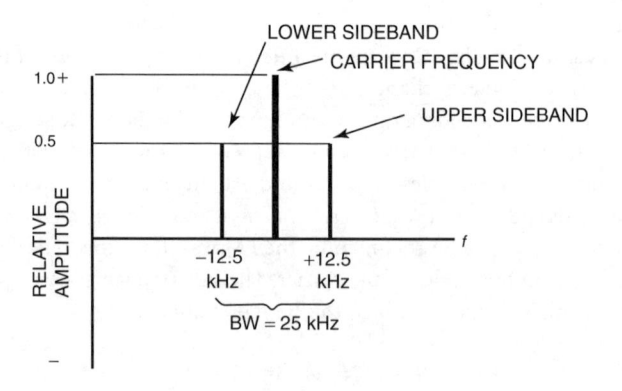

FIGURE 4.4 Example 4.3(1) AM spectral diagram.

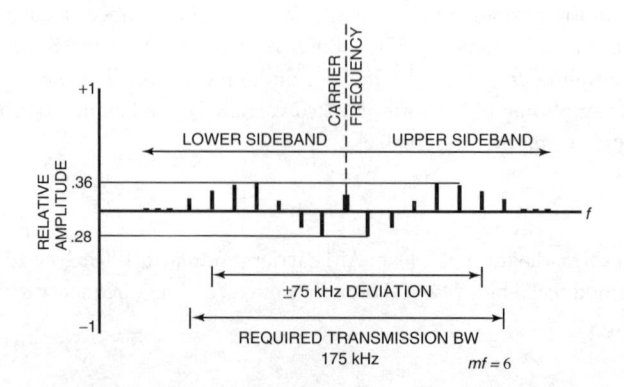

FIGURE 4.5 Example 4.3(2) FM spectral diagram.

Frequency Modulation

There also may be upper and lower sideband frequencies that extend beyond the allowable deviation in an FM signal, in this example beyond the +/−75-kHz limit. The total number of sideband frequencies that are produced is different for each different value of modulation index. The greater the modulation index is, the greater the number of sideband components.

4.7 Narrow-Band vs. Wide-Band FM

The *narrow-band* (NBFM) and *wide-band* (WBFM) terms refer to the amount of frequency deviation (Δf) that is present for a specific transmitted FM signal. This directly correlates to the amount of spectral bandwidth that the transmission occupies. NBFM is typically used for communications services that occupy less spectral bandwidth. This service is used for two-way voice communications, amateurs, and by governmental agencies. The FCC limits the frequency deviation to less than +/−15 kHz for these services. WBFM, on the other hand, typically uses a frequency deviation greater than +/−15 kHz.

4.8 Phase Modulation

So far, we have seen how FM can be produced by shifting the frequency of the carrier above and below a resting frequency as determined by the amplitude of the modulation signal. Frequency modulation can also be produced by shifting the phase of the carrier relative to an arbitrary reference point. This is known as **phase modulation** (PM). When a carrier is phase modulated, the input signal is designed to alter the phase of the carrier. When the amplitude of the modulating signal swings positive, the greater is the phase shift of the carrier results as it advances or leads in phase. This results in a greater frequency swing. As the amplitude of the modulating signal goes negative, the carrier will lag in phase. This method of frequency modulation is often referred to as *indirect FM*.

4.9 FM Transmission Principles

The basic transmission principles used today for frequency modulation fall into two categories. One type, *direct FM*, is the modulation process where the frequency of the transmitter oscillator varies in accordance with the amplitude of the modulating signal. The other type, *indirect FM*, obtains a frequency modulated waveform by phase modulating the carrier.

The method of modulating an FM transmitter vs. modulating an AM transmitter differs significantly. With AM, modulation generally takes place in a higher level stage of the transmitter at the final RF power amplifier stages. With FM, modulation takes place in a much lower stage of the transmitter, usually the master oscillator itself. The lower level modulation of FM is generally performed before any frequency multiplication occurs to minimize increasing phase shift and frequency deviation.

To frequency modulate a carrier directly, an active device is used in conjunction with the input signal to produce a variable reactance (capacitive or inductive) across its output. If the variable reactance is placed across a tuned tank circuit, the effective capacitance or inductance of the tank will change. This, in turn, will change the resonant frequency of the circuit. This methodology was primarily used to generate FM up until the past 15 or so years. Terms, such as *reactance tube, reactance transistor*, and *diode modulators*, were common. Today, *voltage controlled oscillators* are commonly used, and circuits built around variactor diodes are found.

Direct FM Modulators

Over the years, many types of circuits have been used to produce direct FM. In each case, a reactance device is used to shunt capacitive or inductive reactance across an oscillator. The value of capacitive or inductive reactance is made to vary as the amplitude of the modulating signal varies. Since the reactive load is placed across an oscillator tuned circuit, the frequency of the oscillator will therefore shift by a predetermined amount, thereby creating an FM signal.

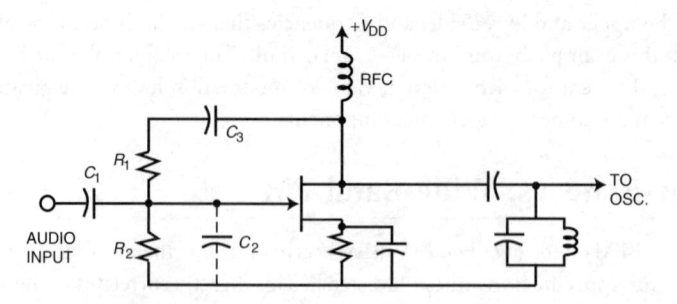

FIGURE 4.6 Simplified reactance modulator.

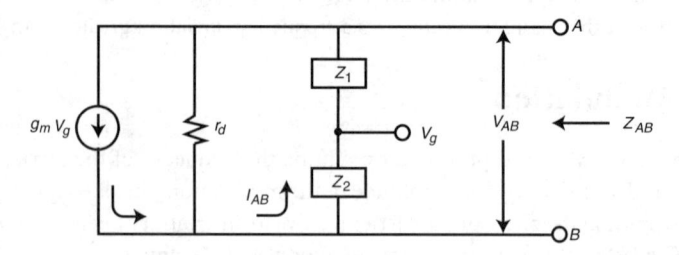

FIGURE 4.7 Equivalent circuit of the reactance modulator.

A typical example of a reactance modulator is shown in Fig. 4.6. The circuit uses a field effect transistor (FET) where the modulating signal applies to the modulator through C_1. The actual components that affect the overall reactance consist of R_1 and C_2. Typically, the value of C_2 is small as this is the input capacitance to the FET. This may only be a few picofarads. However, this capacitance will generally be much larger by a significant amount due to the *Miller effect*. Capacitor C_3 has no significant effect on the reactance of the modulator. It is strictly a blocking capacitor, which keeps DC from affecting the gate bias of the FET.

To further understand the performance of the reactance modulator, an equivalent circuit of Fig. 4.6 is represented in Fig. 4.7. The FET is represented as a current source $g_m V_g$, with the internal drain resistance r_d. The impedances Z_1 and Z_2 are a combination of resistance and capacitive reactance, which are designed to provide a 90° phase shift. This will be evident as the analysis of the reactance FET proceeds.

In an FET, the internal drain resistance r_d is typically very high. Therefore, we can neglect it in our analysis. Looking into the model, the impedance between points A and B is designated Z_{AB} and the voltage across A and B is V_{AB}. We are interested in Z_{AB}, because the value of this impedance will indicate the value of added reactance to the master frequency oscillator. Looking at Z_{AB}, it appears that the output impedance is the series combination of Z_1 and Z_2 (neglecting r_d). This is not exactly true, because one of the components (Z_1 or Z_2) is reactive and is a variable factor that depends on the drain current flowing through the FET and the operating frequency. Since the drain current depends upon the transconductance g_m of the FET, the impedance injected also depends upon the g_m. This is shown as follows.

The impedance is defined as

$$Z_{AB} = V_{AB}/I_{AB} \qquad (4.1)$$

By definition, we also know that the circuit current is (r_d is neglected),

$$I_{AB} = g_m V_g \qquad (4.2)$$

Substituting terms, Eq. (4.2) into Eq. (4.1)

$$Z_{AB} = V_{AB}/g_m V_g \qquad (4.3)$$

Frequency Modulation

Referring to Fig. 4.7, V_g is derived using the voltage divider principle

$$V_g = V_{AB}(Z_2/(Z_1 + Z_2)) \tag{4.4}$$

Substituting terms, Eq. (4.4) into Eq. (4.3)

$$Z_{AB} = \frac{V_{AB}}{g_m V_{AB}(Z_2/(Z_1 + Z_2))} \tag{4.5}$$

$$Z_{AB} = ((Z_1 + Z_2)/Z_1)/(g_m) \tag{4.6}$$

$$Z_{AB} = (1/g_m)((Z_1 + Z_2)/Z_2) \tag{4.7}$$

$$Z_{AB} = (1/g_m)(1 + Z_1/Z_2) \tag{4.8}$$

$$Z_{AB} = (1/g_m) + (1/g_m)(Z_1/Z_2) \tag{4.9}$$

Equation (4.9) represents the impedance seen when looking into the reactive FET circuit, at points A and B. Since the unit for transconductance is given in mho, the term $1/g_m$ will therefore be resistive in ohms. The equation then states that the impedance across points A and B consists of a *resistance in series with a reactance*. In this case, the reactance Z_1 is R_1 (purely resistive) and Z_2 is C_2. As Eq. (4.9) also illustrates, the transconductance value of the FET is a key term and plays an important role in determining the overall added reactance. Since the transconductance of the FET is dependent on the gate voltage,[1] it is apparent that when a modulating signal is applied to C_1, the g_m of the FET will vary as the audio voltage varies. This, in turn, varies the reactance applied to the master oscillator tank circuit.

Using vector diagrams, we can also analyze the phase relationship of the reactance modulator. Referring back to Figs. 4.6 and 4.7, the resistance of R_1 is typically very high compared to the capacitive reactance of C_2. The R_1C_2 circuit is then resistive. Since this circuit is resistive, the current I_{AB} that flows through it is in phase with the voltage V_{AB}. Voltage V_{AB} is also across R_1C_2 (or Z_1Z_2 in Fig. 4.7). This is true because current and voltage tend to be in phase in a resistive network. However, voltage V_{C2}, which is across C_2, is out of phase with I_{AB}. This is because the voltage that is across a capacitor lags behind its current by 90°. This is illustrated in the vector diagram of Fig. 4.8.

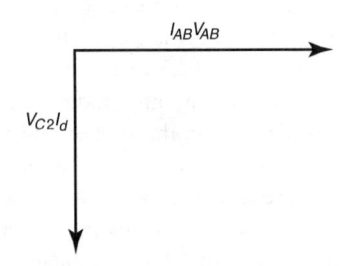

FIGURE 4.8 Vector diagram of reactance modulator producing FM. Note: $V_g = V_{C2}$.

By design, we also see that V_{C2} is also the voltage applied to the gate of the FET, V_g. Since the drain current variations in the FET are a direct result of variations in gate voltage, the drain current I_d is in phase with the gate voltage V_g. This is shown in Fig. 4.8 with I_d next to V_{C2}. As previously shown, we can put V_{AB} next to I_{AB} since they are in phase. The vector diagram shows that I_d, the drain current of the FET, is 90° behind V_{AB}. Notice that V_{AB} is also across the oscillator tuned circuit. Thus, we now have a circuit where the drain current lags the voltage across the oscillator tank by 90°. AC analysis has proven that the current through an inductance lags behind the voltage across the inductor by 90°. The voltage, V_{AB} therefore behaves the same as inductance. Again, as this inductance varies, so too will the total impedance of the oscillator tank circuit, which in turn results in FM.

---

[1] $g_m = \left(\dfrac{T_d}{V_{gs}}\right)\bigg|V_{DS}$

Voltage Controlled Oscillator (VCO) Direct-FM Modulators

Another one of the more common direct-FM modulation techniques in use today uses an analog voltage controlled oscillator (VCO) in a phase locked loop (PLL) arrangement. This is shown in Fig. 4.9. In this configuration, a VCO produces a desired carrier frequency, which is in turn modulated by applying the audio signal to the VCO input via a variactor diode. A variactor diode is generally used to vary the capacitance of a circuit. Therefore, the variactor behaves as a variable capacitor whose capacitance changes as the input voltage across it changes. As the input capacitance of the VCO is changed by the variactor, the output frequency of the VCO is shifted, which produces a direct-FM modulated signal.

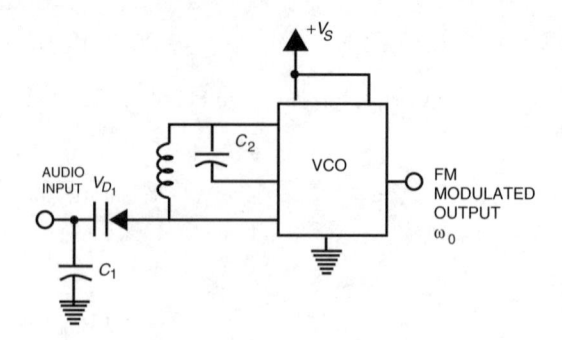

FIGURE 4.9 Voltage controlled direct-FM modulator.

Indirect-FM Modulators

Thus far we have only discussed the direct method of producing FM and just briefly mentioned the indirect method. Historically, the indirect method, or phase modulation, was originally developed by Armstrong in the early 1930s and the results were published in 1936 [Armstrong, 1936]. His was the first method to provide a practical system for producing an FM signal as many of the early FM transmitters used his method of modulation.

Simply stated, in phase modulation, the phase of the carrier signal deviates away from its resting position as modulation is applied. This is accomplished by passing a fixed RF carrier through a time-delay network, which makes the carrier change in phase. If the time-delay network is made to vary in accordance with the amplitude of an input signal, the delay network will change the phase of the carrier in accordance with the applied audio. The resultant output from the time-delay network will then be a phase modulated signal. The center frequency of the carrier is typically produced by using a stable oscillator circuit, such as a crystal, that is resonate at some frequency lower than that of the final desired output frequency. For this reason, there are typically many stages of frequency multiplication that are used following the phase modulator.

The basic principle of phase modulation can easily be illustrated in the following example. Figure 4.10 shows a simple phase modulated circuit consisting of a crystal oscillator (providing a center frequency) and a series RC network. The output of this circuit is taken across the variable resistor R whose resistance varies according to the applied audio signal. The vector diagram illustrates the impedance of the RC circuit for three different values of resistance. With the resistance variable, the resultant phase of the oscillator frequency will be variable and the modulated output voltage V_{AB} varies in phase. Resistance R_2 represents the circuit resistance with no modulation present. Resistance R_1 represents the circuit resistance

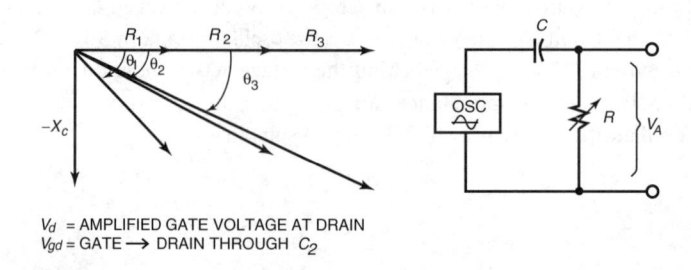

V_d = AMPLIFIED GATE VOLTAGE AT DRAIN
V_{gd} = GATE $\rightarrow$ DRAIN THROUGH C_2

FIGURE 4.10 Phase modulator principle and vector diagram.

Frequency Modulation 4-11

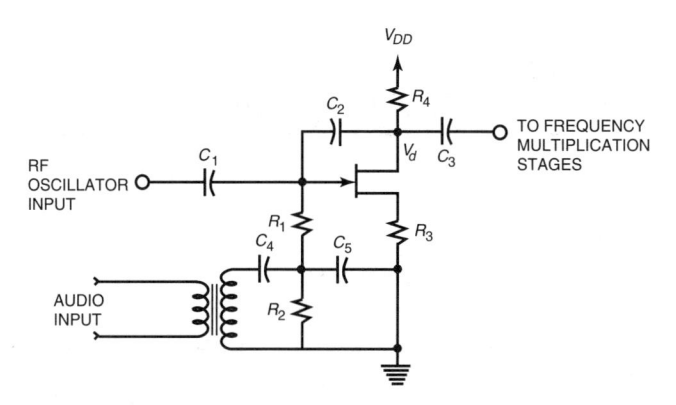

FIGURE 4.11 Simple FET phase modulator.

when the amplitude of the modulating signal swings positive, and resistance R_3 represents the circuit resistance as the amplitude of the modulating signal swings negative. The component $-X_C$ is the capacitive reactance of C.

When the modulating signal is applied to the circuit, it alternately causes the resistance to decrease R_1 and to increase R_3. The phase of the current through the RC circuit varies as long as the phase angle of the impedance varies. As we have shown, as R varies, the resultant phase angle θ varies from θ_1 to θ_3. The output voltage V_{AB} also follows the phase change in the current and impedance, and the result is a phase modulated signal.

In a real application, the variable resistance R in the example can be replaced with an active device such as an FET. For an FET to function properly, it must act like a variable resistance. This happens in the FET, for example, as the dynamic drain resistance is placed in parallel with the load resistor R_4 as illustrated in Fig. 4.11. The capacitive reactance of C_2 determines the amount of phase shift that occurs with the overall variable resistance of the FET. This example is one of the most simple types of phase modulators.

4.10 FM Reception Principles

Regardless of frequency, the receivers for FM communications are similar to those for AM. The functional layout for both types of receivers are similar in that a superheterodyne circuit is used. Both receiver types contain RF amplification, mixing, a local oscillator, IF amplification, detection, and amplification. There are, however, a few important design differences. Retrieval of the intelligence of an FM signal requires a slightly different detector circuit and some form of signal limiting. FM requires a detector that is designed to discriminate between a positive and negative frequency deviation. In addition, any variation in the amplitude of the carrier represents undesirable noise. This is removed by a **limiter** stage before demodulation occurs. A functional comparison between an AM receiver and two types of FM receivers is illustrated in Fig. 4.12.

As is apparent in the illustration, the major circuit differences between the AM and FM receivers consist of the limiter, **discriminator**, and **ratio-detector** stages. There are other minor differences, but these will not be discussed in detail since they are not unique to FM receiver design. These include differences in the receiver tuning and RF amplification stages and final amplification. The tuning range and RF amplification stages are different only in the design of the received frequency.

Limiters

The limiter stage of an FM receiver is basically an IF amplifier that is designed to saturate and clip off undesired AM and noise components from a signal prior to FM detection. With proper design, the output

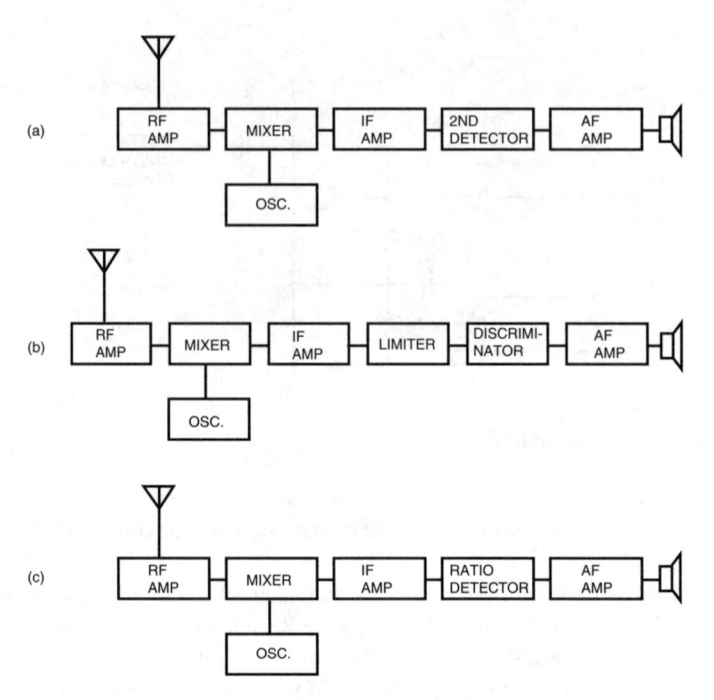

FIGURE 4.12 Functional block diagrams of typical AM and FM receivers: (a) AM receiver, (b) FM receiver using a Foster–Seeley discriminator and limiter, and (c) FM receiver using a ratio detector.

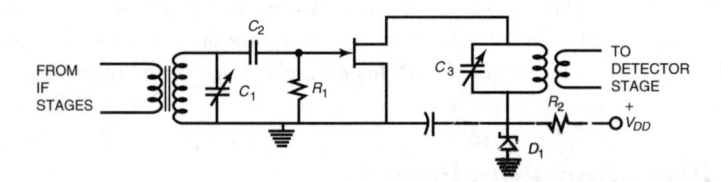

FIGURE 4.13 Typical FET limiter.

of the limiter will have a constant amplitude as the output signal is applied to the following detector stages. Figure 4.13 shows a limiter circuit designed with an FET.

When the noise levels on the positive peaks exceed a specified design level, limiting action is produced by driving the FET out of the active region into saturation. To help ensure saturation, the supply voltage is reduced to a level that still enables a small input signal to drive the FET into saturation. Lowering the drain voltage also reduces the overall gain of the limiter. For this reason, the majority of FM receivers typically have more IF stages than their AM counterparts. As the input signal swings negative, the extreme negative peaks are removed by driving the FET into cutoff. Notice the flattening of the upper portion of the output waveform shown in Fig. 4.14. This is caused by driving the FET into saturation. As the figure shows, an input signal is the voltage applied to the gate of the FET and the output current is the current through the drain.

Operation of the limiter is very straightforward. The input signal to the FET is coupled to the gate through C_2 via the tuned input tank circuit. Gate leak bias is used, which stabilizes the output signal and improves the limiter response time. The time constant of the bias network (C_2R_1) is typically in the range of 1–15 μs. For the limiter to operate properly, the input signal must have sufficient amplitude to drive the gate positive. As the gate signal goes positive, the gate current flows and charges C_2 to a value almost equal to the peak amplitude of the input signal. This causes the drain current to flow, and the FET acts like a normal RF amplifier. If the signal is strong, the FET will be driven into saturation and the drain

Frequency Modulation 4-13

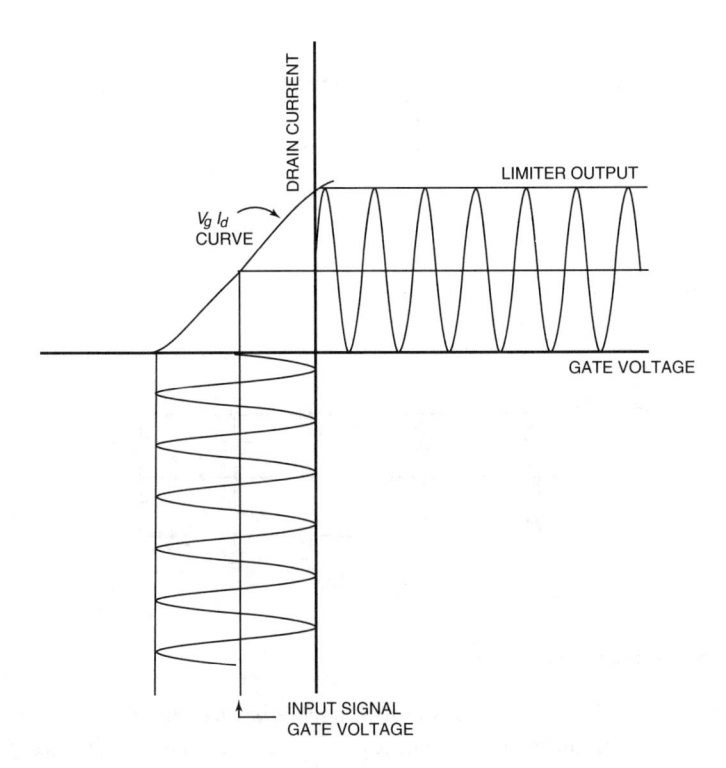

FIGURE 4.14 Dynamic transfer characteristics of an FM limiter (I_d vs. V_g).

current reaches its peak level. As the signal drops, the capacitor discharges through R_1 and produces a negative voltage, which reduces the drain current. The drain voltage is reduced to a level below that of other stages in the receiver by R_2 and zener D_1.

FM Detectors

The two most popular classic detector circuits for FM are the Foster–Seeley discriminator and the ratio detector. Other types of FM detection used today include phase-locked loop circuitry. The basic function of the discriminator is to convert the frequency swings in the FM signal back into amplitude variations for further audio processing. The discriminator is, therefore, susceptible to both amplitude and frequency variations. For this reason, the Foster–Seeley detector is always preceded by a limiter stage. The ratio detector, on the other hand, acts like a limiter, and so there is no need for the use of a separate limiter stage.

Discriminators and Ratio Detectors

Perhaps the most frequently used FM detector is the Foster–Seeley discriminator. The design is simple and its operational characteristics are less critical than other types. A typical Foster–Seeley discriminator circuit is illustrated in Fig. 4.15. The tuned circuit L_1C_1 is tuned to the center resting frequency of the IF stage. With no modulation present, the voltage developed across the center tap of L_1 (L_2 and L_3) is out of phase with the voltage across L_1. When this occurs, the current through D_1 and D_2 are equal, and the voltage developed across R_1 and R_2 will be equal and opposite in polarity. The output voltage from the circuit will, therefore, be zero.

When modulation occurs, the frequency varies above and below the carrier resting frequency. The tuned tank circuit, L_1C_1, becomes inductive or capacitive, depending on which way the signal shifts from the resting frequency. As the tuned circuit turns reactive, the phase angle changes between voltage and current. The resulting current flow is different flowing through D_1 from that flowing through D_2. As a

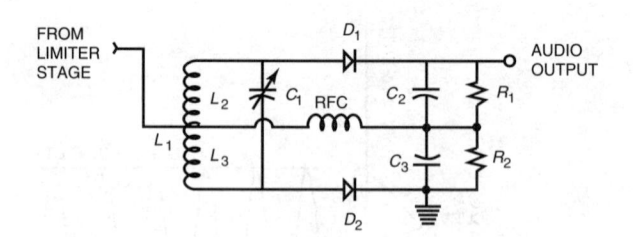

FIGURE 4.15 A simplified Foster–Seeley discriminator.

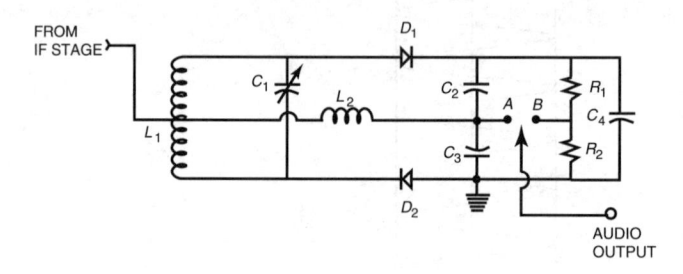

FIGURE 4.16 A simplified ratio detector.

rectified voltage is developed across R_1 and R_2, the difference between these two voltages will no longer be zero. This is, then, the reproduced output voltage that is produced as the modulation frequency varies.

The ratio detector is another commonly used FM detector. Notice that there are some close similarities between this and the Foster–Seeley discriminator. The major difference, as illustrated in Fig. 4.16, is that one of the diodes is reversed, there is a capacitor (C_4) across the other two capacitors, and the output voltage is taken between the junction of R_1 and R_2 and the taped junction of capacitors C_2 and C_3 at terminals A and B.

The basic operation of the ratio detector is similar to that of the Foster–Seeley discriminator. The input is coupled to the tuned circuit consisting of L_1 and C_1. At the resting frequency, diodes D_1 and D_2 conduct equally. As the frequency of the input signal swings above and below the resting frequency, either diode D_1 or D_2 will conduct more heavily. Since one of the diodes is reversed, current also flows through the entire circuit path of D_2 through the time constant of $C_4R_1R_2$, through D_1 and back into the top of L_1. After several RF cycles, C_4 charges to the peak value of the voltage across L_2. Variations in amplitude of the incoming signal have little effect on the charging of C_4, and the voltage across the capacitor remains fairly constant (due to a long time constant). This results in a practically constant voltage across $R_1 + R_2$ and $C_2 + C_3$. This, then, results in a constant output voltage. Since the output voltage was not effected by amplitude variations, the need for limiting is reduced.

With no modulation applied to the circuit, both diodes conduct equally and C_4 charges up. This results in a constant output voltage of zero at terminals AB. When the frequency changes as a result of modulation on the carrier, diode D_1 or D_2 will conduct more heavily. This makes the voltage ratio charged on C_2 and C_3 unequal, which gives the name ratio detector. The voltage at terminals AB is, therefore, the output signal of the demodulated FM signal.

Defining Terms

Direct FM: Frequency modulation produced by changing the frequency of a carrier as a result of applying a modulating signal.
Discriminator: A detector used in an FM receiver to demodulate the FM signal.
Frequency deviation: The peak difference between modulated wave and the carrier frequency.

Frequency modulation: A system of modulation where the instantaneous radio frequency varies in proportion to the instantaneous amplitude of the modulating signal and the instantaneous radio frequency is independent of the frequency of the modulating signal.

Limiter: A stage of an FM receiver that is designed to saturate and clip off undesired AM and noise components from a signal prior to FM detection.

Modulation index: The ratio of the frequency deviation to the frequency of the modulating signal.

Percentage modulation: The ratio of the actual frequency deviation to the frequency deviation defined as 100% modulation, expressed in percentage. For FM broadcast stations, a frequency deviation of +/−75 kHz is defined as 100% modulation.

Phase modulation: Frequency modulation produced by shifting the phase of the carrier relative to an arbitrary reference point. This method of frequency modulation is often referred to as indirect FM.

Ratio detector: A detector used in an FM receiver to demodulate the FM signal.

References

Armstrong, E.H. 1936. A method of reducing disturbances in radio signaling by a system of frequency modulation. *Proc. of IRE* 24(5):689–740.

Code of Federal Regulations. 1993. Vol. 47, Section 73.310.

Crutchfield, E.B., ed. 1985. *National Association of Broadcasters Engineering Handbook,* 17th ed., pp. 3.3-63–3.3-68. Washington, D.C.

Inglis, A.F. 1988. *Electronic Communications Handbook.* McGraw-Hill, New York.

Klapper, J., ed. 1970. *Selected Papers on Frequency Modulation.* Dover, New York.

Lewis, T. 1991. *Empire of the Air — The Men Who Made Radio.* Harper Collins, New York.

Rohde, L.U. and Bucher, T.T. 1988. *Communications Receivers Principles & Design.* McGraw-Hill, New York.

5
Pulse Modulation

Rodger E. Ziemer
*University of Colorado,
Colorado Springs*

5.1 Introduction 5-1
5.2 The Sampling Theorem 5-1
5.3 Analog-to-Digital Conversion 5-4
5.4 Baseband Digital Pulse Modulation 5-5
5.5 Detection of Pulse Modulation Formats 5-10
5.6 Analog Pulse Modulation 5-10

5.1 Introduction

Pulse modulation is important for many applications including telephone call transmission, compact disks for music, airline passenger communication systems, and digital control systems, among others. The reasons for the use of pulse modulation, even in cases where the information is analog in form, are many. For example, in telephone call transmission, the overloading of cable trays by copper wire in the late 1960s in part led to the development of the T carrier system, which employs **time-division multiplexing** (TDM). Thus, several separate telephone calls can be carried by the same transmission line. Airline passenger communication systems were similarly developed based on TDM to save in weight of the transmission lines required to address several functions at each passenger's seat, including the attendant calling function and several channels of entertainment media. In the late 1980s compact disks and the associated recording and playback technology were developed to provide high-fidelity music with little danger of degradation due to damage of the medium (i.e., the compact disk). Control systems in aircraft, particularly high-performance military aircraft, rely on the servomechanism commands and responses being conveyed by transmission lines to the various control points with these commands being represented in digital format. Spacecraft similarly utilize baseband digital pulse modulation to distribute commands and acquire data from the various sensors and systems on board.

5.2 The Sampling Theorem

The basis for pulse modulation applications is representing analog signals as properly spaced samples. The theoretical justification for this is Shannon's **sampling theorem** [Ziemer and Tranter, 1995], which may be stated succinctly as follows:

Lowpass Uniform Sampling Theorem: A signal with no frequency components above W Hz can be uniquely represented by uniformly spaced samples taken at intervals of no greater than $1/2W$ seconds apart.

Various forms of this sampling theorem are often used, including a bandpass uniform sampling version and quadrature sampling for bandpass signals.

The proof of the low-pass uniform sampling theorem (hereafter referred to simply as the sampling theorem) is based on the following Fourier transform theorems [Ziemer and Tranter, 1995]:

Multiplication Theorem: Given two signals, $x(t)$ and $y(t)$, with Fourier transforms $X(f)$ and $Y(f)$, where f is frequency in hertz, the Fourier transform of their product is the convolution of their Fourier transforms:

$$x(t)y(t) \leftrightarrow X(f)*Y(f) = \int_{-\infty}^{\infty} X(\lambda)Y(f-\lambda)d\lambda \tag{5.1}$$

where the double-headed arrow denotes a Fourier transform pair.

Fourier Transform of the Ideal Sampling Waveform: The Fourier transform of a doubly infinite train of uniformly spaced impulses is itself a doubly-infinite train of uniformly spaced impulses, or

$$\sum_{n=-\infty}^{\infty} \delta(t-nT_s) \leftrightarrow f_s \sum_{n=-\infty}^{\infty} \delta(f-nf_s), \; f_s = \frac{1}{T_s} \tag{5.2}$$

Consider a low-pass signal $x(t)$ with Fourier transform $X(f)$ bandlimited so that none of its spectral components lie above W Hz. Ideal impulse sampling of $x(t)$ gives

$$x_\delta(t) = x(t) \sum_{n=-\infty}^{\infty} \delta(t-nT_s) = \sum_{n=-\infty}^{\infty} x(nT_s)\delta(t-nT_s) \tag{5.3}$$

Use of Eqs. (5.1) and (5.2) to obtain the Fourier transform of $x_\delta(t)$ gives

$$X_\delta(f) = X(f)*f_s \sum_{n=-\infty}^{\infty} \delta(f-nf_s) = f_s \sum_{n=-\infty}^{\infty} X(f-nf_s) \tag{5.4}$$

where $\delta(f-nf_s) * X(f) = X(f-nf_s)$ has been used and the asterisk denotes convolution. Equation (5.4) is sketched in Fig. 5.1(b) for an assumed $X(f)$ shown in Fig. 5.1(a). It is seen that if the sampling frequency f_s is greater than $2W$, then the baseband portion of the sampled signal spectrum can be separated from the translated spectra centered around nonzero multiples of f_s, and the result is an undistorted version of the original signal spectrum $X(f)$. If the signal to be sampled, $x(t)$, is bandlimited but $f_s < 2W$, the translates of $X(f)$ centered at $f = \pm f_s$ will overlap with the component centered at $f = 0$. The resulting error is referred to as **aliasing**. If the signal to be sampled, $x(t)$, is not strictly bandlimited, then the sampling frequency cannot be chosen large enough to prevent overlap of the translates making up the sampled signal spectrum. Perfect distortionless recovery is impossible, and distortion results, due to the filtering operation used for signal and recovery, as does aliasing. These non-ideal recovery effects are illustrated in Fig. 5.2.

Clearly, sampling by means of impulses is impossible from a practical standpoint. A more practical waveform is the train of square pulses illustrated in Fig. 5.3(a). The sampling operation can be done in two ways with such a pulse train. The first, referred to as **natural sampling**, consists of multiplication of the signal to be sampled, $x(t)$, by the pulse train such as that shown in Fig. 5.3(a) as illustrated in Fig. 5.3(b). The second, called **flat top sampling**, is accomplished by sampling the waveform $x(t)$ at the instants nT_s and holding that value for the duration of the pulses making up the pulse train as illustrated in Fig. 5.3(c). For the former, the difference in the spectrum from ideal sampling is provided by the transform pair

$$p_\tau(t) = \begin{cases} 1, & 0 \leq t \leq \tau \\ 0, & \text{otherwise} \end{cases} \leftrightarrow \tau \operatorname{sinc}(\tau f)e^{-j\pi\tau f} \tag{5.5}$$

Pulse Modulation

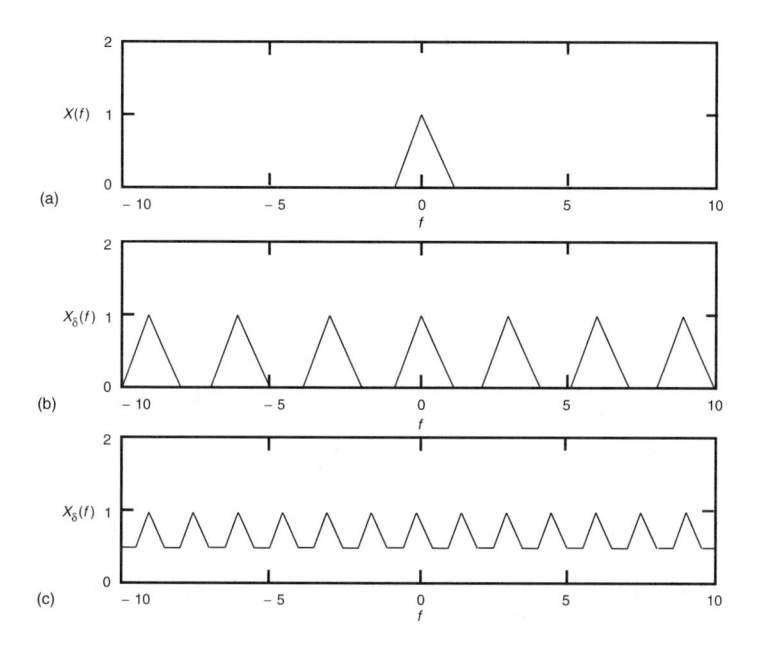

FIGURE 5.1 Spectra for sampling of a strictly bandlimited signal: (a) spectrum of signal to be sampled, (b) sampled signal spectrum of $f_s > 2W$, and (c) sampled signal spectrum for $f_s < 2W$.

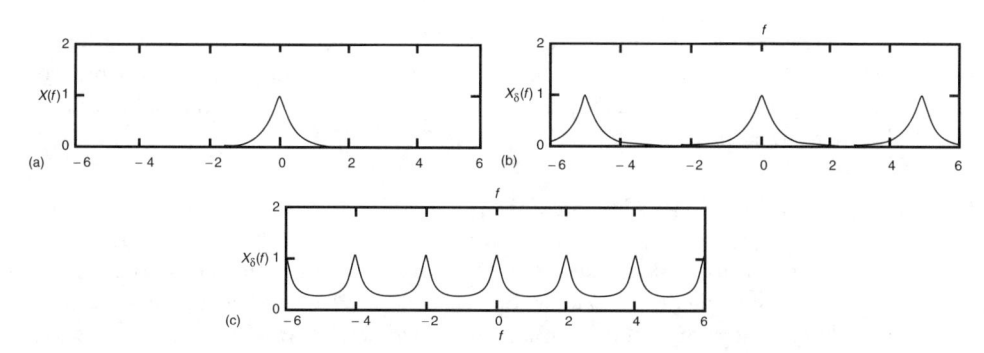

FIGURE 5.2 Sampling of a nonbandlimited signal spectrum: (a) spectrum of signal, (b) spectrum of sampled signal for large f_s (minimum aliasing), and (c) spectrum of sampled signal for moderate f_s (significant aliasing).

where τ is the pulse width and $\text{sinc}(u) = \sin(\pi u)/(\pi u)$. The ideal sampling waveform [Eq. (5.2)] can be converted to the natural sampling waveform shown in Fig. 5.3(a) by convolving $p_\tau(t)$ with Eq. (5.2). The Fourier transform of the natural sampled waveform is, therefore, the product of the transform in Eq. (5.2) and the transform in Eq. (5.5), and the overall effect on the spectrum is to multiply each term of Eq. (5.4) by $\tau \text{sinc}(n\tau f_s) e^{-jn\pi\tau f_s}$. Thus, the spectrum is not distorted, but each term in the spectrum is multiplied by the weighting function $\tau \text{sinc}(n\tau f_s)$. On the other hand, flat top sampling can be represented as

$$\begin{aligned}
x_{sh}(t) &= \sum_{n=-\infty}^{\infty} x(nT_s) p_\tau(t - nT_s) \\
&= p_\tau(t) * \sum_{n=-\infty}^{\infty} x(nT_s) \delta(t - nT_s) \\
&= p_\tau(t) * x_\delta(t)
\end{aligned} \qquad (5.6)$$

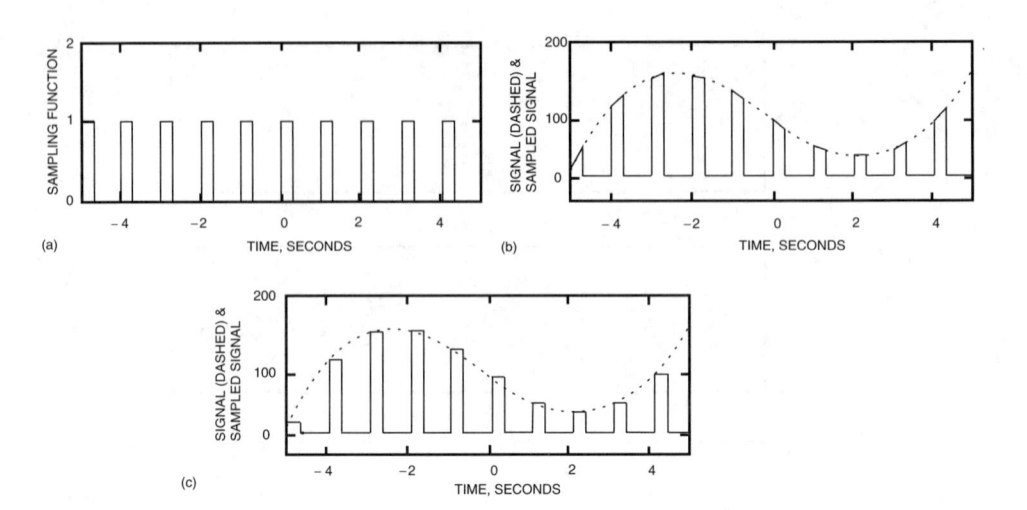

FIGURE 5.3 Illustration of (a) sampling function, (b) natural sampling, and (c) flat top sampling.

In analogy to Eq. (5.1), the Fourier transform of the convolution of two signals is the product of their respective Fourier transforms. Thus, the spectrum of Eq. (5.6) is

$$X_{sh}(f) = \tau f_s \, \text{sinc}(\tau f) \sum_{n=-\infty}^{\infty} X(f - nf_s) \tag{5.7}$$

It is seen that in the case of flat top sampling, the sampled signal spectrum is a distorted version of $X(f)$ because of the factor $\tau f_s \, \text{sinc}(\tau f)$. If $\tau \ll T_s$, this distortion is small and the original signal can be recovered almost exactly. If not, the factor $\tau f_s \, \text{sinc}(\tau f)$ must be compensated for by an inverse filter of the form $[\text{sinc}(\tau f)]^{-1}$ before recovery of $x(t)$.

5.3 Analog-to-Digital Conversion

Although analog samples can be transmitted from the source to destination, and historically this was done, it is now more usual to digitize the samples of an analog signal and transmit them in digital form. To do so, the additional steps of quantization and encoding are required first. This process is called **analog-to-digital (A/D) conversion.** This quantization is usually accomplished in terms of a binary code for each sample. Two methods for such representation will be discussed here. First, the base two representation for any set of positive integers is of the form

$$x(nT_s) = b_{m-1} \times 2^{m-1} + \cdots + b_2 \times 2^2 + b_1 \times 2^1 + b_0 \times 2^0 \tag{5.8}$$

where m is the number of bits b_i in the representation, also called the **wordlength**. As an example, consider the sample values of Fig. 5.3, which are given in Table 5.1, where the sample values are rounded to the nearest integer value. A disadvantage of the binary code is that several bits can change for a one-unit change in the sample value. For example, the **binary representation** of 15_{10} is 01111 whereas the binary representation of 16_{10} is 10000. An alternative representation is the **Gray code**, which can be found from the binary code by means of the following algorithm:

$$\begin{aligned} g_{m-1} &= b_{m-1} \\ g_i &= b_i \oplus b_{i+1}, \quad i < m-1 \end{aligned} \tag{5.9}$$

The Gray code representation of the samples shown in Fig. 5.1 is given in the fourth column of Table 5.1.

Pulse Modulation

Representation of the sample values by a binary code, although straightforward and intuitively satisfying, is not always the best course to follow. For example, large values are represented by the same number of bits as smaller values, and this may not be the best road to follow since smaller values may be more probable than the larger values. Two methods for representing more probable values of the samples may be employed: (1) nonuniform **quantization** and (2) compression of the signal to be sampled and quantized followed by subsequent expansion at the receiver, referred to as **companding**. The objective is to compress the signal before sampling and quantization such that the dynamic range is increased with very little loss in terms of signal of quantization noise power ratio. One such compressor characteristic is called the µ-**law compressor**, which is described by the input–output characteristic [Couch, 1990]

$$y(x) = V \operatorname{sgn}(x) \frac{\ln\left[1 + \frac{\mu |x|}{V}\right]}{\ln(1+\mu)} \qquad (5.10)$$

where V is the maximum input signal amplitude, μ is a parameter, and

$$\operatorname{sgn}(x) = \begin{cases} 1, & x > 0 \\ -1, & x < 0 \end{cases} \qquad (5.11)$$

The value $\mu = 255$ is used in the telephone system in the U.S. At the receiving end, the samples are put through a nonlinearity which is the inverse of Eq. (5.10).

Companders are an attempt at minimizing the error in representing a waveform in terms of quantized samples based on a fixed wordlength. Another approach at minimizing the average number of bits per sample is to use a variable length code. The optimum variable wordlength code for independent samples is called a **Huffman code** after its inventor [Blahut, 1990]. It is an algorithm for obtaining the minimum average number of bits per sample based on the probability distribution of the samples, with the lower probability samples being assigned the most number of bits and the most probable sample values being assigned the least number of bits. The process of encoding samples at a source is called **source encoding** and can be viewed as removing redundancy from the source output. Another type of encoding, called **channel encoding** or *error correction encoding*, adds redundancy in terms of extra bits appended to each encoded sample so that errors can be corrected at the reception point.

TABLE 5.1 Sample Values and Binary Representations for the Waveform of Fig. 5.3(c)

Time, nT_s	Sample Value	Binary Code	Gray Code
−5	18.75	00010011	00011010
−4	121.76	01111010	01000111
−3	163.57	10100100	11110110
−2	163.68	10100100	11110110
−1	137.99	10001010	11001111
0	100	01100100	01010110
1	62.01	00111110	00100001
2	35.32	00100011	00110010
3	36.43	00100100	00110110
4	78.24	01001110	01101001
5	181.25	10110101	11101111

5.4 Baseband Digital Pulse Modulation

After the samples have been quantized, they are transmitted through a channel, received, and converted back to their approximate original form. The reason for the modifier approximate is that they will invariably suffer some degradation from noise and channel-induced distortion. This will be explored shortly. It is first important to point out that transmission does not necessarily mean to a remote location. For example, storage of data on a computer hard disk and retrieval may happen in the same physical location. It is true that the storage of Beethoven's Fifth Symphony on a compact disk and the subsequent playing back will usually happen at remote locations.

The question at hand is in what format, hereafter called modulation, should the quantized samples be placed for faithful transmission through the channel and subsequent reproduction? The answer depends on the channel. In the case of magnetic recording media, low-frequency response of the medium and recording and pickup heads is poor, so a type of modulation must be used that minimizes the low-frequency content of the signal being recorded. A myriad of formats is possible. We will discuss only a few. In Fig. 5.4, several possible formats are shown. The first is called **nonreturn-to-zero** (NRZ) polar because the waveform does not return to zero during each signaling interval, but switches from +V to −V, or vice versa, at the end of each signaling interval (NRZ unipolar uses the levels V and 0). On the other hand, **unipolar return-to-zero** (RZ) format, shown in Fig. 5.4(b) returns to zero in each signaling interval. Since bandwidth is inversely proportional to pulse duration, it is apparent that RZ requires twice the bandwidth that NRZ does. Another is that RZ has a nonzero DC component, whereas NRZ does not have to unless there are more 1s than 0s or vice versa. An advantage of RZ over NRZ is that a pulse transition is guaranteed in each signaling interval whereas this is not the case for NRZ. Thus, in cases where there are long strings of 1s or 0s, it may be difficult to synchronize the receiver to the starting and stopping times of each pulse in the case of NRZ. A very important modulation format from synchronization considerations is NRZ-mark, also known as **differential encoding**, where an initial reference bit is chosen and a subsequent 1 is encoded as a change from the reference and a 0 is encoded as no change. After the initial reference bit, the current bits serves as a reference for the next bit, etc. An example of this modulation format is shown in Fig. 5.4(c).

Another baseband data modulation format that guarantees a transition in each signaling interval and does not have a DC component is known as Manchester, biphase, or **split phase**, which is illustrated in Fig. 5.4(d). It is produced by ORing the data clock with an NRZ-formatted signal. The result is a + to − transition for a logic 1, and a − to + zero crossing for a logic 0.

Several other data formats have been proposed and employed in the past, but we will consider mainly these. An important consideration of any data format is its **bandwidth occupancy**. It can be shown [Haykin, 1988] that NRZ polar has the power spectral density

$$S_{\text{NRZ}}(f) = V^2 T_b \operatorname{sinc}^2(fT_b) \tag{5.12}$$

whereas unipolar RZ format has the power spectrum

$$S_{\text{URZ}}(f) = \frac{V^2 T_b}{16} \operatorname{sinc}^2\left(\frac{fT_b}{2}\right) + \frac{V^2}{16} \sum_{\substack{m=-\infty \\ m \text{ odd}}}^{\infty} \frac{4}{(\pi m)^2} \delta\left(f - \frac{m}{T_b}\right) + \frac{V^2}{16} \delta(f) \tag{5.13}$$

where the data function at $f = 0$ reflects the nonzero DC level of the unipolar return-to-zero format. On the other hand, differential encoded (NRZ-mark) and split phase are formats with no DC level. The power spectral density of the former is the same as for NRZ, and that of the latter is

$$S_{\text{SP}}(f) = V^2 T_b \operatorname{sinc}^2\left(\frac{fT_b}{2}\right) \sin^2\left(\frac{\pi f T_b}{2}\right) \tag{5.14}$$

The total average power in each case is obtained by integrating the power spectrum over $-\infty < f < \infty$. These four spectra are plotted in Fig. 5.5, where it is noted that the nonreturn-to-zero occupies half the bandwidth of unipolar RZ and split phase.

Of the four modulation formats discussed, the first two assumed zero memory between pulses, and the latter two had memory imposed between pulses. Split phase has zero power density at $f = 0$ with the

Pulse Modulation 5-7

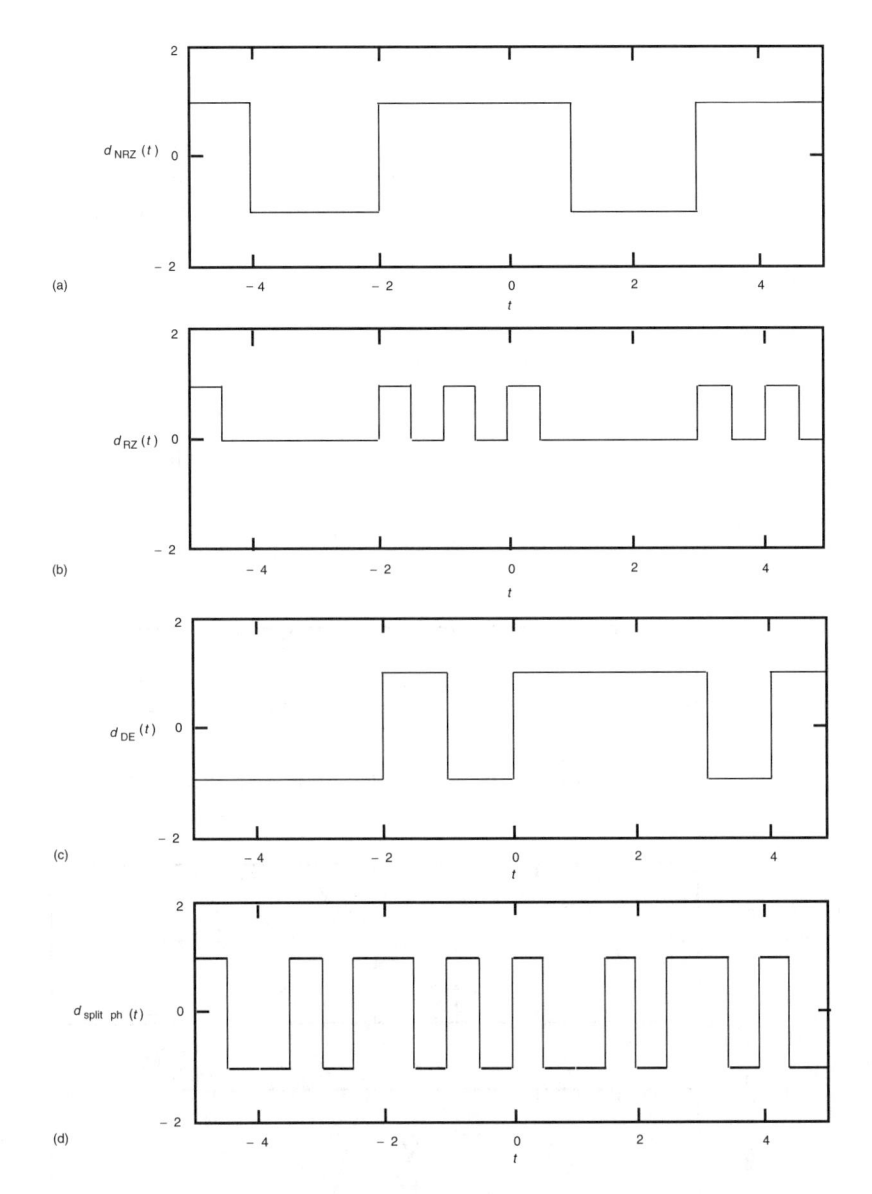

FIGURE 5.4 Various baseband modulation formats: (a) nonreturn-to-zero, (b) unipolar return-to-zero, (c) differential encoded (NRZ-mark), and (d) split phase.

result that its bandwidth is double that of nonreturn-to-zero. In a sense, the zero power density at $f = 0$ was obtained in the case of split phase by imposing a particular type of memory between pulses. More general memory structures are used between pulses for applications such as magnetic recording. These can be classified as **line codes**. It is beyond the scope of this chapter to go into this subject here. A simple example is provided by assuming a square pulse function of width T_b for each bit, but with successive pulse multipliers related by

$$a_k = A_k - A_{k-1} \tag{5.15}$$

where $A_k = \pm 1$ represent the bit value in signaling interval k. Thus the multiplier for pulse k can assume the values 2 ($A_k = 1$ and $A_{k-1} = -1$), 0 ($A_k = 1$ and $A_{k-1} = 1$), or -2 ($A_k = -1$ and $A_{k-1} = 1$). The power spectral density of this pulse modulation format can be shown to be [Ziemer and Tranter, 1995]

$$S_{DC}(f) = 4V^2 T_b \, \text{sinc}^2(fT_b) \sin^2(\pi f T_b) \tag{5.16}$$

Such pulse modulation is referred to as *dicode*. If the original bit stream is precoded with differential encoding, it is referred to as *duobinary*, a modulation scheme that was invented by Lender [1966].

Before leaving the subject of pulse modulation formats, we discuss one more principle involved in choosing pulse shapes with bandwidth occupancy in mind, known as **Nyquist's pulse-shaping criterion**. The idea is to find pulse shapes $p(t)$ with bandlimited spectra that have the zero **intersymbol-interference** property given by

$$p(nT_b) = \begin{cases} 1, & n = 0 \\ 0, & n \neq 0 \end{cases} \tag{5.17}$$

The condition (5.17) says that if the output of the transmitter/channel/receiver filter cascade is of the form

$$x_r(t) = V \sum_{k=-\infty}^{\infty} a_k p(t - kT_b) \tag{5.18}$$

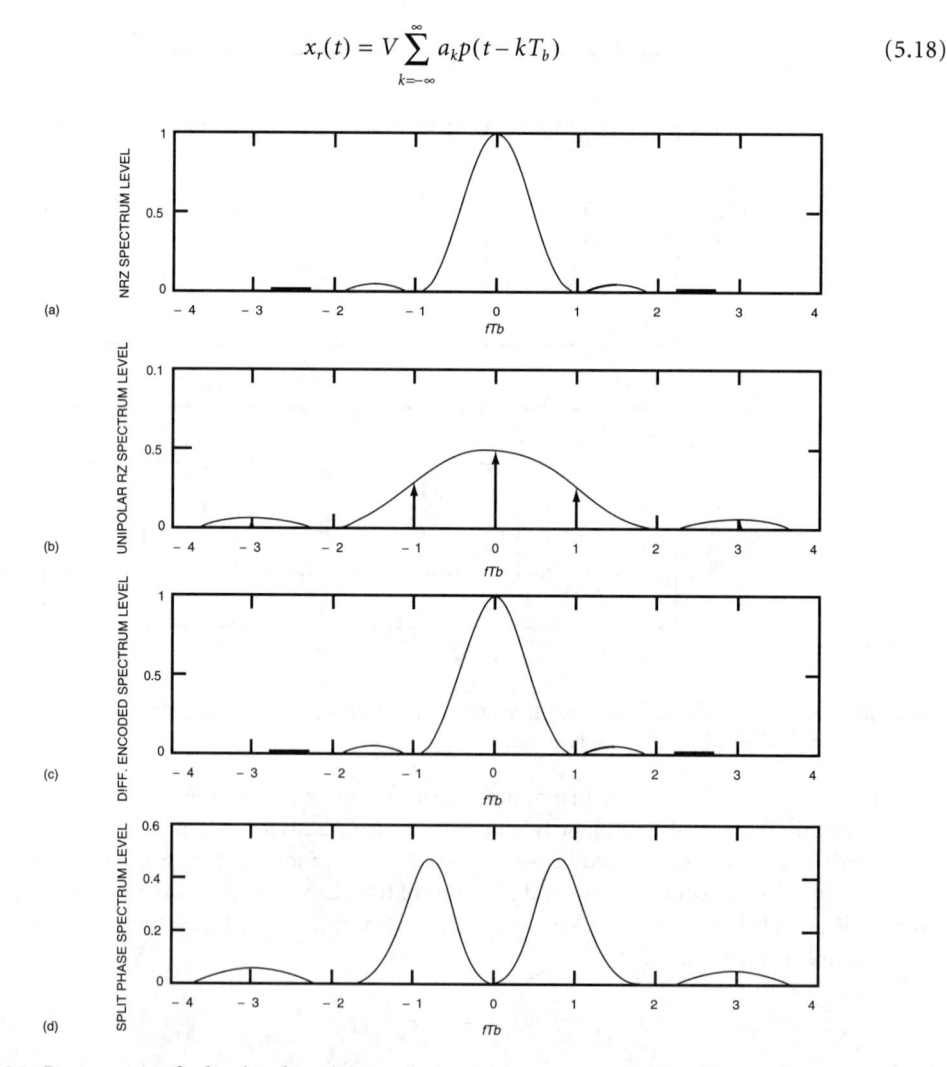

FIGURE 5.5 Power spectra for baseband modulation formats: (a) nonreturn-to-zero, (b) unipolar return-to-zero, (c) differential encoded, and (d) split phase.

Pulse Modulation

where the a_k represent the bit value (± 1), then sampling at intervals of T_b ensures that a given pulse sample is not influenced by preceeding or following pulses. This is zero intersymbol interference. Nyquist's pulse shaping criterion states that Eq. (5.18) holds if the Fourier transform $P(f) = \mathcal{F}[p(t)]$ of $p(t)$ satisfies

$$\sum_{k=-\infty}^{\infty} P\left(f + \frac{k}{T_b}\right) = T_b, \quad |f| \leq \frac{1}{2T_b} \tag{5.19}$$

One pulse family that has this property is that family having **raised-cosine spectra.** This pulse shape family is given by

$$p(t) = \frac{\cos 2\pi\beta t}{1 - (4\beta t)^2} \operatorname{sinc}(t/T_b) \tag{5.20}$$

which has the spectrum

$$P(f) = \begin{cases} T_b, & |f| \leq \left(\frac{1}{2T_b}\right) - \beta \\ \frac{T_b}{2}\left[1 + \cos\left(\frac{\pi(|f| - 1/(2T_b) + \beta)}{2\beta}\right)\right], & \frac{1}{2T_b} - \beta < |f| \leq \frac{1}{2T_b} + \beta \\ 0, & |f| > \left(\frac{1}{2T_b}\right) + \beta \end{cases} \tag{5.21}$$

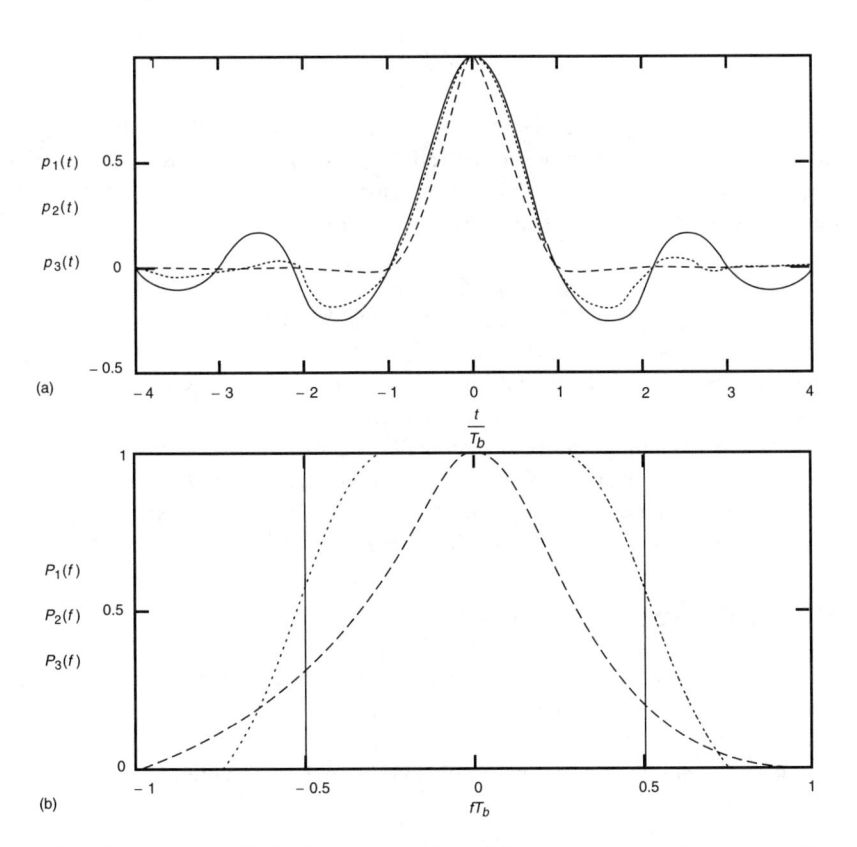

FIGURE 5.6 The raised cosine family for $\beta = 0$, 0.25, and 0.5; (a) pulse shapes and (b) corresponding spectra.

The parameter β determines the bandwidth of the pulse spectrum and its rate of decrease to zero. This pulse shape family and the corresponding spectra are shown in Fig. 5.6.

5.5 Detection of Pulse Modulation Formats

With no dependency between signaling pulses, the optimum detection scheme for a digitally modulated pulse train in additive white Gaussian noise (AWGN) is to pass the received pulse train plus noise through a filter matched to the basic signaling pulse, sample the output at the time of the peak output signal, and compare this sample with a suitably chosen threshold. A **matched filter** has impulse response proportional to the time reverse of the signal to which it is matched. It can be shown that this procedure is equivalent to correlating the received signal plus noise with a replica of the basic pulse shape, sampling, and threshold comparison. These two basic detector structures are shown in Figs. 5.7(a) and 5.7(b). For a rectangular pulse shape, the **correlation receiver** is equivalent to integrating the pulse being detected over its width, sampling at the end of the pulse, and comparing threshold. This receiver structure, called an **integrate-and-dump detector**, is shown in Fig. 5.7(c). NRZ pulse modulation fits this description, and it can be shown that the average probability of making an error in detecting NRZ using this procedure is given by

$$P_E = Q\left(\sqrt{\frac{2E_b}{2N_0}}\right), \quad \text{NRZ} \tag{5.22}$$

where

$$Q(x) = \int_x^\infty \frac{e^{-u^2/2}}{\sqrt{2\pi}} du \tag{5.23}$$

is the Q function, N_0 is the one-sided noise power spectrum level, and $E_b = V^2 T_b$ is the energy in the pulse.

Since unipolar RZ uses half the pulse width of NRZ, and since zero level is used to represent a logic 0 (as opposed to $-V$ for NRZ), its performance is a factor of 4 worse than that of NRZ, with the result for the **probability of error** being (E_b is now the average bit energy)

$$P_E = Q\left(\sqrt{\frac{E_b}{2N_0}}\right), \quad \text{unipolar RZ} \tag{5.24}$$

For differential encoding, essentially two bit errors result each time there is a bit error (the present bit serves as a reference for a succeeding bit) so that for a given E_b/N_0 the probability of error is double that of NRZ. This amounts to about a 0.8-dB shift on the E_b/N_0 axis in a plot of P_E vs. E_b/N_0. The P_E results for these pulse modulation formats are plotted in Fig. 5.8 as a function of E_b/N_0.

Finally, detection of split phase is accomplished by the same detector as for NRZ with a multiplication by the clock proceeding the detector. The probability of error is the same as for NRZ.

5.6 Analog Pulse Modulation

The concentration in this section has been on digital pulse modulation methods. Analog pulse modulation, whereby some attribute of a pulse is made to vary in a one-to-one correspondence with the signal samples, is not as important now as it once was. The main types of analog pulse modulation are pulse amplitude, pulse width, and pulse position modulation [Ziemer and Tranter, 1995].

Pulse Modulation

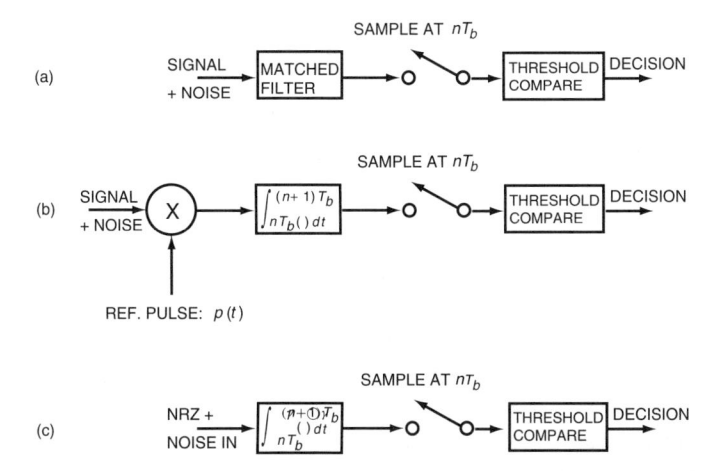

FIGURE 5.7 Probability of error curves for (a) NRZ, (b) unipolar RZ, and (c) differential encoding.

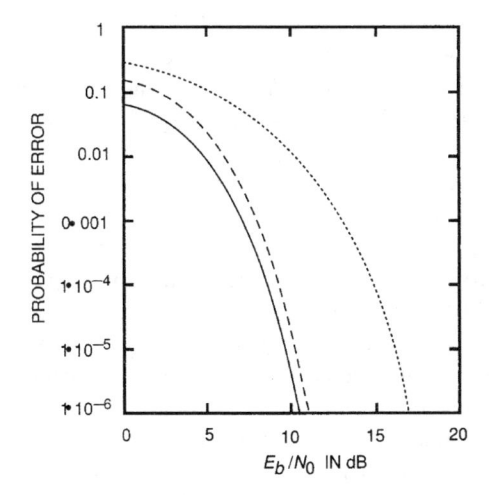

FIGURE 5.8 Detectors for pulse modulated signals: —matched filer detector, – – –correlation detector, - - -integrate-and-dump detector for rectangular pulse signals.

Defining Terms

Aliasing: Distortion arising from the representation of a signal by its samples due to too low a sampling rate relative to the signal bandwidth.

Analog-to-digital (A/D) conversion: The process of approximating an analog sample in terms of a finite-precision number, usually in binary form.

Bandwidth occupancy: The amount of bandwidth in hertz occupied by a signal. Quite often, an approximate measure must be used because the total bandwidth extent of typical signals is infinite.

Binary representation: Representation of sample values in terms of a binary number.

Channel encoding: The process of appending redundant bits onto the digital information sent through a channel with the goal of being able to correct errors at the receiver.

Companding: The process of compressing a signal in amplitude before A/D conversion and then expanding it after digital-to-analog conversion. The purpose is to get more resolution for small and moderate sample values and less at large amplitudes. Without quantization, companding would result in no waveform distortion.

Correlation receiver: A receiver structure that correlates received signal plus noise with replicas of the possible received signal shapes. It is completely equivalent in terms of performance to the matched filter receiver.

Differential encoding (NRZ-mark): The process of encoding a bit stream using the present bit as a reference for the following bit, with a 1 encoded as a change from the reference and a 0 encoded as no change.

Flat top sampling: Sampling in which the sampling pulses hold the value of the signal being sampled for a short period of time.

Gray code: A particular binary number representation in which only one bit changes in going from one level to an adjacent level.

Huffman code: A variable-length source code that optimally represents lower probability samples with long codewords and higher probability samples with short codewords so as to minimize overall average codeword length.

Integrate-and-dump detector: A form of the matched filter or correlation detector that is specialized to rectangular signal shapes. Thus, the correlation operation reduces to an integration of the signal interval.

Intersymbol interference: Interference from preceeding or succeeding pulses being smeared into a pulse of interest during a sampling process, usually at the receiver.

Line codes: Codes that utilize memory between pulses to control spectral shape, among other reasons. *Dicode* and *duobinary* are early examples of such encoding techniques.

Matched filter: A fixed, linear filter that maximizes peak signal-to-rms noise ratio. Its impulse response is the time reverse of the signal.

μ-Law compressor: A particular compressor characteristic (see *companding*) used in telephone representation of speech signals.

Natural sampling: Sampling that can be modeled as multiplication of periodic rectangular-pulse sampling waveform by the signal to be sampled. During the sampling times, the sample values follow the shape of the signal being sampled.

Nonreturn-to-zero: A pulse modulation format where ones are represented by a constant positive level for a time period called the bit period and zeros by minus that constant level for the same period. Thus, the pulse representation never returns to zero.

Nyquist's pulse-shaping criterion: A condition placed on the spectrum of a pulse shape function that guarantees zero intersymbol interference when samples are taken at a proper sampling rate.

Probability of error: A measure of performance for a digital receiver. Over a long string of received symbols (signals), it is approximately the ratio of the number of symbols received erroneously divided by the total number.

Pulse modulation: A process whereby information is impressed on a pulse train carrier for transmission through a channel.

Quantization: One of the steps in A/D conversion whereby samples assuming a continuum of values are approximated by finite-precision values.

Raised-cosine spectra: The spectrum of one family of pulses that obeys Nyquist's pulse-shaping criterion.

Sampling Theorem: One of several theorems, the most common of which is called the low-pass sampling theorem and says that a low-pass bandlimited signal of bandwidth W hertz may be represented in terms of samples taken periodically at a minimum rate of $2W$ per second.

Source code: Any coding technique to represent the output of a source, usually with the objective of minimizing the average wordlength.

Split phase: A pulse modulation format that amounts to nonreturn-to-zero multiplied by the data clock, which assumes ±1 values (also called Manchester or biphase).

Time-division multiplexing: Interlacing pulse samples from different sources in time so that they can be transmitted through a common channel.

Unipolar return-to-zero: A pulse modulation format wherein ones are represented by a positive pulse level during the first-half of the bit period and zero during the last-half and zeros are represented by a zero level.

Wordlength: As pertaining to A/D conversion, the number of digits, usually bits, used to represent a sample.

References

Blahut, R.E. 1990. *Digital Transmission of Information.* Addison-Wesley, Reading, MA.
Couch, L.W. 1990. *Digital and Analog Communication Systems.* 3rd ed. Wiley, New York.
Haykin, S. 1988. *Digital Communications.* Wiley, New York.
Lender, A. 1966. Correlative level encoding for binary data transmission. *IEEE Spectrum* (Feb.):104–115.
Ziemer, R.E. and Tranter, W.H. 1995. *Principles of Communications: Systems, Modulation, and Noise,* 4th ed. Wiley, New York.

6
Digital Modulation

	6.1	Introduction 6-1
	6.2	Detection of Binary Signals in Additive White Gaussian Noise 6-1
		Binary, Coherent Modulation Schemes • Binary, Noncoherent Modulation Schemes • Bandwidth Efficiency
	6.3	Detection of M-ary Signals in Additive White Gaussian Noise 6-8
		Signal Detection in Geometric Terms • The Gram–Schmidt Procedure • Geometric View of Signal Detection
	6.4	Comparison of Modulation Schemes.......................... 6-15
		Bandwidth Efficiency • Power Efficiency • Other Important Types of Digital Modulation Schemes
Rodger E. Ziemer	6.5	Higher Order Modulation Schemes............................ 6-20
University of Colorado, Colorado Springs		Error Correction Coding Fundamentals • Trellis-Coded Modulation

6.1 Introduction

Digital modulation is necessary before digital data can be transmitted through a bandpass channel. Examples of such channels are microwave line of sight, satellite, optical fiber, and cellular mobile radio. Modulation is the process of varying some attribute of a carrier waveform, such as amplitude, phase, or frequency, in accordance with the message to be transmitted. In the case of digital modulation, the message sequence is a stream of digits, quite often binary valued. In the simplest case, this parameter variation is on a symbol-by-symbol basis (zero memory), and the carrier parameters that can be varied are amplitude [**amplitude-shift keying (ASK)**], phase [**phase-shift keying (PSK)**], or frequency [**frequency-shift keying (FSK)**]. So-called higher order modulation schemes impose memory over several symbol periods. Modulation techniques can be classified as **binary** or **M-ary** depending on whether one of two possible **signals** or $M > 2$ signals per signaling interval can be sent. If the latter, and if the source digits are binary, it is clear that several bits must be grouped together in order to make up an M-ary word. Another classification for digital modulation techniques is **coherent** vs. **noncoherent**, depending on whether a reference carrier at the receiver coherent with the received carrier is necessary for demodulation (coherent) or not (noncoherent).

6.2 Detection of Binary Signals in Additive White Gaussian Noise

Binary, Coherent Modulation Schemes

The simplest possible digital communications system is one which transmits a sequence of binary symbols represented for convenience by {0, 1} from a transmitter to a receiver over a channel that degrades the transmitted signal with **additive white Gaussian noise (AWGN)** of two-sided spectral density $N_0/2$. The transmitted binary symbols are associated with two signaling waveforms, denoted as $s_1(t)$ and $s_2(t)$, defined to exist over the time interval $(0, T_b)$. (The symbol T_b will be used to denote **bit period**. Later, T_s will be used to denote **symbol period** when signaling schemes that select from more than two possible transmitted signals are discussed.) One of these signals is transmitted each T_b seconds so that the information transmission rate is $R_b = 1/T_b$ binary symbols (bits) per second (b/s). During signaling interval k, the transmitter associates a symbol, such as a 1, with $s_1(t - kT_b)$ and the other symbol, a 0, with $s_2(t - kT_b)$. The receiver is assumed to have perfect knowledge of both $s_1(t)$ and $s_2(t)$, including the precise time at which they could be received and the probability that they were transmitted, assumed to be equally likely, but the receiver does not know which signal was, in fact, transmitted. During each T_b-signaling interval, the receiver observes the signaling waveform contaminated in AWGN and processes this information so as to minimize the probability of making an error.

It can be shown [Ziemer and Tranter, 1995; Wozencraft and Jacobs, 1965; Blahut, 1990; Proakis, 1995] that the minimum probability of error is achieved when the receiver guesses the transmitted signal to be that signal which, given the received signal plus noise waveform, was most likely to have been transmitted. Such a receiver is called a **maximum-likelihood receiver**. For equally likely binary symbols transmitted in AWGN, it can be shown [Ziemer and Tranter, 1995] that the minimum **probability of error** is

$$P_E = Q[\sqrt{z(1 - R_{12})}] \tag{6.1}$$

where

$$Q(x) = \int_x^\infty \frac{e^{-u^2/2}}{\sqrt{2\pi}} du \tag{6.2}$$

is the Q function. The quantities z and R_{12} are defined as

$$z = \frac{E_1 + E_2}{2N_0} = \frac{E_b}{N_0} \tag{6.3}$$

and

$$R_{12} = \frac{\sqrt{E_1 E_2}}{E_b} \rho_{12} \tag{6.4}$$

in which E_i, $i = 1, 2$, is the energy of signal i defined as

$$E_i = \int_0^{T_b} |s_i(t)|^2 dt \tag{6.5}$$

and $E_b = (E_1 + E_2)/2$ is the average signal energy. The parameter ρ_{12} is the normalized **correlation coefficient** between signals, which is given by

$$\rho_{12} = \frac{1}{\sqrt{E_1 E_2}} \int_0^{T_b} s_1(t) s_2(t) dt \tag{6.6}$$

If $R_{12} = 0$, the signaling scheme is said to be **orthogonal**, whereas if $R_{12} = -1$, the signaling scheme is said to be **antipodal**. An example of the former is coherent, binary FSK and an example of the latter is binary PSK. The probability of error is shown in Fig. 6.1 as a function of $z = E_b/N_0$ in decibel for these two cases.

The receiver for these binary signaling schemes can have one of two equivalently performing structures—a **matched filter implementation** and a **correlation implementation**. A block diagram for the

Digital Modulation

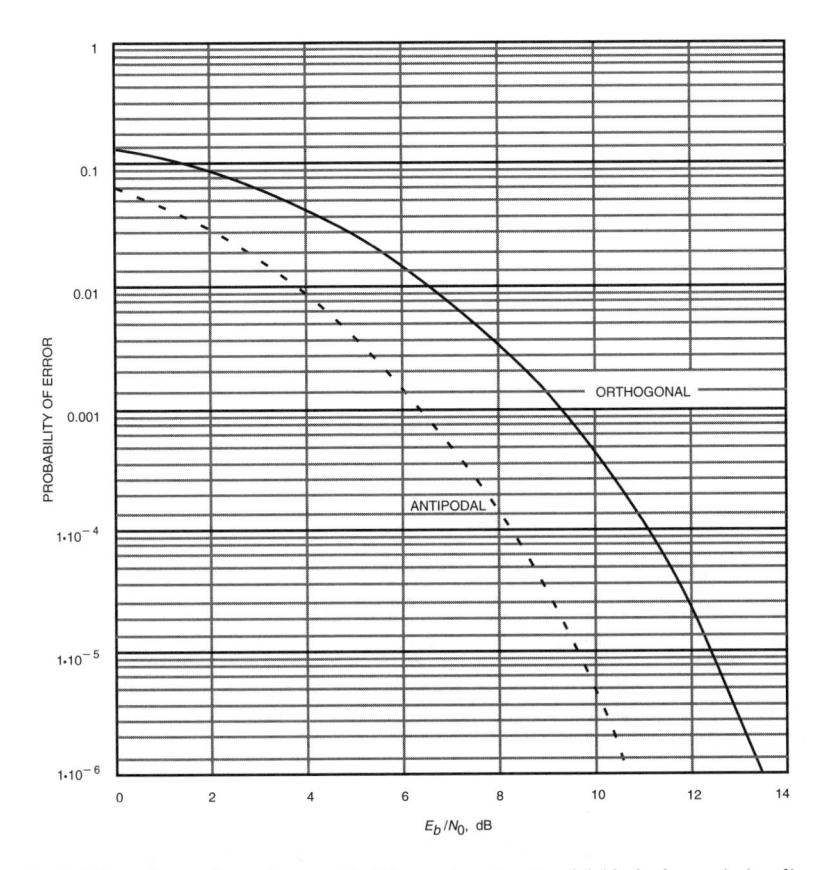

FIGURE 6.1 Probability of error for orthogonal (solid curve) and antipodal (dashed curve) signaling.

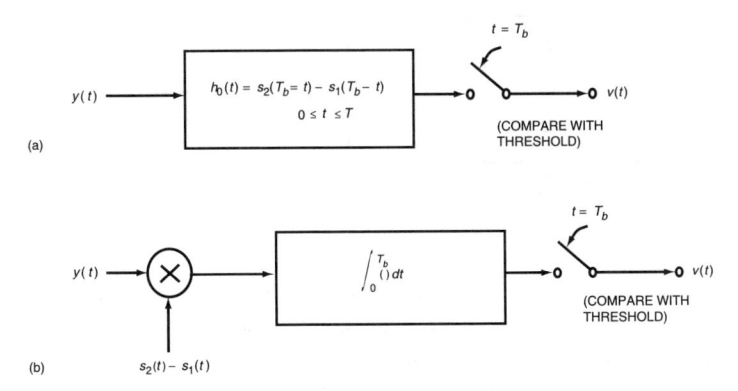

FIGURE 6.2 Implementations of the minimum probability of error receiver for binary signal reception: (a) matched filter, and (b) correlator.

matched filter receiver is shown in Fig. 6.2(a), and is seen to consist of a matched filter followed by a sampler, which samples the output of the matched filter at the end of each T_b-s signaling interval, and a threshold comparator. A matched filter for any signal has an impulse response that is the shifted time reverse of the signal. Since we are dealing with two signals in this case, the matched filter is matched to the *difference* of the two signals and has impulse response

$$h(t) = s_2(T_b - t) - s_1(T_b - t), \quad 0 \leq t \leq T_b \tag{6.7}$$

TABLE 6.1 Characteristics of Coherent Binary Digital Modulation Schemes

Name	Signal	Threshold Eq. (6.8)	Signal Corr. Coeff., R_{12}
Antipodal baseband[a] signaling	$s_{1,2}(t) = \pm\sqrt{\dfrac{E_b}{T_b}}, \quad 0 \le t \le T_b$	0	−1
Amplitude-shift keying	$\left. \begin{array}{l} s_{1,2}(t) = 0 \\ s_2(t) = \sqrt{\dfrac{4E_b}{T_b}}\cos(\omega_c t) \end{array} \right\} \; 0 \le t \le T_t$	$E_2/2$	0
(Binary) phase-shift keying (PSK)	$s_{1,2}(t) = \sqrt{\dfrac{2E_b}{T_b}}(\sin\omega_c t \pm \cos^{-1} m), \quad 0 \le t \le T_b$ where $|m| \le 1$ is the modulation index	0	$2m^2 - 1$
Biphase-shift keying (BPSK)	$s_{1,2}(t) = \pm\sqrt{\dfrac{2E_b}{T_b}}\cos(\omega_c t), \quad 0 \le t \le T_b$	0	−1
Frequency-shift keying (FSK)	$\left. \begin{array}{l} s_1(t) = \sqrt{2E_b/T_b}\cos(\omega_c t) \\ s_2(t) = \sqrt{2E_b/T_b}\cos[(\omega_c + \Delta\omega)t] \end{array} \right\} \; 0 \le t \le T_b$ $\Delta\omega = \pi \times \text{integer } T_b$	0	0

[a] In all cases, E_b is the average signal energy per bit, E_1 is the energy of signal 1, and E_2 is the energy of signal 2. All signalling schemes except antipodal baseband are referred to as coherent, because the carrier phase must be known at the receiver to implement the matched or correlator detector.

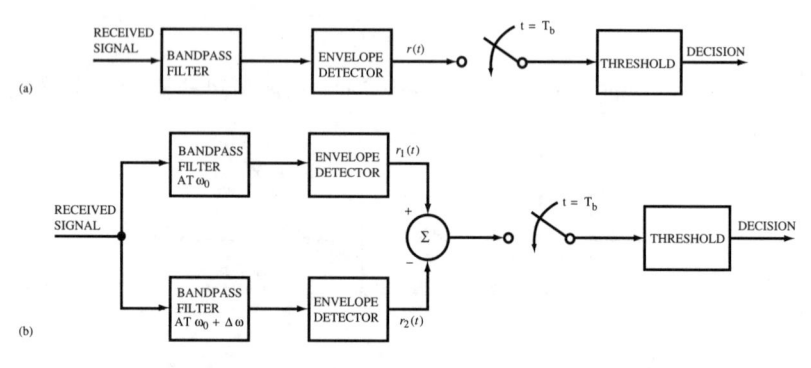

FIGURE 6.3 Receivers for noncoherent detection of binary signals: (a) ASK and (b) FSK.

A block diagram for the correlator receiver is shown in Fig. 6.2(b), and consists of a correlation operation with the difference of the two signals, followed by a sampler and threshold comparison. The correlator consists of the multiplier and integrator cascade.

For equally probable signals, the comparator threshold is set at

$$k = \frac{1}{2}[s_{o1}(T_b) + s_{o2}(T_b)] \tag{6.8}$$

where $s_{o1}(T_b)$ and $s_{o2}(T_b)$ are the output signals from the matched filter at the sampling instant corresponding to $s_1(t)$ and $s_2(t)$, respectively, at its input. The threshold simplifies to [Ziemer and Tranter, 1995]

$$k = \frac{1}{2}(E_2 - E_1) \tag{6.9}$$

Digital Modulation

Several special cases of interest in the binary signaling hierarchy are listed in Table 6.1, along with the thresholds and correlation coefficients in each case.

Binary, Noncoherent Modulation Schemes

In situations where it is difficult to maintain phase stability, for example in fading channels, it is useful to employ modulation schemes that do not require the acquisition of a reference signal at the receiver that is in phase coherence with the received carrier. ASK and FSK are two modulation schemes that lend themselves well to noncoherent detection. Receivers for detection of ASK and FSK noncoherently are shown in Fig. 6.3.

For noncoherent reception of binary ASK, the error probability for large signal-to-noise ratios is well approximated by [Ziemer and Tranter, 1995]

$$P_E \cong \frac{1}{2} e^{-z/2}, \quad z \gg 1 \text{ (noncoherent ASK)} \tag{6.10}$$

where $z = E_b/N_0$ as before. For noncoherent detection of binary FSK, the probability of error is exactly given by [Ziemer and Tranter 1995]

$$P_E = \frac{1}{2} e^{-z/2} \quad \text{(noncoherent FSK)} \tag{6.11}$$

Thus, both perform the same for large signal-to-noise ratios. To compare this with coherent detection of FSK, the asympotic approximation for the Q function given by

$$Q(x) = \frac{e^{-x^2/2}}{\sqrt{2\pi}x}, \quad x \gg 1 \tag{6.12}$$

is employed. Application of this to Eq. (6.1) with $R_{12} = 0$ gives

$$P_E = \frac{e^{-z/2}}{\sqrt{2\pi z}}, \quad z \gg 1 \text{ (coherent FSK)} \tag{6.13}$$

Since the dominant behavior comes through the exponent in Eq. (6.13) it follows that coherent and noncoherent FSK have very nearly the same error probability performance at large signal-to-noise ratios, with coherent FSK slightly better due to the $z^{-1/2}$ in the denominator of Eq. (6.13).

There is one other binary modulation scheme which is, in a sense, noncoherent. It is **differentially coherent PSK (DPSK)**, in which the phase of the preceding bit interval is used as a reference for the current bit interval. This technique depends on the channel being stable enough so that phase changes due to channel pertubations from a given bit interval to the succeeding one are inconsequential. It also depends on there being a known phase relationship from one bit interval to the next. This is ensured by differentially encoding the bits before phase modulation at the transmitter. **Differential encoding** is illustrated in Table 6.2. An arbitrary reference bit is chosen to start the process off. In Table 6.2 a 1 has been chosen. For each bit of the encoded sequence, the present bit is used as a reference for the following bit in the sequence. A 0 in the message sequence is encoded as a transition from the state of the reference

TABLE 6.2 An Example Illustrating the Differential Encoding Process

Message sequence:		1	0	0	1	1	1	0
Encoded sequence:	1	1	0	1	1	1	1	0
Transmitted phase radians:	0	0	π	0	0	0	0	π

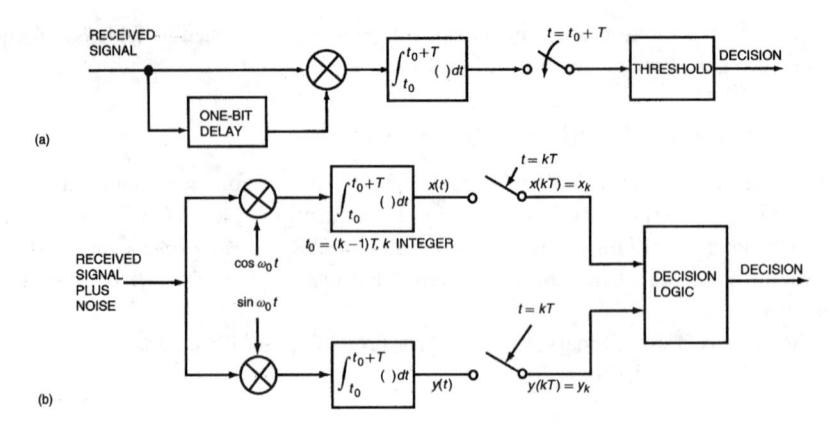

FIGURE 6.4 Receiver block diagrams for detection of DPSK: (a) suboptimum and (b) optimum.

bit to the opposite state in the encoded message sequence; a 1 is encoded as no change of state. Using these rules, it is seen that the encoded sequence shown in Table 6.2 results.

Block diagrams of two possible receiver structures for DPSK are shown in Fig. 6.3. The first is suboptimum, but relatively simple to implement. The second is the optimum receiver for DPSK in AWGN. Its probability of error performance can be shown to be [Ziemer and Tranter, 1995]

$$P_E = \frac{1}{2}e^{-z} \text{ (DPSK)} \tag{6.14}$$

This can be compared with BPSK by again making use of the asymptotic approximation for the Q function given by Eq. (6.12) in Eq. (6.1) with $R_{12} = -1$ to give the following approximate result for BPSK for large signal-to-noise ratios:

$$P_E = \frac{e^{-z}}{2\sqrt{\pi z}}, \quad z \gg 1 \text{ (BPSK)} \tag{6.15}$$

The exponential behavior for DPSK and BPSK is the same for large z; BPSK is slightly better due to the factor $z^{-1/2}$ in Eq. (6.15). Error probabilities for BPSK, DPSK, coherent binary FSK, and noncoherent binary FSK are compared in Fig. 6.5. It is seen that less than 1 dB of degradation results in going from coherent to noncoherent detection at $P_E = 10^{-6}$.

Bandwidth Efficiency

The error probability as a function of E_b/N_0 for a given modulation scheme tells only half of the story, and is often referred to as a measure of its **power efficiency**. Also important is its **bandwidth efficiency**, defined to be the ratio of the bandwidth required to accept a given data rate divided by the data rate. For example, it is well known from Fourier theory that the spectrum of a rectangular pulse of duration T is

$$S(f) = AT \operatorname{sinc}(fT) \tag{6.16}$$

where $\operatorname{sinc}(x) = \sin(\pi x)/(\pi x)$.

When used to modulate a cosinusoid of frequency f_c, the spectrum of the rectangular pulse is centered around the carrier frequency, f_c

$$S_m(f) = \frac{AT}{2}\{\operatorname{sinc}[(f-f_c)T] + \operatorname{sinc}[(f+f_c)T]\} \tag{6.17}$$

Digital Modulation

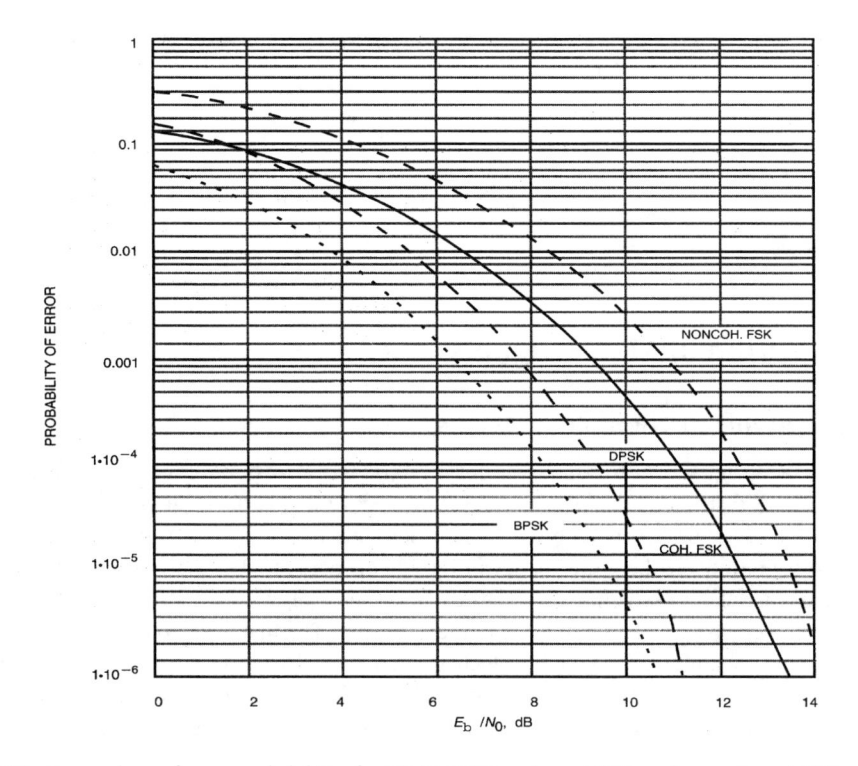

FIGURE 6.5 Comparison of error probabilities for BPSK, DPSK, coherent FSK, and noncoherent FSK.

The bandwidth of the main lobe of the magnitude of this spectrum is

$$B_{RF} = \frac{2}{T} \text{ Hz} \tag{6.18}$$

Since ASK and PSK involve square-pulse modulated sinusoidal carriers, this is the null-to-null main lobe bandwidth for these modulation schemes with T replaced by T_b. For coherent FSK, note that the minimum frequency spacing between cosinusoidal bursts at frequencies f_c and $f_c + \Delta f$ is $1/2T_b$ Hz in order to maintain orthogonality of the two signals. The first null of the sinc function at frequency f_c must have $1/T_b$ Hz below it, and the one at frequency $f_c + \Delta f$ must have $1/T_b$ Hz above it, giving a total bandwidth for FSK of

$$B_{FSK} = \frac{2.5}{T_b} \text{ Hz} \tag{6.19}$$

For noncoherent FSK, the frequency spacing between tones should be $2/T_b$ to allow separation at the receiver by filtering. Allowing the additional $1/T_b$ on either side due to the sinc-function spectrum for each tone, the total bandwidth required for noncoherent FSK is

$$B_{NFSK} = \frac{4}{T_b} \text{ Hz} \tag{6.20}$$

Since $1/T_b$ is the data rate R_b in b/s, the bandwidth efficiencies, defined as the b/s per hertz of bandwidth, of the various modulation schemes considered are as given in Table 6.3.

TABLE 6.3 Bandwidth Efficiencies for Binary Modulation Schemes

Modulation Type	Bandwidth Efficiency, b/s/Hz
Rectangular pulse baseband	1
ASK, PSK, BPSK, DPSK	0.5
Coherent FSK	0.4
Noncoherent FSK	0.25

6.3 Detection of *M*-ary Signals in Additive White Gaussian Noise

Signal Detection in Geometric Terms

It is useful to view digital data transmission in geometric terms for several reasons. First, it provides a general framework that makes the analysis of several types of digital data transmission methods easier, particularly *M*-ary systems. Second, it provides an insight into the digital data transmission problem that allows one to see intuitively the power-bandwidth tradeoffs possible. Third, it suggests ways to improve upon standard modulation schemes. The mathematical basis for the geometric approach is known as signal space (Hilbert space is mathematical literature) theory. The first book to use this approach in the U.S. was Wozencraft and Jacobs [1965], which was based, in part, on earlier work by the Russian Kotelnikov [1960]. An early paper in the literature approaching signal detection from a geometric standpoint is Arthurs and Dym [1962]. An overview of signal space concepts is given next.

The Gram-Schmidt Procedure

Given a finite set of signals, denoted $s_i(t)$, $i = 1, 2, \ldots, M$ for $0 \le t \le T_s$, it is possible to find an orthonormal basis set in terms of which all signals in the set can be represented. (Because more than two possible signals can be sent during each signaling interval, the parameter T_s will now be used to denote the signaling, or *symbol, interval*.) The procedure is known as the **Gram-Schmidt procedure**, and is easy to describe once some notation is defined. (For the most part, signals will be real. However, it is sometimes convenient to represent signals as phasors or complex exponentials.) The **scalar product** of two signals, u and v, defined over the interval $[0, T_s]$, is defined as

$$(u, v) = \int_0^{T_s} u(t) v^*(t) dt \qquad (6.21)$$

and the **norm** of a signal is defined as

$$\|u\| = \sqrt{(u, u)} \qquad (6.22)$$

In terms of this notation, the Gram-Schmidt procedure is as follows:

1. Set $v_1(t) = s_1(t)$ and define the first orthonormal basis function as

$$\phi_1(t) = \frac{v_1(t)}{\|v_1\|} \qquad (6.23)$$

2. Set $v_2(t) = s_2(t) - (s_2, \phi_1) \phi_1(t)$ and let the second orthonormal basis function be

$$\phi_2(t) = \frac{v_2(t)}{\|v_2\|} \qquad (6.24)$$

Digital Modulation

3. Set $v_3(t) = s_3(t) - (s_3, \phi_2)\phi_2(t) - (s_3, \phi_1)\phi_1(t)$ and let the next orthonormal basis function be

$$\phi_3(t) = \frac{v_3(t)}{\|v_3\|} \tag{6.25}$$

4. Continue until all signals have been used. If one or more of the steps yield $v_j(t)$ for which $\|v_j(t)\| = 0$, omit these from consideration so that a set of $K \le M$ orthonormal functions is obtained. This is called a *basis set*.

Using the orthonormal basis set thus obtained, an arbitrary signal in the original set of signals can be represented as

$$s_j(t) = \sum_{i=1}^{K} S_{ij}\phi_i(t), \quad j = 1, 2, \ldots, M \tag{6.26}$$

where

$$S_{ij} = (s_j, \phi_i) = \int_0^{T_s} s_j(t)\phi_i^*(t)\,dt \tag{6.27}$$

With this procedure, any signal of the set can be represented as a point in a signal space [the coordinates of $s_j(t)$ are $S_{1j}, S_{2j}, \ldots, S_{Kj}$]. The representation of the signal in this space will be referred to as the signal vector. Thus, the signal detection problem can be viewed geometrically. This is discussed in the following subsection.

Geometric View of Signal Detection

Given a set of M signals as discussed, an M-ary digital communication system selects one of them with equal likelihood each contiguous T_s-s interval and sends it through a channel in which AWGN of two-sided spectral density $N_0/2$ is added. Letting the noise be represented by $n(t)$ and supposing that the jth signal is transmitted, the coordinates of the noisy received signal, here called the components of the data vector, are

$$Z_i = S_{ij} + N_i, \quad i = 1, 2, \ldots, K \quad (\text{signal } j \text{ transmitted}) \tag{6.28}$$

where

$$N_i = (n, \phi_i) = \int_0^{T_s} n(t)\phi_i^*(t)\,dt \tag{6.29}$$

Schematic diagrams of two receiver front ends that can be used to compute the coordinates of the data vector are shown in Fig. 6.6. The first is called the **correlator implementation**, and the second is called the **matched filter implementation**. Because the noise components are linear transformations of a Gaussian random process, they are also Gaussian, and can be shown to have zero means and covariances

$$\text{cov}(N_i, N_j) = E\{N_i N_j\} = \frac{N_0}{2}\delta_{ij} \tag{6.30}$$

where δ_{ij} is the Kronecker delta, which is zero for the indices equal and zero otherwise. Consequently, the signal coordinates (6.26) are Gaussian with means S_{ij}, zero covariances, and variances of $N_0/2$. Thus, given signal $s_j(t)$ was sent, the joint conditional probability density function of the received data vector components is

$$p[z_1, z_2, \ldots, z_k | s_j(t)] = (\pi N_0)^{-K/2} \exp\left\{-\frac{1}{N_0}\sum_{i=1}^{K}(z_i - s_{ij})^2\right\}, \quad j = 1, 2, \ldots, M \quad (6.31)$$

A reasonable strategy for deciding on the signal that was sent is to choose the most likely, that is, if each signal is transmitted with equal probability, maximize the conditional probability density function (6.31) by choosing the appropriate signal vector, which can also be shown equivalent to minimizing the average probability of error [Wozencraft and Jacobs, 1965]. Given the form of Eq. (6.31), this is accomplished by minimizing its exponent. Minimizing the exponent is equivalent to minimizing the sum of the squares of the differences between the components of the received data vector and those of the signal vector, that is, choosing the signal point that is closest in Euclidian distance to the received data point. This is illustrated in the following specific examples.

M-ary Phase-Shift Keying (MPSK)

Consider MPSK for which the signal set is

$$s_i(t) = \sqrt{\frac{2E_s}{T_s}} \cos\left[2\pi\left(f_c t + \frac{i-1}{M}\right)\right], \quad 0 \le t \le T_s; \, i = 1, 2, \ldots, M \quad (6.32)$$

where E_s is the signal energy, T_s is the signal duration, and f_c is the carrier frequency in hertz. An $M = 4$ MPSK system is referred to as **quadriphase-shift keying (QPSK)**. The Gram–Schmidt procedure could

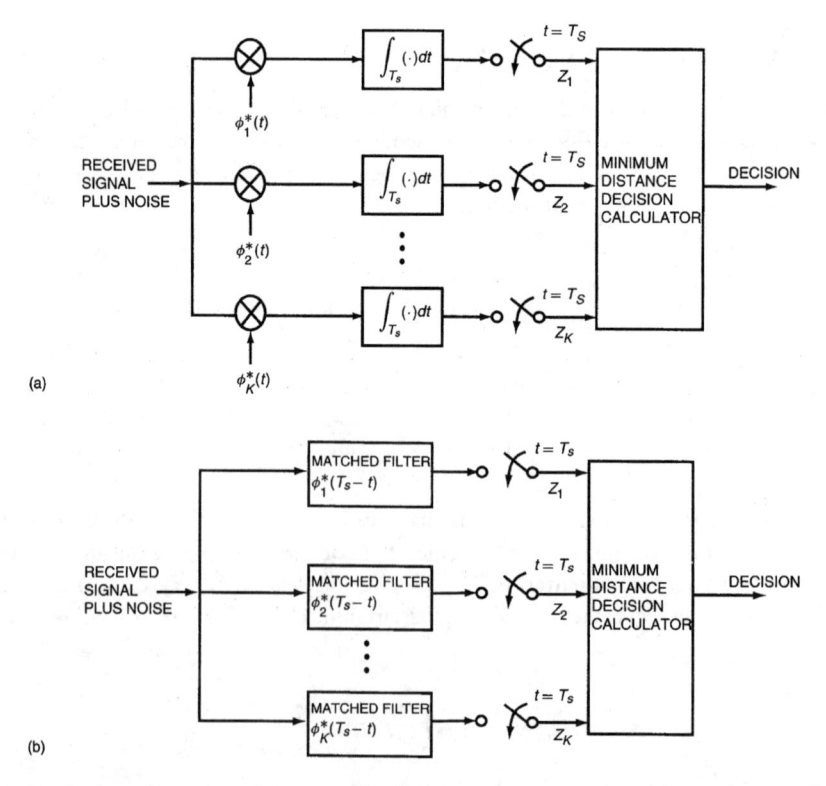

FIGURE 6.6 Receiver configurations for computing data vector components: (a) correlator realization, and (b) matched filter realization.

Digital Modulation

be used to find the orthonormal basis set for expanding this signal set, but it is easier to expand Eq. (6.31) using trigonometric identities as

$$s_i(t) = \sqrt{E_s}\left[\cos\left(\frac{2\pi(i-1)}{M}\right)\sqrt{\frac{2}{T_s}}\cos\omega_c t - \sin\left(\frac{2\pi(i-1)}{M}\right)\sqrt{\frac{2}{T_s}}\sin\omega_c t\right]$$

$$= \sqrt{E_s}\left[\cos\left(\frac{2\pi(i-1)}{M}\right)\phi_1(t) - \sin\left(\frac{2\pi(i-1)}{M}\right)\phi_2(t)\right] \quad (6.33)$$

$$0 \le t \le T_s; \ i = 1, 2, \ldots, M$$

where $\omega_c = 2\pi f_c$, and it follows that

$$\phi_1(t) = \sqrt{\frac{2}{T_s}}\cos\omega_c t, \quad 0 \le t \le T_s \quad \text{and} \quad \phi_2(t) = \sqrt{\frac{2}{T_s}}\sin\omega_c t, \quad 0 \le t \le T_s \quad (6.34)$$

Note that in this case $K = 2 \le M$. A signal space diagram is shown in Fig. 6.7(a) (only the ith signal point is shown). The best decision strategy, as discussed, chooses the signal point in signal space closest in **Euclidian distance** to the received data point. This is accomplished in Fig. 6.7(a) by dividing the signal space up into pie-shaped decision regions associated with each signal point. If the received data point lands in a given region, the decision is made that the corresponding signal was transmitted. The probability of error is the probability that, given a certain signal was transmitted, the noise causes the data vector to land outside of the corresponding decision region. The **probability of *symbol* error** P_S can be upper and lower bounded by [Peterson, Ziemer, and Borth 1995]

$$Q\left(\sqrt{\frac{2E_s}{N_0}}\sin\frac{\pi}{M}\right) \le P_S \le 2Q\left(\sqrt{\frac{2E_s}{N_0}}\sin\frac{\pi}{M}\right) \quad (6.35)$$

which is obtained by considering two half-planes above and below the wedge corresponding to the ith signal as shown in Fig. 6.7(b) and 6.7(c). The upper bound is very tight for M moderately large, a fact which follows by comparing the areas of the plane with one wedge excluded in Fig. 6.7(a) with the area of the two half-planes in Figs. 6.7(b) and 6.7(c). Probability of error curves will be shown later.

(a) DECISION REGION (UNSHADED) FOR iTH TRANSMITTED PHASE

(b) INTEGRATION REGION FOR COMPUTING LOWER BOUND ON $P_b(\varepsilon)$

(c) INTEGRATION REGION FOR COMPUTING UPPER BOUND ON $P_b(\varepsilon)$

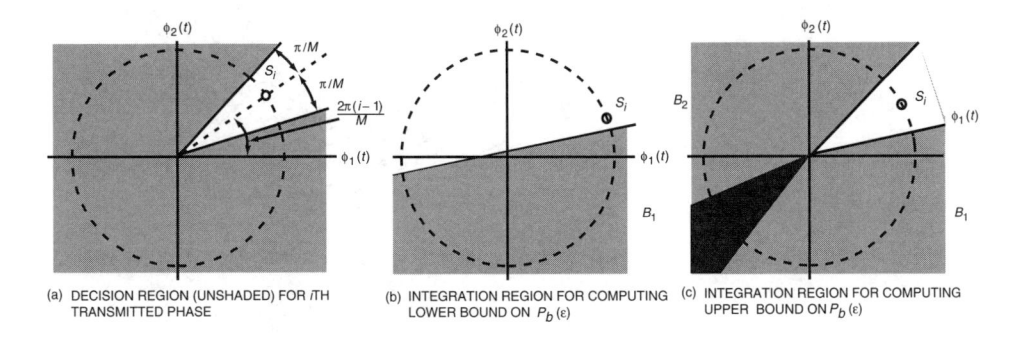

FIGURE 6.7 Signal space diagrams for MPSK: (a) diagram showing typical transmitted signal point with decision region; (b) and (c) half-planes for bounding the error probability. The probability of error is greater than the probability of the received data vector falling into one half-plane, but less than the probability of it falling into either one of both half-planes.

Coherent *M*-ary Frequency-Shift Keying (CMFSK)

The signal set for this modulation scheme is given by

$$s_i(t) = \sqrt{\frac{2E_s}{T_s}} \cos\{2\pi[f_c + (i-1)\Delta f)t], \quad 0 \leq t \leq T_s, \, i = 1, 2, \ldots, M \quad (6.36)$$

where f_c is the lowest tone frequency and Δf is the frequency spacing between tones, both in hertz. The orthonormal basis function set for this modulation scheme is

$$s_i(t) = \sqrt{\frac{2}{T_s}} \cos\{2\pi[f_c + (i-1)\Delta f)t]\}, \quad 0 \leq t \leq T_s, \, i = 1, 2, \ldots, M \quad (6.37)$$

Note that the signal space is M dimensional as opposed to two dimensional for the case of MPSK. A signal space diagram for $M = 3$ is shown in Fig. 6.8. For moderately large M, the symbol error probability can be tightly upper bounded by [Peterson, Ziemer, and Borth, 1995]

$$P_s \leq (M-1) Q\left(\sqrt{\frac{E_s}{N_0}}\right) \quad (6.38)$$

Curves showing bit error probability as a function of signal-to-noise ratio will be presented later.

M-ary Quadrature-Amplitude-Shift Keying (MQASK)

The **M-ary quadrature-amplitude-shift keying** (MQASK) modulation scheme uses the two-dimensional space of MPSK, but with multiple amplitudes. Many such two-dimensional configurations have been considered, but only a simple rectangular grid of signal points will be considered here. The signal set can be expressed as

$$s_i(t) = \sqrt{\frac{2}{T_s}} (A_i \cos\omega_c t + B_i \sin\omega_c t), \quad 0 \leq t \leq T_s \quad (6.39)$$

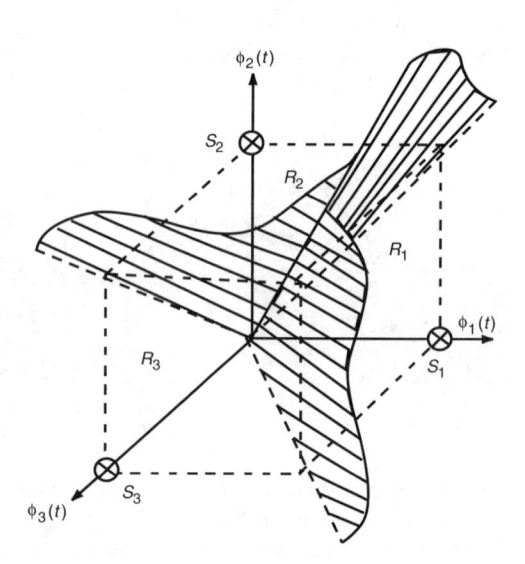

FIGURE 6.8 Signal space showing the possible transmitted signal points and decision boundaries for coherent 3-ary FSK.

Digital Modulation

where ω_c is the carrier frequency in radians per second, and A_i and B_i are amplitudes taking on the values

$$A_i, B_i, = \pm a, \pm 3a, \ldots, \pm \sqrt{M}a \tag{6.40}$$

where M is assumed to be a power of 4. The parameter a can be related to the average signal energy by

$$a = \sqrt{\frac{3E_s}{2(M-1)}} \tag{6.41}$$

A signal space diagram is shown in Fig. 6.9 for $M = 16$ with optimum partitioning of the decision regions. Each signal point is labeled with a Roman numeral I, II, or III. For the type I regions, the probability of correct reception is

$$P(C|I) = \left[1 - 2Q\left(\sqrt{\frac{2a^2}{N_0}}\right)\right]^2 \tag{6.42}$$

For the type II regions, the probability of correct reception is

$$P(C|II) = \left[1 - 2Q\left(\sqrt{\frac{2a^2}{N_0}}\right)\right]\left[1 - Q\left(\sqrt{\frac{2a^2}{N_0}}\right)\right] \tag{6.43}$$

and for the type III regions, it is

$$P(C|III) = \left[1 - Q\left(\sqrt{\frac{2a^2}{N_0}}\right)\right]^2 \tag{6.44}$$

In terms of these probabilities, the probability of symbol error is [Peterson, Ziemar, and Both, 1995]

$$P_s = 1 - \frac{1}{M}[(\sqrt{M} - 2)^2 P(C|I) + 4(\sqrt{M} - 2)P(C|II) + 4P(C|III)] \tag{6.45}$$

Bit error probability plots will be provided later.

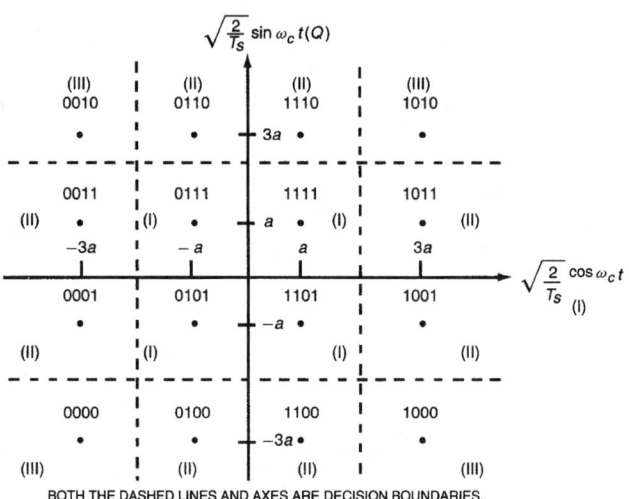

FIGURE 6.9 Signal space diagram for 16-ary QASK.

BOTH THE DASHED LINES AND AXES ARE DECISION BOUNDARIES.
ROMAN NUMERALS SHOW DECISION REGION TYPE.

Differentially Coherent Phase-Shift Keying (DPSK)

DPSK discussed earlier can be generalized to more than two phases. Suppose that the transmitted carrier phase for symbol interval $n-1$ is α_{n-1} and the desired symbol phase for interval n is β_n, which is assumed to take on a multiple of $2\pi/M$ rad. If it is desired to send the particular symbol (phase) $\beta_n = \Phi$, then the *transmitted phase* at interval n, α_n, is

$$\alpha_n = \alpha_{n-1} + \Phi \tag{6.46}$$

where α_{n-1} is the transmitted phase in interval $n-1$. Suppose that the phases detected corresponding to α_{n-1} and α_n are $\theta_{n-1} = \alpha_{n-1} + \gamma$ and $\theta_n = \alpha_n + \gamma$, respectively, where γ is the unknown phase shift introduced by the channel. The first stage of the receiver is a phase discriminator, which detects θ_n. From the previous decision interval, it is assumed that θ_{n-1} is available, so the receiver forms the difference

$$\theta_n - \theta_{n-1} = (\alpha_n + \gamma) - (\alpha_{n-1} + \gamma) = \alpha_n - \alpha_{n-1}$$
$$= \alpha_{n-1} + \Phi - \alpha_{n-1} = \Phi \tag{6.47}$$

where no noise is assumed. In the presence of noise, the receiver must decide which $2\pi/M$ region $\theta_n - \theta_{n-1}$ falls into; hopefully, in this example, this is the region centered on Φ. Thus, a correct decision will be made at the receiver when

$$\Phi - \frac{\pi}{M} < \theta_n - \theta_{n-1} \leq \Phi + \frac{\pi}{M} \tag{6.48}$$

A receiver block diagram implementing this decision strategy is shown in Fig. 6.10. The error probability can be bounded by [Prabhu, 1982]

$$P_1 \leq P_s \leq 2P_1 \tag{6.49}$$

where P_1 is upper and lower bounded by

$$P_1 \leq \frac{\pi}{2} \frac{\cos\left(\frac{\pi}{2M}\right)}{\sqrt{\cos(\pi/M)}} Q\left[2\sqrt{\frac{E_s}{N_0}} \sin\left(\frac{\pi}{2M}\right)\right] \tag{6.50}$$

$$P_1 \geq \frac{1}{2} \frac{\cos\left(\frac{\pi}{2M}\right)}{\sqrt{\cos(\pi/M)}} \left[1 - 2Q\left(\pi\sqrt{\frac{E_s}{N_0}} \sin\left(\frac{\pi}{M}\right)\right)\right] Q\left[2\sqrt{\frac{E_s}{N_0}} \sin\left(\frac{\pi}{2M}\right)\right]$$

Noncoherent *M*-ary FSK

The signal set for noncoherent MFSK can be expressed as

$$s_i(t) = \sqrt{\frac{2E_s}{T_s}} \cos(\omega_i t + \alpha), \quad 0 \leq t \leq T_s; i = 1, 2, \ldots, M \tag{6.51}$$

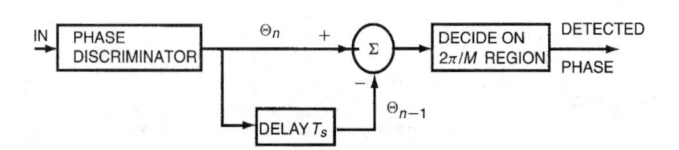

FIGURE 6.10 Block diagram for an *M*-ary differential PSK receiver.

Digital Modulation

where ω_i is the radian frequency of the ith signal, E_s is the symbol energy, T_s is the symbol duration, and α is the unknown phase, which is modeled as a uniformly distributed random variable in $[0, 2\pi]$. The signal space is $2M$ dimensional and can be defined by the basis functions

$$\left. \begin{array}{l} \phi_{xi}(t) = \sqrt{\dfrac{2}{T_s}} \cos \omega_i t \\ \phi_{yi}(t) = \sqrt{\dfrac{2}{T_s}} \sin \omega_i t \end{array} \right\}, \quad 0 \le t \le T_s; \, i = 1, 2, \ldots, M \tag{6.52}$$

A fairly lengthy derivation [Peterson, Ziemer, and Borth, 1995] results in the symbol error probability expression

$$P_s = \sum_{k=1}^{M-1} \binom{M-1}{k} \frac{(-1)^k}{k+1} \exp\left(-\frac{k}{k+1} \frac{E_s}{N_0}\right) \tag{6.53}$$

which reduces to the result for binary noncoherent FSK for $M = 2$. The optimum receiver is an extension of Fig. 6.3 with a parallel set of in-phase and quadrature filters and envelope detectors for each possible transmitted signal.

6.4 Comparison of Modulation Schemes

Bandwidth Efficiency

There are several ways to compare bandwidth of M-ary digital modulation schemes. One way is to compute out-of-band power as a function of bandwidth of an ideal brick wall filter. This requires integration of the power spectrum of the various modulation schemes being compared. The basis for bandwidth comparison used here will be the bandwidth required for the mainlobe of the signal spectrum, which makes for somewhat simpler computation without undue loss of accuracy. For example, replacing T by the symbol duration, T_s, in Eq. (6.17), the radio-frequency bandwidth of the mainlobe of a modulation scheme, which uses a single modulated frequency to transmit the information, such as M-ary PSK, M-ary DPSK, or QASK, is

$$B_{RF} = \frac{2}{T_s} = 2R_s \text{ Hz} \tag{6.54}$$

where R_s is the symbol rate. However, for an M-ary modulation scheme, the symbol duration is related to the bit duration by

$$T_s = T_b \log_2 M = \frac{\log_2 M}{R_b} \tag{6.55}$$

Thus, the bandwidth in terms of bit rate for such modulation schemes is

$$B_{RF} = \frac{2R_b}{\log_2 M} \quad \text{(MPSK, MDPSK, MQASK)} \tag{6.56}$$

Now the ratio of bit rate to required bandwidth is called the bandwidth efficiency of a modulation scheme. In the case at hand, the bandwidth efficiency is

$$\text{bandwidth efficiency} = \frac{R_b}{B_{RF}} = 0.5 \log_2 M \quad \text{(MPSK, MDPSK, MQASK)} \tag{6.57}$$

For schemes using multiple frequencies to transmit the information such as *M*-ary FSK, a more general approach is used. Each symbol is represented by a different frequency. For coherent *M*-ary FSK, the minimum separation per frequency required to maintain orthogonality is $1/2T_s$ Hz. The two outside frequencies use $1/T_s$ Hz for the half of the mainlobe on the left and right of the composite spectrum. The $M - 2$ interior frequencies require a minimum separation of $1/2T_s$ Hz (there are actually $M - 1$ interior slots) for a total RF bandwidth of

$$B_{RF} = \frac{1}{T_s} + \frac{M-1}{2T_s} + \frac{1}{T_s} = \frac{M+3}{2T_s} \text{(coherent MFSK)} \quad (6.58)$$

Substitution of Eq. (6.55) gives a bandwidth efficiency of

$$\text{bandwidth efficiency} = \frac{R_b}{B_{RF}} = \frac{2 \log_2 M}{M+3} \text{(coherent MFSK)} \quad (6.59)$$

For noncoherent MFSK, the minimum separation of the frequencies used to represent the symbols is taken as $2/T_s$ Hz for a total RF bandwidth of

$$B_{RF} = \frac{1}{T_s} + \frac{2(M-1)}{T_s} + \frac{1}{T_s} = \frac{2M}{T_s} \text{(noncoherent MFSK)} \quad (6.60)$$

with a bandwidth efficiency of

$$\text{bandwidth efficiency} = \frac{R_b}{B_{RF}} = \frac{\log_2 M}{2M} \text{(noncoherent MFSK)} \quad (6.61)$$

A comparison of Eq. (6.57) with Eq. (6.59) and (6.61) shows that the bandwidth efficiency of MPSK and MQASK *increases* with *M* whereas the bandwidth efficiency of MFSK *decreases* with *M*. This can be attributed to the dimensionality of the signal space staying constant with *M* for the former modulation schemes, whereas it increases with *M* for the latter. A comparison of bandwidth efficiencies for the various modulation schemes considered in this chapter is given in Table 6.4.

TABLE 6.4 Comparison of Bandwidth Efficiencies for Various Modulation Methods

M	MPSK, MDPSK, MQASK[a]	Coherent MFSK	Noncoherent MFSK
2	0.5	0.400	0.250
4	1.0	0.571	0.250
8	1.5	0.545	0.188
16	2.0	0.421	0.125
32	2.5	0.286	0.078
64	3.0	0.179	0.047

[a] For MQASK, only values of *M* of a power of four are applicable.

Power Efficiency

The power efficiency of a modulation method is indicated by the value of E_b/N_0 required to yield a desired bit error probability, such as 10^{-6}. This being the case, it is necessary to convert from symbol error probability to bit error probability and from E_s/N_0 to E_b/N_0 for *M*-ary modulation schemes. The latter is straightforward since the difference between symbol energy and bit energy is symbol duration vs. bit duration, which are related by

$$T_s = T_b \log_2 M \quad (6.62)$$

giving

$$\frac{E_b}{N_0} = \frac{1}{\log_2 M}\frac{E_s}{N_0} \qquad (6.63)$$

Conversion between symbol error probability and bit error probability is somewhat more complicated. First, considering two-dimensional modulation schemes such as MPSK and MQASK, it is assumed that the most probable errors are those in favor of adjacent signal points and that encoding is used that results in a single bit change in going from one signal point to an adjacent one (i.e., Gray encoding). Since there are $\log_2 M$ bits per symbol, the result is that bit error probability is approximately related to symbol error probability by

$$P_s = \frac{P_b}{\log_2 M} \qquad (6.64)$$

Finally, consider MFSK for which each symbol occupies a separate dimension in the signal space. Thus, all symbol errors are equally probable, which means that each symbol error occurs with probability $P_s/(M-1)$. Suppose that for a given symbol error k bits are in error. There are

$$\binom{\log_2 M}{k}$$

ways that this can happen, since each symbol represents $\log_2 M$ bits. This gives the average number of bit errors per symbol error as

$$\text{Ave. no. of bit errors per symbol error} = \sum_{k=1}^{\log_2 M} k \binom{\log_2 M}{k}\frac{P_s}{M-1} = \frac{M\log_2 M}{2(M-1)} \qquad (6.65)$$

Since there are a total of $\log_2 M$ bits per symbol, the average bit error probability in terms of symbol error probability is

$$P_b = \frac{M}{2(M-1)}P_s \qquad (6.66)$$

The various modulation schemes considered in this chapter are compared on the basis of bit error probability vs. E_b/N_0 in Figs. 6.11–6.15.

Other Important Types of Digital Modulation Schemes

There are many other important digital modulation formats that have not been discussed here. In this section, several of these will be mentioned with appropriate references given so that the reader will have a place to start to learn more about them.

A modulation scheme related to QPSK is **offset QPSK** (OQPSK) [Ziemer and Tranter, 1995]. This modulation format is produced by allowing only ±90° phase changes in a QPSK format. Furthermore, the phase changes can take place at multiples of a half-symbol interval, or a bit period. The reason for limiting phase changes to ±90° is to prevent the large envelope deviations that occur when QPSK is filtered to limit sidelobe power, and then regrowth of the sidelobes after amplitude limiting is used to produce a constant-envelope signal. This is typically encountered in satellite communications where, due to power efficiency considerations, hard limiting repeaters are used in the satellite communications system.

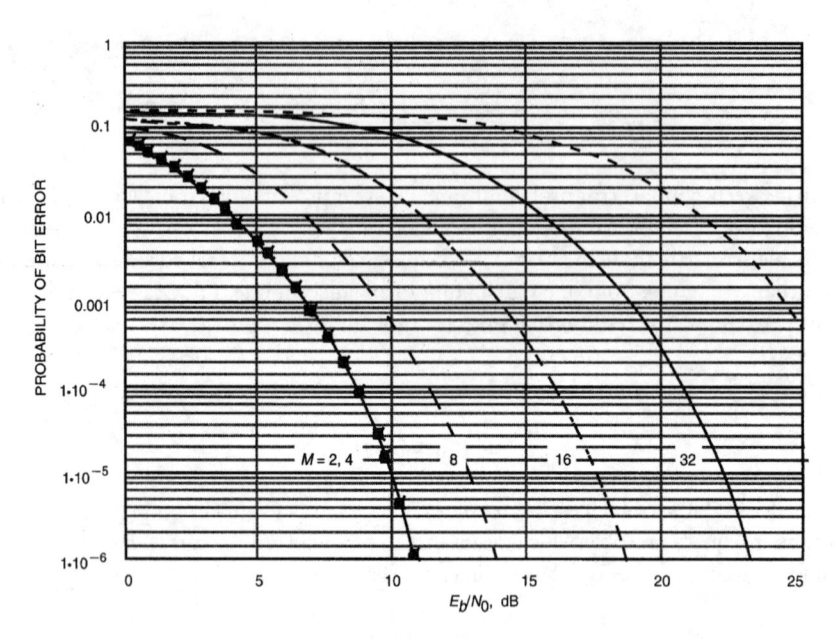

FIGURE 6.11 Bit error probability vs. E_b/N_0 for coherent M-ary PSK.

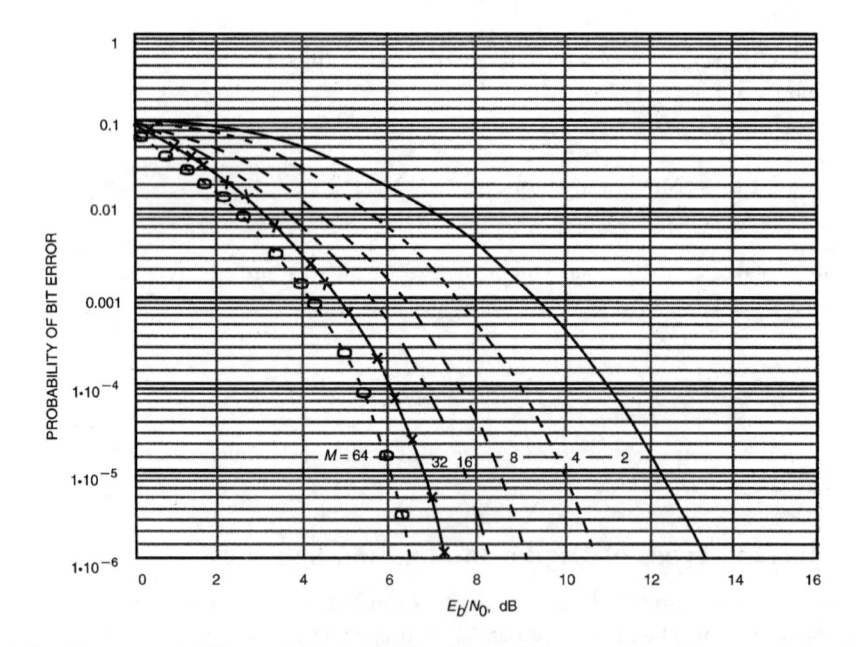

FIGURE 6.12 Bit error probability vs. E_b/N_0 for coherent M-ary FSK.

Another modulation scheme closely related to QPSK and OQPSK is **minimum shift keying (MSK)** [Ziemer and Tranter, 1995], which can be produced from OQPSK by weighting the inphase and quadrature components of the baseband OQPSK signal with half-sinusoids. As with OQPSK, this is a further attempt at producing a modulated signal with spectrum that has less sidelobe power than BPSK, QPSK, or OQPSK and behaves well when filtered and limited. With MSK, there are no abrupt phase changes, but rather, the phase changes linearly over the symbol interval.

Many different forms of MSK have been proposed and investigated over the years. One form is **Gaussian-filtered MSK (GMSK)**, where the baseband data is filtered by a filter having a Gaussian-shaped

Digital Modulation

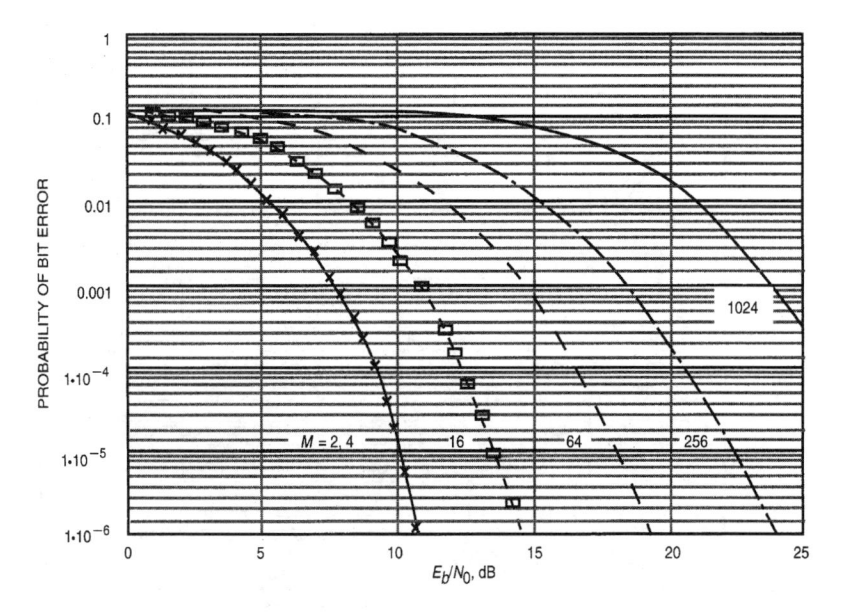

FIGURE 6.13 Bit error probability vs. E_b/N_0 for coherent M-ary QASK.

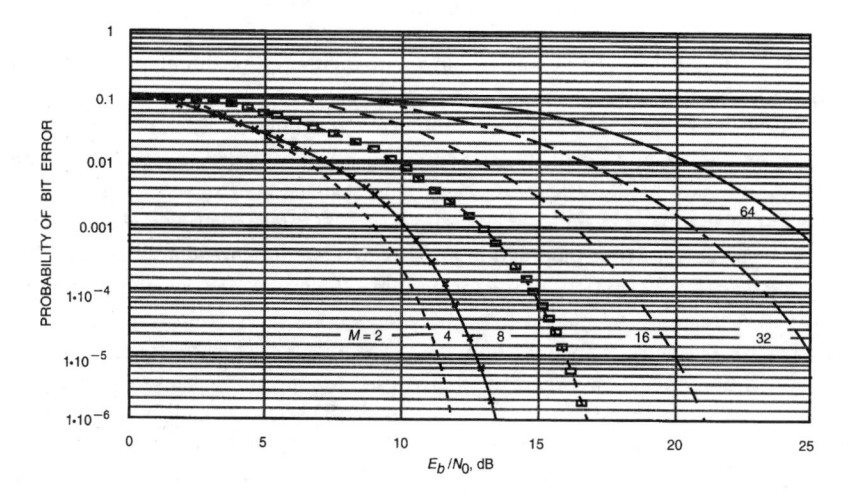

FIGURE 6.14 Bit error probability vs. E_b/N_0 for M-ary DPSK.

transfer function at the baseband before modulating the carrier [Murota, 1981]. GMSK is another attempt at producing a signal spectrum that is compact around the carrier and having low envelope deviation of the modulated carrier. It has been adopted by the European community for second-generation cellular-mobile communication systems.

A modulation scheme that is related to 8-PSK is $\pi/4$-**differential QPSK** ($\pi/4$-DQPSK) [Peterson, Ziemer, and Borth, 1995]. It is essentially an 8-PSK format with differential encoding where, from a given phase state, only phase shifts of $\pm\pi/4$ or $\pm 3\pi/4$ are allowed. It has been adopted as the modulation format for one North American standard for second-generation cellular-mobile radio. The other adopted standard uses direct-sequence spread-spectrum modulation.

A whole host of modulation schemes may be grouped under the heading **continuous-phase modulation (CPM)** [Sklar, 1988]. These modulation formats employ continuous phase trajectories over one or more symbols to get from one phase to the next in response to input data changes. MSK and GMSK

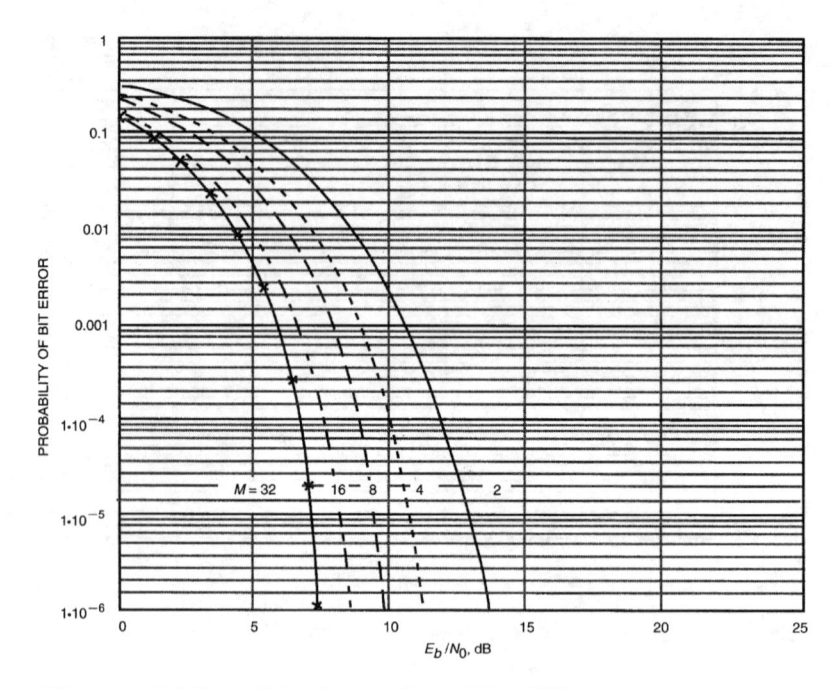

FIGURE 6.15 Bit error probability vs. E_b/N_0 for noncoherent M-ary FSK.

are special cases of CPM. CPM schemes are employed in an attempt to simultaneously improve power and bandwidth efficiency.

A class of modulation that is important for military and cellular radio communications is spread spectrum [Peterson, Ziemer, and Borth, 1995]. Spread spectrum modulation is any modulation format that utilizes a transmission bandwidth much wider than that required to transmit the message signal itself, independent of the message signal bandwidth. It can be categorized several ways. The two most common types of spread spectrum modulation are **direct-sequence spread spectrum (DSSS)** and **frequency-hop spread spectrum (FHSS)**. In the former type, the modulated signal spectrum is spread by multiplication with a pseudonoise binary code, which changes state several times during a symbol interval. In the latter type, the spectrum is spread by hopping the modulated signal spectrum about in a pseudorandom manner. The receiver knows the code that is used for spreading at the transmitter. Thus, it is possible to despread the received signal at the receiver once the receiver's code is synchronized with the spreading coding on the received signal. Spread spectrum modulation is used for several reasons. Among these are to hide the modulated signal from an enemy interceptor, to lower the susceptibility of jamming by an unfriendly or unintentional source, to combat multipath, to provide multiple access capability for the modulation scheme, and to provide a means for range measurement. An example of the latter application is the **global positioning system (GPS)**. An example of the fourth application is the second North American digital cellular-radio standard mentioned earlier.

6.5 Higher Order Modulation Schemes

Error Correction Coding Fundamentals

The previously discussed modulation schemes had dependency between signaling elements only over one signaling element. There are advantages to providing memory over several signaling elements from the standpoint of error correction. Historically, the first way that this was accomplished was to encode the data (usually binary) by adding redundant symbols for error correction and use the encoded symbol stream to modulate the carrier. The ratio of information symbols to total encoded symbols is referred

Digital Modulation

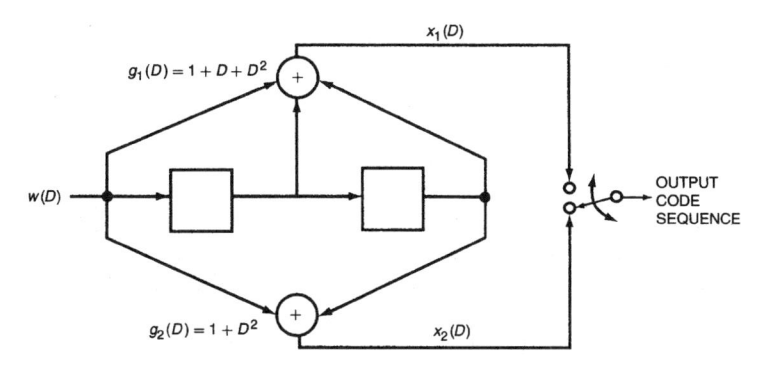

FIGURE 6.16 A rate $-\frac{1}{2}$, constraint length 3 convolutional encoder.

to as the **code rate**. At the receiver, demodulation was accomplished followed by decoding. The latter operation allowed some channel-induced errors to be corrected. Since redundant symbols are added at the transmitter, it is necessary to use a larger bandwidth by a factor of one over the code rate than if no encoding is employed in order to keep the same information rate through the channel with coding as without coding. Thus, the signal-to-noise ratio is smaller by a factor of the code rate with coding than without coding, and the raw bit error probability is higher. Since the code can correct errors, however, this compensates partially for the higher raw bit error probability to a degree. If the code is powerful enough, the overall bit error probability through the channel is lower with coding than without. Usually, this depends on the signal-to-noise ratio: for small values of the signal-to-noise ratio, the compensation by coding is not enough, and the overall bit error probability is higher with coding than without; for sufficiently large values of the signal-to-noise ratio, the error correction capability of the code more than makes up for the higher raw bit error probability, and a lower overall error probability is obtained. The **coding gain** is defined as the ratio of the signal-to-noise ratios without and with coding, or if they are expressed in decibels, the difference between the signal-to-noise ratios without and with coding, at a given bit error probability.

There are two widely used types of coding: **block** and **convolutional**. We will discuss these assuming that binary data is used. For block coding, the information bits are taken a block at a time (k of them) and correction bits are added to the block to make an overall block of length n. The ratio k/n is the code rate in this case. The trick is to introduce dependencies between the bits in the blocks of length n so that error correction can be accomplished at the receiver. The field of block coding dates back to work by Shannon in the late 1940s, and many powerful block codes are now available with efficient decoding techniques to make them implementable. Examples are Hamming, Bose-Chaudhuri–Hocquenghem (BCH), and Golay codes for binary codes; a nonbinary family of codes that is very powerful is the Reed–Solomon codes, which are used in compact disc audio technology. Depending on the code rate and the code used, block codes can provide coding gains from less than 1 to 10 or more dB at a probability of bit error of 10^{-6}.

A second class of error correction codes are convolutional codes. A simple convolutional encoder is shown in block diagram form in Fig. 6.16. It consists of a shift register and two or more modulo-2 adders to combine the input bit and bits residing in the shift register. It is assumed that a clock synchronizes the shifting of the bits into and down the shift register. When a new bit is shifted into the first stage of the shift register, all former bits in the shift register ripple to the right. In this case, the raw data is fed into the shift register bit-by-bit and the outputs of the modulo-2 adders are sampled in turn after each bit is fed in. Since two encoded bits are obtained for each input bit, this is a rate $-\frac{1}{2}$ convolutional encoder. If there were three modulo-2 adders for the bits in the shift register, a rate $-\frac{1}{3}$ code would result. Code rates greater than $\frac{1}{2}$ can be achieved by deleting bits periodically at the output of the encoder. This process is called **puncturing**. At the decoder, arbitrary symbols are inserted at each puncturing time (the period of the puncturing is known at the decoder, but the values of the punctured symbols are, of course, not known) and the decoder treats the arbitrarily inserted bit as a possible error. Convolutional codes have been in existence since the 1960s. A very efficient decoding method for convolutional codes is called the **Viterbi algorithm**, which was invented by Andrew Viterbi in the late 1960s. We will not present this algorithm in detail here, but the reader is referred to the references [Peterson, Ziemer, and Borth, 1995]. The performance of convolutional codes depends on the number of stages in the shift register (the number of stages plus 1 is called the **constraint length** of the code), the code rate, and whether hard or soft decisions are fed into the decoder. A **hard decision** is a definite 1–0 decision at the output of the detector; a **soft decision** is a quantized version of the detector output before a hard decision is made. Coding gains for convolutional codes range from a few decibels to 6 or 7 dB at a probability of error of 10^{-6} for soft decisions, constraint lengths of 8 or 9, and rates of $\frac{1}{2}$ or $\frac{1}{3}$. The higher the rate of the punctured code (i.e., the closer the rate is to 1), the less coding gain yielded by puncturing. Code rates close to 1 can be achieved by methods other than puncturing.

In closing, we note that coding used in conjunction with modulation always expands the required transmission bandwidth by the inverse of the code rate, assuming the overall bit rate is kept constant. In other words, power efficiency goes up but bandwidth efficiency goes down with use of a well-designed code.

Trellis-Coded Modulation

In attempting to simultaneously conserve power and bandwidth, Ungerboeck [1987] in the 1970s began to look at ways to combine coding and modulation. His solution was to use coding in conjunction with M-ary modulation to increase the minimum Euclidian distance between those signal points that are most likely to be confused without increasing the average power or bandwidth over an uncoded scheme transmitting the same number of bits per second. The technique is called **trellis-coded modulation** (**TCM**). An example will illustrate the procedure.

In the example, we wish to compare a TCM system and a QPSK system operating at the same data rates and, hence, the same bandwidths. Since the QPSK system transmits 2 b per signal phase, we can keep that same data rate with the TCM system by employing an 8-PSK modulator, which carries 3 b per signal phase, in conjunction with a convolutional encoder that produces three encoded symbols for every two input data bits (i.e., a rate $\frac{2}{3}$ encoder). Figure 6.17 shows a way in which code symbols can be associated with signal phases. This technique, called **set partitioning**, is carried out according to the following rules: (1) All code triples from the encoder differing by one code symbol are assigned the maximum possible Euclidian distance in encoded phase points; (2) all other possibilities are assigned the next to largest possible Euclidian distance.

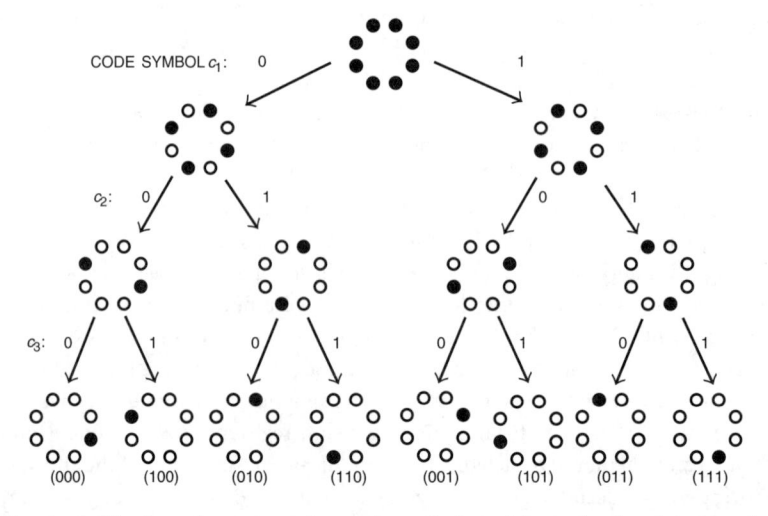

FIGURE 6.17 A set partitioning scheme to assign code symbols to M-ary signal points in order to maximize Euclidian distance between signal points.

Digital Modulation

To decode the TCM signal, the received signal plus noise in each signaling interval is correlated with each possible signal/code combination; a search for the closest symbol combination is made by means of a Viterbi algorithm using the sums of these correlations as the metric in the algorithm (i.e., Euclidian distance). With the level of this chapter, the reader is encouraged to consult the references for the details of the decoding procedure [Zeimer and Tranter, 1995; Blahut, 1990; Proakis, 1995]. Using this procedure, it can be shown that a coding gain results in going from the QPSK system to the combined coded/8-PSK system at *no increase in bandwidth*. In fact, for the system considered here, Ungerboeck has shown that an asymptotic (i.e., in the limit of small bit error probabilities) coding gain in excess of 3.5 dB results. More powerful codes and larger phase shifts result in asymptotic coding gains of over 5.5 dB. Viterbi et al. [1989] has discussed the use of rate $\frac{1}{2}$, constraint length 7 convolutional codes, for which VLSI circuit implementations are available at a low cost, to provide various coding gains using puncturing and MPSK.

Defining Terms

Additive white Gaussian noise (AWGN): Noise that is additive to the signal having a constant power spectrum over all frequencies and Gaussian amplitude distribution.

Amplitude-shift keying (ASK): A signaling scheme whereby the digital data modulates the amplitude of a carrier.

Antipodal signaling: A signaling scheme which represents the digital data as plus (logic 1) or minus (logic 0) a basic pulse shape.

Bandwidth efficiency: The ratio of bit rate to bandwidth occupied for a digital modulation scheme. Technically, it is dimensionless, but it is usually given the dimensions of bits/second/hertz.

Binary signaling: Any signaling scheme where the number of possible signals sent during any given signaling interval is two.

Bit period: The symbol duration for a binary signaling scheme.

Block code: Any encoding scheme that encodes the information symbols block-by-block by adding a fixed number of error correction symbols to a fixed block length of information symbols.

Code rate: The ratio of the number of information bits into to encoded symbols out of a coder corresponding to those input bits.

Coding gain: The amount of improvement in E_b/N_0 at a specified probability of bit error provided by a coding scheme.

Coherent modulation: Any modulation technique that requires a reference at the receiver which is phase coherent with the received modulated carrier signal for demodulation.

Constraint length: The number of shift register stages plus one for a convolutional encoder. It is also the span of information symbols that determines the error correction symbols for a convolutional code.

Continuous-phase modulation (CPM): Any phase modulation scheme where the phase changes continuously over one or more symbol intervals in response to a change of state of the input data.

Convolutional code: Any encoding scheme that encodes a sliding window of information symbols by means of a shift register and two or more modulo-2 adders for the bits in the shift register that are sampled to produce the encoded output.

Correlation coefficient: A measure of the similarity between two signals normalized to a maximum absolute value of 1. If the correlation coefficient of two signals is zero, they are called orthogonal; if −1, they are called antipodal.

Correlator implementation: A receiver implementation that multiplies the incoming signal plus noise during any given signaling interval by a replica of each possible signal transmitted, integrates over the signaling interval, and uses these data for determining the transmitted signal.

Differential encoding: A coding scheme that compares the present differentially encoded bit with the next data bit; if a 1, the next encoded bit is the same as the reference, whereas if a 0, the next encoded bit is opposite the reference.

Differentially coherent PSK (DPSK): The result of using a differentially encoded bit stream to phase-shift key a carrier.

Digital modulation: The variation of some attribute of a carrier, such as amplitude, phase, or frequency, in a one-to-one correspondence with a message taking on a discrete set of values.

Direct-sequence spread-spectrum (DSSS): A modulation scheme employing a transmission bandwidth much wider than that required for the message signal, where the additional spectral spreading is achieved by multiplication with a binary pseudonoise code having many state changes per data symbol.

Euclidian distance: The length of the vector that is the difference of two other vectors in signal space; forms the Euclidian distance between the points defined by these vectors.

Frequency-hop spread-spectrum (FHSS): A modulation scheme employing a transmission bandwidth much wider than that required for the message signal, where the additional spectral spreading is achieved by hopping the modulated signal spectrum pseudorandomly in frequency.

Frequency-shift keying (FSK): A signaling scheme whereby the digital data modulates the frequency of a carrier.

Gaussian MSK (GMSK): MSK in which the baseband signal has been filtered with a filter having a Gaussian-shaped amplitude response before carrier modulation.

Global positioning system: A system consisting of 24 satellites spaced around the Earth with precise timing sources and emitting DSSS signals, at least four of which can be acquired by a receiver simultaneously in order to determine the position of the platform on which the receiver is located.

Gram-Schmidt procedure: An algorithm for obtaining a set of orthonormal functions from a set of signals.

Hard decision: The process of making a definite 1–0 decision at the output of a detector, usually in conjunction with coding. Thus information is discarded before the decoding process.

M-ary signaling: Any signaling scheme where the number of possible signals sent during any given signaling interval is M; binary signaling is a special case with $M = 2$. MPSK uses M phases of a sinusoidal carrier.

Matched filter implementation: A receiver implementation that has a parallel bank of matched filters as the first stage, one matched to each possible transmitted signal. A matched filter has an impulse response that is the time reverse of the signal to which it is matched.

Maximum-likelihood receiver: A receiver that bases its decisions on the probabilities of the received waveform given each possible transmitted signal.

Minimum-shift keying (MSK): OQPSK with the in-phase and quadrature baseband signal components weighted by half-sinusoids. The phase changes linearly over a bit interval.

Noncoherent modulation: Any modulation scheme not requiring a reference at the receiver in phase coherence with the received modulated signal in order to perform demodulation.

Norm: A function of a signal that is analogous to its length.

Offset quadriphase-shift keying (OQPSK): A QPSK scheme where phase shifts can be only $\pm\pi/2$ and can occur at half-symbol intervals.

Orthogonal signaling: Any signaling scheme that uses a signal set that have zero correlation coefficient between any pair of signals.

Phase-shift keying (PSK): A signaling scheme whereby the digital data modulates the phase of a carrier.

$\pi/4$-differential quadriphase-shift keyed signaling ($\pi/4$-DQPSK): A differentially coherent phase-shift keyed modulation scheme in which one of eight possible phases is transmitted each signaling interval, and in which the possible phase changes from the current phase is limited to $\pm\pi/4$ and $\pm 3\pi/4$.

Power efficiency: The energy-per-bit over noise power spectral density (E_b/N_0) required to provide a given probability of bit error for a digital modulation scheme.

Probability of (bit) (symbol) error: The average number of symbol errors over a long string (ideally infinite) of transmitted signals for any digital modulation scheme. If expressed in terms of the number of bits per symbol, the probability of bit error results.

Puncturing: A method to increase the rate of a code, usually convolutional, by periodically deleting symbols from the encoded output. At the decoder, arbitrary symbols are inserted in the places the deleted symbols would have occupied, and the decoder corrects a large majority of these if necessary.

Quadrature-amplitude-shift keying (QASK): A digital modulation scheme in which both the phase and amplitude of the carrier takes on a set of values in one-to-one correspondence with the digital data to be transmitted.

Quadriphase-shift keying (QPSK): An MPSK (see M-ary signaling) system that utilizes four signals in the signal set distinguished by four phases 90° apart.

Scalar product: A function of two signals that is analogous to their inner or dot product in ordinary geometric terms.

Set partitioning: A procedure for implementing TCM encoding rules in order to maximize Euclidian distance between signal points.

Soft decisions: The use of detector outputs quantized to more than two levels (i.e., a 1 or a 0) to feed into a decoder with the intention of improving the coding gain over what is achievable with hard decisions.

Symbol period: The duration of the signaling element for an M-ary modulation scheme.

Trellis-coded modulation (TCM): The process of using convolutional codes of suitable rates in conjunction with M-ary modulation to achieve coding gain without bandwidth expansion.

Viterbi algorithm: An efficient algorithm for decoding convolutional codes, among other applications, that implements maximum likelihood estimation.

References

Arthurs, A. and Dim, H. 1962. On the optimum detection of digital signals in the presence of white Gaussian noise — A geometric interpretation and a study of three basic data transmission systems. *IRE Trans. on Comm. Systems* CS-10(Dec.).

Blahut, R.E. 1990. *Digital Transformation of Information.* Addison-Wesley, Reading, MA.

Kotelnikov, V.A. 1960. *The Theory of Optimum Noise Immunity.* Dover, New York.

Murota, K. 1981. GMSK modulation for digital mobile radio telephony. *IEEE Trans. on Comm.* COM-29(July): 1044–1050.

Peterson, R.L., Ziemer, R.E., and Borth, D.E. 1995. *Introduction to Spread Spectrum Communications.* Prentice-Hall, Englewood Cliffs, NJ.

Prabhu, V.K. 1982. Error rate bounds for differential PSK. *IEEE Trans. on Comm.* COM-30(Dec.):2547–2550.

Proakis, J.G. 1995. *Digital Communications,* 3rd ed., McGraw-Hill, New York.

Sklar, B. 1988. *Digital Communications: Fundamentals and Applications.* Prentice-Hall, Englewood Cliffs, NJ.

Ungerboeck, G. 1987. Trellis-coded modulation with redundant signal sets, part I: Introduction and trellis-coded modulation with redundant signal sets, Part II: State of the art. *IEEE Comm. Mag.* 25(Feb.):5–11 and 12–21.

Viterbi, A.J., Wlolf, J.K., Zehavi, E., and Padovani, R. 1989. A pragmatic approach to trellis-coded modulation. *IEEE Comm. Mag.* 27(July):11–19.

Wozencraft, J.M. and Jacobs, I.M. 1965. *Principles of Communications Engineering.* Wiley, New York (out of print, but available from Waveland Press, Prospect Heights, IL).

Ziemer, R.E. and Tranter, W.H. 1995. *Principles of Communications: Systems, Modulation, and Noise,* 4th ed. Wiley, New York.

7
High-Power Vacuum Devices

Jerry C. Whitaker
Editor

7.1 Introduction 7-1
  Characteristics of Electrons
7.2 Electron Optics 7-2
  Magnetic Field Effects · Thermal Emission from Metals
  · Secondary Emission
7.3 Diode 7-5
7.4 Triode 7-6
7.5 Tetrode 7-9
  Application Example
7.6 Pentode 7-12
7.7 High-Frequency Operating Limits 7-13
  Transit-Time Effects
7.8 Device Cooling 7-15
  Air Cooling · Water Cooling · Vapor-Phase Cooling
  · Special Applications

7.1 Introduction

A power grid tube is a device using the flow of free electrons in a vacuum to produce useful work.[1] It has an emitting surface (the cathode), one or more grids that control the flow of electrons, and an element that collects the electrons (the anode). Power tubes can be separated into groups according to the number of electrodes (grids) they contain. The physical shape and location of the grids relative to the plate and cathode are the main factors that determine the amplification factor (μ) and other parameters of the device. The physical size and types of material used to construct the individual elements determine the power capability of the tube. A wide variety of tube designs are available to commercial and industrial users. By far the most common are triodes and tetrodes.

Characteristics of Electrons

Electrons are minute, negatively charged particles that are constituents of all matter. They have a mass of 9×10^{-28} g ($\frac{1}{1840}$ that of a hydrogen atom) and a charge of 1.59×10^{-19} coulomb. Electrons are always identical, irrespective of their source. Atoms are composed of one or more such electrons associated with a much heavier nucleus, which has a positive charge equal to the number of the negatively charged electrons contained in the atom; an atom with a full quota of electrons is electrically neutral. The differences in chemical elements arise from differences in the nucleus and in the number of associated electrons.

Free electrons can be produced in a number of ways. *Thermonic emission* is the method normally employed in vacuum tubes. The principle of thermonic emission states that if a solid body is heated sufficiently, some of the electrons that it contains will escape from the surface into the surrounding space. Electrons are also ejected from solid materials as a result of the impact of rapidly moving electrons or ions. This phenomenon is referred to as *secondary electron emission* because it is necessary to have a primary source of electrons (or ions) before the secondary emission can be obtained. Finally, it is possible to pull electrons directly out of solid substances by an intense electrostatic field at the surface of the material.

Positive ions represent atoms or molecules that have lost one or more electrons and so have become charged bodies having the weight of the atom or molecule concerned, and a positive charge equal to the negative charge of the lost electrons. Unlike electrons, positive ions are not all alike and may differ in charge or weight, or both. They are much heavier than electrons and resemble the molecule or atom from which they are derived. Ions are designated according to their origin, such as mercury ions or hydrogen ions.

7.2 Electron Optics

Electrons and ions are charged particles and, as such, have forces exerted upon them by an electrostatic field in the same way as other charged bodies. Electrons, being negatively charged, tend to travel toward the positive electrode (or anode), while the positively charged ions travel in the opposite direction (toward the negative electrode or cathode). The force F exerted on a charged particle by an electrostatic field is proportional to the product of the charge e of the particle and the voltage gradient G of the electrostatic field:[1]

$$F = G \times e \times 10^7 \tag{7.1}$$

where: F is the force in dynes, G is the voltage gradient in volts per centimeter, and e is the charge in coulombs. This force upon the ion or electron is exerted in the direction of the electrostatic flux lines at the point where the charge is located. The force acts toward or away from the positive terminal, depending on whether a negative or positive charge, respectively, is involved.

The force that the field exerts on the charged particle causes an acceleration in the direction of the field at a rate that can be calculated by the laws of mechanics where the velocity does not approach that of light:

$$A = \frac{F}{M} \tag{7.2}$$

where A is the acceleration in centimeters per second per second, F is the force in dynes, and M is the mass in grams.

The velocity an electron or ion acquires in being acted upon by an electrostatic field can be expressed in terms of the voltage through which the electron (or ion) has fallen in acquiring the velocity. For velocities well below the speed of light, the relationship between velocity and the acceleration voltage is:

$$v = \sqrt{\frac{2 \times V \times e \times 10^7}{M}} \tag{7.3}$$

where v is the velocity in centimeters per second corresponding to V, V is the accelerating voltage, e is the charge in coulombs, and M is the mass in grams. Electrons and ions move at great velocities in even moderate-strength fields. For example, an electron dropping through a potential difference of 2500 V will achieve a velocity of approximately one-tenth the speed of light.

High-Power Vacuum Devices

Electron optics, as discussed in this section, rely on the principles of classical physics. While modern tube design uses computer simulation almost exclusively, the preceding information remains valid and provides a basis for understanding electron motion within a vacuum tube device.

Magnetic Field Effects

An electron in motion represents an electric current of magnitude ev, where e is the magnitude of the charge on the electron and v is its velocity. A magnetic field accordingly exerts a force on a moving electron exactly as it exerts a force on an electric current in a wire. The magnitude of the force is proportional to the product of the equivalent current ev represented by the moving electron and the strength of the component of the magnetic field in a direction at right angles to the motion of the electron. The resulting force is, then, in a direction at right angles both to the direction of motion of the electron and to the component of the magnetic field that is producing the force. As a result, an electron entering a magnetic field with a high velocity will follow a curved path. Because the acceleration of the electron that the force of the magnetic field produces is always at right angles to the direction in which the electron is traveling, an electron moving in a uniform magnetic field will follow a circular path. The radius of this circle is determined by the strength of the magnetic field and the speed of the electron moving through the field.

When an electron is subjected to the simultaneous action of both electric and magnetic fields, the resulting force acting on the electron is the vector sum of the force resulting from the electric field and the force resulting from the magnetic field, each considered separately.

Magnetic fields are not used for conventional power grid tubes. Microwave power tubes, on the other hand, use magnetic fields to confine and focus the electron stream.

Thermal Emission from Metals

Thermonic emission is the phenomenon of an electric current leaving the surface of a material as the result of thermal activation. Electrons with sufficient thermal energy to overcome the surface-potential barrier escape from the surface of the material. This thermally emitted electron current increases with temperature because more electrons have sufficient energy to leave the material.

The number of electrons released per unit area of an emitting surface is related to the absolute temperature of the emitting material and a quantity b that is a measure of the work an electron must perform in escaping through the surface, according to Eq. (7.4):[1]

$$I = AT^2 \varepsilon^{-b/T} \qquad (7.4)$$

where T is the absolute temperature of the emitting material, b is the work an electron must perform in escaping the emitter surface, I is the electron current in amperes per square centimeter, and A is a constant (value varies with type of emitter). The exponential term in the equation accounts for most of the variation in emission with temperature. The temperature at which the electron current becomes appreciable is accordingly determined almost solely by the quantity b. Figure 7.1 plots the emission resulting from a cathode operated at various temperatures.

Thermal electron emission can be increased by applying an electric field to the cathode. This field lowers the surface-potential barrier, enabling more electrons to escape. This field-assisted emission is known as the *Schottky effect*.

Figure 7.2 illustrates common heater-cathode structures for power tubes.

Secondary Emission

Almost all metals and some insulators will emit low-energy electrons (secondary electrons) when bombarded by other energetic electrons. The number of secondary electrons emitted per primary electron is determined by the velocity of the primary bombarding electrons and the nature and condition of the

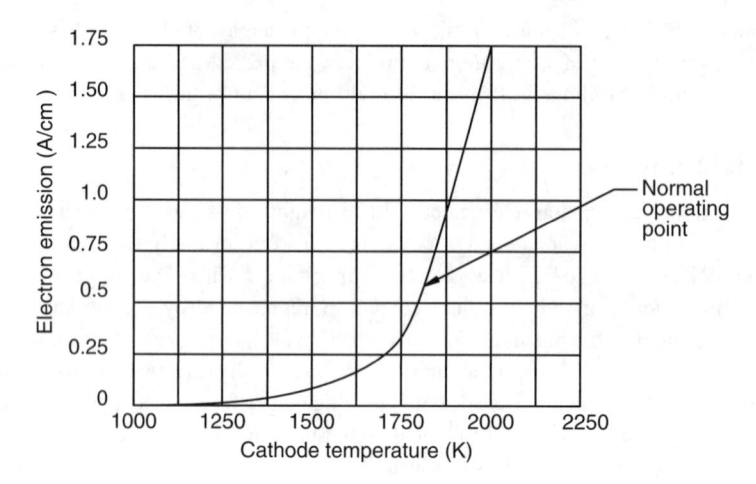

FIGURE 7.1 Variation of electron emission as a function of absolute temperature for a thoriated-tungsten emitter.

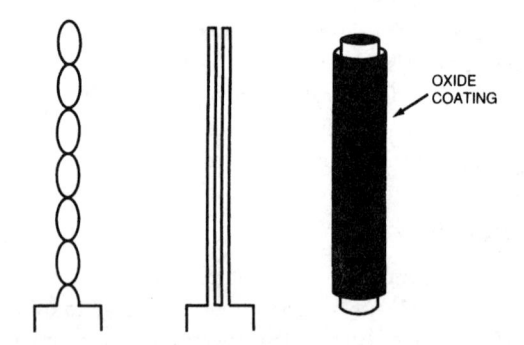

FIGURE 7.2 Common types of heater and cathode structures. (Adapted from Ferris, C. D., *The Electronics Handbook*, J. C. Whitaker, Ed., CRC Press, Boca Raton, FL, 1996, 295–305.)

material composing the surface being bombarded. Figure 7.3 illustrates a typical relationship for two types of surfaces. As shown in the figure, no secondary electrons are produced when the primary velocity is low. However, with increasing potential (and consequently higher velocity), the ratio of secondary to primary electrons increases, reaching a maximum and then decreasing. With pure metal surfaces, the maximum ratio of secondary to primary electrons ranges from less than 1 to approximately 3. Some complex surfaces based on alkali metal compounds yield ratios of secondary to primary electrons as high as 5 to 10.

The majority of secondary electrons emitted from a conductive surface have relatively low velocity. However, a few secondary electrons are usually emitted with a velocity nearly equal to the velocity of the bombarding primary electrons.

For insulators, the ratio of secondary to primary electrons as a function of primary electron potential follows along the same lines as for metals. The net potential of the insulating surface being bombarded is affected by the bombardment. If the ratio of secondary to primary current is less than unity, the insulator acquires a net negative charge because more electrons arrive than depart. This causes the insulator to become more negative and, finally, to repel most of the primary electrons, resulting in a blocking action. In the opposite extreme, when the ratio of secondary to primary electrons exceeds unity, the insulating surface loses electrons through secondary emission faster than they arrive; the surface becomes increasingly positive. This action continues until the surface is sufficiently positive that the ratio of secondary to primary electrons decreases to unity as a result of the increase in the velocity of the bombarding electrons, or until the surface is sufficiently positive that it attracts back into itself a significant

High-Power Vacuum Devices

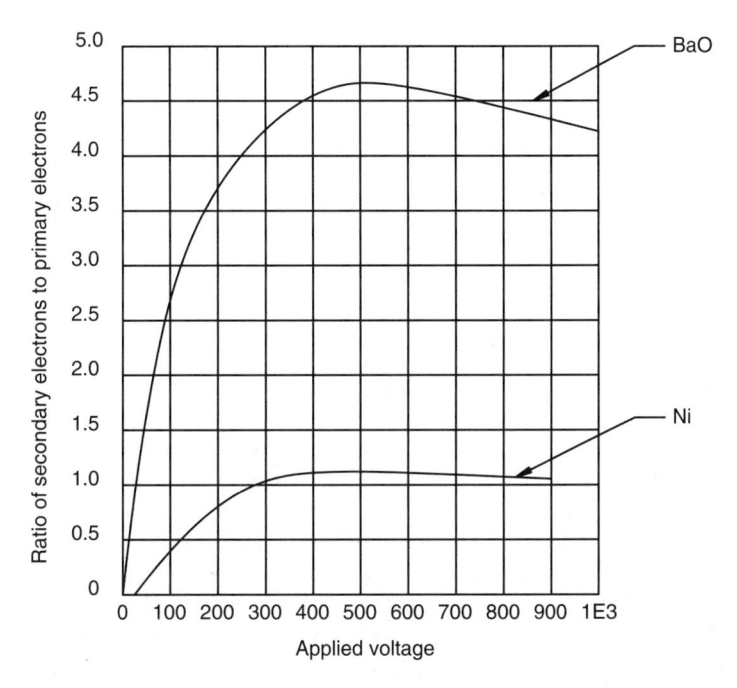

FIGURE 7.3 Ratio of secondary emission current to primary current as a function of primary electron velocity.

number of secondary electrons. This process makes the number of electrons gained from all sources equal to the number of secondary electrons emitted.

7.3 Diode

A diode is a two-electrode vacuum tube containing a cathode, which emits electrons by thermionic emission, surrounded by an anode (or plate) (see Fig. 7.4). Such a tube is inherently a rectifier because when the anode is positive, it attracts electrons; current therefore passes through the tube. When the anode is negative, it repels the electrons and no current flows.

The typical relationship between anode voltage and current flowing to the positive anode is shown in Fig. 7.5. When the anode voltage is sufficiently high, electrons are drawn from the cathode as rapidly as they are emitted. The anode current is then limited by the electron emission of the cathode and, therefore, depends on cathode temperature rather than anode voltage.

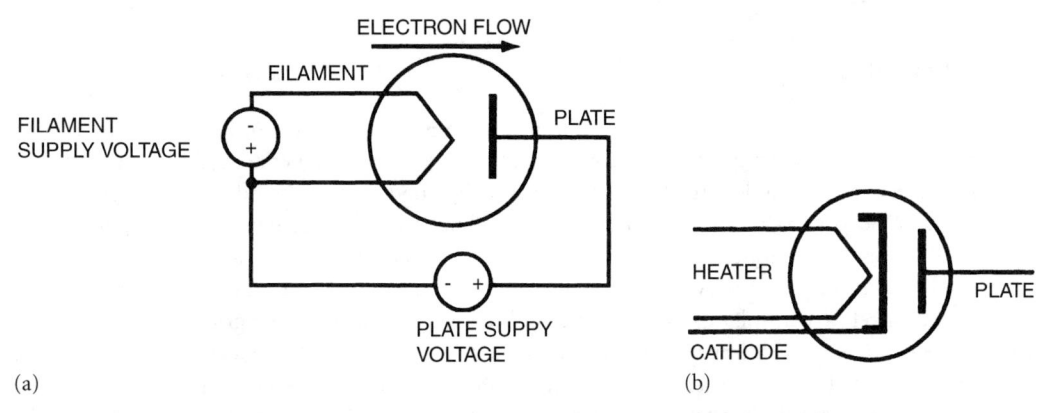

FIGURE 7.4 Vacuum diode: (a) directly heated cathode, and (b) indirectly heated cathode.

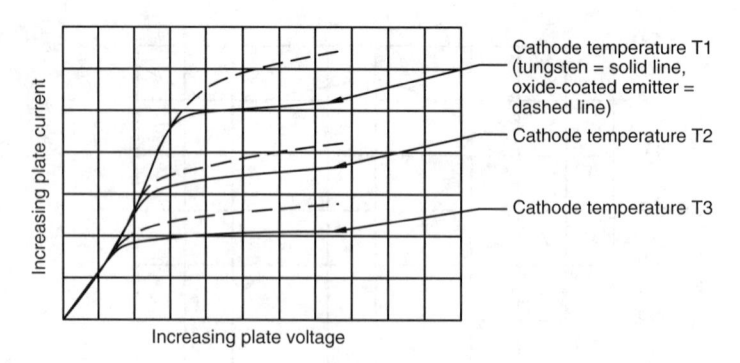

FIGURE 7.5 Anode current as a function of anode voltage in a two-electrode tube for three cathode temperatures.

At low anode voltages, however, plate current is less than the emission of which the cathode is capable. This occurs because the number of electrons in transit between the cathode and plate at any instant cannot exceed the number that will produce a *negative space charge*, which completely neutralizes the attraction of the positive plate upon the electrons just leaving the cathode. All electrons in excess of the number necessary to neutralize the effects of the plate voltage are repelled into the cathode by the negative space charge of the electrons in transit; this situation applies irrespective of how many excess electrons the cathode emits. When the plate current is limited in this way by space charge, plate current is determined by plate potential and is substantially independent of the electron emission of the cathode.

Detailed examination of the space-charge situation will reveal that the negative charge of the electrons in transit between the cathode and the plate is sufficient to give the space in the immediate vicinity of the cathode a slight negative potential with respect to the cathode. The electrons emitted from the cathode are projected out into this field with varying emission velocities. The negative field next to the cathode causes the emitted electrons to slow as they move away from the cathode, and those having a low velocity of emission are driven back into the cathode. Only those electrons having the highest velocities of emission will penetrate the negative field near the cathode and reach the region where they are drawn toward the positive plate. The remainder (those electrons having low emission velocities) will be brought to a stop by the negative field adjacent to the cathode and will fall back into the cathode.

The energy that is delivered to the tube by the source of anode voltage is first expended in accelerating the electrons traveling from the cathode to the anode; it is converted into kinetic energy. When these swiftly moving electrons strike the anode, this kinetic energy is then transformed into heat as a result of the impact and appears at the anode in the form of heat that must be radiated to the walls of the tube.

The basic function of a vacuum tube diode — to rectify an ac voltage — has been superseded by solid-state devices. An understanding of how the diode operates, however, is important in understanding the operation of triodes, tetrodes, and pentodes.

7.4 Triode

The power triode is a three-element device commonly used in a wide variety of RF generators. Triodes have three internal elements: the cathode, control grid, and plate. Most tubes are cylindrically symmetrical. The filament or cathode structure, the grid, and the anode are all cylindrical in shape and are mounted with the axis of each cylinder along the center line of the tube, as illustrated in Figure 7.6.

The grid is normally operated at a negative potential with respect to the cathode, and thus attracts no electrons. However, the extent to which it is negative affects the electrostatic field in the vicinity of the cathode and therefore controls the number of electrons that pass between the grid and the plate. The grid, in effect, functions as an imperfect electrostatic shield. It allows some, but not all, of the electrostatic flux from the anode to leak between its wires. The number of electrons that reach the anode in a triode tube under space-charge-limited conditions is determined almost solely by the electrostatic field near the cathode; the field in the remainder of the interelectrode space has little effect. This phenomenon results

High-Power Vacuum Devices

because the electrons near the cathode are moving slowly compared with the electrons that have traveled some distance toward the plate. The result of this condition is that the volume density of electrons in proportion to the rate of flow is large near the cathode and low in the remainder of the interelectrode space. The total space charge of the electrons in transit toward the plate therefore consists almost solely of the electrons in the immediate vicinity of the cathode. After an electron has traveled beyond this region, it reaches the plate so quickly as to contribute to the space charge for only a brief additional time interval. The result is that the space current in a three-electrode vacuum tube is, for all practical purposes, determined by the electrostatic field that the combined action of the grid and plate potentials produces near the cathode.

When the grid structure is symmetrical, the field E at the surface of the cathode is proportional to the quantity:

$$\frac{E_c + E_b}{\mu} \quad (7.5)$$

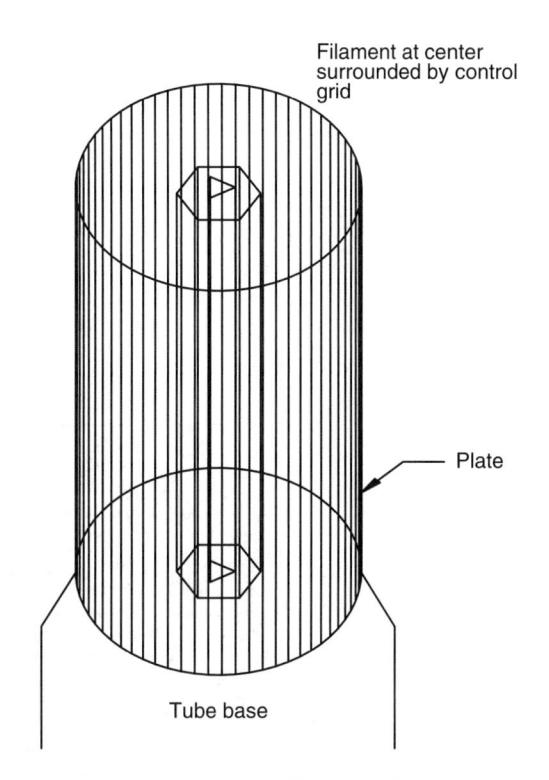

FIGURE 7.6 Mechanical configuration of a power triode.

where E_c is the control grid voltage (with respect to cathode), E_b is the anode voltage (with respect to cathode), and μ is a constant determined by the geometry of the tube. The constant μ, the amplification factor, is independent of the grid and plate voltages. It is a measure of the relative effectiveness of grid and plate voltages in producing electrostatic fields at the surfaces of the cathode. Placement of the control grid relative to the cathode and plate determines the amplification factor. The μ values of triodes generally range from 5 to 200. Key mathematical relationships include the following:

$$\mu = \frac{\Delta E_b}{\Delta E_{c1}} \quad (7.6)$$

$$R_p = \frac{\Delta E_b}{\Delta I_b} \quad (7.7)$$

$$S_m = \frac{\Delta I_b}{\Delta E_{c1}} \quad (7.8)$$

where μ is the amplification factor (with plate current held constant), R_p is the dynamic plate resistance, S_m is the transconductance (also may be denoted G_m), E_b is the total instantaneous plate voltage, and E_{c1} is the total instantaneous control grid voltage, and I_b is the total instantaneous plate current. The total cathode current of an ideal triode can be determined from the equation:

$$I_k = \left\{ E_c + \frac{E_b}{\mu} \right\}^{3/2} \quad (7.9)$$

where I_k is the cathode current, E_c is the grid voltage, E_b is the plate voltage, and μ is the amplification factor.

Figure 7.7 plots plate and grid current as a function of plate voltage at various grid voltages for a triode with a μ of 12. The tube, a 304TL, is a classic design and, while not used in new equipment, provides a common example of the relationship between the parameters plotted. Figure 7.8 plots the same parameters for a tube with a μ of 20. Observe how much more plate current at a given plate voltage can be obtained from the 304TL (μ = 12) without driving the grid into the positive grid region. Note also how much more bias voltage is required for the 304TL to cut the plate current off at some given plate voltage. With this increased bias, there is a corresponding increase in grid voltage swing to drive up the zero grid voltage point on the curve. Low-μ tubes have lower voltage gain by definition. This fact can be seen by comparing Figures 7.7 and 7.8.

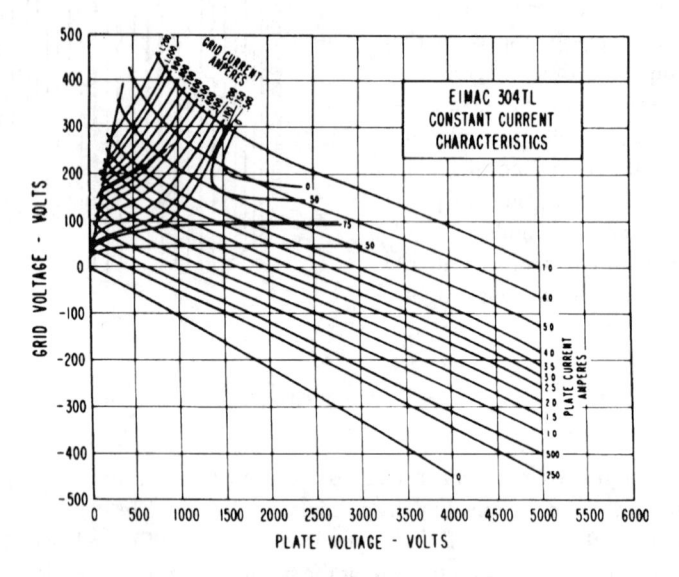

FIGURE 7.7 Constant-current characteristics of a triode with a μ of 12.

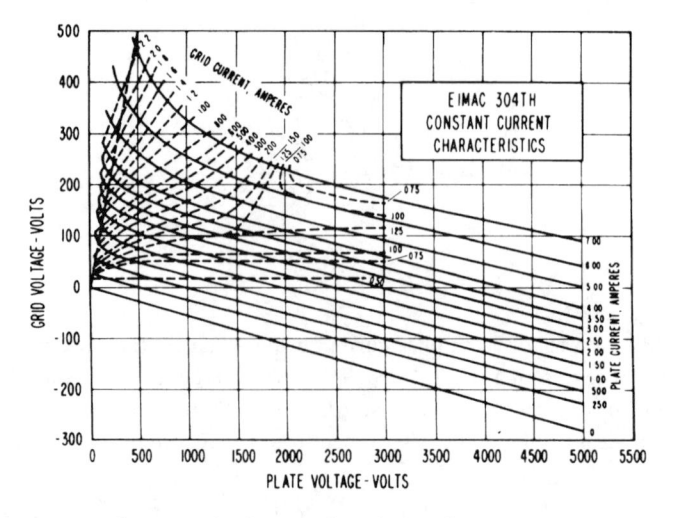

FIGURE 7.8 Constant-current characteristics for a triode with a μ of 20.

High-Power Vacuum Devices

Triodes with μ between 20 and 50 are generally used in conventional RF amplifiers and oscillators. High-μ triodes (≥200) can be designed so that the operating bias is zero, as depicted in Fig. 7.9. These zero-bias triodes are available with plate dissipation ratings of 400 W to 10 kW or more. The zero-bias triode is commonly used in grounded-grid amplification. The tube offers good power gain and circuit simplicity. No bias power source is required. Furthermore, no protection circuits for loss of bias or drive are needed. Despite these attributes, present-day use of the zero-bias triode is limited.

Low- and medium-μ devices are usually preferred for induction heating applications because of the wide variations in load that an induction or dielectric heating oscillator normally works into. Such tubes exhibit lower grid-current variation with a changing load. The grid current of a triode with a μ of 20 will rise substantially less under a light- or no-load condition than a triode with a μ of 40. High-μ triode oscillators can be designed for heating applications, but extra considerations must be given to current rise under no-load conditions.

Vacuum tubes specifically designed for induction heating are available, intended for operation under adverse conditions. The grid structure is ruggedized with ample dissipation capability to deal with wide variations in load. As the load decreases, grid dissipation increases.

Triodes also are manufactured with the cathode, grid, and anode in the shape of a flat surface, as shown in Fig. 7.10. Tubes so constructed are called *planar triodes*. This construction technique permits operation at high frequencies. The close spacing reduces electron transit time, allowing the tube to be used at high frequencies (up to 3 GHz or so). The physical construction of planar triodes results in short lead lengths, which reduces lead inductance. Planar triodes are used in both continuous-wave (CW) and pulsed modes. The contacting surfaces of the planar triode are arranged for easy integration into coaxial and waveguide resonators.

7.5 Tetrode

The tetrode is a four-element tube with two grids. The control grid serves the same purpose as the grid in a triode, while a second (screen) grid with the same number of vertical elements (bars) as the control grid is mounted between the control grid and the anode. The grid bars of the screen grid are mounted directly behind the control-grid bars, as observed from the cathode surface, and serve as a shield or screen

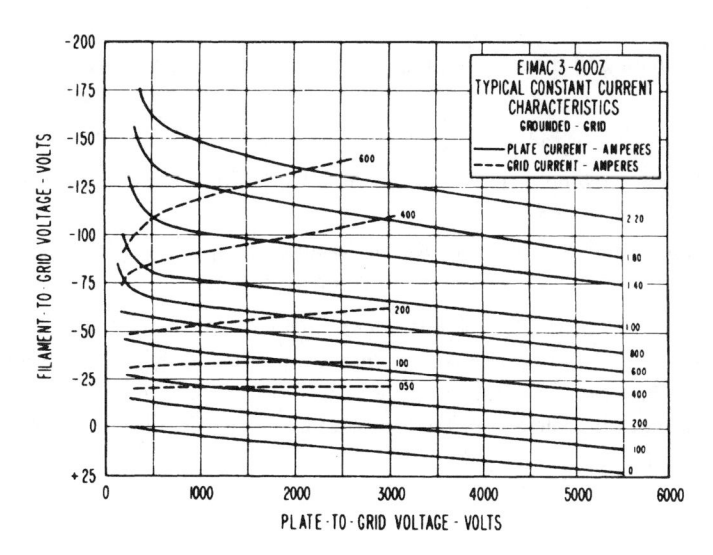

FIGURE 7.9 Grounded-grid, constant-current characteristics for a zero-bias triode with a μ of 200. (Courtesy of Varian/Eimac.)

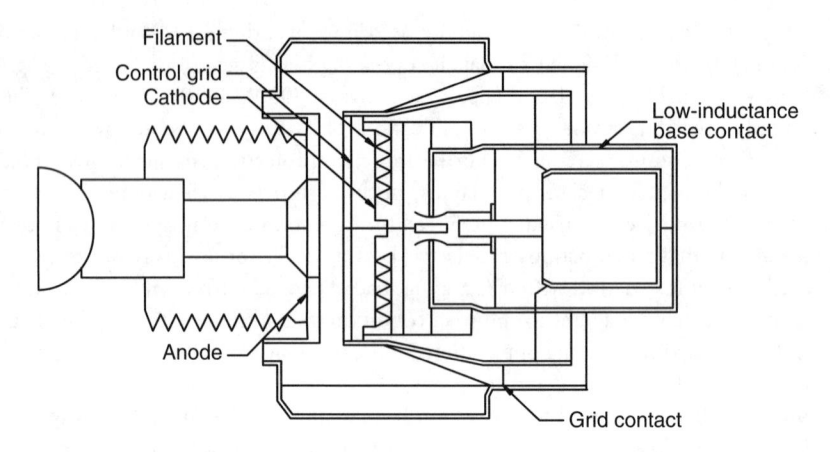

FIGURE 7.10 Internal configuration of a planar triode.

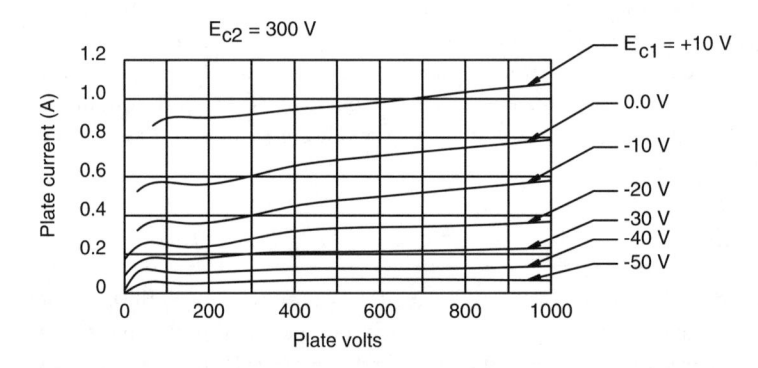

FIGURE 7.11 Tetrode plate current characteristics. Plate current is plotted as a function of plate voltage, with grid voltages as shown.

between the input circuit and the output circuit of the tetrode. The principal advantages of a tetrode over a triode include:

- Lower internal plate-to-grid feedback
- Lower drive power requirements; in most cases, the driving circuit need supply only 1% of the output power
- More efficient operation; tetrodes allow the design of compact, simple, flexible equipment with little spurious radiation and low intermodulation distortion

Plate current is almost independent of plate voltage in a tetrode. Figure 7.11 plots plate current as a function of plate voltage at a fixed screen voltage and various grid voltages. In an ideal tetrode, a change in plate current does not cause a change in plate voltage. The tetrode, therefore, can be considered a constant-current device. The voltages on the screen and control grids determine the amount of plate current.

The total cathode current of an ideal tetrode is determined by:

$$I_k = K \left\{ E_{c1} + \frac{E_{c2}}{\mu_s} + \frac{E_b}{\mu_p} \right\}^{3/2} \tag{7.10}$$

High-Power Vacuum Devices

where I_k is the cathode current, K is a constant determined by tube dimensions, E_{c1} is the control grid voltage, E_{c2} is the screen grid voltage, μ_s is the screen amplification factor, μ_p is the plate amplification factor, and E_b is the plate voltage. The arithmetic value of the screen μ is not generally used in the design of an RF amplifier. In most tetrode applications, the screen amplification factor is useful for roughly categorizing the performance to be expected.

Application Example

Figure 7.12 shows a radial beam power tetrode (4CX15000A) designed for class A_{B1} or class C power amplification. The device is particularly well suited for RF linear power amplifier service. The tube has a directly heated thoriated-tungsten mesh filament for mechanical ruggedness and high efficiency. The maximum rated plate dissipation of the tube is 15 kW using air cooling.

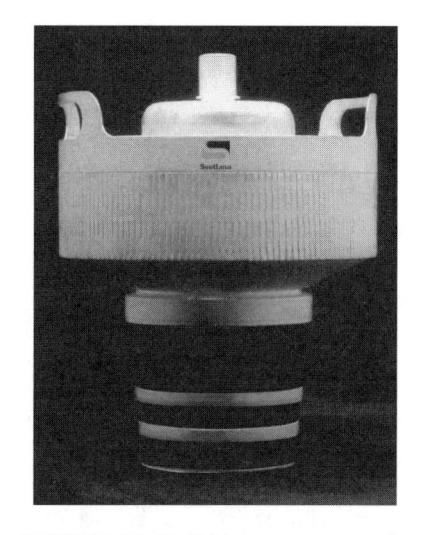

FIGURE 7.12 Radial beam power tetrode (4CX15000A).

The tube must be protected from damage that might result from an internal arc occurring at high plate voltage. A protective resistance is typically inserted in series with the tube anode to help absorb stored power supply energy in case an internal arc occurs.

The maximum control grid dissipation is 200 W, determined (approximately) by the product of the dc grid current and the peak positive grid voltage.

Screen grid maximum dissipation is 450 W. With no ac applied to the screen grid, dissipation is the product of dc screen voltage and dc screen current. Plate voltage, plate loading, and/or bias voltage must never be removed while filament and screen voltages are present.

The 4CX15000A must be mounted vertically, base up or down. The tube requires forced-air cooling in all applications. The tube socket is mounted in a pressurized compartment where cooling air passes through the socket and is guided to the anode cooling fins by an air chimney. Adequate movement of cooling air around the base of the tube keeps the base and socket contact fingers at a safe operating temperature. Although the maximum temperature rating for seals and the anode is 250°C, good engineering practice dictates that a safety factor be provided. Table 7.1 lists cooling parameters for the tube with the cooling air at 50°C and a maximum anode temperature of 225°C. The figures given in the table apply to designs in which air passes in the base-to-anode direction. Pressure drop values shown are approximate and apply to the tube/socket/chimney combination.

At altitudes significantly above sea level, the flow rate must be increased for equivalent cooling. At 5000 ft above sea level, both the flow rate and the pressure drop are increased by a factor of 1.20; at 10,000 ft, both the flow rate and pressure drop are increased by 1.46.

Anode and base cooling is applied before or simultaneously with filament voltage turn-on, and normally should continue for a brief period after shutdown to allow the tube to cool properly.

TABLE 7.1 Minimum Cooling Airflow Requirements for the 4CX15000A Power Tetrode at Sea Level

Plate Dissipation (W)	Airflow (CFM)	Pressure Drop (inches of water)
7,500	230	0.7
12,500	490	2.7
15,000	645	4.6

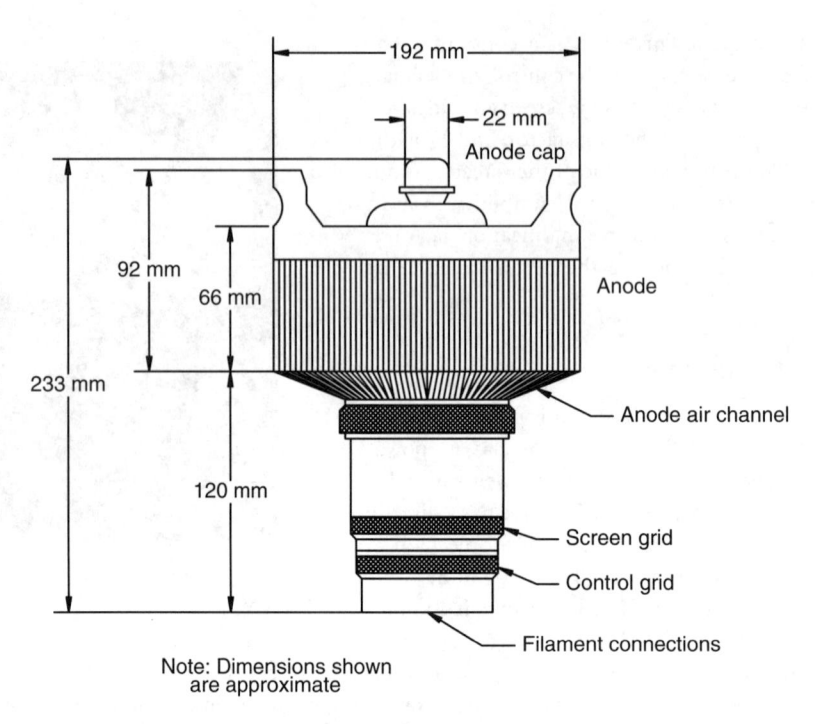

FIGURE 7.13 Principal dimensions of the 4CX15000A tetrode.

An outline of the principal tube dimensions is given in Fig. 7.13. General specifications are listed in Table 7.2.

7.6 Pentode

The pentode is a five-electrode tube incorporating three grids. The control and screen grids perform the same function as in a tetrode. The third grid, the suppressor grid, is mounted in the region between the screen grid and the anode. The suppressor grid produces a potential minimum, which prevents secondary electrons from being interchanged between the screen and plate. The pentode's main advantages over the tetrode include:

- Reduced secondary emission effects
- Good linearity
- Ability to let plate voltage swing below the screen voltage without excessive screen dissipation; this allows slightly higher power output for a given operating plate voltage

Because of the design of the pentode, plate voltage has even less effect on plate current than in the tetrode. The same total space-current equation applies to the pentode as with the tetrode:

$$I_k = K \left\{ E_{c1} + \frac{E_{c2}}{\mu_s} + \frac{E_b}{\mu_p} \right\}^{3/2} \tag{7.11}$$

where I_k is the cathode current, K is a constant determined by tube dimensions, E_{c1} is the control grid voltage, E_{c2} is the screen grid voltage, μ_s is the screen amplification factor, μ_p is the plate amplification factor, and E_b is the plate voltage.

The suppressor grid may be operated negative or positive with respect to the cathode. It also may be operated at cathode potential. It is possible to control plate current by varying the potential on the

High-Power Vacuum Devices

TABLE 7.2 General Characteristics of the 4CX15000A Power Tetrode

Electrical Characteristics		
Filament type		Thoriated-tungsten mesh
Filament voltage		6.3 ± 0.3 V
Filament current		164 A (at 6.3 V)
Amplification factor (average), grid to screen		4.5
Direct interelectrode capacitance (grounded cathode)	C_{in}	158 pF
	C_{out}	25.8 pF
	C_{pk}	1.3 pF
Direct interelectrode capacitance (grounded grid)	C_{in}	67 pF
	C_{out}	25.6 pF
	C_{pk}	0.21 pF
Maximum frequency for full ratings (CW)		110 MHz

Mechanical Characteristics	
Length	238 mm (9.38 in.)
Diameter	193 mm (7.58 in.)
Net weight	5.8 kg (12.8 lb)
Operating position	Axis vertical, base up or down
Maximum operating temperature (seals/envelope)	250°C
Cooling method	Forced air
Base type	Coaxial

Radio Frequency Power Amplifier (class C FM) (absolute maximum ratings)	
DC plate voltage	10,000 V
DC screen voltage	2,000 V
DC grid voltage	–750 V
DC plate current	5.0 A
Plate dissipation	15 kW
Screen dissipation	450 W
Grid dissipation	200 W

Typical Operation (frequencies up to 110 MHz)		
DC plate voltage	7.5 kV dc	10.0 kV dc
DC screen voltage	750 V dc	750 V dc
DC grid voltage	–510 V dc	–550 V dc
DC plate current	4.65 A dc	4.55 A dc
DC screen current	0.59 A dc	0.54 A dc
DC grid current	0.30 A dc	0.27 A dc
Peak RF grid voltage	730 V	790 V
Calculated driving power	220 W	220 W
Plate dissipation	8.1	9.0 kW
Plate output power	26.7	36.5 kW

Source: Courtesy of Svetlana Electron Devices, Palo Alto, California.

suppressor grid. Because of this ability, a modulating voltage can be applied to the suppressor to achieve amplitude modulation. The required modulating power is low because of the low electron interception of the suppressor.

7.7 High-Frequency Operating Limits

As with most active devices, performance of a given vacuum tube deteriorates as the operating frequency is increased beyond its designed limit. Electron *transit time* is a significant factor in the upper-frequency limitation of electron tubes. A finite time is taken by electrons to traverse the space from the cathode, through the grid, and travel on to the plate. As the operating frequency increases, a point is reached at

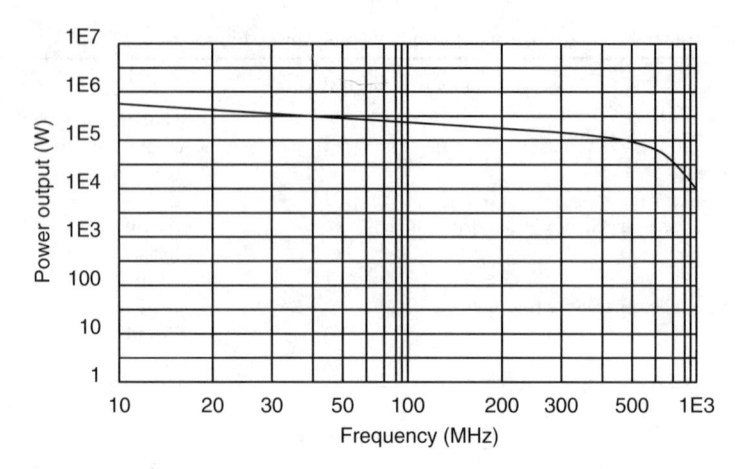

FIGURE 7.14 Continuous-wave output power capability of a gridded vacuum tube.

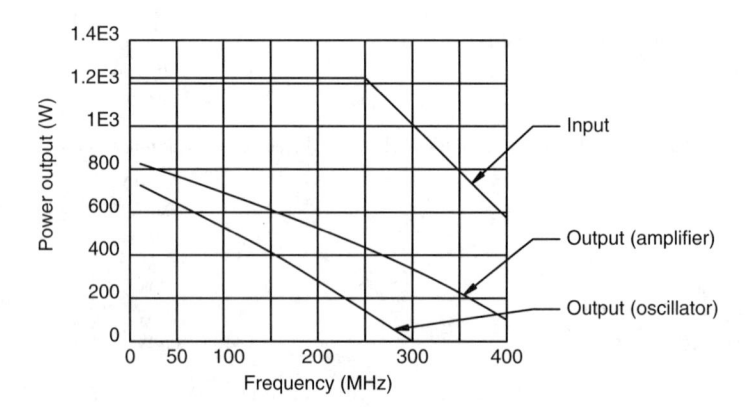

FIGURE 7.15 Performance of a class C amplifier as the operating frequency is increased beyond the design limits of the vacuum tube.

which the electron transit-time effects become significant. This point is a function of the accelerating voltages at the grid and anode and their respective spacings. Tubes with reduced spacing in the grid-to-cathode region exhibit reduced transit-time effects.

A power limitation is also interrelated with the high-frequency limit of a device. As the operating frequency is increased, closer spacing and smaller-sized electrodes must be used. This reduces the power-handling capability of the tube. Figure 7.14 illustrates the relationship.

Gridded tubes at all power levels for frequencies up to about 1 GHz are invariably cylindrical in form. At higher frequencies, planar construction is almost universal. As the operating frequency is increased beyond design limits, output power and efficiency both decrease. Figure 7.15 illustrates the relationship.

Transit time typically is not a problem for power grid tubes operating below 30 MHz. Depending on the application, power grid tubes can be used at 100 MHz and above without serious consideration of transit-time effects.

Transit-Time Effects

When class C, class B, or similar amplifier operations are carried out at frequencies sufficiently high that the transit time of the electrons is not a negligible fraction of the waveform cycle, the following complications are observed in grid-based vacuum tubes:

High-Power Vacuum Devices

- Back-heating of the cathode
- Loading of the control grid circuit as a result of energy transferred to electrons that do not necessarily reach the grid to produce a dc grid current
- Debunching of plate current pulses
- Phase differences between the plate current and the exciting voltage applied to the control grid

Back-heating of the cathode occurs when the transit time in the grid-cathode space is sufficiently great to cause an appreciable number of electrons to be in transit at the instant the plate current pulse would be cut off in the case of low-frequency operation. A considerable fraction of the electrons thus trapped in the interelectrode space are returned to the cathode by the negative field existing in the grid-cathode space during the cutoff portion of the cycle. These returning electrons act to heat the cathode. At very high frequencies, this back-heating is sufficient to supply a considerable fraction of the total cathode heating required for normal operation. Back-heating may reduce the life of the cathode as a result of electron bombardment of the emitting surface. It also causes the required filament current to depend upon the conditions of operation within the tube.

Energy absorbed by the control grid as a result of input loading is transferred directly to the electron stream in the tube. Part of this stream acts to produce back-heating of the cathode. The remainder affects the velocity of the electrons as they arrive at the anode of the tube. This portion of the energy is not necessarily all wasted. In fact, a considerable percentage of it may, under favorable conditions, appear as useful output in the tube. To the extent that this is the case, the energy supplied by the exciting voltage to the electron stream is simply transferred directly from the control grid to the output circuits of the tube without amplification.

An examination of the total time required by electrons to travel from the cathode to the anode in a triode, tetrode, or pentode operated as a class C amplifier reveals that the resulting transit times for electrons at the beginning, middle, and end of the current pulse will differ as the operating frequency is increased. In general, electrons traversing the distance during the first segment of the pulse will have the shortest transit time, while those near the middle and end of the pulse will have the longest transit times, as illustrated in Fig. 7.16. The first electrons in the pulse have a short transit time because they approach the plate before the plate potential is at its minimum value. Electrons near the middle of the pulse approach the plate with the instantaneous plate potential at or near minimum and, consequently, travel less rapidly in the grid-plate space. Finally, those electrons that leave the cathode late in the current pulse (those just able to escape being trapped in the control grid-cathode space and returned toward the cathode) will be slowed as they approach the grid, and thus have a large transit time. The net effect is to cause the pulse of plate current to be longer than it would be in operation at a low frequency. This causes the efficiency of the amplifier to drop at high frequencies because a longer plate current pulse increases plate losses.

7.8 Device Cooling

The main factor that separates tube types is the method of cooling used: air, water, or vapor. Air-cooled tubes are common at power levels below 50 kW. A water cooling system, although more complicated, is more effective than air cooling — by a factor of 5 to 10 or more — in transferring heat from the device. Air cooling at the 100-kW level is virtually impossible because it is difficult to physically move enough air through the device (if the tube is to be of reasonable size) to keep the anode sufficiently cool. Vapor cooling provides an even more efficient method of cooling a power amplifier (PA) tube than water cooling, for a given water flow and a given power dissipation. Naturally, the complexity of the external blowers, fans, ducts, plumbing, heat exchangers, and other hardware must be taken into consideration in selecting a cooling method. Figure 7.17 shows how the choice of cooling method is related to anode dissipation.

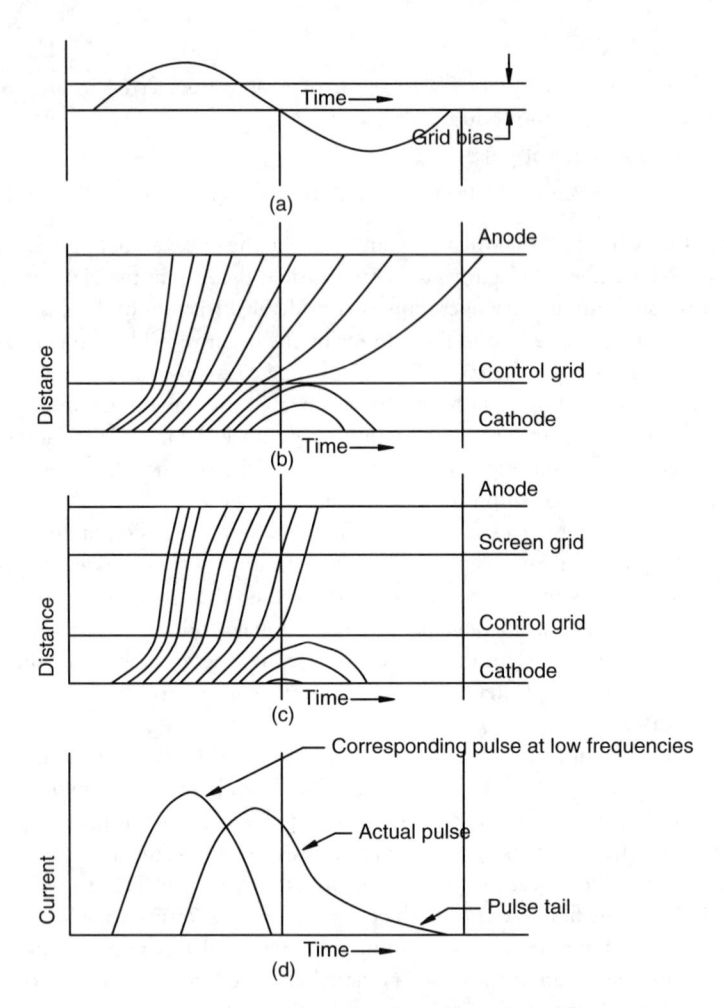

FIGURE 7.16 Transit time in a class C amplifier: (a) control grid voltage; (b) electron position as a function of time, triode case; (c) electron position as a function of time, tetrode case; and (d) plate current, triode case.

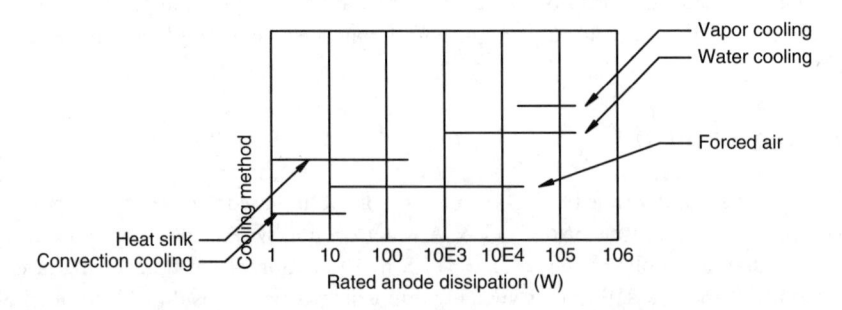

FIGURE 7.17 The relationship between anode dissipation and cooling method.

Air Cooling

A typical air cooling system for a transmitter is shown in Fig. 7.18. Cooling system performance for an air-cooled device is not necessarily related to airflow volume. The cooling capability of air is a function of its mass, not its volume. An appropriate airflow rate within the equipment is established by the manufacturer, resulting in a given resistance to air movement.

High-Power Vacuum Devices

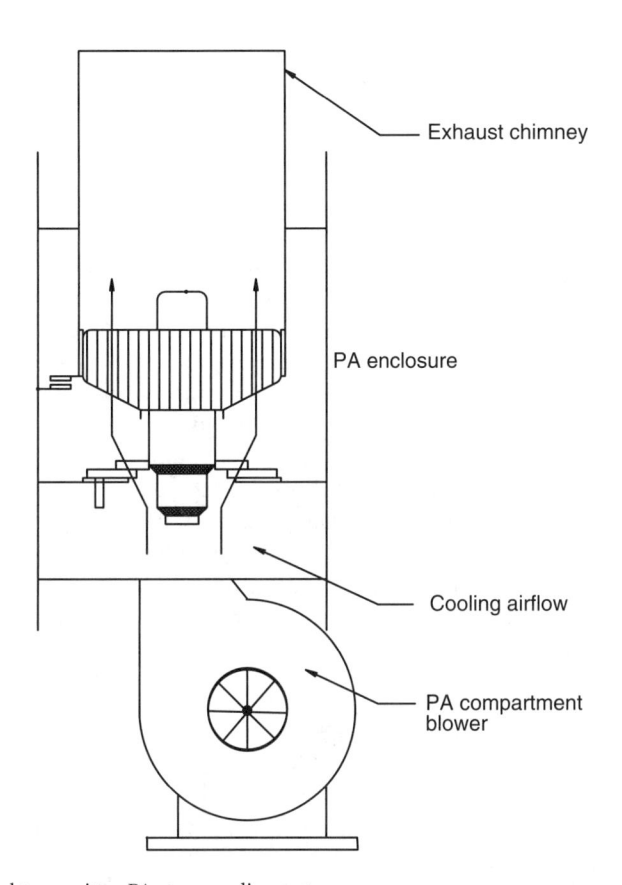

FIGURE 7.18 A typical transmitter PA stage cooling system.

The altitude of operation is also a consideration in cooling system design. As altitude increases, the density (and cooling capability) of air decreases. To maintain the same cooling effectiveness, increased airflow must be provided.

Water Cooling

Water cooling is usually preferred over air cooling for power outputs above about 50 kW. Multiple grooves on the outside of the anode, in conjunction with a cylindrical jacket, force the cooling water to flow over the surface of the anode, as illustrated in Fig. 7.19.

Because the water is in contact with the outer surface of the anode, a high degree of purity must be maintained. A resistivity of 1 mΩ-cm (at 25°C) is typically specified by tube manufacturers. Circulating water can remove about 1 kW/cm^2 of effective internal anode area. In practice, the temperature of water leaving the tube must be limited to 70°C to prevent the possibility of spot boiling.

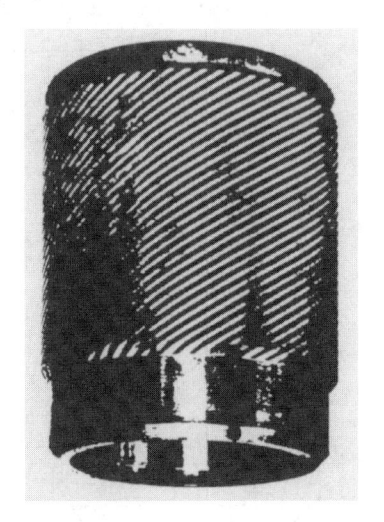

FIGURE 7.19 Water-cooled anode with grooves for controlled water flow.

After leaving the anode, the heated water is passed through a heat exchanger where it is cooled to 30 to 40°C before being pumped back to the tube.

In typical modern water-cooled tubes, the dissipation ranges from 500 W to 1 kW per square centimeter. Water-cooled anodes have deen designed that can operate at 10 kW/cm^2 and above. However, with the exception of "big science" applications, dissipations this high are seldom required.

Vapor-Phase Cooling

Vapor cooling allows the permissible output temperature of the water to rise to the boiling point, giving higher cooling efficiency compared with water cooling. The benefits of vapor-phase cooling are the result of the physics of boiling water. Increasing the temperature of 1 g (1 gram) of water from 40 to 70°C requires 30 calories of energy. However, transforming 1 g of water at 100°C into steam vapor requires 540 calories. Thus, a vapor-phase cooling system permits essentially the same cooling capacity as water cooling, but with greatly reduced water flow. Viewed from another perspective, for the same water flow, the dissipation of the tube can be increased significantly (all other considerations being the same).

A typical vapor-phase cooling system is shown in Fig. 7.20. A tube incorporating a specially designed anode is immersed in a boiler filled with distilled water. When power is applied to the tube, anode dissipation heats the water to the boiling point, converting the water to steam vapor. The vapor passes to a condenser, where it gives up its energy and reverts to a liquid state. The condensate is then returned to the boiler, completing the cycle. Electric valves and interlocks are included in the system to provide for operating safety and maintenance. A vapor-phase cooling system for a transmitter with multiple PA tubes is shown in Fig. 7.21.

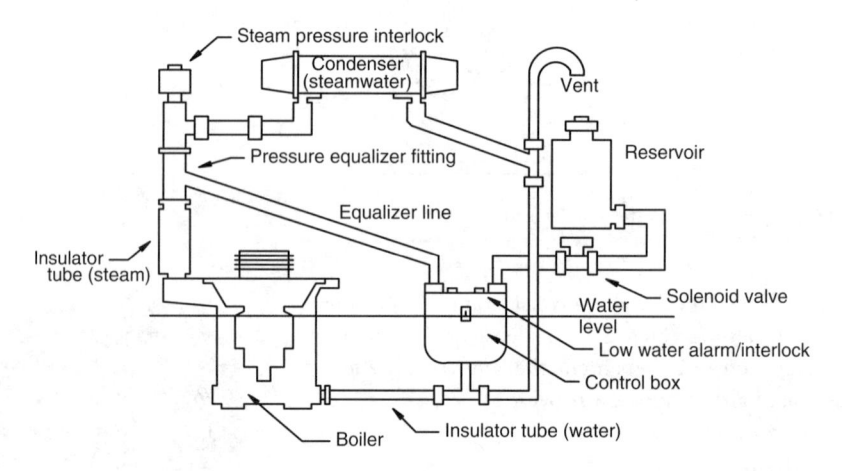

FIGURE 7.20 Typical vapor-phase cooling system.

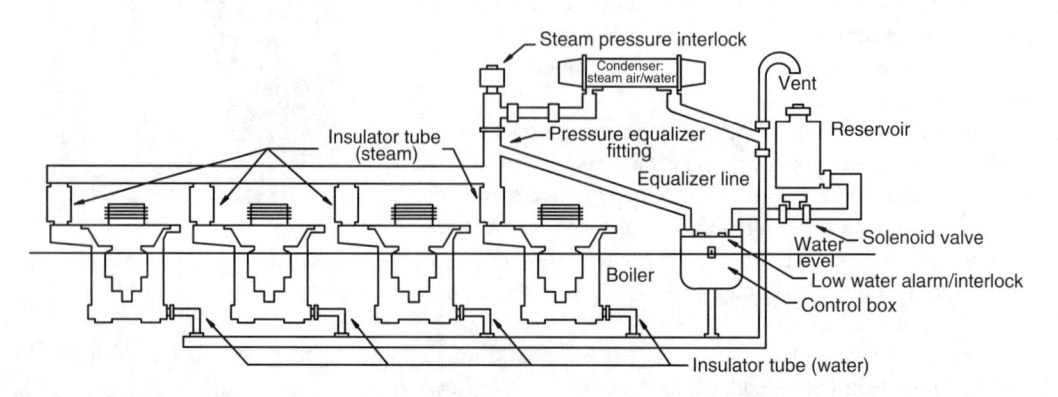

FIGURE 7.21 Vapor-phase cooling system for a four-tube transmitter using a common water supply.

Special Applications

Power devices used for research applications must be designed for transient overloading, requiring special considerations with regard to cooling. Oil, heat pipes, refrigerants (such as Freon), and, where high-voltage hold-off is a problem, gases (such as sulfahexafluoride) are sometimes used to cool the anode of a power tube.

References

1. Terman, F. E., *Radio Engineering*, 3rd ed., McGraw-Hill, New York, 1947.
2. Ferris, Clifford D., Electron tube fundamentals, in *The Electronics Handbook*, Jerry C. Whitaker, Ed., CRC Press, Boca Raton, FL, 1996, 295–305.

Bibliography

Birdsall, C. K., *Plasma Physics via Computer Simulation*, Adam Hilger, 1991.
Block, R., CPS microengineers new breed of materials, *Ceramic Ind.*, April, 51–53, 1988.
Buchanan, R. C., *Ceramic Materials for Electronics*, Marcel Dekker, New York, 1986.
Ceramic Products, Brush Wellman, Cleveland, OH.
Chaffee, E. L., *Theory of Thermionic Vacuum Tubes*, McGraw-Hill, New York, 1939.
Combat Boron Nitride, Solids, Powders, Coatings, Carborundum Product Literature, form A-14, 011, September 1984.
Coors Ceramics — Materials for Tough Jobs, Coors Data Sheet K.P.G.-2500-2/87 6429.
Cote, R. E. and R. J. Bouchard, Thick Film Technology, in *Electronic Ceramics*, L. M. Levinson, Ed., Marcel Dekker, New York, 1988, 307–370.
Dettmer, E. S. and H. K. Charles, Jr., AlNi and SiC substrate properties and processing characteristics, in *Advances in Ceramics*, Vol. 31, American Ceramic Society, Columbus, OH, 1989.
Dettmer, E. S., H. K. Charles, Jr., S. J. Mobley, and B. M. Romenesko, Hybrid design and processing using aluminum nitride substrates, *ISHM 88 Proc.*, 1988, 545–553.
Eastman, Austin V., *Fundamentals of Vacuum Tubes*, McGraw-Hill, New York, 1941.
Fink, D. and D. Christiansen, Eds., *Electronics Engineers' Handbook*, 3rd ed., McGraw-Hill, New York, 1989.
Floyd, J. R., How to Tailor High-Alumina Ceramics for Electrical Applications, *Ceramic Ind.*, February 1969, 44–47; March 1969, 46–49.
Gray, Truman S., *Applied Electronics*, John Wiley & Sons, New York, 1954.
Harper, C. A., *Electronic Packaging and Interconnection Handbook*, McGraw-Hill, New York, 1991.
High Power Transmitting Tubes for Broadcasting and Research, Philips Technical Publication, Eindhoven, The Netherlands, 1988.
Iwase, N. and K. Shinozaki, Aluminum nitride substrates having high thermal conductivity, *Solid State Technology*, October, 135–137, 1986.
Jordan, Edward C., Ed., *Reference Data for Engineers: Radio, Electronics, Computer and Communications*, 7th ed., Howard W. Sams, Indianapolis, IN, 1985.
Kingery, W. D., H. K. Bowen, and D. R. Uhlmann, *Introduction to Ceramics*, John Wiley & Sons, New York, 1976, 637.
Kohl, Walter, *Materials Technology for Electron Tubes*, Reinhold, New York.
Laboratory Staff, *The Care and Feeding of Power Grid Tubes*, Varian Eimac, San Carlos, CA, 1984.
Lafferty, J. M., *Vacuum Arcs*, John Wiley & Sons, New York, 1980.
Mattox, D. M. and H. D. Smith, The role of manganese in the metallization of high alumina ceramics, *J. Am. Ceram. Soc.*, 64, 1363–1369, 1985.
Mistler, R. E., D. J. Shanefield, and R. B. Runk, Tape casting of ceramics, in G. Y. Onoda, Jr. and L. L. Hench, Eds., *Ceramic Processing Before Firing*, John Wiley & Sons, New York, 1978, 411–448.
Muller, J. J., Cathode excited linear amplifiers, *Electrical Communications*, Vol. 23, 1946.

Powers, M. B., Potential Beryllium Exposure While Processing Beryllia Ceramics for Electronics Applications, Brush Wellman, Cleveland, OH.

Reich, Herbert J., *Theory and Application of Electronic Tubes*, McGraw-Hill, New York, 1939.

Roth, A., *Vacuum Technology*, 3rd ed., Elsevier Science Publishers B. V., 1990.

Sawhill, H. T., A. L. Eustice, S. J. Horowitz, J. Gar-El, and A. R. Travis, Low temperature co-fireable ceramics with co-fired resistors, in *Proc. Int. Symp. on Microelectronics*, 1986, 173–180.

Schwartz, B., Ceramic packaging of integrated circuits, in *Electronic Ceramics*, L. M. Levinson, Ed., Marcel Dekker, New York, 1988, 1–44.

Strong, C. E., The inverted amplifier, *Electrical Communications*, 19(3), 1941.

Whitaker, J. C., *Radio Frequency Transmission Systems: Design and Operation*, McGraw-Hill, New York, 1991.

8
Microwave Vacuum Devices

Jerry C. Whitaker
Editor

8.1 Introduction 8-1
Linear-Beam Tubes · Crossed-Field Tubes

8.2 Grid Vacuum Tubes 8-5
Planar Triode · High-Power UHF Tetrode · Diacrode

8.3 Klystron 8-9
Reflex Klystron · The Two-Cavity Klystron · The Two-Cavity Klystron Oscillator · The Two-Cavity Klystron Amplifier · The Multicavity Klystron

8.4 Traveling Wave Tube 8-14
Theory of Operation · Pulse Modulation · Electron Gun · Beam Focusing · Collector Assembly · Operating Efficiency

8.5 Crossed-Field Tubes 8-20
Magnetron · Operating Principles · Coaxial Magnetron · Frequency-Agile Magnetron · Linear Magnetron · Backward Wave Oscillator · Strap-Fed Devices · Gyrotron · Gyrotron Design Variations

8.1 Introduction

Microwave power tubes span a wide range of applications, operating at frequencies from 300 MHz to 300 GHz with output powers from a few hundred watts to more than 10 MW. Applications range from the familiar to the exotic. The following devices are included under the general description of microwave power tubes:

- Klystron, including the reflex and multicavity klystron
- Multistage depressed collector (MSDC) klystron
- Inductive output tube (IOT)
- Traveling wave tube (TWT)
- Crossed-field tube
- Coaxial magnetron
- Gyrotron
- Planar triode
- High-frequency tetrode
- Diacrode

This wide variety of microwave devices has been developed to meet a broad range of applications. Some common uses include:

- UHF-TV transmission
- Shipboard and ground-based radar
- Weapons guidance systems
- Electronic countermeasure (ECM) systems
- Satellite communications
- Tropospheric scatter communications
- Fusion research

As new applications are identified, improved devices are designed to meet the needs. Microwave power tube manufacturers continue to push the limits of frequency, operating power, and efficiency. Microwave technology, therefore, is an evolving science. Figure 8.1 charts device type as a function of operating frequency and power output.

Two principal classes of microwave vacuum devices are in common use today: linear-beam tubes and crossed-field tubes. Each class serves a specific range of applications. In addition to these primary classes, some power grid tubes are also used at microwave frequencies.

Linear-Beam Tubes

In a linear-beam tube, as the name implies, the electron beam and the circuit elements with which it interacts are arranged linearly. The major classifications of linear-beam tubes are shown in Fig. 8.2. In such a device, a voltage applied to an anode accelerates electrons drawn from a cathode, creating a beam of kinetic energy. Power supply potential energy is converted to kinetic energy in the electron beam as it travels toward the microwave circuit. A portion of this kinetic energy is transferred to microwave energy as RF waves slow down the electrons. The remaining beam energy is either dissipated as heat or returned to the power supply at the collector. Because electrons will repel one another, there is usually an applied magnetic focusing field to maintain the beam during the interaction process. The magnetic field is supplied either by a solenoid or permanent magnets. Figure 8.3 shows a simplified schematic of a linear-beam tube.

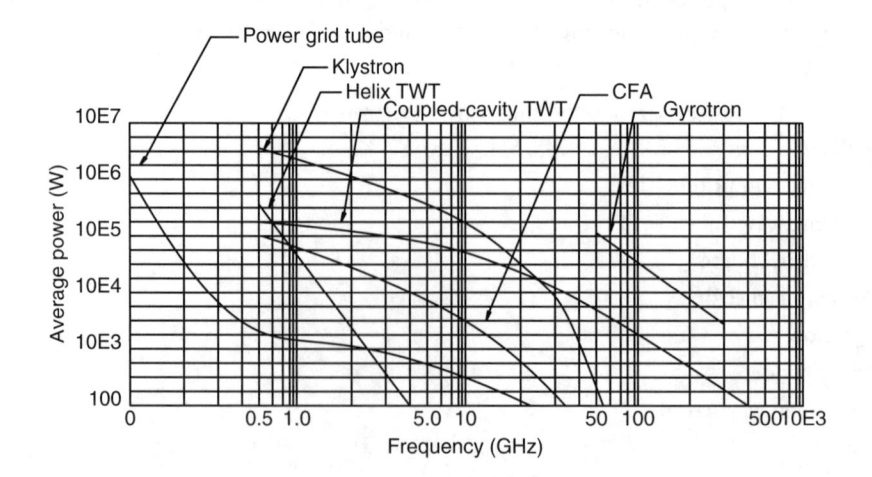

FIGURE 8.1 Microwave power tube type as a function of frequency and output power.

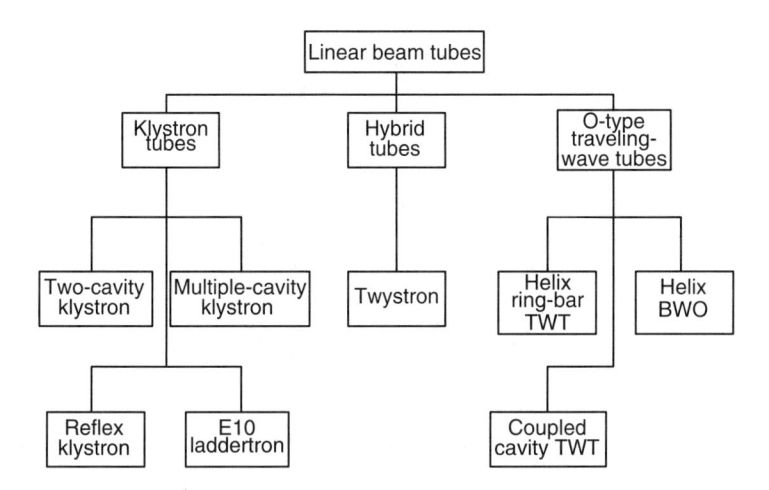

FIGURE 8.2 Types of linear-beam microwave tubes.

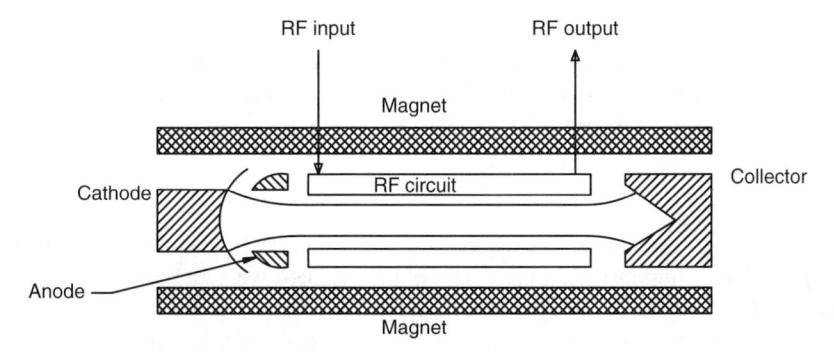

FIGURE 8.3 Schematic diagram of a linear-beam tube.

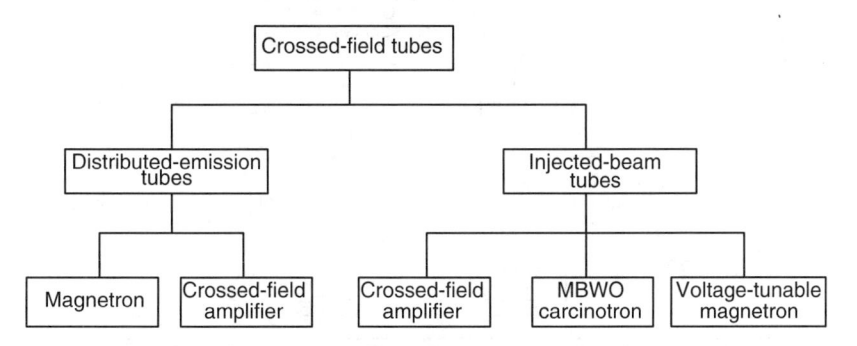

FIGURE 8.4 Types of crossed-field microwave tubes.

Crossed-Field Tubes

The magnetron is the pioneering device of the family of crossed-field tubes. The family tree of this class of devices is shown in Fig. 8.4. Although the physical appearance differs from that of linear-beam tubes, which are usually circular in format, the major difference is in the interaction physics that requires a

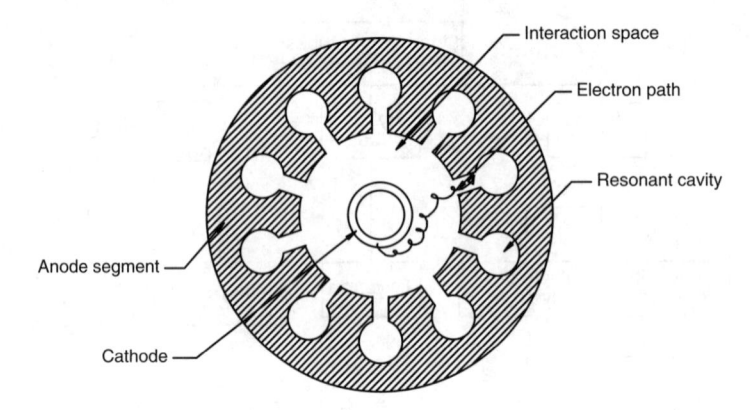

FIGURE 8.5 Magnetron electron path looking down into the cavity with the magnetic field applied.

magnetic field at right angles to the applied electric field. Whereas the linear-beam tube sometimes requires a magnetic field to maintain the beam, the crossed-field tube always requires a magnetic focusing field.

Figure 8.5 shows a cross-section of the magnetron, including the magnetic field applied perpendicular to the cathode–anode plane. The device is basically a diode, with the anode composed of a plurality of resonant cavities. The interaction between the electrons emitted from the cathode and the crossed electric and magnetic fields produces a series of space-charge spokes that travel around the anode–cathode space in a manner that transfers energy to the RF signal supported by the multicavity circuit. The mechanism is highly efficient.

Crossed-Field Amplifiers

Figure 8.6 shows the family tree of the *crossed-field amplifier* (CFA). The configuration of a typical present-day distributed emission amplifier is similar to that of the magnetron except that the device has an input for the introduction of RF energy into the circuit. Current is obtained primarily by secondary emission from the negative electrode that serves as a cathode throughout all or most of the interaction space. The earliest versions of this tube type were called *amplitrons*.

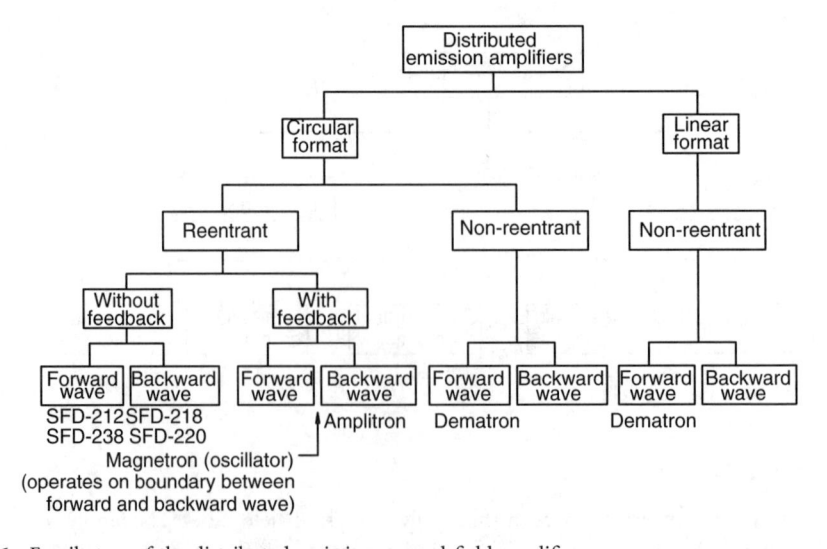

FIGURE 8.6 Family tree of the distributed emission crossed-field amplifier.

Microwave Vacuum Devices

The CFA is deployed in radar systems operating from UHF to the Ku-band, and at power levels up to several megawatts. In general, bandwidth ranges from a few percent to as much as 25% of the center frequency.

8.2 Grid Vacuum Tubes

The physical construction of a vacuum tube causes the output power and available gain to decrease with increasing frequency. The principal limitations faced by grid-based devices include the following:

- *Physical size*. Ideally, the RF voltages between electrodes should be uniform but this condition cannot be realized unless the major electrode dimensions are significantly less than 1/4-wavelength at the operating frequency. This restriction presents no problems at VHF, but as the operating frequency increases into the microwave range, severe restrictions are placed on the physical size of individual tube elements.
- *Electron transit time*. Interelectrode spacing, principally between the grid and the cathode, must be scaled inversely with frequency to avoid problems associated with electron transit time. Possible adverse conditions include (1) excessive loading of the drive source, (2) reduction in power gain, (3) back-heating of the cathode as a result of electron bombardment, and (4) reduced conversion efficiency.
- *Voltage standoff*. High-power tubes operate at high voltages. This presents significant problems for microwave vacuum tubes. For example, at 1 GHz, the grid–cathode spacing must not exceed a few mils. This places restrictions on the operating voltages that may be applied to the individual elements.
- *Circulating currents*. Substantial RF currents may develop as a result of the inherent interelectrode capacitances and stray inductances/capacitances of the device. Significant heating of the grid, connecting leads, and vacuum seals may result.
- *Heat dissipation*. Because the elements of a microwave grid tube must be kept small, power dissipation is limited.

Still, a number of grid-based vacuum tubes find applications at high frequencies. For example, planar triodes are available that operate at several gigahertz, with output powers of 1 to 2 kW in pulsed service. Efficiency (again for pulsed applications) ranges from 30 to 60%, depending on the frequency.

Planar Triode

A cross-sectional diagram of a planar triode is shown in Fig. 8.7. The envelope is made of ceramic, with metal members penetrating the ceramic to provide for connection points. The metal members are shaped either as disks or as disks with cylindrical projections.

The cathode is typically oxide-coated and indirectly heated. The key design objective for a cathode is high emission density and long tube life. Low-temperature emitters are preferred because high cathode temperatures typically result in more evaporation and shorter life.

The grid of the planar triode is perhaps the greatest design challenge for tube manufacturers. Close spacing of small-sized elements is needed, at tight tolerances. Good thermal stability is also required because the grid is subjected to heating from currents in the element itself, plus heating from the cathode and bombardment of electrons from the cathode.

The anode, usually made of copper, conducts the heat of electron bombardment to an external heat sink. Most planar triodes are air-cooled.

Planar triodes designed for operation at 1 GHz and above are used in a variety of circuits. The grounded-grid configuration is most common. The plate resonant circuit is cavity based, using waveguide,

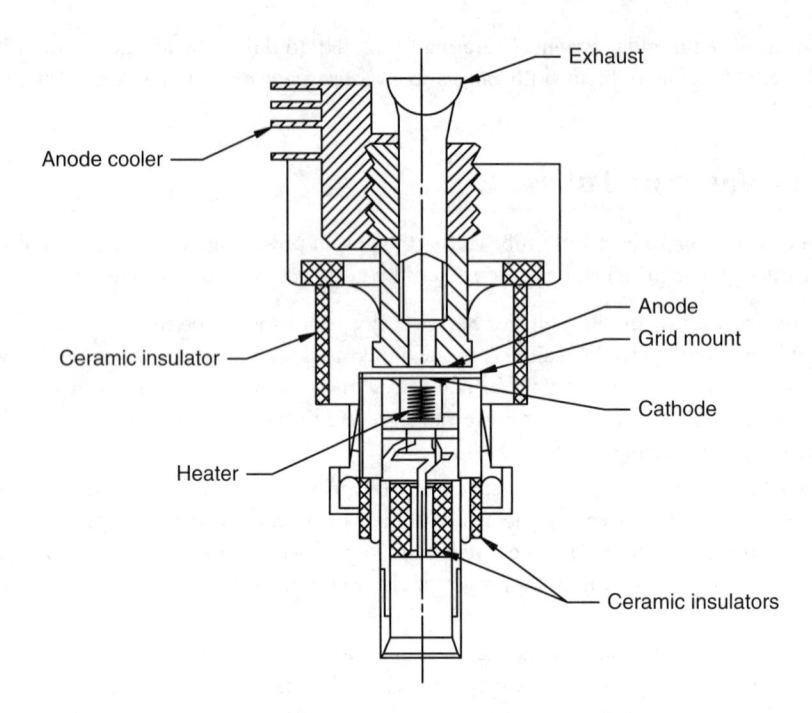

FIGURE 8.7 Cross-section of a 7289 planar triode.

coaxial line, or stripline. Electrically, the operation of the planar triode is much more complicated at microwave frequencies than at low frequencies. Figure 8.8a compares the elements at work for a grounded-grid amplifier operating at low frequencies and Fig. 8.8b compares the situation at microwave frequencies. The equivalent circuit is made more complex by:

- Stray inductance and capacitance of the tube elements
- Effects of the tube contact rings and socket elements
- Distributed reactance of cavity resonators and the device itself
- Electron transit-time effects, which result in resistive loading and phase shifts

Reasonable gains of 5 to 10 dB can be achieved with a planar triode. Increased gain is available by cascading stages. Interstage coupling may consist of waveguide or coaxial-line elements. Tuning is accomplished by varying the cavity inductance or capacitance. Additional bandwidth is possible by stagger tuning of cascaded stages.

High-Power UHF Tetrode

New advancements in vacuum tube technology have permitted the construction of high-power UHF transmitters based on tetrodes. Such devices are attractive because they inherently operate in a relatively efficient class AB mode. UHF tetrodes operating at high power levels provide essentially the same specifications, gain, and efficiency as tubes operating at lower powers. The anode power supply is much lower in voltage than the collector potential of a klystron- or IOT-based system (8 kV is common). Also, the tetrode does not require a focusing magnet system.

Efficient removal of heat is the key to making a tetrode practical at high power levels. Such devices typically use water or vapor-phase cooling. Air cooling at such levels is impractical because of the fin size that would be required. Also, the blower for the tube would have to be quite large, reducing the overall transmitter ac-to-RF efficiency.

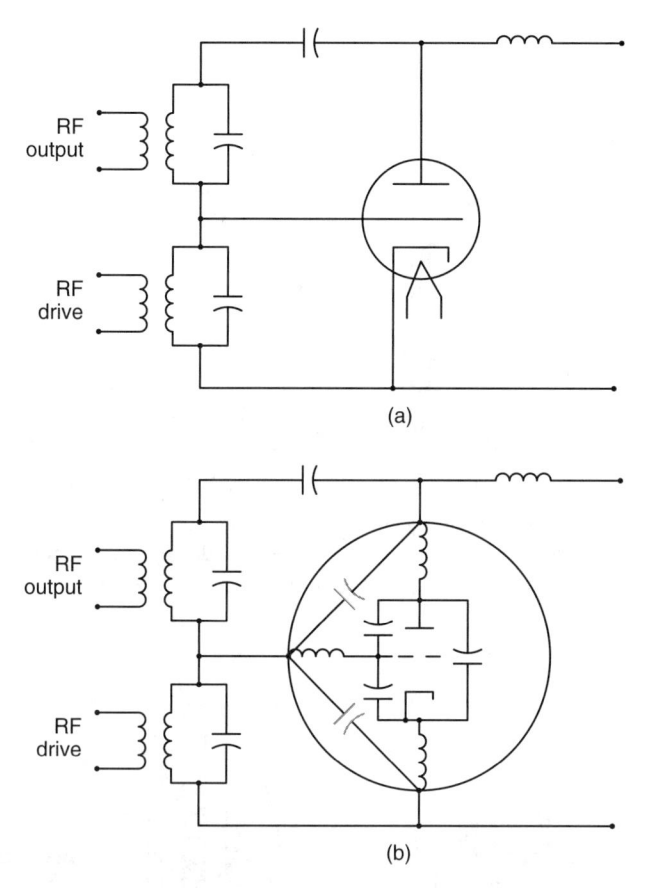

FIGURE 8.8 Grounded-grid equivalent circuits: (a) low-frequency operation, and (b) microwave-frequency operation. The cathode-heater and grid-bias circuits are not shown.

Another drawback inherent in tetrode operation is that the output circuit of the device appears electrically in series with the input circuit and the load.[1] The parasitic reactance of the tube elements, therefore, is a part of the input and output tuned circuits. It follows, then, that any change in the operating parameters of the tube as it ages can affect tuning. More importantly, the series nature of the tetrode places stringent limitations on internal element spacings and the physical size of those elements in order to minimize the electron transit time through the tube vacuum space. It is also fair to point out, however, that the tetrode's input-to-output circuit characteristic has at least one advantage: power delivered to the input passes through the tube and contributes to the total power output of the transmitter. Because tetrodes typically exhibit low gain compared to klystron-based devices, significant power can be required at the input circuit. The pass-through effect therefore contributes to the overall operating efficiency of the transmitter.

The expected lifetime of a tetrode in UHF service is usually shorter than that of a klystron of the same power level. Typical lifetimes of 8000 to 15,000 hours have been reported. Intensive work, however, has led to products that offer higher output powers and extended operating lifetime, while retaining the benefits inherent in tetrode devices.

Diacrode

The Diacrode (Thomson) is a promising adaptation of the high-power UHF tetrode (Fig. 8.9). The operating principle of the Diacrode is basically the same as that of the tetrode. The anode current is modulated by an RF drive voltage applied between the cathode and the grid. The main difference is in

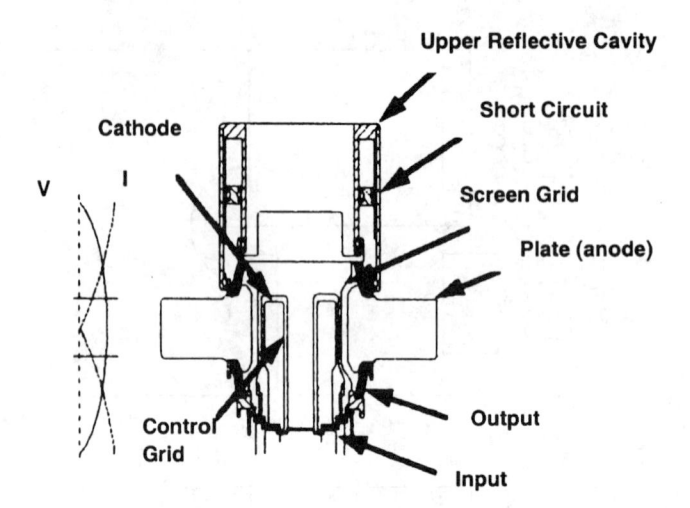

FIGURE 8.9 The elements of the Diacrode, including the upper cavity. Double current, and consequently double power, are achieved with this device because of the current peaks at the top and bottom of the device, as shown. (Adapted from Hulick, T. P., *Proc. 1996 NAB Broadcast Engineering Conf.*, National Association of Broadcasters, Washington, D.C., 1996, 442.)

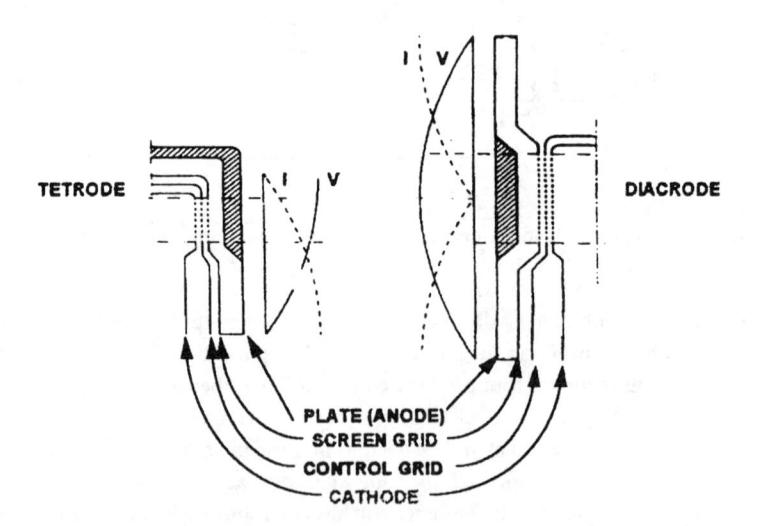

FIGURE 8.10 Cutaway view of the tetrode (left) and the Diacrode (right). Note that the RF current peaks above and below the Diacrode center, while on the tetrode there is only one peak at the bottom. (Adapted from Hulick, T. P., *Proc. 1996 NAB Broadcast Engineering Conf.*, National Association of Broadcasters, Washington, D.C., 1996, 442.)

the position of the active zones of the tube in the resonant coaxial circuits, resulting in improved reactive current distribution in the electrodes of the device.

Figure 8.10 compares the conventional tetrode with the Diacrode. The Diacrode includes an electrical extension of the output circuit structure to an external cavity.[2] The small dc-blocked cavity rests on top of the tube, as illustrated in the figure.

The cavity is a quarter-wave transmission line, as measured from the top of the cavity to the vertical center of the tube. The cavity is short-circuited at the top, reflecting an open circuit (current minimum) at the vertical center of the tube and a current maximum at the base of the tube, like the conventional tetrode, and a second current maximum above the tube at the cavity short-circuit.

Microwave Vacuum Devices

With two current maximums, the RF power capability of the Diacrode is double that of the equivalent tetrode, while the element voltages remain the same. All other properties and aspects of the Diacrode are basically identical to the TH563 high-power UHF tetrode (Thomson), upon which the Diacrode is patterned.

Some of the benefits of such a device, in addition to the robust power output available, include the low high-voltage requirements (low relative to a klystron-based system, that is), small size, and simple replacement procedures.

8.3 Klystron

The klystron is a linear-beam device that overcomes the transit-time limitations of a grid-controlled tube by accelerating an electron stream to a high velocity before it is modulated. Modulation is accomplished by varying the velocity of the beam, which causes the drifting of electrons into bunches to produce RF space current. One or more cavities reinforce this action at the operating frequency. The output cavity acts as a transformer to couple the high-impedance beam to a low-impedance transmission line. The frequency response of a klystron is limited by the impedance-bandwidth product of the cavities, but may be extended through stagger tuning or the use of multiple-resonance filter-type cavities.

The klystron is one of the primary means of generating high power at UHF and above. Output powers for multicavity devices range from a few thousand watts to 10 MW or more. The klystron provides high gain and requires little external support circuitry. Mechanically, the klystron is relatively simple. It offers long life and requires minimal routine maintenance.

Reflex Klystron

The reflex klystron uses a single-cavity resonator to modulate the RF beam and extract energy from it. The construction of a reflex klystron is shown in Fig. 8.11. In its basic form, the tube consists of the following elements:

- A cathode
- Focusing electrode at cathode potential
- Coaxial line or reentrant-type cavity resonator, which also serves as an anode
- Repeller or reflector electrode, which is operated at a moderately negative potential with respect to the cathode

The cathode is shaped so that, in relation to the focusing electrode and anode, an electron beam is formed that passes through a gap in the resonator, as shown in Fig. 8.11, and travels toward the repeller. Because the repeller has a negative potential with respect to the cathode, it turns the electrons back toward the anode, where they pass through the anode gap a second time. By varying the applied voltage on the reflector electrode, phasing of the beam can be varied to produce the desired oscillating mode and to control the frequency of oscillation.

The variation of position with time for electrons in the anode–repeller space is illustrated in Fig. 8.12. Path *a* corresponds to an electron that emerges from the anode with a velocity corresponding to the anode voltage. This electron follows a parabolic path, as shown, determined by the electric field in the anode–repeller space.

Operation of the reflex klystron can best be understood by examining the movement of electrons inside the device. Assume that oscillations exist in the resonator so that an alternating voltage develops across the gap. Assume further that the electron corresponding to path *a* passed through the gap at the instant that this alternating voltage across the gap was zero and becoming negative. An electron passing through the gap just before electron *a* will encounter an accelerating voltage across the gap and therefore will emerge from the anode with greater velocity than the first or reference electron. This second electron, accordingly, penetrates farther toward the repeller against the retarding field and, as a result, takes longer

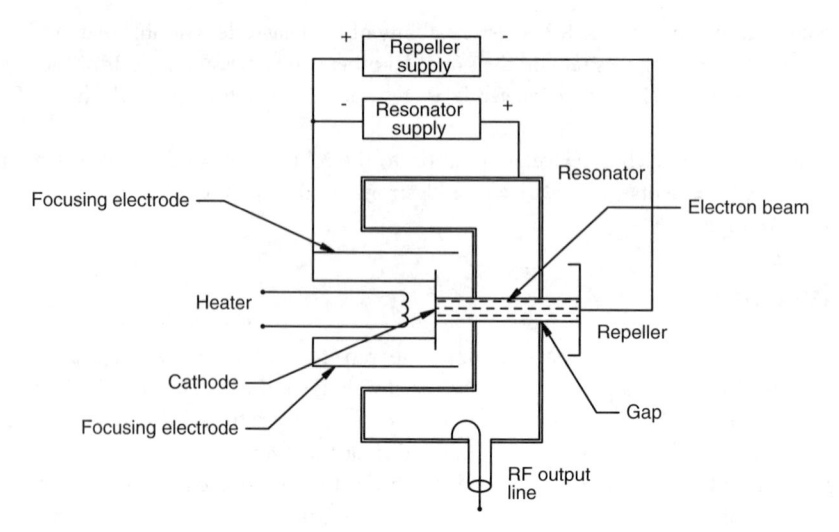

FIGURE 8.11 Schematic representation of a reflex klystron.

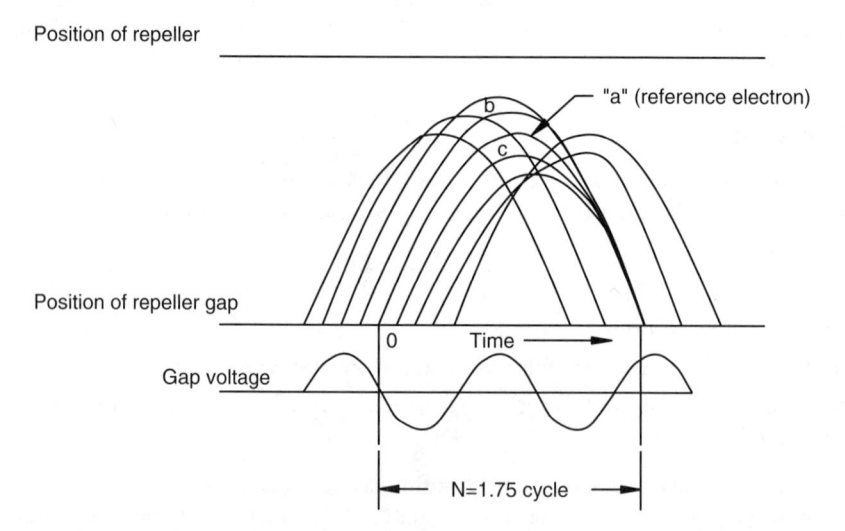

FIGURE 8.12 Position–time curves of electrons in the anode–repeller space, showing the tendency of the electrons to bunch around the electron passing through the anode at the time when the alternating gap voltage is zero and becoming negative.

to return to the anode. Consequently, this electron follows path *b*, as shown in Fig. 8.12, and tends to arrive at the anode on its return path at the same time as the reference electron because its earlier start is more or less compensated for by increased transit time. In a similar manner, an electron passing through the anode gap slightly later than the reference electron will encounter a negative or retarding field across the gap, and thus will emerge from the anode with less velocity than the electron that follows path *a*. This third electron will then follow trajectory *c* and return to the anode more quickly than electron *a*. Electron *c*, therefore, tends to return to the anode at about the same time as electron *a* because the later start of electron *c* is compensated for by the reduced transit time.

This variation with time of the velocity of electrons emerging from the anode is termed *velocity modulation*. The effect of this phenomenon can be seen in Fig. 8.12 to cause a bunching of electrons about the electron that passed through the gap when the resonator voltage was zero and becoming negative. This bunching causes the electrons that are returned toward the anode by the repeller to pass

Microwave Vacuum Devices 8-11

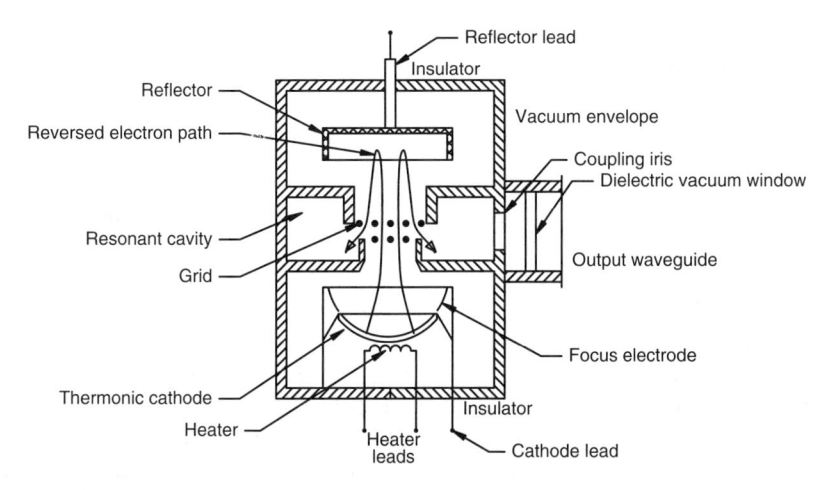

FIGURE 8.13 Schematic cross-section of a reflex klystron oscillator.

through the anode gap in bursts or pulses, one each cycle. When these pulses pass through the gap at such a time that the electrons in the pulse are slowed as a result of the alternating voltage existing across the gap at the instant of their return passage, energy will be delivered to the oscillations in the resonator, thereby assisting in maintaining the oscillations. This condition corresponds to a transit time N from the resonator toward the repeller and back to the resonator of approximately

$$N = n + \frac{3}{4} \tag{8.1}$$

where n is an integer (including zero). The transit time in the anode–repeller space in any particular case depends on the following:

- The anode voltage
- Repeller voltage
- Geometry of the anode-repeller space

The extent of the bunching action that takes place when the transit time of the reference electron has the correct value for sustaining oscillations is determined by the following:

- The amplitude of the alternating voltage across the resonator gap in relation to the anode and repeller voltage
- The geometry of the repeller space

The reflex klystron typically includes a grid to concentrate the electric field so that it can efficiently couple to the electron beam. Such a device is illustrated in Fig. 8.13.

The reflex klystron can be used as a local oscillator, low-power FM transmitter, or test signal source. Reflex tubes are used primarily from 4 to 40 GHz. Power outputs of 1 W or less are common.

The reflex tube is the only klystron in which beam feedback is used to produce output energy. In klystrons with more than one cavity, the electron beam passes through each cavity in succession.

The Two-Cavity Klystron

The two-cavity klystron operates on the same bunching principle as the reflex klystron but incorporates two cavities connected by a drift tube. Figure 8.14 shows a cross-section of a classic device. The heater/cathode element (shown as A in the figure) produces an electron beam in conjunction with a focusing electrode. The route taken by the electrons is as follows:

- The beam passes through grid elements D in the side of a reentrant cavity resonator (the buncher).
- The beam then passes through the drift tube, which is at the same electrical potential as the buncher.
- Finally, the beam enters a second resonator, termed the collector, which is provided with grid E.

The cathode and its associated focusing electrode are maintained at a high negative potential with respect to the remaining part of the structure, all of which is at the same dc potential. The entire arrangement illustrated in Fig. 8.14 is enclosed in a vacuum.

The operational principles of the two-cavity klystron are similar in nature to those of the reflex klystron. Assume, first, that oscillations exist in the buncher so that an alternating voltage is present across the gap D. When this voltage is zero but just becoming positive, an electron passing through the buncher travels through the grids D, down the drift tube, and into the collector resonator with unchanged velocity. However, an electron that passes through the buncher slightly later receives acceleration while passing through, because of the positive alternating field that it encounters between grids D, and enters the drift tube with increased velocity. This later electron therefore tends to overtake the earlier electron. Similarly, an electron that arrives at the buncher slightly earlier than the first (reference electron) encounters a field between grids D that opposes its motion. Hence, this early electron enters the drift tube with reduced velocity and tends to drop back and be overtaken by the reference electron.

As a result of these actions, the electrons bunch together as they travel down the drift tube. This effect is more pronounced at certain distances from the buncher. If the collector is located at a distance where the bunching is pronounced, the electrons enter the element in pulses, one pulse per cycle.

With proper adjustment, the amount of power required to produce the bunching effect is relatively small compared with the amount of energy delivered by the electron beam to the collector. As a result, the klystron operates as an amplifying device.

The Two-Cavity Klystron Oscillator

The two-cavity klystron oscillator is designed for applications requiring moderate power (up to 100 W), stable frequency output, and low sideband noise. The device has a coupling iris on the wall between the

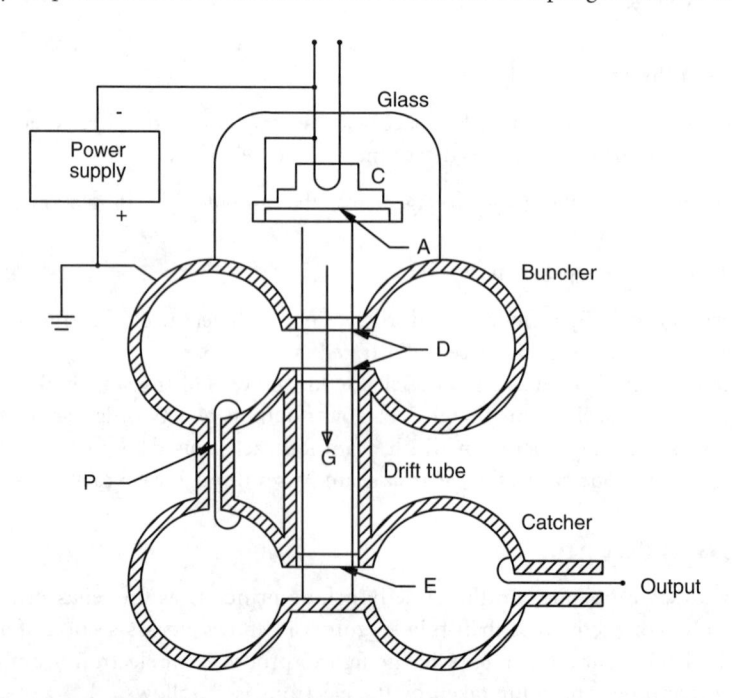

FIGURE 8.14 Cross-section of a classic two-cavity klystron oscillator.

Microwave Vacuum Devices   8-13

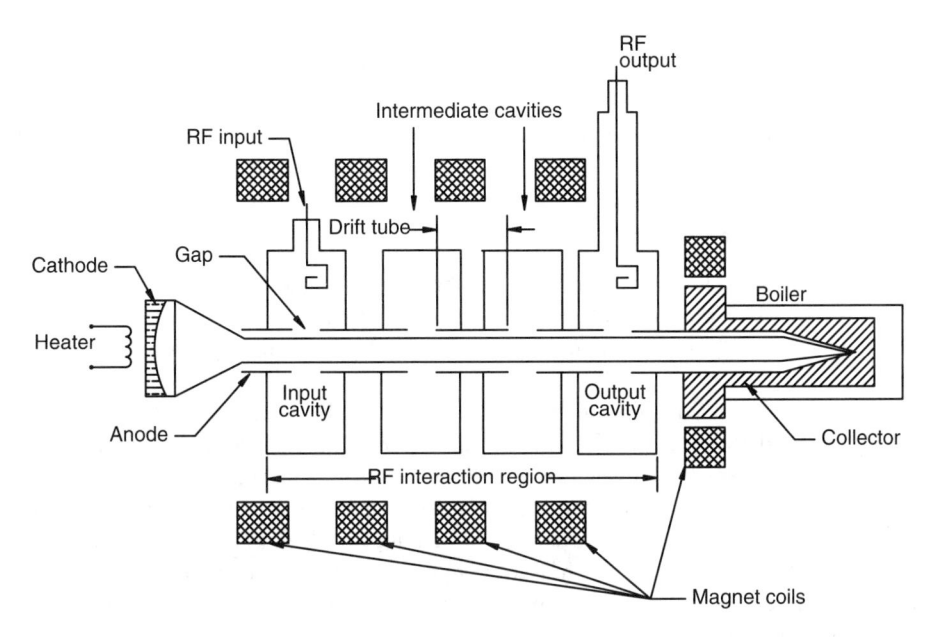

FIGURE 8.15 Principal elements of a multistage klystron.

two cavities. The tube can be frequency-modulated by varying the cathode voltage about the center of the oscillating mode. Although it is more efficient and powerful than the reflex klystron, the two-cavity klystron requires more modulator power. The two-cavity klystron is typically used in Doppler radar systems.

The Two-Cavity Klystron Amplifier

Similar in design to the two-cavity oscillator, the two-cavity klystron amplifier provides limited power output (10 W or less) and moderate gain (about 10 dB). A driving signal is coupled into the input cavity, which produces velocity modulation of the beam. After the drift space, the density-modulated beam induces current in the output resonator. Electrostatic focusing of the beam is common.

The two-cavity klystron amplifier finds only limited applications because of its restrictions on output power and gain. For many applications, solid-state amplifiers are a better choice.

The Multicavity Klystron

The multicavity klystron is an important device for amplifying signals to high power levels at microwave frequencies.[3] Each cavity tuned to the operating frequency adds about 20-dB gain to the 10-dB gain offered by the basic two-cavity klystron amplifier. Overall gains of 60 dB are practical. Cavities can be tuned to either side of resonance to broaden the operating bandwidth of the device. Klystrons with up to eight cavities have been produced. Operating power for continuous-wave klystrons ranges up to 1 MW per device, and as much as 50 MW per device for pulsed applications.

The primary physical advantage of the klystron over a grid-based power tube is that the cathode-to-collector structure is virtually independent of transit-time effects (see Fig. 8.15). Therefore, the cathode can be made large and the electron beam density kept low.

The operating frequency of a klystron may be fixed — determined by the mechanical characteristics of the tube and its cavities — or tunable. Cavities are tuned mechanically using one of several methods, depending on the operating power and frequency. Tuning is accomplished by changing the physical dimensions of the cavities using one or more of the following techniques:

- Cavity wall deformation, in which one wall of the cavity consists of a thin diaphragm that is moved in and out by a tuning mechanism. About 3% frequency shift can be accomplished using this method, which varies the inductance of the cavity.

- Movable cavity wall, in which one wall of the cavity is moved in or out by a tuning mechanism. About 10% frequency shift is possible with this approach, which varies the inductance of the cavity.
- Paddle element, in which an element inside the cavity moves perpendicularly to the beam and adds capacitance across the interaction gap. A tuning range of about 25% is provided by this approach.
- Combined inductive-capacitive tuning, which uses a combination of the previous methods. Tuning variations of 35% are possible.

Each of these tuning methods can be used whether the cavity is inside or outside the vacuum envelope of the tube. Generally speaking, however, tubes that use external cavities provide more adjustment range, usually on the order of 35%. Bandwidth can be increased by stagger tuning of the cavities, at the expense of gain.

High conversion efficiency requires the formation of electron bunches, which occupy a small region in velocity space, and the formation of interbunch regions with low electron density. The latter is particularly important because these electrons are phased to be accelerated into the collector at the expense of the RF field. Studies show that the energy loss as a result of an electron accelerated into the collector may exceed the energy delivered to the field by an equal but properly phased electron. Therein lies a key to improving the efficiency of the klystron: recover a portion of this wasted energy.

Klystrons are cooled by air or liquid for powers up to 5 kW. Tubes operating in excess of 5 kW are usually water- or vapor-cooled.

8.4 Traveling Wave Tube

The traveling wave tube (TWT) is a linear-beam device finding extensive applications in communications and research. Power levels range from a few watts to 10 MW. Gain ranges from 40 to 70 dB for small drive signals. The TWT consists of four basic elements:

- *Electron gun*. The gun forms a high-current-density beam of electrons that interact with a wave traveling along the RF circuit to increase the amplitude of the RF signal. In a typical application, electrons are emitted from a cathode and converged to the proper beam size by focusing electrodes.
- *RF interaction circuit*. The RF wave is increased in amplitude as a result of interaction with the electron beam from the gun. The fundamental principle on which the TWT operates is that an electron beam, moving at approximately the same velocity as an RF wave traveling along a circuit, gives up energy to the RF wave.
- *Magnetic electron beam focusing system*. The beam size is maintained at the proper dimensions through the interaction structure by the focusing system. This can be accomplished using either a permanent magnet or an electromagnetic focusing element.
- *Collector*. The electron beam is received at the collector after it has passed through the interaction structure. The remaining beam energy is dissipated in the collector.

Figure 8.16 shows the basic elements of a TWT.

The primary differences between types of TWT devices involve the RF interaction structure employed. In Fig. 8.16, the interaction structure is a helix. A variety of other structures can be employed, depending on the operating power and frequency. Three common approaches are used to provide the needed magnetic beam focusing. Illustrated in Fig. 8.17, they are:

- Electromagnetic focusing, used primarily on high-power tubes, where tight beam focusing is required
- Permanent-magnet focusing, used where the interaction structure is short
- Periodic permanent-magnet focusing, used on most helix TWT and coupled-cavity tubes; the magnets are arranged with alternate polarity in successive cells along the interaction region

Microwave Vacuum Devices 8-15

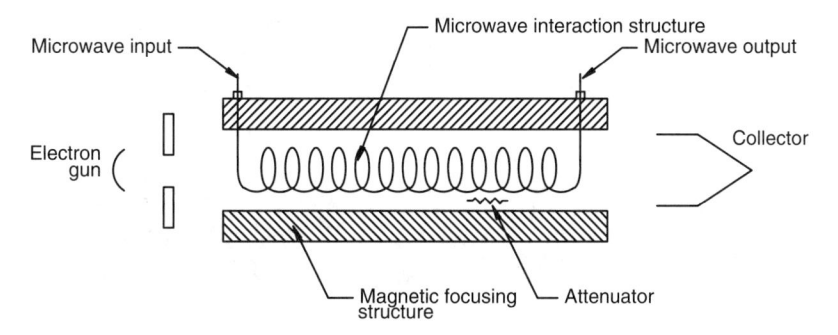

FIGURE 8.16 Basic elements of a traveling wave tube.

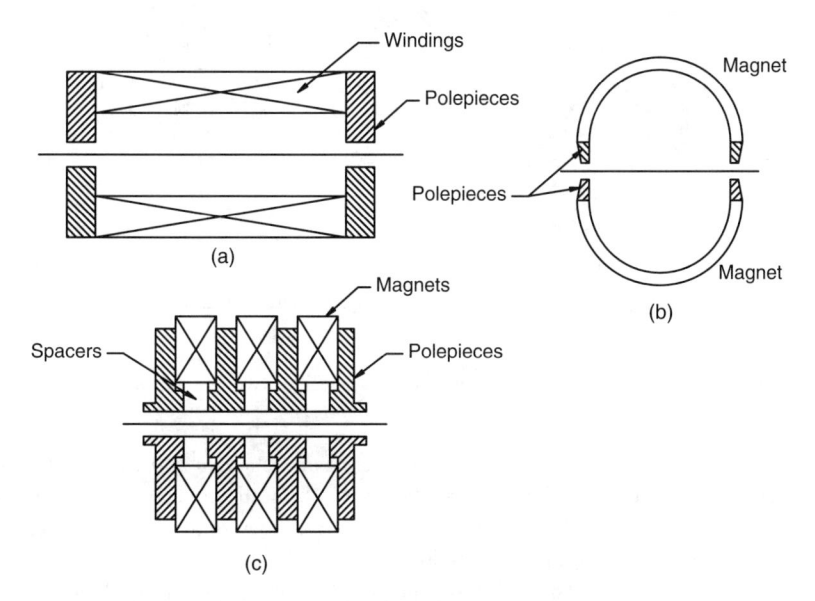

FIGURE 8.17 Magnetic focusing for a TWT: (a) solenoid type, (b) permanent magnet type, and (c) periodic permanent-magnet structure.

Theory of Operation

The interaction structure acts to slow the RF signal so that it travels at approximately the same speed as the electron beam. Electrons enter the structure during both positive and negative portions of the RF cycle. Electrons entering during the positive portion are accelerated; those entering during the negative portion are decelerated. The result is the creation of electron bunches that produce an alternating current superimposed on the dc beam current. This alternating current induces the growth of an RF *circuit wave* that encourages even tighter electron bunching.

One or more *severs* are included to absorb reflected power that travels in a backward direction on the interaction circuit. This reflected power is the result of a mismatch between the output port and the load. Without the sever, regenerative oscillations could occur.

At a given frequency, a particular level of drive power will result in maximum bunching and power output. This operating point is referred to as *saturation*.

Interaction Circuit

The key to TWT operation lies in the interaction element. Because RF waves travel at the speed of light, a method must be provided to slow down the forward progress of the wave to approximately the same velocity as the electron beam from the cathode. The beam speed of a TWT is typically 10 to 50% of the

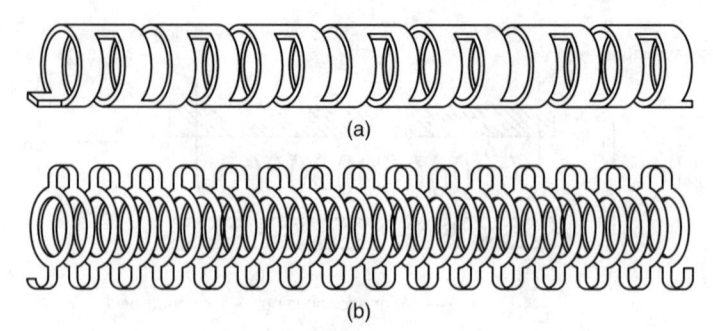

FIGURE 8.18 Helix structures for a TWT: (a) ring-loop circuit, and (b) ring-bar circuit.

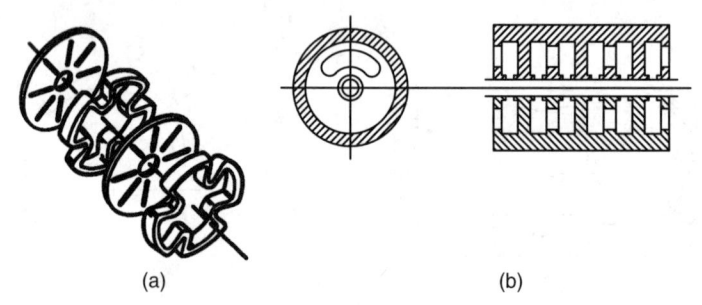

FIGURE 8.19 Coupled-cavity interaction structures: (a) forward fundamental circuit or "cloverleaf," and (b) single-slot space harmonic circuit.

speed of light, corresponding to cathode voltages of 4 to 120 kV. Two mechanical structures commonly are used to slow the RF wave:

- *Helix circuit*. The helix is used where bandwidths of an octave or more are required. Over this range the velocity of the signal carried by the helix is basically constant with frequency. Typical operating frequencies range from 500 MHz to 40 GHz. Operating power, however, is limited to a few hundred watts. TWTs intended for higher-frequency operation may use a variation of the helix, as shown in Fig. 8.18. The ring-loop and ring-bar designs permit peak powers of hundreds of kilowatts. The average power, however, is about the same as that of a conventional helix because the structure used to support the interaction circuit is the same.
- *Coupled-cavity circuit*. The coupled-cavity interaction structure permits operation at high peak and average power levels, and moderate bandwidth (10% being typical). TWTs using coupled-cavity structures are available at operating frequencies from 2 to 100 GHz. The basic design of a coupled-cavity interaction circuit is shown in Fig. 8.19. Resonant cavities, coupled through slots cut in the cavity end walls, resemble a folded waveguide. Two basic schemes are used: the *cloverleaf* and the *single-slot space harmonic* circuit.

The cloverleaf, also known as the *forward fundamental circuit*, is used primarily on high-power tubes. The cloverleaf provides operation at up to 3 MW peak power and 5 kW average at S-band frequencies. The single-slot space harmonic interaction circuit is more common than the cloverleaf. The mechanical design is simple, as shown in Fig. 8.19. The single-slot space harmonic structure typically provides peak power of up to 50 kW and average power of 5 kW at X-band frequencies.

Pulse Modulation

The electron beam from the gun can be pulse-modulated using one of four methods:

- *Cathode pulsing*. The cathode is pulsed in a negative direction with respect to the grounded anode. This approach requires the full beam voltage and current to be switched.

Microwave Vacuum Devices

- *Anode pulsing*. This approach is similar to cathode pulsing, except that the full beam voltage is switched between cathode potential and ground. The current switched, however, is only that value intercepted on the anode. Typically, the intercepted current is a few percent of the full beam potential.
- *Focus electrode pulsing*. If the focus electrode, which normally operates at or near cathode potential, is biased negatively with respect to the cathode, the beam will be turned off. The voltage swing required is typically one third of the full cathode voltage. This approach is attractive because the focus electrode draws essentially no current, making implementation of a switching modulator relatively easy.
- *Grid pulsing*. The addition of a grid to the cathode region permits control of beam intensity. The voltage swing required for the grid, placed directly in front of the cathode, is typically 5% of the full beam potential.

Electron Gun

The electron gun of a TWT is a device that supplies the electron beam to the tube.[4] A schematic diagram of a generic electron gun is given in Fig. 8.20. The device consists of a hot cathode heated by an electric heater, a negatively biased focusing electrode or focuser, and a positively biased accelerating anode. The cross-sectional view given in the figure can be a two-dimensional or three-dimensional coaxial structure.[5,6]

An axially symmetrical solid cylindrical electron beam is produced by the gun structure shown in Fig. 8.20 if the structure is axially cylindrically symmetrical. If the middle of the hot cathode is made nonemitting and only the edge of the cathode is emitting, the cathode becomes an *annular cathode*.[4] The annular cathode produces a hollow beam. The annular electron beam can be used to reduce beam current for a given microwave output power.

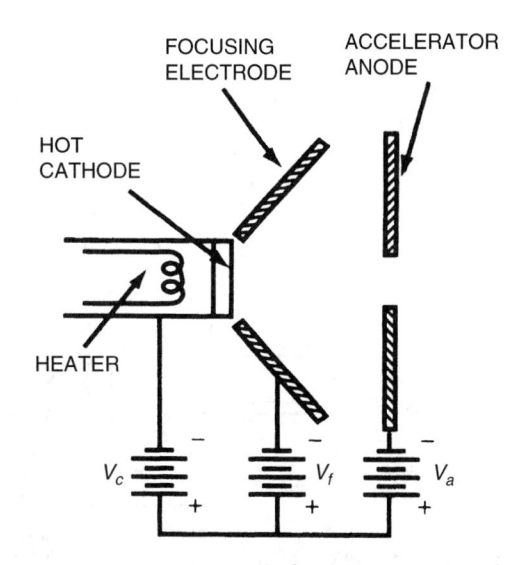

FIGURE 8.20 Generic TWT electron gun structure. (From Ishii, T. K., *The Electronics Handbook*, Whitaker, J. C., Ed., CRC Press, Boca Raton, FL, 1996, 428–443.)

If the gun structure shown in Fig. 8.20 is two dimensional, then a ribbon-shaped electron beam is produced. A ribbon-shaped beam is used for a TWT of a two-dimensional structure.

If the angle of the focusing electrode against the axis of the electron beam is 67.5° and the anode is also tilted forward to produce a *rectilinear flow* (electron flow parallel to the z-axis in Fig. 8.20), then such an electron gun is termed the *Pierce gun*.

In practice, the hot cathode surface is curved as shown in Fig. 8.21 to increase the electron emitting surface and to obtain a high-density electron beam.

Beam Focusing

Electrons in an electron beam mutually repel each other by the electron's own coulomb force because of their negative charge.[4] In addition, the electron beam usually exists in proximity to the positively biased slow-wave structure, as shown in Fig. 8.22. Therefore, the electron beam tends to diverge. The process of confining the electron beam within the desired trajectory against the mutual repulsion and diverging force from the slow-wave structure is termed electron beam focusing.

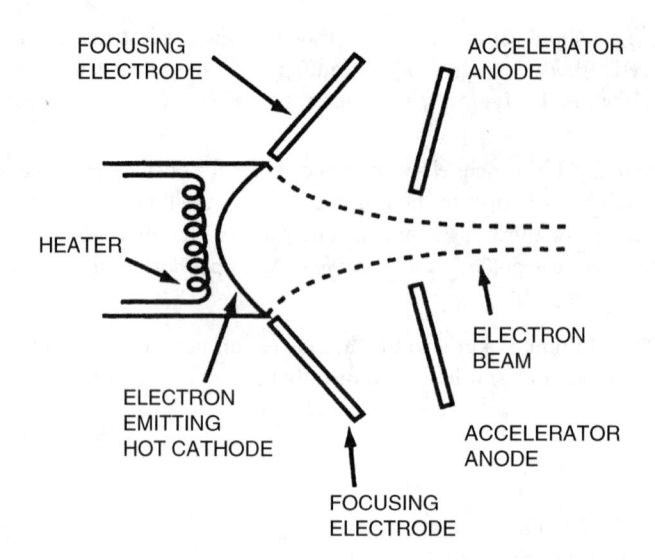

FIGURE 8.21 Cross-sectional view of a TWT electron gun with a curved hot cathode. (From Ishii, T. K., *The Electronics Handbook*, Whitaker, J. C., Ed., CRC Press, Boca Raton, FL, 1996, 428–443.)

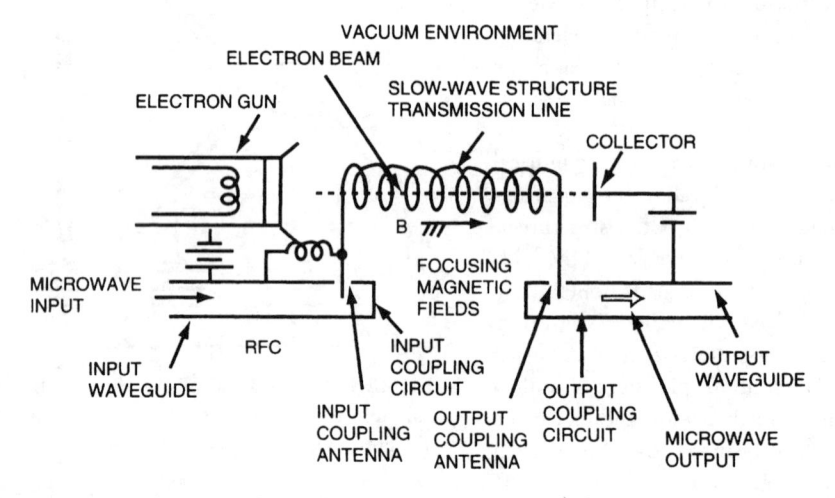

FIGURE 8.22 Generic configuration of a traveling wave tube. (From Ishii, T. K., *The Electronics Handbook*, Whitaker, J. C., Ed., CRC Press, Boca Raton, FL, 1996, 428–443.)

The electron beam in a TWT is usually focused by a dc magnetic flux applied parallel to the direction of the electron beam, which is coaxial to the slow-wave transmission line. Variations on this basic technique include:

- *Brillouin flow*, where the output of the electron gun is not exposed to the focusing magnetic flux[7]
- *Immersed flow*, where the electron gun itself is exposed to and unshielded from the focusing para-axial longitudinal magnetic flux[7]
- *Generic flow*, where the electron gun is not shielded from the focusing magnetic flux, and focusing flux is not para-axia (i.e., neither Brillouin nor immersed flow)

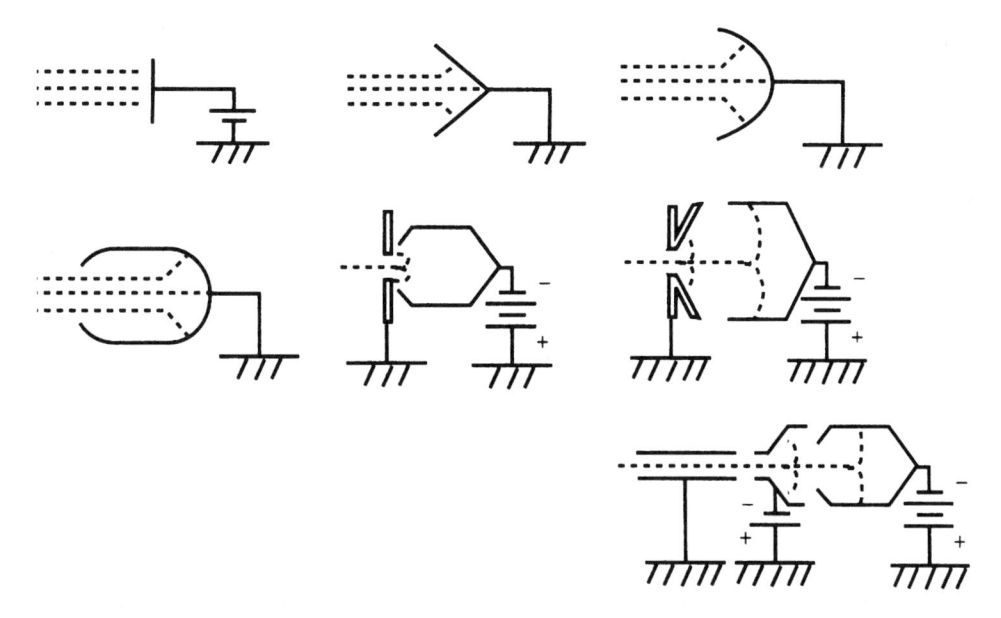

FIGURE 8.23 Cross-sectional view of various collector configurations for a TWT. (From Ishii, T. K., *The Electronics Handbook*, Whitaker, J. C., Ed., CRC Press, Boca Raton, FL, 1996, 428–443.)

Collector Assembly

Various configurations are used for the collector assembly of a TWT. Figure 8.23 shows a selection of the more common, including:[4]

- Plate collector
- Cone collector
- Curved cone
- Cylinder collector
- Depressed potential cylinder
- Two-stage collector
- Three-stage collector

Cooling options include conduction, air, and water.

Cooling of a low-power TWT is accomplished by clamping the tube to a metal baseplate, mounted in turn on an air- or liquid-cooled heat sink. Coupled-cavity tubes below 1 kW average power are convection-cooled by circulating air over the entire length of the device. Higher-power coupled-cavity tubes are cooled by circulating liquid over the tube body and collector.

Operating Efficiency

Efficiency is not one of the TWT's strong points. Early traveling wave tubes offered only about 10% dc-to-RF efficiency. Wide bandwidth and power output are where the TWT shines. TWT efficiency can be increased in two basic ways: (1) *collector depression* for a single-stage collector, or (2) use of a multistage collector.

Collector depression refers to the practice of operating the collector at a voltage lower than the full beam voltage. This introduces a potential difference between the interaction structure and the collector, through which electrons pass. The amount by which a single-stage collector can be depressed is limited

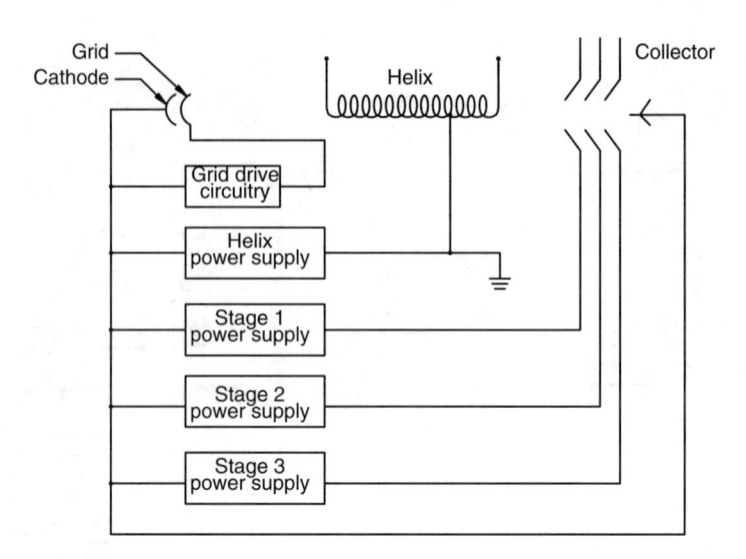

FIGURE 8.24 Power supply configuration for a multistage depressed collector TWT.

by the remaining energy of the slowest electrons. That is, the potential reduction can be no greater than the amount of energy of the slowest electrons or they will turn around and reenter the interaction structure, causing oscillations.

By introducing multiple depressed collector stages, still greater efficiency can be realized. This method provides for the collection of the slowest electrons at one collector potential, while allowing those with more energy to be collected on other stages that are depressed still further. This approach is illustrated in Fig. 8.24.

8.5 Crossed-Field Tubes

A crossed-field microwave tube is a device that converts dc into microwave energy using an electronic energy-conversion process. These devices differ from beam tubes in that they are potential-energy converters, rather than kinetic-energy converters. The term "crossed field" is derived from the orthogonal characteristics of the dc electric field supplied by the power source and the magnetic field required for beam focusing in the interaction region. This magnetic field is typically supplied by a permanent-magnet structure. Such devices also are referred to as *M-tubes*.

Practical devices based on the crossed-field principles fall into two broad categories:

- *Injected-beam crossed-field tubes.* The electron stream is produced by an electron gun located external to the interaction region, similar to a TWT. The concept is illustrated in Fig. 8.25.
- *Emitting-sole tubes.* The electron current for interaction is produced directly within the interaction region by secondary electron emissions, which result when some electrons are driven to the negative electrode and allowed to strike it. The negative electrode is formed using a material capable of producing significant numbers of secondary-emission electrons. The concept is illustrated in Fig. 8.26.

Magnetron

The magnetron encompasses a class of devices finding a wide variety of applications. Pulsed magnetrons have been developed that cover frequency ranges from the low UHF band to 100 GHz. Peak power from a few kilowatts to several megawatts has been obtained. Typical overall efficiencies of 30 to 40% can be realized, depending on the power level and operating frequency. CW magnetrons have also been developed, with power levels of a few hundred watts in a tunable tube, and up to 25 kW or more in a fixed-frequency device. Efficiencies range from 30% to as much as 70%.

Microwave Vacuum Devices 8-21

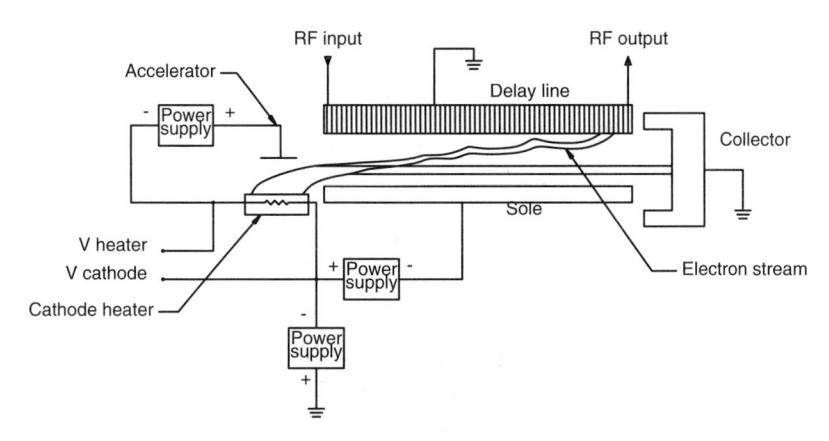

FIGURE 8.25 Linear injected-beam microwave tube.

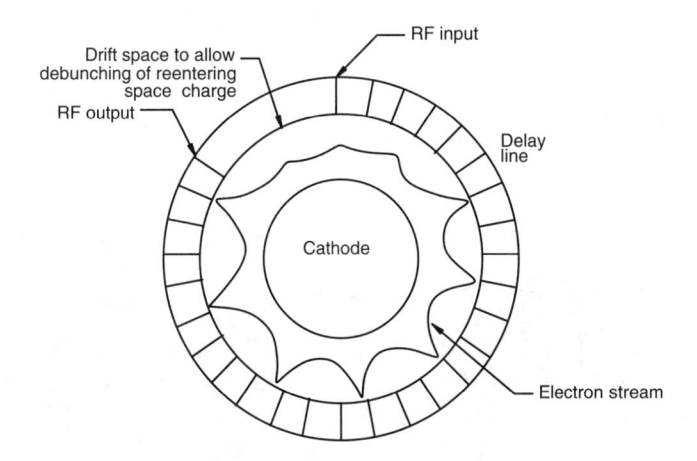

FIGURE 8.26 Reentrant emitting-sole crossed-field amplifier tube.

The magnetron operates electrically as a simple diode. Pulsed modulation is obtained by applying a negative rectangular voltage waveform to the cathode with the anode at ground potential. Operating voltages are less critical than for beam tubes; line-type modulators are often used to supply pulsed electric power. The physical structure of a conventional magnetron is shown in Fig. 8.27.

High-power pulsed magnetrons are used primarily in radar systems. Low-power pulsed devices find applications as beacons. Tunable CW magnetrons are used in ECM (electronic countermeasures) applications. Fixed-frequency devices are used as microwave heating sources.

Tuning of conventional magnetrons is accomplished by moving capacitive tuners or by inserting symmetrical arrays of plungers into the inductive portions of the device. Tuner motion is produced by a mechanical connection through flexible bellows in the vacuum wall. Tuning ranges of 10 to 12% of bandwidth are possible for pulsed tubes, and as much as 20% for CW tubes.

Operating Principles

Most magnetrons are built around a cavity structure of the type shown in Fig. 8.28. The device consists of a cylindrical cathode and anode, with cavities in the anode that open into the cathode–anode space — the interaction space — as shown. Power can be coupled out of the cavities by means of a loop or a tapered waveguide.

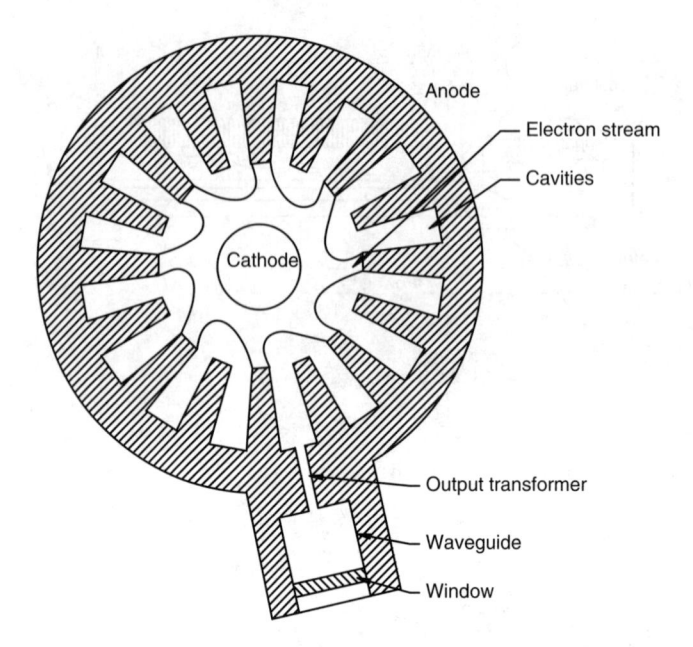

FIGURE 8.27 Conventional magnetron structure.

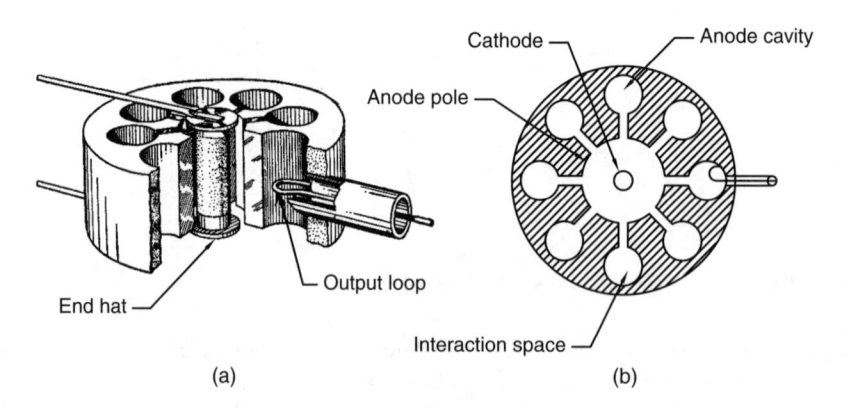

FIGURE 8.28 Cavity magnetron oscillator: (a) cutaway view, and (b) cross-sectional view perpendicular to the axis of the cathode.

Cavities, together with the spaces at the ends of the anode block, form the resonant system that determines the frequency of the generated oscillations. The actual shape of the cavity is not particularly important, and various types are used, as illustrated in Fig. 8.29. The oscillations associated with the cavities are of such a nature that alternating magnetic flux lines pass through the cavities parallel to the cathode axis, while the alternating electric fields are confined largely to the region where the cavities open into the interaction space. The most important factors determining the resonant frequency of the system are the dimensions and shape of the cavities in a plane perpendicular to the axis of the cathode. Frequency is also affected by other factors such as the end space and the axial length of the anode block, but to a lesser degree.

The magnetron requires an external magnetic field with flux lines parallel to the axis of the cathode. This field usually is provided by a permanent magnet or electromagnet.

The cathode is commonly constructed as a cylindrical disk.

Microwave Vacuum Devices 8-23

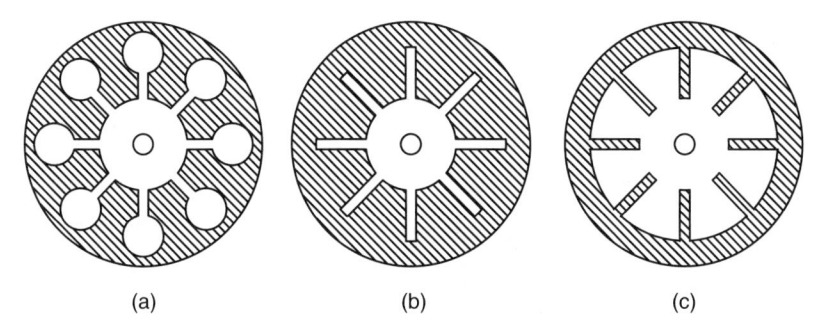

FIGURE 8.29 Cavity magnetron oscillator anode: (a) hole-and-slot type, (b) slot type, and (c) vane type.

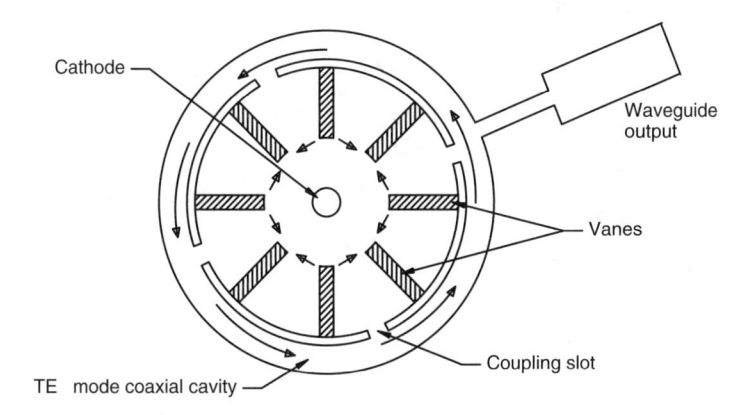

FIGURE 8.30 Structure of a coaxial magnetron.

Coaxial Magnetron

The frequency stability of a conventional magnetron is affected by variations in the load impedance and by cathode-current fluctuations. Depending on the extent of these two influences, the magnetron occasionally may fail to produce a pulse. The coaxial magnetron minimizes these effects by using the anode geometry shown in Fig. 8.30. Alternate cavities are slotted to provide coupling to a surrounding coaxial cavity.

The oscillating frequency is controlled by the combined vane system and the resonant cavity. Tuning can be accomplished through the addition of a movable end plate in the cavity, as shown in Fig. 8.31.

Frequency-Agile Magnetron

Tubes developed for specialized radar and ECM applications permit rapid tuning of the magnetron. A conventional device can be tuned using one of the following methods:

- *A rapidly rotating capacitive element*. Tubes of this type are referred to as *spin-tuned magnetrons*.
- *A hydraulic-driven tuning mechanism*. Tubes of this type are referred to as *mechanically tuned magnetrons*.

Electronic tuning of magnetrons is also possible, with tuning rates as high as several megahertz per microsecond.

Linear Magnetron

Although the most common types of magnetrons are radial in nature, the linear magnetron and inverted magnetron can also be used, depending on the application.[8] A cross-sectional view of a linear magnetron

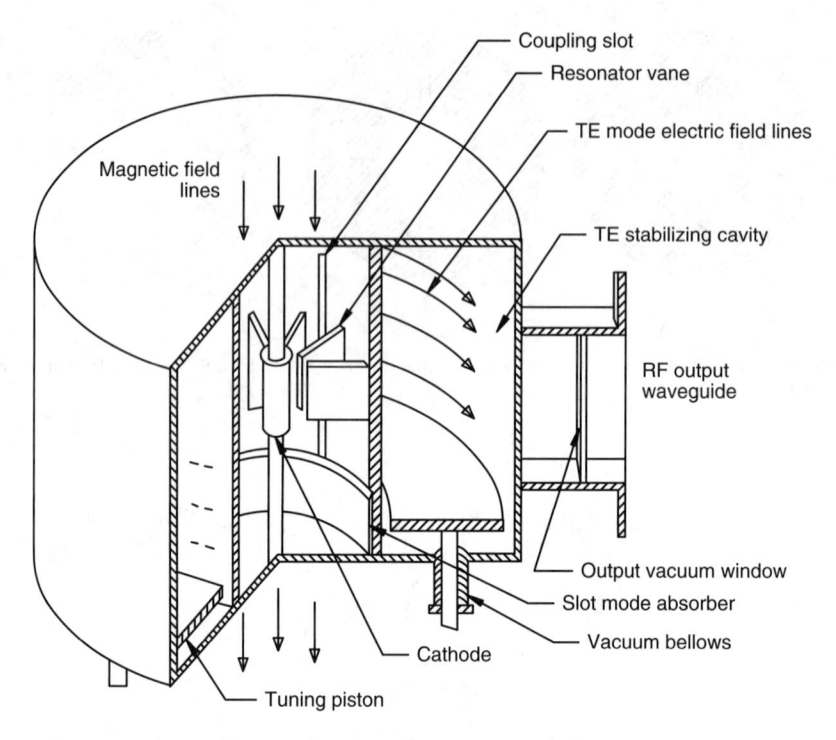

FIGURE 8.31 Structure of a tunable coaxial magnetron.

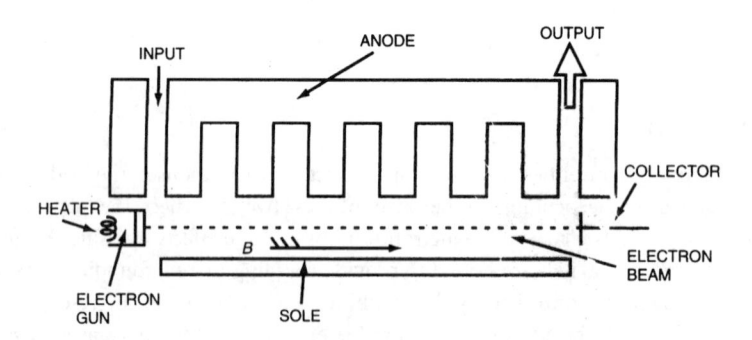

FIGURE 8.32 Cross-sectional view of a linear magnetron. (From Ishii, T. K., *The Electronics Handbook*, Whitaker, J. C., Ed., CRC Press, Boca Raton, FL, 1996, 428–443.)

is given in Fig. 8.32. Shown in the figure is the O-type linear magnetron, in which the electron beam emitted from the electron gun is focused by a longitudinally applied dc magnetic flux density (*B*), as in the case of the traveling wave tube.

As shown in the figure, a number of slots are included in the basic structure. These slots are cut 1/4-wavelength deep, functioning as quarter-wave cavity resonators. This structure forms a series of microwave cavity resonators coupling to an electron beam, in a similar manner to the multicavity klystron. The velocity-modulated electrons are bunched and the tightly bunched electrons produce amplified microwave energy at the output cavity, which is coupled to an external circuit. The linear magnetron typically offers high gain but narrow frequency bandwidth.

Backward Wave Oscillator

In a traveling wave tube, if the microwave signal to be amplified is propagating in the slow-wave structure backwardly to the direction of the electron beam, the device is termed a backward wave oscillator (BWO).[8]

Microwave Vacuum Devices

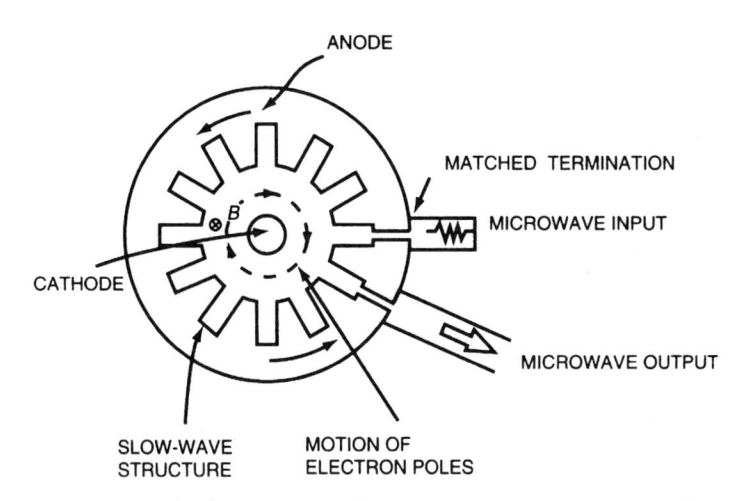

FIGURE 8.33 Functional schematic of the M-type radial BWO. (From Ishii, T. K., *The Electronics Handbook*, Whitaker, J. C., Ed., CRC Press, Boca Raton, FL, 1996, 428–443.)

Microwaves traveling in a backward direction carry positive feedback energy toward the electron gun and yield stronger velocity modulation and bunching. Thus, the system is inherently an oscillator rather than a stable amplifier. The input is typically terminated by an impedance-matched reflectionless termination device. The oscillation frequency is a function of the speed of the electrons and the time constant of the feedback mechanism. The speed of electron motion is controlled by the anode voltage.

An M-type radial BWO is shown in Fig. 8.33. The direction of electron pole motion and the direction of microwave propagation along the annular reentrant type slow-wave structure are opposite each other. It should be noted that the depths of the slits cut in the inner surface of the anode is very shallow — much less than 1/4-wavelength deep. In other words, the slits are not in resonance; they are not cavity resonators, as is the case of a magnetron. Rather, the slits are nonresonating, as in the case of a TWT. In the M-type radial BWO, the electron beam is focused by a magnetic flux density applied perpendicular to the beam, as seen from Fig. 8.33.

An M-type radial BWO is sometimes termed the Carcinotron, a trade name. A key feature of the Carcinotron is its wide voltage tunability over a broad frequency range.

Strap-Fed Devices

A radial magnetron can be configured so that every other pole of the anode resonators are conductively tied for microwave potential equalization, as shown in Fig. 8.34a.[8] These conducting tie rings are termed *straps*; the technique of using strap rings is termed *strapping*. Strapping ensures good synchronization of microwaves in the magnetron resonators with the rotation of electron poles.

The technique of strapping is extended and modified for an M-type radial BWO, as shown in Fig. 8.34b. Strapping rings tie every other pole of the radial slow-wave structure, as in the case of a strapped radial magnetron, but the strapping rings are no longer reentrant. Microwave energy to be amplified is fed to the strap at one end and the amplified output is extracted from the other end. This type of electron tube is termed a *strap-fed device*.

If an M-type radial BWO is strapped, it usually does not start oscillation by itself. But, if microwave energy is fed through the strap from the outside using an external microwave power source to the microwave input, then the oscillation starts — and even if the exciter source is turned off, the oscillation continues. This type of M-type radial BWO is termed the *platinotron*.[9]

In a platinotron, if the output of the tube is fed back to the input through a high-Q cavity resonator, it becomes a self-starting oscillator. The oscillation frequency is stabilized by the high-Q cavity resonator. This type of high-power, frequency-stabilized strapped radial BWO is termed the *stabilotron*.[9] The operating powers are at kilowatt and megawatt levels.

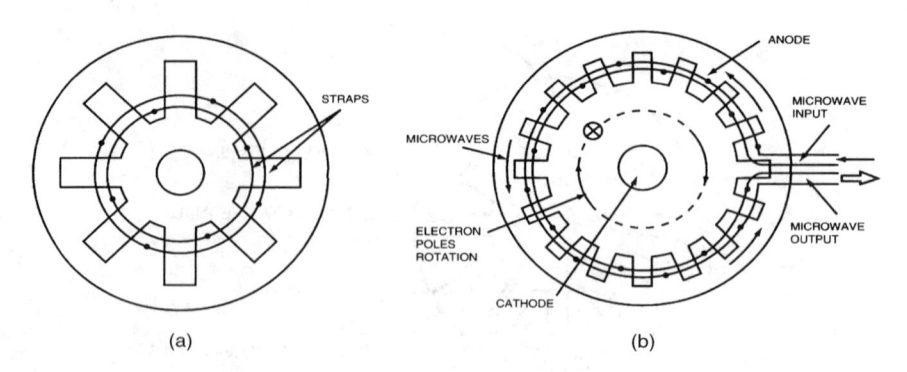

FIGURE 8.34 Strap-fed devices: (a) strapped radial magnetron, and (b) nonreentrant strapping of a BWO. (From Ishii, T. K., *The Electronics Handbook*, Whitaker, J. C., Ed., CRC Press, Boca Raton, FL, 1996, 428–443.)

Performance of the platinotron depends on, among other things, the design of the slow-wave structure. For example, the interdigital slow-wave structure as shown in Fig. 8.34 has limited power handling capability and frequency bandwidth. Design of a slow-wave structure with greater power handling capacity and stability, with broader frequency bandwidth, is possible. For example, instead of an anode with an interdigital slow-wave structure, the anode could be made of an annular open conducting duct, loaded with a number of pairs of conducting posts across the open duct. Strapping is done at every other tip of the pairs of conducting posts. This type of strapping loads the slow-wave structure, stabilizing it and preventing oscillation. The structure of the anode with an annular duct and pairs of posts increases the power handling capability. This type of loaded radial BWO is termed the *amplitron*.[5,9] The amplitron is capable of amplifying high-power microwave signals with pulses and continuous waves. It is used for long-range pulsed radar transmitter amplifiers and industrial microwave heating generators. The operating power levels range from kilowatt to megawatt levels.

Gyrotron

The gyrotron is a cyclotron resonance maser. The device includes a cathode, collector, and circular waveguide of gradually varying diameter. Electrons are emitted at the cathode with small variations in speed. The electrons are then accelerated by an electric field and guided by a static magnetic field through the device. The nonuniform induction field causes the rotational speed of the electrons to increase. The linear velocity of the electrons, as a result, decreases. The interaction of the microwave field within the waveguide and the rotating (helical) electrons causes bunching similar to the bunching within a klystron. A decompression zone at the end of the device permits decompression and collection of the electrons.

The power available from a gyrotron is 100 times greater than that possible from a classic microwave tube at the same frequency. Power output of 100 kW is possible at 100 GHz, with 30% efficiency. At 300 GHz, up to 1.5 kW can be realized, but with only 6% efficiency.

Theory of Operation

The trajectory of an electron in an electron beam focused by a longitudinally applied magnetic field is a helix.[8] If the electron velocity, electron injection angle, and applied longitudinal magnetic flux density are varied, then an electron beam of helical form with different size and pitch will be formed. A coil-shaped electron beam will be produced by adjusting the acceleration voltage, applied magnetic flux density, and the electron injection angle to the focusing magnetic field. The coil of the electron beam can be a simple single coil, or — depending on the adjustment of the aforementioned three parameters — it can be an electron beam of a double coil or a large coil made of thin small coils. In the case of the double-coil trajectory, the large coil-shaped trajectory is termed the major orbit and the smaller coil trajectory is termed the minor orbit.

Microwave Vacuum Devices 8-27

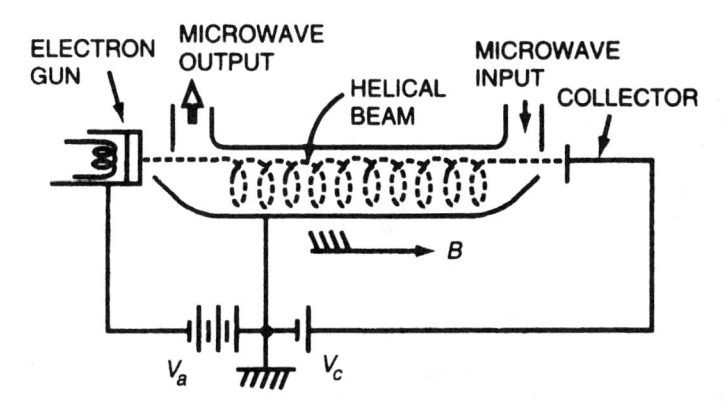

FIGURE 8.35 Basic structure of a helical beam tube. (From Ishii, T. K., *The Electronics Handbook*, Whitaker, J. C., Ed., CRC Press, Boca Raton, FL, 1996, 428–443.)

If a single coil-shaped electron beam is launched in a waveguide, as shown in Fig. 8.35, then microwaves in the waveguide will interact with the helical beam. This type of vacuum tube is termed the helical beam tube.[8] In this class of device, a single-coil helical beam is launched into a TE_{10} mode rectangular waveguide. Inside the waveguide, microwaves travel from right to left and the helical beam travels in an opposite direction. Therefore, the microwave–electron beam interaction is of the backward wave type. If the microwave frequency, the focusing magnetic flux density B, and the acceleration voltage V_a are properly adjusted, this device will function as a backward wave amplifier. Electrons in the helical beam interact with the transverse microwave electric fields and are velocity modulated at the left-hand side of the waveguide as the beam enters into the waveguide. The velocity modulated electrons in the helical beam are bunched as they travel toward the right. If the alternating microwave electric field synchronizes its period and phase with the helical motion of bunched electrons so that the electrons always receive retardation from microwave transverse electric fields, then the electrons lose their kinetic energy and the microwave signal gains in electric field energy according to the principle of kinetic energy conservation. Thus, the amplified microwave power emerges at the waveguide output at the left (because the microwaves travel backward).

In Fig. 8.35, if the microwave input port and the output port are interchanged with each other, then the system becomes a forward wave amplifier. Such a forward wave amplifier is termed a *peniotron*.[8]

If the electron gun is modified to incorporate a side-emitting cathode and the waveguide is changed to TE_{11} mode oversized circular waveguide, as shown in Fig. 8.36a, the gyrotron is formed. In this device, both ends of the waveguide are open and there are sufficient reflections in the waveguide for positive feedback. The gyrotron is thus a forward wave oscillator.

A double-coil helical beam gyrotron is shown in Fig. 8.36b. The device is formed by readjusting the anode voltage and the focusing flux density so that the electron beam is made into a double helical coil (as shown in the figure), and operating the oversized circular waveguide in the TE_{01} mode. In the TE_{01} mode, the microwave transverse electric fields exist as concentric circles. Therefore, the tangential electric fields interact with electrons in the small coil trajectory. The alternating tangential microwave electric fields are made to synchronize with the tangential motion of electrons in the minor coil-shaped trajectory. Thus, electron velocity modulation takes place near the cathode and bunching takes place in the middle of the tube. Microwave kinetic energy transfer takes place as the beam approaches the right. The focusing magnetic flux density B is applied only in the interaction region. Therefore, if the electron beam comes out of the interaction region, it is defocused and collected by the anode waveguide (as depicted in Fig. 8.36b). If the circular waveguide is operated in an oversized TE_{11} mode, with the double-coil helical beam, then the device is referred to as a *tornadotron*.[8] Microwave–electron interaction occurs between the parallel component of tangential motion of the small helical trajectory and the TE_{11} mode microwave electric field. If the phase of the microwave electric field decelerates bunched electrons, then the lost kinetic energy of the bunched electrons is transferred to the microwave signal and oscillation begins.

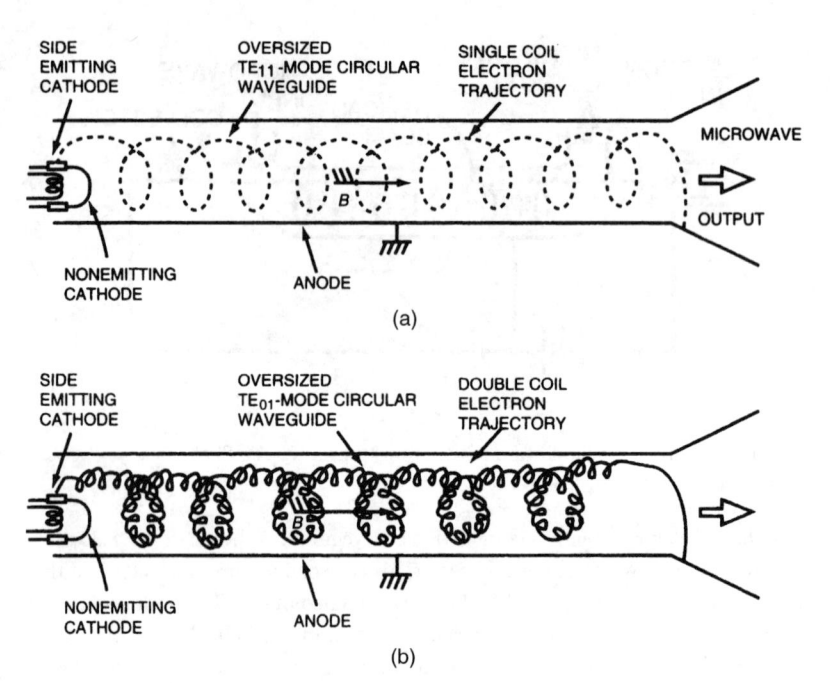

FIGURE 8.36 Functional schematic diagram of the gyrotron: (a) single coiled helical beam gyrotron, and (b) electron trajectory of double-coil helical beam gyrotron. (From Ishii, T. K., *The Electronics Handbook*, Whitaker, J. C., Ed., CRC Press, Boca Raton, FL, 1996, 428–443.)

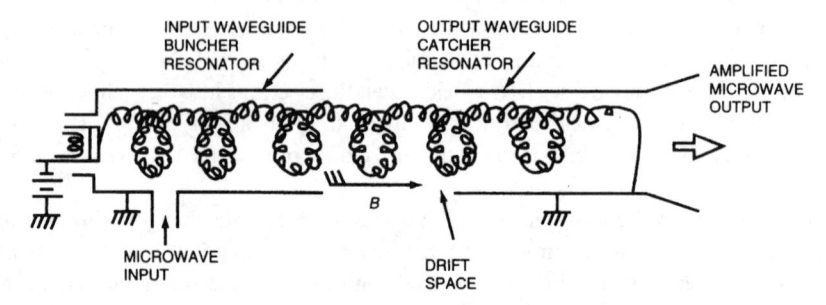

FIGURE 8.37 Basic structure of the gyroklystron amplifier. (From Ishii, T. K., *The Electronics Handbook*, Whitaker, J. C., Ed., CRC Press, Boca Raton, FL, 1996, 428–443.)

Gyrotron Design Variations

The gyrotron exists in a number of design variations, each optimized for a particular feature or application.[8,10,11]

When the gyrotron circular waveguide is split as shown in Fig. 8.37, the tube is termed the *gyroklystron amplifier*.[8] Both waveguides resonate to the input frequency and there are strong standing waves in both waveguide resonators. The input microwave signal to be amplified is fed through a side opening to the input waveguide resonator. This is the buncher resonator, which functions in a manner similar to the klystron. The buncher resonator imparts velocity modulation to gyrating electrons in the double helical coil-shaped electron beam. There is a drift space between the buncher resonator and the catcher resonator at the output. While drifting electrons bunch and bunched electrons enter into the output waveguide catcher resonator, electron speed is adjusted in such a manner that electrons are decelerated by the resonating microwave electric field. This lost kinetic energy in bunched electrons is transformed into microwave energy and microwaves in the catcher resonator are thus amplified. The amplified power appears at the output of the tube.

Microwave Vacuum Devices 8-29

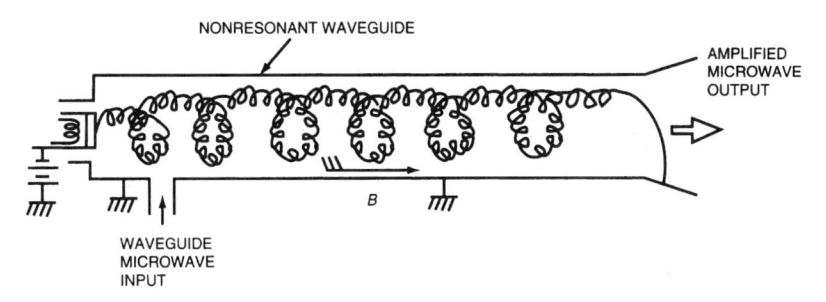

FIGURE 8.38 Basic structure of the gyroklystron traveling wave tube amplifier. (From Ishii, T. K., *The Electronics Handbook*, Whitaker, J. C., Ed., CRC Press, Boca Raton, FL, 1996, 428–443.)

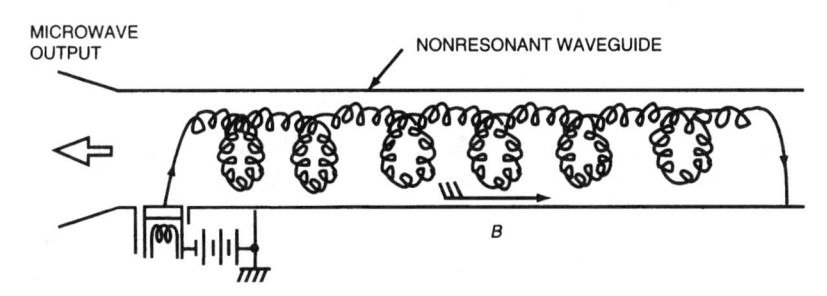

FIGURE 8.39 Basic structure of the gyrotron backward oscillator. (From Ishii, T. K., *The Electronics Handbook*, Whitaker, J. C., Ed., CRC Press, Boca Raton, FL, 1996, 428–443.)

If the gyrotron waveguide is an unsplit one-piece waveguide that is impedance-matched and not resonating, as shown in Fig. 8.38, the tube is termed the *gyrotron traveling wave tube amplifier*.[8] In this tube, the input microwaves are fed through an opening in the waveguide near the electron gun. Microwaves in the waveguide are amplified gradually as they travel toward the output port by interacting with the double-coiled helical electron beam, which is velocity-modulated and bunched. There are no significant standing waves in the waveguide. Microwaves grow gradually in the waveguide as they travel toward the output port as a result of interaction with electrons.

If the electron gun of the gyrotron is moved to the side of the waveguide and microwave power is extracted from the waveguide opening in proximity to the electron gun, as shown in Fig. 8.39, then the device is termed a *gyrotron backward oscillator*.[9] The principle involved is similar to the backward wave oscillator, and the process of velocity modulation, drifting, bunching, and catching is similar to that of the klystron. Microwave energy induced in the waveguide travels in both directions but the circuit is adjusted to emphasize the waves traveling in a backward direction. The backward waves become the output of the tube and, at the same time, carry the positive feedback energy to the electrons just emitted and to be velocity-modulated. The system thus goes into oscillation.

If the gyrotron waveguide is split into two again, but this time the input side waveguide is short and the output side waveguide is long, as shown in Fig. 8.40, then the tube is termed a *gyrotwystron amplifier*.[8] This device is a combination of the gyroklystron and gyrotron traveling wave tube amplifier; thus the name gyrotwystron amplifier. The input side waveguide resonator is the same as the input resonator waveguide of a gyroklystron. There are strong standing waves in the input bunched-waveguide resonator. There is no drift space between the two waveguides. The output side waveguide is a long impedance-matched waveguide and there is no microwave standing wave in the waveguide (a traveling-wave waveguide). As microwaves travel in the waveguide, they interact with bunched electrons and the microwaves grow as they move toward the output port.

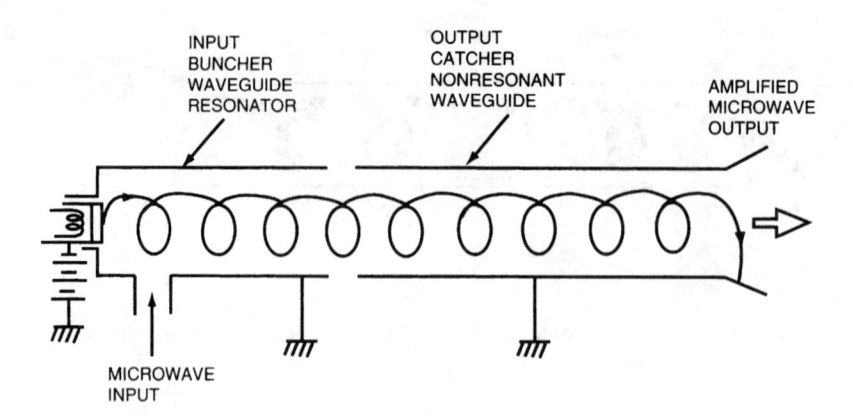

FIGURE 8.40 Basic structure of the gyrotwystron amplifier. (From Ishii, T. K., *The Electronics Handbook*, Whitaker, J. C., Ed., CRC Press, Boca Raton, FL, 1996, 428–443.)

References

1. Ostroff, Nat S., A unique solution to the design of an ATV transmitter, *Proceedings of the 1996 NAB Broadcast Engineering Conference*, National Association of Broadcasters, Washington, D.C., 1996, 144.
2. Hulick, Timothy P., 60 kW Diacrode UHF TV transmitter design, performance and field report, *Proceedings of the 1996 NAB Broadcast Engineering Conference*, National Association of Broadcasters, Washington, D.C., 1996, 442.
3. Integral Cavity Klystrons for UHF-TV Transmitters, Varian Associates, Palo Alto, CA.
4. Ishii, T. K., Traveling wave tubes, in *The Electronics Handbook*, Jerry C. Whitaker, Ed., CRC Press, Boca Raton, FL, 1996, 428–443.
5. Liao, S. Y., *Microwave Electron Tube Devices*, Prentice-Hall, Englewood Cliffs, NJ, 1988.
6. Sims, G. D. and I. M Stephenson, *Microwave Tubes and Semiconductor Devices*, Interscience, London, 1963.
7. Hutter, R. G. E., *Beam and Wave Electronics in Microwave Tubes*, Interscience, London, 1960.
8. Ishiik, T. K., Other microwave vacuum devices, in *The Electronics Handbook*, Jerry C. Whitaker, Ed., CRC Press, Boca Raton, FL, 1996, 444–457.
9. Ishii, T. K., *Microwave Engineering*, Harcourt-Brace-Jovanovich, San Diego, CA, 1989.
10. Coleman, J. T., *Microwave Devices*, Reston Publishing, Reston, VA, 1982.
11. McCune, E. W., Fision plasma heating with high-power microwave and millimeter wave tubes, in *Journal of Microwave Power*, 20, 131–136, 1985.

Bibliography

Badger, George, The klystrode: a new high-efficiency UHF-TV power amplifier, *Proceedings of the NAB Engineering Conference*, National Association of Broadcasters, Washington, D.C., 1986.
Collins, G. B., *Radar System Engineering*, McGraw-Hill, New York, 1947.
Crutchfield, E. B., Ed., *NAB Engineering Handbook*, 8th ed., National Association of Broadcasters, Washington, D.C., 1992.
Dick, Bradley, New developments in RF technology, *Broadcast Engineering*, Intertec Publishing, Overland Park, KS, May 1986.
Fink, D. and D. Christiansen, Eds., *Electronics Engineers' Handbook*, 3rd ed., McGraw-Hill, New York, 1989.
Fisk, J. B., H. D. Hagstrum, and P. L. Hartman, The magnetron as a generator of centimeter waves, *Bell System Tech. J.*, 25, 167, 1946.
Ginzton, E. L. and A. E. Harrison, Reflex klystron oscillators, *Proc. IRE*, 34, 97, March 1946.

IEEE Standard Dictionary of Electrical and Electronics Terms, Institute of Electrical and Electronics Engineers, Inc., New York, 1984.

McCune, Earl, Final Report: The Multi-Stage Depressed Collector Project, *Proceedings of the NAB Engineering Conference*, National Association of Broadcasters, Washington, D.C., 1988.

Ostroff, N., A. Kiesel, A. Whiteside, and A. See, Klystrode-equipped UHF-TV transmitters: report on the initial full service station installations, *Proceedings of the NAB Engineering Conference*, National Association of Broadcasters, Washington, D.C., 1989.

Ostroff, N., A. Whiteside, A. See, and A. Kiesel, A 120 kW klystrode transmitter for full broadcast service, *Proceedings of the NAB Engineering Conference*, National Association of Broadcasters, Washington, D.C., 1988.

Ostroff, N., A. Whiteside, and L. Howard, An integrated exciter/pulser system for ultra high-efficiency klystron operation, *Proceedings of the NAB Engineering Conference*, National Association of Broadcasters, Washington, D.C., 1985.

Pierce, J. R., Reflex oscillators, *Proc. IRE*, 33, 112, February 1945.

Pierce, J. R., Theory of the beam-type traveling wave tube, *Proc. IRE*, 35, 111, February 1947.

Pierce, J. R. and L. M. Field, Traveling-wave tubes, *Proc. IRE*, 35, 108, February 1947.

Pond, N. H. and C. G. Lob, Fifty years ago today or on choosing a microwave tube, *Microwave Journal*, 226–238, September 1988.

Priest, D. and M. Shrader, The klystrode — An unusual transmitting tube with potential for UHF-TV, *Proceedings of the IEEE*, Vol. 70, no. 11, IEEE, New York, November 1982.

Shrader, Merrald B., Klystrode technology update, *Proceedings of the NAB Engineering Conference*, National Association of Broadcasters, Washington, D.C., 1988.

Spangenberg, Karl, *Vacuum Tubes*, McGraw-Hill, New York, 1947.

Terman, F. E., *Radio Engineering*, 3rd ed., McGraw-Hill, New York, 1947.

Varian, R. and S. Varian, A high-frequency oscillator and amplifier, *J. Applied Phys.*, 10, 321, May 1939.

Webster, D. L., Cathode bunching, *J. Applied Physics*, 10, 501, July 1939.

Whitaker, Jerry C. and T. Blankenship, Comparing integral and external cavity klystrons, *Broadcast Engineering*, Intertec Publishing, Overland Park, KS, November 1988.

Whitaker, Jerry C., *Radio Frequency Transmission Systems: Design and Operation*, McGraw-Hill, New York, 1991.

Whitaker, Jerry C., Ed., *NAB Engineering Handbook*, 9th ed., National Association of Broadcasters, Washington, D.C., 1998.

9
Bipolar Junction and Junction Field-Effect Transistors

Sidney Soclof
*California State University,
Los Angeles*

9.1 Bipolar Junction Transistors..9-1
9.2 Amplifier Configurations..9-3
9.3 Junction Field-Effect Transistors.......................................9-5
   JFET as an Amplifier: Small-Signal AC Voltage Gain • JFET as a Constant Current Source • Operation of a JFET as a Voltage-Variable Resistor • Voltage-Variable Resistor Applications

9.1 Bipolar Junction Transistors

A basic diagram of the bipolar junction transistor (BJT) is shown in Fig. 9.1. Whereas the diode has one PN junction, the BJT has two PN junctions. The three regions of the BJT are the emitter, base, and collector. The middle, or base region, is very thin, generally less than 1 μm wide. This middle electrode, or base, can be considered to be the control electrode that controls the current flow through the device between emitter and collector. A small voltage applied to the base (i.e., between base and emitter) can produce a large change in the current flow through the BJT.

BJTs are often used for the amplification of electrical signals. In these applications the emitter-base PN junction is turned on (forward biased) and the collector-base PN junction is off (reverse biased). For the NPN BJT as shown in Fig. 9.1, the emitter will emit electrons into the base region. Since the P-type base region is so thin, most of these electrons will survive the trip across the base and reach the collector-base junction. When the electrons reach the collector-base junction they will roll downhill into the collector, and thus be collected by the collector to become the collector current I_C. The emitter and collector currents will be approximately equal, so $I_C \cong I_E$. There will be a small base current, I_B, resulting from the emission of holes from the base across the emitter-base junction into the emitter. There will also be a small component of the base current due to the recombination of electrons and holes in the base. The ratio of collector current to base current is given by the parameter β or h_{FE}, is $\beta = I_C/I_B$, and will be very large, generally up in the range of 50–300 for most BJTs.

In Fig. 9.2(a) the circuit schematic symbol for the NPN transistor is shown, and in Fig. 9.2(b) the corresponding symbol for the PNP transistor is given. The basic operation of the PNP transistor is similar to that of the NPN, except for a reversal of the polarity of the algebraic signs of all DC currents and voltages.

In Fig. 9.3 the operation of a BJT as an amplifier is shown. When the BJT is operated as an amplifier the emitter-base PN junction is turned on (forward biased) and the collector-base PN junction is off (reverse biased). An AC input voltage applied between base and emitter, $v_{in} = v_{be}$, will produce an AC

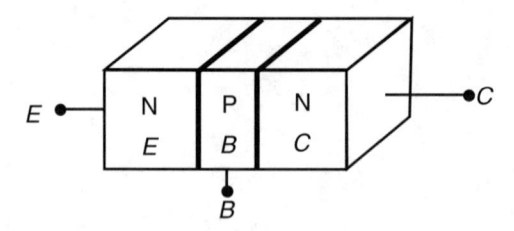

FIGURE 9.1 Bipolar junction transistor.

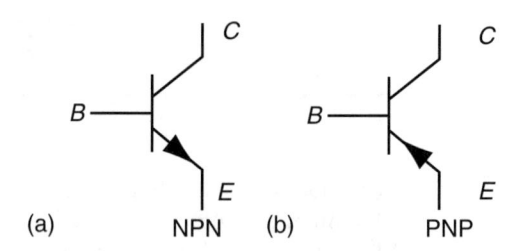

FIGURE 9.2 BJT schematic symbols: (a) NPN BJT and (b) PNP BJT.

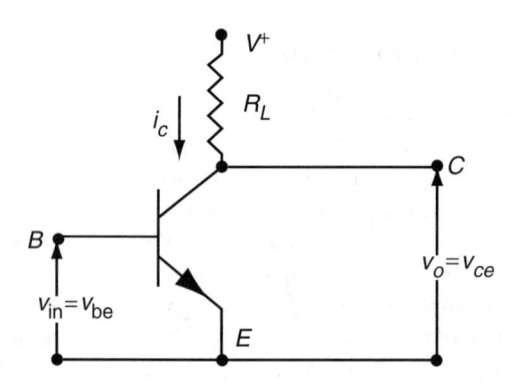

FIGURE 9.3 A BJT amplifier.

component, i_c, of the collector current. Since i_c flows through a load resistor, R_L, an AC voltage, $v_o = v_{ce} = -i_c \cdot R_L$ will be produced at the collector. The AC small-signal voltage gain is $A_V = v_o/v_{in} = v_{ce}/v_{be}$.

The collector current I_C of a BJT when operated as an amplifier is related to the base-to-emitter voltage V_{BE} by the exponential relationship $I_C = I_{CO} \cdot \exp(V_{BE}/V_T)$, where I_{CO} is a constant, and V_T = thermal voltage = 25 mV. The rate of change of I_C with respect to V_{BE} is given by the *transfer conductance*, $g_m = dI_C/dV_{BE} = I_C/V_T$. If the net load driven by the collector of the transistor is R_L, the AC small-signal voltage gain is $A_V = v_{ce}/v_{be} = -g_m \cdot R_L$. The negative sign indicates that the output voltage will be an amplified, but inverted, replica of the input signal. If, for example, the transistor is biased at a DC collector current level of $I_C = 1$ mA and drives a net load of $R_L = 10$ kΩ, then $g_m = I_C/V_T = 1$ mA/25 mV = 40 mS, and $A_V = v_c/v_{be} = -g_m \cdot R_L = -40$ mS $\cdot$ 10 k$\Omega = -400$. Thus we see that the voltage gain of a single BJT amplifier stage can be very large, often up in the range of 100, or more.

The BJT is a three electrode or **triode** electron device. When connected in a circuit it is usually operated as a two-port, or two-terminal, pair device as shown in Fig. 9.4. Therefore, one of the three electrodes of the BJT must be common to both the input and output ports. Thus, there are three basic BJT configurations, common emitter (CE), common base (CB), and common collector (CC), as shown in

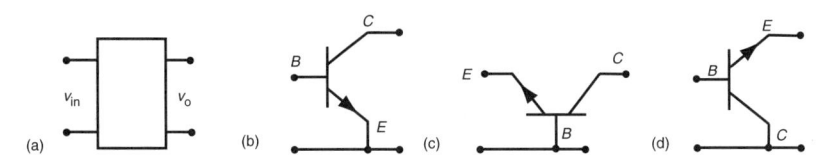

FIGURE 9.4 The BJT as a two-port device: (a) block representation, (b) common emitter, (c) common base, and (d) common collector.

Fig. 9.4. The most often used configuration, especially for amplifiers, is the common-emitter (CE), although the other two configurations are used in some applications.

9.2 Amplifier Configurations

We will first compare the common-emitter circuit of Fig. 9.4(b) to the common-base circuit of Fig. 9.4(c). The AC small-signal voltage gain of the common-emitter circuit is given by $A_V = -g_m R_{NET}$ where g_m is the dynamic forward transfer conductance as given by $g_m = I_C/V_T$, and R_{NET} is the net load resistance driven by the collector of the transistor. Note that the common-emitter circuit is an inverting amplifier, in that the output voltage is an amplified, but inverted, replica of the input voltage. The AC small-signal voltage gain of the common-base circuit is given by $A_V = g_m R_{NET}$, so we see that the common-base circuit is a noninverting amplifier, in that the output voltage is an amplified replica of the input voltage. Note that the magnitude of the gain is given by the same expression for both amplifier circuits.

The big difference between the two amplifier configurations is in the input resistance. For the common-emitter circuit the AC small-signal input resistance is given by $r_{IN} = n\beta V_T/I_C$, where n is the ideality factor, which is a dimensionless factor between 1 and 2, and is typically around 1.5 for silicon transistors operating at moderate current levels, in the 1–10 mA range. For the common-base circuit the AC small-signal input resistance is given by $r_{IN} = V_T/I_E \cong V_T/I_C$. The input resistance of the common-base circuit is smaller than that of the common-emitter circuit by a factor of approximately $n\beta$. For example, taking $n = 1.5$ and $\beta = 100$ as representative values, at $I_C = 1.0$ mA we get for the common-emitter case $r_{IN} = n\beta V_T/I_C = 1.5 \cdot 100 \cdot 25$ mV/1 mA $= 3750 \,\Omega$; whereas for the common-base case, we get $r_{IN} \cong V_T/I_C = 25$ mV/1 mA $= 25 \,\Omega$. We see that r_{IN} for the common emitter is 150 times larger than for the common base. The small input resistance of the common-base case will severely load most signal sources. Indeed, if we consider cascaded common-base stages with one common-base stage driving another operating at the same quiescent collector current level, then we get $A_V = g_m R_{NET} \cong g_m \cdot r_{IN} \cong (I_C/V_T) \cdot (V_T/I_C) = 1$, so that no net voltage gain is obtained from a common-base stage driving another common-base stage under these conditions. For the cascaded common-emitter case, we have $A_V = -g_m R_{NET} \cong -g_m r_{IN} = (I_C/V_T) \cdot (n\beta V_T/I_C) = n\beta$, so that if $n = 1.5$ and $\beta = 100$, a gain of about 150 can be achieved. It is for this reason that the common-emitter stage is usually chosen.

The common-base stage is used primarily in high-frequency applications due to the fact that there is no direct capacitive feedback from output (collector) to input (emitter) as a result of the common or grounded base terminal. A circuit configuration that is often used to take advantage of this, and at the same time to have the higher input impedance of the common-emitter circuit is the *cascode* configuration, as shown in Fig. 9.5. The cascode circuit is a combination of the common-emitter stage directly coupled to a common-base stage. The input impedance is that of the common-emitter stage, and the grounded base of the common-base stage blocks the capacitive feedback from output to input.

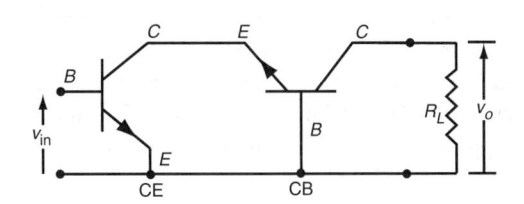

FIGURE 9.5 A cascode circuit.

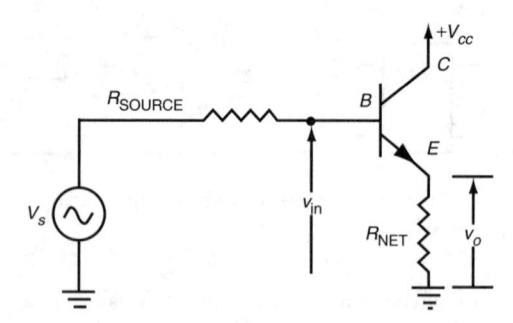

FIGURE 9.6 Emitter-follower circuit.

The common-collector circuit of Fig. 9.6 will now be considered. The common-collector circuit has an AC small-signal voltage gain given by

$$A_V = \frac{g_m R_{NET}}{[1 + g_m R_{NET}]} = \frac{R_{NET}}{[R_{NET} + V_T/I_C]}$$

The voltage gain is positive, but will always be less than unity, although it will usually be close to unity. For example if $I_C = 10$ mA and $R_{NET} = 50\ \Omega$ we obtain

$$A_V = \frac{R_{NET}}{\left[R_{NET} + \frac{V_T}{I_C}\right]} = \frac{50}{\left[50 + \frac{25\ \text{mV}}{10\ \text{mA}}\right]} = \frac{50}{50 + 2.5} = \frac{50}{52.5} = 0.952$$

Since the voltage gain for the common-collector stage is positive and usually close to unity, the AC voltage at the emitter will rather closely follow the voltage at the base, hence, the name *emitter-follower* that is usually used to describe this circuit.

We have seen that the common-collector or emitter-follower stage will always have a voltage gain that is less than unity. The emitter follower is nevertheless a very important circuit because of its impedance transforming properties. The AC small-signal input resistance is given by $r_{IN} = (\beta + 1)\,[(V_T/I_C) + R_{NET}]$, where R_{NET} is the net AC load driven by the emitter of the emitter follower. We see that looking into the base, the load resistance R_{NET} is transformed up in value by a factor of $\beta + 1$. Looking from the load back into the emitter, the AC small-signal output resistance is given by

$$r_O = \frac{1}{g_m} + \frac{R_{SOURCE}}{(\beta + 1)} = \frac{V_T}{I_C} + \frac{R_{SOURCE}}{(\beta + 1)}$$

where R_{SOURCE} is the net AC resistance that is seen looking out from the base toward the signal source. Thus, as seen from the load looking back into the emitter follower, the source resistance is transformed down by a factor of $\beta + 1$. This impedance transforming property of the emitter follower is useful for coupling high-impedance sources to low-impedance loads. For example, if a 1-kΩ source is coupled directly to a 50-Ω load, the transfer ratio will be $T = R_{LOAD}/[R_{LOAD} + R_{SOURCE}] = 50/1050 \cong 0.05$. If an emitter follower with $I_C = 10$ mA and $\beta = 200$ is interposed between the signal source and the load, the input resistance of the emitter follower will be

$$r_{IN} = (\beta + 1)\left[\frac{V_T}{I_C} + R_{NET}\right] = 201 \cdot [2.5\ \Omega + 50\ \Omega] = 201 \cdot 52.5\ \Omega = 10.55\ \text{k}\Omega$$

The signal transfer ratio from the signal source to the base of the emitter follower is now $T = r_{IN}/[r_{IN} + R_{SOURCE}] = 10.55\ \text{k}\Omega/11.55\ \text{k}\Omega = 0.913$. The voltage gain through the emitter follower from base to emitter is

$$A_V = \frac{R_{\text{NET}}}{\left[R_{\text{NET}} + \frac{V_T}{I_C}\right]} = \frac{50}{\left[50 + \frac{25 \text{ mV}}{10 \text{ mA}}\right]} = \frac{50}{[50 + 2.5]} = \frac{50}{52.5} = 0.952$$

and so the overall transfer ratio is $T_{\text{NET}} = 0.913 \cdot 0.952 = 0.87$. Thus, there is a very large improvement in the transfer ratio.

As a second example of the usefulness of the emitter follower, consider a common-emitter stage operating at $I_C = 1.0$ mA and driving a 50 Ω load. We have

$$A_V = -g_m R_{\text{NET}} = -\frac{1 \text{ mA}}{25 \text{ mV}} \cdot 50 \text{ Ω} = -40 \text{ mS} \cdot 50 \text{ Ω} = -2$$

If an emitter follower operating at $I_C = 10$ mA is interposed between the common-emitter stage and the load, we now have for the common-emitter stage a gain of

$$A_V = -g_m R_{\text{NET}} = -\frac{1 \text{ mA}}{25 \text{ mV}} \cdot r_{\text{IN}} = -40 \text{ mS} \cdot 10.55 \text{ kΩ} = -422$$

The voltage gain of the emitter follower from base to emitter is 0.952, so the overall gain is now $-422 \cdot 0.952 = -402$, as compared to the gain of only -2 that was available without the emitter follower.

The BJT is often used as a switching device, especially in digital circuits, and in high-power applications. When used as a switching device, the transistor is switched between the *cutoff region* in which both junctions are off, and the *saturation region* in which both junctions are on. In the *cutoff region* the collector current is reduced to a small value, down in the low nanoampere range, and so the transistor looks essentially like an open circuit. In the saturation region the voltage drop between collector and emitter becomes small, usually less than 0.1 V, and the transistor looks like a small resistance.

9.3 Junction Field-Effect Transistors

A junction field-effect transistor or JFET is a type of transistor in which the current flow through the device between the drain and source electrodes is controlled by the voltage applied to the gate electrode. A simple physical model of the JFET is shown in Fig. 9.7. In this JFET an N-type conducting channel exists between *drain* and *source*. The gate is a heavily doped P-type region (designated as P+), that surrounds the N-type channel. The gate-to-channel PN junction is normally kept reverse biased. As the reverse bias voltage between gate and channel increases, the depletion region width increases, as shown in Fig. 9.8. The depletion region extends mostly into the N-type channel because of the heavy doping on the P+ side. The depletion region is depleted of mobile charge carriers and, thus, cannot contribute to the conduction of current between drain and source. Thus, as the gate voltage increases, the cross-sectional area of the N-type channel available for current flow decreases. This reduces the current flow between drain and source. As the gate voltage increases, the channel becomes further constricted, and the current flow gets smaller. Finally, when the depletion regions meet in the middle of the channel, as shown in Fig. 9.9, the channel is pinched off in its entirety, all of the way between the source and the drain. At this point the current flow between drain and source is reduced to essentially zero. This voltage is called the **pinch-off voltage** V_P. The pinch-off voltage is also represented as V_{GS} (OFF), as being the gate-to-source voltage that turns the drain-to-source current I_{DS} off. We have been considering here an N-channel JFET. The complementary device is the P-channel JFET, which has a heavily doped N-type (N+) gate region surrounding a P-type channel. The operation of a P-channel JFET is the same as for an N-channel device, except the algebraic signs of all DC voltages and currents are reversed.

We have been considering the case for V_{DS} small compared to the pinch-off voltage such that the channel is essentially uniform from drain to source, as shown in Fig. 9.10(a). Now let us see what happens as V_{DS} increases. As an example, assume an N-channel JFET with a pinch-off voltage of $V_P = -4$ V. We

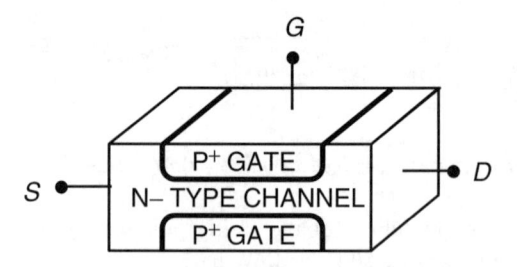

FIGURE 9.7 Model of a JFET device.

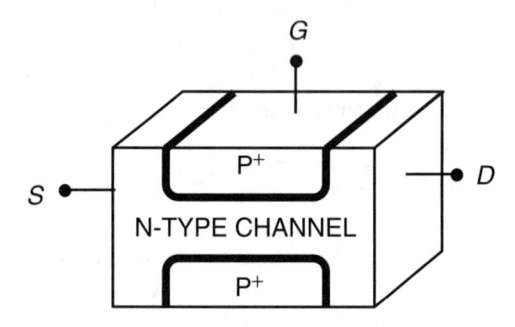

FIGURE 9.8 JFET with increased gate voltage.

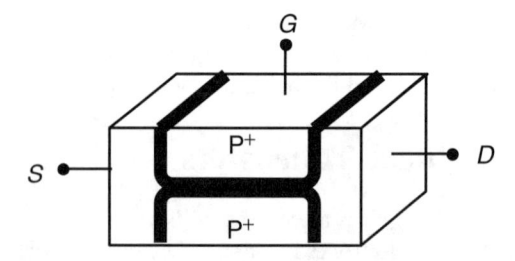

FIGURE 9.9 JFET with pinched-off channel.

will see what happens for the case of $V_{GS} = 0$ as V_{DS} increases. In Fig. 9.10(a) the situation is shown for the case of $V_{DS} = 0$ in which the JFET is fully on and there is a uniform channel from source to drain. This is at point A on the I_{DS} vs. V_{DS} curve of Fig. 9.11. The drain-to-source conductance is at its maximum value of g_{ds} (ON), and the drain-to-source resistance is correspondingly at its minimum value of r_{ds} (ON). Now, consider the case of $V_{DS} = +1$ V as shown in Fig. 9.10(b). The gate-to-channel bias voltage at the source end is still $V_{GS} = 0$. The gate-to-channel bias voltage at the drain end is $V_{GD} = V_{GS} - V_{DS} = -1$ V, so the depletion region will be wider at the drain end of the channel than at the source end. The channel will, thus, be narrower at the drain end than at the source end, and this will result in a decrease in the channel conductance g_{ds}, and correspondingly, an increase in the channel resistance r_{ds}. Thus, the slope of the I_{DS} vs. V_{DS} curve, which corresponds to the channel conductance, will be smaller at $V_{DS} = 1$ V than it was at $V_{DS} = 0$, as shown at point B on the I_{DS} vs. V_{DS} curve of Fig. 9.11.

In Fig. 9.10(c) the situation for $V_{DS} = +2$ V is shown. The gate-to-channel bias voltage at the source end is still $V_{GS} = 0$, but the gate-to-channel bias voltage at the drain end is now $V_{GD} = V_{GS} - V_{DS} = -2$ V, so the depletion region will be substantially wider at the drain end of the channel than at the source end. This leads to a further constriction of the channel at the drain end, and this will again result in a decrease in the channel conductance g_{ds}, and correspondingly, an increase in the channel resistance r_{ds}. Thus the slope of the I_{DS} vs. V_{DS} curve will be smaller at $V_{DS} = 2$ V than it was at $V_{DS} = 1$ V, as shown at point C on the I_{DS} vs. V_{DS} curve of Fig. 9.11.

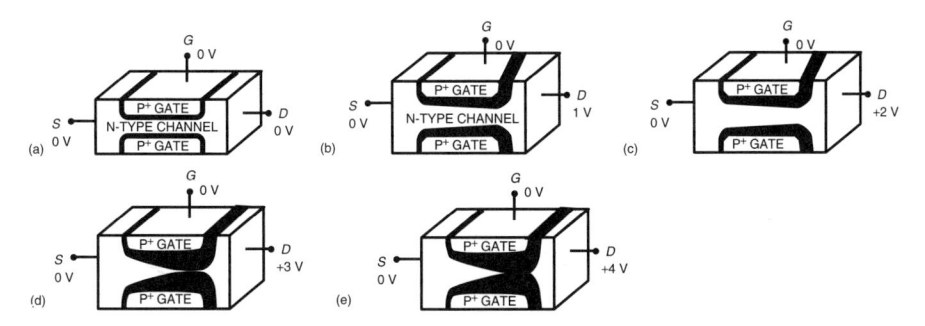

FIGURE 9.10 JFET operational characteristics: (a) uniform channel from drain to source, (b) depletion region wider at the drain end, (c) depletion region significantly wider at the drain, (d) channel near pinchoff, and (e) channel at pinchoff.

In Fig. 9.10(d) the situation for $V_{DS} = +3$ V is shown, and this corresponds to point D on the I_{DS} vs. V_{DS} curve of Fig. 9.11.

When $V_{DS} = +4$ V the gate-to-channel bias voltage will be $V_{GD} = V_{GS} - V_{DS} = 0 - 4\text{ V} = -4\text{ V} = V_P$. As a result the channel is now pinched off at the drain end, but is still wide open at the source end since $V_{GS} = 0$, as shown in Fig. 9.10(e). It is important to note that channel is pinched off just for a very short distance at the drain end, so that the drain-to-source current I_{DS} can still continue to flow. This is not at all the same situation as for the case of $V_{GS} = V_P$ wherein the channel is pinched off in its entirety, all of the way from source to drain. When this happens, it is like having a big block of insulator the entire distance between source and drain, and I_{DS} is reduced to essentially zero. The situation for $V_{DS} = +4\text{ V} = -V_P$ is shown at point E on the I_{DS} vs. V_{DS} curve of Fig. 9.11.

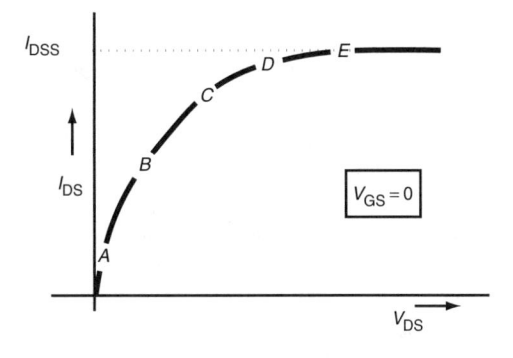

FIGURE 9.11 I_{DS} vs. V_{DS} curve.

For $V_{DS} > +4$ V, the current essentially saturates, and does not increase much with further increases in V_{DS}. As V_{DS} increases above +4 V, the pinched-off region at the drain end of the channel gets wider, which increases r_{ds}. This increase in r_{ds} essentially counterbalances the increase in V_{DS} such that I_{DS} does not increase much. This region of the I_{DS} vs. V_{DS} curve in which the channel is pinched off at the drain end is called the **active region**, also known as the *saturated region*. It is called the active region because when the JFET is to be used as an amplifier it should be biased and operated in this region. The saturated value of drain current up in the active region for the case of $V_{GS} = 0$ is called I_{DSS}. Since there is not really a true saturation of current in the active region, I_{DSS} is usually specified at some value of V_{DS}. For most JFETs, the values of I_{DSS} fall in the range of 1–30 mA. In the current specification, I_{DSS}, the third subscript S refers to I_{DS} under the condition of the gate *shorted* to the source.

The region below the active region where $V_{DS} < +4\text{ V} = -V_P$ has several names. It is called the **nonsaturated region**, the **triode region**, and the **ohmic region**. The term triode region apparently originates from the similarity of the shape of the curves to that of the vacuum tube triode. The term ohmic region is due to the variation of I_{DS} with V_{DS} as in Ohm's law, although this variation is nonlinear except for the region of V_{DS}, which is small compared to the pinch-off voltage, where I_{DS} will have an approximately linear variation with V_{DS}.

The upper limit of the active region is marked by the onset of the breakdown of the gate-to-channel PN junction. This will occur at the drain end at a voltage designated as BV_{DG}, BV_{DS}, since $V_{GS} = 0$. This breakdown voltage is generally in the 30–150 V range for most JFETs.

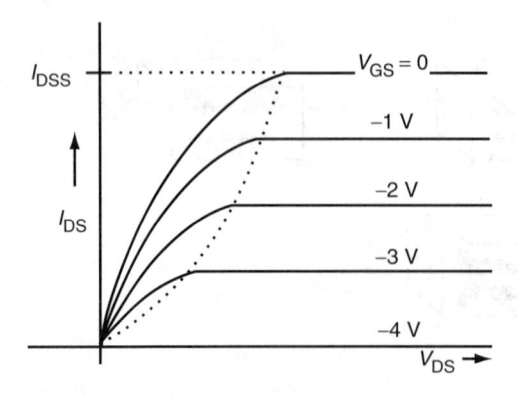

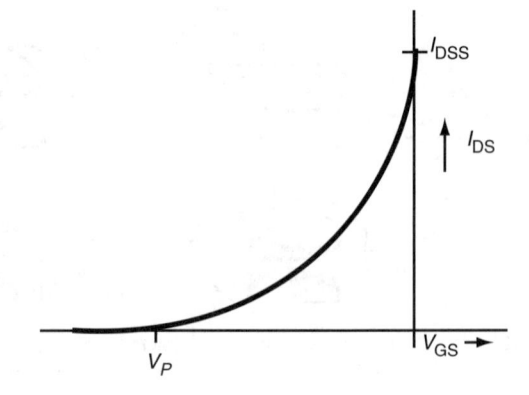

FIGURE 9.12 JFET drain characteristics.

FIGURE 9.13 JFET transfer characteristics.

Thus far we have looked at the I_{DS} vs. V_{DS} curve only for the case of $V_{GS} = 0$. In Fig. 9.12 a family of curves of I_{DS} vs. V_{DS} for various constant values of V_{GS} is presented. This is called the *drain characteristics*, and is also known as the **output characteristics**, since the output side of the JFET is usually the drain side. In the active region where I_{DS} is relatively independent of V_{DS}, there is a simple approximate equation relating I_{DS} to V_{GS}. This is the square law **transfer equation** as given by $I_{DS} = I_{DSS} [1 - (V_{GS}/V_P)]^2$. In Fig. 9.13 a graph of the I_{DS} vs. V_{GS} *transfer characteristics* for the JFET is presented. When $V_{GS} = 0$, $I_{DS} = I_{DSS}$ as expected, and as $V_{GS} \rightarrow V_P$, $I_{DS} \rightarrow 0$. The lower boundary of the active region is controlled by the condition that the channel be pinched off at the drain end. To meet this condition the basic requirement is that the gate-to-channel bias voltage at the drain end of the channel, V_{GD}, be greater than the pinch-off voltage V_P. For the example under consideration with $V_P = -4$ V, this means that $V_{GD} = V_{GS} - V_{DS}$ must be more negative than –4 V. Therefore, $V_{DS} - V_{GS} \geq +4$ V. Thus, for $V_{GS} = 0$, the active region will begin at $V_{DS} = +4$ V. When $V_{GS} = -1$ V, the active region will begin at $V_{DS} = +3$ V, for now $V_{GD} = -4$ V. When $V_{GS} = -2$ V, the active region begins at $V_{DS} = +2$ V, and when $V_{GS} = -3$ V, the active region begins at $V_{DS} = +1$ V. The dotted line in Fig. 9.12 marks the boundary between the nonsaturated and active regions. The upper boundary of the active region is marked by the onset of the avalanche breakdown of the gate-to-channel PN junction. When $V_{GS} = 0$, this occurs at $V_{DS} = BV_{DS} = BV_{DG}$. Since $V_{DG} = V_{DS} - V_{GS}$, and breakdown occurs when $V_{DG} = BV_{DG}$, as V_{GS} increases the breakdown voltages decrease as given by $BV_{DG} = BV_{DS} - V_{GS}$. Thus, $BV_{DS} = BV_{DG} + V_{GS}$. For example, if the gate-to-channel breakdown voltage is 50 V, the V_{DS} breakdown voltage will start off at 50 V when $V_{GS} = 0$, but decreases to 46 V when $V_{GS} = -4$ V.

In the nonsaturated region I_{DS} is a function of both V_{GS} and I_{DS}, and in the lower portion of the nonsaturated region where V_{DS} is small compared to V_P, I_{DS} becomes an approximately linear function of V_{DS}. This linear portion of the nonsaturated region is called the *voltage-variable resistance* (VVR) region, for in this region the JFET acts like a linear resistance element between source and drain. The resistance is variable in that it is controlled by the gate voltage.

JFET as an Amplifier: Small-Signal AC Voltage Gain

Consider the common-source amplifier circuit of Fig. 9.14. The input AC signal is applied between gate and source, and the output AC voltage is taken between drain and source. Thus the source electrode of this triode device is common to input and output, hence the designation of this JFET as a common-source (CS) amplifier.

A good choice of the DC operating point or quiescent point (Q-point) for an amplifier is in the middle of the activate region at $I_{DS} = I_{DSS}/2$. This allows for the maximum symmetrical drain current swing, from the quiescent level of $I_{DSQ} = I_{DSS}/2$, down to a minimum of $I_{DS} \cong 0$ and up to a maximum of $I_{DS} = I_{DSS}$. This choice for the Q-point is also a good one from the standpoint of allowing for an adequate safety margin for the location of the actual Q-point due to the inevitable variations in device and

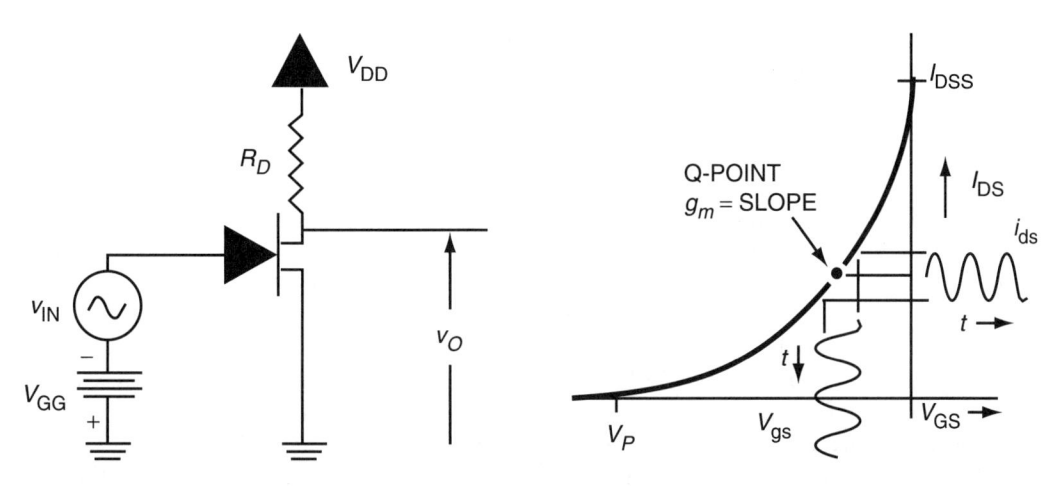

FIGURE 9.14 A common source amplifier.   **FIGURE 9.15** JFET transfer characteristic.

component characteristics and values. This safety margin should keep the Q-point well away from the extreme limits of the active region, and thus ensure operation of the JFET in the active region under most conditions. If $I_{DSS} = +10$ mA, then a good choice for the Q-point would thus be around $+5.0$ mA. The AC component of the drain current, i_{ds} is related to the AC component of the gate voltage, v_{gs} by $i_{ds} = g_m \cdot v_{gs}$, where g_m is the *dynamic transfer conductance*, and is given by

$$g_m = \frac{2\sqrt{I_{DS} \cdot I_{DSS}}}{-V_P}$$

If $V_p = -4$ V, then

$$g_m = \frac{2\sqrt{5 \text{ mA} \cdot 10 \text{ mA}}}{4 \text{ V}} = \frac{3.54 \text{ mA}}{V} = 3.54 \text{ mS}$$

If a small AC signal voltage v_{gs} is superimposed on the quiescent DC gate bias voltage $V_{GSQ} = V_{GG}$, only a small segment of the transfer characteristic adjacent to the Q-point will be traversed, as shown in Fig. 9.15. This small segment will be close to a straight line, and as a result the AC drain current i_{ds}, will have a waveform close to that of the AC voltage applied to the gate. The ratio of i_{ds} to v_{gs} will be the slope of the transfer curve as given by

$$\frac{i_{ds}}{v_{gs}} \cong \frac{dI_{DS}}{dV_{GS}} = g_m$$

Thus $i_{ds} \cong g_m \cdot v_{gs}$. If the net load driven by the drain of the JFET is the drain load resistor R_D, as shown in Fig. 9.14, then the AC drain current i_{ds} will produce an AC drain voltage of $v_{ds} = -i_{ds} \cdot R_D$. Since $i_{ds} = g_m \cdot v_{gs}$, this becomes $v_{ds} = -g_m v_{gs} \cdot R_D$. The AC small-signal voltage gain from gate to drain thus becomes

$$A_V = \frac{v_O}{v_{IN}} = \frac{v_{ds}}{v_{gs}} = -g_m \cdot R_D$$

The negative sign indicates signal inversion as is the case for a common-source amplifier.

If the DC drain supply voltage is $V_{DD} = +20$ V, a quiescent drain-to-source voltage of $V_{DSQ} = V_{DD}/2 = +10$ V will result in the JFET being biased in the middle of the active region. Since $I_{DSQ} = 5$ mA, in the example under consideration, the voltage drop across the drain load resistor R_D is 10 V. Thus $R_D = 10$ V/5 mA = 2 kΩ. The AC small-signal voltage gain, A_V, thus becomes

$$A_V = -g_m \cdot R_D = -3.54 \text{ mS} \cdot 2 \text{ k}\Omega = -7.07$$

Note that the voltage gain is relatively modest, as compared to the much larger voltage gains that can obtain with the bipolar-junction transistor common-emitter amplifier. This is due to the lower transfer conductance of both JFETs and metal-oxide-semiconductor field-effect transistors (MOSFETs) as compared to BJTs. For a BJT the transfer conductance is given by $g_m = I_C/V_T$ where I_C is the quiescent collector current and $V_T = kT/q \cong 25$ mV is the **thermal voltage**. At $I_C = 5$ mA, $g_m = 5$ mA/25 mV = 200 mS for the BJT, as compared to only 3.5 mS for the JFET in this example. With a net load of 2 kΩ, the BJT voltage gain will be –400 as compared to the JFET voltage gain of only 7.1. Thus FETs do have the disadvantage of a much lower transfer conductance and, therefore, lower voltage gain than BJTs operating under similar quiescent current levels; but they do have the major advantage of a much higher input impedance and a much lower input current. In the case of a JFET the input signal is applied to the *reverse-biased* gate-to-channel PN junction, and thus sees a very high impedance. In the case of a common-emitter BJT amplifier, the input signal is applied to the *forward-biased* base-emitter junction and the input impedance is given approximately by $r_{IN} = r_{BE} \cong 1.5 \cdot \beta \cdot V_T/I_C$. If $I_C = 5$ mA and $\beta = 200$, for example, then $r_{IN} \cong 1500$ Ω. This moderate input resistance value of 1.5 kΩ is certainly no problem if the signal source resistance is less than around 100 Ω. However, if the source resistance is above 1 kΩ, there will be a substantial signal loss in the coupling of the signal from the signal source to the base of the transistor. If the source resistance is in the range of above 100 kΩ, and certainly if it is above 1 MΩ, then there will be severe signal attenuation due to the BJT input impedance, and an FET amplifier will probably offer a greater overall voltage gain. Indeed, when high impedance signal sources are encountered, a multistage amplifier with an FET input stage, followed by cascaded BJT stages is often used.

JFET as a Constant Current Source

An important application of a JFET is as a constant current source or as a current regulator diode. When a JFET is operating in the active region, the drain current I_{DS} is relatively independent of the drain voltage V_{DS}. The JFET does not, however, act as an ideal constant current source since I_{DS} does increase slowly with increases in V_{DS}. The rate of change of I_{DS} with V_{DS} is given by the drain-to-source conductance $g_{ds} = dI_{DS}/dV_{DS}$. Since I_{DS} is related to the channel length L by $I_{DS} \propto 1/L$, the drain-to-source conductance g_{ds} can be expressed as

$$g_{ds} = \frac{dI_{DS}}{dV_{DS}} = \frac{dI_{DS}}{dL} \cdot \frac{dL}{dV_{DS}}$$

$$= \frac{-I_{DS}}{L} \cdot \frac{dL}{dV_{DS}} = I_{DS}\left(\frac{-1}{L}\right)\left(\frac{dL}{dV_{DS}}\right)$$

The channel length modulation coefficient is defined as

$$\text{channel length modulation coefficient} = \frac{1}{V_A} = \frac{-1}{L}\left(\frac{dL}{dV_{DS}}\right)$$

where V_A is the JFET *early voltage*. Thus we have that $g_{ds} = I_{DS}/V_A$. The early voltage V_A for JFETs is generally in the range of 20–200 V.

The *current regulation* of the JFET acting as a constant current source can be expressed in terms of the fractional change in current with voltage as given by

$$\text{current regulation} = \left(\frac{1}{I_{DS}}\right)\frac{dI_{DS}}{dV_{DS}} = \frac{g_{ds}}{I_{ds}} = \frac{1}{V_A}$$

For example, if $V_A = 100$ V, the current regulation will be $1/(100 \text{ V}) = 0.01/\text{V} = 1\%/\text{V}$, so I_{DS} changes by only 1% for every 1 V change in V_{DS}.

In Fig. 9.16 a diode-connected JFET or *current regulator diode* is shown. Since $V_{GS} = 0$, $I_{DS} = I_{DSS}$. The current regulator diode can be modeled as an ideal constant current source in parallel with a resistance r_O as shown in Fig. 9.17. The *voltage compliance range* is the voltage range over which a device or system acts as a good approximation to the ideal constant current source. For the JFET this will be the extent of the active region. The lower limit is the point where the channel just becomes pinched off at the drain end. The *voltage compliance range* is the voltage range over which a device or system acts as a good approximation to the ideal constant current source. For the JFET this will be the extent of the active region. The lower limit is the point where the channel just becomes pinched off at the drain end. The requirement is, thus, that $V_{DG} = V_{DS} - V_{GS} > -V_P$. For the case of $V_{GS} = 0$, this occurs at $V_{DS} = -V_P$. The upper limit of the voltage compliance range is set by the breakdown voltage of the gate-to-channel PN junction, BV_{DG}. Since V_P is typically in the 2–5 V range, and BV_{DG} is typically >30 V, this means that the voltage compliance range will be relatively large. For example, if $V_P = -3$ V and $BV_{DG} = +50$ V, the voltage compliance range will extend from $V_{DS} = +3$ V up to $V_{DS} = +50$ V. If $V_A = 100$ V and $I_{DSS} = 10$ mA, the current regulator dynamic output conductance will be

$$g_O = \frac{dI_O}{dV_O} = \frac{dI_{ds}}{dV_{ds}} = g_{ds} = \frac{10 \text{ mA}}{100 \text{V}} = 0.1 \text{ mA}/V = 0.1 \text{ mS}$$

The current regulator dynamic output resistance will be $r_O = r_{ds} = 1/g_{ds} = 10$ kΩ. Thus, the current regulator diode can be represented as a 10-mA constant current source in parallel with a 10-kΩ dynamic resistance.

In Fig. 9.18 a current regulator diode is shown in which a resistor R_S is placed in series with the JFET source in order to reduce I_{DS} below I_{DSS}. The current I_{DS} flowing through R_S produces a voltage drop $V_{SG} = I_{DS} \cdot R_S$. This results in a gate-to-source bias voltage of $V_{GS} = -V_{SG} = -I_{DS} \cdot R_S$. From the JFET transfer equation, $I_{DS} = I_{DSS}[1 - (V_{GS}/V_P)]^2$ we have that $V_{GS} = V_P [1 - \sqrt{I_{DS}/I_{DSS}}]$. From the required value of I_{DS} the corresponding value of V_{GS} can be determined, and from that the value of R_S can be found.

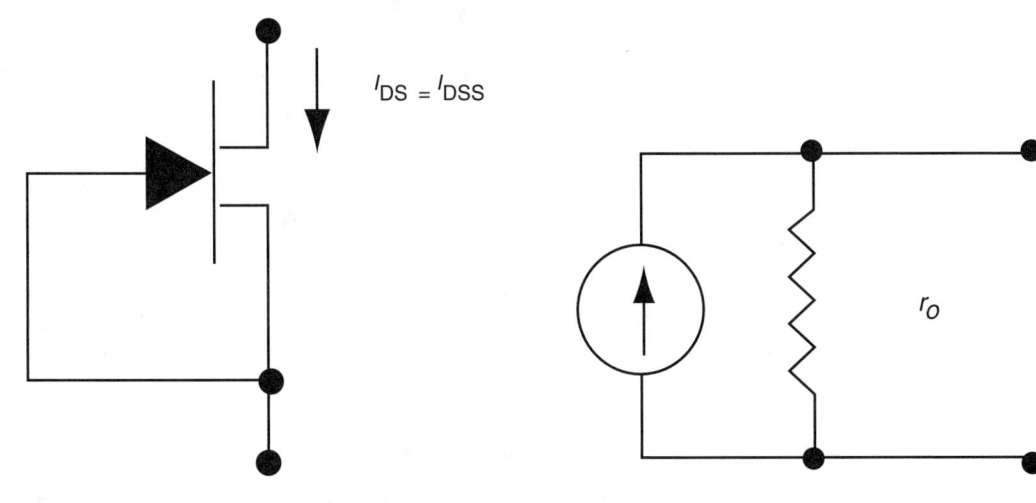

FIGURE 9.16 A current regulator diode.

FIGURE 9.17 Model of the current regulator diode.

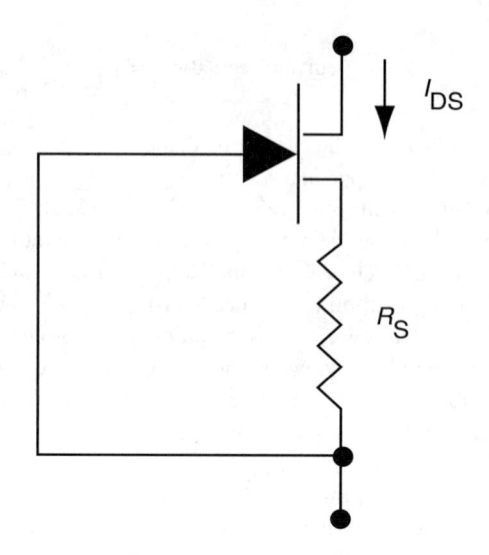

FIGURE 9.18 A current regulator diode for $I_{DS} < I_{DSS}$.

With R_S present, the dynamic output conductance, $g_O = dI_O/dV_O$, becomes $g_O = g_{ds}/(1 + g_m R_S)$. The current regulation now given as

$$\text{current regulation} = \left(\frac{1}{I_O}\right)\frac{dI_O}{dV_O} = \frac{g_O}{I_O}$$

Thus R_S can have a beneficial effect in reducing g_O and improving the current regulation. For example, let $V_P = -3V$, $V_A = 100$ V, $I_{DSS} = 10$ mA, and $I_O = I_{DS} = 1$ mA. We now have that

$$V_{GS} = V_{GS} = V_P\left[1 - \sqrt{\frac{I_{DS}}{I_{DSS}}}\right] = -3\text{ V}\left[1 - \sqrt{\frac{1\text{ mA}}{10\text{ mA}}}\right] = -2.05\text{ V}$$

and so $R_S = 2.05$ V/1 mA = 2.05 kΩ. The transfer conductance g_m is given by

$$g_m = \frac{2\sqrt{I_{DS} \cdot I_{DSS}}}{-V_P} = \frac{2\sqrt{1\text{ mA} \cdot 10\text{ mA}}}{3\text{ V}} = 2.1\text{ mS}$$

and so $g_m R_S$ = 2.1 mS · 2.05 kΩ = 4.32. Since g_{ds} = 1 mA/100 V = 10 μS, we have

$$g_O = \frac{g_{ds}}{1 + g_m R_S} = \frac{10\text{ }\mu\text{S}}{5.32} = 1.9\text{ }\mu\text{S}$$

The current regulation is thus g_O/I_O = 1.9 μS/1 mA = 0.0019/V = 0.19%/V. This is to be compared to the current regulation of 1%/V obtained for the case of $I_{DS} = I_{DSS}$.

Any JFET can be used as a current regulating diode. There are, however, JFETs that are especially made for this application. These JFETs have an extra long channel length, which reduces the channel length modulation effect and, hence, results in a large value for V_A. This in turn leads to a small g_{ds} and, hence, a small g_O and, thus, good current regulation.

Operation of a JFET as a Voltage-Variable Resistor

A JFET can be used as voltage-variable resistor in which the drain-to-source resistance r_{ds} of the JFET can be varied by variation of V_{GS}. For values of $V_{DS} \ll V_P$ the I_{DS} vs. V_{DS} characteristics are approximately linear, and so the JFET looks like a resistor, the resistance value of which can be varied by the gate voltage.

The channel conductance in the region where $V_{DS} \ll V_P$ is given by $g_{ds} = A\sigma/L = WH\sigma/L$, where the channel height H is given by $H = H_0 - 2W_D$. In this equation W_D is the depletion region width and H_0 is the value of H as $W_D \to 0$. The depletion region width is given by $W_D = K\sqrt{V_J} = K\sqrt{V_{GS} + \phi}$ where K is a constant, V_J is the junction voltage, and ϕ is the PN junction **contact potential** (typical around 0.8–1.0 V). As V_{GS} increases, W_D increases and the channel height H decreases as given by $H = H_0 - 2K\sqrt{V_{GS} + \phi}$. When $V_{GS} = V_P$, the channel is completely pinched off, so $H = 0$.

The drain-to-source resistance r_{ds} is given approximately by $r_{ds} \cong r_{ds}(ON)/[1 - \sqrt{V_{GS}/V_P}]$.

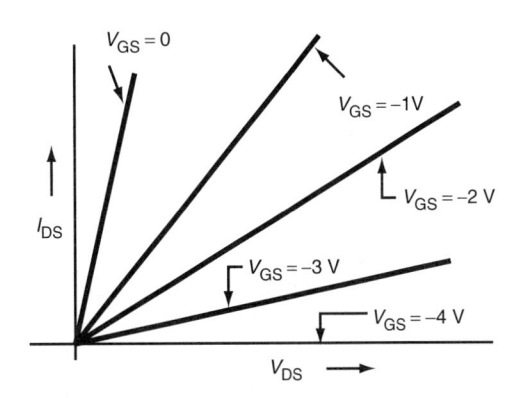

FIGURE 9.19 Voltage-variable resistor characteristics of the JFET.

As $V_{GS} \to 0$, $r_{ds} \to r_{ds}(ON)$ and as $V_{GS} \to V_P$, $r_{ds} \to \infty$. This latter condition corresponds to the channel being pinched off in its entirety, all of the way from source to drain. This is like having big block of insulator (i.e., the depletion region) between source and drain. When $V_{GS} = 0$, r_{ds} is reduced to its minimum value of $r_{ds}(ON)$, which for most JFETs is in the 20–4000 Ω range. At the other extreme, when $V_{GS} > V_P$, the drain-to-source current I_{DS} is reduced to a small value, generally down into the low nanoampere, or even picoampere range. The corresponding value of r_{ds} is not really infinite, but is very large, generally well up into the gigaohm (1000 MΩ) range. Thus, by variation of V_{GS}, the drain-to-source resistance can be varied over a very wide range. As long as the gate-to-channel junction is reverse biased, the gate current will be very small, generally down in the low nanoampere, or even picoampere range, so that the gate as a control electrode draws little current. Since V_P is generally in the 2–5 V range for most JFETs, the V_{DS} values required to operate the JFET in the VVR range is generally < 0.1 V. In Fig. 9.19 the VVR region of the JFET I_{DS} vs. V_{DS} characteristics is shown.

Voltage-Variable Resistor Applications

Applications of VVRs include automatic gain control (AGC) circuits, electronic attenuators, electronically variable filters, and oscillator amplitude control circuits.

When using a JFET as a VVR it is necessary to limit V_{DS} to values that are small compared to V_P to maintain good linearity. In addition, V_{GS} should preferably not exceed $0.8V_P$ for good linearity, control, and stability. This limitation corresponds to an r_{ds} resistance ratio of about 10:1. As V_{GS} approaches V_P, a small change in V_P can produce a large change in r_{ds}. Thus, unit-to-unit variations in V_P as well as changes in V_P with temperature can result in large changes in r_{ds} as V_{GS} approaches V_P.

The drain-to-source resistance r_{ds} will have a temperature coefficient TC due to two causes: (1) the variation of the channel resistivity with temperature and (2) the temperature variation of V_P. The TC of the channel resistivity is positive, whereas the TC of V_P is positive due to the negative TC of the contact potential ϕ. The positive TC of the channel resistivity will contribute to a positive TC or r_{ds}. The negative TC of VP will contribute to a negative TC of r_{ds}. At small values of V_{GS}, the dominant contribution to the TC is the positive TC of the channel resistivity, and so r_{ds} will have a positive TC. As V_{GS} gets larger, the negative TC contribution of V_P becomes increasingly important, and there will be a value of V_{GS} at which the net TC of r_{ds} is zero and above this value of V_{GS} the TC will be negative. The TC of r_{ds} (ON) is typically +0.3%/°C for N-channel JFETs, and +0.7%/°C for P-channel JFETs. For example, for a typical JFET with an r_{ds} (ON) = 500 Ω at 25°C and V_P = 2.6 V, the zero temperature coefficient point will occur at V_{GS} = 2.0 V. Any JFET can be used as a VVR, although there are JFETs that are specifically made for this application.

Example of VVR Application

A simple example of a VVR application is the electronic gain control circuit of Fig. 9.20. The voltage gain is given by $A_V = 1 + (R_F/r_{ds})$. If, for example, $R_F = 19$ kΩ and r_{ds} (ON) = 1 kΩ, then the maximum gain will be $A_{VMAX} = 1 + [R_F/r_{ds}$ (ON)] = 20. As V_{GS} approaches V_P, the r_{ds} will increase and become very large such that $r_{ds} \gg R_F$, so that A_V will decrease to a minimum value of close to unity. Thus, the gain can be varied over a 20:1 ratio. Note that $V_{DS} \cong V_{IN}$, and so to minimize distortion, the input signal amplitude should be small compared to V_P.

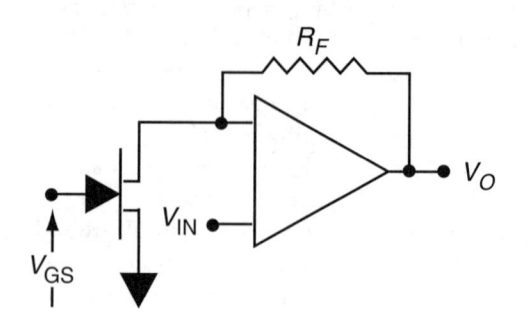

FIGURE 9.20 An electronic gain control circuit.

Defining Terms

Active region: The region of transistor operation in which the output current is relatively independent of the output voltage. For the BJT this corresponds to the condition that the emitter-base junction is on, and the collector-base junction is off. For the FETs this corresponds to the condition that the channel is on, or open, at the source end, and pinched off at the drain end.

Contact potential: The internal voltage that exists across a PN junction under thermal equilibrium conditions, when no external bias voltage is applied.

Ohmic, nonsaturated, or triode region: These three terms all refer to the region of FET operation in which a conducting channel exists all of the way between source and drain. In this region the drain current varies with both the gate voltage and the drain voltage.

Output characteristics: The family of curves of output current vs. output voltage. For the BJT this will be curves of collector current vs. collector voltage for various constant values of base current or voltage, and is also called the collector characteristics. For FETs this will be curves of drain current vs. drain voltage for various constant values of gate voltage, and is also called the drain characteristics.

Pinch-off voltage, V_P: The voltage that when applied across the gate-to-channel PN junction will cause the conducting channel between drain and source to become pinched off. This is also represented as $V_{GS(OFF)}$.

Thermal voltage: The quantity kT/q where k is Boltzmann's constant, T is absolute temperature, and q is electron charge. The thermal voltage has units of volts, and is a function only of temperature, being approximately 25 mV at room temperature.

Transfer conductance: The AC or dynamic parameter of a device that is the ratio of the AC output current to the AC input voltage. The transfer conductance is also called the *mutual transconductance*, and is usually designated by the symbol g_m.

Transfer equation: The equation that relates the output current (collector or drain current) to the input voltage (base-to-emitter or gate-to-source voltage).

Triode: A three-terminal electron device, such as a bipolar junction transistor or a field-effect transistor.

References

Mauro, R. 1989. *Engineering Electronics.* Prentice-Hall, Englewood Cliffs, NJ.
Millman, J. and Grabel, A. 1987. *Microelectronics,* 2nd ed. McGraw-Hill, New York.
Mitchell, F.H., Jr. and Mitchell, F.H., Sr. 1992. *Introduction to Electronics Design,* 2nd ed. Prentice-Hall, Englewood Cliffs, NJ.
Savant, C.J., Roden, M.S., and Carpenter, G.L. 1991. *Electronic Design,* 2nd ed. Benjamin-Cummings, Menlo Park, CA.
Sedra, A.S. and Smith, K.C. 1991. *Microelectronics Circuits,* 3rd ed. Saunders, Philadelphia, PA.

10
Metal-Oxide-Semiconductor Field-Effect Transistor

John R. Brews
The University of Arizona, Tucson

10.1 Introduction 10-1
10.2 Current-Voltage Characteristics 10-3
     Strong-Inversion Characteristics • Subthreshold Characteristics
10.3 Important Device Parameters 10-4
     Threshold Voltage • Driving Ability and $I_{D,\text{sat}}$
     • Transconductance • Output Resistance and Drain Conductance
10.4 Limitations on Miniaturization 10-11
     Subthreshold Control • Hot-Electron Effects • Thin Oxides
     • Dopant-Ion Control • Other Limitations

10.1 Introduction

The metal-oxide-semiconductor field-effect transistor (MOSFET) is a transistor that uses a control electrode, the **gate**, to capacitively modulate the conductance of a surface **channel** joining two end contacts, the **source** and the **drain**. The gate is separated from the semiconductor **body** underlying the gate by a thin *gate insulator,* usually silicon dioxide. The surface channel is formed at the interface between the semiconductor body and the gate insulator, see Fig. 10.1.

The MOSFET can be understood by contrast with other field-effect devices, like the junction field-effect transistor (JFET) and the metal-semiconductor field-effect transistor (MESFET) [Hollis and Murphy, 1990]. These other transistors modulate the conductance of a *majority-carrier* path between two *ohmic* contacts by capacitive control of its cross-section. (Majority carriers are those in greatest abundance in field-free semiconductor, electrons in *n*-type material and holes in *p*-type material.) This modulation of the cross-section can take place at any point along the length of the channel, and so the gate electrode can be positioned anywhere and need not extend the entire length of the channel.

Analogous to these field-effect devices is the *buried-channel, depletion-mode,* or *normally on* MOSFET, which contains a surface layer of the same doping type as the source and drain (opposite type to the semiconductor body of the device). As a result, it has a built-in or normally on channel from source to drain with a conductance that is reduced when the gate depletes the majority carriers.

In contrast, the true MOSFET is an *enhancement-mode* or *normally off* device. The device is normally off because the body forms *p–n* junctions with both the source and the drain, so no majority-carrier current can flow between them. Instead, *minority-carrier* current can flow, provided minority carriers are available. As discussed later, for gate biases that are sufficiently attractive, above **threshold,** minority

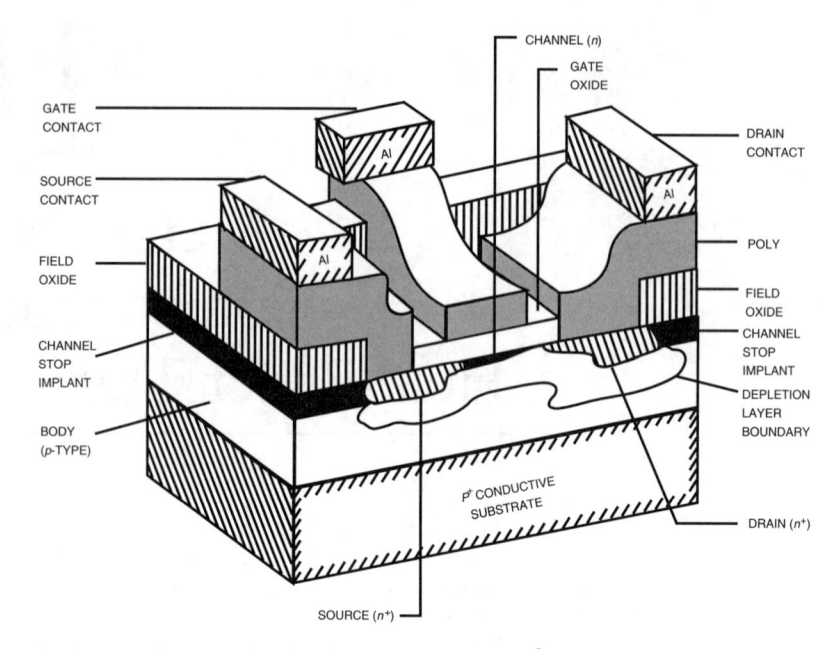

FIGURE 10.1 A high-performance *n*-channel MOSFET. The device is isolated from its neighbors by a surrounding thick *field oxide* under which is a heavily doped *channel stop* implant intended to suppress accidental channel formation that could couple the device to its neighbors. The drain contacts are placed over the field oxide to reduce the capacitance to the body, a parasitic that slows response times. These structural details are described later. (After Brews, J.R. 1990. The submicron MOSFET. In *High-Speed Semiconductor Devices,* ed. S.M. Sze, pp. 139–210. Wiley, New York.)

carriers are drawn into a surface channel, forming a conducting path from source to drain. The gate and channel then form two sides of a capacitor separated by the gate insulator. As additional attractive charges are placed on the gate side, the channel side of the capacitor draws a balancing charge of minority carriers from the source and the drain. The more charges on the gate, the more populated the channel, and the larger the conductance. Because the gate *creates* the channel, to ensure electrical continuity the gate must extend over the entire length of the separation between source and drain.

The MOSFET channel is created by attraction to the gate and relies on the insulating layer between the channel and the gate to prevent leakage of minority carriers to the gate. As a result, MOSFETs can be made only in material systems that provide very good gate insulators, and the best system known is the silicon–silicon dioxide combination. This requirement for a good gate insulator is not as important for JFETs and MESFETs where the role of the gate is to *push away* majority carriers, rather than to *attract* minority carriers. Thus, in GaAs systems where good insulators are incompatible with other device or fabricational requirements, MESFETs are used.

A more recent development in GaAs systems is the heterostructure field-effect transistor (HFET) [Pearton and Shah, 1990] made up of layers of varying compositions of Al, Ga, and As or In, Ga, P, and As. These devices are made using molecular beam epitaxy or by organometallic vapor phase epitaxy, expensive methods still being refined for manufacture. HFETs include a variety of structures, the best known of which is the modulation doped FET (MODFET). HFETs are field-effect devices, not MOSFETs, because the gate simply modulates the carrier density in a pre-existent channel between ohmic contacts. The channel is formed spontaneously, regardless of the quality of the gate insulator, as a condition of equilibrium between the layers, just as a depletion layer is formed in a *p–n* junction. The resulting channel is created very near to the gate electrode, resulting in gate control as effective as in a MOSFET.

The silicon-based MOSFET has been successful primarily because the silicon–silicon dioxide system provides a stable interface with low trap densities and because the oxide is impermeable to many environmental contaminants, has a high breakdown strength, and is easy to grow uniformly and reproducibly

Metal-Oxide-Semiconductor Field-Effect Transistor

[Nicollian and Brews 1982]. These attributes allow easy fabrication using lithographic processes, resulting in integrated circuits (ICs) with very small devices, very large device counts, and very high reliability at low cost. Because the importance of the MOSFET lies in this relationship to high-density manufacture, an emphasis of this chapter is to describe the issues involved in continuing miniaturization.

An additional advantage of the MOSFET is that it can be made using either electrons or holes as channel carrier. Using both types of devices in so-called complementary MOS (CMOS) technology allows circuits that draw no DC power if current paths include at least one series connection of both types of devices because, in steady state, only one or the other type conducts, not both at once. Of course, in exercising the circuit, power is drawn during switching of the devices. The flexibility in choosing n- or p-channel devices has enabled large circuits to be made that use low-power levels. Hence, complex systems can be manufactured without expensive packaging or cooling requirements.

10.2 Current-Voltage Characteristics

The derivation of the current-voltage characteristics of the MOSFET can be found in [Annaratone, 1986; Brews, 1981; and Pierret, 1990]. Here a qualitative discussion is provided.

Strong-Inversion Characteristics

In Fig. 10.2 the source-drain current I_D is plotted vs. drain-to-source voltage V_D (the $I - V$ curves for the MOSFET). At low V_D the current increases approximately linearly with increased V_D, behaving like a simple resistor with a resistance that is controlled by the gate voltage V_G: as the gate voltage is made more attractive for channel carriers, the channel becomes stronger, more carriers are contained in the channel, and its resistance R_{ch} drops. Hence, at larger V_G the current is larger.

At large V_D the curves flatten out, and the current is less sensitive to drain bias. The MOSFET is said to be in *saturation*. There are different reasons for this behavior, depending on the field along the channel caused by the drain voltage. If the source-drain separation is short, near or below a micrometer, the usual drain voltage is sufficient to create fields along the channel of more than a few $\times$ 10^4 V/cm. In this case the carrier energy is sufficient for carriers to lose energy by causing vibrations of the silicon atoms composing the crystal (optical phonon emission). Consequently, the carrier velocity does not increase much with increased field, saturating at a value $v_{sat} \approx 10^7$ cm/s in silicon MOSFETs. Because the carriers do not move faster with increased V_D, the current also saturates.

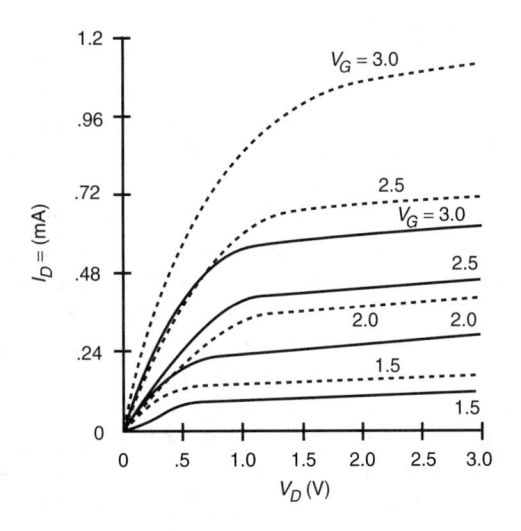

FIGURE 10.2 Drain current I_D vs. drain voltage V_D for various choices of gate bias V_G. The dashed-line curves are for a *long-channel* device for which the current in saturation increases quadratically with gate bias. The solid-line curves are for a *short-channel* device that is approaching *velocity saturation* and thus exhibits a more linear increase in saturation current with gate bias, as discussed in the text.

For longer devices the current-voltage curves saturate for a different reason. Consider the potential along the insulator–channel interface, the surface potential. Whatever the surface potential is at the source end of the channel, it varies from the source end to a value larger at the drain end by V_D because the drain potential is V_D higher than the source. The gate, on the other hand, is at the same potential everywhere. Thus, the difference in potential between the gate and the source is larger than that between the gate and drain. Correspondingly, the oxide field at the source is larger than that at the drain and, as a result, less charge can be supported at the drain. This reduction in attractive power of the gate reduces the number of carriers in the channel at the drain end, increasing channel resistance. In short, we have $I_D \approx V_D/R_{ch}$, but the channel resistance $R_{ch} = R_{ch}(V_D)$ is increasing with V_D. As a result, the current-voltage curves do not continue along the initial straight line, but bend over and saturate.

Another difference between the current-voltage curves for short devices and those for long devices is the dependence on gate voltage. For long devices, the current level in saturation $I_{D,\text{sat}}$ increases quadratically with gate bias. The reason is that the number of carriers in the channel is proportional to $V_G - V_{TH}$ (where V_{TH} is the *threshold voltage*) as is discussed later, the channel resistance $R_{ch} \propto 1/(V_G - V_{TH})$, and the drain bias in saturation is approximately V_G. Thus $I_{D,\text{sat}} = V_D/R_{ch} \propto (V_G - V_{TH})^2$, and we have quadratic dependence. When the carrier velocity is saturated, however, the dependence of the current on drain bias is suppressed because the speed of the carriers is fixed at v_{sat}, and $I_{D,\text{sat}} \propto v_{\text{sat}}/R_{ch} \propto (V_G - V_{TH})v_{\text{sat}}$, a linear gate-voltage dependence. As a result, the current available from a short device is not as large as would be expected if we assumed it behaved like a long device.

Subthreshold Characteristics

Quite different current-voltage behavior is seen in **subthreshold,** that is, for gate biases so low that the channel is in *weak inversion*. In this case the number of carriers in the channel is so small that their charge does not affect the potential, and channel carriers simply must adapt to the potential set up by the electrodes and the dopant ions. Likewise, in subthreshold any flow of current is so small that it causes no potential drop along the interface, which becomes an equipotential.

As there is no lateral field to move the channel carriers, they move by diffusion only, driven by a gradient in carrier density setup because the drain is effective in reducing the carrier density at the drain end of the channel. In subthreshold the current is then independent of drain bias once this bias exceeds a few tens of millivolts, enough to reduce the carrier density at the drain end of the channel to near zero.

In short devices, however, the source and drain are close enough together to begin to share control of the potential with the gate. If this effect is too strong, a drain-voltage dependence of the subthreshold characteristic then occurs, which is undesirable because it increases the MOSFET off current and can cause a drain-bias dependent threshold voltage.

Although for a well-designed device there is no drain-voltage dependence in subthreshold, gate-bias dependence is exponential. The surface is lowered in energy relative to the semiconductor body by the action of the gate. If this *surface potential* is ϕ_S below that of the body, the carrier density is enhanced by a Boltzmann factor $\exp(q\,\phi_S/kT)$ relative to the body concentration, where kT/q is the thermal voltage, ≈ 25 mV at 290 K. As ϕ_S is roughly proportional to V_G, this exponential dependence on ϕ_S leads to an exponential dependence on V_G for the carrier density and, hence, for the current in subthreshold.

10.3 Important Device Parameters

A number of MOSFET parameters are important to the performance of a MOSFET. In this section some of these parameters are discussed, particularly from the viewpoint of digital ICs.

Threshold Voltage

The threshold voltage is vaguely defined as the gate voltage V_{TH} at which the channel begins to form. At this voltage devices begin to switch from off to on, and circuits depend on a voltage swing that straddles this value. Thus, threshold voltage helps in deciding the necessary supply voltage for circuit operation and it also helps in determining the leakage or off current that flows when the device is in the off state.

We now will make the definition of threshold voltage precise and relate its magnitude to the doping profile inside the device, as well as other device parameters such as oxide thickness and flatband voltage.

Threshold voltage is controlled by oxide thickness d and by body doping. To control the body doping, ion implantation is used, so that the dopant-ion density is not simply a uniform extension of the bulk, background level N_B ions/unit volume, but has superposed on it an implanted ion density. To estimate the threshold voltage, we need a picture of what happens in the semiconductor under the gate as the gate voltage is changed from its off level toward threshold.

If we imagine changing the gate bias from its off condition toward threshold, at first the result is to repel majority carriers, forming a surface *depletion layer*, refer to Fig. 10.1. In the depletion layer there are almost no carriers present, but there are dopant ions. In *n*-type material these dopant ions are positive donor impurities that cannot move under fields because they are locked in the silicon lattice, where they have been deliberately introduced to replace silicon atoms. In *p*-type material these dopant ions are negative acceptors. Thus, each charge added to the gate electrode to bring the gate voltage closer to threshold causes an increase in the depletion-layer width sufficient to balance the gate charge by an equal but opposite charge of dopant ions in the silicon depletion layer.

This expansion of the depletion layer continues to balance the addition of gate charge until threshold is reached. Then this charge response changes: above threshold any additional gate charge is balanced by an increasingly strong inversion layer or channel. The border between a depletion-layer and an inversion-layer response, threshold, should occur when

$$\frac{dqN_{\text{inv}}}{d\phi_S} = \frac{dQ_D}{d\phi_S} \tag{10.1}$$

where $d\phi_S$ is the small change in surface potential that corresponds to our incremental change in gate charge, qN_{inv} is the inversion-layer charge/unit area, and Q_D the depletion-layer charge/unit area. According to Eq. (10.1), the two types of responses are equal at threshold, so that one is larger than the other on either side of this condition. To be more quantitative, the rate of increase in qN_{inv} is exponential, that is, its rate of change is proportional to qN_{inv}, and so as qN_{inv} increases, so does the left side of Eq. (10.1). On the other hand, Q_D has a square-root dependence on ϕ_S, which means its rate of change becomes smaller as Q_D increases. Thus, as surface potential is increased, the left side of Eq. (10.1) increases $\propto qN_{\text{inv}}$ until, at threshold, Eq. (10.1) is satisfied. Then, beyond threshold, the exponential increase in qN_{inv} with ϕ_S swamps Q_D, making change in qN_{inv} the dominant response. Likewise, below threshold, the exponential decrease in qN_{inv} with decreasing ϕ_S makes qN_{inv} negligible and change in Q_D becomes the dominant response. The abruptness of this change in behavior is the reason for the term threshold to describe MOSFET switching.

To use Eq. (10.1) to find a formula for threshold voltage, we need expressions for N_{inv} and Q_D. Assuming the interface is held at a lower energy than the bulk due to the charge on the gate, the minority-carrier density at the interface is larger than in the bulk semiconductor, even below threshold. Below threshold and even up to the threshold of Eq. (10.1), the number of charges in the channel/unit area N_{inv} is given for *n*-channel devices approximately by [Brews, 1981]

$$N_{\text{inv}} \approx d_{\text{INV}} \frac{n_i^2}{N_B} e^{q(\phi_S - V_S)/kT} \tag{10.2}$$

where the various symbols are defined as follows: n_i is the intrinsic carrier density/unit volume $\approx 10^{10}/\text{cm}^3$ in silicon at 290 K and V_S is the body reverse bias, if any. The first factor, d_{INV}, is an effective depth of minority carriers from the interface given by

$$d_{\text{INV}} = \frac{\varepsilon_S kT/q}{Q_D} \tag{10.3}$$

where Q_D is the depletion-layer charge/unit area due to charged dopant ions in the region where there are no carriers and ε_S is the dielectric permittivity of the semiconductor.

Equation (10.2) expresses the net minority-carrier density/unit area as the product of the bulk minority-carrier density/unit volume n_i^2/N_B, with the depth of the minority-carrier distribution d_{INV} multiplied in turn by the customary Boltzmann factor $\exp[q(\phi_S - V_S)/kT]$ expressing the enhancement of the interface density over the bulk due to lower energy at the interface. The depth d_{INV} is related to the carrier distribution near the interface using the approximation (valid in weak inversion) that the minority-carrier density decays exponentially with distance from the oxide–silicon surface. In this approximation, d_{INV} is the *centroid* of the minority-carrier density. For example, for a uniform bulk doping of 10^{16} dopant ions/cm^3 at 290K, using Eq. (10.2) and the surface potential at threshold from Eq. (10.7) ($\phi_{TH} = 0.69$ V), there are $Q_D/q = 3 \times 10^{11}$ charges/cm^2 in the depletion layer at threshold. This Q_D corresponds to a $d_{INV} = 5.4$ nm and a carrier density at threshold of $N_{inv} = 5.4 \times 10^9$ charges/cm^2.

The next step in using the definition of threshold, Eq. (10.1), is to introduce the depletion-layer charge/unit area Q_D. For the ion-implanted case, Q_D is made up of two terms [Brews, 1981]

$$Q_D = qN_B L_B [2(q\phi_{TH}/kT - m_1 - 1)]^{\frac{1}{2}} + qD_I \tag{10.4}$$

where the first term is Q_B, the depletion-layer charge from bulk dopant atoms in the depletion layer with a width that has been reduced by the first moment of the implant, namely, m_1 given in terms of the centroid of the implant x_C by

$$m_1 = \frac{D_I x_C}{N_B L_B^2} \tag{10.5}$$

The second term is the additional charge due to the implanted-ion density within the depletion layer, D_I/unit area. The Debye length L_B is defined as

$$L_B^2 \equiv \left[\frac{kT}{q}\right]\left[\frac{\varepsilon_S}{qN_B}\right] \tag{10.6}$$

where ε_S is the dielectric permittivity of the semiconductor. The Debye length is a measure of how deeply a variation of surface potential penetrates into the body when $D_I = 0$ and the depletion layer is of zero width.

Approximating qN_{inv} by Eq. (10.2) and Q_D by Eq. (10.4), Eq. (10.1) determines the surface potential at threshold ϕ_{TH} to be

$$\phi_{TH} = 2(kT/q)\ln(N_B/n_i) + (kT/q)\ln\left[1 + \frac{qD_I}{Q_B}\right] \tag{10.7}$$

where the new symbols are defined as follows: Q_B is the depletion-layer charge/unit area due to bulk body dopant N_B in the depletion layer, and qD_I is the depletion-layer charge/unit area due to implanted ions in the depletion layer between the inversion-layer edge and the depletion-layer edge. Because even a small increase in ϕ_S above ϕ_{TH} causes a large increase in qN_{inv}, which can balance a rather large change in gate charge or gate voltage, ϕ_S does not increase much as $V_G - V_{TH}$ increases. Nonetheless, in strong inversion $N_{inv} \approx 10^{12}$ charges/cm^2, and so in **strong inversion** ϕ_S will be about 7 kT/q larger than ϕ_{TH}.

Equation (10.7) indicates for uniform doping (no implant, $D_I = 0$) that threshold occurs approximately for $\phi_S = \phi_{TH} = 2(kT/q)\ln(N_B/n_i) \equiv 2\phi_B$, but for the nonuniformly doped case, a larger surface potential is needed, assuming the case of a normal implant where D_I is positive, increasing the dopant density. The implant increases the required surface potential because the field at the surface is larger, narrowing the inversion layer and reducing the channel strength for $\phi_S = 2\phi_B$. Hence, a somewhat larger surface potential is needed to increase qN_{inv} to the point that Eq. (10.1) is satisfied. Equation (10.7) would not apply if a significant fraction of the implant were confined to lie within the inversion layer itself. No realistic implant can be confined within a distance comparable to an inversion-layer thickness (a few tens of nanometers), however, and so Eq. (10.7) covers practical cases.

With the surface potential ϕ_{TH} known, the potential on the gate at threshold Φ_{TH} can be found if we know the oxide field F_{ox} by simply adding the potential drop across the semiconductor to that across the oxide. That is, $\Phi_{TH} = \phi_{TH} + F_{ox} d$, where d is the oxide thickness and F_{ox} is given by Gauss' law as

$$\varepsilon_{ox} F_{ox} = Q_D \tag{10.8}$$

There are two more complications in finding the threshold voltage. First, the gate *voltage* V_{TH} usually differs from the gate *potential* Φ_{TH} at threshold because of a work-function difference between the body and the gate material. This difference causes a spontaneous charge exchange between the two materials as soon as the MOSFET is placed in a circuit allowing charge transfer to occur. Thus, even before any voltage is applied to the device, a potential difference exists between the gate and the body due to spontaneous charge transfer. The second complication affecting threshold voltage is the existence of charges in the insulator and at the insulator–semiconductor interface. These nonideal contributions to the overall charge balance are due to traps and fixed charges incorporated during the device processing.

Ordinarily interface-trap charge is negligible (<10^{10}/cm^2 in silicon MOSFETs) and the other nonideal effects on threshold voltage are accounted for by introducing the *flatband voltage* V_{FB}, which corrects the gate bias for these contributions. Then, using Eq. (10.8) with $F_{OX} = (V_{TH} - V_{FB} - \phi_{TH})/d$ we find

$$V_{TH} = V_{FB} + \phi_{TH} + Q_D \frac{d}{\varepsilon_{ox}} \tag{10.9}$$

which determines V_{TH} even for the nonuniformly doped case, using Eq. (10.7) for ϕ_{TH} and Q_D at threshold from Eq. (10.4). If interface-trap charge/unit area is not negligible, then terms in the interface-trap charge/unit area Q_{IT} must be added to Q_D in Eq. (10.9).

From Eqs. (10.4) and (10.7), the threshold voltage depends on the implanted dopant-ion profile only through two parameters, the net charge introduced by the implant in the region between the inversion layer and the depletion-layer edge qD_I, and the centroid of this portion of the implanted charge x_C. As a result, a variety of implants can result in the same threshold, ranging from the extreme of a δ-function spike implant of dose D_I/unit area located at the centroid x_C, to a box type rectangular distribution with the same dose and centroid, namely, a rectangular distribution of width $x_W = 2x_C$ and volume density D_I/x_W. (Of course, x_W must be no larger than the depletion-layer width at threshold for this equivalence to hold true, and x_C must not lie within the inversion layer.) This weak dependence on the details of the profile leaves flexibility to satisfy other requirements, such as control of off current.

As already stated, for gate biases $V_G > V_{TH}$, any gate charge above the threshold value is balanced mainly by inversion-layer charge. Thus, the additional oxide field, given by $(V_G - V_{TH})/d$, is related by Gauss' law to the inversion-layer carrier density approximately by

$$\varepsilon_{ox}(V_G - V_{TH})/d \approx qN_{inv} \tag{10.10}$$

which shows that channel strength above threshold is proportional to $V_G - V_{TH}$, an approximation often used in this chapter. Thus, the switch in balancing gate charge from the depletion layer to the inversion layer causes N_{inv} to switch from an exponential gate-voltage dependence in subthreshold to a liner dependence above threshold.

For circuit analysis Eq. (10.10) is a convenient *definition* of V_{TH} because it fits current-voltage curves. If this definition is chosen instead of the charge-balance definition Eq. (10.1), then Eqs. (10.1) and (10.7) result in an *approximation* to ϕ_{TH}.

Driving Ability and $I_{D, sat}$

The driving ability of the MOSFET is proportional to the current it can provide at a given gate bias. One might anticipate that the larger this current, the faster the circuit. Here this current is used to find some response times governing MOSFET circuits.

MOSFET current is dependent on the carrier density in the channel, or on $V_G - V_{TH}$, see Eq. (10.10). For a long-channel device, driving ability depends also on channel length. The shorter the channel length L, the greater the driving ability, because the channel resistance is directly proportional to the channel length. Although it is an oversimplification, let us suppose that the MOSFET is primarily in saturation during the driving of its load. This simplification will allow a clear discussion of the issues involved in making faster MOSFETs without complicated mathematics. Assuming the MOSFET to be saturated over most of the switching period, driving ability is proportional to current in saturation, or to

$$I_{D,\text{sat}} = \frac{\varepsilon_{\text{ox}} Z \mu}{2dL}(V_G - V_{TH})^2 \tag{10.11}$$

where the factor of two results from the saturating behavior of the $I - V$ curves at large drain biases, and Z is the width of the channel normal to the direction of current flow. Evidently, for long devices driving ability is quadratic in $V_G - V_{TH}$ and inversely proportional to d.

The result of Eq. (10.11) holds for long devices. For short devices, as explained for Fig. 10.2, the larger fields exerted by the drain electrode cause *velocity saturation* and, as a result, $I_{D,\text{sat}}$ is given roughly by [Einspruch and Gildenblat, 1989]

$$I_{D,\text{sat}} \approx \frac{\varepsilon_{\text{ox}} Z v_{\text{sat}}}{d} \frac{(V_G - V_{TH})^2}{V_G - V_{TH} + F_{\text{sat}} L} \tag{10.12}$$

where v_{sat} is the carrier saturation velocity, about 10^7 cm/s for silicon at 290 K, and F_{sat} is the field at which velocity saturation sets in, about 5×10^4 V/cm for electrons and not well established as $\gtrsim 10^5$ V/cm for holes in silicon MOSFETs. For Eq. (10.12) to agree with Eq. (10.11) at long L, we need $\mu \approx 2v_{\text{sat}}/F_{\text{sat}} \approx 400$ cm²/V · s for electrons in silicon MOSFETs, which is only roughly correct. Nonetheless, we can see that for devices in the submicron channel length regime, $I_{D,\text{sat}}$ tends to become independent of channel length L and becomes more linear with $V_G - V_{TH}$ and less quadratic, see Fig. 10.2. Equation (10.12) shows that velocity saturation is significant when $(V_G - V_{TH})/L \gtrsim F_{\text{sat}}$, for example, when $L \lesssim 0.5$ μm if $V_G - V_{TH} = 2.3$V.

To relate $I_{D,\text{sat}}$ to a gate response time τ_G, consider one MOSFET driving an identical MOSFET as load capacitance. Then the current from (Eq. 10.12) charges this capacitance to a voltage V_G in a gate response time τ_G given by [Shoji, 1988]

$$\tau_G = C_G V_G / I_{D,\text{sat}}$$
$$= \left[\frac{L}{v_{\text{sat}}}\right]\left[1 + \frac{C_{\text{par}}}{C_{\text{ox}}}\right]\frac{V_G(V_G - V_{TH} + F_{\text{sat}} L)}{(V_G - V_{TH})^2} \tag{10.13}$$

where C_G is the MOSFET gate capacitance $C_G = C_{\text{ox}} + C_{\text{par}}$, with $C_{\text{ox}} = \varepsilon_{\text{ox}} ZL/d$ the MOSFET oxide capacitance, and C_{par} the parasitic component of the gate capacitance [Chen, 1990]. The parasitic capacitance C_{par} is due mainly to overlap of the gate electrode over the source and drain and partly to fringing-field and channel-edge capacitances. For short-channel lengths, C_{par} is a significant part of C_G, and keeping C_{par} under control as L is reduced is an objective of gate-drain alignment technology. Typically, $V_{TH} \approx V_G/4$, so that

$$\tau_G = \left[\frac{L}{v_{\text{sat}}}\right]\left[1 + \frac{C_{\text{par}}}{C_{\text{ox}}}\right]\left[1.3 + 1.8\frac{F_{\text{sat}} L}{V_G}\right] \tag{10.14}$$

Thus, on an intrinsic level, the gate response time is a multiple of the transit time of an electron from source to drain, which is L/v_{sat} in velocity saturation. At shorter L, a linear reduction in delay with L is predicted, whereas for longer devices the improvement can be quadratic in L, depending on how V_G is scaled as L is reduced.

The gate response time is not the only delay in device switching, because the drain-body p–n junction also must charge or discharge for the MOSFET to change state [Shoji, 1988]. Hence, we must also consider a *drain* response time τ_D. Following Eq. (10.13), we suppose that the drain capacitance C_D is charged by the supply voltage through a MOSFET in saturation so that

$$\tau_D = C_D V_G / I_{D,\text{sat}} = \left[\frac{C_D}{C_G}\right] \tau_G \qquad (10.15)$$

Equation (10.15) suggests that τ_D will show a similar improvement to τ_G as L is reduced, provided that C_D/C_G does not increase as L is reduced. However, $C_{\text{ox}} \propto L/d$, and the major component of C_{par}, namely, the overlap capacitance contribution, leads to $C_{\text{par}} \propto L_{\text{ovlp}}/d$ where L_{ovlp} is roughly three times the length of overlap of the gate over the source or drain [Chen, 1990]. Then $C_G \propto (L + L_{\text{ovlp}})/d$ and, to keep the C_D/C_G ratio from increasing as L is reduced, either C_D or oxide thickness d must be reduced along with L.

Clever design can reduce C_D. For example, various *raised-drain* designs reduce the drain-to-body capacitance by separating much of the drain area from the body using a thick oxide layer. The contribution to drain capacitance stemming from the sidewall depletion-layer width next to the channel region is more difficult to handle, because the sidewall depletion layer is deliberately reduced during miniaturization to avoid *short-channel* effects, that is, drain influence on the channel in competition with gate control. As a result, this sidewall contribution to the drain capacitance tends to increase with miniaturization unless junction depth can be shrunk.

Equations (10.14) and (10.15) predict reduction of response times by reduction in channel length L. Decreasing oxide thickness leads to no improvement in τ_G, but Eq. (10.15) shows a possibility of improvement in τ_D. The *ring oscillator*, a closed loop of an odd number of inverters, is a test circuit whose performance depends primarily on τ_G and τ_D. Gate delay/state for ring oscillators is found to be near 12 ps/stage at 0.1-μm channel length, and 60 ps/stage at 0.5 μm.

For circuits, interconnection capacitances and fan out (multiple MOSFET loads) will increase response times beyond the device response time, even when parasitics are taken into account. Thus, we are led to consider interconnection delay τ_{INT}. Although a lumped model suggests, as with Eq. (10.15), that $\tau_{\text{INT}} \approx (C_{\text{INT}}/C_G)\tau_G$, the length of interconnections requires a *distributed* model. Interconnection delay is then

$$\tau_{\text{INT}} = R_{\text{INT}} C_{\text{INT}} / 2 + R_{\text{INT}} C_G + (1 + C_{\text{INT}}/C_G) \tau_G \qquad (10.16)$$

where R_{INT} is the interconnection resistance, C_{INT} is the interconnection capacitance, and we have assumed that the interconnection joins a MOSFET driver in saturation to a MOSFET load, C_G. For small R_{INT}, the τ_{INT} is dominated by the last term, which resembles Eqs. (10.13) and (10.15). Unlike the ratio C_D/C_G in Eq. (10.15), however, it is difficult to reduce or even maintain the ratio C_{INT}/C_G in Eq. (10.16) as L is reduced. Remember, $C_G \propto Z(L + L_{\text{ovlp}})/d$. Reduction of L, therefore, tends to increase C_{INT}/C_G, especially because interconnect cross-sections cannot be reduced without impractical increases in R_{INT}. What is worse, along with reduction in L, chip sizes usually increase, making line lengths longer, increasing R_{INT} even at constant cross-section. As a result, interconnection delay becomes a major problem as L is reduced. The obvious way to keep C_{INT}/C_G under control is to increase the device width Z so that $C_G \propto Z(L + L_{\text{ovlp}})/d$ remains constant as L is reduced. A better way is to cascade drivers of increasing Z [Chen, 1990; Shoji, 1988]. Either solution requires extra area, however, reducing the packing density that is a major objective in decreasing L in the first place. An alternative is to reduce the oxide thickness d, a major technology objective today.

Transconductance

Another important device parameter is the small-signal transconductance g_m [Malik, 1995; Sedra and Smith, 1991; Haznedar, 1991] that determines the amount of output current swing at the drain that results from a given input voltage variation at the gate, that is, the small-signal gain,

$$g_m = \left.\frac{\partial I_D}{\partial V_G}\right|_{V_D = \text{const}} \tag{10.17}$$

Using the chain rule of differentiation, the transconductance in saturation can be related to the small-signal *transition* or *unity-gain frequency* that determines at how high a frequency ω the small-signal current gain $|i_{\text{out}}/i_{\text{in}}| = g_m/(\omega C_G)$ drops to unity. Using the chain rule,

$$g_m = \frac{\partial I_{D,\text{sat}}}{\partial Q_G}\frac{\partial Q_G}{\partial V_G} = \omega_T C_G \tag{10.18}$$

where C_G is the oxide capacitance of the device, $C_G = \partial Q_G/\partial V_G|_{V_D}$ where Q_G is the charge on the gate electrode. The frequency ω_T is a measure of the small-signal, high-frequency speed of the device, neglecting parasitic resistances. Using Eq. (10.12) in Eq. (10.18) we find that the transition frequency also is related to the transit time L/v_{sat} of Eq. (10.14), so that both the digital and small-signal circuit speeds are related to this parameter.

Output Resistance and Drain Conductance

For small-signal circuits the output resistance r_o of the MOSFET [Malik, 1995; Sedra and Smith, 1991] is important in limiting the gain of amplifiers. This resistance is related to the small-signal drain conductance in saturation by

$$r_o = \frac{1}{g_D} = \left.\frac{\partial V_D}{\partial I_{D,\text{sat}}}\right|_{V_G=\text{const}} \tag{10.19}$$

If the MOSFET is used alone as a simple amplifier with a load line set by a resistor R_L, the gain becomes

$$\left|\frac{v_o}{v_{\text{in}}}\right| = g_m\frac{R_L r_o}{R_L + r_o} \leq g_m R_L \tag{10.20}$$

showing how gain is reduced if r_o is reduced to a value approaching R_L.

As devices are miniaturized, r_o is decreased and g_D increased, due to several factors. At moderate drain biases, the main factor is channel-length modulation, the reduction of the channel length with increasing drain voltage that results when the depletion region around the drain expands toward the source, causing L to become drain-bias dependent. At larger drain biases, a second factor is drain control of the inversion-layer charge density that can compete with gate control in short devices. This is the same mechanism discussed later in the context of subthreshold behavior. At rather high drain bias, carrier multiplication further lowers r_o.

In a digital inverter, a lower r_o widens the voltage swing needed to cause a transition in output voltage. This widening increases power loss due to current spiking during the transition and reduces noise margins [Annaratone, 1986]. It is not, however, a first-order concern in device miniaturization for digital applications. Because small-signal circuits are more sensitive to r_o than digital circuits, MOSFETs designed for small-signal applications cannot be made as small as those for digital applications.

10.4 Limitations on Miniaturization

A major factor in the success of the MOSFET has been its compatibility with processing useful down to very small dimensions. Today channel lengths (source-to-drain spacings) of 0.5 μm are manufacturable, and further reduction to 0.1 μm has been achieved for limited numbers of devices in test circuits, such as ring oscillators. In this section some of the limits that must be considered in miniaturization are outlined [Brews, 1990].

Subthreshold Control

When a MOSFET is in the off condition, that is, when the MOSFET is in subthreshold, the off current drawn with the drain at supply voltage must not be too large in order to avoid power consumption and discharge of ostensibly isolated nodes [Shoji, 1988]. In small devices, however, the source and drain are closely spaced, and so there exists a danger of direct interaction of the drain with the source, rather than an interaction mediated by the gate and channel. In an extreme case, the drain may draw current directly from the source, even though the gate is off (*punchthrough*). A less extreme but also undesirable case occurs when the drain and gate jointly control the carrier density in the channel (*drain-induced barrier lowering*, or drain control of threshold voltage). In such a case, the on–off behavior of the MOSFET is not controlled by the gate alone, and switching can occur over a range of gate voltages dependent on the drain voltage. Reliable circuit design under these circumstances is very complicated, and testing for design errors is prohibitive. Hence, in designing MOSFETs, a drain-bias independent subthreshold behavior is necessary.

A measure of the range of influence of the source and drain is the depletion-layer width of the associated *p–n* junctions. The depletion layer of such a junction is the region in which all carriers have been depleted, or pushed away, due to the potential drop across the junction. This potential drop includes the applied bias across the junction and a spontaneous built-in potential drop induced by spontaneous charge exchange when *p*- and *n*-regions are brought into contact. The depletion-layer width W of an abrupt junction is related to potential drop V and dopant-ion concentration/unit volume N by

$$W = \left[\frac{2\varepsilon_s V}{qN}\right]^{\frac{1}{2}} \tag{10.21}$$

To avoid subthreshold problems, a commonly used rule of thumb is to make sure that the channel length is longer than a minimum length L_{min} related to the junction depth r_j, the oxide thickness d, and the depletion-layer widths of the source and drain W_S and W_D by [Brews, 1990]

$$L_{min} = A[r_j d(W_S W_D)^2]^{\frac{1}{3}} \tag{10.22}$$

where the empirical constant $A = 0.88$ nm$^{-1/3}$ if r_j, W_S, and W_D are in micrometers and d is in nanometers.

Equation (10.22) shows that smaller devices require shallower junctions (smaller r_j), or thinner oxides (smaller d), or smaller depletion-layer widths (smaller voltage levels or heavier doping). These requirements introduce side effects that are difficult to control. For example, if the oxide is made thinner while voltages are not reduced proportionately, then oxide fields increase, requiring better oxides. If junction depths are reduced, better control of processing is required, and the junction resistance is increased due to smaller cross-sections. To control this resistance, various *self-aligned contact* schemes have been developed to bring the source and drain contacts closer to the gate [Brews, 1990: Einspruch and Gildenblat, 1989], reducing the resistance of these connections. If depletion-layer widths are reduced by increasing the dopant-ion density, the *driving ability* of the MOSFET suffers because the threshold voltage increases. That is, Q_D increases in Eq. (10.9), reducing $V_G - V_{TH}$. Thus, increasing V_{TH} results in slower circuits.

As secondary consequences of increasing dopant-ion density, channel conductance is further reduced due to the combined effects of increased scattering of electrons from the dopant atoms and increased oxide fields that pin carriers in the inversion layer closer to the insulator–semiconductor interface, increasing scattering at the interface. These effects also reduce driving ability, although for shorter devices they are important only in the linear region (that is, below saturation), assuming that mobility μ is more strongly affected than saturation velocity v_{sat}.

Hot-Electron Effects

Another limit on how small a MOSFET can be made is a direct result of the larger fields in small devices. Let us digress to consider why proportionately larger voltages, and thus larger fields, are used in smaller devices. First, according to Eq. (10.14), τ_G is shortened if voltages are increased, at least so long as $V_G/L \lesssim F_{sat} \approx 5 \times 10^4$ V/cm. If τ_G is shortened this way, then so are τ_D and τ_{INT}, Eqs. (10.15) and (10.16). Thus, faster response is gained by increasing voltages into the velocity saturation region. Second, the fabricational control of smaller devices has not improved proportionately as L has shrunk, and so there is a larger percentage variation in device parameters with smaller devices. Thus, disproportionately larger voltages are needed to ensure all devices operate in the circuit, to overcome this increased fabricational *noise*. Thus, to increase speed and to cope with fabricational variations, fields get larger in smaller devices.

As a result of these larger fields along the channel direction, a small fraction of the channel carriers have enough energy to enter the insulating layer near the drain. In silicon-based *p*-channel MOSFETs, energetic holes can become trapped in the oxide, leading to a positive oxide charge near the drain that reduces the strength of the channel, degrading device behavior. In *n*-channel MOSFETs, energetic electrons entering the oxide create interface traps and oxide wear out, eventually leading to gate-to-drain shorts [Pimbley et al., 1989].

To cope with these problems *drain engineering* has been tried, the most common solution being the *lightly doped drain* [Chen, 1990; Einspruch and Gildenblat, 1989; Pimbley et al., 1989; Wolf, 1995]. In this design, a lightly doped extension of the drain is inserted between the channel and the drain proper. To keep the field moderate and reduce any peaks in the field, the lightly doped drain extension is designed to spread the drain-to-channel voltage drop as evenly as possible. The aim is to smooth out the field at a value close to F_{sat} so that energetic carriers are kept to a minimum. The expense of this solution is an increase in drain resistance and a decreased gain. To increase packing density, this lightly doped drain extension can be stacked vertically alongside the gate, rather than laterally under the gate, to control the overall device area.

Thin Oxides

According to Eq. (10.22), thinner oxides allow shorter devices and, therefore, higher packing densities for devices. In addition, driving ability is increased, shortening response times for capacitive loads, and output resistance and transconductance are increased. There are some basic limitations on how thin the oxide can be made. For instance, there is a maximum oxide field that the insulator can withstand. It is thought that the intrinsic breakdown voltage of SiO_2 is of the order of 10^7 V/cm, a field that can support $\approx 2 \times 10^{13}$ charges/cm^2, a large enough value to make this field limitation secondary. Unfortunately, as they are presently manufactured, the intrinsic breakdown of MOSFET oxides is much less likely to limit fields than defect-related leakage or breakdown, and control of these defects has limited reduction of oxide thicknesses in manufacture to about 5 nm to date.

If defect-related problems could be avoided, the thinnest useful oxide would probably be about 3 nm, limited by direct tunneling of channel carriers to the gate. This tunneling limit is not well established and also is subject to oxide-defect enhancement due to tunneling through intermediate defect levels. Thus, defect-free manufacture of thin oxides is a very active area of exploration.

Dopant-Ion Control

As devices are made smaller, the precise positioning of dopant inside the device is critical. At high temperatures during processing, dopant ions can move. For example, source and drain dopants can enter the channel region, causing position-dependence of threshold voltage. Similar problems occur in isolation structures that separate one device from another [Primbley et al., 1989; Einspruch and Gildenblat, 1989; Wolf, 1995].

To control these thermal effects, process sequences are carefully designed to limit high-temperature steps. This design effort is shortened and improved by the use of computer modeling of the processes. Dopant-ion movement is complex, however, and its theory is made more difficult by the growing trend to use *rapid thermal processing* that involves short-time heat treatments. As a result, dopant response is not steady state, but transient. Computer models of transient response are primitive, forcing further advance in small device design to be more empirical.

Other Limitations

Besides limitations directly related to the MOSFET, there are some broader difficulties in using MOSFETs of smaller dimension in chips involving even greater numbers of devices. Already mentioned is the increased delay due to interconnections that are lengthening due to increasing chip area and increasing complexity of connection. The capacitive loading of MOSFETs that must drive signals down these lines can slow circuit response, requiring extra circuitry to compensate.

Another limitation is the need to isolate devices from each other [Brews, 1990; Chen, 1990; Einspruch and Gildenblat, 1989; Pimbley et al., 1989; Wolf, 1995], so that their actions remain uncoupled by parasitics. As isolation structures are reduced in size to increase device densities, new parasitics are discovered. A developing solution to this problem is the manufacture of circuits on insulating substrates, *silicon-on-insulator* technology [Colinge, 1991]. To succeed, this approach must deal with new problems, such as the electrical quality of the underlying silicon-insulator interface, and the defect densities in the silicon layer on top of this insulator.

Acknowledgments

The author is pleased to thank R.D. Schrimpf and especially S.L. Gilbert for suggestions that clarified the manuscript.

Defining Terms

Channel: The conducting region in a MOSFET between source and drain. In an *enhancement*-mode, or normally off MOSFET the channel is an inversion layer formed by attraction of minority carriers toward the gate. These carriers form a thin conducting layer that is prevented from reaching the gate by a thin *gate-oxide* insulating layer when the gate bias exceeds *threshold*. In a *buried-channel* or *depletion*-mode, or normally on MOSFET, the channel is present even at zero gate bias, and the gate serves to increase the channel resistance when its bias is nonzero. Thus, this device is based on majority-carrier modulation, like a MESFET.

Gate: The control electrode of a MOSFET. The voltage on the gate capacitively modulates the resistance of the connecting *channel* between the *source* and *drain*.

Source, drain: The two output contacts of a MOSFET, usually formed as *p–n* junctions with the *substrate* or *body* of the device.

Strong inversion: The range of gate biases corresponding to the on condition of the MOSFET. At a fixed gate bias in this region, for low drain-to-source biases the MOSFET behaves as a simple gate-controlled resistor. At larger drain biases, the channel resistance can increase with drain bias, even to the point that the current *saturates*, or becomes independent of drain bias.

Substrate or body: The portion of the MOSFET that lies between the *source* and *drain* and under the *gate*. The gate is separated from the body by a thin *gate insulator*, usually silicon dioxide. The gate modulates the conductivity of the body, providing a gate-controlled resistance between the source and drain. The body is sometimes DC biased to adjust overall circuit operation. In some circuits the body voltage can swing up and down as a result of input signals, leading to body-effect or back-gate bias effects that must be controlled for reliable circuit response.

Subthreshold: The range of gate biases corresponding to the off condition of the MOSFET. In this regime the MOSFET is not perfectly off, but conducts a leakage current that must be controlled to avoid circuit errors and power consumption.

Threshold: The gate bias of a MOSFET that marks the boundary between on and off conditions.

References

The following references are not to the original sources of the ideas discussed in this chapter, but have been chosen to be generally useful to the reader.

Annaratone, M. 1986. *Digital CMOS Circuit Design.* Kluwer Academic, Boston, MA.

Brews, J.R. 1981. Physics of the MOS Transistor. In *Applied Solid States Science, Supplement 2A,* ed. D. Kahng, pp. 1–20. Academic Press, New York.

Brews, J.R. 1990. The Submicron MOSFET. In *High-Speed Semiconductor Devices,* ed. S.M. Sze, pp. 139–210. Wiley, New York.

Chen, J.Y. 1990. *CMOS Devices and Technology for VLSI.* Prentice-Hall, Englewood Cliffs, NJ.

Colinge, J.-P. 1991. *Silicon-on-Insulator Technology: Materials to VLSI.* Kluwer Academic, Boston, MA.

Einspruch, N.G. and Gildenblat, G.S., eds. 1989. *Advanced MOS Devices Physics,* Vol. 18, *VLSI Microstructure Science.* Academic, New York.

Haznedar, H. 1991. *Digital Microelectronics.* Benjamin/Cummings, Redwood City, CA.

Hollis, M.A. and Murphy, R.A. 1990. Homogeneous Field-Effect Transistors. In *High-Speed Semiconductor Devices,* ed. S.M. Sze, pp. 211–282. Wiley, New York.

Malik, N.R. 1995. *Electronic Circuits: Analysis, Simulation, and Design.* Prentice-Hall, Englewood Cliffs, NJ.

Nicollian, E.H. and Brews, J.R. 1982. *MOS Physics and Technology,* Chap. 1. Wiley, New York.

Pearton, S.J. and Shah, N.J. 1990. Heterostructure Field-Effect Transistors. In *High-Speed Semiconductor Devices,* ed. S.M. Sze, pp. 283–334. Wiley, New York.

Pierret, R.F. 1990. *Field Effect Devices,* 2nd ed., Vol. 4, *Modular Series on Solid State Devices.* Addison-Wesley, Reading, MA.

Pimbley, J.M., Ghezzo, M., Parks, H.G., and Brown, D.M. 1989. *Advanced CMOS Process Technology,* ed. N.G. Einspruch, Vol. 19, *VLSI Electronics Microstructure Science.* Academic Press, New York.

Sedra, S.S. and Smith, K.C. 1991. *Microelectronic Circuits,* 3rd ed. Saunders College Publishing, Philadelphia, PA.

Shoji, M. 1988. *CMOS Digital Circuit Technology.* Prentice-Hall, Englewood Cliffs, NJ.

Wolf, S. 1995. *Silicon Processing for the VLSI Era: Volume 3—The Submicron MOSFET.* Lattice Press, Sunset Beach, CA.

11
Solid-State Amplifiers

Timothy P. Hulick
Electrical Engineering Consultant

11.1 Linear Amplifiers and Characterizing Distortion **11**-2
   Compression • Odd Order Intermodulation Distortion
   • Other Distortions
11.2 Nonlinear Amplifiers and Characterizing Distortion **11**-8
   Distortions
11.3 Linear Amplifier Classes of Operation **11**-10
   Class A Amplifier • Class B Amplifier
11.4 Nonlinear Amplifier Classes of Operation **11**-17
   Nonsaturated Class C Amplifier • Saturated Class C Amplifier
   • Class D Voltage Switch-Mode Amplifier

Amplifiers are used in all types of electronic equipment and form the basis from which other active circuits are derived. Designing an oscillator would be impossible without a built-in amplifier. Without amplifiers, any generated signal (if a meaningful signal could be generated without one) would only suffer loss after loss as it passed through a circuit to the point where it would become immeasurably small and useless. In fact, amplifiers are designed and built to counter loss so that deterioration of a signal as it passes through a medium is restored to its former or even greater level while allowing it to do something to the circuit that it just passed through. An amplified signal does not have to be a greater reproduction of itself, although this may be desirable, but only has to be bigger in some way than the one it came from. The form of an amplified signal may not resemble the signal from which it came, but must be controlled by it. The controlling of a big signal by a smaller one defines the notion of **gain**. All useful amplifiers have a gain magnitude equal to or greater than one which counters the effect of gain less than one (or loss) by passive elements of a circuit. Gain of an amplifier is defined as the ratio of output power level to a load, divided by input power level driving the amplifier. In general, those devices that have a gain magnitude greater than one are considered to be *amplifiers*, whereas those of the other case are considered to be *attenuators*. The universally accepted electronic symbol for an amplifier is the triangle, however oriented, and it may or may not have more than one input or output or other connection (s) made to it. A simple amplifier is shown in Fig. 11.1.

Technically speaking, active logic gates found in the various types of logic families are also amplifiers since they offer power gain in their ability to drive relatively low impedance loads, while not acting as a significant load themselves. These are special cases discussed elsewhere in this book.

Amplifiers discussed in this chapter are limited to those that use solid-state devices as the active element. In addition, amplifiers discussed in this chapter are presented in a way leading the reader to think that it is a chapter on *RF* amplifiers. This could not be further from the truth. Examples and presentations made here are meant to be *general*, using the general case where carrier frequencies are not zero. The same rules apply to amplifiers, whether they are used for baseband purposes, in which case the carrier frequency is zero, or are used as RF amplifiers, in which case the carrier frequency is not zero.

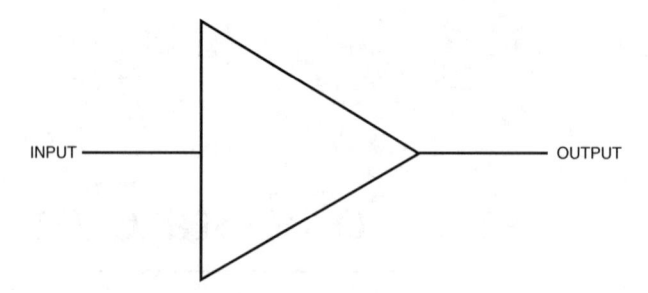

FIGURE 11.1 Any amplifier is universally symbolically represented as a triangle.

11.1 Linear Amplifiers and Characterizing Distortion

Amplifiers overcome loss and may or may not preserve the amplitude–time–frequency attributes of the signal to be amplified. Those that do come under the broad category known as **linear** amplifiers, whereas those that do not come under the category called *nonlinear* amplifiers. A linear amplifier has a frequency domain transfer function so that each point along the time-domain curve that defines the input signal perfectly maps into the output signal in direct proportion with constant time delay. This can only happen if a plot of output vs. input is a straight line, hence, the term linear. Since a straight line function represents only the ideal case, there are degrees of linearity that define the usefulness of such a thing in a particular application. What is considered linear in one application may not be linear enough in another. Linearity is measured in the amount of distortion of many different kinds, again, according to the application.

Compression

For an amplifier to be considered linear, the amplified signal output must be a faithful reproduction of the input signal. Without any *fixes* applied, such as negative feedback, the basis linear amplifier has limitations that are functions of its load impedance, supply voltage, input bias current (or voltage depending on device type), and its ability to just make the power asked of it. With sufficiently large drive, any device can be driven beyond its limits to the point of burnout, and so it is important to know the limitations of any amplifier. The onset of output signal compression is the first sign that a linear amplifier is being driven beyond its linear limits, that is, the transfer function curve is beginning to slope away from a straight line such that the output *compresses* compared to what it should be. Figure 11.2 illustrates the normal straight line path compared to the **compression** taking place when the amplifier is being driven too hard.

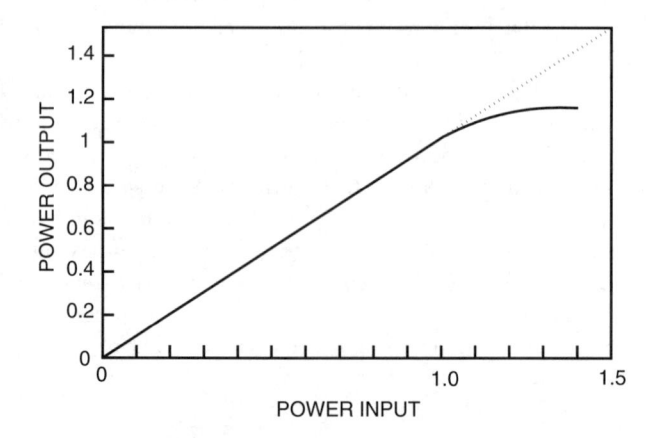

FIGURE 11.2 In the normalized plot of power output vs. power input, the amplifier is seen to be linear in the region from 0 to 1.0. Beyond this point, the output becomes compressed and finally flattens.

Linear amplifiers are characterized according to their 1-dB power gain compression point. Simply put, the 1-dB compression point is that power output from which a 9-dB decrease is realized when the drive level is decreased by 10 dB. A single-frequency continuous wave (CW) signal is usually used as the driving signal. When performing this test it is best to begin with a low-level signal far away from the compression region. The drive level is increased in 1-dB steps while monitoring the linear tracking of input and output powers. Steps taken in 1-dB increments are adequate and beneficial when looking for less than a 1-dB increase in output power as the compression region is reached. At the onset of compression, the drive should be increased by 1 dB more. At this point drive may be decreased by 10 dB. If the output power decreases by more than 9 dB, the 1-dB compression point is still higher. Increase drive by a few tenths of a decibel and try again. Finding the 1-dB compression point should be done by decreasing drive, not increasing it. Not knowing where it is, an increase this large could put the amplifier well into saturation (compression) and perhaps permanent damage may result.

The 1-dB compression point should be noted. Generally excellent linear operation will be 10 dB or more below this point, but more power may be gotten from an amplifier if the compression point is approached more closely and compensated for by a low-level precorrector (expander) circuit before the input to the amplifier.

Odd Order Intermodulation Distortion

Another very important measure of linearity is the relative level of generated odd order intermodulation products in the amplifier. For intermodulation to take place, two or more signals must exist at the same time as a combined input signal. They must be at different frequencies, but close enough to each other so that they both fit within the passband of the amplifier. A suitable example input signal is of the form of Eq. (11.1)

$$S_i(t) = A\cos(2\pi f_1 t) + B\cos(2\pi f_2 t) \tag{11.1}$$

where A and B are the amplitudes of the independent input signals of frequencies f_1 and f_2, respectively.

In the absence of any nonlinearity in the amplifier transfer function, the output is of the same form as the input, and there is no change except that A and B are multiplied by the gain of the amplifier G

$$S_o(t) = GA\cos(2\pi f_1 t) + GB\cos(2\pi f_2 t) \tag{11.2}$$

If the transfer function is not perfectly linear, the input components are also multiplied together within the amplifier and appear at the output. The amplifier actually behaves like an ideal amplifier shunted (input to output) by a mixer, as shown in Fig. 11.3. The expression showing all frequency components out of the amplifier is

$$S_o(t) = GA\cos(2\pi f_1 t) + GB\cos(2\pi f_2 t) + \sum_{m=\text{int}}\sum_{n=\text{int}} c_{mn} AB\cos(2\pi m f_1 t)\cos(2\pi n f_2 t)$$

$$S_o(t) = GA\cos(2\pi f_1 t) + GB\cos(2\pi f_2 t)$$
$$+ \sum_{m=\text{int}}\sum_{n=\text{int}}\left[\frac{1}{2}c_{mn} AB\cos 2\pi(mf_1 + nf_2)t + \frac{1}{2}c_{mn} AB\cos 2\pi(mf_1 - nf_2)t\right] \tag{11.3}$$

where c_{mn} is the amplitude of the product component at frequency $f_{mn} = (mf_1 + nf_2)$ and is determined by the degree of nonlinearity.

Not that m and n can be any positive or negative integer and that the frequency components f_{mn} closest to the original signal frequencies are for m or n negative, but not both, $|m| + |n| =$ odd and m and n differ only by one. Generally, the amplitudes of the multiplication (intermodulation product) frequency components at f_{mn} are greater for $|m| + |n| = 3$ than for $|m| + |n| = 5$. Also the magnitudes of the $|m| + |n| = 5$ components are greater than the $|m| + |n| = 7$ components. The difference frequencies for the sum of m

and n being odd are special because these frequency components may be difficult to filter out. Even order components, that is, when $|m| + |n|$ = even and the sum frequencies are far away in frequency, therefore, are easy to filter and generally are of no consequence. Because of this, odd order **intermodulation distortion** components are of importance to the amplifier linearity whereas even order components are not. Specifically, the third-order components are of concern because they are the closest in frequency to the original signals and are of the greatest amplitude. Figure 11.4 is a view of the relative positions and amplitudes of odd order intermodulation products around the original signals; their frequencies are listed in Table 11.1.

The CW signal used to find the 1-dB compression point sets the maximum **peak envelope power (PEP)** with which the amplifier should be driven. The PEP of Eq. (11.1) is $(A + B)^2$, whereas the PEP of

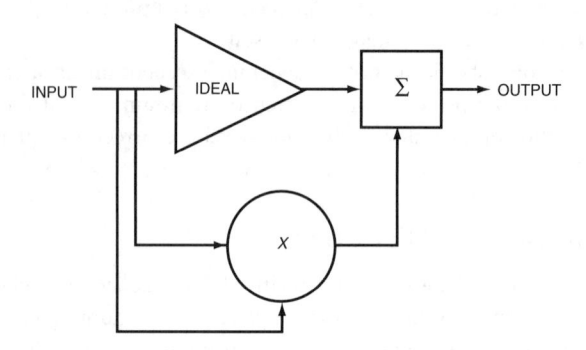

FIGURE 11.3 For intermodulation distortion analysis, the amplifier under test may be thought of as an ideal amplifier shunted by a mixer that creates the intermodulation product components. Multiple frequency signal components make up the inputs to the mixer.

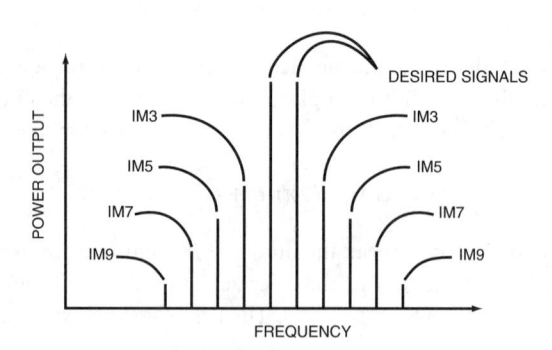

FIGURE 11.4 Example output spectral components are shown for a real linear amplifier with odd order intermodulation distortion. A two-frequency (two-tone) input signal is shown amplified at the output along with odd order difference frequencies generated within the amplifier.

TABLE 11.1 Some Odd Order Intermodulation Distortion Components to the Ninth Order

Third Order	Fifth Order	Seventh Order	Ninth Order	Comment
$2f_1 - f_2$	$3f_1 - 2f_2$	$4f_1 - 3f_2$	$5f_1 - 4f_2$	Near
$2f_2 - f_1$	$3f_2 - 2f_1$	$4f_2 - 3f_1$	$5f_2 - 4f_1$	Near
$f_1 - 2f_2$	$2f_1 - 3f_2$	$3f_1 - 4f_2$	$4f_1 - 5f_2$	Negative
$f_2 - 2f_1$	$2f_2 - 3f_1$	$3f_2 - 4f_1$	$4f_2 - 5f_1$	Negative
$f_1 + 2f_2$	$2f_1 + 3f_2$	$3f_1 + 4f_2$	$4f_1 + 5f_2$	Far away
$f_2 + 2f_1$	$2f_2 + 3f_1$	$3f_2 + 4f_1$	$4f_2 + 5f_1$	Far away

each frequency component is A^2 or B^2. The CW signal used in the compression test would be, for example, C^2. Using a *two-tone* signal with equal amplitudes such that $A = B$, the PEP becomes $4A^2$ or $4B^2$. The PEP of the two-tone signal should not exceed the PEP of the CW signal, or the output will enter the compressed region more than anticipated. If $4A^2 = C^2$, then the PEP needed in each of the two tones is one-fourth of the equivalent CW PEP of the single-tone compression test signal, or A^2. For this reason the two-tone test signals are individually 6 dB below the compression test limit chosen.

The frequencies of the two tones are spaced such that the third-order intermodulation products commonly known as the IM3 products fall within the passband of the amplifier. They are not to be suppressed by means of any filters or bandwidth shaping circuits. As the amplifier is driven by the two-tone test signal, the output is observed on a spectrum analyzer. The transfer function of one of the desired signals is plotted, whereas the level of one of the IM3 signals is also plotted directly below that. The IM3 point plotted in the *power output vs. power input* space indicates that it could have come from an input signal that already had the IM3 component in it and it was just amplified and that the amplifier was not the source of the IM3 component at all. Of course, this is not true, but the IM3 curve, as it is derived along with the real transfer function as points are plotted at other power levels, begins to show itself as an imaginary transfer function. At increasing power levels, the loci of the two curves may be linearly extended to a crossover point where the desired signal output power equals the IM3 level. An example plot is shown in Fig. 11.5. Linear amplifiers are never driven this hard, but this point is specified in decibels relative to a milliwatt or dBm for every linear amplifier and used along with the gain (slope of the real transfer function) to recreate a plot such as that of Fig. 11.5.

The level of IM3 may be related to the PEP of the two tones, in which case it is referred to as (X) dB_{PEP}, or it may be related to the PEP of either of the two tone levels that are, of course, 6 dB lower than the PEP value. When expressed this way, it is referred to as (X) dB_c, referenced to one of the carrier or tone levels. It is important to know which is being used.

Although the two-tone test has become a standard to determine the third-order intercept point, it does not tell the whole story. The two-tone test applies where only two tones are expected in the signal to be amplified. In the case of a modern television signal, for example, there are three tones of constant amplitude: the picture carrier, commonly called the *visual carrier*; the **color subcarrier**; and the sound carrier, commonly called the **aural subcarrier**. Their respective equivalent CW amplitudes in most television formats around the world are −8 dB, −17 dB and −10 dB, respectively, referenced to the peak of the synchronization (sync) level. The PEP level is the PEP of the peak of sync and is used as the reference, but for common visual/aural amplification, the combined peak of sync value is an additional 73% greater.

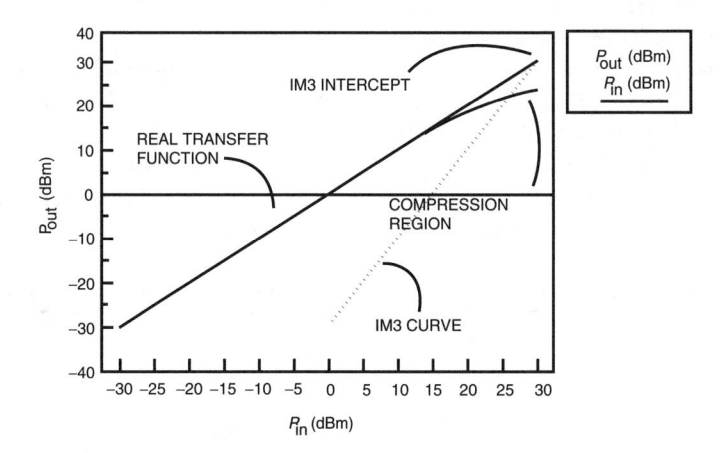

FIGURE 11.5 The output power where the IM3 line and the real transfer function line intersect is called the IM3 intercept point. The slope of the IM3 line is twice that of the transfer function so that for every decibel increase in output power, the IM3 level increases by 3 dB.

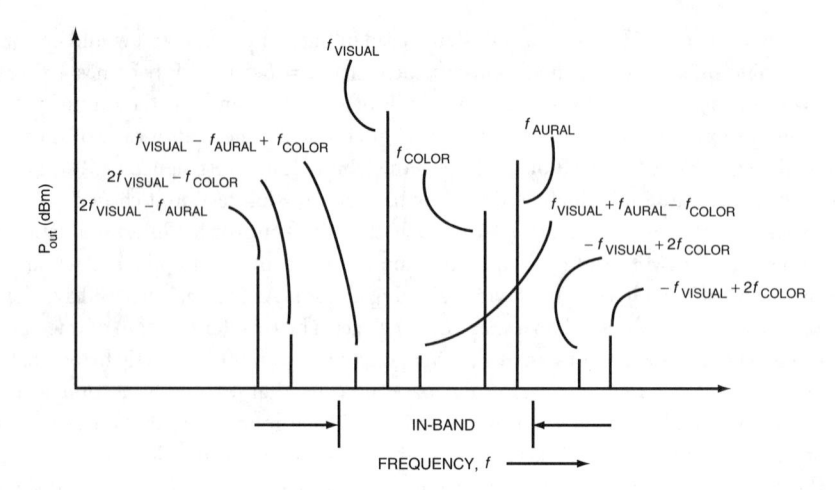

FIGURE 11.6 A three-tone test is used when evaluating a linear amplifier for television applications. The tones are set to the level of the carriers making up the signal, for example, –8-dB visual, –17-dB color, and –10-dB aural, relative to peak of sync. Some IM3 products fall in between the carriers and, therefore, are in-band and impossible to filter out.

If the peak of sync level is considered to be normalized to one, and the aural subcarrier is –10 dBm down from that, it follows that

$$P_{PEP} = (\sqrt{P_{PEP\ VISUAL}} + \sqrt{P_{AURAL}})^2$$
$$P_{PEP} = (\sqrt{1} + \sqrt{0.1})^2 = 1.73 \qquad (11.4)$$

The additional PEP of the color subcarrier is insignificant at 17 dBm down from the peak of sync value. In television, however, the level of intermodulation is related to the peak of sync value as if the aural subcarrier were not present. A three-tone test is commonly used in this industry with the tone levels set to their respective values and placed at the proper frequencies. The visual carrier frequency is taken as the reference, whereas the color subcarrier is about 3.579 MHz higher and the aural subcarrier is 4.5 MHz above the visual carrier. When three tones are used for an intermodulation test, it is important to realize that some of the IM products will fall in between the original three signals. This is not the case when only two tones are used. For IM distortion products in between desired signals, filtering these components to an acceptable level is not possible since they fall on top of sideband information of one or more subcarriers. Concentrating on enhanced linearity and precorrection offers the only possible acceptable performance solutions. Generally, in-band IM products, as they are called, are considered acceptable when they are at least –52 dB lower than the peak of sync level using an average gray picture format. Figure 11.6 shows the placement of IM3 products when three tones are present.

Table 11.2 lists the in-band and nearby IM3 products for an amplified three-tone signal where the three carriers (tones) are designated f_V, f_C and f_A for *visual, color,* and *aural*. One peculiarity is that

TABLE 11.2 Some Third-Order Intermodulation Products for a Three-Tone Signal

IM3 In-band	IM3 Out-of-Band	Out-of-Band Comment
$f_V + f_A - f_C$	$f_V + 2f_A$	Far away
$f_V - f_A + f_C$	$f_V + 2f_C$	Far away
	$f_V - 2f_A$	Negative, only exists mathematically
	$f_V - 2f_C$	Negative, only exists mathematically
	$2f_V - f_A$	Sideband Regeneration
	$2f_V - f_C$	Sideband Regeneration

Solid-State Amplifiers

$2f_v - f_a$ and $2f_v - f_c$ are also IM3 products, but appear to look as if they were lower sideband images of the legitimate signals at f_a and f_c, respectively. Although these components arise due to intermodulation, they were originally present when the color and aural subcarriers modulated the visual carrier before they were filtered out by a vestigial sideband filter to conform to television standards. This reappearance of the lower sidebands is sometimes called *sideband regeneration* because of where they are placed even though they are third-order intermodulation products.

Modern communications systems, such as those used by the telephone industry, make use of large numbers of carriers through one amplifier placing very demanding requirements on intermodulation performance. For a two-tone application, it was shown that the PEP of the individual carriers is 6 dB down from the PEP of the combined carriers. IM3 levels are 6 dB higher when referenced to either of these two carriers compared to the combined PEP of the two tones. For n equal amplitude carriers of power P, the PEP of the combined carriers is shown to be

$$\text{PEP} = n^2 P$$

or that each carrier's peak envelope power contribution to the total PEP becomes

$$P = \frac{\text{PEP}}{n^2} \qquad (11.5)$$

If 16 carriers are used to drive an amplifier capable of 100-W PEP (50 dBm), the individual carriers cannot exceed about 0.4 W (26 dBm). This is a 24-dB degradation in IM3 level relative to the individual carriers. If Δ is the relative increase in IM3 in dB due to a multicarrier signal and n is the number of carriers, it follows that

$$\Delta = 20 \log n$$

For this amplifier passing 16 carriers, requiring the IM3 levels to be 60 dB$_c$ to prevent cross talk, the IM3 level relative to total PEP must be $60 + 20 \log 16 = 84$ dB$_{\text{PEP}}$.

Other Distortions

Odd order intermodulation distortion and 1-dB gain compression go a long way to characterizing the linearity of a linear amplifier, but they are by no means the only ones. Other distortion types tend to be application specific relating to their importance. For video signals other distortion types include the following:

- Differential gain
- Differential phase
- High-frequency to low-frequency delay
- Group delay
- Incidental carrier phase modulation
- Cross modulation (one carrier modulation to another)
- RF harmonic distortion
- Total harmonic distortion

Each of these is briefly discussed, but for a more in-depth understanding, the reader is referred to more specialized sources of information [Craig, 1994].

Differential gain distortion is a distortion of the amplitude information on a subcarrier in the presence of amplitude modulation of another subcarrier or main carrier. *Differential phase distortion* is a distortion of the phase information on a subcarrier in the presence of amplitude modulation of another subcarrier or main carrier. The former is measured in percent whereas the latter in degrees.

High-frequency to low-frequency delay is a distortion caused by the difference in time delay presented to high-frequency sidebands compared to low-frequency sidebands when it is important that their time relationship not be changed. A distortion of this type causes misalignment of subcarrier information with main carrier or other subcarriers and is measured in time.

Group delay distortion has the same effect as high-frequency to low-frequency delay, but is caused by associated filters (notch, bandpass, etc.) that may be built into an amplifier. Sudden changes of transfer function of a filter such as that near a notch filter frequency will cause sudden changes in group delay for amplifier frequency components nearby. It is also measured in time.

Incidental carrier phase modulation (ICPM) could be characterized as AM to PM conversion taking place within an amplifier. As an amplifier passes an amplitude modulated signal, dynamic capacitances within the amplifying device itself act in conjunction with fixed impedances in the amplifier circuit to cause dynamic time constants leading to carrier phase shift with amplitude modulation. ICPM is measured in degrees relative to some clamped amplitude level.

Cross modulation is a distortion caused by the modulation of one carrier through the amplifier onto another subcarrier or carrier. A cause of cross modulation is the mixer or multiplier behavior of real amplifiers modeled in Fig. 11.3. If two or more carriers are multiplied together, at least one of which is modulated with AM, its modulation sidebands may appear on the other. This is different from intermodulation distortion where sum and difference carrier frequencies arise whether they are modulated or not, but the mechanism is the same. Cross modulation occurs in linear amplifiers when the signal levels at the input are simply too high in amplitude for the amplifier to pass without *mixing*. Cross modulation is also caused when the amplifier modulates the power supply voltage as supply current changes with amplitude modulated signal level. If an AM signal and a frequency modulated (FM) or phase modulated (PM) signal are being amplified simultaneously, the AM signal causes a modulated current draw from the supply in certain classes of linear amplifiers (see Class B and Class AB amplifiers later in this chapter). This in turn causes a corresponding modulated voltage drop across the internal resistance of the supply to the amplifier causing synchronous amplitude modulation of the FM or PM signal. Cross modulation is measured in terms of percent AM in the case where AM is the result, or, frequency deviation in the case where FM is the result.

When the peaks of the instantaneous RF carrier enter the compression region, the RF carrier sinusoidal waveform is compressed to something other than a perfect sinusoid. When this happens, a Fourier series of harmonic components is generated leading to RF *harmonic distortion*. These are far away in frequency and easy to filter out, but may cause significant degradation of power efficiency since it is power that cannot be used and frequently is as high as 10% of the total RF power coming out of the amplifier. With no harmonic filter, a frequency insensitive power meter may indicate 100 W, but only 90 of these 100 W may be at the fundamental frequency.

In RF amplifiers, harmonics of the carrier frequencies are located at integer multiples of those frequencies and are easy to filter out. Harmonics of the modulating (baseband) signals may fall within the passband of the amplifier. Distortion by the appearance of baseband harmonics will arise out of a real amplifier when they did not exist going in. For example, a typical AM radio amplifier needs a passband of 30 kHz to faithfully pass all of the sidebands out to 15 kHz for quality amplification of a music modulated carrier. If just a single note at 1 kHz is the modulating signal, there is room for 15 harmonics to be passed along should the harmonic distortion create them. What an amplifier does to baseband signals is characterized by *total harmonic distortion*, which is expressed as a percent of total harmonic power to fundamental power. It is an envelope or baseband distortion, not an RF distortion.

11.2 Nonlinear Amplifiers and Characterizing Distortion

Any amplifier that significantly degrades amplitude modulation is a nonlinear amplifier. Linear amplifiers are supposed to be perfect, whereas nonlinear amplifiers are deliberately not this way. Usually a nonlinear amplifier is optimized in one or more of its characteristics, such as efficiency or ability to make power no matter how distorted. Nonlinear amplifiers have their place where a single tone input signal or one that is not amplitude modulated (except for the case where the modulation is a full on or full off signal) is the signal to be amplified. An example of this is a signal that is only frequency modulated. Specifically, an RF carrier that is deviated in frequency does not require *amplitude* linearity. Even though an FM signal may be carrying beautiful music, it is always only a single carrier that is moving around in frequency, hopefully symmetrically about the carrier frequency or *rest point*. Instantaneously it is only at one frequency. This is important to understand and often thought of only casually. The instantaneous frequency of a deviated carrier is singular even if there are subcarriers as part of the baseband signal providing the modulation. The fact remains that instantaneously, there is simply no other carrier to intermodulate with. For this reason, intermodulation cannot exist and so there is no intermodulation distortion. There is no distortion of the modulating signal because of amplitude compression. There is even no distortion of the frequency modulation if there is simultaneous amplitude modulation as long as the AM is not near 100% where the carrier is made to *pinch off* or disappear at times. Phase modulation amplification is affected in the same way as frequency modulation amplification through nonlinear amplifiers. Nonlinear amplifiers are also very good at making high power much more efficiently than linear amplifiers. Industrial RF heating applications also do not care about linearity since the purpose is to make high-power RF at a particular frequency in the most efficient way possible. If all these good things are attributable to nonlinear amplifiers amplifying frequency or phase modulated signals why are they not used universally and forget about AM? References on the subject of modulation methods will indicate that theoretically FM and PM require infinite bandwidth and practically five or six times the bandwidth of an equivalent AM signal carrying the same information. Amplitude modulation methods have their place, with their linearity requirements, whereas nonamplitude modulation methods have theirs.

Distortions

Just as there are those conditions that cause distortion in linear amplifiers, there are other conditions that cause distortion in nonlinear amplifiers. For example, it was stated that the bandwidth requirement for FM or PM is much greater than for AM. In fact, infinite bandwidth is the ideal case, when in practice, of course, this is not possible. The bandwidth, however, must be wide enough to pass enough of the sideband components so that harmonic distortion of the recovered baseband signal at the receiving end is below a specified limit. Therefore, distortion of the modulated signal can only be avoided by ensuring that the bandwidth of the nonlinear amplifier is sufficient. The bandwidth required is estimated by the well-known *Carson's Rule* [Taub and Schilling, 1986] which states that

$$B = 2(\Delta f + f_m) \qquad (11.6)$$

where B is the bandwidth that will carry 98% of the sideband power, Δf is the frequency deviation of the carrier, and f_m is the highest frequency component of the modulating signal.

Other distortions that affect an FM or PM signal take place in the modulator or source of the modulated signal to be amplified, not in the amplifier, thus they will not be discussed in this chapter.

Linear amplifiers are used wherever the integrity of the signal to be amplified must be maintained to the greatest extent possible. Minimizing the distortion types presented in Sec. 11.1 will optimize a linear amplifier so that it can be used for its intended purpose, but an amplifier must be classified as a linear amplifier in the first place or it will be impossible to achieve *acceptable linear performance*. Linear is a relative term implying that distortion is imperceivable for the application. Relative linearity is classified alphabetically from A to Z with A being the most linear. Not all letters are used, thus there are enough of them available for future classification as new amplifier types are discovered. The most widely used linear classes of amplifiers are Class A, Class AB, and Class B. Other classes are BD, G, H, and S.

Often while perusing through the literature, two opposing phrases may appear even within the same paragraph. They are *small signal* and *large signal*. Nebulous words such as these are used among amplifier design engineers and appear to be unscientific in a most unlikely place, but they do have definition. These terms are used because active solid-state amplifying devices have a small signal behavior and a large signal behavior. *Small signal* behavior means simply that the signal going through a transistor is small enough in magnitude so that the transistor may be modeled in terms of equivalent linear elements. The signal is not large enough to be noticed by the modeled elements of the transistor to cause them to change value. Using the model provides predictable behavior. In addition, the signal is not large enough to cause departure from the user's definition of what is linear enough. Best possible linearity requires that the compression region shown in Fig. 11.2 or Fig. 11.5 be avoided, but if it is acceptable, the small signal domain may be extended into this region of the transfer function. The term small signal is confined to Class A amplifiers and has no meaning in the other classes. The dividing line between small and large signals is at the power level below which the transistor equivalent elements are linear components and the amplifier is still considered linear according to the needs of the application.

11.3 Linear Amplifier Classes of Operation

Class A Amplifier

The term Class A does not refer to whether an amplifier is used in a small or large signal application even though all small signal amplifiers are Class A amplifiers. A Class A amplifier could also be used for large signals if the linear model no longer applies. Class A refers to the conduction angle of the amplifying device with respect to the driving signal and how it is biased to achieve this. For an amplifying device to be operating in Class A, it must be biased so that it is conducting current throughout the entire swing of the signal to be amplified. In other words, it is biased so that extremes of the amplitude of the driving signal never cause the device to go to zero **current** draw (cut off) and never cause the voltage across the output of the device to go to zero (**saturation**). In this manner, the device is always active to follow the waveform of the driving signal. Intuitively, it has the best chance at being linear compared to other bias conditions to be presented later. Notice that the word *device* is used when addressing the part that does the amplifying. **Class of operation** has nothing to do with the device type. The requirements for Class A operation are the same whether the device is a tetrode vacuum tube, bipolar junction transistor, field effect transistor, klystron, or anything else, and the mathematical behavior and theoretical limits of a device operated in Class A are the same regardless of the device type. It is true, however, that some devices are better than others in practical linearity, which allows variations in power efficiency and other performance parameters, but not in theoretical limits. The more general statement is that class of operation has nothing to do with device type, but only how it is biased, which, in turn, determines the **conduction angle**.

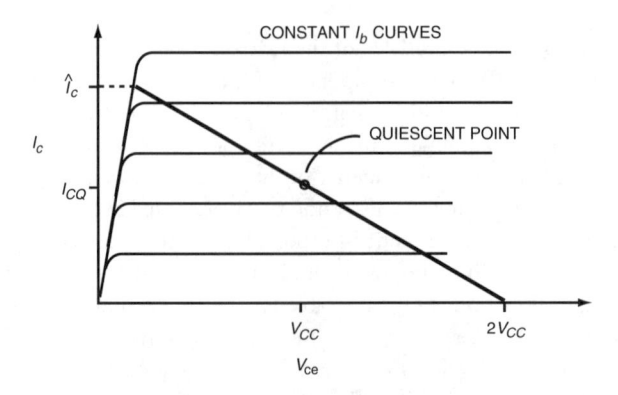

FIGURE 11.7 The I_c vs. V_{ce} plot for a bipolar junction transistor biased for Class A operation is shown with the static quiescent point halfway between voltage and current limits for symmetrical excursions along the loadline. Strict Class A operation requires that the ends of the loadline not be exceeded.

The operation of the Class A amplifier is easily understood if it is analyzed in graphical form. A bipolar junction NPN transistor is assumed. Figure 11.7 is a plot of collector current I_c vs. collector to emitter voltage V_{ce} for various values of constant base current I_b for a common emitter configuration. There is no resistor in the collector bias circuit so that the transistor quiesces at the supply voltage V_{CC}. To allow a symmetrical current and voltage swing about the quiescent point, the quiescent point is chosen at $\frac{1}{2}$ ($2V_{CC}$) or V_{CC} and $\frac{1}{2}\hat{I}_c$, the peak value of collector current. I_{CQ} is the quiescent collector current. Collector–emitter saturation voltage is assumed to be zero. The negative reciprocal of the correct *load resistance* is the slope of the loadline plotted to allow maximum power transfer to the load. The loadline must, therefore, connect $2V_{CC}$ at $I_c = 0$ with $\hat{I}_c$ at $V_{ce} = 0$. Class A operation dictates that the ends of the loadline cannot be exceeded, but approaching the ends while maintaining linearity is possible if the harmonic power generated is filtered out so that it does not reach the load. If the peak of the AC component of the collector–emitter voltage is $\hat{V}_{ac}$ and the load resistance is R, then the output power P_{out}, may be calculated to be

$$P_{out} = \frac{\hat{V}_{ac}^2}{2R} \tag{11.7}$$

The DC input power from the power supply P_{in}, is calculated from

$$P_{in} = V_{CC}I_{CQ} = \frac{V_{CC}^2}{R} \tag{11.8}$$

Comparing Eq. (11.7) with Eq. (11.8), it is obvious that

$$P_{out} = \frac{1}{2}P_{in} \tag{11.9}$$

when $\hat{V}_{ac} = V_{CC}$. This occurs when the collector current and collector–emitter voltage are driven to the limits. The maximum theoretical power efficiency η_{max} is

$$\eta_{max} = \frac{P_{out}}{P_{in}} = \frac{1}{2} \tag{11.10}$$

And, of course, this number decreases from the maximum value as $\hat{V}_{ac}$ is decreased from the end limits of the loadline.

In practice, the limits imposed on $\hat{V}_{ac}$ are usually much less than that to provide an efficiency of 50%. A Class A amplifier used as a power amplifier typically operates at 10–20% efficiency in order to maintain third-order intermodulation distortion down to required levels for most communications applications. Since the Class A stage is usually used as the driver in a cascaded series of power amplifiers, the class A stage must be very clean. The IM3 distortion products that the class A stage produces are amplified by follow-on stages and the distortion only gets worse as more IM3 is produced. At UHF, as of this writing, it is practical to design and construct Class A power amplifiers using one bipolar junction NPN transistor that will provide 13 dB of gain at 40-W PEP with −30 dB IM3 performance using the two-tone test. Of course, any number of these may be power combined to produce hundreds of watts. After combining four or eight Class A stages, it is no longer practical to continue, since power combining losses become significant and adding stages only serves to cancel out losses.

Looking at the amplifier as a whole and not just the transistor, the input and output impedances are designed to be some value: 50 Ω for power RF amplifiers, 75 Ω for video, 600 Ω for audio or some other standard value for interfacing with the outside world. Whatever value the amplifier is designed for, it must be a good resistive load to whatever is driving it, and it must work into a load designed to be its load resistance. The actual impedances of the transistor input and the load presented to its output usually

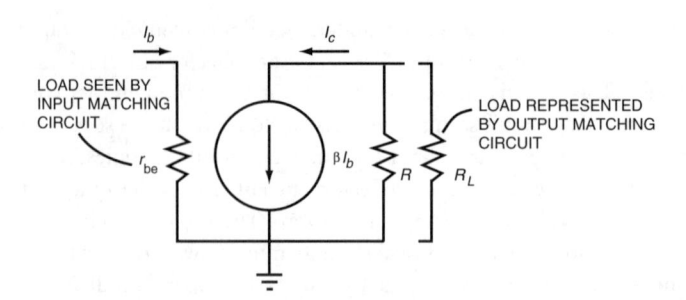

FIGURE 11.8 The Class A amplifier transistor can be modeled as a current controlled current source with internal output resistance R. When $R = R_L$, half of the collector power is dissipated in R and the other half in R_L, providing a maximum efficiency η_{max} of 50%. This agrees with Eq. (11.9). Not shown are the matching circuits transforming r_{be} and R_L to a standard impedance, such as 50 Ω.

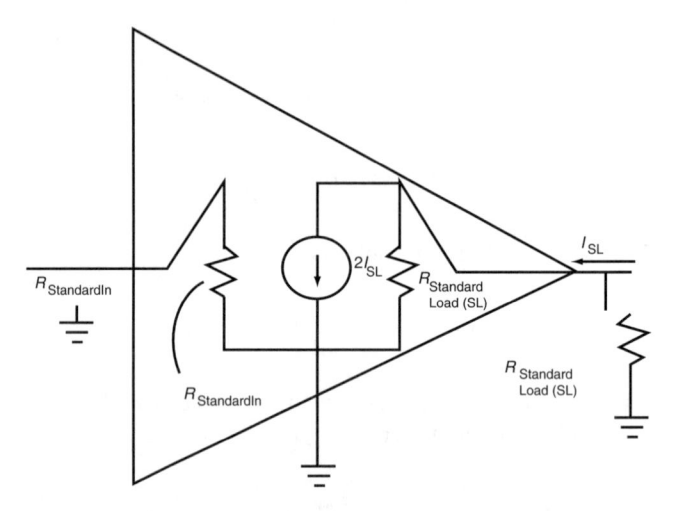

FIGURE 11.9 The Class A transistor amplifier is modeled as a resistive load to its driver whereas the output impedance seen at the output port is equal to the standard load resistance. The input and output impedances may be measured from the outside looking in with a network analyzer or impedance bridge.

are much different than the external impedances, called the **standard**. In fact, for a power amplifier these actual impedances could be a fraction of an ohm. Input and output matching circuits connect these very low-impedance values to the standard. Looking into either port of a Class A amplifier with an impedance measuring device such as an impedance bridge or network analyzer will allow the impedance standard to be measured when the amplifier supply voltages are applied and the transistor is operating at its quiescent point. A return loss of 20 dB or more across the frequency band of interest signifies a *good* amplifier. In the case of Class A, the entire amplifier may be modeled to be a resistor of the chosen standard value at the input $R_{StandardIn}$ and a current controlled current source shunted by the standard resistance value $R_{StandardLoad(SL)}$ at the output. These are measured values. A model of the Class A transistor (neglecting the effects of capacitances and stray inductances that are *tuned out* by the matching circuits) is presented in Fig 11.8, whereas the model for the transistor amplifier with internal impedance matching circuits modeled is shown in Fig. 11.9.

Class B Amplifier

At high-power levels not practical for Class A, the Class B amplifier is used. Class B is more power efficient than Class A and unlike Class A, a transistor (or other amplifying device) conducts for precisely half of

Solid-State Amplifiers

the drive cycle of π radians. It is biased precisely at collector current cutoff. As stated previously, the class of operation has nothing to do with the device type, but rather its bias condition and its conduction angle. Since Class B operation means that the device is only conducting half of the time, a single ended amplifier may only be used at narrow bandwidths where the tuned output network provides the missing half-cycle by storing the energy presented to it by the active half-cycle, then returning it to the circuit when the device is off. For untuned applications, such as audio amplifiers and wideband RF amplifiers, a push–pull pair of devices operate during their respective half-cycles causing the amplifier to be active all of the time. In this situation, the point on the loadline halfway between the ends is at zero supply current for both devices and driven away from zero with drive. Figure 11.10 shows a graphical presentation of the Class B push–pull configuration.

Because the phase of one transistor is of the opposite phase of the other, a center tapped transformer is needed to combine the two halves properly. This may be done without the use of a transformer if complementary pairs of transistors are available (one NPN and one PNP), but rarely is this the case at frequencies above the audio range. Both cases are presented in Fig. 11.11.

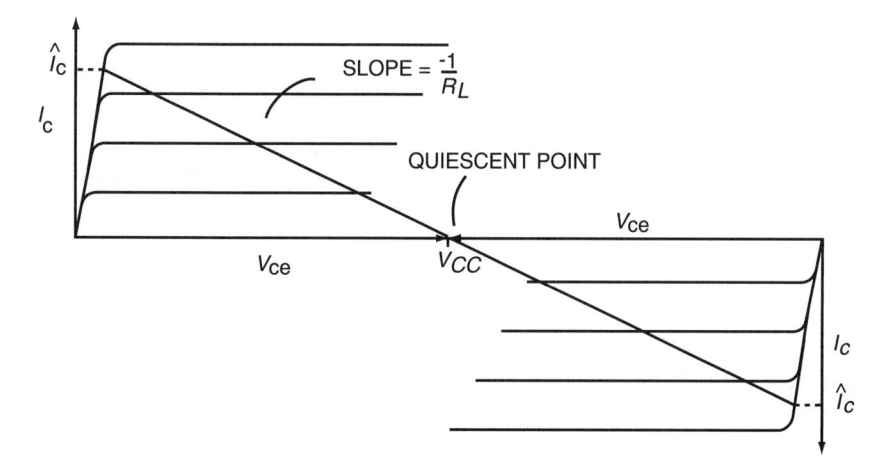

FIGURE 11.10 The graphical solution to the push–pull Class B amplifier configuration is shown. The quiescent point is midway between the ends of the loadline as in Class A, but unlike Class A, the quiescent current is zero. Each transistor output sees R_L at the collector since the center-tapped primary transformer has a 1:1 turns ratio from R_L to each transistor even though it has a 2:1 primary (end-to-end) to secondary turns ratio.

*Signifies complex conjugate

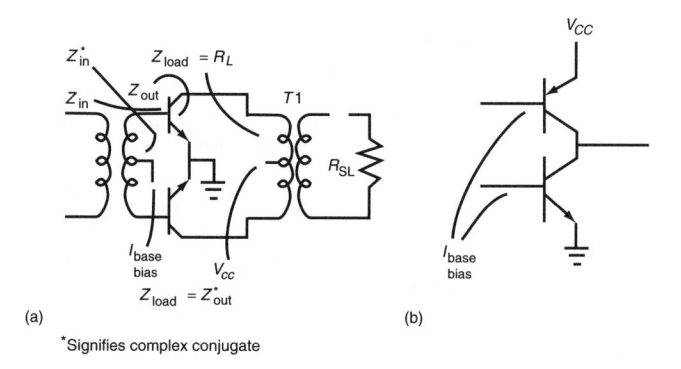

FIGURE 11.11 Push–pull amplifier: (a) like transistors are configured in push–pull Class B. An input and an output transformer are needed to invert the phase of one transistor with respect to the other. The turns ratio of T1 is chosen so that the standard load impedance R_{SL} is transformed to the load impedance R_L presented to each collector. Their graphical solution is shown in Fig. 11.10. In (b) the phase of one transistor with respect to the other is inverted without transformers since one transistor is NPN whereas the other is PNP.

To find the maximum linear output power and maximum theoretical power efficiency of a Class B push–pull amplifier, the loadline in Fig. 11.10 provides the information for the following analysis. Let

$\overset{\diamond}{V}_{ac}$ = maximum peak-to-peak voltage swing along the loadline, $\leq 2V_{CC}$
$\overset{\diamond}{I}_{ac}$ = maximum peak-to-peak current swing along the loadline, $\leq 2\hat{I}_c$
$\hat{V}_{ac}$ = maximum peak voltage swing along the loadline from the quiescent point, $\leq V_{CC}$
$\hat{I}_{ac}$ = maximum peak current swing along the loadine from the quiescent point, $\leq \hat{I}_c$

It follows that

$$2\hat{V}_{ac} = \overset{\diamond}{V}_{ac} \leq 2V_{CC}$$
$$2\hat{I}_{ac} = \overset{\diamond}{I}_{ac} \leq 2\hat{I}_{ac}$$

For a sinusoidal signal

$$P_{out_{max}} = \left(\frac{\hat{V}_{ac}}{\sqrt{2}}\right)\left(\frac{\hat{I}_{ac}}{\sqrt{2}}\right) = \left(\frac{\hat{I}_{ac}}{2}\right)\hat{V}_{ac} \tag{11.11}$$

For simplicity, assume that the output transformer primary to secondary turns ratio is 2:1 so that either active-half has a 1:1 turns ratio with the secondary. For a load at the secondary winding R_L, the primary end-to-end load is $4R_L$, but each transistor works into R_L for each half-cycle. The load presented to each transistor is, therefore, R_L. It follows that

$$\overset{\diamond}{I}_{ac} = 2\hat{I}_{ac} = \frac{\overset{\diamond}{V}_{ac}}{R_L} = \frac{2\hat{V}_{ac}}{R_L} \Rightarrow \left(\frac{\hat{I}_{ac}}{2}\right) = \frac{\overset{\diamond}{V}_{ac}}{4R_L} = \frac{2\hat{V}_{ac}}{4R_L} = \frac{\hat{V}_{ac}}{2R_L}$$

Substituting for $\hat{I}_{ac}/2$ into Eq. (11.11) gives

$$P_{out_{max}} = \frac{\hat{V}_{ac}^2}{2R_L} \leq \frac{V_{CC}^2}{2R_L} \tag{11.12}$$

The center tap current of the transformer is half-wave DC at twice the frequency of the drive signal so that

$$I_{ct}(\theta) = \hat{I}_{ac}|\sin\theta|$$

The DC component of I_{ct} is found from

$$I_{DC} = \frac{1}{2\pi}\int_0^{2\pi}\hat{I}_{ac}|\sin\theta|d\theta = \frac{1}{2\pi}\int_0^{\pi}\hat{I}_{ac}\sin\theta d\theta - \frac{1}{2\pi}\int_{\pi}^{2\pi}\hat{I}_{ac}\sin\theta d\theta = \frac{2\hat{I}_{ac}}{\pi}$$

and

$$I_{DC} = \frac{2\hat{V}_{ac}}{\pi R_L} \leq \frac{2V_{CC}}{\pi R_L}$$

The DC input power is found from

$$P_{in} = I_{DC}V_{CC} = \frac{2\hat{V}_{ac}V_{CC}}{\pi R_L} \leq \frac{2V_{CC}^2}{\pi R_L} \tag{11.13}$$

Solid-State Amplifiers

The power efficiency becomes [from Eqs. (11.12) and (11.13)]

$$\eta = \frac{P_{out}}{P_{in}} = \frac{\hat{V}_{ac}^2}{2R_L} \frac{\pi R_L}{2\hat{V}_{ac} V_{CC}} = \frac{\pi \hat{V}_{ac}}{4 V_{CC}} \qquad (11.14)$$

whereas the maximum theoretical power efficiency (when $\hat{V}_{ac} = V_{CC}$) becomes

$$\eta_{max} = \frac{\pi}{4} = 78.5\% \qquad (11.15)$$

The upper limit of Class B power efficiency is considerably higher than that of Class A at 50%. The power dissipated in each of the two transistors P_{Dis}, is half of the total. The total power dissipated in the case of maximum efficiency is

$$2P_{Dis} = P_{in} - P_{out}$$
$$= \frac{2V_{CC}^2}{\pi R_L} - \frac{V_{CC}^2}{2R_L} \qquad (11.16)$$
$$2P_{Dis} = \frac{V_{CC}^2}{R_L}\left(\frac{2}{\pi} - \frac{1}{2}\right)$$

and

$$P_{Dis} = \frac{V_{CC}^2}{R_L}\left(\frac{1}{\pi} - \frac{1}{4}\right) \qquad (11.17)$$

In Fig. 11.10 the loadline passes through the zero collector current point of each transistor where one device is turning off and the other is turning on. At this precise point both amplifier devices could be *off* or *on*. To avoid this *glitch* or *discontinuity*, strict class B is never really used, but instead, each transistor is biased on to a small value of current to avoid this *crossover distortion*. The choice of bias current has an effect on IM3 so that it may be adjusted for best IM performance under a multitone test condition. The quiescent bias current is usually set for about 1–10% of the highest DC collector current under drive. Biased under this condition, the active devices are not operating in strict Class B, but rather, in a direction toward Class A, and is usually referred to as Class AB.

Since the quiescent point of a Class AB or B amplifier is near or at zero, looking into the input port or output port expecting to measure the standard impedance to which the amplifier was designed is not possible. Large signal behavior comes into play in power amplifiers other than Class A and only exists under drive. Without drive, the amplifier is cut off in the case of Class B or near cutoff in the case of Class AB. With drive, the input return loss may be measured as in the case of Class A, but a directional coupler is used to sense the incident and reflected power. Forward output power is also measured with a directional coupler and compared with theoretical efficiency to determine the worth of the output matching circuit. Efficiency should be in the 40–60% range for good linearity, staying away from the ends of the loadline. Frequently, impedance information about the input and output of a transistor used in Class AB or B is for large signal (power) applications and arrived at experimentally by beginning with a best guess set of parameters about the transistor, then designing a matching circuit and optimizing it for best performance under power. When the goal is reached, the transistor is removed, and the input and output impedances of the circuit are measured at the points where the transistor was connected. These points may then be plotted on a Smith chart. If the input impedance to the transistor is called Z_{in}, the complex conjugate of what the circuit measured is Z_{in} of the transistor. If the output impedance of the transistor is called Z_{out}, then this is the complex conjugate of that measured for the output circuit. If, however, the output impedance is called Z_{load}, then it is not the complex conjugate of what was

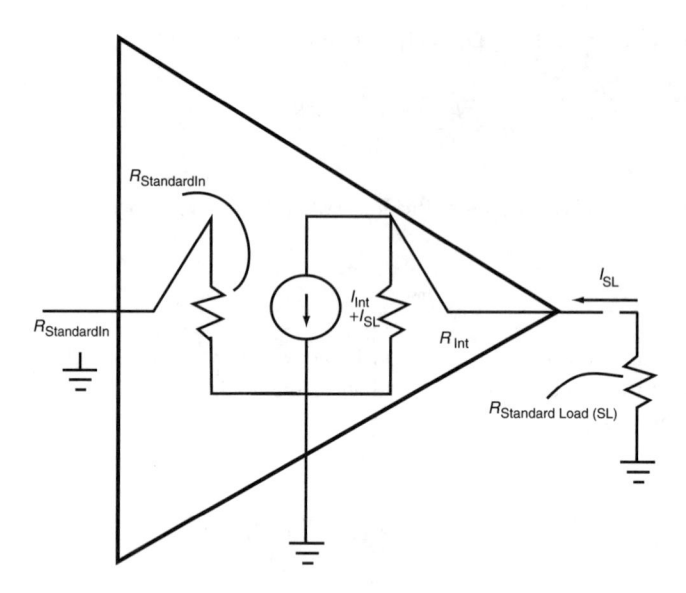

FIGURE 11.12 The Class AB or B amplifier may be modeled as having a measurable input resistance when driven and an internal output resistance that maintains the power efficiency at or below the theoretical maximum of 78.5%. For this, $R_{\text{Int}} \leq 3.6512 R_{\text{SL}}$.

measured, but the load impedance that the transistor is working into. There is some ambiguity about the output impedance among the stated parameters of many manufacturers of devices. Some call it Z_{out} when they really mean Z_{load}. A general guide is that the load that a power transistor wants to work into for best overall performance is capacitive, so if Z_{load} is specified to be $R + jX$, this is probably Z_{out} and not Z_{load}. Z_{load} would be the complex conjugate of this or $R - jX$.

Third-order intermodulation performance for Class AB or B is generally worse than Class A, and harmonic distortion is better in the sense that even harmonics are canceled to a large degree in the primary of the output transformer. These currents flow in opposite directions at the same time and net to zero, whereas odd order harmonic currents do not. An amplifier model for a Class AB or B amplifier is shown in Fig. 11.12. Its topology is the same as that for the Class A in Fig. 11.9 except that $R_{\text{StandardLoad(SL)}}$ *inside* the amplifier is replaced by R_{Internal} or R_{Int}. For a maximum power efficiency, $\eta_{\text{max}} = 78.5\%$,

$$\eta_{\text{max}} = \frac{P_{\text{out}_{\text{max}}}}{P_{\text{in}}} = \frac{R_{\text{Int max}}}{R_{\text{Int max}} + R_{\text{SL}}}$$
$$= 0.785 \tag{11.18}$$

and

$$R_{\text{Int max}} = 3.6512 R_{\text{SL}}$$

so that for efficiency figures less than the theoretical maximum,

$$R_{\text{Int}} \leq 3.6512 R_{\text{SL}} \tag{11.19}$$

Deleting max from Eq. (11.18),

$$\eta = \frac{R_{\text{Int}}}{R_{\text{Int}} + R_{\text{SL}}} \tag{11.20}$$

Solid-State Amplifiers

R_{Int} may be measured indirectly by knowing the efficiency of the amplifier and the standard load resistance,

$$R_{Int} = R_{SL}\left(\frac{\eta}{1-\eta}\right) \tag{11.21}$$

An enhancement to the efficiency of the Class B amplifier makes use of the addition of a third harmonic component of the right amount to the collector–emitter voltage waveform to cause near square-wave flattening when it is near zero (where the collector current is greatest). This modification alters the amplifier enough so that it enjoys a different classification called Class F. Flattening enhances efficiency by as much as one-eighth so that

$$\eta_{ClassF} = \frac{9}{8}\eta_{ClassB}$$

From Eq. (11.15)

$$\eta_{ClassFmax} = \frac{9}{8}\left(\frac{\pi}{4}\right) = 0.884$$

11.4 Nonlinear Amplifier Classes of Operation

If the Class A amplifier is biased to cause device current for the entire drive cycle, and Class B is biased to cause device current for precisely half of the cycle, then it is logical to assign the Class C designation to that bias condition that causes less than half-cycle device current. There are two types of Class C amplifiers: nonsaturated and saturated.

Nonsaturated Class C Amplifier

It is not convenient to analyze the Class C amplifier by means of the loadline argument as in the cases of Classes A or B, but rather, a time- or phase-domain analysis proves to be more workable. The reason is that in the other classes conduction time is fixed at full time or half. In Class C, the conduction time is variable so that a general analysis must be performed where time or phase is the independent variable, not collector–emitter voltage. A time or phase analysis for Classes A or B could have been done, but then only one method would be presented here and the intention is to present another way. The results are the same and comparisons are presented at the end of this section.

A phase domain diagram of the nonsaturated Class C amplifier is shown in Fig. 11.13 along with a representative schematic diagram. The term nonsaturated means that the active device is not driven to the point where the collector–emitter voltage is at the lowest possible value. The curve of $V_{ce}(\theta)$ in Fig. 11.13(a) does not quite touch the zero line. Another way of saying this is that the transistor is always active or never conducts like a switch. It is active in the same way as Classes A and B, and because of this similarity, it may be modeled in the same way, that is, as a dependent current source.

In an amplifier where the drive signal causes the onset of conduction so that the conduction angle is less than π, it is important to realize the following.

1. The transistor will not conduct until its base-emitter junction self-reverse bias voltage of about 0.7 V is overcome. With no external base bias applied, the conduction angle will be less than π and the quiescent collector current fixed at a negative default value I_{CQ}. If this default value is acceptable in the outcome, then no external bias needs to be applied and the design becomes simpler. Applying a DC base current will allow control over I_{CQ}.
2. The collector current $I_c(\theta)$, is sinusoidal in shape when it exists.
3. The collector–emitter voltage $V_{ce}(\theta)$, is sinusoidal, but π radians out of phase with the collector current.

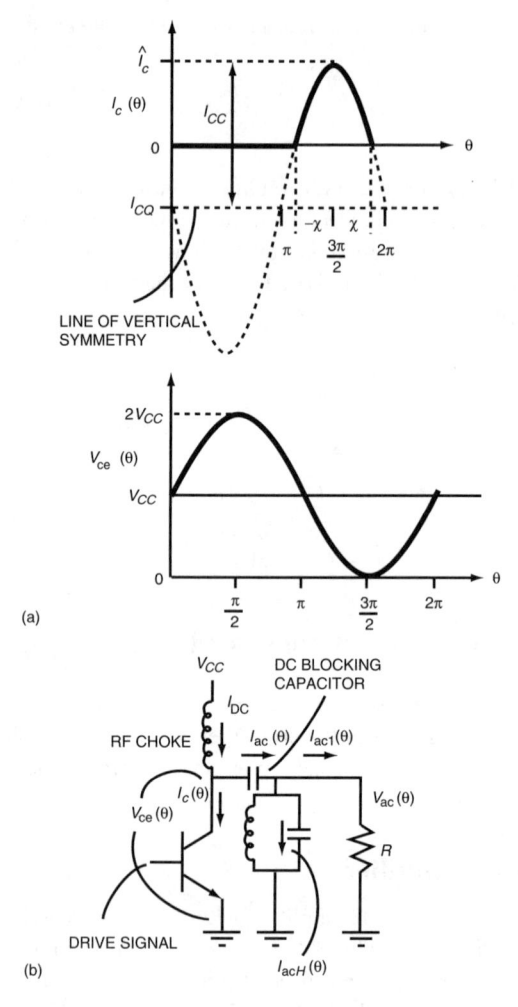

FIGURE 11.13 The phase- or time-domain diagrams of all significant voltages and currents of the nonsaturated Class C amplifier are shown: (a) a representative NPN transistor, and (b) schematic diagram.

Some definitions are in order.

$I_c(\theta)$ = instantaneous collector current
$\hat{I}_c$ = peak value of collector current
I_{CQ} = negative self bias collector current
$I_{CC}(\theta)$ = positive net value of $I_c(\theta)$ and I_{CQ}
$I_c(\theta) = I_{CQ} - I_{CC} \sin\theta$ when the net collector current $I_c(\theta) \geq 0$; it is zero elsewhere
I_{DC} = DC current draw from the collector power supply
$I_{acH}(\theta)$ = harmonic current shunted to ground by the tank circuit capacitor
$\hat{I}_{ac1}$ = peak current amplitude of the fundamental frequency component in the load; it is the first harmonic amplitude coefficient of the Fourier series of the collector current with the conduction angle establishing the limits of integration.
$I_{ac1}(\theta) = \hat{I}_{ac1} \sin\theta$
$V_{ce}(\theta)$ = collector–emitter voltage
V_{CC} = collector supply voltage
2χ = conduction angle of the transistor collector current
$\hat{V}_{ac1}$ = peak voltage amplitude across the load
$V_{ac1}(\theta) = \hat{V}_{ac1} \sin\theta$

By the definition of the Class C amplifier and analysis of Fig. 11.13, $I_c(\theta)$ is found to be

$$I_c(\theta) = I_{CQ} - I_{CC}\sin\theta \geq 0$$
$$= 0, \text{ elsewhere} \tag{11.22}$$

and

$$I_{CQ} = -I_{CC}\cos\chi \tag{11.23}$$

The DC current from the power supply is found by averaging $I_c(\theta)$,

$$I_{DC} = \frac{1}{2\pi}\int_0^{2\pi} I_c(\theta)d\theta$$

$$I_{DC} = \frac{1}{2\pi}\int_{\frac{3\pi}{2}-\chi}^{\frac{3\pi}{2}+\chi} (I_{CQ} - I_{CC}\sin\theta)d\theta$$

$$I_{DC} = \frac{2\chi}{2\pi}I_{CQ} + \frac{I_{CC}}{2\pi}\left[\cos\left(\frac{3\pi}{2}+\chi\right) - \cos\left(\frac{3\pi}{2}-\chi\right)\right]$$

However,

$$\cos\left(\frac{3\pi}{2}+\chi\right) = \sin\chi \quad \text{and} \quad -\cos\left(\frac{3\pi}{2}-\chi\right) = \sin\chi$$

so that

$$I_{DC} = \frac{\chi}{\pi}(-I_{CC}\cos\chi) + \frac{I_{CC}}{\pi}\sin\chi$$

and

$$I_{DC} = \frac{I_{CC}}{\pi}(\sin\chi - \chi\cos\chi) \tag{11.24}$$

Also,

$$I_{CC} = \frac{\pi I_{DC}}{\sin\chi - \chi\cos\chi} \tag{11.25}$$

DC power supplied to the collector is found to be

$$P_{in} = I_{DC}V_{CC} = \frac{I_{CC}V_{CC}}{\pi}(\sin\chi - \chi\cos\chi) \tag{11.26}$$

The peak current value of the fundamental frequency component in the load, $\hat{I}_{acl}$ is the coefficient of the fundamental frequency component of the Fourier series of the collector current $I_c(\theta)$. All harmonic current is shunted to ground by the tank circuit capacitor and assumed to be zero. The tank circuit is resonant at the fundamental frequency.

$\hat{I}_{ac1}$ is found from

$$\hat{I}_{ac1} = -\frac{1}{\pi}\int_0^{2\pi} I_c(\theta)\sin\theta\, d\theta$$

$$\hat{I}_{ac1} = -\frac{1}{\pi}\int_{\frac{3\pi}{2}-\chi}^{\frac{3\pi}{2}+\chi}(I_{CQ}-I_{CC}\sin\theta)\sin\theta\, d\theta$$

$$\hat{I}_{ac1} = -\frac{I_{CQ}}{\pi}\int_{\frac{3\pi}{2}-\chi}^{\frac{3\pi}{2}+\chi}\sin\theta\, d\theta + \frac{I_{CC}}{\pi}\int_{\frac{3\pi}{2}-\chi}^{\frac{3\pi}{2}+\chi}\sin^2\theta\, d\theta$$

$$\hat{I}_{ac1} = \frac{I_{CQ}}{\pi}\cos\theta\bigg|_{\frac{3\pi}{2}-\chi}^{\frac{3\pi}{2}+\chi} + \frac{I_{CC}}{\pi}\left(\frac{\theta}{2}+\frac{\sin 2\theta}{4}\right)\bigg|_{\frac{3\pi}{2}-\chi}^{\frac{3\pi}{2}+\chi}$$

$$\hat{I}_{ac1} = \frac{I_{CQ}}{\pi}\left[\cos\left(\frac{3\pi}{2}+\chi\right)-\cos\left(\frac{3\pi}{2}-\chi\right)\right] + \frac{I_{CC}}{\pi}\left[\chi - \frac{-\sin 2\chi - \sin 2\chi}{4}\right]$$

$$\hat{I}_{ac1} = \frac{2I_{CQ}}{\pi}\sin\chi + \frac{I_{CC}}{\pi}\chi + \frac{I_{CC}}{2\pi}\sin 2\chi$$

Since $I_{CQ} = -I_{CC}\cos\chi$,

$$\hat{I}_{ac1} = -\frac{2I_{CC}}{\pi}\sin\chi\cos\chi + \frac{I_{CC}}{\pi}\chi + \frac{I_{CC}}{2\pi}\sin 2\chi$$

$$\hat{I}_{ac1} = \frac{I_{CC}}{2\pi}(-4\sin\chi\cos\chi + 2\chi + \sin 2\chi)$$

rearranging and making use of the identity, $\sin 2x = 2\sin x \cos x$,

$$\hat{I}_{ac1} = \frac{I_{CC}}{\pi}(\chi - \sin\chi\cos\chi) \tag{11.27}$$

The peak voltage across the load becomes

$$\hat{V}_{ac1} = \frac{RI_{CC}}{\pi}(\chi - \sin\chi\cos\chi) \tag{11.28}$$

The output power in the load is

$$P_{out} = P_{ac1} = \frac{\hat{V}_{ac1}^2}{2R} = \frac{RI_{CC}^2}{2\pi^2}(\chi - \sin\chi\cos\chi)^2 \tag{11.29}$$

Substituting Eq. (11.25) for I_{CC},

$$P_{out} = \frac{RI_{DC}^2}{2}\left(\frac{\chi - \sin\chi\cos\chi}{\sin\chi - \chi\cos\chi}\right)^2 \tag{11.30}$$

Dividing Eq. (11.29) by Eq. (11.26), the power efficiency η is

$$\eta = \frac{P_{out}}{P_{in}} = \frac{RI_{CC}^2}{2\pi^2} \frac{\pi}{I_{CC}V_{CC}} \frac{(\chi - \sin\chi\cos\chi)^2}{(\sin\chi - \chi\cos\chi)}$$

$$\eta = \frac{R}{2\pi} \frac{I_{CC}}{V_{CC}} \frac{(\chi - \sin\chi\cos\chi)^2}{(\sin\chi - \chi\cos\chi)} \tag{11.31}$$

Here, η_{max} occur when $\hat{V}_{ac1} = V_{CC}$. Setting Eq. (11.28) equal to V_{CC}, solving for I_{CC}, and substituting into Eq. (11.31), $\eta \to \eta_{max}$

$$\eta_{max} = \frac{\chi - \sin\chi\cos\chi}{2(\sin\chi - \chi\cos\chi)} \tag{11.32}$$

$$V_{ce}(\theta) = V_{CC} + \hat{V}_{ac1}\sin\theta$$

and in the case of maximum efficiency where $\hat{V}_{ac1} = V_{CC}$,

$$V_{ce}(\theta) = V_{CC}(1 + \sin\theta)|_{\eta=\eta_{max}} \tag{11.33}$$

The collector–emitter voltage $V_{ce}(\theta)$ swings from $2V_{CC}$ to zero and differs from the voltage across the load R by the DC component V_{CC}. Stated before, $I_{acH}(\theta)$, the harmonic currents, do not enter the equations because these currents do not appear in the load.

The first thing to note about Eq. (11.32) is that the theoretical maximum efficiency is not limited to a value less than one. The second thing is that it is only a function of conduction angle, 2χ. Here, $\eta_{max} \to 1$ as $\chi \to 0$. For $\chi \to 0$, Eq. (11.32) becomes

$$\lim_{x \to 0} \eta_{max} = \frac{\chi - \sin\chi\cos\chi}{2(\sin\chi - \chi\cos\chi)} \tag{11.34}$$

For small χ,

$$\sin\chi \to \chi, \quad \sin 2\chi \to 2\chi, \quad -4\sin\chi \to -4\chi, \quad \text{and} \quad \cos\chi \to (<1)$$

It follows that

$$\lim_{x \to 0} \eta_{max} = \frac{2\chi - 2\chi(<1)}{4\chi - 4\chi(<1)} = 1 \tag{11.35}$$

To achieve 100% power efficiency, the collector current pulse $I_c(\theta)$ would become infinitesimal in width, carry no energy, and make no power. Even though the amplifier would produce no power, it would consume no power, and so it would be 100% efficient. In real applications, power efficiency can be in the 80s of percent.

Power dissipated as heat in the transistor is

$$P_{Dis} = (1 - \eta)P_{in} \tag{11.36}$$

Furthermore, the resistance represented by the transistor collector–emitter path, that is, that which is represented by the internal shunt resistance of the current-dependent current source, is found by dividing

Eq. (11.36) by I_{DC}^2. This resistance may be called R_{Dis} and shares the input power with the load according to Eq. (11.36)

$$R_{Dis} = \frac{P_{Dis}}{I_{DC}^2} = \frac{V_{CC}}{I_{DC}} - \frac{R}{2}\left(\frac{\chi - \sin\chi\cos\chi}{\sin\chi - \chi\cos\chi}\right)^2 \tag{11.37}$$

For the amplifier (not the transistor), the equivalent internal resistance may be found referenced to the standard amplifier load resistance by simply scaling $R \to R_{standardLoad(SL)}$ and

$$R_{int} = R_{Dis}\frac{R_{SL}}{R} \tag{11.38}$$

The nonsaturated Class C amplifier, like the linear classes, may be modeled as a current controlled current source. Figure 11.14 is a representation of the amplifier and is very similar to Fig. 11.12.

Measuring port impedances without drive and making output power is senseless because the transistor is biased off without drive. Input resistance, $R_{StandardIn}$ may only be measured when the transistor is driven to full power. For nonsaturated Class C, $R_{int} \to \infty$ as $\eta \to 1$.

The transistor impedances within the amplifier may be measured by removing the transistor after matching circuit adjustments are made for best specified performance and connecting the impedance measuring instrument, such as a network analyzer, to the input and output connections left open by the transistor. Figure 11.15 illustrates the same concept as that presented in Fig. 11.11(a) for the Class B amplifier except that the latter is for a push–pull configuration and the former is for the single ended.

A measuring instrument will give the complex conjugate of the transistor input impedance Z_{in}. Measuring the impedance seen at the input of the output matching network will give Z_{load}, which will be of the from $R - jX$. R is the same R used in the equations derived for the transistor. There may be a small jX component for best operation of a particular transistor type.

An $I_c(\theta)$ vs. $V_{ce}(\theta)$ loadline representation of the nonsaturated Class C amplifier is shown in Fig. 11.16 and is based on the derived results of the given equations.

The greatest P_{out} is obtained when the conduction angle 2χ is near π in width, that is, where the breakpoint in the loadline is just below V_{CC}. The greatest power efficiency occurs where the breakpoint

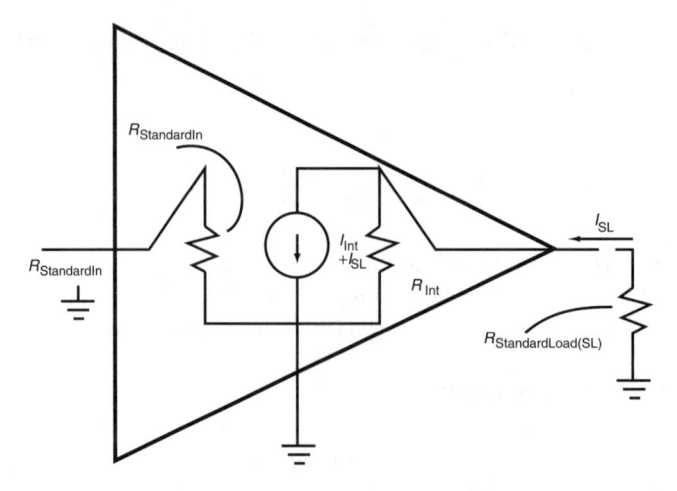

FIGURE 11.14 The nonsaturated Class C amplifier may be modeled as a current controlled current source with a measurable input resistance when driven, and an internal output resistance determined by the power efficiency, η. R_{Int} shares power with R_{SL}, but $R_{Int} \to \infty$ when $\eta \to 1$.

Solid-State Amplifiers

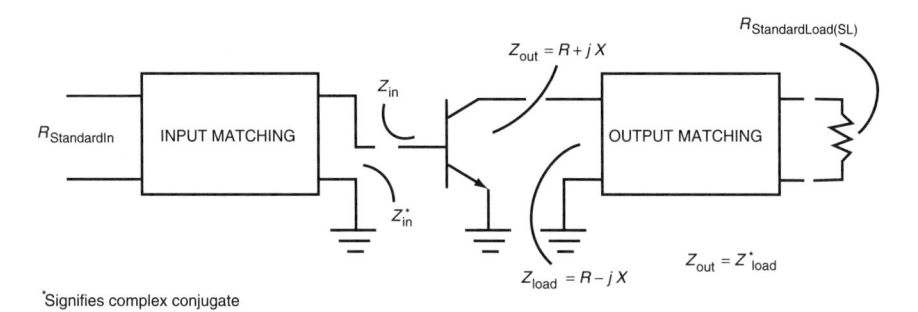

FIGURE 11.15 For the nonsaturated Class C amplifier, input and load impedances seen by the transistor may be measured after its removal. The input and output parts must be terminated in the standard impedance for this static measurement. This figure also applies to the saturated Class C amplifier.

is far to the left near $V_{ce}(\theta) = 0$ and the conduction angle is near zero. The load R presented to a Class C amplifier is usually chosen to be

$$R = \frac{V_{CC}^2}{2P_{out}} \tag{11.39}$$

for a practical design giving a high conduction angle and power making capability using the simple default base bias current. This is where the drive signal must overcome the base-emitter reverse bias voltage of about 0.7 V. A base bias source is not needed and the base is simply grounded for DC.

Saturated Class C Amplifier

The saturated Class C amplifier allows the collector **voltage** (in the case of the bipolar junction transistor) to be driven into **saturation** for a portion of the active time. Figures 11.2 and 11.5 show this as the compression region, but in Class C with no regard to amplitude linearity, power efficiency is increased when the transistor is driven into hard saturation so that further increases in drive yield little, if any, increase in output power. The collector–emitter voltage is as low as it can be. The variable in this approach is the length of time or phase that the transistor is saturated compared to the conduction angle. With the nonsaturated Class C transistor, there is a sequence of three changes of collector current behavior, that is, cutoff, linear, and cutoff. With the saturated transistor, there are five: cutoff, linear, saturated, linear, and cutoff. The mathematical expressions for each change are necessary to completely describe the transistor behavior, and this involves the introduction of another variable, **saturation angle** $2\chi_s$ to those described for the nonsaturated case.

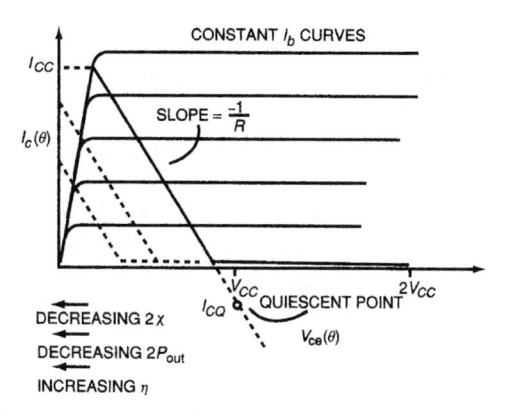

FIGURE 11.16 The nonsaturated Class C amplifier is shown graphically with trends as a function of 2γ, P_{out}, and η. The intersection of the loadline with the $I_c(\theta) = 0$ line at a point lower than V_{CC} accounts for higher efficiency than classes A or B. Quiescent point I_{CQ} is negative.

When using a bipolar junction transistor in the saturated mode, the saturation voltage must be taken into account. It cannot be ignored as in the previously discussed classifications because, in those, V_{Sat}

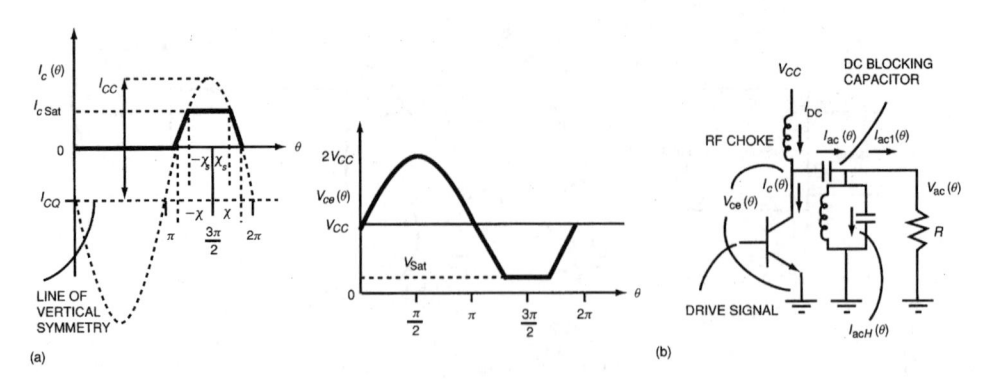

FIGURE 11.17 Phase- or time-domain diagrams of all significant voltages and currents of the saturated Class C amplifier: (a) representative NPN transistor, and (b) circuit diagram.

was never reached. Also when saturated, collector current is not infinite. It is limited by an equivalent **on resistance** so that $I_{cSat} = V_{Sat}/R_{on}$.

In addition to the definitions described for the nonsaturated case, the following are also appropriate for the saturated case:

$2\chi_s$ = saturation angle of the transistor collector current
V_{Sat} = saturation voltage of the transistor collector–emitter path
I_{cSat} = saturated collector current when $V_{ce} = V_{Sat}$
R_{on} = equivalent collector–emitter resistance during saturation

Figure 11.17(a) illustrates the collector current vs. collector–emitter voltage behavior of the saturated transistor. Of course, the saturation angle is always contained within the conduction angle. The circuit diagram in Fig. 11.17(b) is the same as Fig. 11.13(b).

Certain assumptions are made regarding the ideal behavior of a transistor in the saturated Class C mode:

1. The transistor has no dynamic collector–emitter or collector–base capacitances associated with it, which in reality it has.
2. The transistor collector current and collector–emitter voltage waveforms are sinusoidal, which they are not.
3. The saturated collector current is constant, which it is not.
4. Saturation on-resistance is constant, which it is not.
5. There is no parasitic inductance or capacitance associated with the circuit, which there is.
6. There are no harmonic currents flowing in the load, which there are.

The precise saturated Class C amplifier is nearly impossible to analyze for a number of reasons:

1. The transistor has dynamically changing capacitances from collector to base and collector to emitter as functions of dynamic voltages.
2. The current and voltage waveforms are anything but sinusoidal.
3. The saturated collector current has peaks and dips.
4. The saturated on-resistance has peaks and dips.
5. There may be significant parasitic inductances and capacitances at VHF and UHF.
6. There are some harmonic currents in the load.

Because of these things, the assumptions are used in the analysis of the saturated Class C amplifier and in the associated figures to make it manageable.

Solid-State Amplifiers

By definition of saturated Class C transistor operation and analysis of Fig. 11.17, $I_c(\theta)$ is found to be

$$I_c(\theta) = \begin{cases} 0, & \text{when } 0 < \theta < \frac{3\pi}{2} - \chi, \text{ and } \frac{3\pi}{2} + \chi < \theta < 2\pi \\ I_{CQ} - I_{CC}\sin\theta, & \text{when } \frac{3\pi}{2} - \chi \leq \theta \leq \frac{3\pi}{2} - \chi_s, \text{ and } \frac{3\pi}{2} + \chi_s \leq \theta \leq \frac{3\pi}{2} + \chi, \\ \dfrac{V_{Sat}}{R_{on}}, & \text{when } \frac{3\pi}{2} - \chi_s < \theta < \frac{3\pi}{2} + \chi_s \end{cases} \quad (11.40)$$

The onset of saturation occurs when the collector–emitter voltage equals the saturation voltage. R_{on} is finite allowing V_{Sat} to be greater than zero so that the saturated collector current is finite during saturation. Also, as in the nonsaturated case,

$$I_{CQ} = -I_{CC}\cos\chi \quad (11.41)$$

The DC from the power supply is found by averaging $I_c(\theta)$, therefore,

$$I_{DC} = \frac{1}{2\pi}\int_0^{2\pi} I_c(\theta)\,d\theta$$

$$I_{DC} = \frac{1}{2\pi}\int_{\frac{3\pi}{2}-\chi}^{\frac{3\pi}{2}-\chi_s}(I_{CQ} - I_{CC}\sin\theta)\,d\theta + \frac{1}{2\pi}\int_{\frac{3\pi}{2}+\chi_s}^{\frac{3\pi}{2}+\chi}(I_{CQ} - I_{CC}\sin\theta)\,d\theta + \frac{1}{2\pi}\int_{\frac{3\pi}{2}-\chi_s}^{\frac{3\pi}{2}+\chi_s}\frac{V_{Sat}}{R_{on}}\,d\theta$$

$$I_{DC} = \frac{I_{CQ}}{\pi}(\chi - \chi_s) + \frac{I_{CC}}{\pi}(\sin\chi - \sin\chi_s) + \frac{V_{Sat}\chi_s}{\pi R_{on}}$$

Substituting Eq. (11.41) for I_{CQ},

$$I_{DC} = \frac{I_{CC}}{\pi}[(\sin\chi - \sin\chi_s) - (\chi - \chi_s)\cos\chi] + \frac{V_{Sat}\chi_s}{\pi R_{on}} \quad (11.42)$$

Compare Eq. (11.42) with Eq. (11.24). Solving (11.42) for I_{CC},

$$I_{CC} = \frac{\pi I_{DC} - \dfrac{V_{Sat}\chi_s}{\pi R_{on}}}{(\sin\chi - \sin\chi_s) - (\chi - \chi_s)\cos\chi} \quad (11.43)$$

Compare Eq. (11.43) with Eq. (11.25). DC power supplied to the collector is found to be

$$P_{in} = I_{DC}V_{CC} = \left\{\frac{I_{CC}}{\pi}[(\sin\chi - \sin\chi_s) - (\chi - \chi_s)\cos\chi] + \frac{V_{Sat}\chi_s}{\pi R_{on}}\right\}V_{CC} \quad (11.44)$$

Compare Eq. (11.43) with Eq. (11.26).

As in the nonsaturated case, the peak current value of the fundamental frequency component in the load, $\hat{V}_{ac1}$, is the coefficient of the fundamental frequency component of the Fourier series of the collector current $I_c(\theta)$. All harmonic current is shunted to ground by the tank circuit capacitor and *assumed* to be zero. The tank circuit is resonant at the fundamental frequency. $\hat{I}_{ac1}$ is found from

$$\hat{I}_{ac1} = -\frac{1}{\pi}\int_0^{2\pi} I_c(\theta)\sin\theta\,d\theta$$

$$\hat{I}_{ac1} = -\frac{1}{\pi}\int_{\frac{3\pi}{2}-\chi}^{\frac{3\pi}{2}-\chi_s}(I_{CQ} - I_{CC}\sin\theta)\sin\theta\,d\theta$$

$$-\frac{1}{\pi}\int_{\frac{3\pi}{2}-\chi_s}^{\frac{3\pi}{2}+\chi_s}\frac{V_{Sat}}{R_{on}}\sin\theta\,d\theta - \frac{1}{\pi}\int_{\frac{3\pi}{2}+\chi_s}^{\frac{3\pi}{2}+\chi}(I_{CQ} - I_{CC}\sin\theta)\sin\theta\,d\theta$$

After much painful integration,

$$\hat{I}_{ac1} = \frac{2I_{CQ}}{\pi}(\sin\chi - \sin\chi_s) + \frac{I_{CC}}{\pi}(\chi - \chi_s) - \frac{I_{CC}}{2\pi}(\sin 2\chi_s - \sin 2\chi) + \frac{2V_{Sat}}{\pi R_{on}}\sin\chi_s$$

Rearranging,

$$\hat{I}_{ac1} = \frac{2I_{CQ}}{\pi}(\sin\chi - \sin\chi_s) + \frac{I_{CC}}{4\pi}(4\chi - 4\chi_s + 2\sin 2\chi - 2\sin 2\chi_s) + \frac{2V_{Sat}}{\pi R_{on}}\sin\chi_s$$

And again applying Eq. (11.41),

$$\hat{I}_{ac1} = \frac{I_{CC}}{2\pi}[2(\chi - \chi_s) - 4\cos\chi(\sin\chi - \sin\chi_s) + (\sin 2\chi - \sin 2\chi_s)] + \frac{2V_{Sat}}{\pi R_{on}}\sin\chi_s \quad (11.45)$$

Compare Eq. (11.45) with Eq. (11.27).

Hypothetically, if $\chi = \chi_s$, then the transistor is immediately saturated and there is no linear region. This can only happen if V_{ce} never is higher than V_{Sat}. Having no linear region, the peak current in the load becomes

$$\hat{I}_{ac1} = \frac{2V_{Sat}}{\pi R_{on}}\sin\chi_s$$

The entire term within the brackets in Eq. (11.45) is zero, meaning that I_{CC} could be very large, approaching ∞. To find the voltage across the load R, under this condition, $\hat{I}_{ac1}$ is multiplied by R, so that

$$\hat{V}_{ac1} = \frac{2RV_{Sat}}{\pi R_{on}}\sin\chi_s$$

Now, hypnothetically suppose that very hard limiting is maintained to ensure that $\chi = \chi_s$, but that the supply voltage is allowed to increase from V_{Sat} to some higher V_{CC}, so that VSat is replaced by ($V_{CC} - V_{Sat}$). It follows that

$$\hat{V}_{ac1} = \frac{2R(V_{CC} - V_{Sat})}{\pi R_{on}}\sin\chi_s$$

for very hard limitng. The voltage across the load is proportional to supply voltage and can be much greater than it is as a function of R/R_{on} and χ_s, and this is the case as will be shown.

Finding $\hat{V}_{ac1}$ by multiplying Eq. (11.45) by R,

$$\hat{V}_{ac1} = \frac{RI_{CC}}{2\pi}[2(\chi - \chi_s) - 4\cos\chi(\sin\chi - \sin\chi_s) + (\sin 2\chi - \sin 2\chi_s)] + \frac{2RV_{Sat}}{\pi R_{on}}\sin\chi_s \quad (11.46)$$

Solid-State Amplifiers

Compare Eq. (11.46) with Eq. (11.28)
To make things a bit easier, let

$$\Psi = [2(\chi - \chi_s) - 4\cos\chi(\sin\chi - \sin\chi_s) + (\sin 2\chi - \sin 2\chi_s)] \tag{11.47}$$

and it has no unit.
Let

$$\Upsilon = \frac{2V_{Sat}}{\pi R_{on}}\sin\chi_s \tag{11.48}$$

and it has the unit of current.
Let

$$\Gamma = \frac{V_{Sat}}{\pi R_{on}}\chi_s \tag{11.49}$$

and it has the unit of current.
Let

$$\Lambda = [\sin(\chi - \sin\chi_s) - (\chi - \chi_s)\cos\chi] \tag{11.50}$$

and it has no unit.
Rewriting Eq. (11.46),

$$\hat{V}_{ac1} = \frac{RI_{CC}}{2\pi}\Psi + \Upsilon R \tag{11.51}$$

Rewriting Eq. (11.43)

$$I_{CC} = \frac{\pi I_{DC} - \Gamma}{\Lambda} \tag{11.52}$$

Output power is

$$P_{out} = \frac{\hat{V}_{ac1}^2}{2R} = \frac{1}{2R}\left[\frac{RI_{CC}}{2\pi}\Psi + \Upsilon R\right]^2 \tag{11.53}$$

Substituting Eq. (11.52) into Eq. (11.53), we get

$$P_{out} = \frac{R}{2}\left[\frac{1}{2\pi}\left(\frac{\pi I_{DC} - \Gamma}{\Lambda}\right)\Psi + \Upsilon\right]^2 \tag{11.54}$$

Rewriting Eq. (11.44)

$$P_{in} = \left(\frac{I_{CC}}{\pi}\Lambda + \Gamma\right)V_{CC} \tag{11.55}$$

Efficiency η is found by dividing Eq. (11.53) by Eq. (11.55), so that

$$\eta = \frac{P_{out}}{P_{in}} = \frac{\left[\frac{RI_{CC}}{2\pi}\Psi + \Upsilon R\right]^2}{2RV_{CC}\left[\frac{I_{CC}}{\pi}\Lambda + \Gamma\right]} \tag{11.56}$$

Compare Eq. (11.56) with Eq. (11.31).

At this point, η appears to be a function of both I_{CC} and V_{CC}. Equation (11.52) may be used to rid Eq. (11.56) of I_{CC}, but this will only introduce I_{DC} into the equation. Instead, let $\hat{V}_{ac1} = \alpha V_{CC}$, where α is a constant of proportionality greater than one. (This has been shown to be the case for hard saturation.) This will allow expressing Eq. (11.56) only in terms of one voltage αV_{CC}, and no current variable. Being simplified in this way provides for a more meaningful analysis. It also forces the case for hard saturation so that Eq. (11.56) becomes an expression for η_{max}. Setting Eq. (11.51) equal to αV_{CC}, solving for I_{CC}; and substituting into Eq. (11.56) gives

$$\eta_{max} = \frac{1}{4}\frac{\alpha^2 \Psi V_{CC}}{(\alpha V_{CC} - R\Upsilon)\Lambda + \frac{1}{2}R\Gamma\Psi} \tag{11.57}$$

We dare to expand Eq. (11.57) so that

$$\eta_{max} = \frac{1}{4}\frac{\alpha^2 V_{CC}[2(\chi - \chi_s) - 4\cos\chi(\sin\chi - \sin\chi_s) + (\sin 2\chi - \sin 2\chi_s)]}{\alpha V_{CC}[(\sin\chi - \sin\chi_s) - (\chi - \chi_s)\cos\chi] - \left[\frac{R}{R_{ON}}\frac{2V_{Sat}}{\pi}\sin\chi_s\right][X]}$$

$$\overline{\text{Numerator}}$$

$$\frac{[(\sin\chi - \sin\chi_s) - (\chi - \chi_s)\cos\chi] + \frac{R}{R_{ON}}\left[\frac{V_{Sat}}{2\pi}\chi_s\right][X]}{\text{Numerator}}$$

$$\overline{[2(\chi - \chi_s) - 4\cos\chi(\sin\chi - \sin\chi_s) + (\sin 2\chi - \sin 2\chi_s)]} \tag{11.58}$$

Although this equation is lengthy, it offers some intuitive feel for those things that affect efficiency. At first, the casual reader would expect that efficiency would not depend on V_{CC} because the efficiency of the nonsaturating Class C case does not depend on anything but 2χ. There is conduction angle dependency because, in saturation, the transistor collector equivalent circuit is a voltage source, not a current source. Efficiency is high as long as V_{CC} is much larger than V_{Sat}. Intuitively, efficiency is high because the collector current is highest when the collector–emitter is in saturation. As V_{CC} is made to approach V_{Sat}, there is less voltage to appear across the load and more current flows through R_{on}. When V_{CC} is made equal to V_{Sat}, all of the current flows through R_{on}, none to the load, and $\eta \to 0$. Other important conclusions may be drawn by verifying that Eq. (11.58) collapses to η_{max} predicted for the following:

1. Nonsaturated Class C case, where $2\chi_s = 0$ and $\alpha \to 1$.
2. Linear Class B case, where $2\chi = \pi$.
3. Linear Class A case, where $2\chi = 2\pi$.

For the nonsaturated Class C case, Eq. (11.58) reduces to

$$\eta_{max} = \frac{1}{4}\frac{\alpha^2 V_{CC}[2\chi - 4\cos\chi\sin\chi + \sin 2\chi]}{\alpha V_{CC}[\sin\chi - \chi\cos\chi]}$$

and realizing the identity, $\sin 2\chi = 2\sin\chi\cos\chi$,

$$\eta_{max} = \frac{1}{2}\frac{(\chi - \cos\chi\sin\chi)}{(\sin\chi - \chi\cos\chi)} \tag{11.59}$$

which agrees with Eq. (11.32).

For Class B, Eq. (11.59) reduces to

$$\eta_{max} = \frac{\frac{\pi}{2} - \cos\frac{\pi}{2}\sin\frac{\pi}{2}}{2\left(\sin\frac{\pi}{2} - \frac{\pi}{2}\cos\frac{\pi}{2}\right)} = \frac{\frac{\pi}{2} - 0}{2(1-0)} = \frac{\pi}{4} \tag{11.60}$$

which agrees with Eq. (11.15).

For Class A, Eq. (11.59) reduces to

$$\eta_{max} = \frac{\pi - 2\sin\pi\cos\pi}{2(\sin\pi - \pi\cos\pi)} = \frac{\pi - 0}{2(0 + \pi)} = \frac{1}{2} \tag{11.61}$$

which agrees with Eq. (11.10).

In saturated Class C, the voltage across the load is proportional to the supply voltage V_{CC}. Because of this it may be *high-level modulated* by connecting a baseband modulation signal voltage $V_m(\theta)$, in series with the supply line. V_{CC} then becomes $V_{CC}(\varphi)$ so that

$$V_{CC}(\varphi) = V_{CC} + V_m(\varphi)$$

Assuming that $V_m(\varphi)$ is sinusoidal and that $\hat{V}_m$ is made equal to V_{CC}, the amplifier will be modulated 100%, that is, $V_{CC}(\varphi)$ will swing from V_{Sat} to $2V_{CC}$. The amplifier is modeled in Fig. 11.18 and transistor, circuit, and standard impedances are measured in the same way as the nonsaturated amplifier. The loadline representation is given in Fig. 11.19 and should be compared with Fig. 11.16.

Referring to Fig. 11.18, R_{Int} is contained in

$$P_{out} = P_{in}\frac{R_{SL}}{R_{SL} + R_{Int}} \tag{11.62}$$

and along with R_{SL}, determines efficiency. In fact,

$$\eta = \frac{R_{SL}}{R_{SL} + R_{Int}} \tag{11.63}$$

and should be compared with the complexity of Eq. (11.58). It is much easier to measure efficiency than it is to calculate it analytically.

Power dissipated in the transistor P_{Dis}, is found from

$$P_{Dis} = P_{in} - P_{out} = P_{in}\left(1 - \frac{R_{SL}}{R_{SL} + R_{Int}}\right) = P_{in}(1 - \eta) \tag{11.64}$$

Although Class C amplifiers, saturated or not, cannot perform as linear amplifiers, they are almost always used to amplify phase or frequency modulated signals to higher levels where high-power efficiency is important. Since the saturated Class C amplifier may be V_{CC} modulated, it is often used as the output stage of high-power high-level modulated full carrier AM transmitters. And since the Class C amplifier output tank circuit is high in harmonic current, the amplifier is useful as a frequency multiplier by

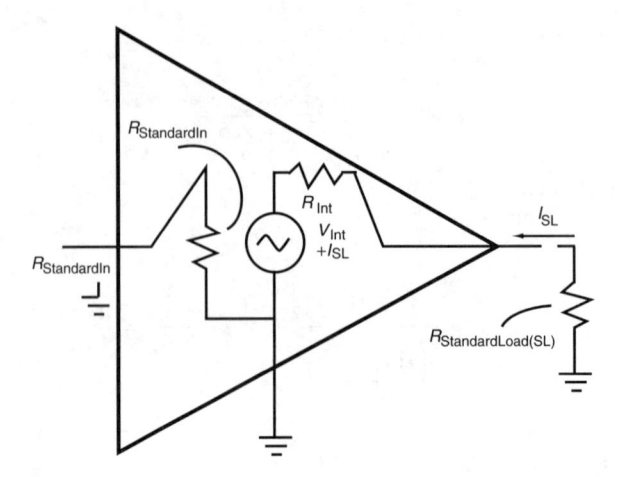

FIGURE 11.18 The saturated Class C amplifier may be modeled as a current controlled voltage source with a measurable input resistance when driven, and an internal output resistance determined by the power efficiency η. R_{Int} shares power with R_{SL}, but $R_{Int} \to 0$ when $\eta \to 1$.

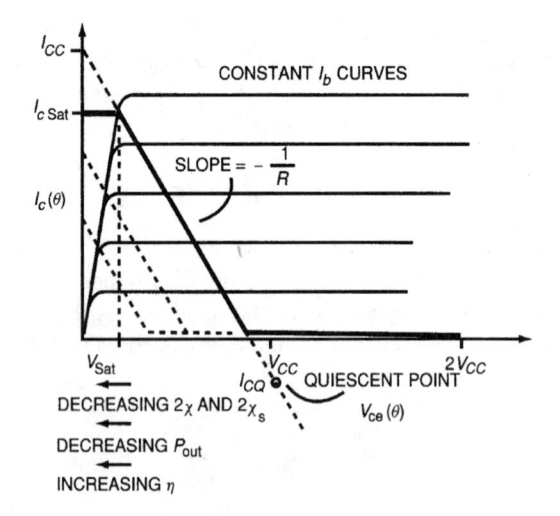

FIGURE 11.19 The saturated Class C amplifier is shown graphically with trends as a function of 2χ, $2\chi_s$, P_{out}, and η. The intersection of the loadline with I_{CQ} is negative current, whereas saturation occurs where the loadline meets the transistor curve.

resonating the tank to a harmonic so that only harmonic energy is passed to the load. An undesirable trait of the Class C is that the transistor collector current *snaps on* as drive appears and cannot be controlled with drive signal if it is saturated. Only V_{CC} will control collector current.

Because of the assumptions made about the saturated Class C amplifier before its mathematical analysis, it must be realized that the final circuit will need *adjusting* to get it to work right. Large signal behavior is only defined for a single set of values, such as supply voltage, load resistance, saturated currents, tank circuit Q, and other parameters. Change one thing and everything else changes! The idealized analysis presented here can only be viewed for basic theoretical understanding and initial design and test. Experience has shown many times that the actual power [O'Reilly, 1975] available from a saturated Class C amplifier is closer to

$$P_{out} \approx 0.625 \frac{(V_{CC} - V_{Sat})^2}{2R} \qquad (11.65)$$

Solid-State Amplifiers   11-31

rather than

$$P_{out} = \frac{(V_{CC} - V_{Sat})^2}{2R}$$

since reality does not agree with the assumptions of this analysis. Particularly troublesome are the parasitic impedances associated with the emitter ground connection where frequently fractions of an ohm are involved and in bypassing supply lines with real capacitors and decoupling chokes that allow more than zero RF current to flow where in the ideal case, should not.

Class D Voltage Switch-Mode Amplifier

Switch-mode amplifiers comprise a special class unlike Classes A, B, or C or their derivatives. Switch mode implies that something is on or off, but not in between. In switch mode there is no linear region as there is in Class A, B, or C. In fact, the active device is modeled as a switch. Insulated gate bipolar junction transistors (IGBJT) are used in switch-mode amplifiers up to a few hundred kilohertz whereas field effect transistors (FETs) are used to about 10 MHz. IGBJTs are advanced in their voltage and current capability, but are slower to switch on and off than FETs. Device development has advanced in these areas over the past few years at such a rapid rate that it is difficult to predict current, voltage, and switching speed limits with any degree of accuracy. For this discussion, the depletion type power FET is used, but it is only representative. Typically, a power FET has several thousand picofarads of gate-source capacitance, a saturation resistance R_{on}, of tenths or hundredths of an ohm, and snaps closed (to R_{on}) whenever gate-source voltage exceeds about +4 V for an N-channel type. Likewise, it snaps open whenever gate-source voltage falls below this same threshold. For the following discussion it is assumed that the time to pass through +4 V in either direction is zero and independent of drive signal shape. Except for R_{on}, it is close to being an ideal switch. In many ways it is better than a real switch because it is faster, does not bounce, and does not wear out like a real mechanical switch. Moreover, for this discussion, drain-source capacitance when the FET switch is open is also considered to be zero.

Figure 11.20(a) shows the schematic diagram of a half-wave depletion type FET Class D amplifier and the drain-source voltage of Q2 is shown in Fig. 11.20(b). The purpose of the amplifier is to produce a sinusoid output voltage across the load R at the switching frequency. The driving signal may be any shape, but a square voltage drive is preferred so that passage of the gate drive voltage through the threshold voltage is fast. As stated, R_{on} is assumed to be zero, as is any drain-source capacitance.

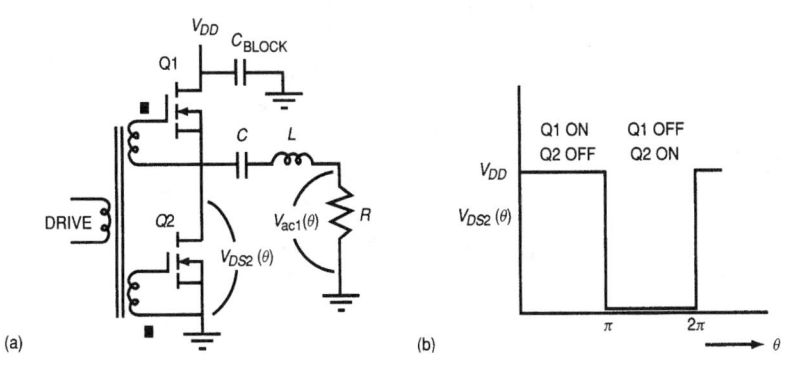

FIGURE 11.20 The half-wave Class D switch-mode pair in (a) produces power at the fundamental component of the switching frequency. Near-100% efficiency may be achieved when the transistors are modeled as ideal switches. The on-resistance of the transistors and drain-source capacitance prevents 100% efficient operation. Drive waveform may be of any shape, but must transition through the on–off threshold region (+4 V) very fast to avoid the active region. (b) The drain-source voltage of Q2.

Because the drain-source voltage of Q2 is square, it contains only odd harmonics according to the nonzero Fourier series coefficients.

$$V_{DS2}(\theta) = V_{DD}\left[\frac{1}{2\pi}\int_0^\pi d\theta + \left(\frac{1}{\pi}\int_0^\pi \sin\theta d\theta\right)\sin\theta + \left(\frac{1}{\pi}\int_0^\pi \sin 2\theta d\theta\right)\sin 2\theta \right.$$
$$\left. + \cdots \left(\frac{1}{\pi}\int_0^\pi \sin 3\theta d\theta\right)\sin 3\theta + \cdots\right] \quad (11.66)$$

$$V_{DS2}(\theta) = V_{DD}\left[\frac{1}{2} - \frac{1}{\pi}\cos\theta\Big|_0^\pi \sin\theta - \frac{1}{2\pi}\cos 2\theta\Big|_0^\pi \sin 2\theta - \frac{1}{3\pi}\cos 3\theta\Big|_0^\pi \sin 3\theta + \cdots\right]$$

$$V_{DS2}(\theta) = V_{DD}\left[\frac{1}{2} + \frac{2}{\pi}\sin\theta + \frac{2}{3\pi}\sin 3\theta + \frac{2}{5\pi}\sin 5\theta + \frac{2}{7\pi}\sin 7\theta + \cdots\right] \quad (11.67)$$

If L and C are resonant at the fundamental frequency of the Fourier series, the load R only sees the fundamental frequency component of the drain-source voltage of Q2, $V_{acl}(\theta)$

$$V_{acl}(\theta) = \frac{2V_{DD}}{\pi}\sin\theta \quad (11.68)$$

and

$$\hat{V}_{acl} = \frac{2V_{DD}}{\pi}$$

Output power is simply

$$P_{out} = \frac{\hat{V}_{acl}^2}{2R} = \frac{4V_{DD}^2}{2\pi^2 R} = \frac{2V_{DD}^2}{\pi^2 R} \quad (11.69)$$

The current through Q1 or Q2 is a half-sinusoid since each transistor conducts for only half of the time. The average current through Q1 is the DC supply current I_{DC}

$$I_{acl}(\theta) = \frac{V_{acl}(\theta)}{R} = \frac{2V_{DD}}{\pi R}\sin\theta$$

$$I_{Q1}(\theta) = \begin{cases} I_{acl}(\theta), & \text{when } \sin\theta > 0 \\ 0, & \text{when } \sin\theta \leq 0 \end{cases}$$

$$I_{DC} = \frac{1}{2\pi}\int_0^\pi I_{Q1}(\theta)d\theta$$

$$I_{DC} = \frac{-V_{DD}}{\pi^2 R}\cos\theta\Big|_0^\pi = \frac{2V_{DD}}{\pi^2 R} \quad (11.70)$$

The DC input power is found from

$$P_{in} = I_{DC}V_{DD} = \frac{2V_{DD}^2}{\pi^2 R} \quad (11.71)$$

Since Eq. (11.71) equals Eq. (11.69), the efficiency η is equal to one.

Solid-State Amplifiers

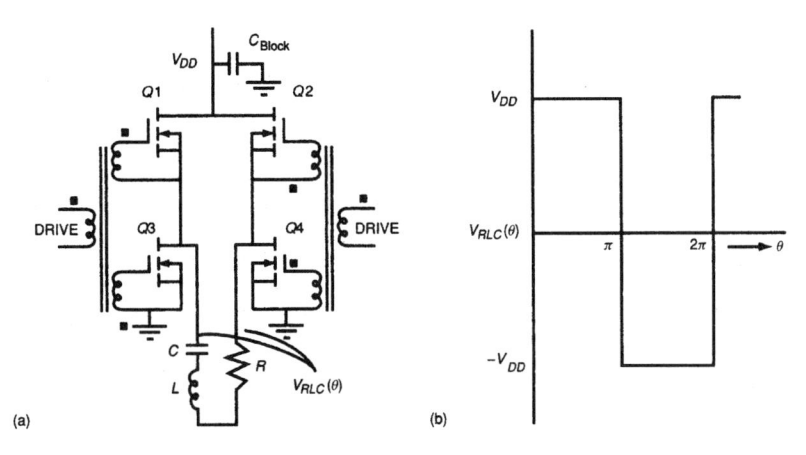

FIGURE 11.21 (a) The H-bridge full-wave Class D amplifier configuration. A diagonal pair of FETs is on, while the opposite pair is off at any instant. The H-bridge provides four times the power as the half-wave configuration. *LC* is a series resonant circuit at the fundamental frequency so that harmonic currents do not flow in the load *R*. (b) The waveform is the bridge voltage across the *RLC* network.

A nonzero R_{on} is the only resistance preventing perfect efficiency if *L* and *C* in Fig. 11.20 are considered ideal. In reality, power efficiency near 98% is possible with very low on-resistance depletion type FETs. C_{block} is necessary because the current through Q1 has a DC component.

The full-wave depletion type FET Class D amplifier may be configured as a pair of transistors such as the push-pull Class B schematic diagram shown in Fig. 11.11(a). Using a center-tapped primary transformer in the output circuit allows both transistors to conduct to the supply instead of one to the supply and one to ground, hence, full wave. A series tuned *L–C* resonant circuit in series with the load is necessary to allow only fundamental frequency load current. Full wave may also be achieved by using the four transistor *H-bridge*. It is the configuration of choice here in order to illustrate another way. The H-bridge name is obvious and is shown in Fig. 11.21. As before, R_{on} and drain-source capacitance are assumed to be zero.

The voltage across the series *RLC* output circuit alternates polarity between $+V_{DD}$ and $-V_{DD}$ so that $V_{ac}(\theta)$ is of the same form as Eq. (11.67). When *L* and *C* are tuned to fundamental resonance,

$$V_{acl}(\theta) = 2V_{DD}\left(\frac{2}{\pi}\sin\theta\right) \tag{11.72}$$

and

$$\hat{V}_{acl} = \frac{4V_{DD}}{\pi}$$

The output power is found from

$$P_{out} = \frac{\hat{V}_{acl}^2}{2R} = \frac{8V_{DD}^2}{\pi^2 R} \tag{11.73}$$

The current through the load is

$$I_{acl}(\theta) = \frac{4V_{DD}}{\pi R}\sin\theta$$

and

$$\hat{I}_{ac1} = \frac{4V_{DD}}{\pi R}$$

The currents through Q1–Q4 are half-sinusoids since complementary pairs (Q1, Q4) and/or (Q2, Q3) only conduct for half of each drive cycle. Over 2π, the pulsating DC from the supply consists of two half-sinusoids so the C_{block} is needed and

$$I_{DC} = 2\left(\frac{1}{2\pi}\int_0^\pi I_{ac1}(\theta)d\theta\right)$$
$$= -\frac{4V_{DD}}{\pi^2 R}\cos\theta\Big|_0^\pi = \frac{8V_{DD}}{\pi^2 R} \qquad (11.74)$$

Input power is

$$P_{in} = I_{DC}V_{DD} = \frac{8V_{DD}^2}{\pi^2 R} \qquad (11.75)$$

Since Eq. (11.73) equals Eq. (11.75), power efficiency is equal to one as long as the transistors act as ideal switches with $R_{on} = 0$ and drain-source capacitance is also equal to zero. Four times the power is available from the full-wave H-bridge compared to the two transistor half-wave circuit of Fig. 11.20 because twice V_{DD} is the voltage swing across RLC.

Either the full-wave or half-wave Class D amplifier may be transformer coupled to R. It is only necessary to consider the turns ratio, but P_{out} and P_{in} remain the same.

Modeling the entire switch mode amplifier in Fig. 11.22 shows that R_{Int} of the output circuit of the amplifier is zero. It must be since power efficiency is 100%, and the output circuit is a voltage source. It may be a current controlled voltage source if the switching devices are bipolar function transistors (BJTs), or voltage controlled if FETs or IGBJTs are used. Although not much attention is given to the drive input circuit, it must be designed to present itself as some standard impedance at the operating frequency and, perhaps, very broadband if the drive signal is square. With no tuned circuits in the way and a very low source impedance, the drive signal should remain square.

In practice, each transistor in the full-wave of half-wave case is shunted by a *fast recovery diode* placed in the reverse direction from drain to source. The diode provides a path for current to flow when the load R is better represented by a complex impedance, $Z = R \pm jX$. A reactive component in the load causes out-of-phase currents to want to flow backwards through any off transistor. Without the diode, the output capacitance of an off transistor will develop a spike of voltage due to the out-of-phase current charging the capacitance, and permanent damage will result. The diodes must be fast enough to conduct or not conduct dictated by the switching phase. Ordinary 60-Hz diodes imitate a resistor at 100 kHz.

As of this writing, Class D amplifiers only find usefulness below carrier frequencies of 10 MHz simply because switching speed and internal capacitances of the transistors prevent *instantaneous* switching. They are used as power amplifiers for RF broadcast transmitters in the medium wave and short wave bands and in switching power supplies.

In the case of the FET, drive power is derived by charging and discharging the gate-source capacitance. The amount of energy required to charge it is

$$E = \frac{1}{2}CV^2 \qquad (11.76)$$

Solid-State Amplifiers

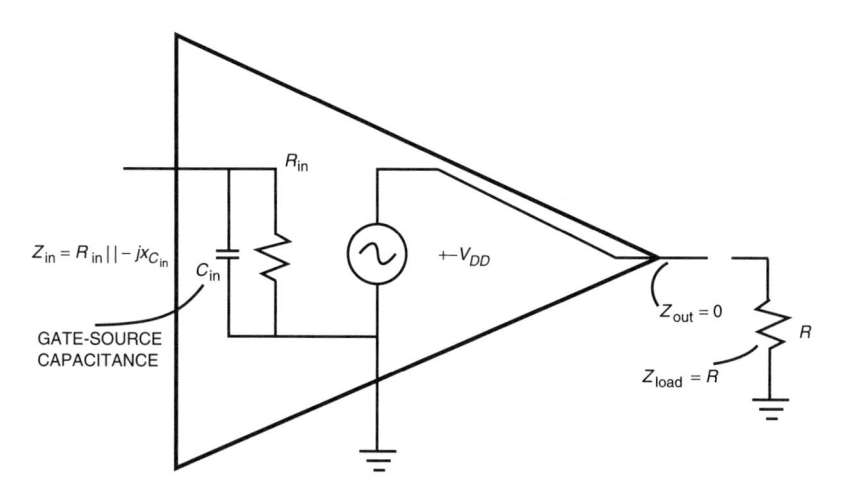

FIGURE 11.22 The full-wave H-bridge is represented by a voltage source with no internal output resistance, assuming the on-resistance and drain-source capacitance are zero. The input circuit must be a good load to the driving source, but usually is driven by a low impedance source capable of driving the capacitive input, R_{in} is high.

For example, at 10 MHz, the period, τ is 10^{-7} s. A half-period (charge time) is $\tau = 0.5 \, (10)^{-7}$ s. If gate capacitance is 2000 pF, $2(10)^{-9}$ F, for each transistor charging to 10 V, then the energy needed to charge each gate becomes

$$E = \frac{1}{2}(2)(10)^{-9}(10)^2 = (10)^{-7} \, J$$

The same energy must be taken out and dissipated (during the off half-period) at a rate of 10 MHz. The power needed to do this is

$$P = \frac{dE}{dt} = \frac{E}{\frac{\tau}{2}} = 2Ef = 2(10)^{-7}(10)^7 = 2 \text{ watts per gate} \qquad (11.77)$$

This may or may not be a large amount of power, but it is not insignificant and must be taken into consideration when designing the overall amplifier. Furthermore, drive power is proportional to carrier frequency.

A potential for disaster exists if a set of on transistors does not turn off as quickly as it should and, instead, remains on while the opposite phase set is turning on. Should this happen, a momentary short circuit appears across the supply, and the transistors are history. A solution is to relatively slowly charge the gates and relatively rapidly discharge them to prevent on-state overlap. The charge path is slow, whereas the discharge path is fast. An effective circuit is shown in Fig. 11.23.

An enhancement to Class D makes use of the real drain-source capacitance previously ignored. It is a source of inefficiency and takes energy to charge, which is dissipated as heat during discharge much like the gate-source capacitance. In the Class E enhancement mode (modified Class D), the capacitance is used as part of the output network, and an additional external capacitance is selected so that the total is the correct amount

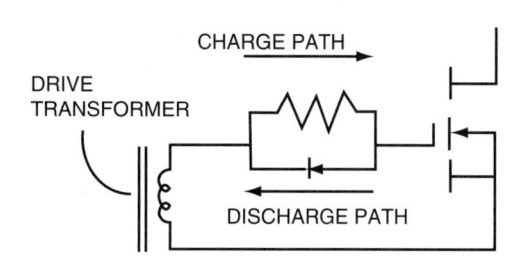

FIGURE 11.23 Slow charge and rapid discharge of the gate-source capacitance prevents possible transistor on-state overlap, which places a short circuit across the supply and destroys transistors.

to cause the drain voltage to reach zero at the instant the transistor turns off. In this way no discharge of the capacitance takes place and efficiency approaches unity.

The switch-mode Class D amplifier can be used as a baseband linear amplifier by applying a square gate signal, the pulse width (duty cycle) of which is proportional to the amplitude of the signal to be amplified.

Assume that a 25% duty cycle represents the halfway point between off and full on (50%) for each transistor set. Modulating up or down from 25% drives the amplifier to zero and 50% duty cycle for each set (100%) total). The average of the output voltage across R will follow the pulse width represented drive signal. The output must be low-pass filtered to obtain the average, which is a reproduction of the signal to be amplified. (Use of the Class D amplifier in this way is sometimes referred to as Class S.)

Applying a pulse width modulated baseband amplifier like this as the power supply for a switch-mode RF amplifier (same configuration) provides a high-level amplitude modulated RF signal. After bandpass filtering at its output, it is suitable for transmission as a full carrier double sideband signal.

There are many variations of the amplifier classes presented in this section and most are *enhancements* of the basic ones.

Defining Terms

Amplifier class of operation: An alphabetical tag associated with an amplifier configuration describing its bias condition (which, in turn, determines its conduction angle) and sometimes its external circuit configuration.

Attenuation: For a signal passing through a circuit, the ratio of the signal power level at the output of the circuit to the signal power level at the input to the circuit when the output signal is of less power than the input. Usually expressed in decibels,

$$\text{attenuation} = -10\log\frac{P_{out}}{P_{in}} \geq 0$$

Aural subcarrier: In a composite television signal, the frequency division multiplexed carrier placed outside the visual passband that carries the audio modulation. In the NTSC (United States) system, it is placed 4.5 MHz higher than the visual carrier.

Color subcarrier: In a composite television signal, the frequency division multiplexed carrier placed within the visual passband which carries the color modulation. In the National Television Systems Committee (NTSC) (United States) system, it is placed approximately 3.579 MHz higher than the visual carrier.

Compression: For the transfer function of an amplifier, the relatively higher power region where the output power is less than proportional to the input power. The 1-dB compression point is that output power level that decreases by 9 dB when the input power decreases by 10 dB.

Conduction angle: The drive or input signal phase angle over which an amplifier conducts current from its output power supply. It is expressed mathematically by 2χ.

Gain: For a signal passing through a circuit, the ratio of the signal power level at the output of the circuit to the signal power level at the input to the circuit when the output signal is of greater power than the input. This can only occur when the circuit is an amplifier. Usually expressed in decibels,

$$\text{gain} = 10\log\frac{P_{out}}{P_{in}} \geq 0$$

Intermodulation distortion: The result of two or more time-domain signals being multiplied together when they were intended to be only added together. Undesirable multiplication takes place in every amplifier amplifying multiple signals yielding sum and difference frequency components that are not present at the input to the amplifier.

Linear: Refers to the ability of an amplifier to maintain the integrity of the signal being amplified. A perfectly linear amplifier causes no distortion to the signal while making it greater in amplitude. The output vs. input transfer function plots as a straight line.

On-resistance: The collector–emitter resistance of a bipolar junction transistor or the drain-source resistance of a field effect transistor when driven to saturation. In the case of the BJT, it is equal to the collector–emitter saturation voltage divided by the collector current and is a function of both. In the case of the FET, it is equal to the saturated drain-source voltage divided by the drain current, but is not a function of either one.

Peak envelope power: The average power of the peak of a signal.

Saturation angle: The driver or input signal phase angle over which an amplifying device is driven into saturation thereby resulting in a flat maximum of amplifying device supply current. Although drive level may continue to increase, output circuit current does not. It is expressed mathematically by $2\chi_s$.

Saturation current: The collector current of a bipolar junction transistor or the drain current of a field effect transistor during the time of the saturation angle.

Standard load: Refers to the resistance of a load that is a standard established by an organization or committee commissioned to decide standards. Input, output, and interstage resistance in RF circuits is established to be 50 Ω unbalanced. Input, output, and interestage resistance in video circuits is established to be 75 Ω unbalanced, and that for audio circuits in 300 Ω balanced and sometimes 600 Ω balanced or 8 Ω unbalanced.

References

Craig, M. 1994. *Television Measurements NTSC Systems.* Tektronix, Inc., Beaverton, OR.
Daley, J.L., ed. 1962. *Principles of Electronics and Electronic Systems.* U.S. Naval Inst., Annapolis, MD.
Hulick, T.P. 1991. Switching power supplies for high voltage. *QEX* (Feb).
Jordan, E.C., ed. 1986. *Reference Data for Engineers: Radio, Electronics, Computer, and Communications,* 7th ed. Howard W. Sams & Co., Indianapolis, IN.
Krauss, H.L., Bostian, C.W., and Raab, F.H. 1980. *Solid State Radio Engineering.* Wiley, New York.
O'Reilly, W.P. 1975. Transmitter power amplifier design II. *Wireless World* 81 (Oct.).
Orr, W.I. 1986. *Radio Handbook,* 23rd ed. Howard W. Sams & Co., Indianapolis, IN.
Schetgen, R., ed. 1995. *The ARRL Handbook,* 72nd ed. American Radio Relay League. Newington, CT.
Taub, H. and Schilling, D.L. 1986. *Principles of Communication Systems,* 2nd ed. McGraw-Hill, New York.

12
Coaxial Transmission Lines

Jerry C. Whitaker
Editor

12.1 Introduction 12-1
   Skin Effect
12.2 Coaxial Transmission Line 12-2
   Electrical Parameters · Transverse Electromagnetic Mode
   · Dielectric · Impedance · Resonant Characteristics
12.3 Electrical Considerations 12-6
12.4 Coaxial Cable Ratings 12-7
   Power Rating · Connector Effects · Attenuation · Phase Stability
   · Mechanical Parameters

12.1 Introduction

The components that connect, interface, transfer, and filter RF energy within a given system — or between systems — are critical elements in the operation of vacuum tube devices. Such hardware, usually passive, determines to a large extent the overall performance of the RF generator. To optimize the performance of power vacuum devices, it is first necessary to understand and optimize the components upon which the tube depends.

The mechanical and electrical characteristics of the transmission line, waveguide, and associated hardware that carry power from a power source (usually a transmitter) to the load (usually an antenna) are critical to proper operation of any RF system. Mechanical considerations determine the ability of the components to withstand temperature extremes, lightning, rain, and wind. That is, they determine the overall reliability of the system.

Skin Effect

The effective resistance offered by a given conductor to radio frequencies is considerably higher than the ohmic resistance measured with direct current. This is because of an action known as the *skin effect*, which causes the currents to be concentrated in certain parts of the conductor and leaves the remainder of the cross-section to contribute little or nothing toward carrying the applied current.

When a conductor carries an alternating current, a magnetic field is produced that surrounds the wire. This field continually expands and contracts as the ac wave increases from zero to its maximum positive value and back to zero, then through its negative half-cycle. The changing magnetic lines of force cutting the conductor induce a voltage in the conductor in a direction that tends to retard the normal flow of current in the wire. This effect is more pronounced at the center of the conductor. Thus, current within the conductor tends to flow more easily toward the surface of the wire. The higher the frequency, the

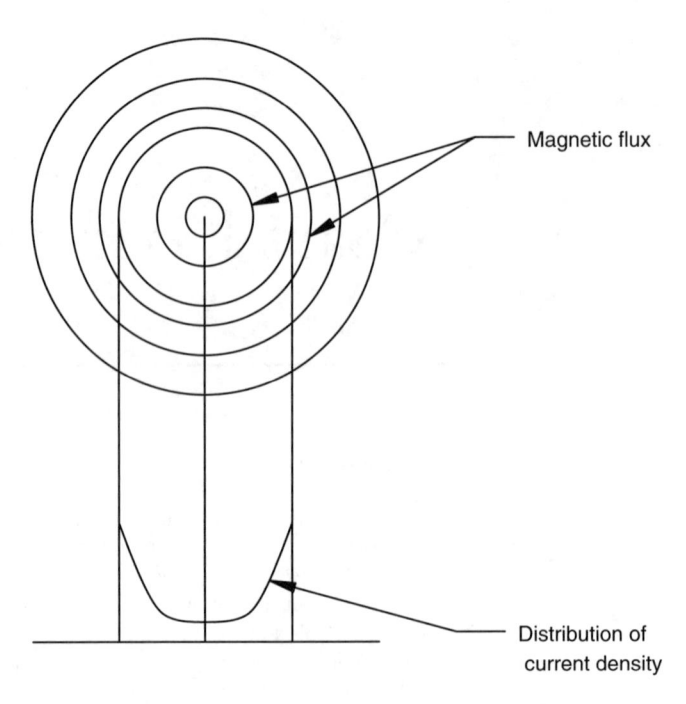

FIGURE 12.1 Skin effect on an isolated round conductor carrying a moderately high frequency signal.

greater the tendency for current to flow at the surface. The depth of current flow d is a function of frequency and is determined from the following equation:

$$d = \frac{2.6}{\sqrt{\mu f}} \tag{12.1}$$

where d is the depth of current in mils, μ is the permeability (copper = 1, steel = 300), and f is the frequency of signal in MHz. It can be calculated that at a frequency of 100 kHz, current flow penetrates a conductor by 8 mils. At 1 MHz, the skin effect causes current to travel in only the top 2.6 mils in copper, and even less in almost all other conductors. Therefore, the series impedance of conductors at high frequencies is significantly higher than at low frequencies. Figure 12.1 shows the distribution of current in a radial conductor.

When a circuit is operating at high frequencies, the skin effect causes the current to be redistributed over the conductor cross-section in such a way as to make most of the current flow where it is encircled by the smallest number of flux lines. This general principle controls the distribution of current, regardless of the shape of the conductor involved. With a flat-strip conductor, the current flows primarily along the edges, where it is surrounded by the smallest amount of flux.

12.2 Coaxial Transmission Line

Two types of coaxial transmission line are in common use today: rigid line and corrugated (semiflexible) line. Rigid coaxial cable is constructed of heavy-wall copper tubes with Teflon or ceramic spacers. (Teflon™ is a registered trademark of DuPont.) Rigid line provides electrical performance approaching an ideal transmission line, including:

- High power-handling capability
- Low loss
- Low VSWR (voltage standing wave ratio)

Coaxial Transmission Lines

Rigid transmission line is, however, expensive to purchase and install.

The primary alternative to rigid coax is semiflexible transmission line made of corrugated outer and inner conductor tubes with a spiral polyethylene (or Teflon) insulator. The internal construction of a semiflexible line is shown in Fig. 12.2. Semiflexible line has four primary benefits:

- It is manufactured in a continuous length, rather than the 20-foot sections typically used for rigid line.
- Because of the corrugated construction, the line can be shaped as required for routing from the transmitter to the antenna.
- The corrugated construction permits differential expansion of the outer and inner conductors.
- Each size of line has a minimum bending radius. For most installations, the flexible nature of corrugated line permits the use of a single piece of cable from the transmitter to the antenna, with no elbows or other transition elements. This speeds installation and provides for a more reliable system.

Electrical Parameters

A signal traveling in free space is unimpeded; it has a free-space velocity equal to the speed of light. In a transmission line, capacitance and inductance slow the signal as it propagates along the line. The degree to which the signal is slowed is represented as a percentage of the free-space velocity. This quantity is called the relative velocity of propagation and is described by

$$V_p = \frac{1}{\sqrt{LC}} \tag{12.2}$$

where L is the inductance in henrys per foot, C is the capacitance in farads per foot, and

$$V_r = \frac{V_p}{c} \times 100\,\% \tag{12.3}$$

where V_p is the velocity of propagation, c is 9.842×10^8 feet per second (free-space velocity), and V_r is the velocity of propagation as a percentage of free-space velocity.

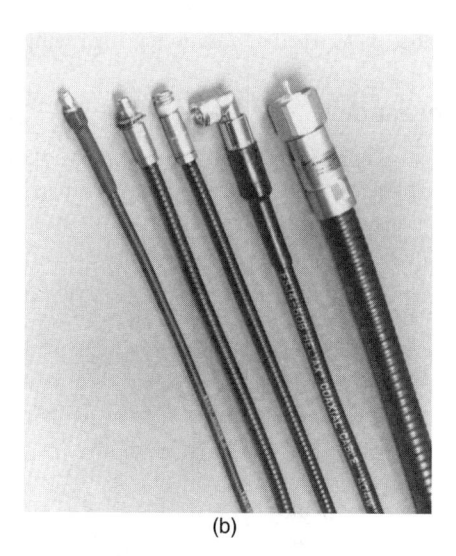

FIGURE 12.2 Semiflexible coaxial cable: (a) section of cable showing the basic construction, and (b) cable with various terminations. (Courtesy of Andrew Corporation, Orland Park, IL.)

Transverse Electromagnetic Mode

The principal mode of propagation in a coaxial line is the *transverse electromagnetic mode* (TEM). This mode will not propagate in a waveguide, and that is why coaxial lines can propagate a broad band of frequencies efficiently. The cutoff frequency for a coaxial transmission line is determined by the line dimensions. Above cutoff, modes other than TEM can exist and the transmission properties are no longer defined. The cutoff frequency is equivalent to

$$F_c = \frac{7.50 \times V_r}{D_i D_o} \qquad (12.4)$$

where F_c is the cutoff frequency in gigahertz, V_r is the velocity (percent), D_i is the inner diameter of outer conductor in inches, and D_o is the outer diameter of inner conductor in inches.

At dc, current in a conductor flows with uniform density over the cross-section of the conductor. At high frequencies, the current is displaced to the conductor surface. The effective cross-section of the conductor decreases and the conductor resistance increases because of the skin effect.

Center conductors are made from copper-clad aluminum or high-purity copper and can be solid, hollow tubular, or corrugated tubular. Solid center conductors are found on semiflexible cable with $1/2$-inch or smaller diameter. Tubular conductors are found in $7/8$-inch or larger-diameter cables. Although the tubular center conductor is used primarily to maintain flexibility, it can also be used to pressurize an antenna through the feeder.

Dielectric

Coaxial lines use two types of dielectric construction to isolate the inner conductor from the outer conductor. The first is an air dielectric, with the inner conductor supported by a dielectric spacer and the remaining volume filled with air or nitrogen gas. The spacer, which may be constructed of spiral or discrete rings, typically is made of Teflon or polyethylene. Air-dielectric cable offers lower attenuation and higher average power ratings than foam-filled cable but requires pressurization to prevent moisture entry.

Foam-dielectric cables are ideal for use as feeders with antennas that do not require pressurization. The center conductor is completely surrounded by foam-dielectric material, resulting in a high dielectric breakdown level. The dielectric materials are polyethylene-based formulations that contain antioxidants to reduce dielectric deterioration at high temperatures.

Impedance

The expression *transmission line impedance* applied to a point on a transmission line signifies the vector ratio of line voltage to line current at that particular point. This is the impedance that would be obtained if the transmission line were cut at the point in question, and the impedance looking toward the receiver were measured.

Because the voltage and current distribution on a line are such that the current tends to be small when the voltage is large (and vice versa), as shown in Fig. 12.3, the impedance will, in general, be oscillatory in the same manner as the voltage (large when the voltage is high and small when the voltage is low). Thus, in the case of a short-circuited receiver, the impedance will be high at distances from the receiving end that are odd multiples of $1/4$-wavelength, and will be low at distances that are even multiples of $1/4$-wavelength.

The extent to which the impedance fluctuates with distance depends on the *standing wave ratio* (ratio of reflected to incident waves), being less as the reflected wave is proportionally smaller than the incident wave. In the particular case where the load impedance equals the characteristic impedance, the impedance of the transmission line is equal to the characteristic impedance at all points along the line.

Coaxial Transmission Lines

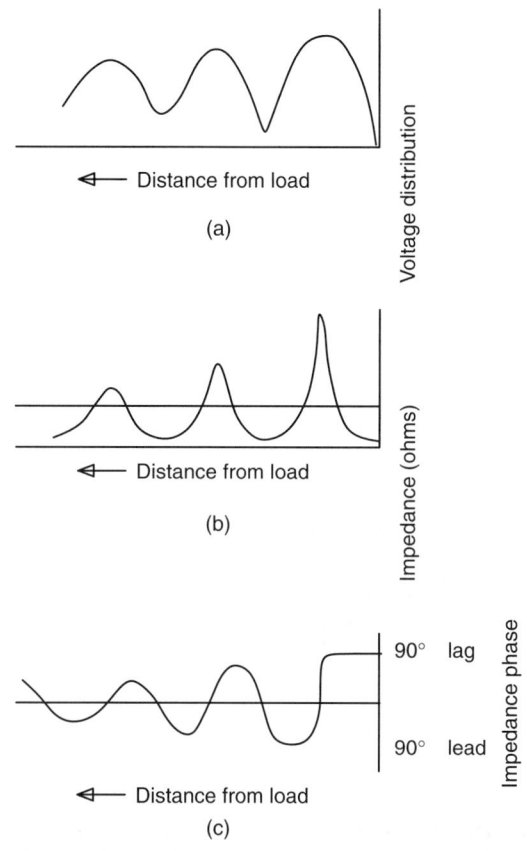

FIGURE 12.3 Magnitude and power factor of line impedance with increasing distance from the load for the case of a short-circuited receiver and a line with moderate attenuation: (a) voltage distribution, (b) impedance magnitude, and (c) impedance phase.

The *power factor* of the impedance of a transmission line varies according to the standing waves present. When the load impedance equals the characteristic impedance, there is no reflected wave and the power factor of the impedance is equal to the power factor of the characteristic impedance. At radio frequencies, the power factor under these conditions is accordingly resistive. However, when a reflected wave is present, the power factor is unity (resistive) only at the points on the line where the voltage passes through a maximum or a minimum. At other points the power factor will be reactive, alternating from leading to lagging at intervals of $1/4$-wavelength. When the line is short-circuited at the receiver, or when it has a resistive load less than the characteristic impedance so that the voltage distribution is of the short-circuit type, the power factor is inductive for lengths corresponding to less than the distance to the first voltage maximum. Thereafter, it alternates between capacitive and inductive at intervals of $1/4$-wavelength. Similarly, with an open-circuited receiver or with a resistive load greater than the characteristic impedance so that the voltage distribution is of the open-circuit type, the power factor is capacitive for lengths corresponding to less than the distance to the first voltage minimum. Thereafter, the power factor alternates between capacitive and inductive at intervals of $1/4$-wavelength, as in the short-circuited case.

Resonant Characteristics

A transmission line can be used to perform the functions of a resonant circuit. For example, if the line is short-circuited at the receiver, at frequencies in the vicinity of a frequency at which the line is an odd number of $1/4$-wavelengths long, the impedance will be high and will vary with frequency in the vicinity of resonance. This characteristic is similar in nature to a conventional parallel resonant circuit. The difference is that with the transmission line, there are a number of resonant frequencies, one for each of the infinite number of frequencies that make the line an odd number of $1/4$-wavelengths long. At VHF, the parallel impedance at resonance and the circuit Q obtainable are far higher than can be realized with lumped circuits. Behavior corresponding to that of a series resonant circuit can be obtained from a transmission line that is an odd number of $1/4$-wavelengths long and open-circuited at the receiver.

Transmission lines can also be used to provide low-loss inductances or capacitances if the proper combination of length, frequency, and termination is employed. Thus, a line short-circuited at the receiver will offer an inductive reactance when less than $1/4$-wavelength, and a capacitive reactance when between $1/4$- and $1/2$-wavelength. With an open-circuited receiver, the conditions for inductive and capacitive reactances are reversed.

12.3 Electrical Considerations

VSWR (voltage standing wave ratio), attenuation, and power-handling capability are key electrical factors in the application of coaxial cable. High VSWR can cause power loss, voltage breakdown, and thermal degradation of the line. High attenuation means less power delivered to the antenna, higher power consumption at the transmitter, and increased heating of the transmission line itself.

VSWR is a common measure of the quality of a coaxial cable. High VSWR indicates nonuniformities in the cable that can be caused by one or more of the following conditions:

- Variations in the dielectric core diameter
- Variations in the outer conductor
- Poor concentricity of the inner conductor
- Nonhomogeneous or periodic dielectric core

Although each of these conditions may contribute only a small reflection, they can add up to a measurable VSWR at a particular frequency.

Rigid transmission line is typically available in a standard length of 20 ft, and in alternative lengths of 19.5 and 19.75 ft. The shorter lines are used to avoid VSWR buildup caused by discontinuities resulting from the physical spacing between line section joints. If the section length selected and the operating frequency have a $1/2$-wave correlation, the connector junction discontinuities will add. This effect is known as flange buildup. The result can be excessive VSWR. The *critical frequency* at which a $1/2$-wave relationship exists is given by

$$F_{cr} = \frac{490.4 \times n}{L} \qquad (12.5)$$

where F_{cr} is the critical frequency, n is any integer, and L is the transmission line length in feet. For most applications, the critical frequency for a chosen line length should not fall closer than ±2 MHz of the passband at the operating frequency.

Attenuation is related to the construction of the cable itself and varies with frequency, product dimensions, and dielectric constant. Larger-diameter cable exhibits lower attenuation than smaller-diameter cable of similar construction when operated at the same frequency. It follows, therefore, that larger-diameter cables should be used for long runs.

Air-dielectric coax exhibits less attenuation than comparable-sized foam-dielectric cable. The attenuation characteristic of a given cable is also affected by standing waves present on the line resulting from an impedance mismatch. Table 12.1 shows a representative sampling of semiflexible coaxial cable specifications for a variety of line sizes.

TABLE 12.1 Representative Specifications for Various Types of Flexible Air-Dielectric Coaxial Cable

Cable Size (in.)	Maximum Frequency (MHz)	Velocity (%)	Peak Power 1 MHz (kW)	Average Power 100 MHz (kW)	Average Power 1 MHz (kW)	Attenuation[a] 100 MHz (dB)	Attenuation[a] 1 MHz (dB)
1 5/8	2.7	92.1	145	145	14.4	0.020	0.207
3	1.64	93.3	320	320	37	0.013	0.14
4	1.22	92	490	490	56	0.010	0.113
5	0.96	93.1	765	765	73	0.007	0.079

[a] Attenuation specified in dB/100 ft.

12.4 Coaxial Cable Ratings

Selection of a type and size of transmission line is determined by a number of parameters, including power-handling capability, attenuation, and phase stability.

Power Rating

Both peak and average power ratings are required to fully describe the capabilities of a transmission line. In most applications, the peak power rating limits the low frequency or pulse energy, and the average power rating limits high-frequency applications, as shown in Fig. 12.4. Peak power ratings are usually stated for the following conditions:

- VSWR = 1.0
- Zero modulation
- One atmosphere of absolute dry air pressure at sea level

The peak power rating of a selected cable must be greater than the following expression, in addition to satisfying the average-power-handling criteria:

$$E_{pk} > P_t \times (1 + M)^2 \times \text{VSWR} \tag{12.6}$$

where E_{pk} is the cable peak power rating in kilowatts, P_t is the transmitter power in kilowatts, M is the amplitude modulation percentage expressed decimally (100% = 1.0), and VSWR is the voltage standing wave ratio. From this equation, it can be seen that 100% amplitude modulation will increase the peak power in the transmission line by a factor of 4. Furthermore, the peak power in the transmission line increases directly with VSWR.

The peak power rating is limited by the voltage breakdown potential between the inner and outer conductors of the line. The breakdown point is independent of frequency. It varies, however, with the line pressure (for an air-dielectric cable) and the type of pressurizing gas.

The average power rating of a transmission line is limited by the safe, long-term operating temperature of the inner conductor and the dielectric. Excessive temperatures on the inner conductor will cause the dielectric material to soften, leading to mechanical instability inside the line.

The primary purpose of pressurization of an air-dielectric cable is to prevent the ingress of moisture. Moisture, if allowed to accumulate in the line, can increase attenuation and reduce the breakdown voltage between the inner and outer conductors. Pressurization with high-density gases can be used to increase both the average power and the peak power ratings of a transmission line. For a given line pressure, the increased power rating is more significant for peak power than for average power. High-density gases used for such applications include Freon 116 and sulfur hexafluoride. Figure 12.5 illustrates the effects of pressurization on cable power rating.

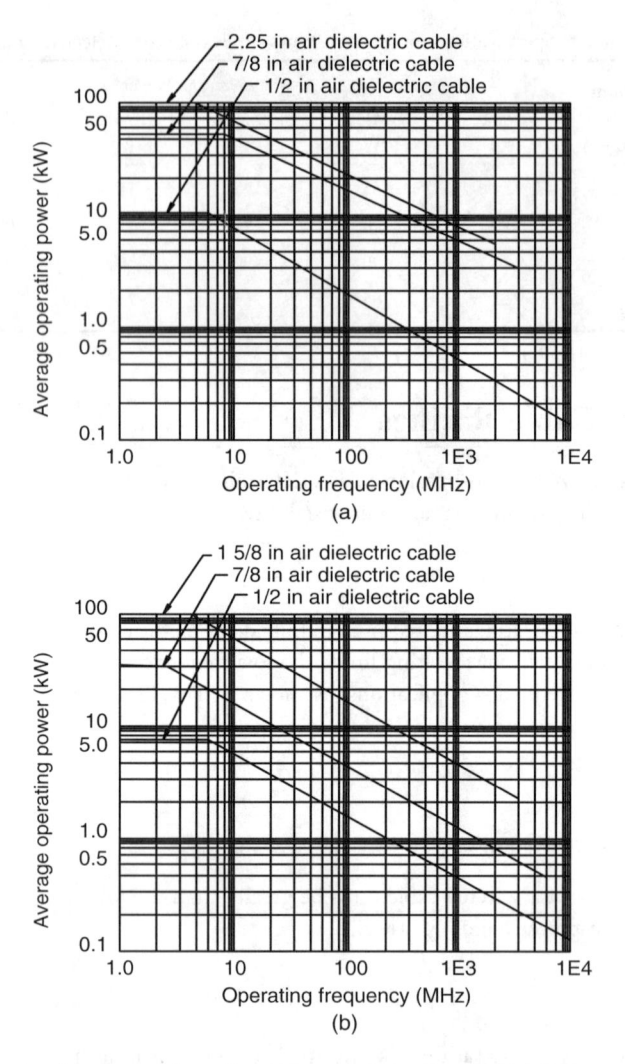

FIGURE 12.4 Power rating data for a variety of coaxial transmission lines: (a) 50 Ω line, and (b) 75 Ω line.

An adequate safety factor is necessary for peak and average power ratings. Most transmission lines are tested at two or more times their rated peak power before shipment to the customer. This safety factor is intended as a provision for transmitter transients, lightning-induced effects, and high-voltage excursions resulting from unforeseen operating conditions.

Connector Effects

Foam-dielectric cables typically have a greater dielectric strength than air-dielectric cables of similar size. For this reason, foam cables would be expected to exhibit higher peak power ratings than air lines. Higher values, however, usually cannot be realized in practice because the connectors commonly used for foam cables have air spaces at the cable/connector interface that limit the allowable RF voltage to "air cable" values.

The peak-power-handling capability of a transmission line is the smaller of the values for the cable and the connectors attached to it. Table 12.2 lists the peak power ratings of several common RF connectors at standard conditions (as defined in the previous section).

Coaxial Transmission Lines

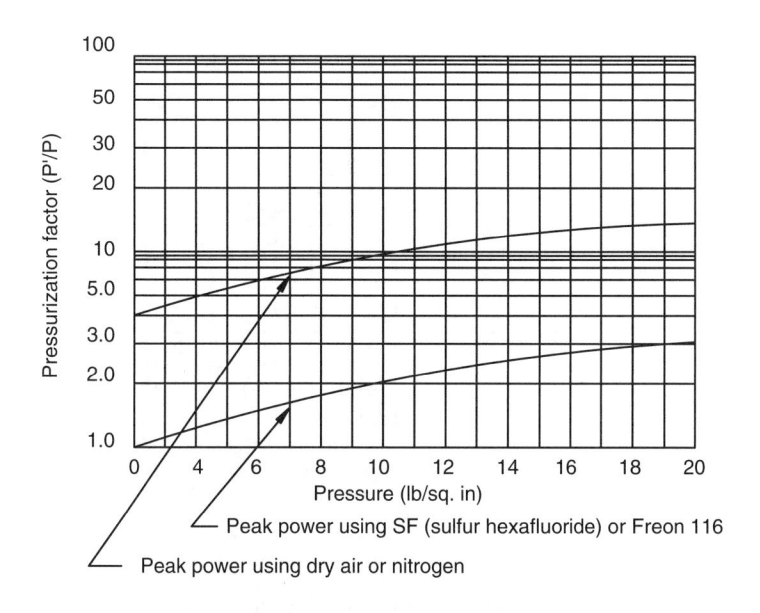

FIGURE 12.5 Effects of transmission line pressurization on peak power rating. Note that P' = rating of the line at the increased pressure and P = rating of the line at atmospheric pressure.

TABLE 12.2 Electrical Characteristics of Common RF Connectors

Connector Type	DC Test Voltage (kW)	Peak Power (kW)
SMA	1.0	1.2
BNC, TNC	1.5	2.8
N, UHF	2.0	4.9
GR	3.0	11
HN, $7/16$	4.0	20
LC	5.0	31
$7/8$ EIA, F Flange	6.0	44
1 $5/8$ EIA	11.0	150
3 $1/8$ EIA	19.0	4

Attenuation

The attenuation characteristics of a transmission line vary as a function of the operating frequency and the size of the line itself. The relationships are shown in Fig. 12.6.

The efficiency of a transmission line dictates how much power output from the transmitter actually reaches the antenna. Efficiency is determined by the length of the line and the attenuation per unit length.

The attenuation of a coaxial transmission line is defined by

$$\alpha = 10 \times \log\left\{\frac{P_1}{P_2}\right\} \quad (12.7)$$

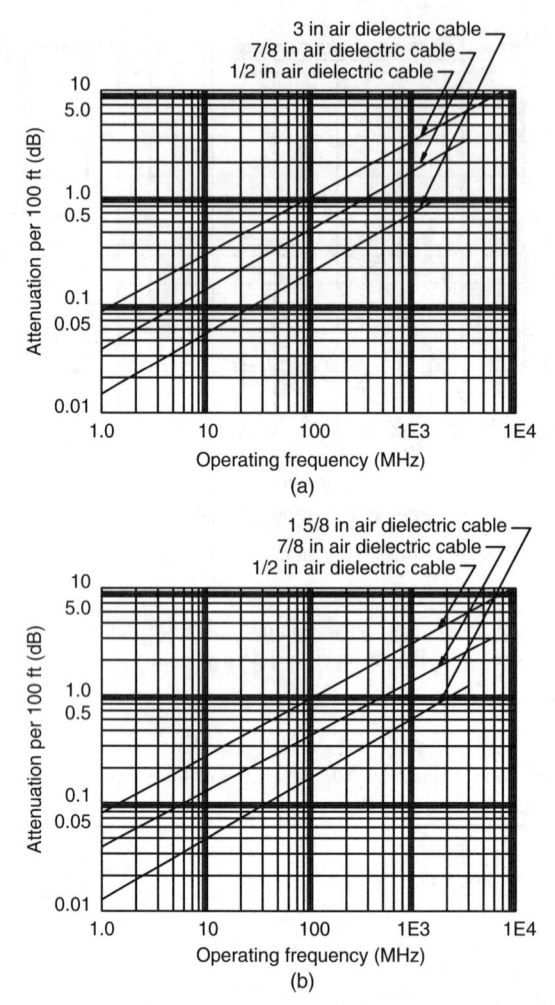

FIGURE 12.6 Attenuation characteristics for a selection of coaxial cables: (a) 50 Ω line, and (b) 75 Ω line.

where α is the attenuation in decibels per 100 meters, P_1 is the input power into a 100-meter line terminated with the nominal value of its characteristic impedance, and P_2 is the power measured at the end of the line. Stated in terms of efficiency (E, percent),

$$E = 100 \times \left\{ \frac{P_o}{P_i} \right\} \tag{12.8}$$

where P_i is the power delivered to the input of the transmission line and P_o is the power delivered to the antenna. The relationship between efficiency and loss in decibels (insertion loss) is illustrated in Fig. 12.7.

Manufacturer-supplied attenuation curves are typically guaranteed to within approximately ±5%. The values given usually are rated at 24°C (75°F) ambient temperature. Attenuation increases slightly with higher temperature or applied power. The effects of ambient temperature on attenuation are illustrated in Fig. 12.8.

Loss in connectors is negligible, except for small (SMA and BNC) connectors at frequencies of several gigahertz and higher. Small connectors used at high frequencies typically add 0.1 dB of loss per connector.

When a transmission line is attached to a load such as an antenna, the VSWR of the load increases the total transmission loss of the system. This effect is small under conditions of low VSWR. Figure 12.9 illustrates the interdependence of these two elements.

Coaxial Transmission Lines

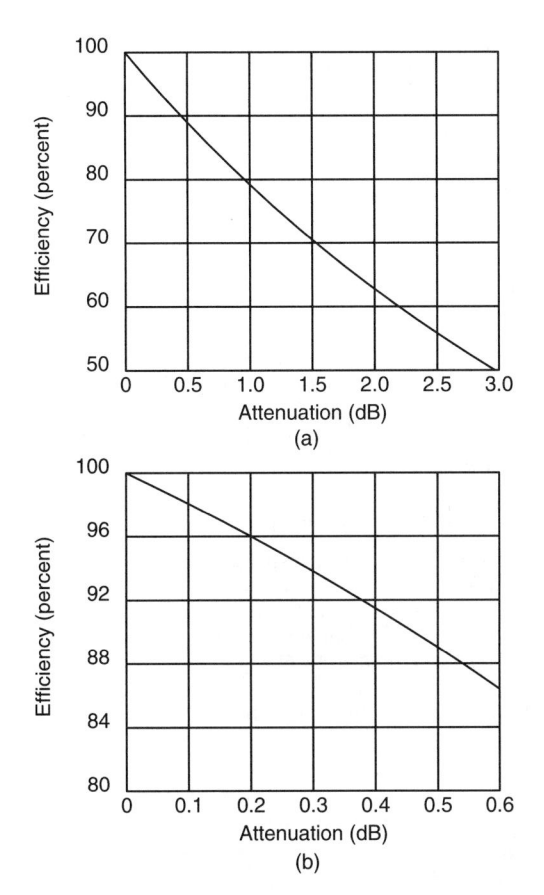

FIGURE 12.7 Conversion chart showing the relationship between decibel loss and efficiency of a transmission line: (a) high-loss line, and (b) low-loss line.

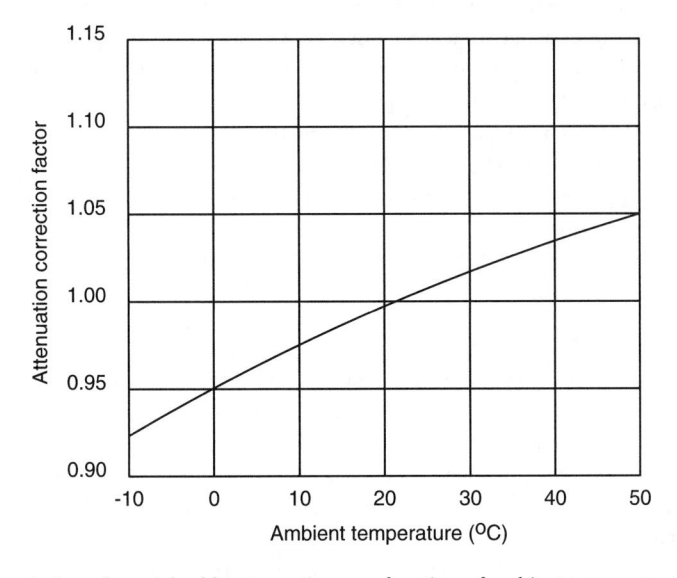

FIGURE 12.8 The variation of coaxial cable attenuation as a function of ambient temperature.

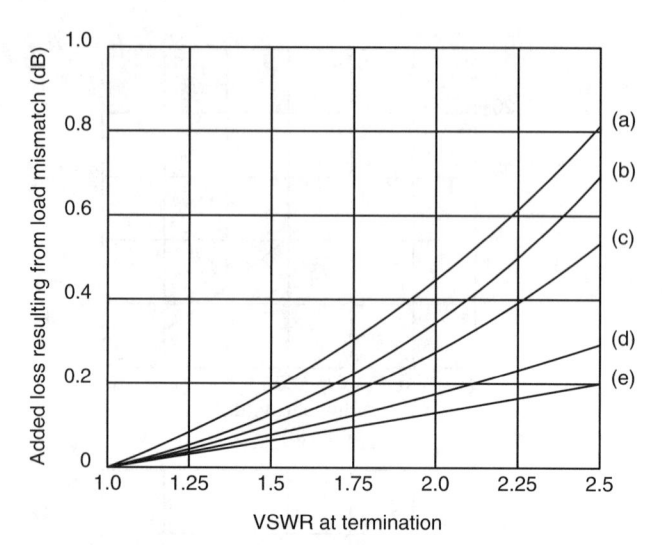

(a) 6 dB normal line attenuation
(b) 3 dB normal line attenuation
(c) 2 dB normal line attenuation
(d) 1 dB normal line attenuation
(e) 0.5 dB normal line attenuation

FIGURE 12.9 The effect of load VSWR on transmission line loss.

Phase Stability

A coaxial cable expands as the temperature of the cable increases, causing the electrical length of the line to increase as well. This factor results in phase changes that are a function of operating temperature. The phase relationship can be described by

$$\theta = 3.66 \times 10^{-7} \times P \times L \times T \times F \qquad (12.9)$$

where θ is the phase change in degrees, P is the phase-temperature coefficient of the cable, L is the length of coax in feet, T is the temperature range (minimum-to-maximum operating temperature), and F is the frequency in MHz. Phase changes that are a function of temperature are important in systems utilizing multiple transmission lines, such as a directional array fed from a single phasing source. To maintain proper operating parameters, the phase changes of the cables must be minimized. Specially designed coaxial cables that offer low-phase-temperature characteristics are available. Two types of coax are commonly used for this purpose: (1) phase-stabilized cables, which have undergone extensive temperature cycling until such time as they exhibit their minimum phase-temperature coefficient, and (2) phase-compensated cables, in which changes in the electrical length have been minimized through adjustment of the mechanical properties of the dielectric and inner/outer conductors.

Mechanical Parameters

Corrugated copper cables are designed to withstand bending with no change in properties. Low-density foam- and air-dielectric cables generally have a minimum bending radius of ten times the cable diameter. Super flexible versions provide a smaller allowable bending radius.

Rigid transmission lines will not tolerate bending. Instead, transition elements (elbows) of various sizes are used. Individual sections of rigid line are secured by multiple bolts around the circumference of a coupling flange.

When a large cable must be used to meet attenuation requirements, short lengths of a smaller cable (jumpers or pigtails) can be used on either end for ease of installation in low-power systems. The trade-off is slightly higher attenuation and some additional cost.

The *tensile strength* of a cable is defined as the axial load that may be applied to the line with no more than 0.2% permanent deformation after the load is released. When divided by the weight per foot of cable, this gives an indication of the maximum length of cable that is self-supporting and therefore can be readily installed on a tower with a single hoisting grip. This consideration usually applies only to long runs of corrugated line; rigid line is installed one section at a time.

The *crush strength* of a cable is defined as the maximum force per linear inch that can be applied by a flat plate without causing more than a 5% deformation of the cable diameter. Crush strength is a good indicator of the ruggedness of a cable and its ability to withstand rough handling during installation.

Cable jacketing affords mechanical protection during installation and service. Semiflexible cables typically are supplied with a jacket consisting of low-density polyethylene blended with 3% carbon black for protection from the sun's ultraviolet rays, which can degrade plastics over time. This approach has proved to be effective, yielding a life expectancy of more than 20 years. Rigid transmission line has no covering over the outer conductor.

For indoor applications, where fire-retardant properties are required, cables can be supplied with a fire-retardant jacket, usually listed by Underwriters Laboratories. Note that under the provisions of the National Electrical Code, outside plant cables such as standard black polyethylene-jacketed coaxial line may be run as far as 50 ft inside a building with no additional protection. The line can also be placed in conduit for longer runs.

Low-density foam cable is designed to prevent water from traveling along its length, should it enter through damage to the connector or the cable sheath. This is accomplished by mechanically locking the outer conductor to the foam dielectric by annular corrugations. Annular or ring corrugations, unlike helical or screw-thread-type corrugations, provide a water block at each corrugation. Closed-cell polyethylene dielectric foam is bonded to the inner conductor, completing the moisture seal.

A coaxial cable line is only as good as the connectors used to tie it together. The connector interface must provide a weatherproof bond with the cable to prevent water from penetrating the connection. This is ensured by the use of O-ring seals. The cable connector interface must also provide a good electrical bond that does not introduce a mismatch and increase VSWR. Good electrical contact between the connector and the cable ensures that proper RF shielding is maintained.

Bibliography

Andrew Corporation, Broadcast Transmission Line Systems, Technical Bulletin 1063H, Orland Park, IL, 1982.

Benson, K. B. and J. C. Whitaker, *Television and Audio Handbook for Technicians and Engineers*, McGraw-Hill, New York, 1989.

Cablewave Systems, The Broadcaster's Guide to Transmission Line Systems, Technical Bulletin 21A, North Haven, CT, 1976.

Cablewave Systems, Rigid Coaxial Transmission Lines, Cablewave Systems Catalog 700, North Haven, CT, 1989.

Crutchfield, E. B., Ed., *NAB Engineering Handbook*, 8th ed., National Association of Broadcasters, Washington, D.C., 1992.

Fink, D. and D. Christiansen, Eds., *Electronics Engineers' Handbook*, 3rd ed., McGraw-Hill, New York, 1989.

Jordan, Edward C., Ed., *Reference Data for Engineers: Radio, Electronics, Computer and Communications*, 7th ed., Howard W. Sams, Indianapolis, IN, 1985.

Perelman, R. and T. Sullivan, Selecting flexible coaxial cable, in *Broadcast Engineering*, Intertec Publishing, Overland Park, KS, May 1988.

Terman, F. E., *Radio Engineering*, 3rd ed., McGraw-Hill, New York, 1947.

Whitaker, Jerry C., G. DeSantis, and C. Paulson, *Interconnecting Electronic Systems*, CRC Press, Boca Raton, FL, 1993.

Whitaker, Jerry C., *Radio Frequency Transmission Systems: Design and Operation*, McGraw-Hill, New York, 1990.

13
Waveguide

Jerry C. Whitaker
Editor

13.1 Introduction 13-1
    Propagation Modes · Dual-Polarity Waveguide · Efficiency
13.2 Ridged Waveguide 13-3
13.3 Circular Waveguide 13-4
    Parasitic Energy
13.4 Doubly Truncated Waveguide 13-4
13.5 Impedance Matching 13-6
    Waveguide Filters · Installation Considerations · Tuning
    · Waveguide Hardware · Cavity Resonators

13.1 Introduction

As the operating frequency of a system reaches into the UHF band, waveguide-based transmission line systems become practical. From a mechanical standpoint, waveguide is simplicity itself. There is no inner conductor; RF energy is launched into the structure and propagates to the load. Several types of waveguide are available, including rectangular, square, circular, and elliptical. Waveguide offers several advantages over coax. First, unlike coax, waveguide can carry more power as the operating frequency increases. Second, efficiency is significantly better with waveguide at higher frequencies.

Rectangular waveguide is commonly used in high-power transmission systems. Circular waveguide may also be used, especially for applications requiring a cylindrical member, such as a rotating joint for an antenna feed. The physical dimensions of the guide are selected to provide for propagation in the *dominant* (lowest-order) mode.

Waveguide is not without its drawbacks, however. Rectangular or square guide constitutes a large windload surface, which places significant structural demands on a tower. Because of the physical configuration of rectangular and square guides, pressurization is limited, depending on the type of waveguide used (0.5 psi is typical). Excessive pressure can deform the guide shape and result in increased VSWR. Wind may also cause deformation and ensuing VSWR problems. These considerations have led to the development of circular and elliptical waveguides.

Propagation Modes

Propagation modes for waveguide fall into two broad categories:

- Transverse-electric (TE) waves
- Transverse-magnetic (TM) waves

With TE waves, the electric vector (E vector) is perpendicular to the direction of propagation. With TM waves, the magnetic vector (H vector) is perpendicular to the direction of propagation. These propagation modes take on integers (from 0 or 1 to infinity) that define field configurations. Only a limited number of these modes can be propagated, depending on the dimensions of the guide and the operating frequency.

Energy cannot propagate in waveguide unless the operating frequency is above the cutoff frequency. The cutoff frequency for rectangular guide is

$$F_c = \frac{c}{2 \times a} \tag{13.1}$$

where F_c is the waveguide cutoff frequency, $c = 1.179 \times 10^{10}$ inches per second (the velocity of light), and a is the wide dimension of the guide.

The cutoff frequency for circular waveguide is defined by

$$F_c = \frac{c}{3.41 \times a'} \tag{13.2}$$

where a' is the radius of the guide.

There are four common propagation modes in waveguide:

- $TE_{0,1}$, the principal mode in rectangular waveguide.
- $TE_{1,0}$, also used in rectangular waveguide.
- $TE_{1,1}$, the principal mode in circular waveguide. $TE_{1,1}$ develops a complex propagation pattern with electric vectors curving inside the guide. This mode exhibits the lowest cutoff frequency of all modes, which allows a smaller guide diameter for a specified operating frequency.
- $TM_{0,1}$, which has a slightly higher cutoff frequency than $TE_{1,1}$ for the same size guide. Developed as a result of discontinuities in the waveguide, such as flanges and transitions, $TM_{0,1}$ energy is not coupled out by either dominant or cross-polar transitions. The parasitic energy must be filtered out, or the waveguide diameter chosen carefully to reduce the unwanted mode.

The field configuration for the dominant mode in rectangular waveguide is illustrated in Fig. 13.1. Note that the electric field is vertical, with intensity maximum at the center of the guide and dropping off sinusoidally to zero intensity at the edges. The magnetic field is in the form of loops that lie in planes that are at right angles to the electric field (parallel to the top and bottom of the guide). The magnetic field distribution is the same for all planes perpendicular to the Y-axis. In the X-direction, the intensity of the component of magnetic field that is transverse to the axis of the waveguide (the component in the direction of X) is at any point in the waveguide directly proportional to the intensity of the electric field at that point. This entire configuration of fields travels in the direction of the waveguide axis (the Z-direction in Fig. 13.1).

The field configuration for the $TE_{1,1}$ mode in circular waveguide is illustrated in Fig. 13.2. The $TE_{1,1}$ mode has the longest cutoff wavelength and is, accordingly, the dominant mode. The next higher mode is $TM_{0,1}$, followed by $TE_{2,1}$.

Dual-Polarity Waveguide

Waveguide will support dual-polarity transmission within a single run of line. A combining element (dual-polarized transition) is used at the beginning of the run, and a splitter (polarized transition) is used at the end of the line. Square waveguide has found numerous applications in such systems. Theoretically, the $TE_{1,0}$ and $TE_{0,1}$ modes are capable of propagation without cross-coupling, at the same frequency, in lossless waveguide of square cross-section. In practice, surface irregularities, manufacturing tolerances, and wall losses give rise to $TE_{1,0}$- and $TE_{0,1}$-mode cross-conversion. Because this conversion occurs continuously along the waveguide, long guide runs usually are avoided in dual-polarity systems.

Waveguide

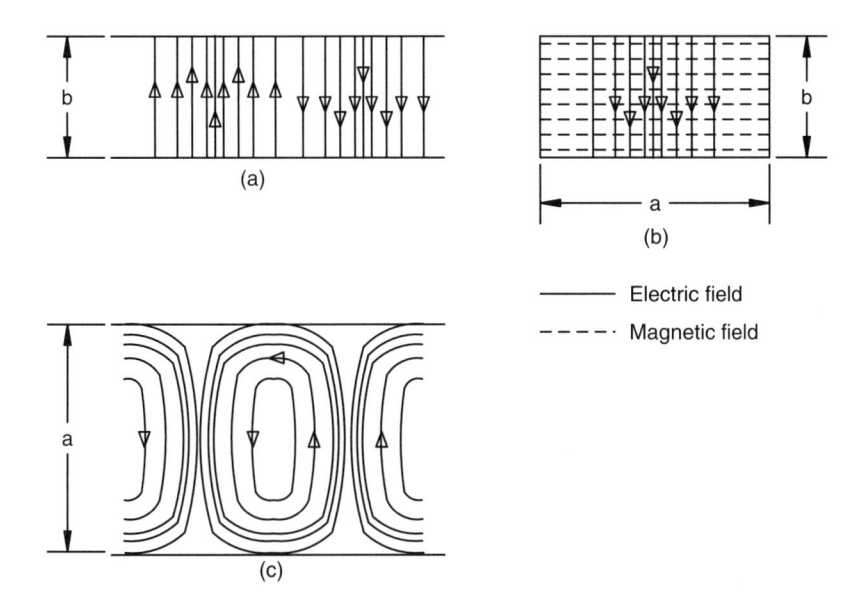

FIGURE 13.1 Field configuration of the dominant or TE$_{1,0}$ mode in a rectangular waveguide: (a) side view, (b) end view, and (c) top view.

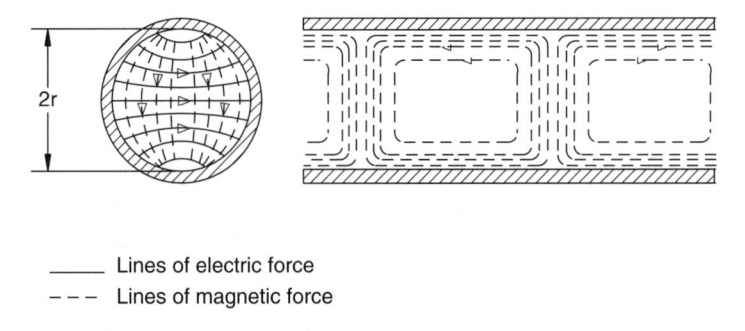

FIGURE 13.2 Field configuration of the dominant mode in circular waveguide.

Efficiency

Waveguide losses result from the following:

- Power dissipation in the waveguide walls and the dielectric material filling the enclosed space
- Leakage through the walls and transition connections of the guide
- Localized power absorption and heating at the connection points

The operating power of waveguide may be increased through pressurization. Sulfur hexafluoride commonly is used as the pressurizing gas.

13.2 Ridged Waveguide

Rectangular waveguide may be ridged to provide a lower cutoff frequency, thereby permitting use over a wider frequency band. As illustrated in Fig. 13.3, one- and two-ridged guides are used. Increased bandwidth comes at the expense of increased attenuation, relative to an equivalent section of rectangular guide.

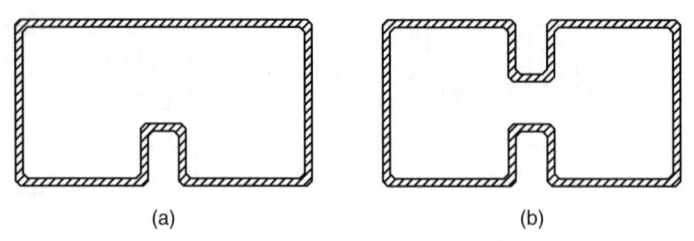

FIGURE 13.3 Ridged waveguide: (a) single-ridged, and (b) double-ridged.

13.3 Circular Waveguide

Circular waveguide offers several mechanical benefits over rectangular or square guide. The windload of circular guide is two-thirds that of an equivalent run of rectangular waveguide. It also presents lower and more uniform windloading than rectangular waveguide, reducing tower structural requirements.

The same physical properties of circular waveguide that give it good power handling and low attenuation also result in electrical complexities. Circular waveguide has two potentially unwanted modes of propagation: the cross-polarized $TE_{1,1}$ and $TM_{0,1}$ modes.

Circular waveguide, by definition, has no short or long dimension and, consequently, no method to prevent the development of cross-polar or orthogonal energy. Cross-polar energy is formed by small ellipticities in the waveguide. If the cross-polar energy is not trapped out, the parasitic energy can recombine with the dominant-mode energy.

Parasitic Energy

Hollow circular waveguide works as a high-Q resonant cavity for some energy and as a transmission medium for the rest. The parasitic energy present in the cavity formed by the guide will appear as increased VSWR if not disposed of. The polarization in the guide meanders and rotates as it propagates from the source to the load. The end pieces of the guide, typically circular-to-rectangular transitions, are polarization sensitive (see Fig. 13.4a). If the polarization of the incidental energy is not matched to the transition, energy will be reflected.

Several factors can cause this undesirable polarization. One cause is out-of-round guides that result from nonstandard manufacturing tolerances. In Fig. 13.4, the solid lines depict the situation at launching: perfectly circular guide with perpendicular polarization. However, certain ellipticities cause polarization rotation into unwanted states, while others have no effect. A 0.2% change in diameter can produce a –40 dB cross-polarization component per wavelength. This is roughly 0.03 in. for 18 in. of guide length.

Other sources of cross-polarization include twisted and bent guides, out-of-roundness, offset flanges, and transitions. Various methods are used to dispose of this energy trapped in the cavity, including absorbing loads placed at the ground and/or antenna level.

13.4 Doubly Truncated Waveguide

The design of doubly truncated waveguide (DTW) is intended to overcome the problems that can result from parasitic energy in a circular waveguide. As shown in Fig. 13.5, DTW consists of an almost elliptical guide inside a circular shell. This guide does not support cross-polarization; tuners and absorbing loads are not required. The low windload of hollow circular guide is maintained, except for the flange area.

Each length of waveguide is actually two separate pieces: a doubly truncated center section and a circular outer skin, joined at the flanges on each end. A large hole in the broadwall serves to pressurize the circular outer skin. Equal pressure inside the DTW and inside the circular skin ensures that the guide will not "breathe" or buckle as a result of rapid temperature changes.

DTW exhibits about 3% higher windloading than an equivalent run of circular waveguide (because of the transition section at the flange joints), and 32% lower loading than comparable rectangular waveguide.

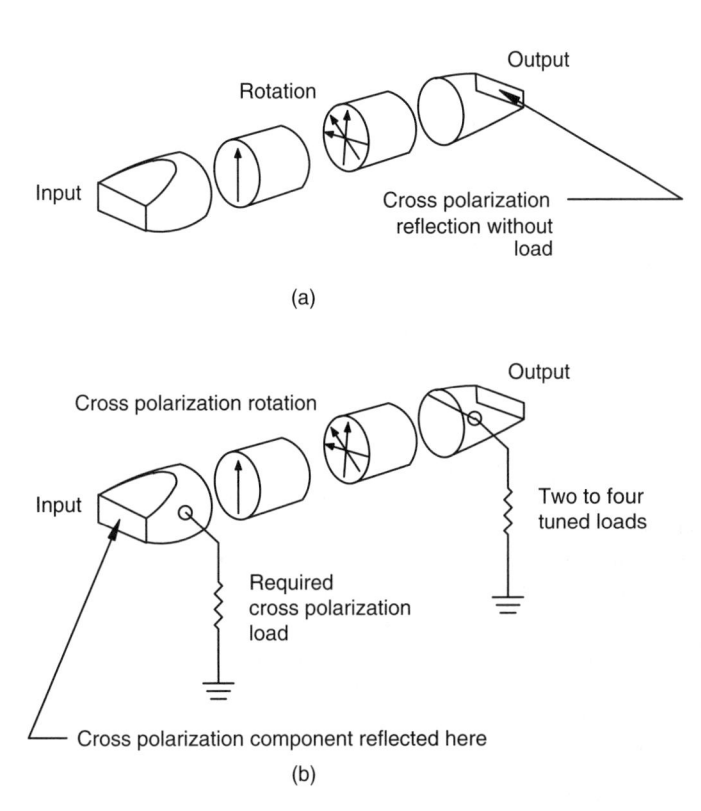

FIGURE 13.4 The effects of parasitic energy in circular waveguide: (a) trapped cross-polarization energy, and (b) delayed transmission of the trapped energy.

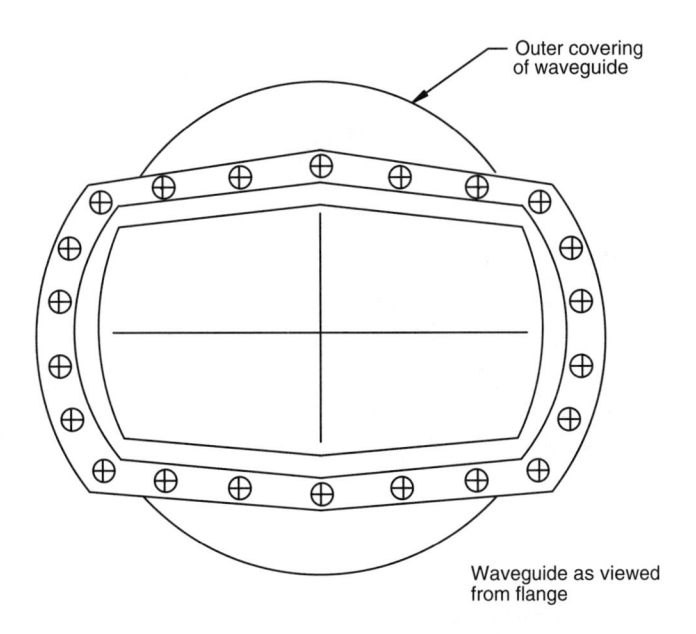

FIGURE 13.5 Physical construction of doubly truncated waveguide.

13.5 Impedance Matching

The efficient flow of power from one type of transmission medium to another requires matching of the field patterns across the boundary to launch the wave into the second medium with a minimum of reflections. Coaxial line is typically matched into rectangular waveguide by extending the center conductor of the coax through the broadwall of the guide, parallel to the electric field lines across the guide. Alternatively, the center conductor can be formed into a loop and oriented to couple the magnetic field to the guide mode.

Standing waves are generally to be avoided in waveguide for the same reasons that they are to be avoided in transmission lines. Accordingly, it is usually necessary to provide impedance-matching systems in waveguides to eliminate standing waves. One approach involves the introduction of a compensating reflection in the vicinity of a load that neutralizes the standing waves that would exist in the system because of an imperfect match. A probe or tuning screw is commonly used to accomplish this, as illustrated in Fig. 13.6. The tuning screw projects into the waveguide in a direction parallel to the electric field. This is equivalent to shunting a capacitive load across the guide. The susceptance of the load increases with extension into the guide up to 1/4 wavelength. When the probe is exactly 1/4 wavelength long, it becomes resonant and causes the guide to behave as though there were an open circuit at the point of the resonant probe. Probes longer than 1/4 wavelength but shorter than 3/4 wavelength introduce inductive loading. The extent to which such a probe projects into the waveguide determines the compensating reflection, and the position of the probe with respect to the standing wave pattern to be eliminated determines the phasing of the reflected wave.

Dielectric slugs produce an effect similar to that of a probe. The magnitude of the effect depends on the following considerations:

- Dielectric constant of the slug
- Thickness of the slug in an axial direction
- Whether the slug extends entirely across the waveguide

The phase of the reflected wave is controlled by varying the axial position of the slug.

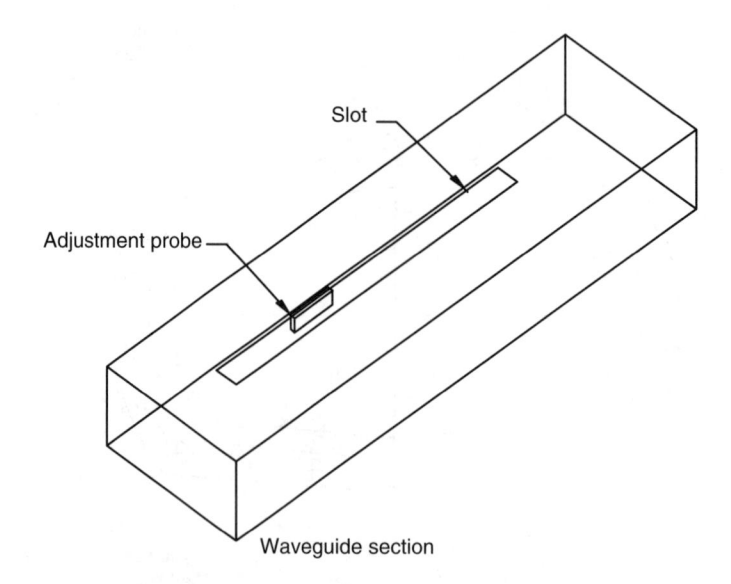

FIGURE 13.6 A probe configured to introduce a reflection in a waveguide that is adjustable in magnitude and phase.

There are several alternatives to the probe and slug for introducing controllable irregularity for impedance matching, including a metallic barrier or window placed at right angles to the axis of the guide, as illustrated in Fig. 13.7. Three configurations are shown:

- The arrangement illustrated in Fig. 13.7(a) produces an effect equivalent to shunting the waveguide with an inductive reactance.
- The arrangement shown in Fig. 13.7(b) produces the effect of a shunt capacitive susceptance.
- The arrangement shown in Fig. 13.7(c) produces an inductive shunt susceptance.

The waveguide equivalent of the coaxial cable tuning stub is a *tee* section, illustrated in Fig. 13.8. The magnitude of the compensating effect is controlled by the position of the short-circuiting plug in the branch. The phase of the compensating reflected wave produced by the branch is determined by the position of the branch in the guide.

Waveguide Filters

A section of waveguide beyond cutoff constitutes a simple high-pass reflective filter. Loading elements in the form of posts or stubs can be employed to supply the reactances required for conventional lumped-constant filter designs.

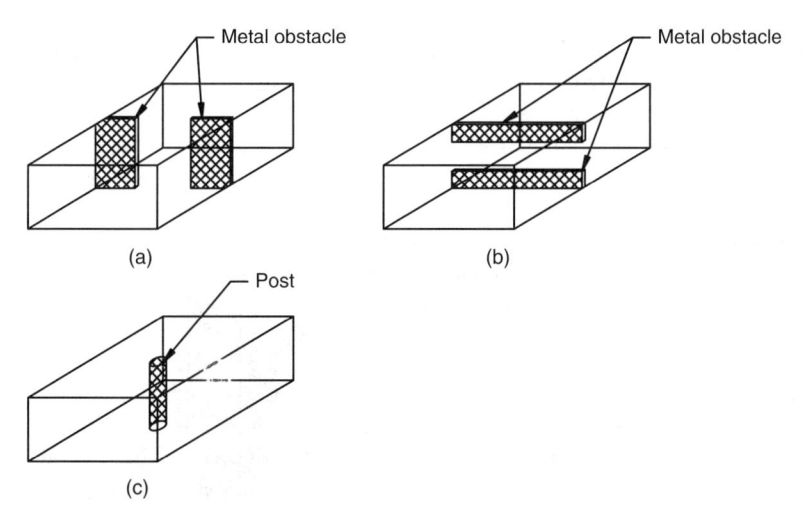

FIGURE 13.7 Waveguide obstructions used to introduce compensating reflections: (a) inductive window, (b) capacitive window, and (c) post (inductive) element.

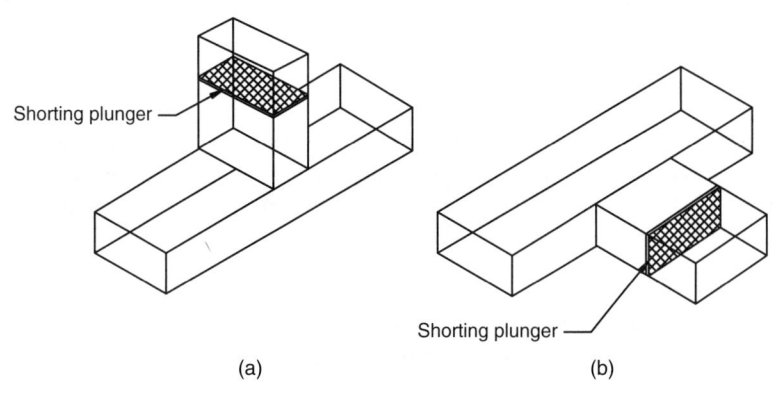

FIGURE 13.8 Waveguide stub elements used to introduce compensating reflections: (a) series *tee* element, and (b) shunt *tee* element.

Absorption filters prevent the reflection of unwanted energy by incorporating lossy material in secondary guides that are coupled through leaky walls (small sections of guide beyond cutoff in the passband). Such filters are typically used to suppress harmonic energy.

Installation Considerations

Waveguide system installation is both easier and more difficult than traditional transmission line installation. There is no inner conductor to align, but alignment pins must be set and more bolts are required per flange. Transition hardware to accommodate loads and coax-to-waveguide interfacing is also required.

Flange reflections can add up in phase at certain frequencies, resulting in high VSWR. The length of the guide must be chosen so that flange reflection buildup does not occur within the operating bandwidth.

Flexible sections of waveguide are used to join rigid sections or components that cannot be aligned otherwise. Flexible sections also permit controlled physical movement resulting from thermal expansion of the line. Such hardware is available in a variety of forms. Corrugated guide is commonly produced by shaping thin-wall seamless rectangular tubing. Flexible waveguide can accommodate only a limited amount of mechanical movement. Depending on the type of link, the manufacturer may specify a maximum number of bends.

Tuning

Circular waveguide must be tuned. This requires a two-step procedure. First, the cross-polar $TE_{1,1}$ component is reduced, primarily through axial ratio compensators or mode optimizers. These devices counteract the net system ellipticity and indirectly minimize cross-polar energy. The cross-polar filters can also be rotated to achieve maximum isolation between the dominant and cross-polar modes. Cross-polar energy manifests itself as a net signal rotation at the end of the waveguide run. A perfect system would have a net rotation of zero.

In the second step, tuning slugs at both the top and bottom of the waveguide run are adjusted to reduce the overall system VSWR. Tuning waveguide can be a complicated and time-consuming procedure. Once set, however, tuning normally does not drift and must be repeated only if major component changes are made.

Waveguide Hardware

Increased use of waveguide has led to the development of waveguide-based hardware for all elements from the output of the RF generator to the load. Waveguide-based filters, elbows, directional couplers, switches, combiners, and diplexers are currently available. The RF performance of a waveguide component is usually better than the same item in coax. This is especially true in the case of diplexers and filters. Waveguide-based hardware provides lower attenuation and greater power-handling capability for a given physical size.

Cavity Resonators

Any space completely enclosed by conducting walls can contain oscillating electromagnetic fields. Such a cavity possesses certain frequencies at which it will resonate when excited by electrical oscillations. These cavity resonators find extensive use as resonant circuits at VHF and above. Advantages of cavity resonators over conventional LC circuits include:

- Simplicity in design
- Relatively large physical size compared with alternative methods of obtaining resonance, an attribute that is important in high-power, high-frequency applications
- High Q
- Capability to configure the cavity to develop an extremely high shunt impedance

Waveguide

Cavity resonators are commonly used at wavelengths on the order of 10 cm or less.

The simplest cavity resonator is a section of waveguide shorted at each end with a length l equal to

$$l = \frac{\lambda_g}{2} \tag{13.3}$$

where $\lambda_g =$ is the guide wavelength. This configuration results in a resonance similar to that of a 1/2-wavelength transmission line short-circuited at the receiving end.

A sphere or any other enclosed surface (irrespective of how irregular the outline) can also be used to form a cavity resonator.

Any given cavity is resonant at a number of frequencies, corresponding to the different possible field conditions that can exist within the space. The resonance having the longest wavelength (lowest frequency) is termed the dominant or fundamental resonance. The resonant wavelength is proportional to the size of the resonator. If all dimensions are doubled, the wavelength corresponding to resonance will likewise be doubled. The resonant frequency of a cavity can be changed by incorporating one or more of the following mechanisms:

- Altering the mechanical dimensions of the cavity. Small changes can be achieved by flexing walls, but large changes require some form of sliding member.
- Coupling reactance into the resonator through a coupling loop.
- Introducing a movable copper paddle into the cavity. A paddle placed inside the resonator will affect the normal distribution of flux and tend to raise the resonant frequency by an amount determined by the orientation of the paddle.

The Q of a cavity resonator has the same significance as for a conventional resonant circuit. Q can be defined for a cavity by the relationship

$$Q = 2\pi \left(\frac{E_s}{E_l}\right) \tag{13.4}$$

where E_s is the energy stored and E_l is the energy lost per cycle. The energy stored is proportional to the square of the magnetic flux density integrated throughout the volume of the resonator. The energy lost in the walls is proportional to the square of the magnetic flux density integrated over the surface of the cavity. To obtain high Q, the resonator should have a large ratio of volume to surface area because it is the volume that stores energy and the surface area that dissipates energy.

Coupling can be obtained from a resonator by means of a coupling loop or coupling electrode. Magnetic coupling is accomplished through the use of a loop oriented so as to enclose magnetic flux lines existing in the desired mode of operation. This technique is illustrated in Fig. 13.9. A current passed through the loop will excite oscillations of this mode. Conversely, oscillations existing in the resonator

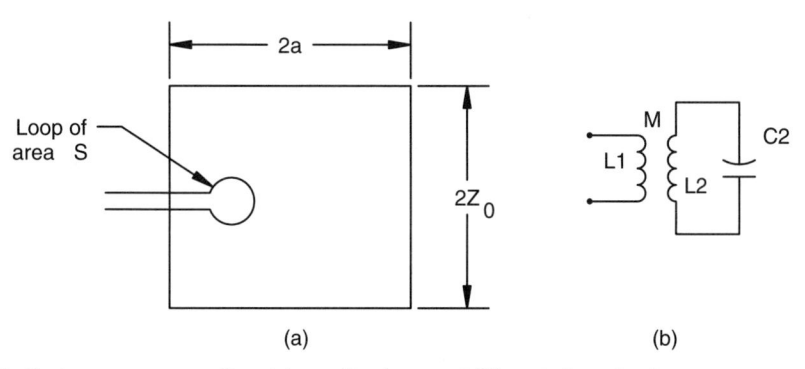

FIGURE 13.9 Cavity resonator coupling: (a) coupling loop, and (b) equivalent circuit.

will induce a voltage in the coupling loop. The magnitude of the coupling can be controlled by rotating the loop; the coupling reduces to zero when the plane of the loop is parallel to the magnetic flux.

Coupling of a resonator also may be accomplished through the use of a probe or opening in one wall of the cavity.

Bibliography

Andrew Corporation, Circular Waveguide: System Planning, Installation and Tuning, Technical Bulletin 1061H, Orland Park, IL, 1980.

Ben-Dov, O. and C. Plummer, Doubly truncated waveguide, in *Broadcast Engineering*, Intertec Publishing, Overland Park, KS, January 1989.

Benson, K. B. and J. C. Whitaker, *Television and Audio Handbook for Technicians and Engineers*, McGraw-Hill, New York, 1989.

Crutchfield, E. B., Ed., *NAB Engineering Handbook*, 8th ed., National Association of Broadcasters, Washington, D.C., 1992.

Fink, D. and D. Christiansen, Eds., *Electronics Engineers' Handbook*, 3rd ed., McGraw-Hill, New York, 1989.

Jordan, Edward C., Ed., *Reference Data for Engineers: Radio, Electronics, Computer and Communications*, 7th ed., Howard W. Sams, Indianapolis, IN, 1985.

Krohe, Gary L., Using circular waveguide, in *Broadcast Engineering*, Intertec Publishing, Overland Park, KS, May 1986.

Terman, F. E., *Radio Engineering*, 3rd ed., McGraw-Hill, New York, 1947.

Whitaker, Jerry C., G. DeSantis, and C. Paulson, *Interconnecting Electronic Systems*, CRC Press, Boca Raton, FL, 1993.

Whitaker, Jerry C., *Radio Frequency Transmission Systems: Design and Operation*, McGraw-Hill, New York, 1990.

14
RF Combiner and Diplexer Systems

Jerry C. Whitaker
Editor

14.1	Introduction 14-1
14.2	Passive Filters 14-2
	Filter Type · Filter Alignment · Filter Order
14.3	Four-Port Hybrid Combiner 14-4
14.4	Nonconstant-Impedance Diplexer 14-6
14.5	Constant-Impedance Diplexer 14-8
	Band-Stop Diplexer · Bandpass Constant-Impedance Diplexer · Intermodulation Products · Group Delay
14.6	Microwave Combiners 14-12
14.7	Hot-Switching Combiners 14-13
	Phase Relationships
14.8	High-Power Isolators 14-17
	Theory of Operation · Applications

14.1 Introduction

The basic purpose of an RF combiner is to add two or more signals to produce an output signal that is a composite of the inputs. The combiner performs this signal addition while providing isolation between inputs. Combiners perform other functions as well, and can be found in a wide variety of RF equipment utilizing solid-state devices and power vacuum tubes. Combiners are valuable devices because they permit multiple amplifiers to drive a single load. The isolation provided by the combiner permits tuning adjustments to be made on one amplifier — including turning it on or off — without significantly affecting the operation of the other amplifier. In a typical application, two amplifiers drive the hybrid and provide two output signals:

- A combined output representing the sum of the two input signals, typically directed toward the antenna
- A difference output representing the difference in amplitude and phase between the two input signals, and typically directed toward a dummy (reject) load

For systems in which more than two amplifiers must be combined, two or more combiners can be cascaded.

Diplexers are similar in nature to combiners but permit the summing of output signals from two or more amplifiers operating at different frequencies. This allows, for example, the outputs of several transmitters operating on different frequencies to utilize a single broadband antenna.

14.2 Passive Filters

A *filter* is a multiport network designed specifically to respond differently to signals of different frequency.[1] This definition excludes networks, which incidentally behave as filters, sometimes to the detriment of their main purpose. Passive filters are constructed exclusively with passive elements (i.e., resistors, inductors, and capacitors). Filters are generally categorized by the following general parameters:

- Type
- Alignment (or class)
- Order

Filter Type

Filters are categorized by type, according to the magnitude of the frequency response, as one of the following:[1]

- Low-pass (LP)
- High-pass (HP)
- Bandpass (BP)
- Band-stop (BS)

The terms *band-reject* and *notch* are also used as descriptive of the BS filter. The term *all-pass* is sometimes applied to a filter whose purpose is to alter the phase angle without affecting the magnitude of the frequency response. Ideal and practical interpretations of the types of filters and the associated terminology are illustrated in Fig. 14.1.

In general, the voltage gain of a filter in the stop band (or attenuation band) is less than $\sqrt{2}/2$ (approximately) 0.707 times the maximum voltage gain in the pass band. In logarithmic terms, the gain in the stop band is at least 3.01 dB less than the maximum gain in the pass band. The *cutoff* (break or

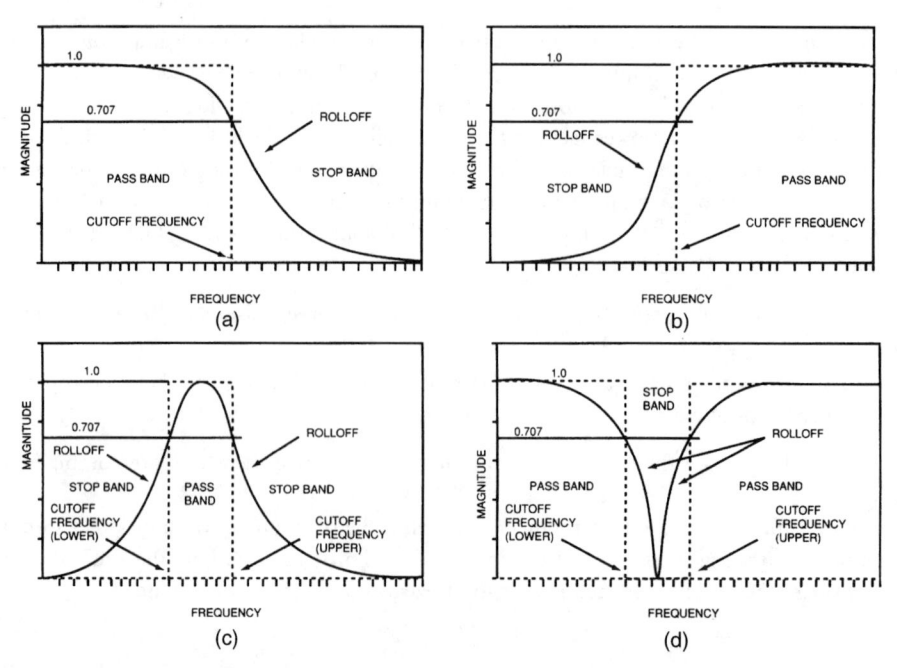

FIGURE 14.1 Filter characteristics by type: (a) low pass, (b) high pass, (c) bandpass, and (d) band stop. (From Harrison, C., Passive filters, in *The Electronics Handbook*, Whitaker, J. C., Ed., CRC Press, Boca Raton, FL, 1996, 279–290.)

corner) frequency separates the pass band from the stop band. In BP and BS filters, there are two cutoff frequencies, sometimes referred to as the lower and upper cutoff frequencies. Another expression for the cutoff frequency is *half-power frequency*, because the power delivered to a resistive load at cutoff frequency is one-half the maximum power delivered to the same load in the pass band. For BP and BS filters, the *center frequency* is the frequency of maximum or minimum response magnitude, respectively, and *band-width* is the difference between the upper and lower cutoff frequencies. *Rolloff* is the transition from pass band to stop band and is specified in gain unit per frequency unit (e.g., gain unit/Hz, dB/decade, dB/octave, etc.).

Filter Alignment

The *alignment* (or class) of a filter refers to the shape of the frequency response.[1] Fundamentally, filter alignment is determined by the coefficients of the filter network transfer function, so there are an indefinite number of filter alignments, some of which may not be realizable. The more common alignments are:

- Butterworth
- Chebyshev
- Bessel
- Inverse Chebyshev
- Elliptic (or Cauer)

Each filter alignment has a frequency response with a characteristic shape, which provides some particular advantage (see Fig. 14.2). Filters with Butterworth, Chebyshev, or Bessel alignment are called *all-pole filters* because their low-pass transfer functions have no zeros. Table 14.1 summarizes the characteristics of the standard filter alignments.

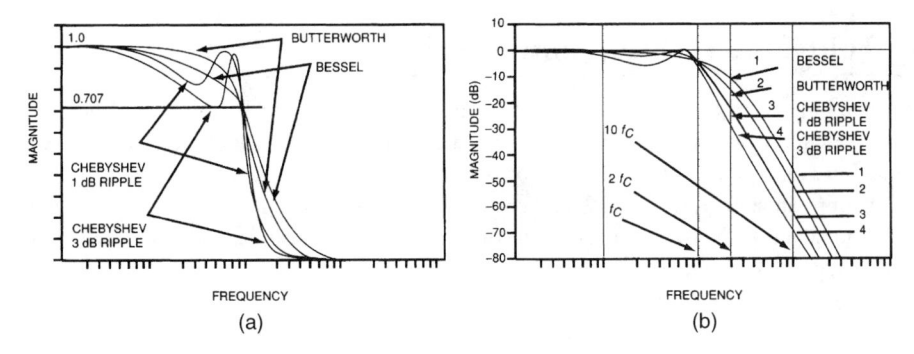

FIGURE 14.2 Filter characteristics by alignment, third-order, all-pole filters: (a) magnitude, and (b) magnitude in decibels. (From Harrison, C., Passive filters, in *The Electronics Handbook*, Whitaker, J. C., Ed., CRC Press, Boca Raton, FL, 1996, 279–290.)

TABLE 14.1 Summary of Standard Filter Alignments

Alignment	Pass-Band Description	Stop-Band Description	Comments
Butterworth	Monotonic	Monotonic	All-pole; maximally flat
Chebyshev	Rippled	Monotonic	All-pole
Bessel	Monotonic	Monotonic	All-pole; constant phase shift
Inverse Chebyshev	Monotonic	Rippled	
Elliptic (or Cauer)	Rippled	Rippled	

Source: Adapted from Harrison, D., Passive filters, in *The Electronics Handbook*, Whitaker, J.C., Ed., CRC Press, Boca Raton, FL, 1996, 279–290.

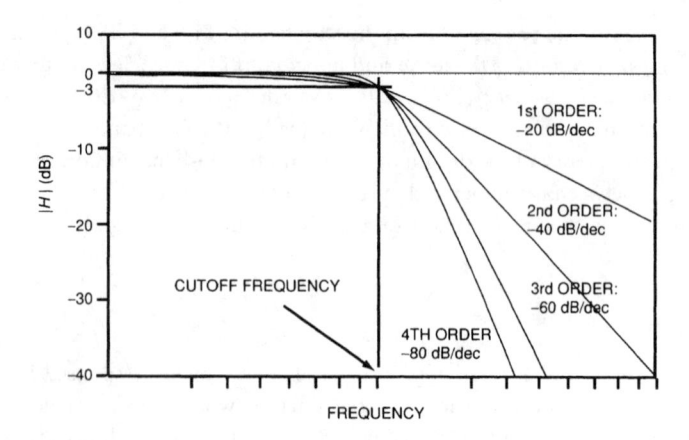

FIGURE 14.3 The effects of filter order on rolloff (Butterworth alignment). (From Harrison, C., Passive filters, in *The Electronics Handbook*, Whitaker, J. C., Ed., CRC Press, Boca Raton, FL, 1996, 279–290.)

Filter Order

The *order* of a filter is equal to the number of poles in the filter network transfer function.[1] For a lossless LC filter with resistive (nonreactive) termination, the number of reactive elements (inductors or capacitors) required to realize an LP or HP filter is equal to the order of the filter. Twice the number of reactive elements are required to realize a BP or a BS filter of the same order. In general, the order of a filter determines the slope of the rolloff: the higher the order, the steeper the rolloff. At frequencies greater than approximately one octave above cutoff (i.e., $f > 2 f_c$), the rolloff for all-pole filters is $20n$ dB/decade (or approximately $6n$ dB/octave), where n is the order of the filter (Fig. 14.3). In the vicinity of f_c, both filter alignment and filter order determine rolloff.

14.3 Four-Port Hybrid Combiner

A hybrid combiner (coupler) is a reciprocal four-port device that can be used for either splitting or combining RF energy over a wide range of frequencies. An exploded view of a typical 3-dB 90° hybrid is illustrated in Fig. 14.4 The device consists of two identical parallel transmission lines coupled over a distance of approximately $1/4$ wavelength and enclosed within a single outer conductor. Ports at the same end of the coupler are in phase, and ports at the opposite end of the coupler are in *quadrature* (90° phase shift) with respect to each other.

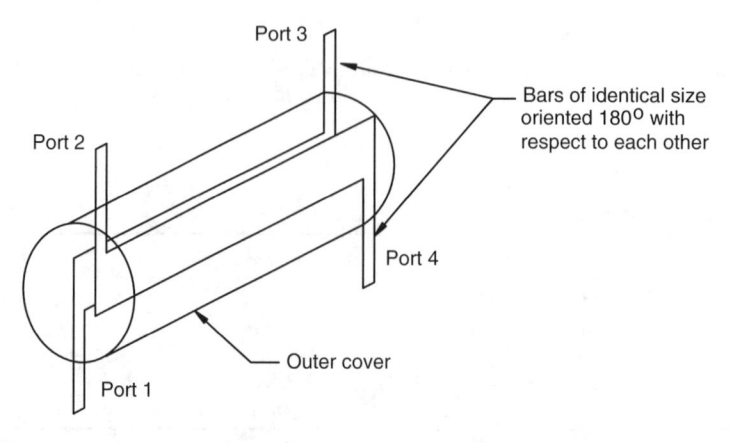

FIGURE 14.4 Physical model of a 90° hybrid combiner.

RF Combiner and Diplexer Systems

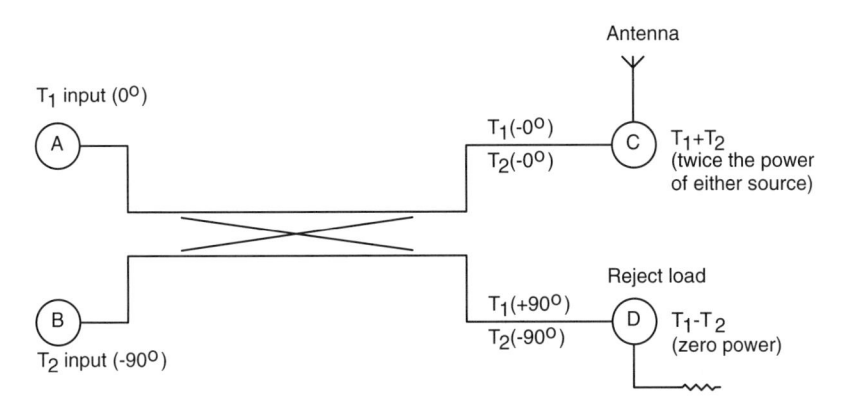

FIGURE 14.5 Operating principles of a hybrid combiner. This circuit is used to add two identical signals at inputs A and B.

The phase shift between the two inputs or outputs is always 90° and is virtually independent of frequency. If the coupler is being used to combine two signals into one output, these two signals must be fed to the hybrid in phase quadrature. When the coupler is used as a power splitter, the division is equal (half-power between the two ports). The hybrid presents a constant impedance to match each source.

Operation of the combiner can best be understood through observation of the device in a practical application. Figure 14.5 shows a four-port hybrid combiner used to add the outputs of two transmitters to feed a single load. The combiner accepts one RF source and splits it equally into two parts. One part arrives at output port C with 0° phase (no phase delay; it is the reference phase). The other part is delayed by 90° at port D. A second RF source connected to input port B, but with a phase delay of 90°, also will split in two but the signal arriving at port C now will be in phase with source 1 and the signal arriving at port D will cancel, as shown in the figure.

Output port C, the summing point of the hybrid, is connected to the load. Output port D is connected to a resistive load to absorb any residual power resulting from slight differences in amplitude and/or phase between the two input sources. If one of the RF inputs fails, half of the remaining transmitter output will be absorbed by the resistive load at port D.

The four-port hybrid works only when the two signals being mixed are identical in frequency and amplitude, and when their relative phase is 90°.

Operation of the hybrid can best be described by a scattering matrix in which vectors are used to show how the device operates. Such a matrix is shown in Table 14.2. In a 3-dB hybrid, two signals are fed to the inputs. An input signal at port 1 with 0° phase will arrive in phase at port 3, and at port 4 with a 90° lag (−90°) referenced to port 1. If the signal at port 2 already contains a 90° lag (−90° referenced to port 1), both input signals will combine in phase at port 4. The signal from port 2 also experiences another 90° change in the hybrid as it reaches port 3. Therefore, the signals from ports 1 and 2 cancel each other at port 3.

If the signal arriving at port 2 leads by 90° (mode 1 in the table), the combined power from ports 1 and 2 appears at port 4. If the two input signals are matched in phase (mode 4), the output ports (3 and 4) contain one-half of the power from each of the inputs.

If one of the inputs is removed, which would occur in a transmitter failure, only one hybrid input receives power (mode 5). Each output port then would receive one-half the input power of the remaining transmitter, as shown.

The input ports present a predictable load to each amplifier with a VSWR that is lower than the VSWR at the output port of the combiner. This characteristic results from the action of the difference port, typically connected to a dummy load. Reflected power coming into the output port will be directed to

TABLE 14.2 Single 90° Hybrid System Operating Modes

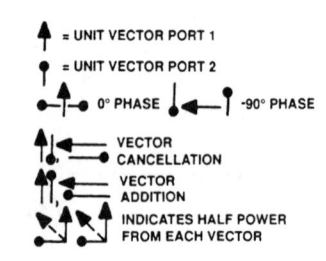

MODE	INPUT 1	INPUT 2	SCHEMATIC	OUTPUT 3	OUTPUT 4
1	$P_1 / 0°$	$P_2 \angle -90°$		0	$P_1 + P_2$
2	$P_1 / 0°$	$P_2 / 90°$		$P_1 + P_2$	0
3	$P_1 / 0°$	$P_2 / 0°$		$P_{½} + P_{2/2}$	$P_{½} + P_{2/2}$
4	$P_1 / 0°$	$P_2 = 0$		$P_{½}$	$P_{½}$
5	$P_1 = 0$	$P_2 / 0°$		$P_{2/2}$	$P_{2/2}$

- ↑ = UNIT VECTOR PORT 1
- ↑ = UNIT VECTOR PORT 2
- 0° PHASE
- -90° PHASE
- VECTOR CANCELLATION
- VECTOR ADDITION
- INDICATES HALF POWER FROM EACH VECTOR

the reject load and only a portion will be fed back to the amplifiers. Figure 14.6 illustrates the effect of output port VSWR on input port VSWR, and on the isolation between ports.

As noted previously, if the two inputs from the separate amplifiers are not equal in amplitude and not exactly in phase quadrature, some power will be dissipated in the difference port reject load. Figure 14.7 plots the effect of power imbalance, and Fig. 14.8 plots the effects of phase imbalance. The power lost in the reject load can be reduced to a negligible value by trimming the amplitude and/or phase of one (or both) amplifiers.

14.4 Nonconstant-Impedance Diplexer

Diplexers are used to combine amplifiers operating on different frequencies (and at different power levels) into a single output. Such systems are typically utilized to sum different transmitter outputs to feed a single broadband antenna.

The *branch diplexer* is the typical configuration for a diplexer that does not exhibit constant-impedance inputs. As shown in Fig. 14.9, the branch diplexer consists of two banks of filters each feeding into a coaxial tee. The electrical length between each filter output and the centerline of the tee is frequency sensitive, but this fact is more of a tuning nuisance than a genuine user concern.

RF Combiner and Diplexer Systems

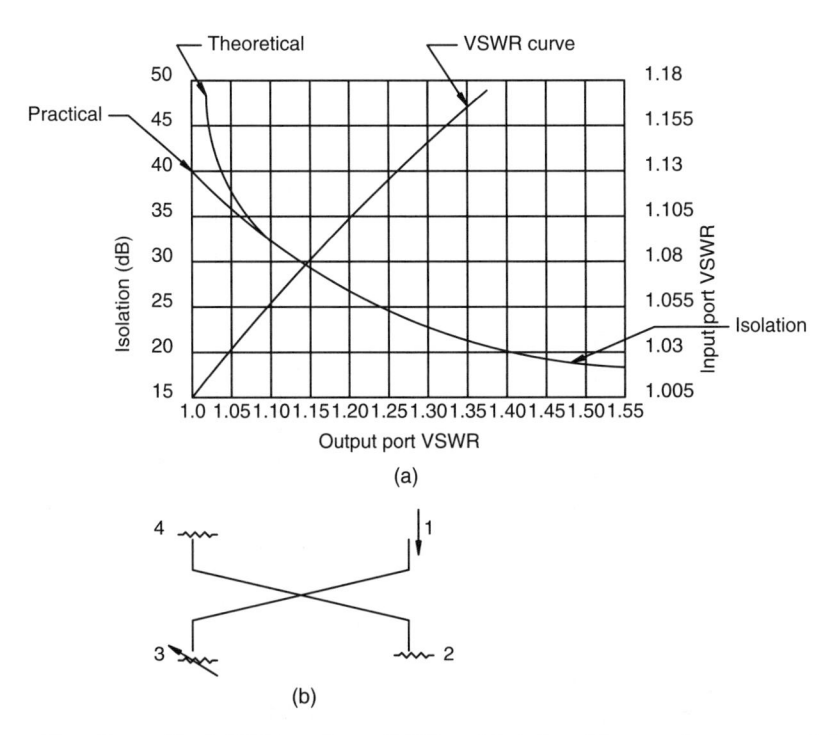

FIGURE 14.6 The effects of load VSWR on input VSWR and isolation: (a) respective curves, and (b) coupler schematic.

$$K = \sqrt{\frac{P_a}{P_b}}$$

P_a = Output power of transmitter 1 (lower power system)
P_b = Output power of transmitter 2

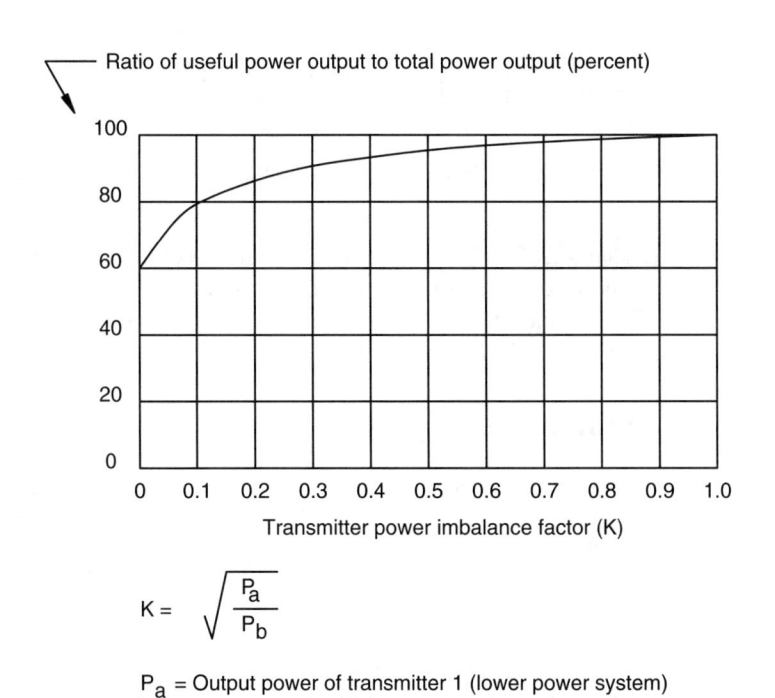

FIGURE 14.7 The effects of power imbalance at the inputs of a hybrid coupler.

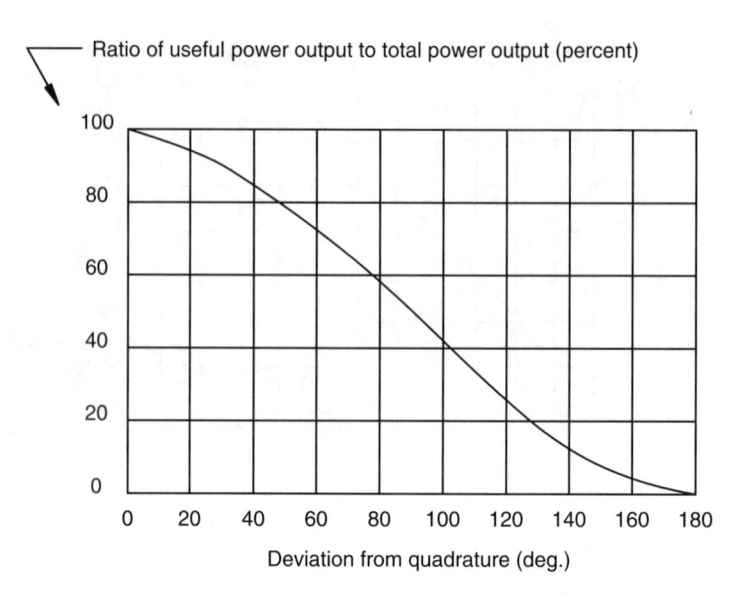

FIGURE 14.8 Phase sensitivity of a hybrid coupler.

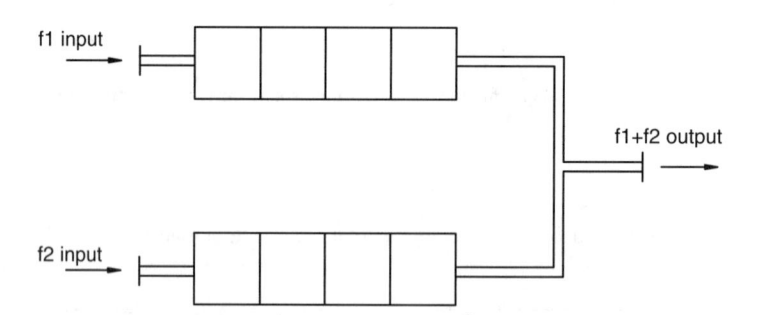

FIGURE 14.9 Non-constant-impedance branch diplexer. In this configuration, two banks of filters feed into a coaxial tee.

For this type of diplexer, all of the electrical parameters are a function of the filter characteristics. The VSWR, insertion loss, group delay, and rejection/isolation will be the same for the overall system as they are for the individual banks of cavities. The major limitation of this type of combiner is the degree of isolation that can be obtained for closely spaced channels.

14.5 Constant-Impedance Diplexer

The *constant-impedance diplexer* employs 3-dB hybrids and filters with a terminating load on the isolated port. The filters in this type of combiner can be either notch type or bandpass type. The performance characteristics are noticeably different for each design.

Band-Stop Diplexer

The *band-stop* (notch) constant-impedance diplexer is configured as shown in Fig. 14.10. For this design, the notch filters must have a high Q response to keep insertion loss low in the passband skirts. The high Q characteristic results in a sharp notch. Depending on the bandwidth required of the diplexer, two or more cavities may be located in each leg of the diplexer. They are typically stagger tuned, one high and

RF Combiner and Diplexer Systems

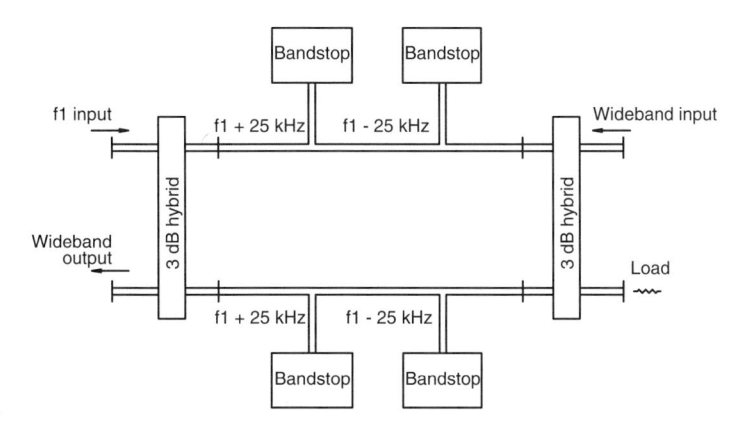

FIGURE 14.10 Band-stop (notch) constant-impedance diplexer module. This design incorporates two 3-dB hybrids and filters, with a terminating load on the isolated port.

one low for the two-cavity case. With this dual-cavity reject response in each leg of the band-stop diplexer system, the following analysis explains the key performance specifications.

If frequency f_1 is fed into the top left port of Fig. 14.10, it will be split equally into the upper and lower legs of the diplexer. Both of these signals will reach the filters in their respective leg and be rejected/reflected back toward the input hybrid, recombine, and emerge through the lower left port, also known as the wideband output.

The VSWR looking into the f_1 input is near 1:1 at all frequencies in the band. Within the bandwidth of the reject skirts, the observed VSWR is equal to the termination of the wideband output. Outside of the passband, the signals will pass by the cavities, enter the rightmost hybrid, recombine, and emerge into the dummy load. Consequently, the out-of-band VSWR is, in fact, the VSWR of the load.

The insertion loss from the f_1 input to the wideband output is low, typically on the order of 0.1 dB at carrier. This insertion loss depends on perfect reflection from the cavities. As the rejection diminishes on the skirts of the filters, the insertion loss from f_1 to the wideband output increases.

The limitation in reject bandwidth of the cavities causes the insertion loss to rise at the edges of the passband. The isolation from f_1 to the wideband input consists of a combination of the reject value of the cavities plus the isolation of the rightmost hybrid.

A signal entering at f_1 splits and proceeds in equal halves rightward through both the upper and lower legs of the diplexer. It is rejected by the filters at carrier and, to a lesser extent, on both sides of the carrier. Any residual signal that gets by the cavities reaches the rightmost hybrid. There it recombines and emerges from the load port. The hybrid provides a specified isolation from the load port to the wideband input port. This hybrid isolation must be added to the filter rejection to obtain the total isolation from the f_1 input to the wideband input.

If a signal is fed into the wideband input of the combiner shown in Fig. 14.10, the energy will split equally and proceed leftward along the upper and lower legs of the diplexer. Normally, f_1 is not fed into the wideband input. If it were, f_1 would be rejected by the notch filters and recombine into the load. All other signals sufficiently removed from f_1 will pass by the cavities with minimal insertion loss and recombine into the wideband output.

The VSWR looking into the wideband input is equal to the VSWR at the output for frequencies other than f_1. If f_1 were fed into the wideband input, the VSWR would be equal to the VSWR of the load.

Isolation from the wideband input to the f_1 input is simply the isolation available in the leftmost 3-dB hybrid. This isolation is usually inadequate for high-power applications. To increase the isolation from the wideband input to the f_1 input, it is necessary to use additional cavities between the f_1 transmitter and the f_1 input that will reject the frequencies fed into the wideband input. Unfortunately, adding these cavities to the input line also cancels the constant-impedance input. Although the notch diplexer as shown in the figure is truly a constant-impedance type of diplexer, the constant impedance is presented to the two inputs by virtue of using the hybrids at the respective inputs.

The hybrids essentially cause the diplexer to act as an absorptive type of filter to all out-of-band signals. The out-of-band signals generated by the transmitter are absorbed by the load rather than reflected to the transmitter.

If a filter is added to the input to supplement isolation, the filter will reflect some out-of-band signals back at the transmitter. The transmitter then will be seeing the passband impedance of this supplemental filter rather than the constant impedance of the notch diplexer. This is a serious deficiency for applications that require a true constant-impedance input.

Bandpass Constant-Impedance Diplexer

The bandpass constant-impedance diplexer is shown in Fig. 14.11. This system takes all of the best features of diplexers and combines them into one unit. It also provides a constant-impedance input that need not be supplemented with input cavities that rob the diplexer of its constant-impedance input.

The bandpass filters exhibit good bandwidth, providing near 1:1 VSWR across the operating bandpass. Insertion is low at carrier (0.28 dB is typical), rising slightly at bandpass extremes. Diplexer rejection, when supplemented by isolation of the hybrids, provides ample transmitter-to-transmitter isolation. Group delay is typically exceptional, providing performance specifications similar to those of a branch-style bandpass system. This configuration has the additional capability of providing high port-to-port isolation between closely spaced operating channels, as well as a true constant-impedance input.

The hybrids shown in Fig. 14.11 work in a manner identical to those described for a band-stop diplexer. However, the bandpass filters cause the system to exhibit performance specifications that exceed the band-stop system in every way. Consider a signal entering at the f_1 input.

Within the pass bands of the filters, which are tuned to f_1, the VSWR will be near 1:1 at carrier, rising slightly at the bandpass extremes. Because of the characteristic of the leftmost hybrid, the VSWR is, in fact, a measure of the similarity of response of the top and bottom bands of filters. The insertion loss looking from the f_1 input to the wideband output will be similar to the insertion loss of the top and bottom filters individually. A value of approximately 0.28 dB at carrier is typical.

Both the insertion loss and group delay can be determined by the design bandwidth of the filters. Increasing bandwidth causes the insertion loss and group delay deviation to decrease. Unfortunately, as the bandwidth increases with a given number of cavities, the isolation suffers for closely spaced channels because the reject skirt of the filter decreases with increasing bandwidth.

Isolation of f_1 to the wideband input is determined as follows. A signal enters at the f_1 input, splits equally into the upper and lower banks of filters, passes with minimal loss through the filters, and recombines into the wideband output of the rightmost hybrid. Both the load and the wideband input ports are isolated by their respective hybrids to some given value below the f_1 input level. Isolation of the f_1 input to the wideband input is supplemented by the reject skirt of the next module.

A signal fed into the wideband input could be any frequency removed from f_1 by some minimal amount. As the wideband signal enters, it will be split into equal halves by the hybrid, then proceed to the left until the two components reach the reject skirts of the filters. The filters will shunt all frequencies

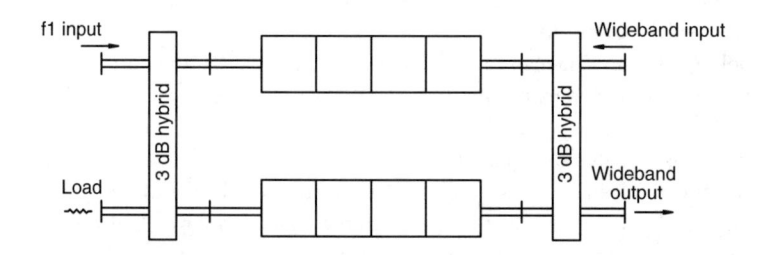

FIGURE 14.11 Bandpass constant-impedance diplexer.

removed from f_1 by some minimal amount. If the shunt energy is in phase for the given frequency when the signal is reflected back to the right hybrid, it will recombine into the wideband output. The VSWR under these conditions will be equal to the termination at the wideband output. If the reject skirts of the filter are sufficient, the insertion loss from wideband input to wideband output will be minimal (0.03 dB is typical).

The isolation from the wideband input to f_1 can be determined as follows. A signal enters at the wideband input, splits equally into upper and lower filters, and is rejected by the filters. Any residual signal that passes through the filters despite the rejection still will be in the proper phase to recombine into the load, producing additional isolation to the f_1 input port. Thus, the isolation of the wideband input to the f_1 input is the sum of the rejection from the filters and from the left hybrid.

Isolation of f_1 to f_2

Extending the use of the diplexer module into a multiplexer application supplements the deficient isolation described previously (narrowband input performance), while maintaining the constant-impedance input. In a multiplexer system, the wideband input of one module is connected to the wideband output of the next module, as illustrated in Fig. 14.12.

It has already been stated that the isolation from the f_1 input to the wideband input is deficient, but additional isolation is provided by the isolation of the wideband output to the f_2 input of the next module. Consider that f_1 has already experienced 30-dB isolation to the wideband input of the same module. When this signal continues to the next module through the wideband output of module 2, it will be split into equal halves and proceed to the left of module 2 until it reaches the reject skirts of the filters in module 2. Assume that these filters are tuned to f_2 and reject f_1 by at least 25 dB. The combined total isolation of f_1 to f_2 is the sum of the 30 dB of the right hybrid in module 1, plus the 25 dB of the reject skirts of module 2, for a total of 55 dB.

Intermodulation Products

The isolation just described is equal in magnitude to that for a band-stop module but provides further protection against the generation of intermodulation products. The most troublesome intermodulation (intermod) products usually occur when an incoming (secondary) signal mixes with the second harmonic of a primary transmitter.

When the primary transmitter is operating on frequency A, the intermod will occur at that frequency. This formula invariably places the intermod from the primary transmitter symmetrically about the operating frequency. By an interesting coincidence, the bandpass filters in the bandpass module also provide symmetrical reject response on both sides of the primary operating frequency.

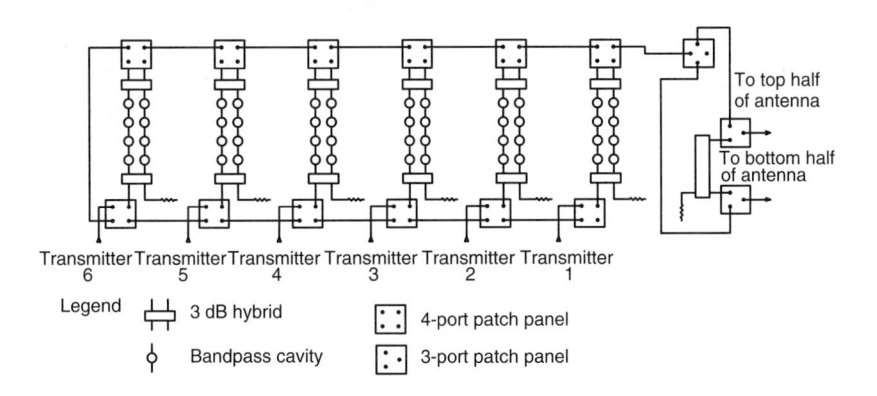

FIGURE 14.12 Schematic diagram of a six-module bandpass multiplexer. This configuration accommodates a split antenna design and incorporates patch panels for bypass purposes.

Assume again that the incoming signal is attenuated by 30 dB in the respective hybrid and by 25 dB in the filter, for a total of 55 dB. If an intermod is still generated despite this isolation, it will emerge on the other skirt of the filter attenuated by 25 dB. In the bandpass system, the incoming signal is attenuated by 55 dB and the resulting outgoing spur by 25 dB, for a total of 80 dB suppression.

Interestingly, the entire 80 dB of attenuation is supplied by the diplexer regardless of the turnaround loss of the transmitter. The tendency toward wideband final stage amplifiers in transmitters requires constant-impedance inputs. The transmitters also require increased isolation because they offer limited turnaround loss.

Group Delay

Group delay in the bandpass multiplexer module is equal to the sum of the narrowband input group delay and the wideband input group delay of all modules between the input and the load (antenna). The narrowband input group delay is a U-shaped response, with minimum at center and rising to a maximum on both sides at the frequency where the reject rises to 3 dB. Group delay then decreases rapidly at first, then more slowly.

If the bandwidth of the pass band is made such that the group delay is ±25 nsec over ±150 kHz (in an example system operating near 100 MHz), the 3 dB points will be at ±400 kHz and the out-of-band group delay will fall rapidly at ±800 kHz and possibly ±1.0 MHz. If there are no frequencies 800 kHz or 1.0 MHz removed upstream in other modules, this poses no problem.

If modules upstream are tuned to 800 kHz or 1.0 MHz on either side, then the group delay (when viewed at the upstream module) will consist of its own narrowband input group delay plus the rapidly falling group delay of the wideband input of the closely spaced downstream module. Under these circumstances, if good group delay is desired, it is possible to utilize a group delay compensation module.

A group delay compensation module consists of a hybrid and two cavities used as notch cavities. It typically provides a group delay response that is inverted, compared with a narrowband input group delay. Because group delay is additive, the inverted response subtracts from the standard response, effectively reducing the group delay deviation.

It should be noted that the improvement in group delay is obtained at a cost of insertion loss. In large systems (eight to ten modules), the insertion loss can be high because of the cumulative total of all wideband losses. Under these conditions, it may be more prudent to accept higher group delay and retain minimal insertion loss.

14.6 Microwave Combiners

Hybrid combiners are typically used in microwave amplifiers to add the output energy of individual power modules to provide the necessary output from an RF generator. Quadrature hybrids effect a VSWR-canceling phenomenon that results in well-matched power amplifier inputs and outputs that can be broadbanded with proper selection of hybrid tees. Several hybrid configurations are possible, including the following:

- Split-tee
- Branch-line
- Magic-tee
- Backward-wave

Key design parameters include coupling bandwidth, isolation, and ease of fabrication. The equal-amplitude, quadrature-phase, reverse-coupled TEM 1/4-wave hybrid is particularly attractive because of its bandwidth and amenability to various physical implementations. Such a device is illustrated in Fig. 14.13.

RF Combiner and Diplexer Systems 14-13

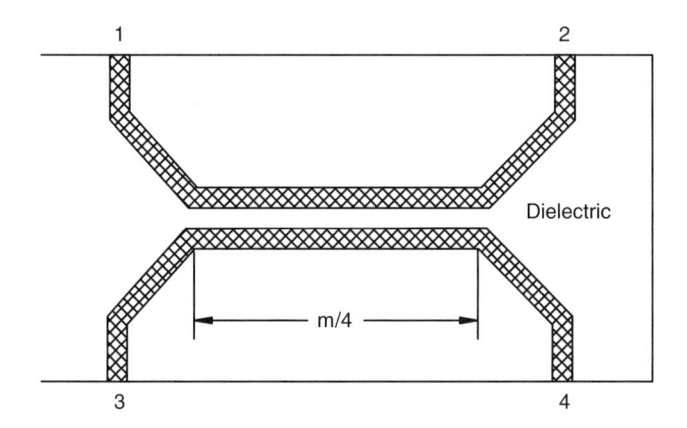

FIGURE 14.13 Reverse-coupled 1/4-wave hybrid coupler.

14.7 Hot-Switching Combiners

Switching RF is nothing new. Typically, the process involves coaxial switches, coupled with the necessary logic to ensure that the "switch" takes place with no RF energy on the contacts. This process usually takes the system off-line for a few seconds while the switch is completed. Through the use of hybrid combiners, however, it is possible to redirect RF signals without turning the carrier off. This process is referred to as *hot-switching*. Figure 14.14 illustrates two of the most common switching functions (SPST and DPDT) available from hot-switchers.

The unique phase-related properties of an RF hybrid make it possible to use the device as a switch. The input signals to the hybrid in Fig. 14.15a are equally powered but differ in phase by 90°. This phase difference results in the combined signals being routed to the output terminal at port 4. If the relative phase between the two input signals is changed by 180°, the summed output then appears on port 3, as shown in Fig. 14.15b. The 3-dB hybrid combiner thus functions as a switch.

This configuration permits the switching of two RF generators to either of two loads. Remember, however, that the switch takes place when the phase difference between the two inputs is 90°. To perform the switch in a useful way requires adding a high-power phase shifter to one input leg of the hybrid. The addition of the phase shifter permits the full power to be combined and switched to either output. This configuration of hybrid and phase shifter, however, will not permit switching a main or standby generator to a main or auxiliary load (DPDT function). To accomplish this additional switch, a second hybrid and phase shifter must be added, as shown in Fig. 14.16. This configuration can then perform the following switching functions:

- RF source 1 routed to output B
- RF source 2 routed to output A
- RF source 1 routed to output A
- RF source 2 routed to output B

The key element in developing such a switch is a high-power phase shifter that does not exhibit reflection characteristics. In this application, the phase shifter allows the line between the hybrids to be electrically lengthened or shortened. The ability to adjust the relative phase between the two input signals to the second hybrid provides the needed control to switch the input signal between the two output ports.

If a continuous analog phase shifter is used, the transfer switch shown in Fig. 14.16 can also act as a hot-switchless combiner where RF generators 1 and 2 can be combined and fed to either output A or B. The switching or combining functions are accomplished by changing the physical position of the phase shifter.

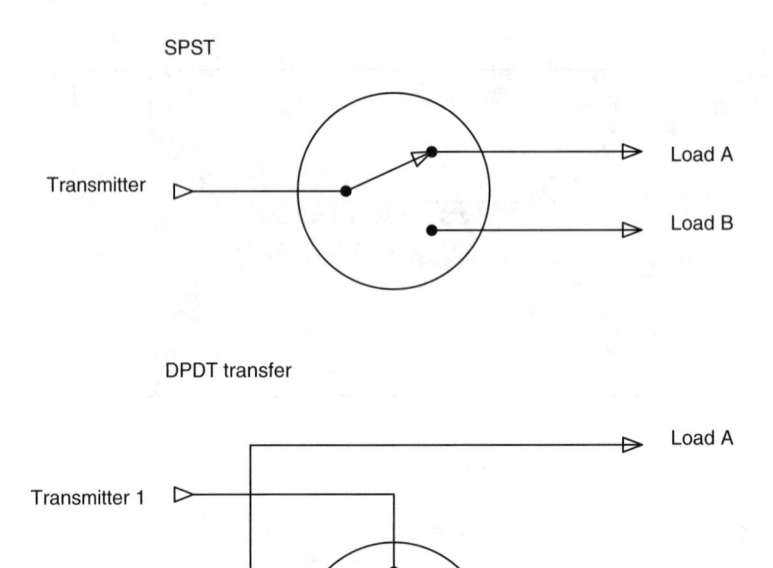

FIGURE 14.14 Common RF switching configurations.

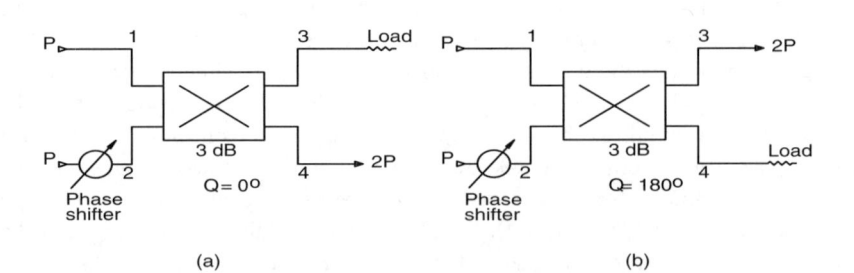

FIGURE 14.15 Hybrid switching configurations: (a) phase set so that the combined energy is delivered to port 4, and (b) phase set so that the combined energy is delivered to port 3.

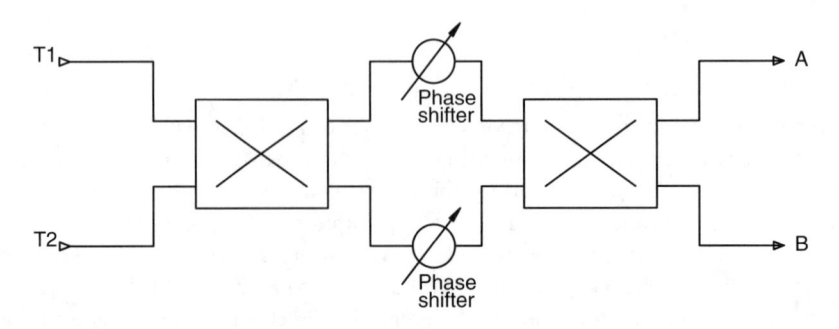

FIGURE 14.16 Additional switching and combining functions enabled by adding a second hybrid and another phase shifter to a hot-switching combiner.

RF Combiner and Diplexer Systems

Note that it does not matter whether the phase shifter is in one or both legs of the system. It is the phase difference ($\theta_1 - \theta_2$) between the two input legs of the second hybrid that is important. With two-phase shifters, dual drives are required. However, the phase shifter needs only two positions. In a one-phase shifter design, only a single drive is required but the phase shifter must have four fixed operating positions.

Phase Relationships

To better understand the dual-hybrid switching and combining process, it is necessary to examine the primary switching combinations. Table 14.3 lists the various combinations of inputs, relative phase, and output configurations that are possible with the single-phase shifter design.

Using vector analysis, note that when two input signals arrive in phase (mode 1) at ports 1 and 2 with the phase shifter set to 0°, the circuit acts like a crossover network with the power from input port 1 routed to output port 4. Power from input port 2 is routed to output port 3. If the phase shifter is set to 180°, the routing changes, with port 1 being routed to port 3 and port 2 being routed to port 4.

TABLE 14.3 Operating Modes of the Dual 90° Hybrid/Single-Phase Shifter Combiner System

MODE	INPUT 1	INPUT 2	∅	INPUT VECTOR	SCHEMATIC	OUTPUT VECTOR	OUTPUT 3	OUTPUT 4
1	$P_1 / 0°$	$P_2 / 0°$	0°				P_2	P_1
1	$P_1 / 0°$	$P_2 / 0°$	180°				P_1	P_2
2	$P_1 / 0°$	$P_2 = 0$	0°				0	P_1
3	$P_1 = 0$	$P_2 / 0°$	0°				P_2	0
4	$P_1 / 0°$	$P_2 = 0$	180°				P_1	0
5	$P_1 = 0$	$P_2 / 0°$	180°				0	P_2
6	$P_1 / 0°$	$P_2 / 0°$	90°				0	$P_1 + P_2$
	$P_1 / 0°$	$P_2 / 0°$	−90°				$P_1 + P_2$	0

↑ = UNIT VECTOR PORT 1
● = UNIT VECTOR PORT 2
↑↑ = PATHS 1-3, 2-4   ↑● = PATHS 1-4, 2-3
↑ OR ↑ 0° PHASE   ●— OR ←− −90° PHASE
—● OR ↓ −180° PHASE

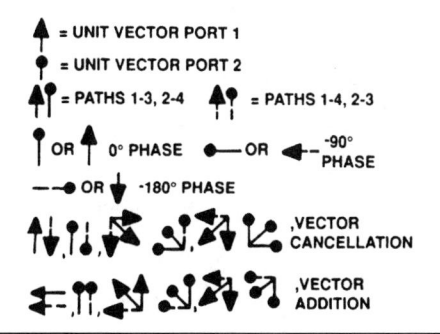

, VECTOR CANCELLATION
, VECTOR ADDITION

Mode 2 represents the case where one of the dual-input RF generators has failed. The output signal from the first hybrid arrives at the input to the second hybrid with a 90° phase difference. Because the second hybrid introduces a 90° phase shift, the vectors add at port 4 and cancel at port 3. This effectively switches the working transmitter connected to port 1 to output port 4, the load.

By introducing a 180° phase shift between the hybrids, as shown in modes 4 and 5, it is possible to reverse the circuit. This allows the outputs to be on the same side of the circuit as the inputs. This configuration might be useful if generator 1 fails and all power from generator 2 is directed to a diplexer connected to output 4.

Normal operating configurations are shown in modes 6 and 7. When both generators are running, it is possible to have the combined power routed to either output port. The switching is accomplished by introducing a ±90° phase shift between the hybrids.

As shown in the Table 14.3, it is possible to operate in all the listed modes through the use of a single-phase shifter. The phase shifter must provide four different phase positions. A similar analysis would show that a two-phase shifter design, with two positions for each shifter, is capable of providing the same operational modes.

The key to making hybrid switches work in the real world lies in the phase shifter. The dual 90° hybrid combiner just discussed requires a phase shifter capable of introducing a fixed phase offset of –90°, 0°, +90°, and +180°. This can be accomplished easily at low power levels through the use of a sliding short-circuit (trombone-type) line stretcher. However, when high-frequency and high-power signals are being used, the sliding short-circuit is not an appropriate design choice. In a typical case, the phase shifter must be able to handle 100 kW or more at UHF. Under these conditions, sliding short-circuit designs are often unreliable. Therefore, three other methods have been developed:

- Variable-dielectric vane
- Dielectric post
- Variable-phase hybrid

Variable-Dielectric Vane

The variable-dielectric vane consists of a long dielectric sheet mounted in a section of rectangular waveguide, as illustrated in Fig. 14.17. The dielectric sheet is long enough to introduce a 270° phase shift when located in the center of the waveguide. As the dielectric sheet is moved toward the wall, into the lower field, the phase shift decreases. A single-sided phase shifter can easily provide the needed four positions. A two-stage 1/4-wave transformer is used on each end of the sheet to maintain a proper match for any position over the desired operating band. The performance of a typical switchless combiner, using the dielectric vane, is given in Table 14.4.

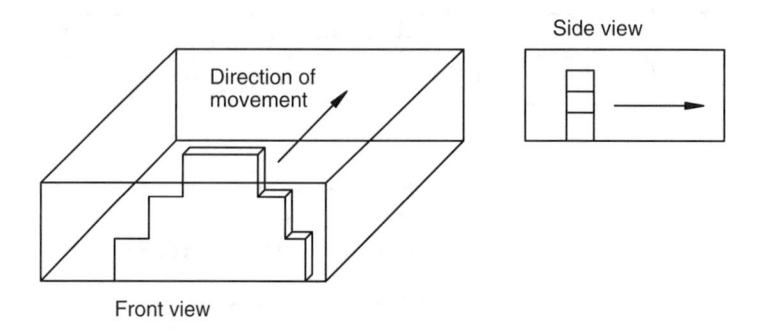

FIGURE 14.17 The dielectric vane switcher, which consists of a long dielectric sheet mounted within a section of rectangular waveguide.

RF Combiner and Diplexer Systems

TABLE 14.4 Typical Performance of a Dielectric Vane Phase Shifter

Type	Input	Phase Change (deg)	VSWR	Input Attenuation 1 (dB)	Input Attenuation 2 (dB)	Output Attenuation 3 (dB)	Output Attenuation 4 (dB)
Single input	T1	180	1.06	—	39	0.1	39
	T1	0	1.05	—	39	39	0.1
	T2	180	1.05	39	—	39	0.1
	T2	0	1.06	39	—	0.1	39
Dual input	T1 + T2	270	1.06	—	—	0.1	36
	T1 + T2	90	1.06	—	—	36	0.1

Dielectric Posts

Dielectric posts, shown in Fig. 14.18, operate on the same principle as the dielectric vane. The dielectric posts are positioned 1/4-wavelength apart from each other to cancel any mismatch, and to maintain minimal VSWR.

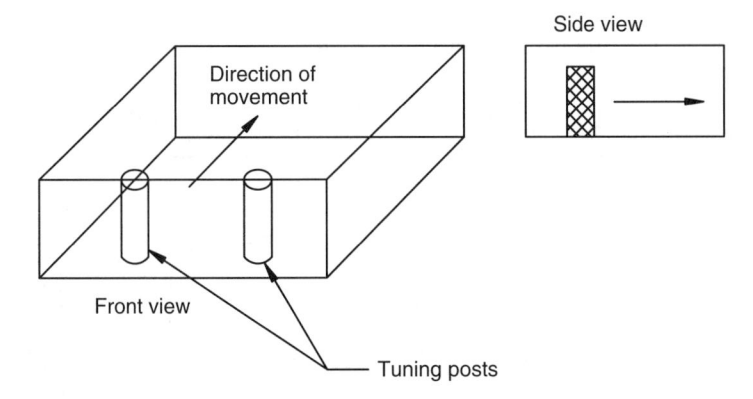

FIGURE 14.18 Dielectric post waveguide phase shifter.

Variable-Phase Hybrid

The variable-phase hybrid, shown in Fig. 14.19, relies on a 90° hybrid, similar to those used in a combiner. With a unit vector incident on port 1, the power is split by the 90° hybrid. The signal at ports 3 and 4 is reflected by the short circuit. These reflected signals are out of phase at port 1 and in phase at port 2. The relative phase of the hybrid can be changed by moving the short circuit.

The variable-phase hybrid is linear with respect to position. Noncontacting choke-type short circuits, with high front-to-back ratios, are typically used in the device. The performance available from a typical high-power, variable-phase switchless combiner is given in Table 14.5.

14.8 High-Power Isolators

The high-power ferrite isolator offers the ability to stabilize impedance, isolate the RF generator from load discontinuities, eliminate reflections from the load, and absorb harmonic and intermodulation products. The isolator can also be used to switch between an antenna or load under full power, or to combine two or more generators into a common load.

Isolators are commonly used in microwave transmitters at low power to protect the output stage from reflections. Until recently, however, the insertion loss of the ferrite made use of isolators impractical at high-power levels (25 kW and above). Ferrite isolators are now available that can handle 500 kW or more of forward power with less than 0.1 dB of forward power loss.

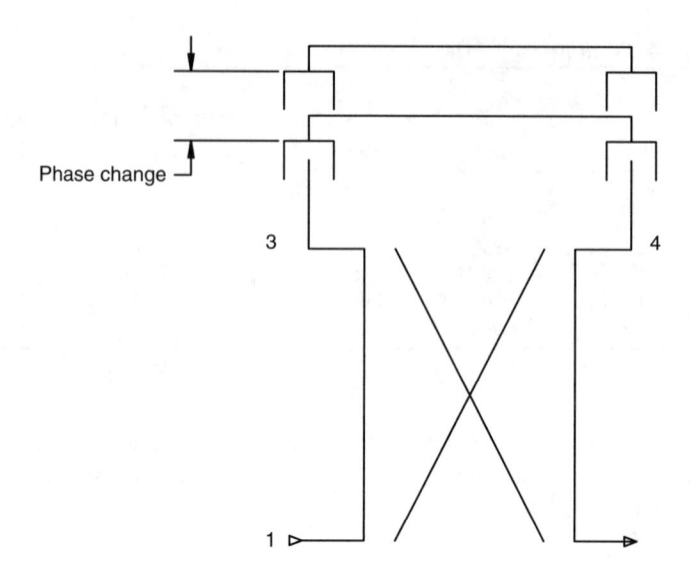

FIGURE 14.19 Variable-phase hybrid phase shifter.

TABLE 14.5 Typical Performance of a Variable-Phase Hybrid Phase Shifter

Type	Input	Phase Change (deg)	VSWR	Input Attenuation 1 (dB)	Input Attenuation 2 (dB)	Output Attenuation 3 (dB)	Output Attenuation 4 (dB)
Single input	T1	180	1.06	—	36	0.1	52
	T1	0	1.04	—	36	50	0.1
	T2	180	1.06	36	—	52	0.1
	T2	0	1.07	36	—	0.1	50
Dual input	T1 + T2	270	1.06	—	—	0.1	36
	T1 + T2	90	1.06	—	—	36	0.1

Theory of Operation

High-power isolators are three-port versions of a family of devices known as *circulators*. The circulator derives its name from the fact that a signal applied to one of the input ports can travel in only one direction, as shown in Fig. 14.20. The input port is isolated from the output port. A signal entering port 1 appears only at port 2; it does not appear at port 3 unless reflected from port 2. An important benefit of this one-way power transfer is that the input VSWR at port 1 is dependent only on the VSWR of the load placed at port 3. In most applications, this load is a resistive (dummy) load that presents a perfect load to the RF generator.

The unidirectional property of the isolator results from magnetization of a ferrite alloy inside the device. Through correct polarization of the magnetic field of the ferrite, RF energy will travel through the element in only one direction (port 1 to 2, port 2 to 3, and port 3 to 1). Reversing the polarity of the magnetic field makes it possible for RF flow in the opposite direction. Recent developments in ferrite technology have resulted in high isolation with low insertion loss.

In the basic design, the ferrite is placed in the center of a Y-junction of three transmission lines, either waveguide or coax. Sections of the material are bonded together to form a thin cylinder perpendicular to the electric field. Although the insertion loss is low, the resulting power dissipated in the cylinder can be as high as 2% of the forward power. Special provisions must be made for heat removal. It is efficient heat-removal capability that makes high-power operation possible.

RF Combiner and Diplexer Systems

The insertion loss of the ferrite must be kept low so that minimal heat is dissipated. Values of ferrite loss on the order of 0.05 dB have been produced. This equates to an efficiency of 98.9%. Additional losses from the transmission line and matching structure contribute slightly to loss. The overall loss is typically less than 0.1 dB, or 98% efficiency. The ferrite element in a high-power system is usually water-cooled in a closed-loop path that uses an external radiator.

The two basic circulator implementations are shown in Figs. 14.20b and 14.20c. These designs consist of Y-shaped conductors sandwiched between magnetized ferrite discs.[2] The final shape, dimensions, and type of material varies according to frequency of operation, power handling requirements, and the method of coupling. The distributed constant circulator is the older design; it is a broadband device, not quite as efficient in terms of insertion loss and leg-to-leg isolation, and considerably more expensive to produce. It is useful, however, in applications where broadband isolation is required. More common today is the lump constant circulator, a less expensive and more efficient, but narrowband, design.

At least one filter is always installed directly after an isolator because the ferrite material of the isolator generates harmonic signals. If an ordinary bandpass or band-reject filter is not to be used, a harmonic filter will be needed.

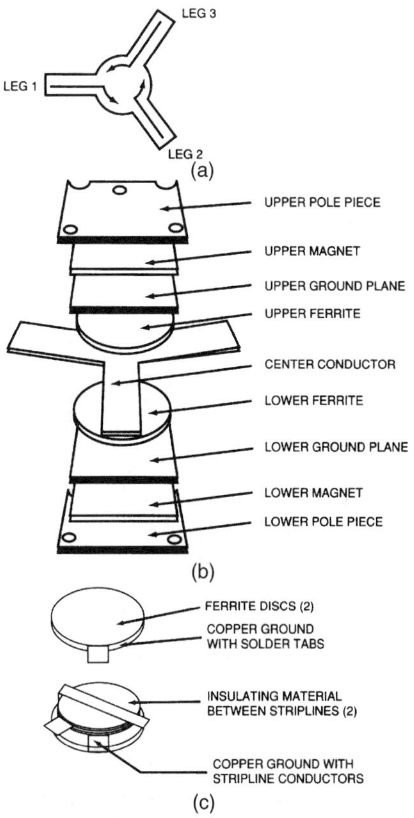

FIGURE 14.20 Basic characteristics of a circulator: (a) operational schematic, (b) distributed constant circulator, and (c) lump constant circulator. (From Surette, R. A., Combiners and combining networks, in *The Electronics Handbook*, Whitaker, J. C., Ed., CRC Press, Boca Raton, FL, 1996, 1368–1381.)

Applications

The high-power isolator permits an RF generator to operate with high performance and reliability despite a load that is less than optimum. The problems presented by ice formations on a transmitting antenna provide a convenient example. Ice buildup will detune an antenna, resulting in reflections back to the transmitter and high VSWR. If the VSWR is severe enough, transmitter power will have to be reduced to keep the system on the air. An isolator, however, permits continued operation with no degradation in signal quality. Power output is affected only to the extent of the reflected energy, which is dissipated in the resistive load.

A high-power isolator can also be used to provide a stable impedance for devices such as klystrons that are sensitive to load variations. This allows the device to be tuned for optimum performance, regardless of the stability of the RF components located after the isolator. Figure 14.21 shows the output of a wideband (6 MHz) klystron operating into a resistive load, and into an antenna system. The power loss is the result of an impedance difference. The periodicity of the ripple shown in the trace is a function of the distance of the reflections from the source.

Hot Switch

The circulator can be made to perform a switching function if a short circuit is placed at the output port. Under this condition, all input power will be reflected back into the third port. The use of a high-power stub on port 2 therefore permits redirecting the output of an RF generator to port 3.

At odd 1/4-wave positions, the stub appears as a high impedance and has no effect on the output port. At even 1/4-wave positions, the stub appears as a short circuit. Switching between the antenna and a test load, for example, can be accomplished by moving the shorting element 1/4-wavelength.

Multiplexer

A multiplexer can be formed by cascading multiple circulators, as illustrated in Fig. 14.22. Filters must be added, as shown. The primary drawback of this approach is the increased power dissipation that occurs in circulators nearest the antenna.

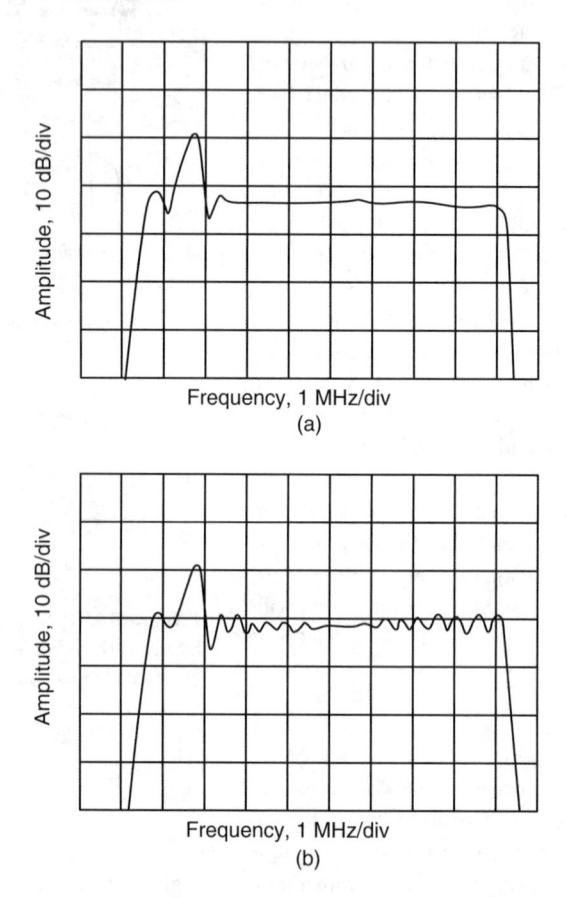

FIGURE 14.21 Output of a klystron operating in different loads through a high-power isolator: (a) resistive load, and (b) antenna system.

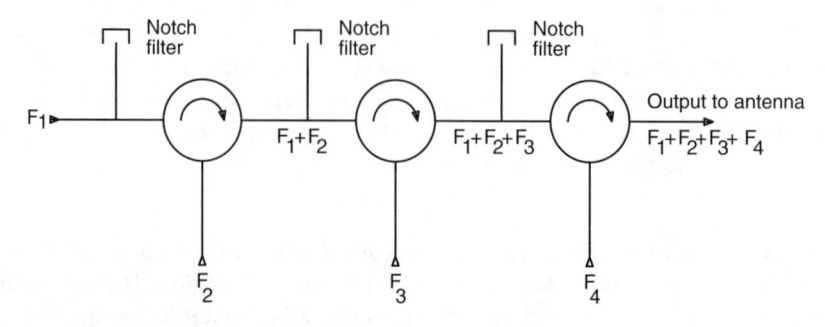

FIGURE 14.22 Using multiple circulators to form a multiplexer.

References

1. Harrison, Cecil, Passive filters, in *The Electronics Handbook*, Jerry C. Whitaker, Ed., CRC Press, Boca Raton, FL, 1996, 279–290.
2. Surette, Robert A., Combiners and combining networks, in *The Electronics Handbook*, Jerry C. Whitaker, Ed., CRC Press, Boca Raton, FL, 1996, 1368–1381.

Bibliography

Benson, K. B. and J. C. Whitaker, *Television and Audio Handbook for Technicians and Engineers*, McGraw-Hill, New York, 1989.

Crutchfield, E. B., Ed., *NAB Engineering Handbook*, 8th ed., National Association of Broadcasters, Washington, D.C., 1992.

DeComier, Bill, Inside FM multiplexer systems, in *Broadcast Engineering*, Intertec Publishing, Overland Park, KS, May 1988.

Fink, D. and D. Christiansen, Eds., *Electronics Engineers' Handbook*, 3rd ed., McGraw-Hill, New York, 1989.

Heymans, Dennis, Hot switches and combiners, in *Broadcast Engineering*, Intertec Publishing, Overland Park, KS, December 1987.

Jordan, Edward C., Ed., *Reference Data for Engineers: Radio, Electronics, Computer and Communications*, 7th ed., Howard W. Sams, Indianapolis, IN, 1985.

Stenberg, James T., Using super power isolators in the broadcast plant, in *Proceedings of the Broadcast Engineering Conference*, Society of Broadcast Engineers, Indianapolis, IN, 1988.

Terman, F. E., *Radio Engineering*, 3rd ed., McGraw-Hill, New York, 1947.

Vaughan, T. and E. Pivit, High power isolator for UHF television, in *Proceedings of the NAB Engineering Conference*, National Association of Broadcasters, Washington, D.C., 1989.

Whitaker, Jerry C., G. DeSantis, and C. Paulson, *Interconnecting Electronic Systems*, CRC Press, Boca Raton, FL, 1993.

Whitaker, Jerry C., *Radio Frequency Transmission Systems: Design and Operation*, McGraw-Hill, New York, 1990.

15
Radio Wave Propagation

15.1	Introduction	15-1
15.2	Radio Wave Basics	15-1
15.3	Free Space Path Loss	15-3
15.4	Reflection, Refraction, and Diffraction	15-4
15.5	Very Low Frequency (VLF), Low Frequency (LF), and Medium Frequency (MF) Propagation	15-7
15.6	HF Propagation	15-9
15.7	VHF and UHF Propagation	15-12
15.8	Microwave Propagation	15-14

Gerhard J. Straub
Hammett & Edison, Inc.

15.1 Introduction

From the sparks of the beginning of radio to the present day congested radio frequency environment, the understanding of radio wave propagation plays a vital role in any communications system. With the ever increasing demand to communicate farther with less power and with less interference comes the demand to know how to design a reliable radio frequency propagation path and how to evaluate the potential for system outages and interference. Without a firm grasp of radio wave propagation principles, the system engineer may be forced to specify equipment with performance in excess of that necessary so as to feel secure that the planned communications system will perform as expected. Use of higher than necessary power to establish the desired communications link may result in better path reliability; however, there is the increased risk of interference to others and, in many instances, a violation of applicable rules and regulations. The basics of radio wave propagation are the same whether one is designing a system for operation at 100 kHz or 1000 MHz, but each frequency range has its own advantages, disadvantages, and peculiarities that must be understood if optimum use of the electromagnetic spectrum is to be achieved.

15.2 Radio Wave Basics

To visualize a radio wave, consider the image of a sine wave being traced across the screen of an oscilloscope. As the image is traced, it sweeps across the screen at a specified rate, constantly changing amplitude and phase with relation to its starting point at the left side of the screen. Consider the left side of the screen to be the antenna, the horizontal axis to be distance instead of time, and the sweep speed to be the speed of light, or at least very close to the speed of light, and the propagation of the radio wave is visualized. To be correct, the traveling, or propagating, radio wave is really a wavefront, as it comprises an electric field component and an orthogonal magnetic field component as shown in Fig. 15.1. The distance between wave crests is defined as the **wavelength** and is calculated by

$$\lambda = \frac{c}{f}$$

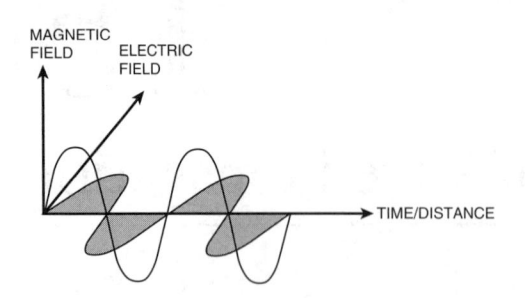

FIGURE 15.1 Propagation of wavefront.

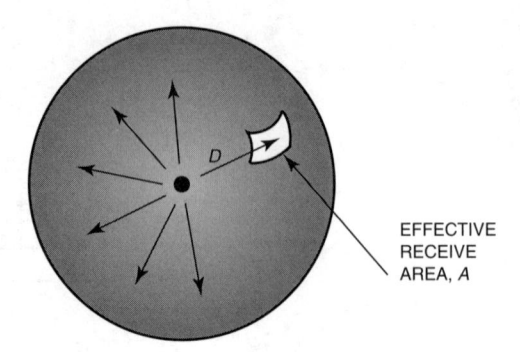

FIGURE 15.2 Power density distribution for an isotropic radiator.

where:

λ = wavelength, m
c = the speed of light, approximately 2.998×10^8 m/s
f = frequency, Hz

At any point in space far away from the antenna, on the order of 10 wavelengths or 10 times the aperture of the antenna to avoid near-field effects, the electric and magnetic fields will be orthogonal and remain constant in amplitude and phase in relation to any other point in space. The polarization of the radio wave is defined by the polarization of the electric field, horizontal if parallel to the Earth's surface and vertical if perpendicular to it. Typically, polarization can be determined by the orientation of the antenna radiating elements.

An **isotropic antenna** is one that radiates equally in all directions. To state this another way, it has a gain of unity. If this isotropic antenna is located in an absolute vacuum and excited with a given amount of power at some frequency, as time progresses the radiated power must be equally distributed along the surface of an ever expanding sphere surrounding the isotropic antenna as in Fig. 15.2. The power density at any point on the surface of this imaginary sphere is simply the radiated power divided by the surface area of the sphere, or,

$$P_d = \frac{P_t}{4\pi D^2}$$

where:

P_d = power density, W/m²
D = distance from antenna, m
P_t = radiated power, W

Since power and voltage, in this case power density and electric field strength, are related by impedance, it is possible to determine the electric field strength as a function of distance given that the impedance of free space is taken to be approximately 377 Ω,

$$E = \sqrt{ZP_d} = 5.48\frac{\sqrt{P_t}}{D}$$

where E is the electric field strength in volts per meter. Converting to units of kilowatts of power, the equation becomes

$$E = 173\frac{\sqrt{P_{t(kW)}}}{D} \text{ V/m}$$

Radio Wave Propagation

which is the form in which the equation is usually seen. Since a half-wave dipole has a gain of 2.15 dB over that of an isotropic radiator (dBi), the equation for the electric field strength from a half-wave dipole is

$$E = 222 \frac{\sqrt{P_{t(kW)}}}{D} \text{ V/m}$$

From these equations it is evident that, for a given radiated power, the electric field strength decreases linearly with the distance from the antenna and power density decreases as the square of the distance from the antenna.

15.3 Free Space Path Loss

A typical problem in the design of a radio frequency communications system requires the calculation of the power available at the output terminals of the receive antenna. Although the gain or loss characteristics of the equipment at the receiver and transmitter sites can be ascertained from manufacturer's data, the effective loss between the two antennas must be stated in a way that allows for the characterization of the transmission path between the antennas. The ratio of the power radiated by the transmit antenna to the power available at the receive antenna is known as the *path loss* and is usually expressed in decibels. The minimum loss on any given path occurs between two antennas when there are no intervening obstructions and no ground losses. In such a case when the receive and transmit antennas are isotropic, the path loss is known as **free space path loss.**

If the transmission path is between isotropic antennas, then the power received by the receive antenna is the power density at the receive antenna multiplied by the effective area of the antenna and is expressed as

$$P_r = \frac{P_t}{4\pi D^2}$$

where A is the effective area of the receive antenna in square meters.

The effective area of an isotropic antenna is defined as $\lambda^2/4\pi$. Note that an isotropic antenna is not a point source, but has a defined area; this is often a misunderstood concept. As a result, the received power is

$$P_r = \frac{P_t}{4\pi D^2} \cdot \frac{\lambda^2}{4\pi} = P_t \left(\frac{\lambda}{4\pi D}\right)^2$$

The term $(\lambda/4\pi D)^2$ is the free space path loss. Expressed in decibels with appropriate constants included for consistency of units, the resulting equation for free space path loss, written in terms of frequency, becomes

$$L_{fs} = 32.5 + 20 \log D + 20 \log f$$

where:

D = distance, km
f = frequency, MHz

The equation for the received power along a path with no obstacles and long enough to be free from any near-field antenna effects, such as that in Fig. 15.3, then becomes

$$P_r = P_t - L_t + G_t - L_{fs} + G_r - L_r$$

where:

P_r = received power, dB
P_t = transmitted power, dB
L_t = transmission line loss, dB
G_t = gain of transmit antenna referenced to an isotropic antenna, dBi
L_{fs} = free space path loss, dB
G_r = gain of receive antenna, dBi
L_r = line loss of receiver download, dB

It should be pointed out that the only frequency-dependent term in the equation for free space path loss occurs in the expression for the power received by an isotropic antenna. This is a function of the antenna area and, as stated previously, the area of an isotropic radiator is defined in terms of wavelength. As a result, the calculated field strength at a given distance from sources with equal radiated powers but on frequencies separated by one octave will be identical, but the free space path loss equation will show 6-dB additional loss for the higher frequency path. To view this another way, for the two paths to have the same calculated loss, the antennas for both paths must have equal effective areas. An antenna with a constant area has higher gain at higher frequencies. As a result, to achieve the same total path loss over these two paths, the higher frequency path requires a higher gain antenna, but the required effective areas of the antennas for the two paths are equal. The most important concept to remember is that the resultant field strength and power density at a given distance for a given radiated power are the same regardless of frequency, as long as the path approximates a free space path, but that the free space path loss increases by 6 dB for a doubling of frequency or distance.

The representation of the radio wave path in Fig. 15.3 and the previous discussion have only considered a direct path between the receiver and transmitter. In reality, there are two major modes of propagation: the **skywave** and the **groundwave.** The skywave refers to propagation via the ionosphere, which consists of several layers of ionized particles in the Earth's atmosphere from approximately 50 to several hundred kilometers in altitude. Some frequencies will be reflected by the ionosphere resulting in potentially long-distance propagation. This propagation mode is discussed in detail in a later section.

The other major mode of propagation is known as groundwave propagation. Groundwave propagation itself consists of two components, the space wave and the surface wave. The space wave also has two components known as the direct path and the reflected path. The direct path is the commonly depicted line-of-sight path that has been previously discussed and is represented in Fig. 15.3. The reflected path is that path that ends at the receiver by way of reflection from the ground or some other object. Note that there may be multiple reflected paths. The surface wave is that portion of the wavefront that interacts with and travels along the surface of the Earth. The surface wave, which will be discussed in more detail in the next section, is commonly incorrectly called the groundwave.

15.4 Reflection, Refraction, and Diffraction

As with light, the direction of propagation of a radio wave may be changed by **reflection, refraction,** or **diffraction.** Like light from a mirror, the propagation of a radio wave may be abruptly changed by

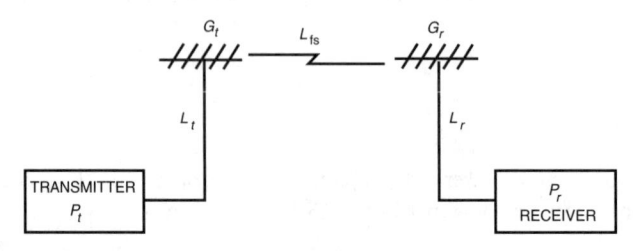

FIGURE 15.3 Path loss variables.

Radio Wave Propagation

reflection from a smooth surface. Smooth in this case is a relative term since the surface must be smooth in terms of wavelength. As a result, although a surface may not appear smooth to the eye of an observer, that is at optical frequencies, the surface may be very smooth and serve as an excellent reflector at the frequencies of interest. Reflection from a perfectly conducting surface results in no energy loss and a complete phase reversal at the reflecting surface. In reality some energy will be lost in the process, as there is no such perfectly conducting surface to be found, but large metal objects or bodies of water may come very close. Perfect reflection results in equal angles of arrival and departure of the direct and reflected wavefronts relative to the reflecting surface.

Assume for a moment that an RF propagation path has been established parallel to and at some height above a perfectly conducting surface. The receive antenna will intercept the transmitted energy by two separate paths: the direct path and the reflected path from the conducting surface. There will be a path height such that the total reflected path length will be 180° longer than the direct path. At this height, the direct and reflected fields will add to result in a 6-dB increase in field strength, since there is a 180° change in phase at the point of reflection. The path height at the point of reflection under these conditions is known as the radius of the first **Fresnel zone**. As the path height at the point of reflection is increased farther, a point will be reached where the direct and reflected paths will be equal in terms of phase. At this point, due to the phase reversal at the reflection point, the fields from the direct and reflected paths will cancel. The height of the path above the reflection point under these conditions is known as the radius of the second Fresnel zone. These conditions repeat themselves as the path height is increased with every odd Fresnel zone radius resulting in a field strength increase and every even Fresnel zone radius resulting in field strength cancellation, or at least a very significant reduction. The Fresnel zone radius increases with increasing distance from the transmitter reaching a maximum at the path midpoint and then decreases with decreasing distance to the receiver. This is a three-dimensional phenomenon, since reflections may occur from a surface on any side of the path. To be more precise, the first Fresnel zone is defined as the locus of all points from which a reflected path will have a pathlength one-half wavelength greater in length. These points form an ellipsoid with the transmitter and receiver antennas as focal points.

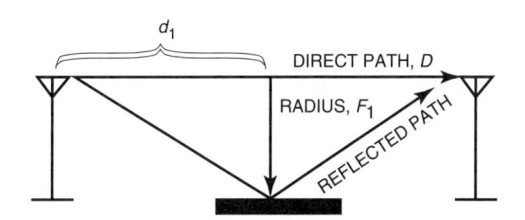

FIGURE 15.4 First Fresnel zone: reflected path length equals direct path length plus 180°.

Although it is not necessarily incorrect, and many times in fact useful, to visualize the path of the radio frequency energy as a ray between the transmitting and receiving antennas, the path actually has dimensions that are important to the path designer. It is generally accepted that an RF path must have at least 0.6 Fresnel zone radius clearance to any obstruction to be considered an unobstructed or free space path for which the free space path loss equation is directly applicable. A formula for calculating the approximate first Fresnel zone radius at any point along the path is

$$F_1 = \frac{\sqrt{(D-d_1)d_1 \lambda}}{D}$$

where:

F_1 = radius of the first Fresnel zone
d_1 = distance from the transmitter to the point of interest

Assuming a 50-km path the radius of the first Fresnel zone at the path midpoint is approximately 61 m at 1 GHz, 194 m at 100 MHz, and 612 m at 10 MHz. From these examples, it is apparent that free space path loss conditions do not often exist for lower frequency systems. If there is less than 0.6 Fresnel zone radius clearance on a given path, then it is presumed that diffraction effects must be considered.

Diffraction may occur on any path that does not meet the requirements for free space propagation. When an obstacle is placed in the path of an electromagnetic wavefront, some energy may be bent around the obstacle such that areas expected to be completely shadowed from the source of the energy may actually receive a signal. This diffraction is really an interference effect between the radiated energy from the source and currents induced in the surface of the obstruction. Any time a path is subject to diffraction effects, losses greater than those predicted by free space path equations will occur. The minimum additional loss occurs when the path is partially obstructed by a sharp edge. This is known as *knife-edge diffraction* and paths may actually be designed to rely on this propagation mode. Where a path is at grazing or near grazing incidence over a gently rolling surface, such as a smooth hill or the surface of the Earth, losses can be very high and may render the path unusable. Several methods for calculating diffraction losses have been developed, including those of the National Bureau of Standards (now National Institute of Standards and Technology) compiled in a long standing reference for non-line-of-sight propagation loss calculation methods [Rice et al., 1967]. It should be understood that although propagation paths using a mode such as knife-edge diffraction are sometimes said to have obstacle gain, the total path loss will never be less than that of a free space path of the same length.

Refraction is a change of direction of the wavefront due to a change of the refractive index of the medium of propagation. As a wavefront, either at light or radio frequencies, passes through the interface between two mediums with different refractive indices, as shown in Fig. 15.5, the velocity of propagation of the portion of the wavefront entering the medium with the smaller refractive index will be increased relative to that in the denser medium. As the entire wavefront enters the new medium, the resulting effect will be to change the direction of propagation of the wavefront. The medium through which terrestrial communications travel is the Earth's atmosphere, which decreases in density, as well as refractive index, as the altitude increases. As a result, as the wavefront increases in altitude, it undergoes a constant bending back toward the Earth.

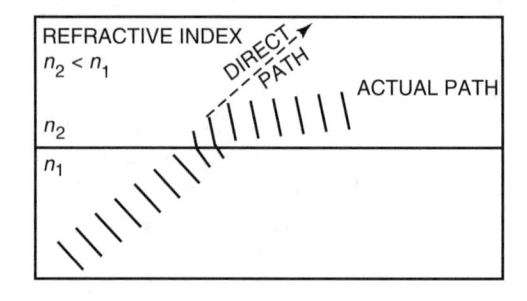

FIGURE 15.5 Effect of refraction: the wavefront is bent toward the denser medium.

An interesting effect of this atmospheric refraction is that, depending on the actual conditions at the time, the radio horizon may be closer or farther away than the true horizon. For an antenna aimed at the horizon, with no intervening terrain or other obstructions, the radiated energy, using a ray analogy, travels toward the horizon. At the horizon, as the Earth's surface curves away, the path of propagation begins to increase in altitude. As the altitude increases, the density of the atmosphere decreases and the propagation path is gradually bent back toward the surface of the Earth, effectively traveling beyond the horizon and somewhat following the curvature of the Earth. As a result, the effective communications range, even for line-of-sight type links, is somewhat beyond the horizon due to the refraction. Stated another way, the distance to the radio horizon, or the **effective Earth radius**, has been increased.

Under typical atmospheric conditions, the effective Earth radius used in propagation planning is 4/3 of the actual Earth radius. Assuming 4/3 Earth radius, the distance to the radio horizon derived from equations in White [1975] can be calculated by

$$d = 3.56\sqrt{Kh}$$

where:

d = distance to the radio horizon, km
h = height of the antenna above ground, m
K = effective Earth radius multiplier, usually 4/3

Radio Wave Propagation

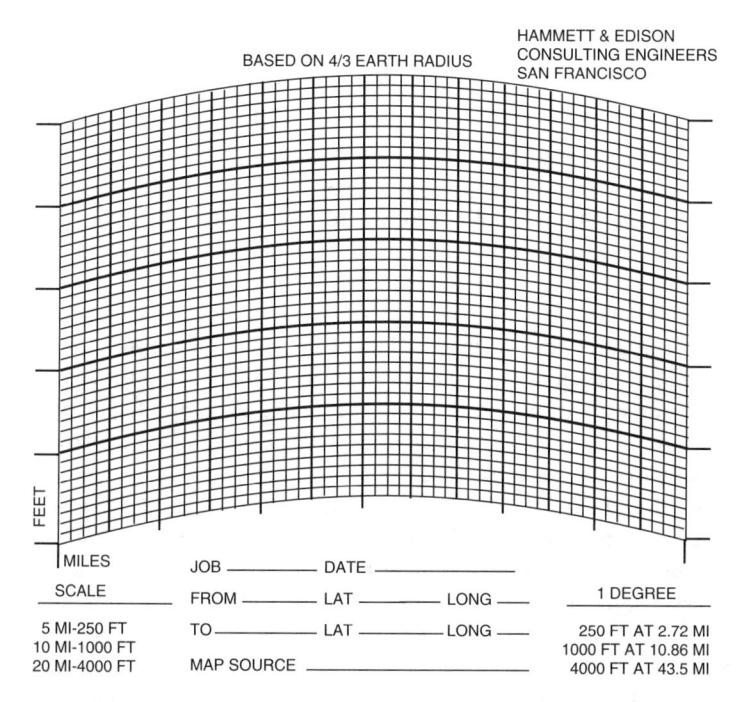

FIGURE 15.6 4/3 Earth paper for path design.

By rearranging this equation, the amount of Earth bulge at any point along a path can be calculated by

$$h = \frac{(D - d_1)d_1}{12.69K}$$

where h is the height of the Earth bulge in meters and D and d_1 are expressed in kilometers.

To aid in planning line-of-sight propagation paths without performing calculations, many times the desired path is plotted on what is known as 4/3 Earth paper as shown in Fig. 15.6. This graph paper is curved with height on the vertical axis and distance on the horizontal axis. The curve of the coordinate system represents the effective curvature of the Earth's surface. The height of the path endpoints as well as the height of significant intervening terrain or manmade obstructions are plotted at the appropriate distances to determine whether or not the path is obstructed. As stated previously, however, clearance of the direct path of transmission based on ray theory is not sufficient, as the radius of the first Fresnel zone must also be considered.

Although the preceding discussion applies generally to all frequencies, each portion of the electromagnetic spectrum has unique qualities that make it suitable for specific tasks. Line-of-sight type links are normally established on VHF and higher frequencies, whereas the lower frequencies are normally suitable for longer distance communications. A brief discussion of the propagation characteristics of various frequency ranges follows.

15.5 Very Low Frequency (VLF), Low Frequency (LF), and Medium Frequency (MF) Propagation

This portion of the frequency spectrum includes frequencies below 3 MHz and has been the mainstay of communications systems since the beginning of wireless communications. Use of higher frequencies has only been a relatively recent occurrence. This portion of the spectrum today is primarily used for relatively low data rate, long distance communications and long range navigation. The Omega and LORAN navigation systems operate in this frequency range. Also found in this portion of the spectrum is the familiar AM or medium wave broadcast band. Frequencies in this range are characterized by very stable communications over a relatively large distance. A major disadvantage to this portion of the spectrum is that due to the very long wavelength, greater than 100 m at the high frequency end of this range and measured in terms of kilometers in the lower portion of this range, efficient antennas or antennas with considerable directional characteristics are very difficult to construct. As a result, communications systems operating in this frequency range typically require acres of antennas and transmitters with very high power. Atmospheric and man-made noise can also cause significant disruption to communications on these frequencies.

At these frequencies, propagation is primarily by surface wave and waveguide modes. The direct and reflected waves of the space wave essentially cancel leaving only the surface wave component. Again, due to the proximity of the ground, at least in typical installations, horizontally polarized antennas experience very high losses and so most antennas in this frequency range are vertically polarized. At the very low end of the frequency range, significant surface penetration is possible so that communications with subsurface facilities, submarines for instance, are possible.

When the distance from the surface of the Earth to the lower boundary of the ionosphere is on the order of one wavelength, long distance propagation by waveguide modes is possible. The surface of the Earth forms one boundary of the waveguide and the ionosphere the other. The energy is essentially continually reflected between these two boundaries. At frequencies supporting this mode of propagation, the surface of the Earth can be thought of as a smooth reflecting plane. Propagation via this mode results in characteristics that are quite stable in terms of phase and amplitude, and is influenced by typical waveguide propagation characteristics, including frequency cutoff and reflections due to changes in properties of the waveguide along the path [Aarons et al., 1984].

As the frequency of operation increases into the MF region, propagation by groundwave, or more correctly surface wave, becomes the dominant mode during daytime hours. Skywave propagation is rare during daylight hours due to almost complete absorption by the lower portion of the ionosphere. In this propagation mode, the wavefront travels along the surface of the Earth. As it does so, currents are generated on the surface of the Earth. If the conductivity of the surface is poor, these currents cause energy to be dissipated as heat and attenuation is increased. As a result, propagation range is heavily dependent on the ground conductivity of the desired path with transmission over high conductivity surfaces, such as sea water, providing the greatest range. In spite of this, since ground conductivity along a particular path remains fairly constant, the characteristics of a given path generally remain quite stable. The exception to this is in areas with a significant change in ground moisture content between various seasons. For example, a large change in a particular path could be expected if the soil is generally dry and has a poor conductivity and then is subject to unusual amounts of rainfall.

Another effect of the ground losses is known as *wave tilt*. The wavefront is not confined to the immediate surface of the Earth but extends significantly upward. The ground losses tend to reduce the velocity of travel of the portion of the wavefront nearest the ground. As a result, the electric field vector, instead of remaining vertical, begins to lean in the direction of travel. This wave tilt becomes more pronounced with increasing frequency, and attenuation increases with increasing wave tilt, as the wavefront alters its polarization from vertical with no horizontal component to having a significant horizontally polarized component, because of the effect of the Earth's short circuiting condition on the horizontally polarized component. This effect and the effect of ground conductivity can be seen in Fig. 15.7 showing FCC propagation curves for the high- and low-frequency limits of the AM broadcasting band for various soil conductivities.

After sunset, when the D layer of the ionosphere (the source of MF skywave attenuation) disappears (see next section), skywave propagation is possible. With this propagation mode, coverage beyond normal surface wave range is possible as the energy directed upward is reflected back to Earth at great distances. This particular mode of propagation is treated in more detail in the next section on HF propagation. It should be noted that not all skywave propagation is beneficial. Energy radiated at high angles can return to Earth within the groundwave coverage area of a station. If the groundwave and skywave field strengths

Radio Wave Propagation

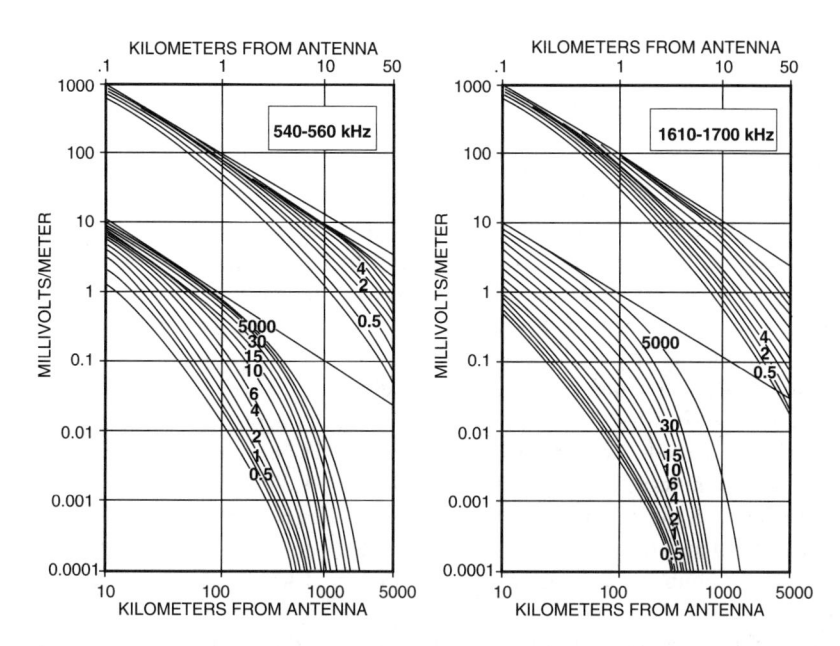

FIGURE 15.7 Ground wave field strength; electric field strength vs. distance for various soil conductivities for the band edges of the AM broadcast band. Conductivities are in millisiemens per meter.

are approximately equal, then severe fading can occur, effectively reducing the groundwave range of that station during daytime hours. Also, skywave interference inside the normal groundwave service area can be experienced from other stations removed by such a distance that groundwave interference is not a concern.

15.6 HF Propagation

The HF portion of the radio spectrum, from about 3 to 30 MHz, has recently experienced renewed interest. Systems operating in this frequency range are capable of worldwide communications without the use of satellites or other types of relay stations. This capability can be enjoyed with relatively low-power levels and unsophisticated antennas with consequently relatively low cost. Although satellite communications systems have overshadowed HF systems for most routine or high-speed and high-quality requirements, long distance communications systems that simply must always be available are sure to have some sort of HF backup due to the total lack of dependence on any outside assistance for the communication to take place. Additionally, amateur radio operators and international broadcasters have used this medium for decades to deliver their intended messages to far corners of the globe.

Beyond horizon propagation on HF circuits is made possible by reflection from the ionosphere. Actually, there are several specific layers that form this region of ionized particles created primarily by solar radiation and occurring from approximately 50 to 350 km above the surface of the Earth. Because of the dependency on solar radiation, the ionization density of these layers, and the resulting propagation characteristics, are highly dependent on not only solar activity variations, but seasonal and diurnal variations as well. As a result, any communications system operating in this region must vary frequency and times of operation for the expected conditions.

HF radio waves interacting with a layer of the ionosphere undergo refraction in the direction of the Earth. If the wave is sufficiently refracted, its path will return to the Earth at some distance from the transmitter, normally well beyond the theoretical radio horizon. This is known as the skywave. It is common to visualize this phenomenon with ray concepts with a reflection from the ionosphere at some virtual height. This virtual height is the height at which a reflection would result in the same downward path as traversed by the refracted wave. Multiple hops are possible as the wave is reflected from the Earth back up to the ionosphere for another pass. The area beyond the usable range of the groundwave and the point of return to Earth of the skywave is known as the skip zone as no usable signal is present in this area. Knowledge of the extent of this zone can be used to establish communications with a low probability of intercept by unwanted receivers. Conversely, inadvertently designing a system with the desired receiver located in the skip zone will yield disappointing results.

The range of single hop propagation and extent of the skip zone are directly related to the virtual height of the ionosphere layer in use and the takeoff angle of the radiated energy from the antenna, the angle between the radiated energy and horizontal, with the longest propagation range and skip zone occurring with the smallest takeoff angle. For some frequency and takeoff angle, there will be insufficient refraction for the wave to be bent far enough to actually return to Earth and the wave penetrates the ionosphere and is lost into space. In areas where the skywave and groundwave are both present at a relatively equivalent field strength, severe distortion and fading can occur as the two waves alternatively cancel and reinforce each other. This is a rather annoying problem for some AM broadcasters, who depend primarily on groundwave coverage, and some work has been done on antiskywave antennas to eliminate, or at least significantly reduce, radiation upward toward the ionosphere at high enough angles to return to Earth within the groundwave coverage area. Figure 15.8 demonstrates the interaction of the virtual heights, takeoff angle, and communications range.

The virtual height of an ionospheric layer can be measured with a device known as an ionosonde. This device radiates a swept frequency signal directly toward the zenith and times the return echo much like a conventional radar system. Half the round-trip travel time multiplied by the speed of light yields the virtual height of the ionosphere layer. The frequency at which no echo is returned is known as the critical frequency for that layer.

A very useful HF propagation mode is known as *near vertical incidence skywave* (NVIS) propagation. As the name implies, this mode utilizes high-angle radiation to provide relatively short range propagation, perhaps on the order of 0 to several hundred kilometers. Energy radiated toward the zenith, below the critical frequency of the pertinent ionospheric layer, is reflected down around the transmitter. As a result, frequencies used for NVIS propagation are usually in the lower-half of the HF spectrum. Note that directional characteristics of the transmitting antenna are not important, or necessarily desirable, as the returned energy from the ionosphere forms an essentially omnidirectional radiation pattern around the transmitter.

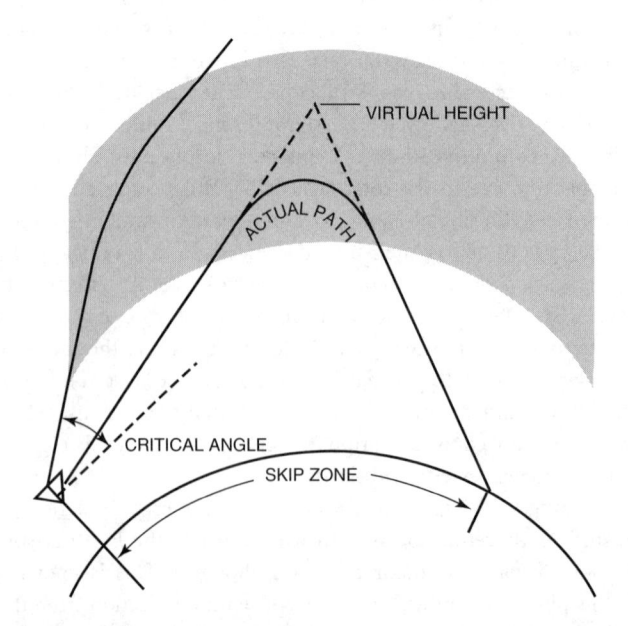

FIGURE 15.8 The mechanism of skywave propagation.

This form of propagation has found followers in the military and in agencies charged with providing emergency communications. Since the transmitted energy is reaching the receiver from such a high arrival angle, it is essentially unaffected by terrain, foliage, or other terrestrial attenuators, resulting in a relatively uniform field strength throughout the coverage area. Using this propagation mode, it is possible to communicate from deep canyons or over local hills and mountains without the use of relays or repeaters that would be required if line-of-sight type propagation modes were being utilized in areas shielded by terrain features. From a security standpoint, this propagation mode is useful as it is extremely difficult to use direction finding techniques to locate the transmitter of a signal that has a very large angle of arrival. For NVIS propagation, it is desirable to have the majority of the radiation aimed at high angles with almost no radiation at the horizon. Specifically, it is important not to have groundwave and skywave propagation paths of similar loss characteristics simultaneously to any given receiver site or signal cancellation or severe fading and distortion could occur. As a result, NVIS antennas are typically horizontally polarized antennas mounted less than one quarter wavelength above ground. Because of the high angle radiation, the resultant antenna pattern and coverage area is essentially omnidirectional.

There are three major layers of the ionosphere that are of importance in HF propagation. The lowest layer of concern is known as the D layer and is only present during daylight hours. Of all of the important ionospheric layers, the D layer is located in the region of highest atmospheric density. Because of this, recombination of ionized particles occurs rather rapidly and constant solar energy is required to sustain ionization in this layer. As the sun sets, this layer recombines and disappears. This layer is primarily responsible for almost complete absorption of frequencies below about 4 MHz. As energy from the electromagnetic wave sets electrons in motion, there is a high probability that the energy will be absorbed in a collision with a neutral particle. More precisely, the electromagnetic energy has been changed to useless, at least for radio wave propagation purposes, kinetic energy. Note that as the angle of incidence of the wave entering the layer is increased, which also means that the takeoff angle from the antenna is increased, the absorption is decreased. This occurs because of the shorter distance traveled in the ionized layer. However, as the angle of incidence is decreased, so is the usable communications range.

The next higher layer of the ionosphere at an altitude of approximately 115 km is the E layer. This is generally considered to be the lowest useful region of the ionosphere. Still low enough so that recombination occurs quickly, this layer is also only present during daylight hours, forming a useful ionized density around midday and disappearing after sunset.

The most useful ionosphere layer for over the horizon propagation is known as the F layer. This layer is located in the least dense portion of the atmosphere at an altitude of approximately 300 km at night. During daylight hours, this layer essentially splits into two separate layers known as the F1 and F2 layers at altitudes of approximately 200 and 300 km, respectively. This layer remains ionized throughout the night with minimum ionization density just before sunrise. Figure 15.9 shows the relative positions of these layers.

In characterizing the propagation characteristics between two geographical locations at a given time of day and time of year with a given level of solar activity, two frequencies are of considerable importance. These are the *maximum usable frequency* (MUF) and the *lowest usable frequency* (LUF). Propagation between the two points much occur on a frequency between these two limits. Lower frequencies require a lower electron density for sufficient refraction for usable service and are consequently reflected from lower ionospheric levels. Also, absorption in the lower levels increases with decreasing frequency. As a result, the LUF is typically determined by the absorption in the lower regions of the ionosphere. The MUF is usually determined by the peak electron density in the ionosphere, which occurs in the F layer, as higher electron density is required for sufficient refraction. The optimum operating frequency, known as the *frequence optimum de travail* (FOT), is typically 75–80% of the MUF where the refracting layer electron density is sufficient for reflection and the absorption in lower ionospheric layers is low. Note that frequencies refracted back to Earth by the F layer experience absorption as they pass through the D and E layers twice, assuming daytime propagation. As a result, the MUF usually gradually increases to its highest point in the afternoon and then gradually decreases, reaching a minimum during the hours of darkness.

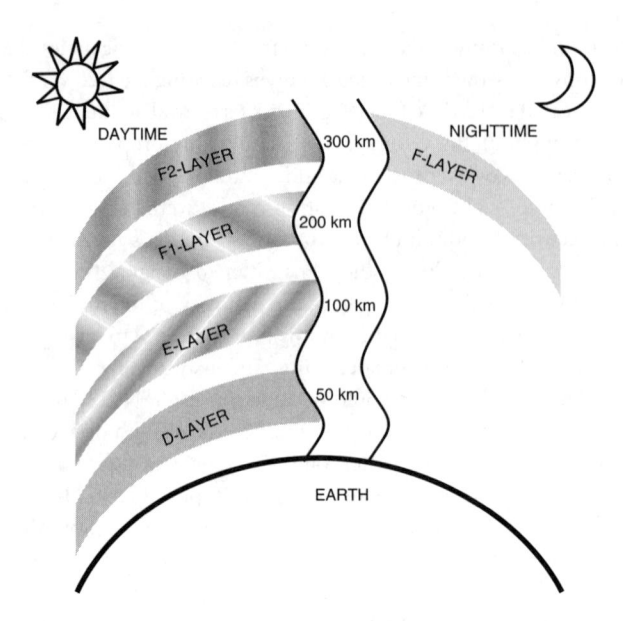

FIGURE 15.9 The ionosphere.

Although HF prediction programs are available that will compute the MUF, LUF, and FOT based on available solar data, HF propagation is subject to some unpredictability, making the choice of frequency and time for a particular propagation path not a precise science. As with line-of-sight-type propagation modes on higher frequencies, HF propagation paths are subject to multipath interference. Ionospheric conditions may support reflection from more than one layer or multiple hop propagation from energy radiated at a higher angle from the transmitter. This type of multipath interference may result in deep fading as the phase of the fields from the two paths alternatively cancel and reinforce each other. Proper system design and choice of frequency can help to reduce this type of interference.

Other HF propagation anomalies are difficult to predict and occur without warning. Many of these anomalies are the result of solar flares. A shortwave fade (SWF) is a sudden and complete absorption of HF radio waves in the D region of the ionosphere. It occurs on the sunlit side of the Earth and is caused by ultraviolet and X-ray emissions from a solar flare. The fade occurs approximately 8 minutes after the solar event and can last from a few minutes to a few hours. Protons entering the ionosphere near the magnetic poles can cause a complete loss of HF propagation in these regions. This effect may occur several hours after the flare. Ionospheric storms are another potential effect that can drastically alter the expected MUF and may occur one to two days after the solar event. Sporadic ionization in the E layer, known as sporadic E, may occur at any time and is not necessarily solar related. This effect can isolate the F layer, altering the path characteristics such that communication is disrupted or perhaps enhanced. Even VHF frequencies may be affected by sporadic E conditions causing a significant increase in effective range.

15.7 VHF and UHF Propagation

VHF and UHF frequencies, considered to be 30–3000 MHz, are used primarily for line-of-sight communications. Aircraft, public service, government, Amateur, and business communications as well as FM and TV broadcasting have allocations in this frequency range. These frequencies are typically used for local area and point-to-point communications. Usable propagation range is usually limited to the radio horizon. At these frequencies, antenna heights that result in sufficient path clearance for free space propagation to be applicable are practical. Ionospheric skywave propagation at these frequencies, at least the higher portion of the band, is essentially nonexistent with the exception of occasional sporadic E

Radio Wave Propagation

propagation. Transmit antennas are located as high above ground as practical in order to have the greatest range to the radio horizon. Communications paths requiring propagation to areas beyond line of sight from the transmitter, either because the receiver is located over the radio horizon or is obstructed by terrain or manmade obstacles, are routinely constructed using repeaters located on mountain tops or tall buildings.

Since they are essentially line-of-sight paths, propagation at VHF and UHF is relatively consistent, although, there are several potential disturbances to reliable propagation at these frequencies. Wavelengths in this frequency range vary from 0.1 to 10 m. At these wavelengths, many natural and manmade objects appear smooth enough to exhibit good reflective properties. As a result, the potential for radiated energy to arrive at the receiver from the direct path to the transmitter as well as various reflected paths is quite high. If the energy from the reflected path and from the direct path reach the receiver in phase, the strength of the received signal is enhanced. If, on the other hand, the energy from the reflected path arrives out of phase with that from the direct path, then there will be complete signal cancellation. This phenomenon is known as *multipath interference* and can be apparent as picket-fence-type noise on a VHF signal, such as that from an FM broadcast station while in a moving vehicle. The picket fencing occurs with an interval related to the wavelength of the operating frequency as the receiver passes through standing waves of destructive interference between the direct and reflected signal. If stopped at a fixed location, moving only a few inches usually can make a significant difference in the strength of the received signal. For analog video signals, such multipath problems may cause "ghosting" in pictures as the reflected signal arrives slightly time delayed from the direct signal.

Although VHF and UHF frequencies are used primarily for line-of-sight paths, propagation beyond the radio horizon is not only possible, but sometimes depended on. Such propagation can also unexpectedly cause disruption to established communications paths and interference to paths in use far beyond the radio horizon. Tropospheric **ducting** is one such phenomena that occurs mostly in warm marine climates. Tropospheric ducting occurs because of a large and rapid change in the atmospheric index of refraction and can result in propagation distances as large as 4000 km [Hutchinson, 1985]. This type of propagation can occur in two different ways, but both are typically caused by temperature inversions where atmospheric temperature increases with altitude. In one type of tropospheric ducting, a well-defined boundary between warm and cool air, and the resulting rapid change in refractive index, bends the traveling wave significantly and essentially traps it between the inversion layer and the surface of the Earth, resulting in greatly enhanced communications range. For this to occur, however, the wave must have a very shallow arrival angle at the inversion layer.

The second type of tropospheric ducting occurs between two layers of air with temperature inversions. In this case the wave is trapped between the two layers and propagates much like a wave in a waveguide as shown in Fig. 15.10. Also, as in a waveguide, the dimensions of the atmospheric duct result in some frequency selectivity for this mode of propagation. For this mode of propagation to be useful, both the transmitter and receiver antennas must be located within the duct. Such ducting can disrupt an existing communications system, especially if highly directive antennas are used, by trapping the transmitted wave inside a duct with the receive antenna located either above or below the duct. This problem can occur with aircraft flying within a duct trying to communicate with a receiver on the ground well below the duct. At higher frequencies, such ducting can cause radar systems to miss close targets or to detect targets at ranges that are theoretically beyond detection.

Another method of propagation beyond the radio horizon is known as tropospheric scatter or *troposcatter* and is shown in Fig. 15.11. In this mode of propagation, energy is reflected and refracted by small changes in the atmospheric index of refraction and by various particulate matter in the troposphere, such as dust. Communications range is limited by the height of the scattering region as both stations must have a line of sight to it. The amount of energy scattered depends on the density of the scattering particles and the power impinging on that area, and so systems regularly using this mode are typically designed with high transmit power levels.

Another mode of propagation that extends the range of VHF systems is known as meteor scatter or *meteor burst* propagation. In this propagation mode, the traveling wave is reflected from the ionized trail

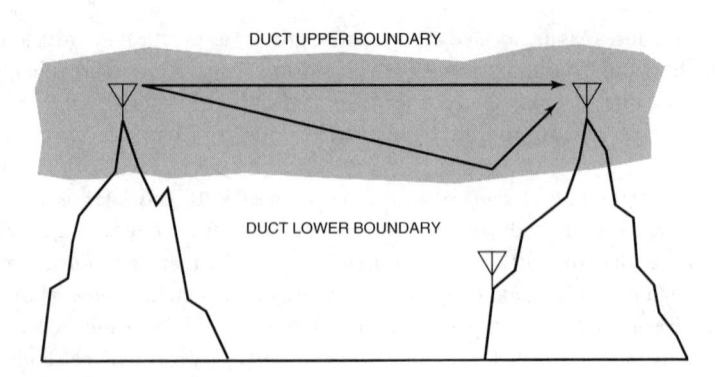

FIGURE 15.10 The mechanism of atmospheric ducting; the propagation path is limited by the boundaries of the duct. Communications with lower receive antenna are not possible.

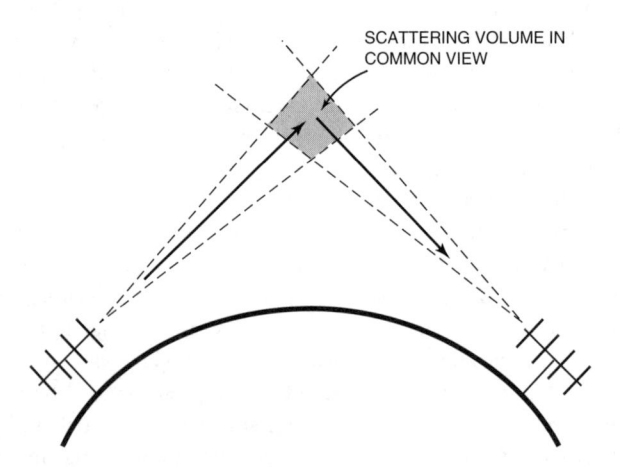

FIGURE 15.11 The principle of tropospheric scattering.

of a meteor passing through the Earth's atmosphere. These ionized trails occur at an altitude of approximately 100 km, resulting in a single-hop communications range of approximately 1500–2000 km. Frequencies in the range of 40–60 MHz are most effective for meteor scatter work, although higher frequencies are sometimes used. While meteor paths are sporadic, they are sufficient in number that digital systems with acceptable throughput can be established. Because of the orbit most meteors and the Earth around the sun, the number of meteors and, hence, meteor scatter performance is typically highest in the early morning hours [Weitzen and Ralston, 1988]. There are seasonal variations as well. Remote sensing systems have been established around this propagation mode. Meteor scatter systems have some distinct advantages besides over the horizon propagation. Meteor scatter systems do not need to change frequency over the course of the day and with changing seasons as HF systems would. Additionally, in order for communication to occur between two stations, both stations must be able to "see" the meteor trail at a proper angle for reflection to occur. This makes meteor scatter systems somewhat secure from unintended reception as well as providing some level of antijam capability. This characteristic also allows simplified frequency sharing or reuse as opposed to other communications systems.

15.8 Microwave Propagation

Propagation at microwave frequencies, which for purposes of this discussion will be considered anything above approximately 3 GHz, is quite similar to that at VHF and UHF frequencies with the addition of a few peculiarities. Because of the very small wavelength at these frequencies, obtaining appropriate Fresnel zone clearance is usually relatively easy. However, also because of the very small wavelength, energy in this frequency range may be readily absorbed by atmospheric gases or scattered by particulate matter in the atmosphere. Vegetation and foliage become obstructions at these frequencies and, if a path is through an area of deciduous trees, the performance between the various seasons can be substantially different. For this reason, propagation paths at microwave frequencies should be completely clear of any type of obstruction, often requiring visual inspection of the path.

Rain attenuation may be significant on paths operating at frequencies above approximately 10 GHz. The actual amount of attenuation is a function of many variables including the size and shape of the drops and the instantaneous intensity of the rainfall as well as the frequency of operation. For a frequency of 13 GHz, rainfall intensity of 1 in/h can result in over 1 dB/km of additional attenuation [White, 1975]. The amount of attenuation increases with rainfall intensity and frequency. Significantly less attenuation is caused by snow and fog as the water density is considerably less. Absorption by atmospheric gases, oxygen to be specific, may be an additional consideration for paths operating in the tens of gegahertz range. As a result, if a path is being planned in areas subject to periods of locally intense rainfall, this additional attenuation must be accounted for in the design of the path fade margin. Other causes of atmospheric absorption are usually secondary considerations for most typical paths.

At these frequencies, diversity receiving systems are often used to combat various forms of multipath fading. Although a path may perform flawlessly most of the time, it may be subject to occasional deep fades. These fades can be very brief or can last for minutes or even hours. The fading is primarily caused by multipath effects. The fading conditions are not constant, however, as changes in the refractive index of the atmosphere effectively change the point of reflection. These fades can also be caused by changes on the ground between the receiver and transmitter. Indeed, these multipath effects can occur with no true reflective surface, but rather from atmospheric effects themselves, such as the already discussed temperature inversion. To minimize these effects, either frequency or space diversity are used.

Diversity systems are based on the premise that fades are a function of frequency and distance from the point of reflection, since the direct and reflected signals must have a phase difference of 180°. Using another frequency, preferably far removed from the primary frequency, usually ensures that both paths will not have simultaneous fades. The drawback to this system is that it is an inefficient use of spectrum. Alternatively, space diversity can be used. With a space diversity system, two receive systems are used with antennas spaced many wavelengths apart, usually vertically. The desired distance is a function of path length and wavelength. With such spaced antennas, the point of reflection will not be the same at both heights so that while one antenna is in a deep fade, the other is not. It is important to note that a diversity system will not improve reliability due to obstructions, such as foliage, insufficient path clearance, or other such nonfrequency specific problems.

At microwave frequencies, non-line-of-sight paths can be established with the use of passive repeaters as well as active repeaters. At the millimeter wavelengths involved, passive reflectors can be constructed with sufficient gain to make their use practical. The most common passive repeater is a large, flat, billboard-looking reflector that acts exactly like an optical mirror. As with an active repeater, the passive repeater must have line of sight to both the receiver and transmitter. A single passive reflector cannot act as a relay when all three components of the link are in-line or are nearly in-line. In such a case, two closely spaced passive reflectors are required.

Communications between Earth and space are also conducted in this frequency range as well as at lower frequencies. The primary concern with space communications is to choose a frequency high enough that will not be refracted back to Earth by the ionosphere and low enough so that absorption by atmospheric gases is not significant. Because of the lower powers and great distances involved, high-gain antennas are typically required and rain and other causes of atmospheric absorption are important considerations. Further complicating space communications is the RF noise associated with thermal radiation from celestial objects, notably the sun. The interference can completely destroy a path when the sun is directly in line with the space borne transmitter.

Defining Terms

Diffraction: Change in direction of propagating energy around an object caused by interference between the radiated energy and induced currents in the object.

Ducting: A change in typical propagation conditions caused by anomalous atmospheric conditions which result in a waveguide effect.

Effective Earth radius: The assumed Earth radius required to result in a distance to the horizon equivalent to the radio horizon.

Free space path loss: The amount of attenuation of RF energy on an unobstructed path between isotropic antennas.

Frequence optimum de travail (FOT): The optimum working frequency.

Fresnel zone: A locus of points along a path where a reflection results in a change of overall path length by $n \times 180°$.

Groundwave: That portion of radiated radio frequency energy that is not influenced by the ionosphere.

Isotropic antenna: An antenna with a gain of unity.

Refraction: Change in direction of propagating radio energy caused by a change in the refractive index, or density, of a medium.

Skywave: That portion of radiated radio frequency energy that propagates skyward and interacts with the ionosphere.

Wavelength: The distance traveled by a radio frequency wavefront in one cycle.

References

Aarons, J., ed. 1984. Ionospheric radio wave propagation. *Handbook of Geophysics and Space Environments,* Chap. 10, 1983 Revision. Environmental Research Paper 879. Air Force Geophys. Lab, Hanscom AFB, MA.

Al'pert, Y.L., 1974. *Radio Wave Propagation and the Ionosphere.* Consultants Bureau, Plenum, New York.

Bekefi, G. and Barrett, A.H. 1977. *Electromagnetic Vibrations, Waves and Radiation.* MIT Press, Cambridge, M.A.

Bothias, L. 1987. *Radio Wave Propagation.* McGraw-Hill, New York.

Department of the Army. 1953. *Antennas and Radio Propagation.* U.S. Government Printing Office, Washington, D.C.

Hutchinson, C.L., ed. 1985. *The ARRL Handbook for the Radio Amateur.* The American Radio Relay League, Newington, CT.

Rice, P.L., Longley, A.G., Norton, K.A., and Barsis, A.P. 1967. *Transmission Loss Predictions for Tropospheric Communication Circuits.* NBS Technical Note 101. U.S. Department of Commerce, Washington, D.C.

Shibuya, Sh. 1987. *A Basic Atlas of Radio-Wave Propagation.* Wiley, New York.

Straw, R.D., ed. 1994. *The ARRL Antenna Book.* American Radio Relay League, Newington, CT.

Watt, A.D. 1967. *VLF Radio Engineering.* Pergamon, New York.

Weitzen, J.A. and Ralston, W.T. 1988. Meteor scatter: an overview. *IEEE Trans. Antennas Propagation.* 36(12): 1813–1819.

White, R.F., ed. 1975. *Engineering Considerations for Microwave Communications Systems.* GTE Lenkurt, San Carlos, CA.

16
Antenna Principles

16.1	Antenna Types	**16**-1
16.2	Antenna Bandwidth	**16**-2
16.3	Antenna Parameters	**16**-3
	Current Distribution • Polarization • Fields • Radiation Patterns • Pattern Multiplication • Impedance • Mutual Impedance • Directivity, Directive Gain, and Power Gain • Efficiency	
16.4	Antenna Characteristics	**16**-13
	Effect of the Earth • Dipole Characteristics • Loop Antennas • Yagi–Uda Arrays	
16.5	Apertures	**16**-26
	Equivalence Theorem • Huygens' Sources • Practical Apertures	
16.6	Wide-Band Antennas	**16**-27
	Frequency-Independent Principles • Log-Periodic Antennas • Frequency Independent Phased Arrays • Self-Similar Fractal Antennas	

Pingjuan L. Werner
Pennsylvania State University

Anthony J. Ferraro
Pennsylvania State University

Douglas H. Werner
Pennsylvania State University

16.1 Antenna Types

Antennas find extensive utilization in communication systems ranging from cellular telephone, television, radio, radar, and numerous other applications. An antenna is designed to launch an electromagnetic signal with desired characteristics; this could be direction of radiation, area of coverage, strength of emission, beamwidth, sidelobe levels, etc. Any metal structure carrying a time varying electrical current will radiate electromagnetic waves as dictated by the well established Maxwell equations. Antennas however are purposefully designed to radiate with certain specified characteristics.

Antenna types are numerous; generally a metallic structure such as wires or metal surfaces serve as the radiator. Simple configurations like a straight length of wire excited from a signal source at the center, for instance, is called a **dipole**. Dipoles can be grouped to form what is known as an *antenna array* for enhancing some characteristic like power gain. A wire curved into a closed circuit, like a circle, forms a simple **loop antenna**. On the other hand, certain useful antennas do not bear any resemblance to the wire types. For instance, the common satellite television receiving antenna is actually a paraboloidal reflector collecting and focusing the received electromagnetic signal to a feed antenna. Such an antenna is called an *aperture*. Antennas can either receive, or transmit, or be designed to do both. The *laws of reciprocity* allow the receiving characteristics to be defined from the transmitting characteristics.

Figure 16.1 depicts several common types of antennas. In Fig. 16.1(a), a half-wave dipole is shown oriented parallel to and above the Earth. The antenna length is adjusted to be near one-half of the emitted wavelength of the electromagnetic signal. If the frequency is changed, then the structure is no longer one half-wave in length, and all of the characteristics will change. One of the most important characteristics is the impedance that the center of the antenna, called the *feed-point*, presents to the signal source. A circular loop antenna is shown in Fig. 16.1(b). Like the dipole, the characteristics of the loop depend on whether the loop circumference is small or large compared to the signal wavelength. Figure 16.1(c) is an

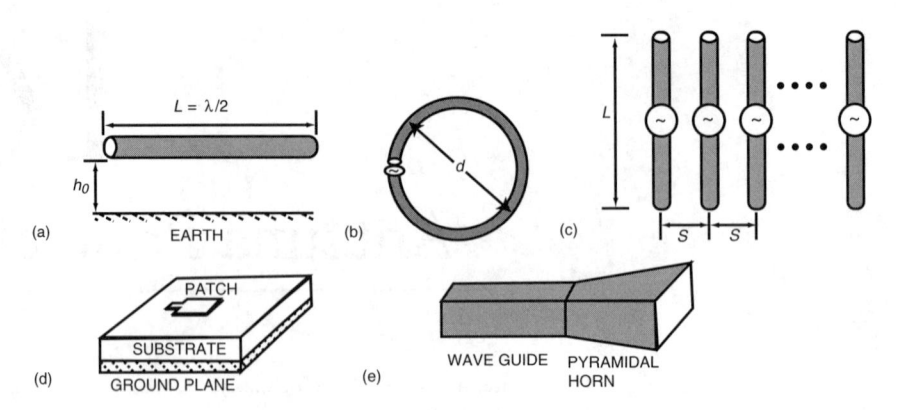

FIGURE 16.1 Examples of different types of antennas: (a) dipole, (b) loop, (c) array, (d) patch, and (e) pyramidal horn.

array of dipoles that are parallel to each other and spaced a fixed distance between individual elements. A typical *patch antenna* is shown in Fig. 16.1(d). The patch antenna is constructed from a rectangular sheet of metal on a substrate. The electrical properties of the substrate influence the characteristics of this antenna. Figure 16.1(e) is a *pyramidal electromagnetic horn* connected to a waveguide for delivering the excitation to the mouth of the horn. This is a type of aperture antenna in which a knowledge of the electromagnetic fields across the face of the mouth are used to calculate the electromagnetic fields radiated by the horn.

16.2 Antenna Bandwidth

Antennas can find use in systems that require narrow or large bandwidths depending on the intended application. Bandwidth is a measure of the frequency range over which a parameter, such as impedance, remains within a given tolerance. Dipoles, for example, by their nature are very narrow band.

For narrow-band antennas, the percent bandwidth can be written as

$$\frac{(f_U - f_L)}{f_C} \times 100 \qquad (16.1)$$

where:

f_L = lowest useable frequency
f_U = highest useable frequency
f_C = center design frequency

In the case of a broadband antenna, it is more convenient to express bandwidth as

$$\frac{f_U}{f_L} \qquad (16.2)$$

One can arbitrarily define an antenna to be *broadband* if the impedance, for instance, does not change significantly over one octave ($f_U/f_L = 2$).

The design of a broadband antenna relies in part on the concept of a **frequency-independent** antenna. This is an idealized concept, but understanding of the theory can lead to practical applications. Broadband antennas are of the helical, biconical, spiral, and **log-periodic** types. Frequency independent antenna concepts are discussed later in this chapter. Some newer concepts employing the idea of fractals are also discussed for a new class of wide-band antennas.

Antenna Principles

Narrow-band antennas can be made to operate over several frequency bands by adding resonant circuits in series with the antenna wire. Such traps allow a dipole to be used at several spot frequencies, but the dipole still has a narrow band around the central operating frequency in each band. Another technique for increasing the bandwidth of narrow-band antennas is to add parasitic elements, such as is done in the case of the open-sleeve antenna [Hall, 1992].

16.3 Antenna Parameters

Current Distribution

To calculate the characteristics of an antenna, the distribution of current on the elements must be known in advance. From the current distribution, the vector potential can be computed, and ultimately the electric field **E** and magnetic field **H** can be found. With this information, the radiation pattern, polarization, **directivity**, and other parameters can be described.

In the case of an aperture such as that shown in Fig. 16.1(e), the primary currents are flowing on the waveguide and horn surfaces and are too difficult to find. In this case, the standard practice is to introduce what are known as *equivalent currents*. For the rectangular aperture at the mouth of the horn, equivalent electric and magnetic currents can be found from the equivalence principle so that the radiation from the equivalents is the same as from the primary currents. In any event, current must be known before proceeding with the analysis of an antenna. Several techniques are available for determining the current distribution including transmission line approximations, the **method of moments (MoM)** and measurements.

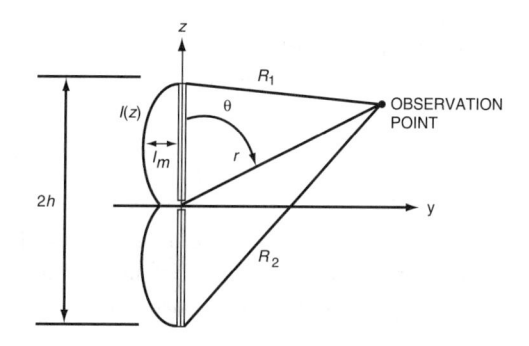

FIGURE 16.2 Transmission line approximation for the current distribution of a center fed dipole.

The transmission line approximation looks at a center fed dipole, as in Fig. 16.1(a), as an open-circuited transmission line that has been opened up or fanned out. This interpretation suggests a sinusoidal current distribution with current nodes at the ends as shown in Fig. 16.2. The current distribution can be written as:

$$I(z) = I_m \sin \beta(h-z) \quad z > 0$$
$$I(z) = I_m \sin \beta(h+z) \quad z < 0 \qquad (16.3)$$

where:

β = propagation constant
h = half-length of the center fed dipole
z = distance measured along the dipole with the center as the origin
I_m = maximum current that occurs along the wire

Exact closed-form expressions for the fields may be derived based on the assumed sinusoidal current distribution of Eq. (16.3). The z component of the free space electric field intensity **E** is given by

$$E_z = -j30 I_m \left(\frac{e^{-j\beta R_1}}{R_1} + \frac{e^{-j\beta R_2}}{R_2} - 2\cos(\beta h)\frac{e^{-j\beta r}}{r} \right) \qquad (16.4)$$

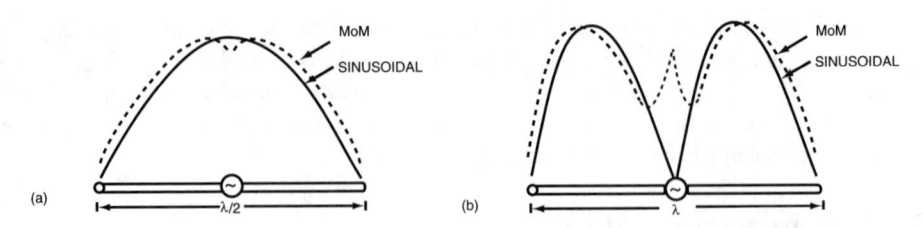

FIGURE 16.3 Comparison of current distributions using the transmission line approximation (sinusoidal) and method of moments: (a) half-wave dipole, and (b) full-wave dipole.

where R_1, R_2, and r are as indicated in Fig. 16.2. The far-field radiation pattern is of primary interest for most practical antenna applications in communications. The far-field approximation is valid for very large values of r. In particular, using the sinusoidal current distribution given in Eq. (16.3), it can be shown that the corresponding far-zone θ component of the electric field is

$$E_\theta = \frac{60 I_m}{r}\left[\frac{\cos(\beta h) - \cos(\beta h \cos\theta)}{\sin\theta}\right] e^{-j\beta r} \qquad (16.5)$$

For preliminary investigations the sinusoidal distribution is useful. For more accurate analysis, however, the MoM is required. The MoM solves an integral equation for the current distribution required to satisfy the boundary conditions on the surface of the antenna wires. There are several user friendly software packages available for the analysis of complex antenna geometries.

The **Numerical Electromagnetic Code** (NEC-2) [Burke and Poggio, 1981] is by far the most powerful antenna simulation program available for modeling wire-based antennas. NEC has many features that include:

- Ability to handle wire arcs
- Coordinate transformations
- Generates cylindrical structures
- Scales all of the units
- Surface patches
- Many different ground types
- Networks integrated into the wires
- Generates patterns
- Computes far and near fields
- Computes efficiency

MININEC [Logan and Rockway, 1986] was originally written in order to allow antenna simulations to run on personal computers. ELNEC [Lewallen, 1993] was developed as a user friendly interface to MININEC. MININEC can perform near-field calculations, whereas ELNEC does not. A comparison of the current distribution for a one-half wave and a full-wave dipole using the MoM and the transmission line approximation is shown in Fig. 16.3. Agreement is quite good for the half-wave dipole; on the other hand, there are significant differences in the case of a full-wave dipole. Hence, the limited usefulness of the transmission line approximation is clearly demonstrated in Fig. 16.3(b).

Polarization

Antennas creates a state of electromagnetic polarization generally described as *linear, circular* or *elliptical*. Figure 16.4(a) depicts a linear polarized electric field (vertical polarization in this case). This would result from one of the dipoles shown in Fig. 16.1(c). The polarization of the wave is related to the orientation

Antenna Principles

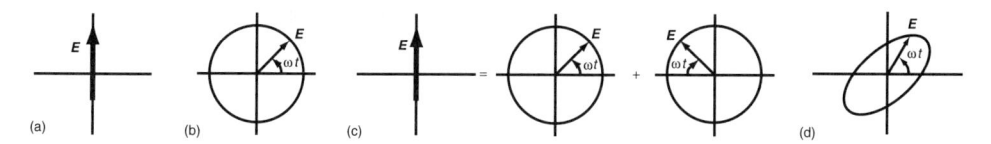

FIGURE 16.4 Various wave polarization states: (a) vertical polarization, (b) right-hand circular polarization, (c) representing linear polarization as the superposition of two circular polarized states, and (d) right-handed elliptical polarization.

of the antenna wire. The electric field in Fig. 16.4(a) oscillates up and down at the angular frequency ω. Circular polarization is shown in Fig. 16.4(b); here the electric vector stays constant in amplitude but rotates at an angular frequency ω. If the wave is approaching the observer, then the rotation direction shown in Fig. 16.4(b) is termed right handed.

A linear polarized wave can be decomposed into two oppositely rotating circular polarized waves having equal magnitudes as shown in Fig. 16.4(c). A circular polarized wave can be generated from two linear polarized waves using two orthogonal dipoles with equal amplitude excitation but 90° electrical phase shift between the excitations. The sense of rotation is controlled by selection of either a leading or lagging 90° phase shift. In general, the polarization of the wave can be elliptical as shown in Fig. 16.4(d). Elliptical polarized waves can be decomposed into two oppositely rotating circular waves of unequal magnitudes and specific electrical phase shift between the circular components.

In a communication system, there can be a polarization mismatch between the transmitting and receiving polarizations. If the transmitted wave is linear polarized, say in the vertical direction, and the receiving antenna is linear polarized in the horizontal direction then, theoretically, there is no far-field received signal at the receiver. If the transmitted wave is linear polarized and the receiving antenna is circular polarized then there is a 3-dB loss in received signal since the linear wave decomposes into two equal amplitude waves as shown in Fig. 16.4(c). This might not be an undesirable situation since the direction of the linear polarized signal could change with time due to propagation environment changes or due to motion of the source (a satellite mounted antenna). In this case, the received signal will be constant in amplitude if received on a circular polarized antenna with a 3-dB signal loss. Polarization mismatch requirements will dictate the type of receiving and transmitting polarizations required in the design of the system.

Fields

The regions around an electrically small radiator contain three types of field terms, that is, electrostatic field terms (field that decreases as $1/r^3$ where r is the distance from the center of the radiator), induction field terms (fields that decrease as $1/r^2$), and the radiation field or **Fraunhofer region** term (fields that decrease as $1/r$). The electrostatic and induction field terms do not contribute to the radiated power and are responsible for the reactive component to the input impedance of the radiator. The complete electric field expression of a very short ideal dipole is given by

$$E = j\omega\mu\frac{I_0 l}{4\pi}\left[1 + \frac{1}{j\beta r} + \frac{1}{(j\beta r)^2}\right]\frac{e^{-j\beta r}}{r}\sin\theta\boldsymbol{\theta} + j\omega\mu\frac{I_0 l}{2\pi}\left[\frac{1}{j\beta r} + \frac{1}{(j\beta r)^2}\right]\frac{e^{-j\beta r}}{r}\cos\theta\boldsymbol{r} \quad (16.6)$$

where:

l = length of the electrically short radiator
I_0 = current in the ideal dipole
μ = magnetic permeability of the medium
$\boldsymbol{\theta}, \boldsymbol{r}$ = unit vectors in the theta and radial directions

For larger radiators the **Fresnel** and Fraunhofer regions are important. The boundary between the two regions has been arbitrarily taken to be

$$\frac{2L^2}{\lambda} \tag{16.7}$$

where L is the largest dimension of the antenna and λ is the wavelength. For accurate radiation patterns the measurements should be in the Fraunhofer region or farther where the shape of the pattern is essentially independent of the distance r from the antenna. In the Fresnel region, however, the pattern is dependent upon distance r.

Radiation Patterns

The radiation pattern of an antenna is a curve plotted in either Cartesian or polar coordinates from which the electric field intensity or a quantity proportional to the electric field can be depicted in various polar directions.

The radiation pattern is a three-dimensional plot, but for simplicity it is usually plotted as a two-dimensional cut along the three-dimensional display. Conventionally, two important cuts are the E-plane and H-plane plots. In the case of a dipole such as that shown in Fig. 16.2, the electric field is in the theta direction, such that it is parallel to any plane that contains the dipole. A plot of the intensity in that plane is called the E-plane radiation pattern. The magnetic field intensity is phi directed and encircles the dipole. The plane that has the dipole element perpendicular to it and also has the magnetic field intensity parallel to it yields the H-plane radiation pattern.

Figure 16.5(a) illustrates both the E- and H-plane patterns for a one-half wavelength dipole oriented along the z coordinate axis. The ELNEC program was used to compute both the current distribution and the radiation pattern. The current distribution, which is graphed using the wire as the axis, is close to what the transmission line approximation would give. The E-plane is shown for the observer in the xz plane. Radiation patterns presume that the observer moves around the antenna while in the far-field region, keeping a constant distance from the antenna. This is a free space pattern with no ground system included. It clearly shows the direction of maximum radiation and that this antenna has a rather broad

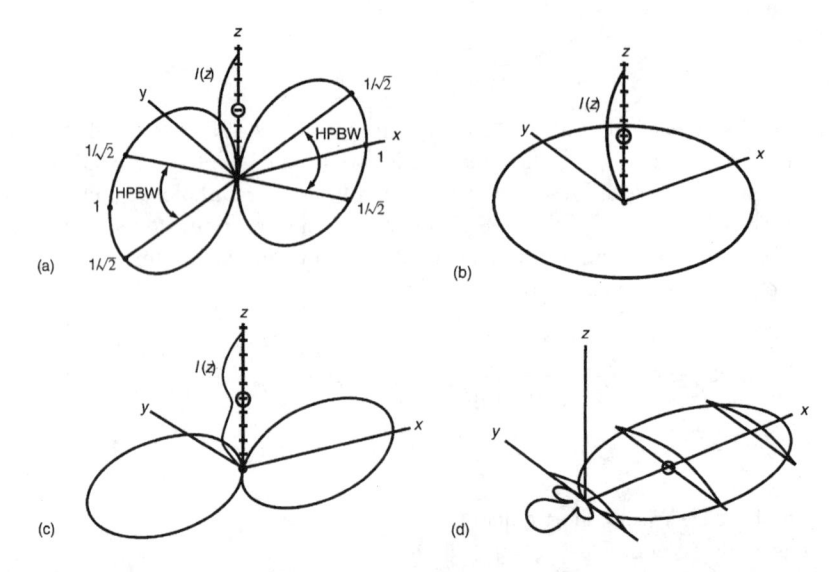

FIGURE 16.5 (a) E-plane plot of the radiation pattern of a half-wave dipole with half power beamwidth shown, (b) H-plane plot of the radiation pattern of a half-wave dipole, (c) E-plane plot of the radiation pattern of a one-wavelength dipole, and (d) E-plane plot of the radiation pattern of a three element Yagi.

pattern. Also displayed are null directions where the fields ideally go to zero. Plotted is a quantity proportional to the field intensity, or one could also choose an option to use the decibel scale where usually the maximum is normalized to 0 dB. The *H*-plane plot for the half-wave dipole is shown in Fig. 16.5(b). Because of symmetry, the dipole is expected to have uniform field intensity as the observer moves around the antenna in the *H*-plane or specifically the xy plane. For a longer dipole of one full wavelength, Fig. 16.5(c) displays the *E*-plane radiation pattern. Clearly the main beam of field intensity is narrower than the previous example, and the current distribution takes on a different form having a minimum of current at the center feed point. This case, which has already been considered in Fig. 16.3(b), is one in which the transmission line approximation would incorrectly predict a null at the feed point. However, using the transmission line current distribution would not show any significant difference in radiation pattern, but would create a sizeable error in the computation of impedance at the center point. A more complex pattern is shown in Fig. 16.5(d), which results from a three element Yagi array. In this type of antenna a central element is fed from a source and two other elements are unfed (parasitic), but due to the mutual coupling with the fed element there is a current induced in the other two. The fed element is close to a half-wavelength long, whereas the element on the right is slightly longer and the one on the left is slightly shorter. Proper pruning of the lengths and spacings results in an *E*-plane pattern that shows the signal is now predominantly directed to the right and that there is a smaller back lobe in the opposite direction. The front-to-back ratio defines the ratio of the field to the right referenced to that directed to the left. The front-to-back ratio is usually measured in decibels. Also evident in fig. 16.5(d) are smaller sidelobes in the back direction.

A radiation pattern provides good visualization of the distribution of the field intensity in different compass directions. More accurate values could be tabulated if desired. A measure of the broadness of the main beam of energy is called the half-power-beamwidth (HPBW). In the linear plots this is the point where the field diminishes to $1/\sqrt{2}$ of the maximum as illustrated in Fig. 16.5(a).

For an antenna that is not straight or one that is not aligned along the z axis as in the previous example, there can be two components of fields, one that is phi and one that is theta directed. In this case, patterns can be found for either component or for the magnitude of the total field. Sometimes a pattern is displayed as the power pattern, in which case the points on the plot are proportional to the square of the field pattern values.

Pattern Multiplication

Pattern multiplication is a procedure that allows the determination of the radiation pattern of an array of identical radiators, spaced equally along a baseline. The method permits the rapid visualization of the effect of changing the spacings, the orientation of the radiators, the type of radiators, and the magnitude and phase shift of excitation to the radiators. Central to the method is to first determine the radiation pattern of an array of **isotropic elements** having the same number, spacings and excitations as the original array. Isotropic elements are fictitious elements that radiate equal intensity in all directions. With information of the isotropic array pattern and knowledge of the element pattern one can construct the resultant pattern of the original array graphically. Use of computers can give the pattern directly, but the multiplication technique does allow better insight into the performance of the array and permits visualization of array parameter changes.

Figure 16.6 illustrates the geometry for an array of *N* isotropic elements arranged along the z axis, having spacings *d* with equal excitation magnitudes *I* and electrical phase shift α between pairs of elements. It can be shown that the radiation pattern for this array may be expressed as

$$AF(\theta) = \frac{1}{N} \frac{\sin\left(\frac{N\psi}{2}\right)}{\sin\left(\frac{\psi}{2}\right)} \qquad (16.8)$$

where

$$\psi = \beta d \cos\theta + \alpha \qquad (16.9)$$

If the isotropic elements are replaced by elements having radiation pattern function $f(\theta)$, then it can be shown that the resultant pattern of the array is given by

$$g(\theta) = f(\theta) \times AF(\theta) \qquad (16.10)$$

Modification of the element type or orientation requires the updated $f(\theta)$ in order to graph the resultant pattern $g(\theta)$. On the other hand, changes in the geometry of the isotropic array require a new function for $AF(\theta)$. Antenna books have numerous graphs of $AF(\theta)$ and $f(\theta)$ for different situations; $g(\theta)$ can be sketched graphically from a knowledge of $f(\theta)$ and $AF(\theta)$. For the case of unequal spacings and nonidentical elements, the pattern multiplication method is not applicable.

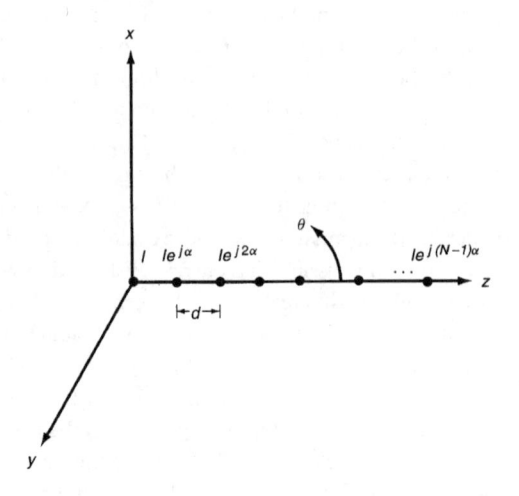

FIGURE 16.6 Geometry for an array of N isotropic elements having spacing d, equal excitation magnitudes I, and phase shift α between pairs of elements.

An example of pattern multiplication is shown in Fig. 16.7. Here a six-element array of half-wave dipoles is collinear and arranged on a baseline with center to center spacings of 0.75 wavelengths. The excitations are equal, and the phasing is set to zero degrees between the individual elements. Figure 16.7(a) is the pattern for the half-wave element observed in the yz-plane, whereas Fig. 16.7(b) is the pattern of the six-element isotropic array in the same plane. Figure 16.7(c) shows the resultant from the multiplication of the $AF(\theta)$ and $f(\theta)$. In Fig. 16.8, the excitation between elements is changed to 180°, and this results in the pattern shown in Fig. 16.8(b) with the final result of multiplication shown in Fig. 16.8(c). The location of nulls in either $f(\theta)$ or $AF(\theta)$ are preserved in the final pattern, as these figures confirm.

Impedance

The input impedance of an antenna plays an important role in the matching of the source to the antenna. Knowledge of the impedance over the operating bandwidth is of concern. The *real* part of the impedance

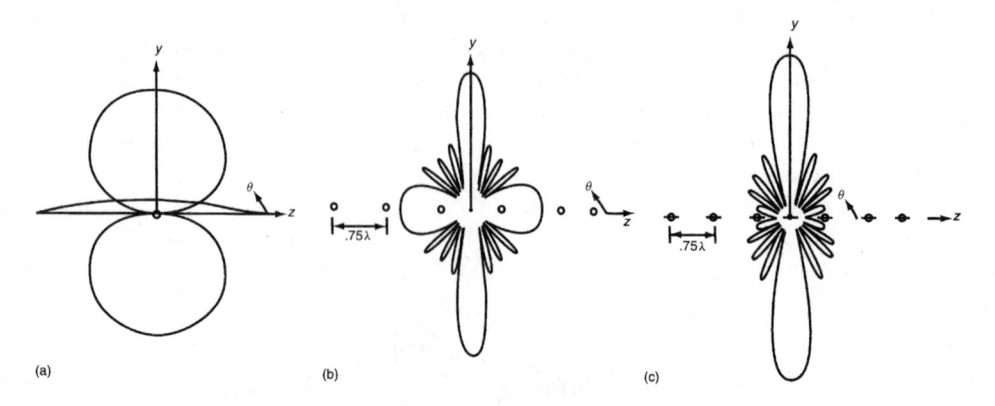

FIGURE 16.7 Example of Pattern multiplication: (a) pattern of a dipole element, (b) pattern of a six-element isotropic array, spacing $d = 0.75\lambda$ and $\alpha = 0°$, and (c) result of multiplications of (a) and (b).

Antenna Principles

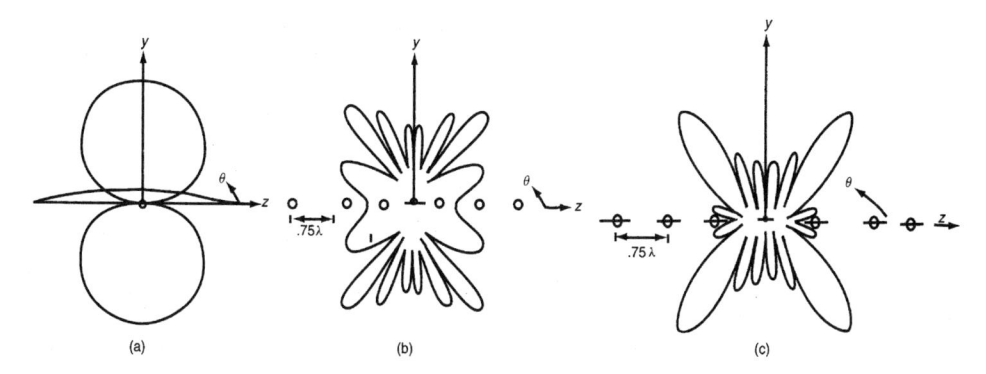

FIGURE 16.8 A second example of pattern multiplication: (a) same as Fig. 16.7(a), (b) pattern of six-element isotropic array with spacing $d = 0.75\lambda$ but α changed to 180°, and (c) result of multiplication of (a) and (b).

is primarily due to the radiation resistance, and in part due to the ohmic loss of the conductors. The radiation resistance is the equivalent resistance, which if connected to the source in place of the antenna absorbs the same power as radiated by the antenna. Impedance can be determined in a number of ways. Use of the method of moments gives the most definitive results subject to modeling limitations. Method of moments software was discussed in the first part of Sec. 16.3.

From the transmission line approximation for the current distribution, the far electric fields can be computed. From this the Poynting vector, (watts per square meter), can be integrated over a large spherical surface to find the total radiated power. The radiation resistance at the feed point is then found from this radiated power and the known current at the feed point. The near fields dictate the *imaginary* part of the impedance. Some typical values of radiation resistance for various elements, which have a small diameter, are given in Table 16.1. The section of dipole characteristics gives the input impedance of dipoles of various lengths computed from the latest moment method formulation described in Sec. 16.4.

TABLE 16.1 Radiation Resistance for Various Elements

Element Type	Element Length, λ	Radiation Resistance, Ω
Short dipole in free space	$L < 0.1$	$R_{rad} = 20\pi^2 (L/\lambda)^2$
Short monopole on a perfectly conducting ground plane	$L < 0.05$	$R_{rad} = 40\pi^2 (L/\lambda)^2$
1/2 wavelength dipole in free space	$L = 0.5$	$R_{rad} = 73$
3/2 wavelength dipole in free space	$L = 1.5$	$R_{rad} = 105.5$

Mutual Impedance

Several elements collected in an array have interactions among themselves. Since in an array environment the elements could be closely spaced in units of wavelengths, the interaction between them can result in significant mutual coupling measured in terms of the mutual impedance. It is convenient to treat the antenna array as an N-port network in order to determine mutual impedances. Hence, for an antenna array the mutual impedances can be computed from knowledge of the total fields acting upon an element.

The port equations for an N-element array with element excitation V_i and element current I_j can be written as

$$V_1 = I_1 Z_{11} + I_2 Z_{12} + \cdots + I_N Z_{1N}$$
$$V_2 = I_1 Z_{21} + I_2 Z_{22} + \cdots + I_N Z_{2N}$$
$$\vdots$$
$$V_N = I_1 Z_{N1} + I_2 Z_{N2} + \cdots + I_N Z_{NN}$$

(16.11)

where Z_{ii} is the self-impedance of the ith element and Z_{ij} is the mutual impedance between the ith and jth elements. These lead to expressions for the driving point or input impedance viewed by each source connected to the elements. Thus, from these port equations one obtains

$$Z_{D1} = \frac{V_1}{I_1} = Z_{11} + \left(\frac{I_2}{I_1}\right)Z_{12} + \cdots + \left(\frac{I_N}{I_1}\right)Z_{1N}$$

$$Z_{D2} = \frac{V_2}{I_2} = \left(\frac{I_1}{I_2}\right)Z_{21} + Z_{22} + \cdots + \left(\frac{I_N}{I_2}\right)Z_{2N} \quad (16.12)$$

$$\vdots$$

$$Z_{DN} = \frac{V_N}{I_N} = \left(\frac{I_1}{I_N}\right)Z_{N1} + \left(\frac{I_2}{I_N}\right)Z_{N2} + \cdots + Z_{NN}$$

Consider the simple case of a two-element array of half-wave dipoles separated by a half-wavelength with elements parallel and side-by-side. The driving point impedances for each element reduces to

$$Z_{D1} = Z_{11} + \left(\frac{I_2}{I_1}\right)Z_{12}$$

$$Z_{D2} = Z_{22} + \left(\frac{I_1}{I_2}\right)Z_{21} \quad (16.13)$$

The known impedance for a thin half-wave dipole is $73 + j42.5$ and available tables [Hickman and Tillman, 1961] for the mutual impedance give a value of $-12.5 - j30$. At this point knowledge of the complex current ratios is required to complete the analysis. Several interesting cases are $I_1/I_2 = 1$ and $I_1/I_2 = -1$. The mutual impedance tables referred to assumed that the array elements had sinusoidal current distributions. Each case gives the following results:

- If $I_1/I_2 = 1$, then $Z_{D1} = Z_{D2} = 60.5 + j12.5\Omega$.
- If $I_1/I_2 = -1$, then $Z_{D1} = Z_{D2} = 85.5 + j72.5\Omega$.

The moment method gives the driving point impedance directly without any need for separate calculations for the self and mutual terms. The same problem repeated with the ELNEC code yields for a wire radius of 1.0 mm the following results:

- If $I_1/I_2 = 1$, then $Z_{D1} = Z_{D2} = 60.08 + j9.38\Omega$.
- If $I_1/I_2 = -1$, then $Z_{D1} = Z_{D2} = 97.43 + j70.66\Omega$.

For the case of a two-element Yagi where only one element is fed and the other obtains excitation from only the mutual coupling, the driving point impedance is derived as

$$Z_{D1} = Z_{11} - Z_{12}^2/Z_{22}$$

since V_2 and, hence, Z_{D2} are zero. The expression shows that the subtraction of the two terms can result in a low-input impedance, which can restrict bandwidth and lower efficiency. Numerical results from ELNEC for the Yagi with two parallel and side-by-side half-wavelength dipoles spaced 0.1 and 0.05 wavelengths are as follows:

- If spacing is 0.1λ, then $Z_{D1} = 22.29 + j59.27\Omega$.
- If spacing is 0.05λ, then $Z_{D1} = 5.50 + j31.15\Omega$.

Generally, if the dipoles in an array are not parallel or collinear, the computation for self and mutual impedances is more difficult, and using the method of moments code is a more suitable approach.

Directivity, Directive Gain, and Power Gain

The **directivity** of an antenna provides a measure of performance compared to a reference antenna. The reference is either the hypothetical isotropic element or the practical half-wave dipole. Specifically, if the antenna under test and the reference are supplied with the same input power, then the directivity would simply compare the ratio of the power densities of the former with the latter. The directivity is used to express the performance in the direction of maximum power density from the antenna under test. It is presumed that all of the power supplied to the antenna is radiated, that is, it is 100% efficient. The *directive gain* is similar to directivity except it can be used for any direction and not necessarily the direction of maximum radiation. The power gain takes into account the efficiency of the antenna system since all of the power supplied is not radiated. This can include the losses in the transmission line feed configuration.

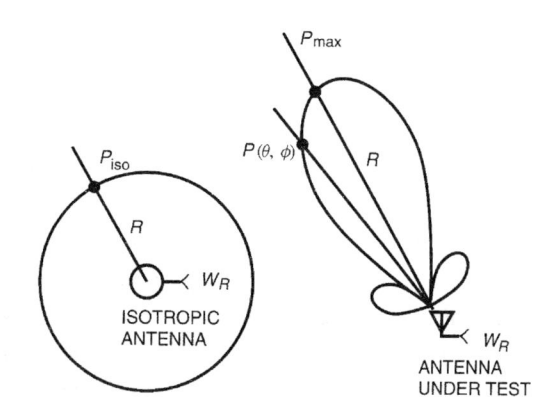

FIGURE 16.9 Comparison of power densities between an antenna and an isotropic reference. W_R is the power radiated, $P(\theta, \phi)$ is the power density, P_{max} is the maximum power density and R is the range at which measurements are made.

Figure 16.9 shows the concept for defining the directive gain and directivity. The antenna under test and the isotropic radiator are supplied with the same power W_R, which is 100% radiated. The power density P is measured from both antennas at the same distance R. The directive gain $D(\theta, \phi)$ is the ratio $P(\theta, \phi)/P_{iso}$. The quantity P_{iso} is the power density from the isotropic antenna that is radiating W_R watts. If P is measured in the direction of maximum P_{max}, then the ratio is called the directivity, otherwise it is known as directive gain. From the three-dimensional radiation pattern $f(\theta, \phi)$, the directive gain and directivity can be computed from the following:

$$D(\theta,\phi) = \frac{4\pi f^2(\theta,\phi)}{\int_0^{2\pi}\int_0^{\pi} f^2(\theta,\phi)\sin\theta d\theta d\phi} \tag{16.14}$$

when $f^2(\theta, \phi) = 1$, that is, in the direction of P_{max}

$$D = \frac{4\pi}{\int_0^{2\pi}\int_0^{\pi} f^2(\theta,\phi)\sin\theta d\theta d\phi} \tag{16.15}$$

where $f^2(\theta, \phi)$ is the power pattern, which is the square of the field pattern.

Table 16.2 gives the values of directivity for some common antenna types. The results are referenced to both the isotropic and half-wave dipole.

The *power gain G* is derived in a similar manner, except the total power input to the antenna is used in place of the radiated power; thus, the relationship between G and D is

$$G = (\text{eff}) \times D \tag{16.16}$$

where eff is the antenna efficiency.

TABLE 16.2 Directivity for Various Antenna Types

Antenna Type	Directivity, dB, Isotropic Reference	Directivity, dB, Half-Wave Dipole Reference
Isotropic	0	−2.15
Free space half-wave dipole	2.15	0
Short dipole	1.76	−0.39
Free space full-wave dipole (from ELNEC)	3.92	1.77
Three element Yagi of Fig. 16.5(d) (from ELNEC)	9.52	7.37

Efficiency

The efficiency of an antenna element depends on the radiation resistance R_{rad} and the ohmic resistance R_Ω. If a dipole, for instance, has peak current I_{in} at the feed point, then the radiated power is $I_{in}^2/2R_{rad}$, whereas the ohmic power loss is dependent on the current values at the various positions along the antenna wire. The ohmic power loss per unit length of the wire is the given by

$$dW_\Omega = \frac{I^2(z)}{2} R_\Omega \tag{16.17}$$

The total ohmic power lost over the entire wire is found from

$$W_\Omega = \int \frac{I^2(z)}{2} R_\Omega dz \tag{16.18}$$

The calculation of radiation resistance has been discussed in an earlier subsection. Here an approximate value of R_Ω will be found. This information may be used to find the efficiency from the expression

$$\text{eff} = \frac{W_{rad}}{W_{rad} + W_\Omega} \tag{16.19}$$

Figure 16.10(a) depicts the geometry used for calculating the surface impedance of a flat conductor 1 × 1 m but very deep. The **surface impedance** [Stutzman and Thiele, 1981] for this case is shown for a good conductor to be

$$Z_s = \sqrt{\frac{j\omega\mu_0}{\sigma}} \tag{16.20}$$

where:

σ = conductivity of the wire
μ_0 = permeability of free space
ω = angular frequency of the antenna source

The real part of Z_s gives the surface resistance R_s to be

$$R_s = \sqrt{\frac{\omega\mu_0}{2\sigma}} \tag{16.21}$$

In the case of a cylindrical wire, as shown in Fig. 16.10(b), the small section with dimensions Δz and Δs can approximate the flat conductor of Fig. 16.10(a). At high frequencies, due to skin effect, the current

Antenna Principles

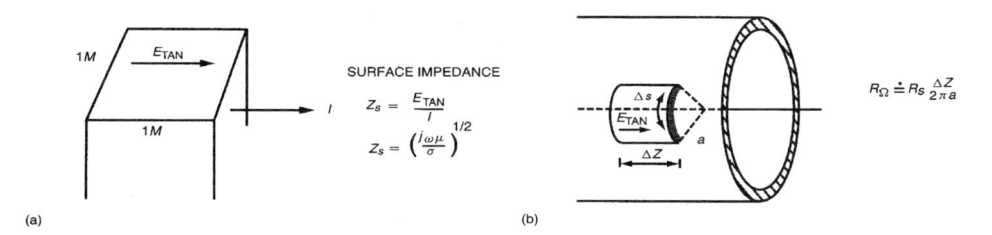

FIGURE 16.10 (a) Surface impedance for a flat thick conductor, and (b) application of the surface impedance for finding the ohmic resistance of a cylindrical conductor at high frequencies.

will be confined to the surface and so a deep section of conductor is not required. Under these circumstances, unfolding the cylindrical surface will approximate a flat conductor of dimensions $2\pi a$ and Δz. This results in an approximate ohmic resistance of:

$$R_\Omega = R_s \frac{\Delta z}{2\pi a} \tag{16.22}$$

From a knowledge of the current distribution discussed earlier, the total ohmic power loss can be computed. A less strenuous method of computing the efficiency is with a method of moments code. Since wire conductivity can be inputted into the code, it is possible to evaluate the directivity (assuming perfectly conducting wires) and then repeat with wire conductivity included. From a knowledge of D and G, efficiency can be inferred.

16.4 Antenna Characteristics

Antennas over ground can be modeled by several methods. The simplest is to assume a perfectly conducting Earth. This would be an excellent approximation for some situations, such as an antenna above sea water. For this case the method of images can be used. If Earth parameters do not warrant the perfectly conducting approximation, then the Fresnel reflection coefficients can be used to compute the ground reflected wave, which is superimposed with the direct wave. For antennas very close to Earth the **Sommerfeld/Norton solution** is used. This method uses the exact solution for fields in the presence of Earth. The NEC code discussed earlier uses this method as an option, although computations require more time to complete.

Effect of the Earth

Perfectly Conducting Ground Plane

If the Earth can be approximated as perfectly conducting, then the method of images can be applied and the solution is straight-forward. Figure 16.11 shows a horizontal and vertical dipole elevated above a perfectly conducting and infinitely large ground plane. For the horizontal case, it can be shown that there is image that can account for the correct amplitude and phase of the reflected wave. The image has the same current distribution as the antenna above ground but is 180° out of phase. The image is the same distance below the ground plane interface as the antenna is above the ground plane interface. The image and original antenna acting together in free space properly accounts for the reflected wave from the ground plane. The combination of the direct and ground reflected waves can be done by using the dipole field equation given earlier. For the vertical dipole, the image has the same current distribution as the original antenna and is in phase. The distance requirements of the image are indicated in Fig. 16.11.

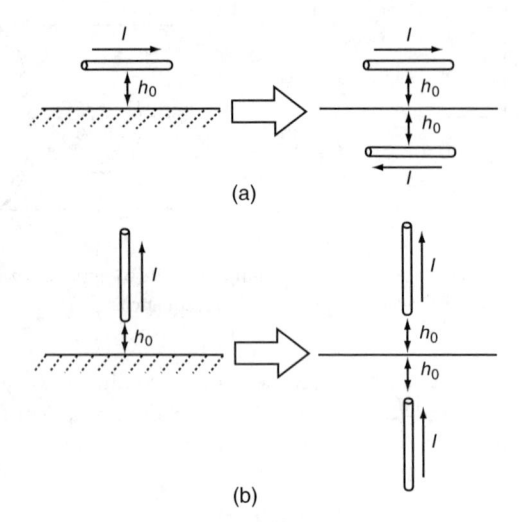

FIGURE 16.11 (a) Horizontal dipole above perfectly conducting ground plane and the image equivalent, and (b) vertical dipole above perfectly conducting ground plane and the image equivalent.

The driving point impedance of the dipole above Earth is easily calculated from the self and mutual impedances already discussed. This gives for the case of a single dipole above a perfectly conducting Earth

$$Z_{D1} = \frac{V_1}{I_1} = Z_{11} = \left(\frac{I_2}{I_1}\right)Z_{12} \tag{16.23}$$

for the horizontal dipole,

$$I_1/I_2 = -1$$

therefore

$$Z_{D1} = Z_{11} - Z_{12}$$

for the vertical dipole,

$$I_1/I_2 = 1 \tag{16.24}$$

therefore

$$Z_{D1} = Z_{11} + Z_{12} \tag{16.25}$$

Imperfectly Conducting Ground Plane

For the case of a ground defined by the electrical permittivity and conductivity, the reflection coefficient can be computed [Kraus and Carver, 1973]. The reflection coefficient defines the amplitude of the reflected wave relative to the direct wave so that the two fields can be added to determine the far-field radiation pattern. The reflection coefficients for horizontal and vertical polarizations are given respectively by

$$\rho_H = \frac{\sin\alpha - \sqrt{\varepsilon_r - \cos^2\alpha}}{\sin\alpha + \sqrt{\varepsilon_r - \cos^2\alpha}} \tag{16.26}$$

$$\rho_V = \frac{\varepsilon_r \sin\alpha - \sqrt{\varepsilon_r - \cos^2\alpha}}{\varepsilon_r \sin\alpha + \sqrt{\varepsilon_r - \cos^2\alpha}} \tag{16.27}$$

Antenna Principles

where:

- α = elevation angle for the reflected ray (see Fig. 16.12)
- ϵ_r = complex permittivity, $e'_r - j\sigma/\omega e_0$
- e'_r, σ = actual ground permittivity and conductivity parameters of the Earth, respectively.

The terminology *horizontal* and *vertical* polarizations refers to the electric field and magnetic field being parallel to the Earth interface, respectively. For example, if the source shown in Fig. 16.12 is the end view of a dipole, then the far electric field is perpendicular to the figure and, consequently, parallel to the interface. The value of the reflected electric field is computed from the quantity ρ_H. If the source in Fig. 16.12 is a vertical dipole in the plane of the figure, then the far magnetic field is parallel to the interface or the far electric field has a component perpendicular to the interface; hence, the ρ_V is used to determine the reflected field.

The resultant far field from a short horizontal dipole of length ds perpendicular to the page would be expressed by

$$E = j30\beta I_0 ds \left(\frac{e^{-j\beta R_1}}{R_1} + \rho_H \frac{e^{-j\beta R_2}}{R_2} \right) \quad (16.28)$$

For a short vertical dipole of length ds parallel to the page, the resultant far field would be expressed by

$$E = j30\beta I_0 ds \cos\alpha \left(\frac{e^{-j\beta R_1}}{R_1} - \rho_V \frac{e^{-j\beta R_2}}{R_2} \right) \quad (16.29)$$

Several examples [Jordan and Balmain, 1968] of the influence of an imperfect ground on an antenna are available in the literature.

The Sommerfeld/Norton Method

The NEC codes provide for several options of modeling a structure over ground. For the perfectly conducting ground the code generates an image as discussed earlier. For a finitely conducting ground plane the code generates an image modified by the Fresnel reflection coefficients as discussed previously. The alternate method uses the Sommerfeld/Norton exact solution for the fields in the presence of Earth and is claimed to be accurate for wire antennas close to the ground. However, computational times are considerably longer. NEC also includes the option to model a radial-wire ground screen as an approximation and a two medium ground approximation, such as a cliff overlooking a different medium.

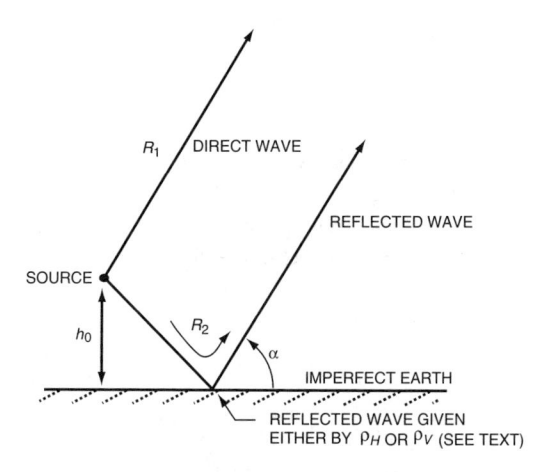

FIGURE 16.12 Direct and reflected waves for an antenna above an imperfect Earth.

Irregular Terrain

The assumption of a flat ground plane is not always realistic because terrain can be very irregular. Some recent work [Young, 1994] has developed a practical computer code for the simulation of antenna patterns over three-dimensional terrain using the **uniform geometrical theory of diffraction (UTD)**. The software is capable of generating a three-dimensional terrain model from existing terrain databases. Using existing diffraction programs, patterns can be obtained. This code is useful for siting antennas and determining optimum locations for high frequency (HF) communications. Various two- and three-dimensional generic shape profiles were analyzed and tabulated as a quick reference guide in selecting an antenna site.

Dipole Characteristics

Consider a center-fed dipole antenna of length $2L$ and diameter $2a$ oriented along the z axis, as illustrated in Fig. 16.13. If it is assumed that the current on the surface of the wire dipole is circumferentially invariant, then it has been shown that the exact form of the current distribution $I(z')$ on the dipole may be determined by solving the following integral equation [Pocklington, 1897]:

$$\int_{-L/2}^{L/2} \left[\frac{\partial^2}{\partial z^2} + \beta^2\right] K(z-z') I(z') dz' = -j4\pi\omega\varepsilon E_z^i(z) \tag{16.30}$$

where E_z^i represents the field incident or impressed on the surface of the wire by a source and

$$K(z-z') = \frac{1}{2\pi} \int_0^{2\pi} \frac{e^{-j\beta R'}}{R'} d\phi' \tag{16.31}$$

is the cylindrical wire kernel, in which

$$R' = \sqrt{(z-z')^2 + \rho^2 + a^2 - 2\rho a \cos(\phi-\phi')} \tag{16.32}$$

This is known as Pocklington's integral equation in which the unknown $I(z')$ appears under the integral.

For modeling of thin-wire antennas, that is, wires with radii $a \lesssim 0.01\lambda$, the reduced kernel approximation of Eq. (16.31) is often used. This approximation is

$$K(z-z') \approx K_0(z-z') = \frac{e^{-j\beta r_0}}{r_0} \quad \text{for} \quad \rho \geq a \tag{16.33}$$

where

$$r_0 = \sqrt{(z-z')^2 + \rho^2} \tag{16.34}$$

It should be noted that Eq. (16.33) is exact for a vanishingly thin wire ($a = 0$) and becomes approximate as soon as the wire radius begins to increase ($a > 0$). On the other hand, for moderately thick wires with radii in the range $0.01\lambda \lesssim a \lesssim 0.1\lambda$, the complete kernel expression of Eq. (16.31) must be considered in order to achieve an accurate solution. Several techniques for evaluating Eq. (16.31) have been discussed in Werner [1995]. These techniques include numerical integration schemes, as well as analytical methods. Among the more useful techniques is a recently found exact representation of Eq. (16.31) given by [Werner, 1993, 1995]

$$K(z-z') = -\frac{e^{-j\beta R}}{R} \sum_{n=0}^{\infty} \sum_{k=0}^{2n} A_{nk} \frac{(\beta^2 \rho a)^{2n}}{(\beta R)^{2n+k}} \tag{16.35}$$

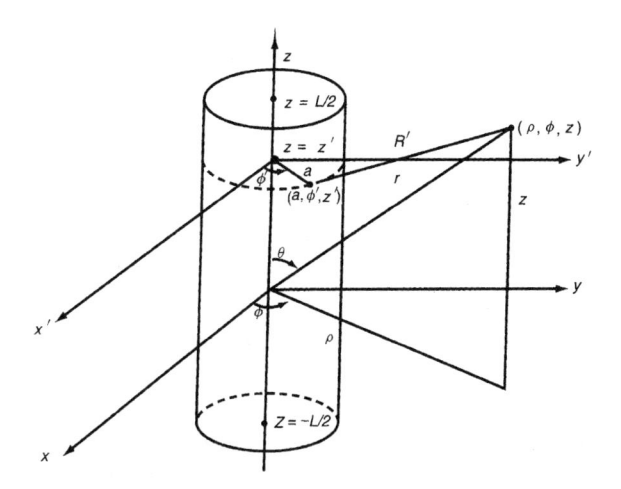

FIGURE 16.13 Dipole antenna geometry.

where

$$R = \sqrt{(z-z')^2 + \rho^2 + a^2} \tag{16.36}$$

and the coefficients A_{nk} may be determined through the following set of recursions:

$$A_{nk} = \frac{(2n+1-k)(2n+k)}{j(2k)} A_{nk-1}, \quad n > 0 \text{ and } k > 0 \tag{16.37}$$

$$A_{n0} = \begin{cases} -1, & n = 0 \\ -\frac{1}{(2n)^2} A_{n-10}, & n > 0 \end{cases} \tag{16.38}$$

A common technique for obtaining solutions to Eq. (16.30) is the method of moments [Harrington, 1968]. The beauty of this technique is that it converts the integral equation to an associated matrix equation, which can then be readily solved using digital computers. In other words, suppose we express Eq. (16.30) in the form

$$\int_{-L/2}^{L/2} I(z')G(z,z')dz' = -E_z^i(z) \tag{16.39}$$

where

$$G(z,z') = \frac{1}{j4\pi\omega\varepsilon}\left[\frac{\partial^2}{\partial z^2} + \beta^2\right] K(z-z') \tag{16.40}$$

Then the MoM may be used to reduced Eq. (16.39) to a system of simultaneous linear algebraic equations of the form

$$\sum_{n=1}^{N} Z_{mn} I_n = V_m \quad \text{for} \quad m = 1,2,\ldots,N \tag{16.41}$$

which may be written in compact matrix notation as

$$[Z_{mn}][I_n] = [V_m] \tag{16.42}$$

where $[Z_{mn}]$, $[I_n]$, and $[V_m]$ are known as the impedance, the current and voltage matrices, respectively. Hence, the problem has been reduced from solving for the unknown current distribution $I(z')$ in Eq. (16.30) to solving for the I_n in Eq. (16.42).

There are several powerful antenna analysis codes available that are based on an MoM formulation. These codes are used by antenna engineers to aid in the design of complex antenna systems where it is important to be able to predict the interaction of antennas with each other and their environment. Some of the more popular antenna modeling codes, which have already been briefly discussed include NEC, ELNEC, and MININEC.

The MoM formulations used in these codes are essentially valid for the analysis of electrically thin wires since they are based on the reduced kernel approximation of Eq. (16.33). A thin-wire antenna analysis capability is sufficient for many practical applications. However, there are some situations that may require the analysis of moderately thick-wire antennas, especially at higher frequencies (100 MHz $\leq f \leq$ 1 GHz). For these cases, an MoM code has been developed at the Pennsylvania State University, Applied Research Laboratory, which is capable of accurately analyzing moderately thick- as well as thin-wire antennas. This code uses a robust treatment of the cylindrical wire kernel Eq. (16.31) in its analysis of thicker wires. Figure 16.14 graphically illustrates the improvements in input impedance predictions for moderately thick wires when using a robust treatment of the kernel as compared to a strictly reduced kernel thin-wire treatment. The code predictions as well as measured values of input resistance and reactance for various moderately thick monopoles are plotted in Figs. 16.14(a) and 16.14(b), respectively.

Figure 16.15 contains design curves of impedance as a function of length for a center fed dipole in free space. The data for these curves was generated using the Pennsylvania State University MoM code. The MoM may also be used to determine radiation patterns and directivity of a particular antenna. The radiation patterns for several popular dipole antennas, including the half-wave and full-wave dipole, have already been discussed in Sec. 16.3 with their corresponding directivities listed in Table 16.2.

Loop Antennas

A thin wire bent into a closed contour is called a loop antenna. The terminals of a loop antenna are formed by a small discontinuity or gap in the conducting wire. Loops have been used in such diverse applications as radio receiving antennas, direction finding, magnetic field-strength probes, as well as array elements. There are various types of loop antennas including rectangular, triangular, rhombic, and circular. In this section we will focus on the thin circular loop antenna, which is one of the most popular and commonly used configurations.

Radiated Fields

Figure 16.16 illustrates the coordinate system that will be adopted for the circular loop antenna of radius a. The source point and field point are designated by the spherical coordinates $(r' = a, \theta' = 90°, \phi')$ and (r, θ, ϕ), respectively. Hence, using the geometry depicted in Fig. 16.16 it may be shown that the distance from the source point on the loop to the field point at some arbitrary location in space is

$$R' = \sqrt{R^2 - 2ar \sin\theta \cos(\phi - \phi')} \tag{16.43}$$

where

$$R = \sqrt{r^2 + a^2} \tag{16.44}$$

Antenna Principles

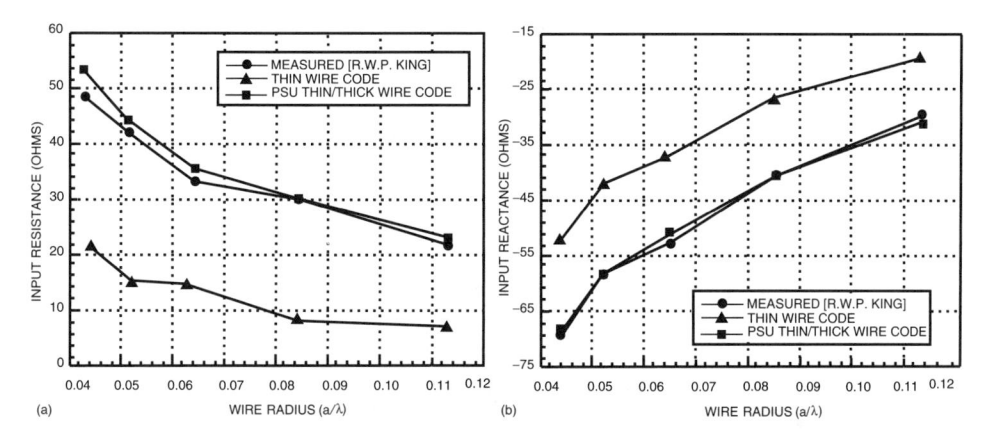

FIGURE 16.14 Code predictions for the input impedance of a 0.4λ monopole compared to measurements: (a) input resistance, and (b) input reactance.

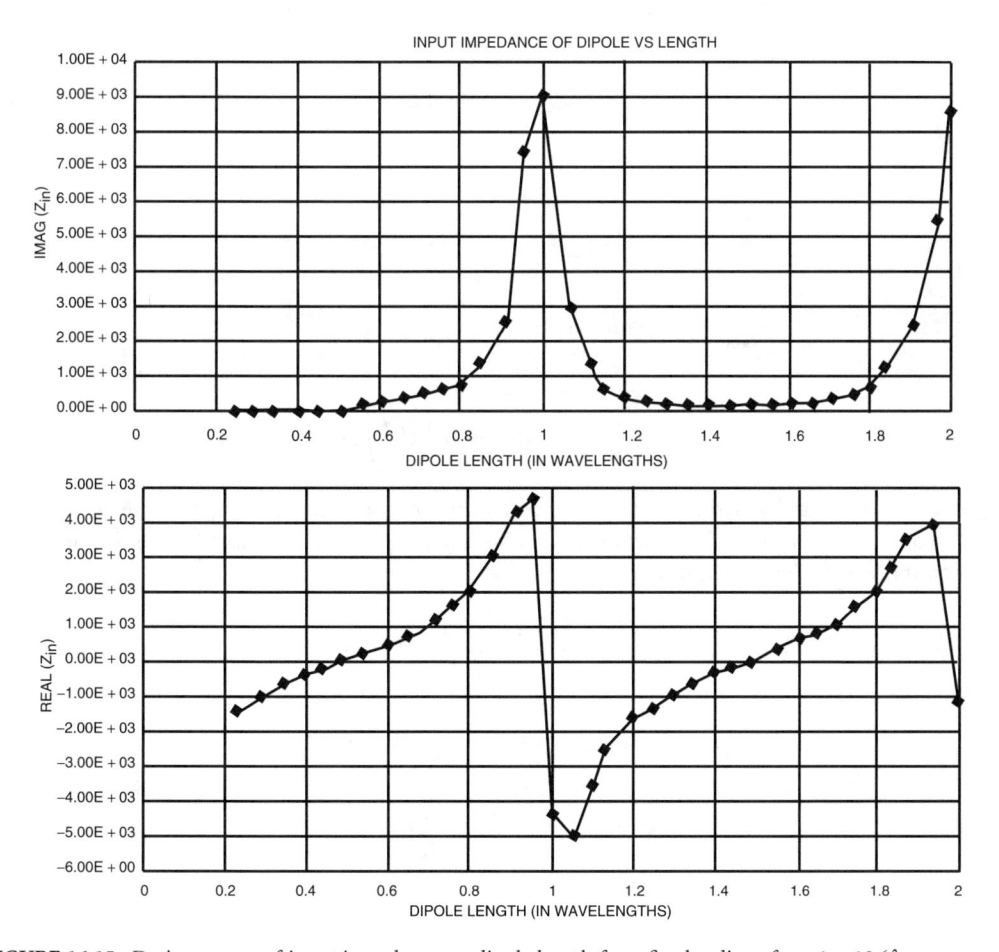

FIGURE 16.15 Design curves of input impedance vs. dipole length for a fixed radius of $a = 1 \times 10^{-6}\lambda$.

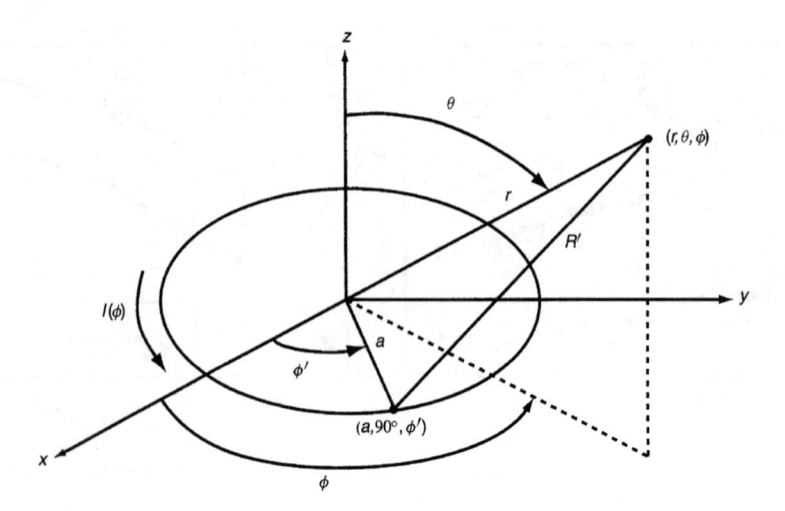

FIGURE 16.16 Circular loop antenna geometry.

The vector potential for the circular loop with a ϕ-directed current $I(\phi)$ may be expressed in the general form [Balanis 1982]

$$A(r,\theta,\phi) = A_r(r,\theta,\phi)r + A_\theta(r,\theta,\phi)\theta + A_\phi(r,\theta,\phi)\phi \qquad (16.45)$$

where

$$A_r(r,\theta,\phi) = \frac{\mu a \sin\theta}{4\pi}\int_0^{2\pi} I(\phi')\sin(\phi-\phi')\frac{e^{-j\beta R'}}{R'}d\phi' \qquad (16.46)$$

$$A_\theta(r,\theta,\phi) = \frac{\mu a \cos\theta}{4\pi}\int_0^{2\pi} I(\phi')\sin(\phi-\phi')\frac{e^{-j\beta R'}}{R'}d\phi' \qquad (16.47)$$

$$A_\phi(r,\theta,\phi) = \frac{\mu a}{4\pi}\int_0^{2\pi} I(\phi')\cos(\phi-\phi')\frac{e^{-j\beta R'}}{R'}d\phi' \qquad (16.48)$$

A general method for evaluating the vector potential integrals defined in Eqs. (16.46–16.48) has been developed in [Werner, 1996]. This exact integration technique is based on the fact that Eqs. (16.46–16.48) may be expressed in the following way:

$$A_r(r,\theta,\phi) = -\frac{\mu}{j2\beta r d\phi}\frac{d}{d\phi}\Im(r,\theta,\phi) \qquad (16.49)$$

$$A_\theta(r,\theta,\phi) = -\frac{\mu}{j2\beta r \tan\theta d\phi}\frac{d}{d\phi}\Im(r,\theta,\phi) \qquad (16.50)$$

$$A_\phi(r,\theta,\phi) = \frac{\mu}{j2\beta r \cos\theta d\phi}\frac{d}{d\phi}\Im(r,\theta,\phi) \qquad (16.51)$$

where the integral

$$\Im(r,\theta,\phi) = \frac{1}{2\pi}\int_0^{2\pi} I(\phi')e^{-j\beta R'}d\phi' \qquad (16.52)$$

Antenna Principles

is common to all three components of the vector potential. This integral may be expressed in the form of an infinite series given by

$$\Im(r,\theta,\phi) = G_0 e^{-j\beta R} + \sum_{m=1}^{\infty} G_m(\phi)\frac{(\beta^2 ar \sin\theta)^m}{m!}\frac{h^2_{m-1}(\beta R)}{(\beta R)^{m-1}} \tag{16.53}$$

where

$$G_0 = \frac{1}{2\pi}\int_0^{2\pi} I(\phi')d\phi' \tag{16.54}$$

$$G_m(\phi) = \frac{1}{2\pi}\int_0^{2\pi} I(\phi+\phi')\cos^m\phi'\, d\phi' \tag{16.55}$$

and $h^{(2)}_{m-1}$ are spherical Hankel functions of the second kind of order $m - 1$. An exact representation for $\Im$ may be obtained from Eq. (16.53), provided closed-form solutions to the integrals Eqs. (16.54) and (16.55) exist for a particular current distribution. Fortunately, it is possible to evaluate these integrals analytically for the majority of commonly assumed current distributions. Once an exact representation for a specific $\Im$ has been found, then it is a straightforward procedure to determine exact expressions for the vector potential components A_r, A_θ, and A_ϕ by substituting the series expansion for $\Im$ into Eqs. (16.46), (16.47), and (16.48), respectively, and performing the necessary differentiation. Finally, expressions for the electric and magnetic fields of the circular loop may be derived from its vector potential by making use of the relationships

$$\mathbf{H} = \frac{1}{\mu}(\nabla \times \mathbf{A}) \tag{16.56}$$

$$\mathbf{E} = \frac{1}{j\omega\varepsilon}\nabla \times \mathbf{H} = \frac{1}{j\omega\mu\varepsilon}[\nabla(\nabla \cdot \mathbf{A}) + \beta^2 \mathbf{A}] \tag{16.57}$$

Scalar equations for the three components of the magnetic field, which are in terms of the vector potential components, follow directly from Eq. (16.56). These equations are given by

$$H_r(r,\theta,\phi) = \frac{1}{\mu r \sin\theta}\left[\frac{\partial}{\partial \theta}(\sin\theta A_\phi) - \frac{\partial}{\partial \phi}A_\theta\right] \tag{16.58}$$

$$H_\theta(r,\theta,\phi) = \frac{1}{\mu r}\left[\frac{1}{\sin\theta}\frac{\partial}{\partial \phi}A_r - \frac{\partial}{\partial r}(rA_\phi)\right] \tag{16.59}$$

$$H_\phi(r,\theta,\phi) = \frac{1}{\mu r}\left[\frac{\partial}{\partial r}(rA_\theta) - \frac{\partial}{\partial \theta}A_r\right] \tag{16.60}$$

Likewise, the vector form of the electric field Eq. (16.57) may be separated into three scalar components, which results in

$$E_r(r,\theta,\phi) = \frac{\eta}{j\beta r \sin\theta}\left[\frac{\partial}{\partial \theta}(\sin\theta H_\phi) - \frac{\partial}{\partial \phi}H_\theta\right] \tag{16.61}$$

$$E_\theta(r,\theta,\phi) = \frac{\eta}{j\beta r}\left[\frac{1}{\sin\theta}\frac{\partial}{\partial \phi}H_r - \frac{\partial}{\partial r}(rH_\phi)\right] \tag{16.62}$$

$$E_\phi(r,\theta,\phi) = \frac{\eta}{j\beta r}\left[\frac{\partial}{\partial r}(rH_\theta) - \frac{\partial}{\partial \theta}H_r\right] \qquad (16.63)$$

where

$$\eta = \sqrt{\frac{\mu}{\varepsilon}} \qquad (16.64)$$

is the characteristic impedance of the medium.

Electrically Small Circular Loop

Circular loop antennas in which the radius of the loop is less than about 0.04λ may be considered electrically small. Electrically small-loop antennas are not usually used in transmitting applications because of their very low-radiation resistance, making them inefficient radiators. On the other hand, small circular loops are frequently employed in receiving applications. The standard approximation for the current distribution on an electrically small circular loop is uniform, that is, $I(\phi) = I_0$ where I_0 is a constant [Balanis, 1982].

The integral form of the vector potential for a uniform current loop is well known and may be expressed as

$$A = A_\phi(r,\theta)\phi \qquad (16.65)$$

where

$$A_\phi(r,\theta) = \frac{a\mu I_0}{2}\frac{1}{\pi}\int_0^\pi \cos\phi' \frac{e^{-j\beta R'}}{R'}d\phi' \qquad (16.66)$$

and R' of Eq. (16.43) reduces to

$$R' = \sqrt{R^2 - 2ar\sin\theta\cos\phi'} \qquad (16.67)$$

The procedure outlined in the "Radiated Fields" subsection may be followed to find an analytical expression for the uniform current vector potential integral of Eq. (16.66). This results in the following exact series representation for the near-zone of the loop:

$$A_\phi(r,\theta) = \frac{\beta a \mu I_0}{2j}\sum_{m=1}^\infty \frac{[(\beta^2 ar\sin\theta)/2]^{2m-1} h^{(2)}_{2m-1}(\beta R)}{m!(m-1)!\,(\beta R)^{2m-1}} \qquad (16.68)$$

The corresponding exact representations of the magnetic and electric field components may be obtained from Eqs. (16.58–16.60) and (16.61–16.63), respectively, by making use of Eq. (16.68). These field expressions are given by

$$H_r(r,\theta) = \frac{\beta(\beta a)^2 I_0 \cos\theta}{2j}\sum_{m=1}^\infty \frac{[(\beta^2 ar\sin\theta)/2]^{2m-2} h^{(2)}_{2m-1}(\beta R)}{[(m-1)!]^2\,(\beta R)^{2m-1}} \qquad (16.69)$$

$$H_\theta(r,\theta) = \frac{\beta(\beta a)^2 I_0 \sin\theta}{2j}\left[\frac{(\beta r)^2}{2}\sum_{m=1}^\infty \frac{[(\beta^2 ar\sin\theta)/2]^{2m-2} h^{(2)}_{2m}(\beta R)}{m!(m-1)!\,(\beta R)^{2m}}\right.$$

$$\left.-\sum_{m=1}^\infty \frac{[(\beta^2 ar\sin\theta)/2]^{2m-2} h^{(2)}_{2m-1}(\beta R)}{[(m-1)!]^2\,(\beta R)^{2m-1}}\right] \qquad (16.70)$$

Antenna Principles

$$H_\phi = 0 \tag{16.71}$$

$$E_r = E_\theta = 0 \tag{16.72}$$

$$E_\phi(r,\theta) = -\frac{\eta\beta(\beta a)I_0}{2} \sum_{m=1}^{\infty} \frac{[(\beta^2 ar\sin\theta)/2]^{2m-1}}{m!(m-1)!} \frac{h_{2m-1}^{(2)}(\beta R)}{(\beta R)^{2m-1}} \tag{16.73}$$

Small-loop approximations to the fields may be obtained from Eqs. (16.69), (16.70), and (16.73) by retaining only the first term ($m = 1$) in each series and recognizing that $R \to r$ as $a \to 0$. Under these conditions, we find that

$$H_r(r,\theta) \approx \frac{j\beta a^2 I_0 \cos\theta}{2r^2}\left[1 + \frac{1}{j\beta r}\right]e^{-j\beta r} \tag{16.74}$$

$$H_\theta(r,\theta) \approx \frac{(\beta a)^2 I_0 \sin\theta}{4r}\left[1 + \frac{1}{j\beta r} - \frac{1}{(\beta r)^2}\right]e^{-j\beta r} \tag{16.75}$$

$$E_\phi(r,\theta) \approx -\frac{(\beta a)^2 \eta I_0 \sin\theta}{4r}\left[1 + \frac{1}{j\beta r}\right]e^{-j\beta r} \tag{16.76}$$

In the far zone of the loop, that is, when $\beta r \gg 1$, expressions (16.74)–(16.76) may be used to show that the fields reduce to the simple form given by

$$H_r \approx H_\phi = E_r = E_\theta = 0 \tag{16.77}$$

$$H_\phi \approx -\frac{(\beta a)^2 I_0 e^{-j\beta r}}{4r}\sin\theta \tag{16.78}$$

$$E_\phi \approx \eta\frac{(\beta a)^2 I_0 e^{-j\beta r}}{4r}\sin\theta \tag{16.79}$$

Hence, the far-field radiation pattern of an electrically small-loop antenna has a $\sin\theta$ variation as demonstrated by Eqs. (16.78) and (16.79). Figure 16.17 illustrates the three-dimensional radiation pattern that would be produced by a small loop. Finally, by using Eq. (16.15) with $f(\theta,\phi) = \sin\theta$, it follows that the directivity of a small loop is

$$D = \frac{3}{2} \text{ or } 1.76 \text{ dB} \tag{16.80}$$

which is the same as that of an infinitesimal or short dipole (see Table 16.2). The corresponding radiation resistance of a small circular loop with circumference $C = 2\pi a$ is given by the formula [Balanis, 1982]

$$R_{\text{rad}} = 20\pi^2\left(\frac{C}{\lambda}\right)^4 \Omega \tag{16.81}$$

Circular Loops with Nonuniformly Distributed Current

The uniform current analysis that was presented earlier in this section is valid for electrically small loops. However, as the radius of the loop increases, the current is no longer uniform as it begins to vary along the circumference of the loop. One common assumption for the variation of the loop current under these conditions is a consinusoidal distribution of the form [Balanis, 1982; Lindsey, 1960]

$$I(\phi) = I_n \cos(n\phi); \quad n = 0, 1, 2, \ldots \tag{16.82}$$

This assumption may be generalized to include loops having an arbitrary current distribution represented by the following Fourier cosine series [Werner, 1996; Storer, 1956]:

$$I(\phi) = \sum_{n=0}^{\infty} I_n \cos(n\phi) \tag{16.83}$$

This form of the current distribution is very general and may be applied to the analysis of the radiation characteristics of loop antennas for a wide variety of circumstances [King and Smith, 1981; King, 1969]. Exact expressions for the vector potential and electromagnetic fields may be derived from the current distributions Eq. (16.82) and (16.83) by following the procedure outlined earlier in this section [Werner, 1996]. The resulting field expressions, although useful, are complicated and beyond the scope of this chapter.

The actual form of the current distribution on a particular loop antenna may be determined by performing a method of moments analysis, as discussed in two earlier sections. It is a straightforward procedure to find the input impedance, radiation pattern, and gain of a loop once the current distribution is known. Method of moments generated plots of the input impedance vs. loop radius may be found in Fig. 5.6 of Harrington [1968].

Yagi–Uda Arrays

A **Yagi–Uda** (or Yagi) antenna is a parasitic linear array that contains one driven element and one or more parasitic elements. A parasitic element is called a *reflector* when the radiation pattern enhancement is from the parasitic to the driver, or it is referred to as a *director* when the radiation enhancement is from the driver to the parasitic. A reflector and director can be obtained by properly adjusting the length of the parasitic. Usually, a reflector is cut longer than the driver and a director is cut shorter than the driver, so that the main beam is enhanced in the same direction by both reflector and director. Numerous curves [Hall, 1992] are available to show the effect of element spacings and lengths on input resistance, input reactance, front-to-back ratio, gain, and radiation patterns. These features are for two, three, and four element Yagis. Generally a Yagi having closely spaced elements has a low-radiation resistance and suffers from having a narrow bandwidth of 3–5% of the center frequency.

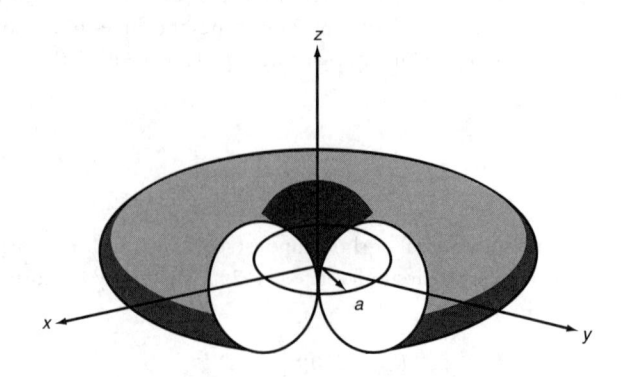

FIGURE 16.17 Three-dimensional radiation pattern produced by an electrically small-loop antenna with a uniform current distribution.

Antenna Principles

The numerical electromagnetic codes described in the first subsection of Sec. 16.3 are suitable for analyzing Yagi arrays. However, these codes can only be used for thin-wire analysis (wire radius less than 0.01λ). They do have limitations on the radius of the elements and are considered to be thin-wire codes. There are many applications where thick wires are used to construct antennas. The same wire radius for relatively lower frequencies is no longer a thin wire at a much higher frequency. Also, thick wires, such as large diameter aluminum or copper tubing, are often used to increase antenna radiation resistance, lower power loss, as well as to increase mechanical stability. In any of these cases, the thin wire codes in the first subsection of Sec. 16.3 can no longer provide accurate results. The electromagnetic code developed at the Pennsylvania State University, discussed in the second subsection of Sec. 16.4, has improved the moment method to allow very thick conductors to be modeled (wire radius from 0.01λ to 0.1λ), and, consequently, can give improved accuracy in the design of a Yagi. Figure 16.18 shows the geometry for a six-element Yagi array where a thicker wire was used for the driven element.

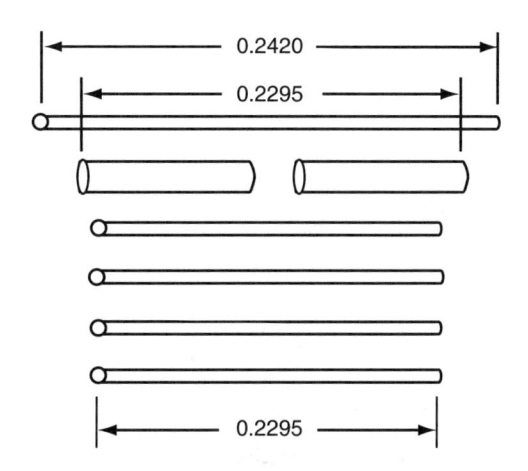

FIGURE 16.18 Geometry for a six-element Yagi array with a thick-wire driven element.

The maximum gain of a Yagi can be improved by increasing the number of directors. However, there is no significant improvement by having more than one reflector [Stutzman and Thiele 1981]. The gain or the front-to-back ratio (front-to-back ratio was discussed in an earlier section) can be controlled by adjusting the spacing between elements and the length of the elements.

Input impedance, which is measured at the feeding point of the driver, is affected by the reflector and directors due to the mutual coupling effect. The amount of effect on resistance and reactance on the driven element by the parasitic depends on the spacing and the length of the elements. A good front-to-back ratio gives maximum forward signal and minimum rearward signal. Normally, the best front-to-back ratio can not be obtained without sacrificing the maximum gain. An optimum design, which gives a maximum front-to-back ratio with a small sacrifice in gain, is obtainable by proper adjustment of the spacing and length of Yagi elements.

Tables 16.3 and 16.4 give design data for thin-wire and thick-wire antennas, respectively. Yagis having 3–7 elements provide the designer with the various options for gain and input impedance. (All calculations were made with the Pennsylvania State University method of moments code discussed earlier.) Note the dramatic increase in input impedance for the case of a six-element Yagi array that was achieved by using a thicker driven element.

TABLE 16.3 Characteristics of Equally Spaced Yagi–Uda Antennas (conductor radii = 0.0015λ)

No. of Elements	Spacing, λ	Reflector, λ	Driver, λ	Directors, λ	Gain, dB	Z_{in}, Ω
3	0.25	0.479	0.453	0.451	9.5	$22.79 + j7.62$
4	0.25	0.486	0.463	0.456	11.03	$11.94 + j13.76$
5	0.25	0.477	0.451	0.442	11.03	$45.87 + j4.07$
6	0.25	0.484	0.459	0.446	12.18	$23.04 + j10.26$
7	0.25	0.477	0.454	0.434	12.06	$53.31 + j2.86$

TABLE 16.4 Characteristics of Equally Spaced Yagi–Uda Antennas (driven element radius = .015λ, all other conductor radii = .0015λ)

No. of Elements	Spacing, λ	Reflector, λ	Driver, λ	Directors, λ	Gain, dB	Z_{in}, Ω
3	0.25	0.479	0.453	0.451	9.54	35.71 + j65.16
4	0.25	0.486	0.463	0.456	11.08	35.70 + j65.16
5	0.25	0.477	0.451	0.442	11.04	17.92 + j59.84
6	0.25	0.484	0.459	0.446	12.20	70.34 + j56.04
7	0.25	0.477	0.454	0.434	12.07	34.84 + j59.43

16.5 Apertures

Equivalence Theorem

The calculation of the characteristics of a wire antenna started from the knowledge of the current distribution. In the case of an aperture antenna, shown in Fig. 16.1(e), there is no easy solution to finding the distribution of current. In this case the currents that flow on the inner surfaces of the waveguide and the horn are the primary currents responsible for radiation. Problems of this type are more easily solved using the *equivalence theorem*. The theorem states that if the electric and magnetic fields are known over a closed surface, then this surface can be replaced by a distribution of magnetic and electric surface currents. Maintaining the inner volume of the closed surface as source free, then the two surface currents can reproduce the fields exterior to the closed surface. This equivalence leads to powerful methods for analyzing apertures. The electric and magnetic surface currents are given by

$$J_s = n \times H_A$$
$$M_s = -n \times E_A \tag{16.84}$$

where

J_s, M_s = electric and magnetic currents
n = unit outward normal to the closed surface
H_A, E_A = magnetic and electric fields over the closed surface

In the case of an aperture, the closed surface would include the open aperture, such as the mouth of the horn. Even in this case, knowledge of the distribution of the electric and magnetic fields over the aperture is now known exactly. One can use, for instance, waveguide theory to estimate the field distribution over the aperture. Or in some instances a good engineering guess can be made. With these approximations in mind, the equivalence theorem still remains a good technique for calculating the radiation characteristics of aperture structures.

Huygens' Sources

Figure 16.19 gives the geometry for an elementary aperture of dimensions $dx \times dy$ having a uniform electric field E_{XA} distribution across it. Visualizing this distribution as part of a wavefront, then the magnetic field distribution H_{YA} is computed based on the electric field and the intrinsic impedance, η_0, of free space; that is, $H_{YA} = E_{XA}/\eta_0$. Applying the equivalence theorem, it can be shown that the far field from this elementary aperture is

$$E_\theta = \frac{jE_{XA}dxdye^{-j\beta r}}{2\lambda r}\cos\phi(1+\cos\theta) \tag{16.85}$$

$$E_\phi = \frac{-jE_{XA}dxdye^{-j\beta r}}{2\lambda r}\sin\phi(1+\cos\theta) \tag{16.86}$$

Antenna Principles

The application of these results to a larger aperture is found from Huygens' principle. This states that each point on a wavefront sends out secondary waves that can be combined to form a new wavefront. Thus by knowing the distribution of amplitude and phase of the fields across the aperture under examination, the superposition of the Huygens' source fields across the aperture can determine the remote field on a wavefront elsewhere.

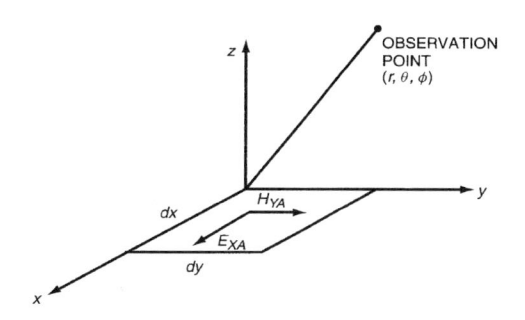

FIGURE 16.19 Geometry of an elementary aperture: a Huygen's source.

Practical Apertures

Figure 16.20 shows several practical aperture antennas. Figure 16.20(a)–16.20(d) pertain to horn types, whereas Fig. 16.20(e) is the paraboloidal antenna, which is popular with satellite television receiving systems for home use. Shown in Figs. 16.20(a)–16.20(c) is a TE_{10} electric field distribution. The terminology TE_{10} implies a transverse electric field, as shown, with one-half of a cosine variation in one direction and no variation in the other direction, hence the subscript 10. In Fig. 16.20(a), it is called a sectorial H-plane because the horn is flared out in the H-field direction; whereas in Fig. 16.20(b), the horn is a sectorial E-plane because the horn is flared out in the E-field direction. Flaring out in both directions gives the pyramidal horn shown in Fig. 16.20(c). Universal radiation patterns and directivity curves [Stutzmam and Thiele, 1981] provide design data.

The paraboloidal antenna derives its name from the paraboloidal shape of the reflecting metallic surface. A source is placed at the feed point, like a horn, and the reflected wave illuminates the circular aperture. For the simplistic case of a uniform field distribution over the circular aperture, the useful parameters for this case are given in Table 16.5.

The quantity J_1 is the Bessel function of the first kind of order unity. Although a uniform distribution is not realistic, it shows the general characteristics of a paraboloidal reflector antenna. For more accuracy, it is usually assumed that there is a parabolic taper [Stutzman and Thiele, 1981] to the field distribution with the maximum being at the center and tapering to smaller values at the edge of the circular aperture. The value of the field at the edges is a function of the beamwidth of the feed antenna.

16.6 Wide-Band Antennas

Antennas can be designed to work over several octaves in frequency. Dipoles or arrays of dipoles as discussed in earlier sections are very frequency dependent. This section discusses some of the principles

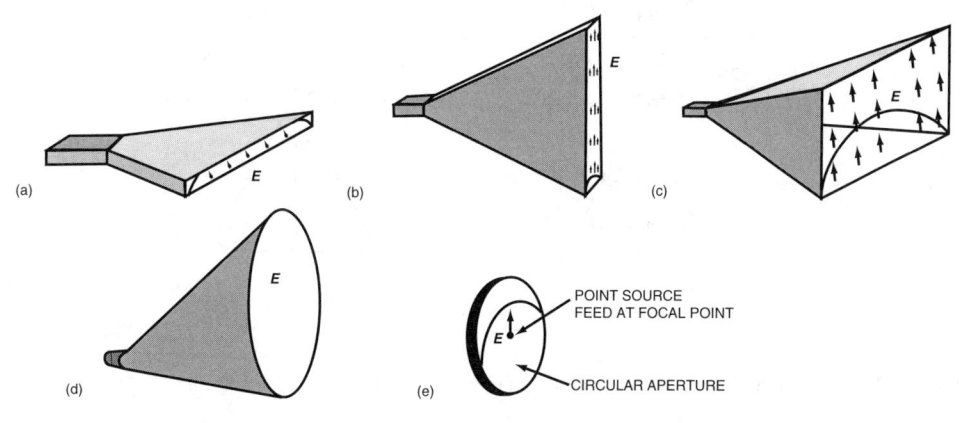

FIGURE 16.20 Some practical aperture antennas: (a) sectoral H-plane, (b) sectoral E-plane, (c) pyramidal, (d) conical, and (e) paraboloid.

TABLE 16.5 Characteristics of a Circular Aperture of Radius a with Uniform Excitation

Parameter	Result
Radiation pattern	$\dfrac{2J_1(\beta a \sin\theta)}{\beta a \sin\theta}$
Half-power beamwidth, rad	$1.02\dfrac{\lambda}{2a}$
Directivity	$\dfrac{4\pi}{\lambda^2}(\pi a^2)$
Side lobe level, dB	-17.6

that in the ideal sense lead to frequency-independent characteristics. Generally, the ideal structure is infinite in extent; hence, it is not practical. However, truncating these ideal structures to finite dimensions can lead to wide-band operation.

Frequency-Independent Principles

There are two basic concepts that are inherent in the conception of a frequency-independent antenna. The first is to arrive at a geometry of conductors that can be defined independent of a dimension. For it is the lengths, spacings, and diameters of wires that appear in the electromagnetic formulas as a ratio to wavelength, thus changing frequency results in nonconstant antenna characteristics. If a structure can be defined only by angles, then one eliminates the characteristic dimensions that are responsible for the frequency dependence of an antenna. Figure 16.21 shows the infinite bow tie and infinite biconical antenna. These structures when truncated in length are no longer frequency independent. They do find usefulness for increasing the bandwidth over a cylindrical wire dipole. This concept of defining a structure entirely by angles is due to Rumsey [1966].

A second concept is centered around a self-complementary planar structure. If the metallic part of the antenna is replaced by free space and the free space is replaced by metal, then this forms the complement of the structure. If the complementing results in the same original structure except for a rotation, then this is a self-complementary structure. In Fig. 16.21, if the infinite bow tie as α equal to

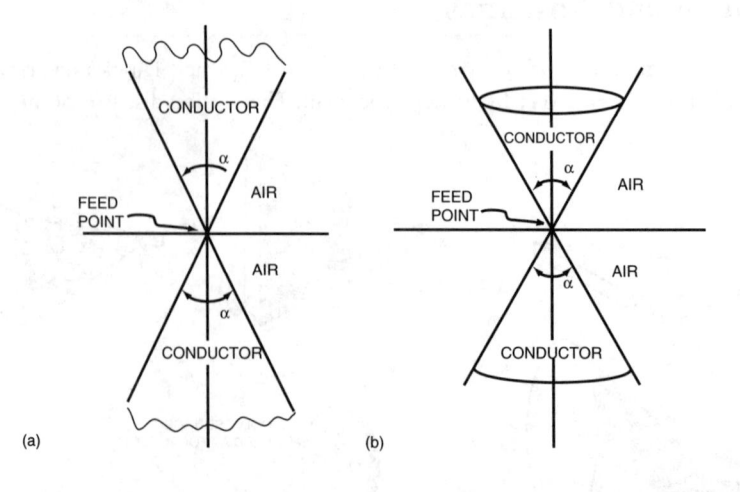

FIGURE 16.21 Two antenna shapes defined by an angle α: (a) bow tie, and (b) biconical.

Antenna Principles

90° for instance, then this can create a self-complementary structure. It has been shown [Stutzman and Thiele 1981] that the input impedance relationship between an antenna and its complement is

$$Z_A(f) \times Z_{AC}(f) = \frac{Z_0^2}{4} \qquad (16.87)$$

where $Z_A(f)$ and $Z_{AC}(f)$ are the input impedances for the antenna and its complement and $Z_0 = 120\pi$ is the characteristic impedance of free space. For a self-complementary structure $Z_A = Z_{AC}$, since it is exactly the original structure, so that

$$Z_A(f) = Z_{AC}(f) = 60\pi \ \Omega \qquad (16.88)$$

From this one infers that the input impedance is independent of frequency and equals $60\pi \ \Omega$.

The spiral shape has been used to develop a series of wide-band antennas. Figure 16.22 shows two arms of an antenna constructed from conductors with edges forming a spiral shape. The equation of a single spiral as shown in Fig. 16.23 is

$$\rho = \rho_0 e^{a\phi} \qquad (16.89)$$

where:

ρ_0 = radius when $\phi = 0$
a = constant that controls the rate of expansion of the spiral

This is an infinite spiral, it continues into the origin as well as out to infinity. It is completely described by an angle and not by any characteristics dimensions. The parameters ρ_0 and a control the specific characteristics of the spiral. The shape in Fig. 16.22 is shown with truncation near the origin. In practice, the feed is connected to the two conductors near the origin. The antenna in Fig. 16.22 can be made self-complementary and, if infinite in extent, will be frequency independent as already discussed. Truncation at the feed point leads to an upper useful frequency limit, whereas truncation of the outer region leads to a lower frequency limit. Wide-band operation is achieved by controlling the dimensions of the truncated regions.

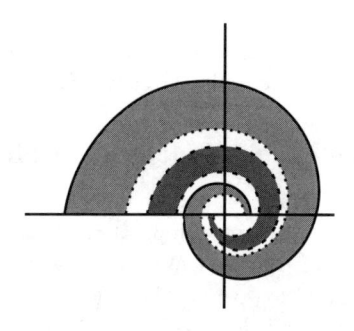

FIGURE 16.22 Spirals for forming the two arms of a self-complementary antenna.

Log-Periodic Antennas

The most practical form of a wide-band antenna originates from the log-periodic dipole array (LPDA) [Stutzman and Thiele, 1981]. The property of a spiral like that shown in Fig. 16.23 is that the ratio of distances from the origin of adjacent arms is a fixed quantity τ. This idea is carried over into the design of the LPDA shown in Fig. 16.24. The dimensions that are scaled by the factor τ are the lengths of the elements L, the diameters of the wire conductors $2a$, and the positions R of the elements measured from the apex.

$$\tau = \frac{R_{n+1}}{R_n} = \frac{L_{n+1}}{L_n} = \frac{a_{n+1}}{a_n} \qquad (16.90)$$

At frequency f_n, the element of length L_n has a certain length in fractional wavelengths. If the frequency is increased to f_n/τ, then the next smallest element of length L_{n+1} has the same fractional length in wavelengths as the previous element. Therefore, the array in total is scaled as the frequency is increased to $f_n/\tau, f_n/\tau^2, f_n/\tau^3$, etc. This relationship allows the array to be scaled to each of these frequencies so that it is expected that the performance of the array at these frequencies should be identical. This would be absolutely true if the LPDA were infinite in both directions toward and away from the origin. Because the relationship of the frequencies is τ, the performance is repeatable in log f, hence the name log periodic. A useful design curve for finding the gain of a LPDA is given in Figs. 6–30 of Stutzman and Thiele [1981]. Note the optimum gain is defined at the point where the parameter τ is the smallest, thus leading to a smaller number of required elements. In practice, the lowest frequency of operation is where the longest element is approximately a half-wave long and the upper frequency of operation is where the shortest element is approximately a half-wave long,

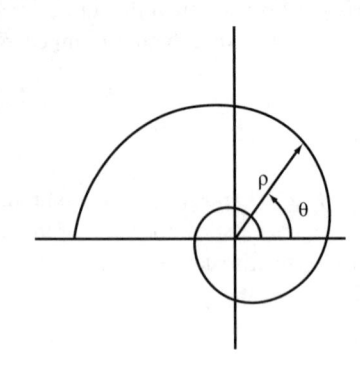

FIGURE 16.23 Basic spiral shape.

$$L_1 \simeq \frac{\lambda_{\text{lower}}}{2} \qquad (16.91)$$

and

$$L_N \simeq \frac{\lambda_{\text{upper}}}{2} \qquad (16.92)$$

The mainbeam of the LPDA is away from the apex, that is, toward the shortest elements. A few extra elements are added onto each end for approximately complying with the idea that the structure should be infinite in the direction of the traveling wave.

Frequency Independent Phased Arrays

If the LPDA is part of an array, for example, of equally spaced elements, such as those discussed in Sec. 16.3 (Pattern Multiplication subsection), then a characteristic dimension is introduced (the spacings between the LPDAs) and the wide-band operation can be seriously affected. A new concept for maintaining log-periodic behavior and still arraying the elements is found in the three-dimensional frequency-independent phased array (FIPA) [Breakall, 1992]. In this array, elements in the log periodic, which make

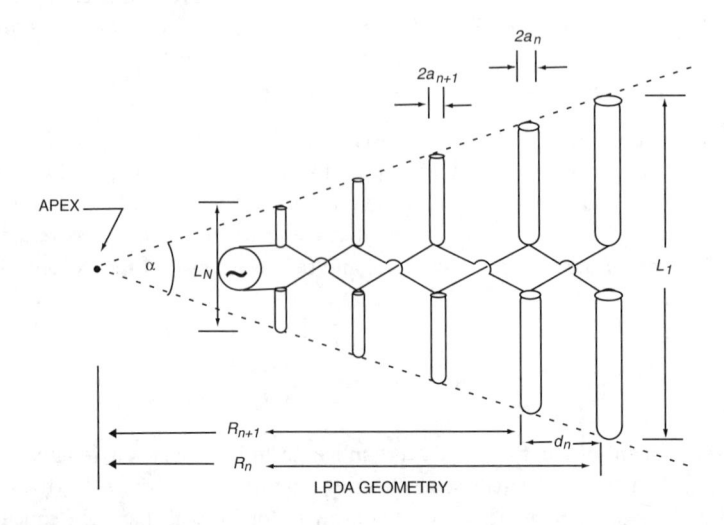

FIGURE 16.24 LPDA geometry.

Antenna Principles　　　**16**-31

up this array, are positioned such that each wire's height above a ground plane is some constant (i.e., 0.25λ) in wavelengths at each wire's resonant frequency. At the prescribed height in wavelengths above the ground plane, each wire's center is also at a constant spacing distance in wavelengths (i.e., 0.60λ) from adjacent wire centers. Scaling of the antenna elements in a log-periodic sense causes the physical wire lengths, heights, and interwire spacings to get progressively smaller as one moves toward the feed end of the element. The scaling factor is the same τ factor as used in the classical log-periodic design.

Figure 16.25 shows an overhead view of a 4 × 4 circularly polarized model of the three-dimensional FIPA. The dark highlighted wires depict the planar nature of the array with the excitation frequency creating large currents at the resonant element frequency. The figure is shown for operation

FIGURE 16.25 A 4 × 4 3D-FIPA (top view) shown excited at the midfrequency.

at a midfrequency. The 4 × 4 planar array maintains fixed lengths, spacings, and heights in wavelengths as the frequency is changed. Therefore the gain, impedance and radiation pattern remain fairly constant making this a respectable wideband array. The feed wires shown in this figure are a single wire for simplicity. The three-dimensional FIPA antenna described overcomes traditional phased-array limitations by using arrays of log-periodic elements.

Self-Similar Fractal Antennas

It has been recognized that one of the fundamental properties of frequency-independent antennas is their ability to retain the same shape under certain scaling transformations. More recently, however, it has been demonstrated that this property of self-similarity is also shared by many fractals [Mandelbrot, 1983]. This commonality has led to the notion that fractal geometric principles be used to provide a natural extension to the traditional approaches for classification, analysis, and design of frequency-independent antennas [D.H. Werner and P.L. Werner]. This new theory allows the classical interpretation of frequency-independent antennas to be generalized to include the radiation from structures that are not only self-similar in the smooth or discrete sense but also in the rough sense.

The standard approach of categorizing frequency-independent antennas is to consider them as being constructed from a multiplicity of adjoining cells. Each cell is identical to the previous cell except for a scaling factor. This periodicity may be characterized in terms of the constant tau [Lo and Lee, 1988],

$$\tau = D_n/D_{n+1} \qquad (16.93)$$

Where D_n represents the dimension of the nth cell and D_{n+1} represents the dimension of the next adjoining cell. In view of the recent introduction of fractal geometry, however, a natural extension to this classical definition of broadband antennas is to consider them as being composed of a sequence of self-similar cells with a similarity factor of τ. For instance, scaling the logarithmic spiral defined in Eq. (16.89) by a factor of τ yields the same spiral rotated by a constant angle $\ell n(\tau)/a$. This suggests that the logarithmic spiral is self-similar, with a similarity factor $\tau = e^{2\pi|a|}$. If rotations are disregarded, however, then the logarithmic spiral can be considered self-similar for any real scaling factor τ. The self-similarity arguments developed for the log spiral are easily generalized to apply to the conical log spiral as well.

Another example of a broadband antenna in common use is the log-periodic dipole antenna discussed earlier in this section. Log-periodic dipole antennas exhibit self-similarity in their geometrical structure

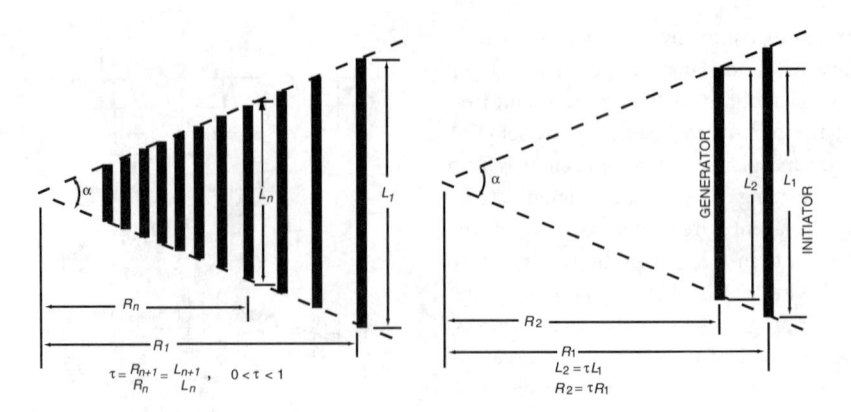

FIGURE 16.26 Fractal geometric description of a log-periodic dipole antenna with initiator and generator.

at discrete frequencies, as illustrated by Fig. 16.26. The lengths and spacings of adjacent dipole elements are scaled by the same similarity factor τ. Figure 16.26 also demonstrates how the log-periodic dipole antenna may be constructed through the use of an initiator and an associated generator.

The scaling properties of log-periodic dipole antennas have also been exploited in the design of a three-dimensional frequency-independent phased array (3D-FIPA) presented earlier in this section. The 3D-FIPA concept can be thought of as having many separate $N \times N$ vertically stacked self-similar planar dipole subarrays. The lengths, heights, and horizontal spacings of the dipoles in each consecutive subarray are physically scaled by a similarity factor of τ. This has the desired effect of ensuring that the dimensions of each dipole subarray remain electrically invariant. Other concepts for broadband arrays of log periodics are discussed in Dunhammel and Berry [1958]. Mei and Moberg [1965], and Johnson and Jasik [1984], which also take advantage of self-similarity in their designs.

A special class of nonuniformly but symmetrically spaced linear arrays, known as Weierstrass arrays, which possess a self-similar geometrical structure, have been investigated in D.H. Werner and P.L. Werner [1996, 1995] and Werner [1994]. It can be shown that the array factor of a Weierstrass array satisfies the self-similarity relation

$$f(\gamma v) = \gamma^{(2-D)} f(v) \qquad (16.94)$$

when

$$v = \cos\theta - \cos\theta_0 \qquad (16.95)$$

and

$$\gamma = \tau^k \text{ for } k = 0, 1, 2, \ldots \qquad (16.96)$$

provided $0 < \tau < 1$ and $1 < D < 2$. The parameters τ and D represent the similarity factor and fractal dimension, respectively. The box-counting definition of fractal dimension for a given fractal F was adopted in this case. In other words

$$D = \lim_{\delta \to 0} \frac{\ell n N_\delta(F)}{\ell n(1/\delta)} \qquad (16.97)$$

where N_δ represents the smallest number of sets of diameter at most δ required to cover the fractal F [Falconer, 1988]. The connection between the box-counting definition of fractal dimension and the intuitive Euclidean concept of dimension is discussed in Voss [1988] and Jaggard [1990].

The self-similarity in the geometrical structure and radiation pattern of infinite Weierstrass arrays suggests that they may be used as multiband arrays that maintain the same radiation characteristics at an infinite number of frequencies. For example, this multiband performance may be achieved by selecting a sequence of discrete frequencies that satisfy the relationship

$$\frac{f_{k+1}}{f_k} = \tau \text{ for } k = 0, 1, \ldots \qquad (16.98)$$

By making use of the multiband properties of Eqs. (16.94), (16.96) and (16.98), it may easily be demonstrated that the directive gain obeys the relationship

$$G(\gamma v) = \frac{|f(\gamma v)|^2}{\frac{1}{2}\int_{-1-u_0}^{1-u_0} |f(\gamma v)|^2 dv} = \frac{|f(v)|^2}{\frac{1}{2}\int_{-1-u_0}^{1-u_0} |f(v)|^2 dv} = G(v) \qquad (16.99)$$

where $u_0 = \cos\theta_0$. This suggests that the directive gain of an infinite Weierstrass array is a log periodic function of frequency with a log period of τ, that is,

$$G(w) = G(w + kT) \text{ for } k = 0, 1, \ldots \qquad (16.100)$$

where

$$w = \log v \qquad (16.101)$$

$$T = \log \tau \qquad (16.102)$$

Defining Terms

Dipole antenna: A straight length of wire excited by a signal source at its center.
Directivity: Provides a measure of performance compared to a reference antenna, such as the isotropic element.
Fraunhofer region: The region that the field from an antenna is a good representation of the far-field expected values. The pattern is essentially independent of the distance from the antenna.
Frequency independent: The term used to define an antenna that has some parameter, such as impedance, independent of frequency. This is an idealization but the concepts can lead to broadband operation.
Fresnel region: In this region the pattern is dependent on the distance from the antenna and does not represent the far-field expected values.
Isotropic source: A fictitious antenna element that radiates equal intensity in all directions. This source is usually the reference antenna element to which gain or directivity of an actual antenna is compared to.
Log periodic: A particular type of broadband antenna in which an antenna parameter, such as impedance, is essentially constant at multiple frequencies that are related logarithmically.
Loop antenna: A type of antenna that is constructed by bending a thin wire into a closed contour.
Method of moments: A numerical method for transforming an integral equation to an associated matrix equation, which can then be readily solved using a computer.
Numerical electromagnetic code: Various antenna software that compute the parameters of an antenna method of moments formulation. NEC, MININEC, and ELNEC are names of some of the popular codes for modeling antennas.
Sommerfeld/Norton solution: A technique that provides the exact solution for fields of wires in the presence of a ground. It is used in antenna modeling software such as NEC for greater accuracy when wires are close to ground.

Surface impedance: The ratio of the tangential electric field on a conductor divided by the resulting linear current density that flows in the conductor.

Uniform theory of diffraction (UTD): Concept allows diffracted fields to be computed for various types of incident waves, such as spherical, plane, and cylindrical.

Yagi–Uda: A form of antenna array in which there is only one driven element. The other elements derive the current excitations from the mutual coupling between elements. Element spacing must be a small fraction of a wavelength in order to have sufficient current excitation.

References

Balanis, C.A. 1982. *Antenna Theory, Analysis, and Design.* Harper & Row, New York.

Breakall, J.K. 1992. Introduction to the three-dimensional frequency-independent phased array (3D-FIPA), A new class of phased array design. *IEEE Antennas and Propagation Society International Symposium Digest,* Vol. III, pp. 1414–1417, July, Chicago.

Burke, G.J. and Poggio, A.J. 1981. Numerical electromagnetic code (NEC), User's Guide. Lawrence Livermore Laboratory, Livermore, CA.

DuHammel, R.H. and Berry, D.G. 1958. Logarithmically periodic antenna arrays. *Wescon Convention Record,* Pt. 1, pp. 161–174.

Falconer, K. 1988. *Fractal Geometry.* Wiley, New York.

Hall, G. 1992. *The ARRL Antenna Book.* The American Radio Relay League, Newington, CT.

Harrington, R.F. 1968. *Field Computation by Moment Methods.* MacMillan, New York.

Hickman, C.E. and Tillman, J.D. 1961. The mutual impedance between identical, parallel dipoles. Bulletin No. 25, Engineering Experiment Station, University of Tennessee, Knoxville.

Jaggard, D.L. 1990. On fractal electrodynamics. In *Recent Advances in Electromagnetic Theory.* Springer-Verlag, New York.

Johnson R.C. and Jasik, H. 1984. *Antenna Engineering Handbook.* McGraw-Hill, New York.

Jordan, E.C. and Balmain, K.G. 1968. *Electromagnetic Waves and Radiating Systems,* 2nd ed., Prentice-Hall, Englewood Cliffs, NJ.

King, R.W.P. 1969. The loop antenna for transmission and reception. In *Antenna Theory,* Pt. I, McGraw-Hill, New York.

King, R.W.P. and Smith, G.S. 1981. *Antennas in Matter: Fundamentals, Theory and Applications.* MIT Press, Cambridge, MA.

Kraus, J.D. and Carver, K.R. 1973. *Electromagnetic,* 2nd ed. McGraw-Hill, New York.

Lewallen, R. 1993. ELNEC antenna analysis program. P.O. Box 6658, Beaverton, OR.

Lindsay, J.E., Jr. 1960. A circular loop antenna with nonuniform current distribution. *IRE Trans. Antennas Propagat.* AP-8(4):439–441.

Lo, Y.T. and Lee, S.W. 1988. *Antennas Handbook.* Van Nostrand Reinhold, New York.

Logan, J.C. and Rockway, J.W. 1986. The new MININEC (version 3): A mini-numerical electromagnetic code. Tech. Doc. 938, Naval Ocean System Center, San Diego, CA.

Mandelbrot, B.B. 1983. *The Fractal Geometry of Nature.* W.H. Freeman, New York.

Mei, K.K., Moberg, M.W., Rumsey, V.H., and Yeh, Y.S. 1965. Directive frequency independent arrays. *IEEE Trans. on Antennas and Propagation* AP-13(5):807–809.

Pocklington, H.C. 1897. Electrical oscillations in wire. *Cambridge Philosophical Soc. Proc.* London, England, 9:324–332.

Rumsey, V. 1966. *Frequency Independent Antennas.* Academic Press, New York.

Storer, J.E. 1956. Impedance of thin-wire loop antennas. *AIEE Trans.* Part I. Communication and Electronics 75(Nov.):606–619.

Stutzman, W.L. and Thiele, G.A. 1981. *Antenna Theory and Design,* 2nd ed. Wiley, New York.

Voss, R.F. 1988. Fractals in nature: From characterization to simulation. In *The Science of Fractal Images.* Springer-Verlag, New York.

Werner, D.H. 1993. An exact formulation for the vector potential of a cylindrical antenna with uniformly distributed current and arbitrary radius. *IEEE Trans. on Antennas and Propagation.* 41(8): 1009–1018.

Werner, D.H. 1994. Fractal radiators. In *Proceedings of IEEE Dual-Use Technologies & Applications Conference,* Vol. I, pp. 478–482. 4th Annual IEEE Mohawk Valley Section, SUNY Inst. of Technology at Utica/Rome, New York.

Werner, D.H. 1995. Analytical and numerical methods for evaluating the electromagnetic field integrals associated with current-carrying wire antennas. In *Advanced Electromagnetism: Foundations, Theory and Applications.* World Scientific, Ltd.

Werner, D.H. 1996. An exact integration procedure for vector potentials of thin circular loop antennas. *IEEE Trans. on Antennas and Propagation.* 44(2): 157–165.

Werner, D.H. and Werner, P.L. 1995. On the synthesis of fractal radiation patterns. *Radio Science* 30(1):29–45.

Werner, D.H. and Werner, P.L. 1996. Frequency independent features of self-similar fractal antennas. *IEEE Trans. on Antennas and Propagation Society International Symposium Digest,* July, Baltimore, MD.

Young, J.S. 1994. Simulation of antenna patterns over 3-dimensional irregular terrain using the uniform geometrical theory of diffraction (UTD) [development of the PAINT system]. Ph.D. thesis in Electrical Engineering, Pennsylvania State University, University Park, PA.

17
Practical Antenna Systems

Jerry C. Whitaker
Editor

17.1 Introduction 17-1
 Operating Characteristics · Antenna Bandwidth · Polarization
 · Antenna Beamwidth · Antenna Gain · Space Regions
 · Impedance Matching
17.2 Antenna Types 17-6
17.3 Antenna Applications 17-11
 AM Broadcast Antenna Systems · FM Broadcast Antenna
 Systems · Television Antenna Systems
17.4 Phased-Array Antenna Systems 17-23
 Phase-Shift Devices · Radar System Duplexer

17.1 Introduction

Transmission is accomplished by the emission of coherent electromagnetic waves in free space from one or more radiating elements that are excited by RF currents. Although, by definition, the radiated energy is composed of mutually dependent magnetic and electric vector fields, it is conventional practice to measure and specify radiation characteristics in terms of the electric field only.

The purpose of an antenna is to efficiently radiate the power supplied to it by the transmitter. A simple antenna, consisting of a single vertical element over a ground plane can do this job quite well at low to medium frequencies. Antenna systems may also be required to concentrate the radiated power in a given direction and minimize radiation in the direction of other stations sharing the same or adjacent frequencies. To achieve such directionality may require a complicated antenna system that incorporates a number of individual elements or towers and matching networks.

As the operating frequency increases into VHF and above, the short wavelengths permit the design of specialized antennas that offer high directivity and gain.

Operating Characteristics

Wavelength is the distance traveled by one cycle of a radiated electric signal. The frequency of the signal is the number of cycles per second. It follows that the frequency is inversely proportional to the wavelength. Both wavelength and frequency are related to the speed of light. Conversion between the two parameters can be accomplished with the formula

$$c = f \times \lambda$$

where c is the speed of light, f is the operating frequency, and λ is the wavelength. The velocity of electric signals in air is essentially the same as that of light in free space (2.9983×10^{10} cm/sec).

The *electrical length* of a radiating element is the most basic parameter of an antenna:

$$H = \frac{H_t \times F_o}{2733}$$

where H is the length of the radiating element in electrical degrees, H_t is the length of the radiating element in feet, and F_o is frequency of operation in kHz. Where the radiating element is measured in meters,

$$H = \frac{H_t \times F_o}{833.23}$$

The *radiation resistance* of an antenna is defined by:

$$R = \frac{P}{I^2}$$

where R is the radiation resistance, P is the power delivered to the antenna, and I is the driving current at the antenna base.

Antenna Bandwidth

Bandwidth is a general classification of the frequencies over which an antenna is effective. This parameter requires specification of acceptable tolerances relating to the uniformity of response over the intended operating band.

Strictly speaking, *antenna bandwidth* is the difference in frequency between two points at which the power output of the transmitter drops to one-half the midrange value. The points are called *half-power points*. A half-power point is equal to a VSWR of 5.83:1, or the point at which the voltage response drops to 0.7071 of the midrange value. In most communications systems, a VSWR of less than 1.2:1 within the occupied bandwidth of the radiated signal is preferable.

Antenna bandwidth depends on the radiating element impedance and the rate at which the reactance of the antenna changes with frequency.

Bandwidth and RF coupling go hand in hand, regardless of the method used to excite the antenna. All elements between the transmitter output circuit and the antenna must be analyzed, first by themselves, and then as part of the overall system bandwidth. In any transmission system, the *composite bandwidth*, not just the bandwidths of individual components, is of primary concern.

Polarization

Polarization is the angle of the radiated electric field vector in the direction of maximum radiation. Antennas may be designed to provide horizontal, vertical, or circular polarization. Horizontal or vertical polarization is determined by the orientation of the radiating element with respect to Earth. If the plane of the radiated field is parallel to the ground, the signal is said to be *horizontally polarized*. If it is at right angles to the ground, it is said to be *vertically polarized*. When the receiving antenna is located in the same plane as the transmitting antenna, the received signal strength will be maximum.

Circular polarization (CP) of the transmitted signal results when equal electrical fields in the vertical and horizontal planes of radiation are out-of-phase by 90° and are rotating a full 360° in one wavelength of the operating frequency. The rotation can be clockwise or counterclockwise, depending on the antenna design. This continuously rotating field gives CP good signal penetration capabilities because it can be received efficiently by an antenna of any random orientation. Figure 17.1 illustrates the principles of circular polarization.

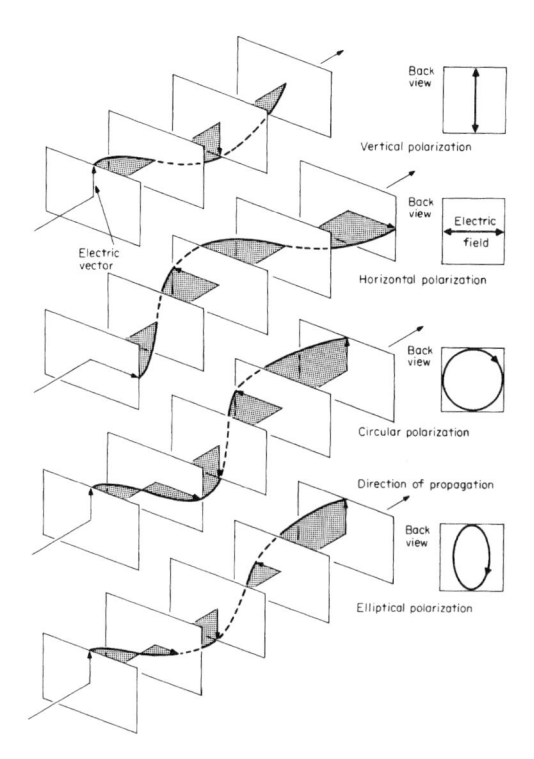

FIGURE 17.1 Polarization of the electric field of a transmitted wave.

Antenna Beamwidth

Beamwidth in the plane of the antenna is the angular width of the directivity pattern where the power level of the received signal is down by 50% (3 dB) from the maximum signal in the desired direction of reception.

Antenna Gain

Directivity and *gain* are measures of how well energy is concentrated in a given direction. Directivity is the ratio of power density in a given direction to the power density that would be produced if the energy were radiated isotropically. The reference can be linearly or circularly polarized. Directivity is usually given in dBi (decibels above isotropic).

Gain is the field intensity produced in a given direction by a fixed input power to the antenna, referenced to a dipole. It is frequently used as a figure of merit. It is closely related to directivity, which in turn is dependent upon the radiation pattern. High values of gain are usually obtained with a reduction in beamwidth.

An antenna is typically configured to exhibit "gain" by narrowing the beamwidth of the radiated signal to concentrate energy toward the intended coverage area. The actual amount of energy being radiated is the same with a unity-gain antenna or a high-gain antenna, but the useful energy (commonly referred to as the *effective radiated power*, or ERP) can be increased significantly.

Electrical *beam tilt* can also be designed into a high-gain antenna. A conventional antenna typically radiates more than half of its energy above the horizon. This energy is lost for practical purposes in most applications. Electrical beam tilt, caused by delaying the RF current to the lower elements of a multi-element antenna, can be used to provide more useful power to the service area.

Pattern optimization is another method that can be used to maximize radiation to the intended service area. The characteristics of the transmitting antenna are affected, sometimes greatly, by the presence of the supporting tower, if side-mounted, or by nearby tall obstructions (such as another transmitting tower) if top-mounted. Antenna manufacturers use various methods to reduce pattern distortions. These generally involve changing the orientation of the radiators with respect to the tower and adding parasitic elements.

Space Regions

Insofar as the transmitting antenna is concerned, space is divided into three regions:

- *Reactive near-field region.* This region is the area of space immediately surrounding the antenna in which the reactive components predominate. The size of the region varies, depending on the antenna design. For most antennas, the reactive near-field region extends 2 wavelengths or less from the radiating elements.
- *Radiating near-field region.* This region is characterized by the predictable distribution of the radiating field. In the near-field region, the relative angular distribution of the field is dependent on the distance from the antenna.
- *Far-field region.* This region is characterized by the independence of the relative angular distribution of the field with varying distance. The pattern is essentially independent of distance.

Impedance Matching

Most practical antennas require some form of impedance matching between the transmission line and the radiating elements. The implementation of a matching network can take on many forms, depending on the operating frequency and output power.

The *negative sign* convention is generally used in impedance matching analysis. That is, if a network delays or retards a signal by θ°, the phase shift across the network is said to be "minus θ degrees." For example, a 1/4-wave length of transmission line, if properly terminated, has a phase shift of –90°. Thus, a *lagging* or low-pass network has a negative phase shift, and a *leading* or high-pass network has a positive phase shift. There are three basic network types that can be used for impedance matching: *L*, *pi*, and *tee*.

L Network

The L network is shown in Fig. 17.2 The loaded Q of the network is determined from Equation 1. Equation 2 defines the shunt leg reactance, which is negative (capacitive) when θ is negative, and positive (inductive) when θ is positive. The series leg reactance is found using Equation 3, the phase shift via Equation 4, and the currents and voltages via Ohm's law. Note that R_2 (the resistance on the shunt leg side of the L network) must always be greater than R_1. An L network cannot be used to match equal resistances, or to adjust phase independently of resistance.

Tee Network

The tee network is shown in Fig. 17.3. This configuration can be used to match unequal resistances. The tee network has the added feature that phase shift is independent of the resistance transformation ratio. A tee network can be considered simply as two L networks back-to-back. Note that there are two loaded Qs associated with a tee network: an input Q and an output Q. To gauge the bandwidth of the tee network, the lower value of Q must be ignored. Note that the Q of a tee network increases with increasing phase shift.

Equations 5 through 14 describe the tee network. It is a simple matter to find the input and output currents via Ohm's law, and the shunt leg current can be found via the Cosine law (Equation 12). Note that this current increases with increasing phase shift. Equation 13 describes the mid-point resistance of a tee network, which is always higher than R_1 or R_2. Equation 14 is useful when designing a *phantom tee network*; that is, where X_2 is made up only of the antenna reactance and there is no physical component in place of X_2. Keep in mind that a tee network is considered as having a lagging or negative phase shift when the shunt leg is capacitive (X_3 negative), and vice versa. The input and output arms can go either negative or positive, depending on the resistance transformation ratio and desired phase shift.

Practical Antenna Systems

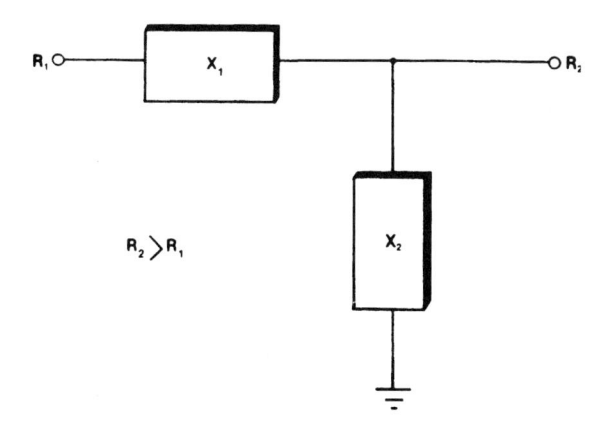

Equation 1:
$$Q = \sqrt{\frac{R_2}{R_1} - 1} = \left|\frac{X_1}{R_1}\right| = \left|\frac{R_2}{X_2}\right|$$

Equation 2:
$$X_2 = \frac{\pm R_2}{Q}$$

Equation 3:
$$X_1 = \frac{-R_1 R_2}{X_2}$$

Equation 4:
$$\Theta = TAN^{-1}\left(\frac{R_2}{X_2}\right)$$

Where: R_1 = L network input resistance (ohms)
R_2 = L network output resistance (ohms)
X_1 = Series leg reactance (ohms)
X_2 = Shunt leg reactance (ohms)
Q = Loaded Q of the L network

FIGURE 17.2 L network parameters.

Pi Network

The pi network is shown in Fig. 17.4. It can also be considered as two L networks back-to-back and, therefore, the same comments about overall loaded Q apply. Note that susceptances have been used in Equations 15 through 19 instead of reactances in order to simplify calculations. The same conventions regarding tee network currents apply to pi network voltages (Equations 20, 21, and 22). The mid-point resistance of a pi network is always less than R_1 or R_2. A pi network is considered as having a negative or lagging phase shift when Y_3 is positive, and vice versa.

Line Stretcher

A *line stretcher* makes a transmission line look longer or shorter in order to produce sideband impedance symmetry at the transmitter PA (see Fig. 17.5). This is done to reduce audio distortion in an envelope detector, the kind of detector that most AM receivers employ. Symmetry is defined as equal sideband resistances, and equal — but opposite sign — sideband reactances.

There are two possible points of symmetry, each 90° from the other. One produces sideband resistances greater than the carrier resistance, and the other produces the opposite effect. One side will create a pre-emphasis effect, and the other a de-emphasis effect.

Depending on the Q of the transmitter output network, one point of symmetry may yield lower sideband VSWR at the PA than the other. This results from the Q of the output network opposing the Q of the antenna in one direction, but aiding the antenna Q in the other direction.

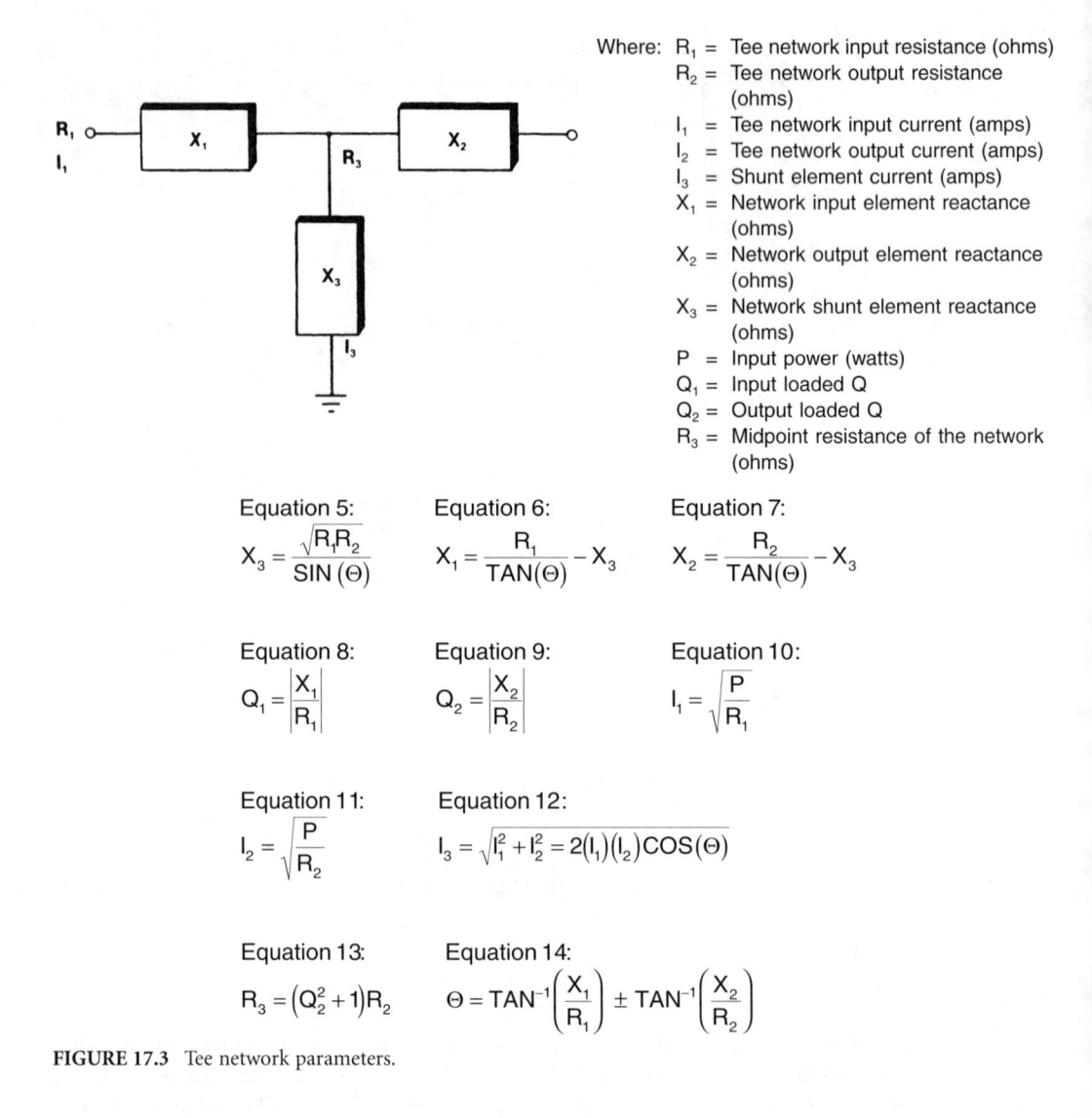

Where: R_1 = Tee network input resistance (ohms)
R_2 = Tee network output resistance (ohms)
I_1 = Tee network input current (amps)
I_2 = Tee network output current (amps)
I_3 = Shunt element current (amps)
X_1 = Network input element reactance (ohms)
X_2 = Network output element reactance (ohms)
X_3 = Network shunt element reactance (ohms)
P = Input power (watts)
Q_1 = Input loaded Q
Q_2 = Output loaded Q
R_3 = Midpoint resistance of the network (ohms)

Equation 5:
$$X_3 = \frac{\sqrt{R_1 R_2}}{\text{SIN}(\Theta)}$$

Equation 6:
$$X_1 = \frac{R_1}{\text{TAN}(\Theta)} - X_3$$

Equation 7:
$$X_2 = \frac{R_2}{\text{TAN}(\Theta)} - X_3$$

Equation 8:
$$Q_1 = \left|\frac{X_1}{R_1}\right|$$

Equation 9:
$$Q_2 = \left|\frac{X_2}{R_2}\right|$$

Equation 10:
$$I_1 = \sqrt{\frac{P}{R_1}}$$

Equation 11:
$$I_2 = \sqrt{\frac{P}{R_2}}$$

Equation 12:
$$I_3 = \sqrt{I_1^2 + I_2^2 = 2(I_1)(I_2)\text{COS}(\Theta)}$$

Equation 13:
$$R_3 = (Q_2^2 + 1)R_2$$

Equation 14:
$$\Theta = \text{TAN}^{-1}\left(\frac{X_1}{R_1}\right) \pm \text{TAN}^{-1}\left(\frac{X_2}{R_2}\right)$$

FIGURE 17.3 Tee network parameters.

17.2 Antenna Types

The *dipole antenna* is simplest of all antennas, and the building block of most other designs. The dipole consists of two in-line rods or wires with a total length equal to 1/2-wave at the operating frequency. Figure 17.6 shows the typical configuration, with two 1/4-wave elements connected to a transmission line. The radiation resistance of a dipole is on the order of 73 Ω. The bandwidth of the antenna may be increased by increasing the diameter of the elements, or by using cones or cylinders rather than wires or rods, as shown in the figure. Such modifications also increase the impedance of the antenna.

The dipole can be straight (in-line) or bent into a V-shape. The impedance of the V-dipole is a function of the V angle. Changing the angle effectively tunes the antenna. The vertical radiation pattern of the V-dipole antenna is similar to the straight dipole for angles of 120° or less.

A *folded dipole* can be fashioned as shown in Fig. 17.7. Such a configuration results in increased bandwidth and impedance. Impedance can be further increased by using rods of different diameter and by varying the spacing of the elements. The 1/4-wave dipole elements connected to the closely-coupled 1/2-wave element act as a matching stub between the transmission line and the single-piece 1/2-wave element. This broadbands the folded dipole antenna by a factor of 2.

Practical Antenna Systems

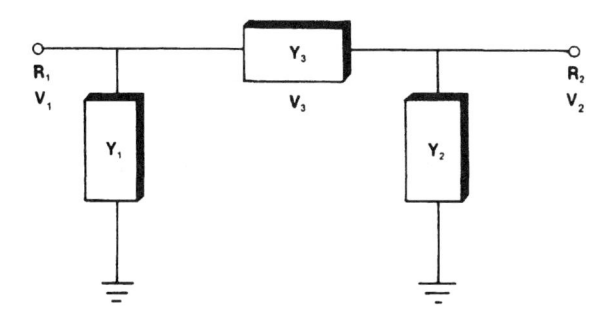

Equation 15:
$$Y_3 = \frac{1}{-\text{SIN}(\Theta)\sqrt{R_1 R_2}}$$

Equation 16:
$$Y_1 = \frac{\text{TAN}(\Theta)}{R_1 - Y_3}$$

Equation 17:
$$Y_2 = \frac{\text{TAN}(\Theta)}{R_2 - Y_3}$$

Equation 18:
$$Q_1 = |R_1 Y_1|$$

Equation 19:
$$Q_2 = |R_2 Y_2|$$

Equation 20:
$$V_1 = \sqrt{R_1 P}$$

Equation 21:
$$V_2 = \sqrt{R_2 P}$$

Equation 22:
$$V_3 \sqrt{V_1^2 + V_2^2 - 2(v_1)(V_2)\text{COS}(\Theta)}$$

Equation 23:
$$R_3 = \frac{Q_2^2 + 1}{R_2}$$

Where: R_1 = Pi network input resistance (ohms)
R_2 = Output resistance (ohms)
V_1 = Input voltage (volts)
V_2 = Output voltage (volts)
V_3 = Voltage across series element (volts)
P = Power input to pi network (watts)
Y_1 = Input shunt element susceptance (mhos)
Y_2 = Output shunt element susceptance (mhos)
Y_3 = Series element susceptance (mhos)
Q_1 = Input loaded Q
Q_2 = Output loaded Q

FIGURE 17.4 Pi network parameters.

A *corner-reflector* antenna may be formed as shown in Fig. 17.8. A ground plane or flat reflecting sheet is placed at a distance of 1/16- to 1/4-wavelengths behind the dipole. Gain in the forward direction can be increased by a factor of 2 with this type of design.

Quarter-Wave Monopole

A conductor placed above a ground plane forms an image in the ground plane such that the resulting pattern is a composite of the *real antenna* and the *image antenna* (see Fig. 17.9). The operating impedance is one-half of the impedance of the antenna and its image when fed as a physical antenna in free space. An example will help illustrate this concept. A 1/4-wave monopole mounted on an infinite ground plane has an impedance equal to one-half the free-space impedance of a 1/4-wave dipole. It follows, then, that the theoretical characteristic resistance of a 1/4-wave monopole with an infinite ground plane is 37 Ω.

For a real-world antenna, an infinite ground plane is neither possible nor required. An antenna mounted on a ground plane that is 2 to 3 times the operating wavelength has about the same impedance as a similar antenna mounted on an infinite ground plane.

Equation 25:

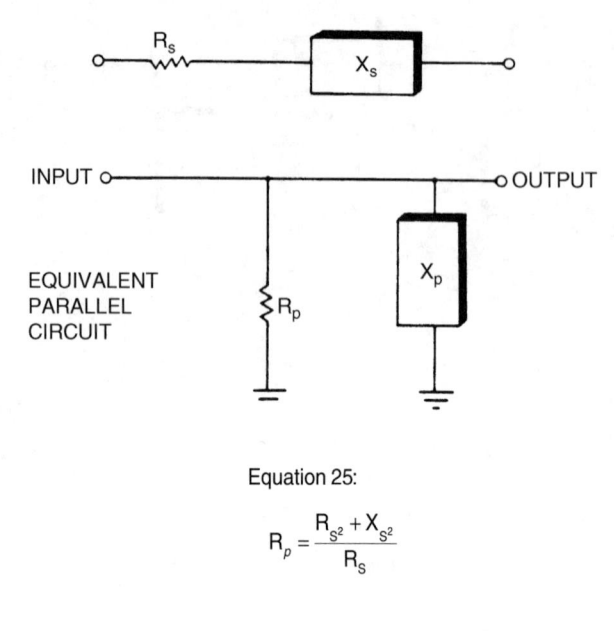

$$R_p = \frac{R_{s^2} + X_{s^2}}{R_s}$$

Equation 26:

$$X_p = \frac{R_{s^2} = X_{s^2}}{X_s}$$

Where: R_s = Series configuration resistance (ohms)
R_p = Parallel configuration resistance (ohms)
X_s = Series reactance (ohms)
X_p = Parallel reactance (ohms)

FIGURE 17.5 Line stretcher configuration.

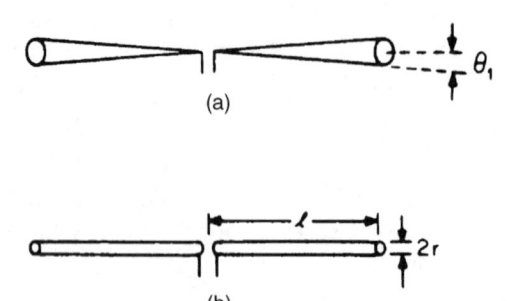

FIGURE 17.6 Half-wave dipole antenna: (a) conical dipole, and (b) conventional dipole.

Log-Periodic Antenna

The log-periodic antenna can take on a number of forms. Typical designs include the *conical log spiral*, *log-periodic V*, and *log-periodic dipole*. The most common of these antennas is the log-periodic dipole. The antenna can be fed either by using alternating connections to a balanced line, or by a coaxial line running through one of the feeders from front to back. In theory, the log-periodic antenna can be designed

Practical Antenna Systems 17-9

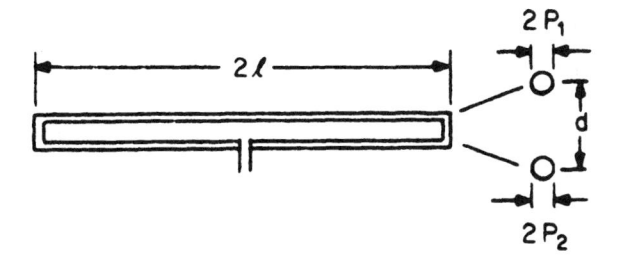

FIGURE 17.7 Folded dipole antenna.

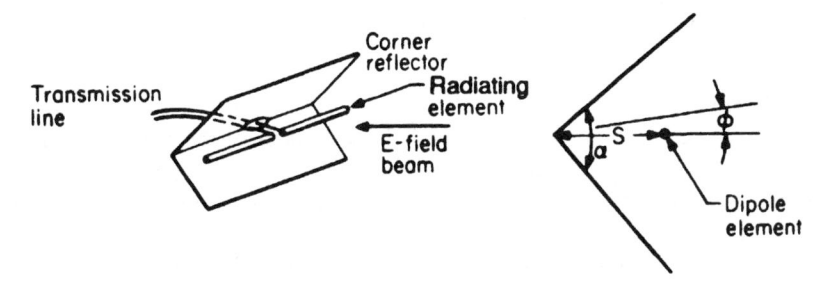

FIGURE 17.8 Corner-reflector antenna.

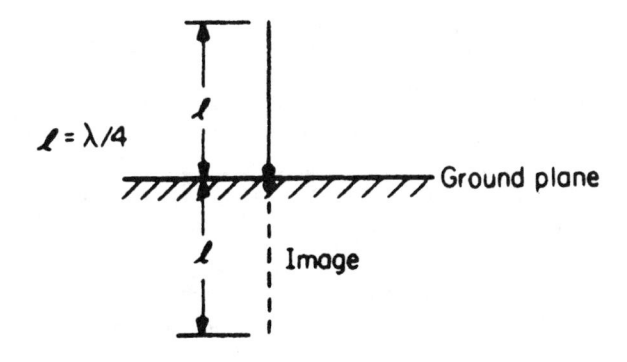

FIGURE 17.9 Vertical monopole mounted above a ground plane.

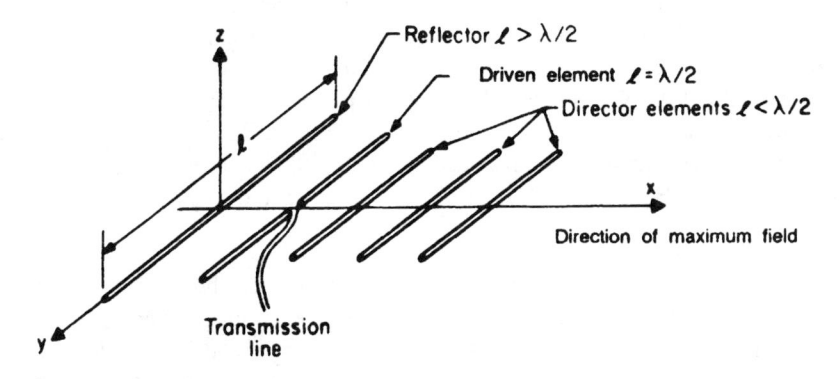

FIGURE 17.10 The Yagi-Uda array.

to operate over many octaves. In practice, however, the upper frequency is limited by the precision required in constructing the small elements, feed lines, and support structure of the antenna.

Yagi-Uda Antenna

The Yagi-Uda is an *end-fire array* consisting typically of a single driven dipole with a *reflector dipole* behind the driven element, and one or more *parasitic director elements* in front (see Fig. 17.10). Common designs use from one to seven director elements. As the number of elements is increased, directivity increases. Bandwidth, however, decreases as the number of elements is increased. Arrays of more than four director elements are typically narrowband.

The radiating element is $1/2$-wavelength at the center of the band covered. The single reflector element is slightly longer, and the director elements are slightly shorter, all spaced approximately 1/4-wavelength from each other.

Table 17.1 demonstrates how the number of elements determines the gain and beamwidth of a Yagi-Uda antenna.

Waveguide Antenna

The waveguide antenna consists of a dominant-mode-fed waveguide opening onto a conducting ground plane. Designs may be based on rectangular, circular, or coaxial waveguide (also called an *annular slot*). The slot antenna is simplicity itself. A number of holes of a given dimension are placed at intervals along

TABLE 17.1 Typical Characteristics of Single-Channel Yagi-Uda Arrays

No. of Elements	Gain (dB)	Beam Width (°)
2	3–4	65
3	6–8	55
4	7–10	50
5	9–11	45
9	12–14	37
15	14–16	30

Source: Benson, B., Ed., *Television Engineering Handbook*, McGraw-Hill, New York, 1986, 388.

a section of waveguide. The radiation characteristics of the antenna are determined by the size, location, and orientation of the slots. The antenna offers optimum reliability because there are no discrete elements, except for the waveguide section itself.

Horn Antenna

The horn antenna may be considered a natural extension of the dominant-mode waveguide feeding the horn in a manner similar to the wire antenna, which is a natural extension of the two-wire transmission line. The most common types of horns are the *E-plane sectoral, H-plane sectoral, pyramidal horn* (formed by expanding the walls of the $TE_{0,1}$-mode-fed rectangular waveguide), and the *conical horn* (formed by expanding the walls of the $TE_{1,1}$-mode-fed circular waveguide). Dielectric-loaded waveguides and horns offer improved pattern performance over unloaded horns. Ridged and tapered horn designs improve the bandwidth characteristics. Horn antennas are available in single and dual polarized configurations.

Reflector Antenna

The reflector antenna is formed by mounting a radiating feed antenna above a reflecting ground plane. The most basic form of reflector is the loop or dipole spaced over a finite ground plane. This concept is the basis for the parabolic or spherical reflector antenna. The *parabolic reflector antenna* may be fed directly or through the use of a subreflector in the focal region of the parabola. In this approach, the subreflector is illuminated from the parabolic surface. The chief disadvantage of this design is the aperture blockage of the subreflector, which restricts its use to large-aperture antennas. The operation of a parabolic or spherical reflector antenna is typically described using physical optics.

Parabolic reflector antennas are usually illuminated by a flared-horn antenna with a flare angle of less than 18°. A rectangular horn with a flare angle less than 18° has approximately the same aperture field as the dominant-mode rectangular waveguide feeding the horn.

Spiral Antenna

The bandwidth limitations of an antenna are based on the natural change in the critical dimensions of the radiating elements caused by differing wavelengths. The spiral antenna overcomes this limitation because the radiating elements are specified only in angles. A two-arm equiangular spiral is shown in Fig. 17.11. This common design gives wideband performance. Circular polarization is inherent in the antenna. Rotation of the pattern corresponds to the direction of the spiral arms. The gain of a spiral antenna is typically slightly higher than a dipole.

The basic spiral antenna radiates on both sides of the arms. Unidirectional radiation is achieved through the addition of a reflector or cavity.

FIGURE 17.11 Two-arm equiangular spiral antenna.

Array Antenna

The term "array antenna" covers a wide variety of physical structures. The most common configuration is the *planar array antenna*, which consists of a number of radiating elements regularly spaced on a rectangular or triangular lattice. The *linear array antenna*, where the radiating elements are placed in a single line, is also common. The pattern of the array is the product of the element pattern and the array configuration. Large array antennas may consist of 20 or more radiating elements.

Correct phasing of the radiating elements is the key to operation of the system. The radiating pattern of the structure, including direction, can be controlled through proper adjustment of the relative phase of the elements.

17.3 Antenna Applications

An analysis of the applications of antennas for commercial and industrial use is beyond the scope of this chapter. It is instructive, however, to examine three of the most obvious antenna applications: AM and FM radio, and television. These applications illustrate antenna technology as it applies to frequencies ranging from the lower end of the MF band to the upper reaches of UHF.

AM Broadcast Antenna Systems

Vertical polarization of the transmitted signal is used for AM broadcast stations because of its superior groundwave propagation, and because of the simple antenna designs that it affords. The Federal Communications Commission (FCC) and licensing authorities in other countries have established classifications of AM stations with specified power levels and hours of operation. Protection requirements set forth by the FCC specify that some AM stations (in the United States) reduce their transmitter power at sunset, and return to full power at sunrise. This method of operation is based on the propagation characteristics of AM band frequencies. AM signals propagate further at nighttime than during the day.

The different day/night operating powers are designed to provide each AM station with a specified coverage area that is free from interference. Theory rarely translates into practice insofar as coverage is concerned, however, because of the increased interference that all AM stations suffer at nighttime.

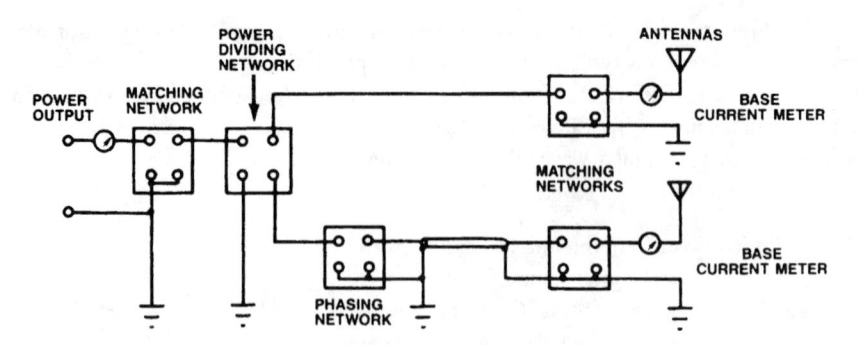

FIGURE 17.12 Block diagram of an AM directional antenna feeder system for a two-tower array.

The tower visible at any AM radio station transmitter site is only half of the antenna system. The second element is a buried ground system. Current on a tower does not simply disappear; rather, it returns to Earth through the capacitance between the Earth and the tower. Ground losses are greatly reduced if the tower has a radial copper ground system. A typical single-tower ground system is made up of 120 radial ground wires, each 140 electrical degrees long (at the operating frequency), equally spaced out from the tower base. This is often augmented with an additional 120 interspersed radials 50 ft long.

Directional AM Antenna Design

When a nondirectional antenna with a given power does not radiate enough energy to serve the station's primary service area, or radiates too much energy toward other radio stations on the same or adjacent frequencies, it is necessary to employ a directional antenna system. Rules set out by the FCC and regulatory agencies in other countries specify the protection requirements to be provided by various classes of stations, for both daytime and nighttime hours. These limits tend to define the shape and size of the most desirable antenna pattern.

A directional antenna functions by carefully controlling the amplitude and phase of the RF currents fed to each tower in the system. The directional pattern is a function of the number and spacing of the towers (vertical radiators), and the relative phase and magnitude of their currents. The number of towers in a directional AM array can range from two to six, or even more in a complex system. One tower is defined as the *reference tower*. The amplitude and phase of the other towers are measured relative to this reference.

A complex network of power splitting, phasing, and antenna coupling elements is required to make a directional system work. Figure 17.12 shows a block diagram of a basic two-tower array. A power divider network controls the relative current amplitude in each tower. A phasing network provides control of the phase of each tower current, relative to the reference tower. Matching networks at the base of each tower couple the transmission line impedance to the base operating impedance of the radiating towers.

In practice, the system shown in the figure would not consist of individual elements. Instead, the matching network, power dividing network, and phasing network would all usually be combined into a single unit, referred to as the *phasor*.

Antenna Pattern Design

The pattern of any AM directional antenna system (array) is determined by a number of factors, including:

- Electrical parameters (phase relationship and current ratio for each tower)
- Height of each tower
- Position of each tower with respect to the other towers (particularly with respect to the reference tower)

Practical Antenna Systems   17-13

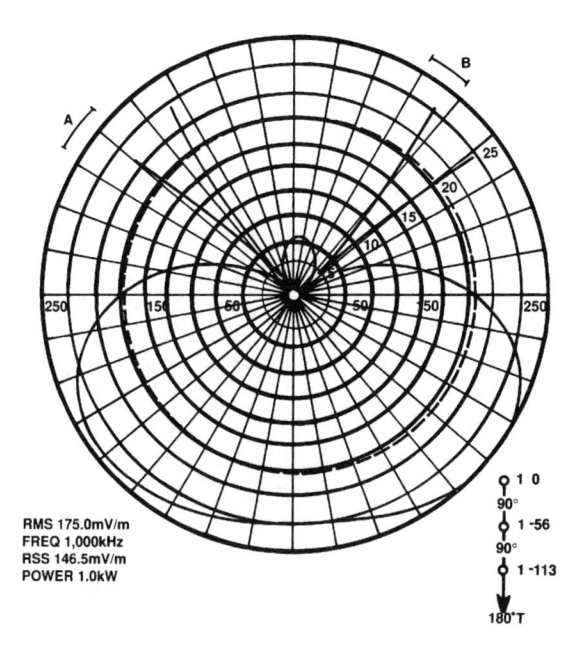

FIGURE 17.13 Radiation pattern generated with a three-tower in-line directional array using the electrical parameters and orientation shown.

A *directional array* consists of two or more towers arranged in a specific manner on a plot of land. Figure 17.13 shows a typical three-tower array, as well as the pattern such an array would produce. This is an *in-line array*, meaning that all the elements (towers) are in line with one another. Notice that the *major lobe* is centered on the same line as the line of towers, and that the *pattern nulls* (*minima*) are positioned symmetrically about the line of towers, protecting co-channel stations A and B at true bearings of 315° and 45°, respectively.

Figure 17.14 shows the same array, except that it has been rotated by 10°. Notice that the pattern shape is not changed, but the position of the major lobe and the nulls follow the line of towers. Also notice that the nulls are no longer pointed at the stations to be protected.

If this directional antenna system were constructed on a gigantic turntable, the pattern could be rotated without affecting the shape. But, to accomplish the required protections and to have the major lobe(s) oriented in the right direction, there is only one correct position. In most cases, the position of the towers will be specified with respect to a single reference tower. The location of the other towers will be given in the form of a distance and bearing from that reference. Occasionally, a reference point, usually the center of the array, will be used for a geographic coordinate point.

Bearing

The bearing or azimuth of the towers from the reference tower or point is specified clockwise in degrees from *true north*. The distinction between true and *magnetic north* is vital. The magnetic North Pole is not at the true or geographic North Pole. (In fact, it is in the vicinity of 74° north, 101° west, in the islands of northern Canada.) The difference between magnetic and true bearings is called variation or *magnetic declination*. Declination, a term generally used by surveyors, varies for different locations. It is not a constant. The Earth's magnetic field is subject to a number of changes in intensity and direction. These changes take place over daily, yearly, and long-term (or *secular*) periods. The secular changes result in a relatively constant increase or decrease in declination over a period of many years.

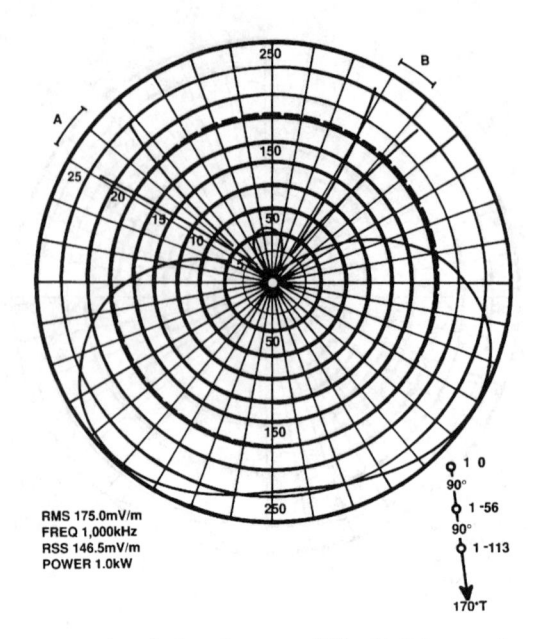

FIGURE 17.14 Radiation pattern produced when the array of Fig. 17.13 is rotated to a new orientation.

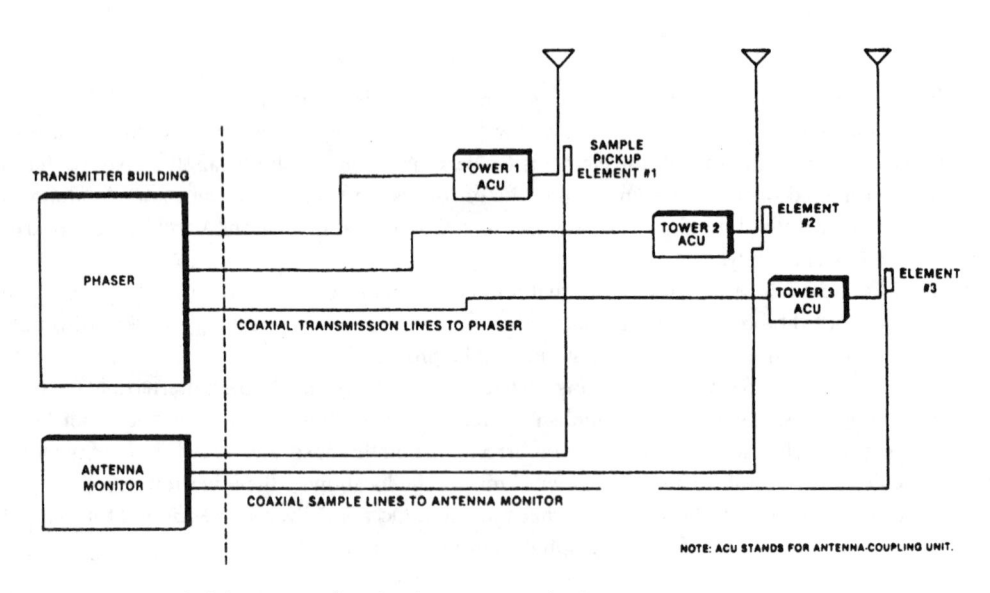

FIGURE 17.15 A typical three-tower directional antenna monitoring system.

Antenna Monitoring System

Monitoring the operation of an AM directional antenna basically involves measuring the power into the system, the relative value of currents into the towers, their phase relationships, and the levels of radiated signal at certain monitoring points some distance from the antenna. Figure 17.15 shows a block diagram of a typical monitoring system for a three-tower array. For systems with additional towers, the basic layout is extended by adding more pickup elements, sample lines, and a monitor with additional inputs.

Phase/Current Sample Loop

Two types of phase/current sample pickup elements are commonly used: the *sample loop* and *torroidal current transformer* (TCT). The sample loop consists of a single turn unshielded loop of rigid construction,

Practical Antenna Systems

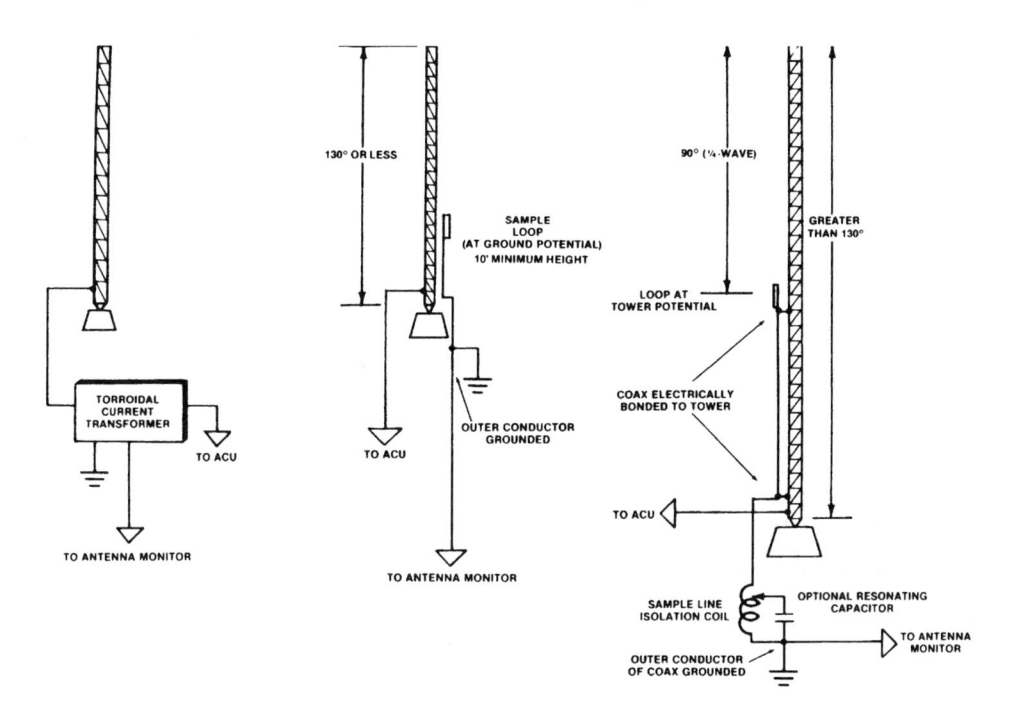

FIGURE 17.16 Three possible circuit configurations for phase sample pickup.

with a fixed gap at the open end for connection of the sample line. The device must be mounted on the tower near the point of maximum current. The loop can be used on towers of both uniform and nonuniform cross-section. It must operate at tower potential, except for towers of less than 130 electrical degrees height, where the loop can be operated at ground potential.

When the sample loop is operated at tower potential, the coax from the loop to the base of the tower is also at tower potential. To bring the sample line across the base of the tower, a sample line isolation coil is used.

A shielded torroidal current transformer can also be used as the phase/current pickup element. Such devices offer several advantages over the sample loop, including greater stability and reliability. Because they are located inside the tuning unit cabinet or house, TCTs are protected from wind, rain, ice, and vandalism.

Unlike the rigid, fixed sample loop, torroidal current transformers are available in several sensitivities, ranging from 0.25 to 1.0 V per ampere of tower current. Tower currents of up to 40 A can be handled, providing a more usable range of voltages for the antenna monitor. Figure 17.16 shows the various arrangements that can be used for phase/current sample pickup elements.

Sample Lines

The selection and installation of the sampling lines for a directional monitoring system are important factors in the ultimate accuracy of the overall array.

With *critical arrays* (antennas requiring operation within tight limits specified in the station license), all sample lines must be of equal electrical length and installed in such a manner that corresponding lengths of all lines are exposed to equal environmental conditions.

While sample lines may be run above ground on supports (if protected and properly grounded), the most desirable arrangement is direct burial using jacketed cable. Burial of sample line cable is almost a standard practice because proper burial offers good protection against physical damage and a more stable temperature environment.

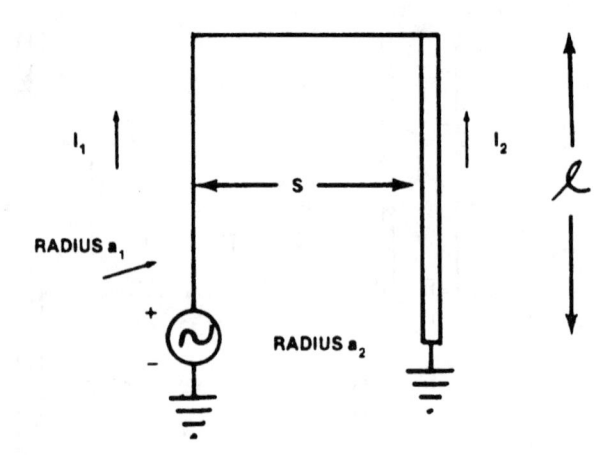

FIGURE 17.17 The folded unipole antenna can be thought of as a 1/2-wave folded dipole antenna perpendicular to the ground and cut in half.

The Common Point

The power input to a directional antenna is measured at the phasor *common point*. Power is determined by the *direct method*:

$$P = I^2 R$$

where P is the power in watts (W), I is the common point current in amperes (A), and R is the common point resistance in ohms (Ω).

Monitor Points

Routine monitoring of a directional antenna involves the measurement of field intensity at certain locations away from the antenna, called *monitor points*. These points are selected and established during the initial tune-up of the antenna system. Measurements at the monitor points should confirm that radiation in prescribed directions does not exceed a value that would cause interference to other stations operating on the same or adjacent frequencies. The field intensity limits at these points are normally specified in the station license. Measurements at the monitor points may be required on a weekly or a monthly basis, depending on several factors and conditions relating to the particular station. If the system is not a critical array, quarterly measurements may be sufficient.

Folded Unipole Antenna

The *folded unipole* antenna consists of a grounded vertical structure with one or more conductors folded back parallel to the side of the structure. It can be visualized as a half-wave folded dipole perpendicular to the ground and cut in half (see Fig. 17.17). This design makes it possible to provide a wide range of resonant radiation resistances by varying the ratio of the diameter of the folded-back conductors in relation to the tower. Top loading can also be used to broaden the antenna bandwidth. A side view of the folded unipole is shown in Fig. 17.18.

The folded unipole antenna can be considered a modification of the standard shunt-fed system. Instead of a slant wire that leaves the tower at an approximate 45° angle (as used for shunt-fed systems), the folded unipole antenna has one or more wires attached to the tower at a predetermined height. The wires are supported by standoff insulators and run parallel to the sides of the tower down to the base.

The tower is grounded at the base. The folds, or wires, are joined together at the base and driven through an impedance matching network. Depending on the tuning requirements of the folded unipole, the wires may be connected to the tower at the top and/or at predetermined levels along the tower with shorting stubs.

Practical Antenna Systems

The folded unipole can be used on tall (130° or greater) towers. However, if the unipole is not divided into two parts, the overall efficiency (unattenuated field intensity) will be considerably lower than the normally expected field for the electrical height of the tower.

FM Broadcast Antenna Systems

The propagation characteristics of VHF FM radio are much different than for MF AM. There is essentially no difference between day and night FM propagation. FM stations have relatively uniform day and night service areas with the same operating power.

A wide variety of antennas is available for use in the FM broadcast band. Nearly all employ circular polarization. Although antenna designs differ from one manufacturer to another, generalizations can be made that apply to most units.

Antenna Types

There are three basic classes of FM broadcast transmitting antennas in use today: *ring stub* and *twisted ring*, *shunt-* and *series-fed slanted dipole*, and *multi-arm short helix*. While each design is unique, all have the following items in common:

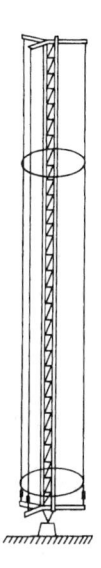

FIGURE 17.18 The folds of the unipole antenna are arranged either near the legs of the tower or near the faces of the tower.

- The antennas are designed for side-mounting to a steel tower or pole.
- Radiating elements are shunted across a common rigid coaxial transmission line.
- Elements are placed along the rigid line every one wavelength.
- Antennas with one to seven bays are fed from the bottom of the coaxial transmission line.
- Antennas with more than seven bays are fed from the center of the array to provide more predictable performance in the field.
- Antennas generally include a means of tuning out reactances after the antenna has been installed through the adjustment of variable capacitive or inductive elements at the feed point.

Figure 17.19 shows a shunt-fed slanted dipole antenna that consists of two half-wave dipoles offset 90°. The two sets of dipoles are rotated 22.5° (from their normal plane) and are *delta-matched* to provide a 50-Ω impedance at the radiator input flange. The lengths of all four dipole arms can be matched to resonance by mechanical adjustment of the end fittings. Shunt-feeding (when properly adjusted) provides equal currents in all four arms.

Wide-band *panel antennas* are a fourth common type of antenna used for FM broadcasting. Panel designs share some of the characteristics listed previously, but are intended primarily for specialized installations in which two or more stations will use the antenna simultaneously. Panel antennas are larger and more complex than other FM antennas, but offer the possibility for shared tower space among several stations and custom coverage patterns that would be difficult or even impossible with more common designs.

The ideal combination of antenna gain and transmitter power for a particular installation involves the analysis of a number of parameters. As shown in Table 17.2, a variety of pairings can be made to achieve the same ERP.

Television Antenna Systems

Television broadcasting uses horizontal polarization for the majority of installations worldwide. More recently, interest in the advantages of circular polarization has resulted in an increase in this form of transmission, particularly for VHF channels.

Both horizontal and circular polarization designs are suitable for tower-top or side-mounted installations. The latter option is dictated primarily by the existence of a previously installed tower-top antenna. On the other hand, in metropolitan areas where several antennas must be located on the same structure, either a stacking or candelabra-type arrangement is feasible. Figure 17.20 shows an example of antenna stacking on the top of the Sears Tower in Chicago, where numerous TV and FM transmitting antennas are located. Figure 17.21 shows a candelabra installation atop the Mt. Sutro tower in San Francisco. The Sutro tower supports eight TV antennas on its uppermost level. A number of FM transmitting antennas and two-way radio antennas are located on lower levels of the structure.

Another approach to TV transmission involves combining the RF outputs of two stations and feeding a single wide-band antenna. This approach is expensive and requires considerable engineering analysis to produce a combiner system that will not degrade the performance of either transmission system. In the Mt. Sutro example (Fig. 17.21), it can be seen that two stations (channels 4 and 5) are combined into a single transmitting antenna.

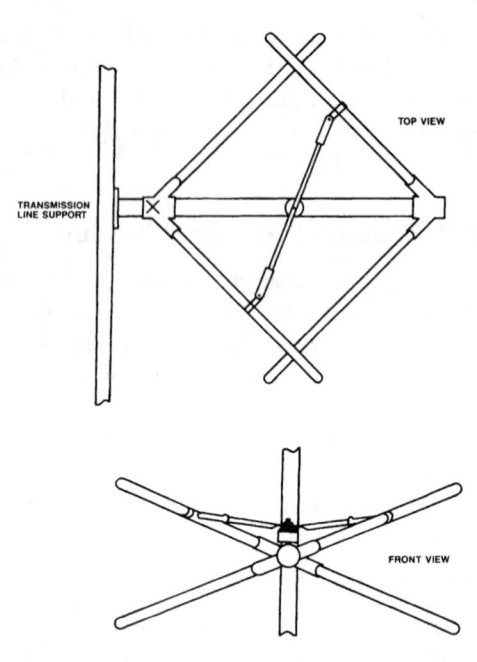

FIGURE 17.19 Mechanical configuration of one bay of a circularly polarized FM transmitting antenna.

Top-Mounted Antenna Types

The typical television broadcast antenna is a broadband radiator operating over a bandwidth of several megahertz with an efficiency of over 95%. Reflections from the antenna and transmission line back to the transmitter must be kept small enough to introduce negligible picture degradation. Furthermore, the gain and pattern characteristics of the antenna must be designed to achieve the desired coverage within acceptable tolerances. Tower-top, pole-type antennas designed to meet these parameters can be classified into two categories: *resonant dipoles* and *multi-wavelength traveling-wave elements*.

The primary considerations in the design of a top-mounted antenna are the achievement of uniform omnidirectional azimuth fields and minimum windloading. A number of different approaches have been tried successfully. Figure 17.22 illustrates the basic mechanical design of the most common antennas.

TABLE 17.2 Various Combinations of Transmitter Power and Antenna Gain that will Produce 100-kW ERP for an FM Station

No. Bays	Antenna Gain	Transmitter Power (kW)
3	1.5888	66.3
4	2.1332	49.3
5	2.7154	38.8
6	3.3028	31.8
7	3.8935	27.0
8	4.4872	23.5
10	5.6800	18.5
12	6.8781	15.3

Practical Antenna Systems **17**-19

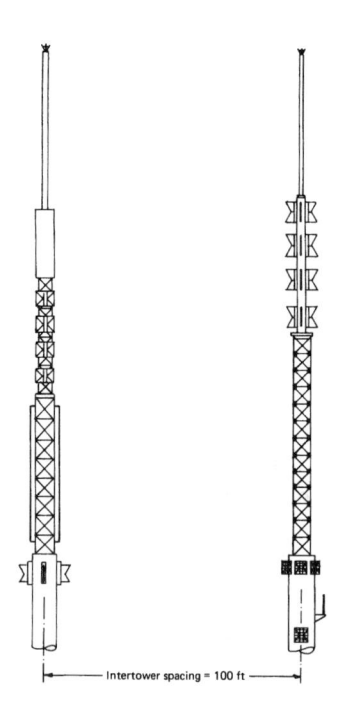

FIGURE 17.20 Twin-tower antenna array atop the Sears Tower in Chicago. Note how antennas have been stacked to overcome space restrictions.

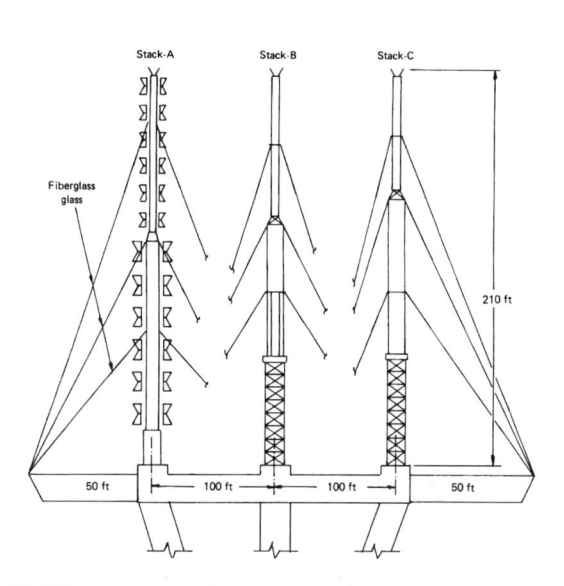

FIGURE 17.21 Installation of TV transmitting antennas on the candelabra structure at the top level of the Mt. Sutro tower in San Francisco. This installation makes extensive use of antenna stacking.

Turnstile Antenna.

The turnstile is the earliest and most popular resonant antenna for VHF broadcasting. The antenna is made up of four *batwing*-shaped elements mounted on a vertical pole in a manner resembling a turnstile. The four batwings are, in effect, two dipoles fed in quadrature phase. The azimuth-field pattern is a function of the diameter of the support mast. The pattern is usually within 10 to 15% of a true circle.

The turnstile antenna is made up of several layers, usually six layers for channels 2 through 6 and twelve layers for channels 7 through 13. The turnstile is not suitable for side-mounting, except for standby applications in which coverage degradation can be tolerated.

Coax Slot Antenna.

Commonly referred as the *pylon antenna*, the coax slot (Fig. 17.22b) is the most popular top-mounted unit for UHF applications. Horizontally polarized radiation is achieved using axial resonant slots on a cylinder to generate RF current around the outer surface of the cylinder. A good omnidirectional pattern is achieved by exciting four columns of slots around the circumference, which is basically just a section of rigid coaxial transmission line.

The slots along the pole are spaced approximately one wavelength per layer, and a suitable number of layers are used to achieve the desired gain. Typical gains range from 20 to 40. By varying the number of slots around the periphery of the cylinder, directional azimuth patterns can be achieved.

Waveguide Slot Antenna.

The UHF waveguide slot (Fig. 17.22c) is a variation on the coax slot antenna. The antenna is simply a section of waveguide with slots cut into the sides. The physics behind the design is long and complicated. However, the end result is the simplest of all antennas. This is a desirable feature in field applications because simple designs translate into long-term reliability.

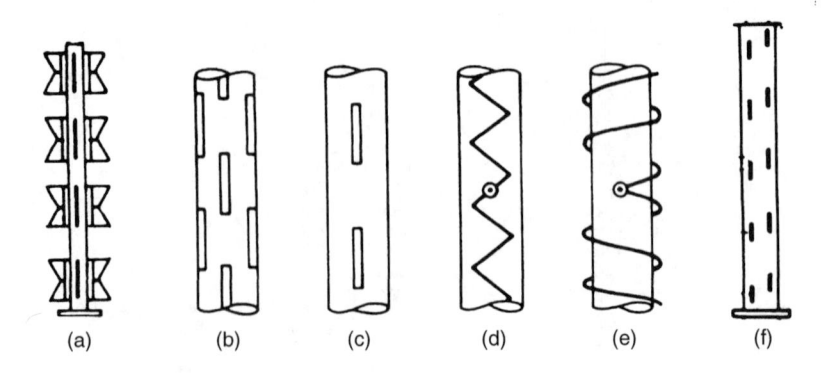

FIGURE 17.22 Various antennas used for VHF and UHF broadcasting. All designs provide linear (horizontal) polarization. Illustrated are (a) turnstile antenna, (b) coax slot antenna, (c) waveguide slot antenna, (d) zigzag antenna, (e) helix antenna, and (f) multi-slot traveling-wave antenna.

Zigzag Antenna.
The zigzag is a panel array design that utilizes a conductor routed up the sides of a three- or four-sided panel antenna in a "zigzag" manner (see Fig. 17.22d). With this design, the vertical current component along the zigzag conductor is mostly canceled out, and the antenna can effectively be considered an array of dipoles. With several such panels mounted around a polygonal periphery, the required azimuth pattern can be shaped by proper selection of feed currents to the various elements.

Helix Antenna.
A variation on the zigzag, the helix antenna (Fig. 17.22e) accomplishes basically the same goal using a different mechanical approach. Note the center feed point shown in the figure.

VHF Multi-slot Antenna.
Mechanically similar to the coax slot antenna, the VHF multi-slot antenna (Fig. 17.22f) consists of an array of axial slots on the outer surface of a coaxial transmission line. The slots are excited by a traveling wave inside the slotted line. The azimuth pattern is typically within 5% of omnidirectional. The antenna is generally about 15 wavelengths long.

Circularly Polarized Antennas

Circular polarization holds the promise of improved penetration into difficult coverage areas. There are a number of points to weigh in the decision to use a circularly polarized (CP) antenna, not the least of which is that a station must double its transmitter power in order to maintain the same ERP if it installs a CP antenna. This assumes equal vertical and horizontal components.

It is possible, and sometimes desirable, to operate with *elliptical polarization*, in which the horizontal and vertical components are not equal. Furthermore, the azimuth patterns for each polarization can be customized to provide the most efficient service area coverage.

Three major antenna types have been developed for CP television applications: the *normal mode helix*, various *panel* antenna designs, and the *interlaced traveling wave array*.

Normal Mode Helix.
The normal mode helix consists of a supporting tube with helical radiators mounted around the tube on insulators. The antenna is broken into sub-arrays, each powered by a divider network. The antenna is called the "normal mode helix" because radiation occurs normal to the axis of the helix, or perpendicular to the support tube. The antenna provides an omnidirectional pattern.

Panel Antennas.
The basic horizontally polarized panel antenna can be modified to produce circular polarization through the addition of vertically polarized radiators. Panel designs offer broad bandwidth and a wide choice of radiation patterns. By selecting the appropriate number of panels located around the tower, and the proper phase and amplitude distribution to the panels, a number of azimuth patterns can be realized. The primary drawback to the panel is the power distribution network required to feed the individual radiating elements.

Interlaced Traveling Wave Array.
As the name implies, the radiating elements of this antenna are interlaced along an array, into which power is supplied. The energy input at the bottom of the antenna is extracted by the radiating elements as it moves toward the top. The antenna consists of a cylindrical tube that supports the radiating elements. The elements are slots for the horizontally polarized component and dipoles for the vertically polarized component. The radiating elements couple RF directly off the main input line; thus, a power dividing network is not required.

Side-Mounted Antenna Types

Television antennas designed for mounting on the faces of a tower must meet the same basic requirements as a top-mounted antenna — wide bandwidth, high efficiency, predictable coverage pattern, high gain, and low windloading — plus the additional challenge that the antenna must work in a less-than-ideal environment. Given the choice, no broadcaster would elect to place its transmitting antenna on the side of a tower instead of at the top. There are, however, a number of ways to solve the problems presented by side-mounting.

Butterfly Antenna.
The butterfly is essentially a batwing panel developed from the turnstile radiator. The butterfly is one of the most popular panel antennas used for tower-face applications. It is suitable for the entire range of VHF applications. A number of variations on the basic batwing theme have been produced, including modifying the shape of the turnstile-type wings to rhombus or diamond shapes. Another version utilizes multiple dipoles in front of a reflecting panel.

For CP applications, two crossed dipoles or a pair of horizontal and vertical dipoles are used. A variety of cavity-backed crossed-dipole radiators are also utilized for CP transmission.

The azimuth pattern of each panel antenna is unidirectional, and three or four such panels are mounted on the sides of a triangular or square tower to achieve an omnidirectional pattern. The panels can be fed in-phase, with each one centered on the face of the tower, or fed in rotating phase with the proper mechanical offset. In the latter case, the input impedance match is considerably better.

Directionalization of the azimuth pattern is realized by proper distribution of the feed currents to individual panels in the same layer. Stacking layers provides gains comparable with top-mounted antennas.

The main drawbacks of panel antennas include (1) high windload, (2) complex feed system inside the antenna, and (3) restrictions on the size of the tower face, which determine to a large extent the omnidirectional pattern of the antenna.

UHF Side-Mounted Antennas

Utilization of panel antennas in a manner similar to those for VHF applications is not always possible at UHF installations. The high gains required for UHF broadcasting (in the range of 20 to 40, compared with gains of 6 to 12 for VHF) require far more panels with an associated complex feed system.

The zigzag panel antenna has been used for special omnidirectional and directional applications. For custom directional patterns, such as a cardioid shape, the pylon antenna can be side-mounted on one of the tower legs. Many stations, in fact, simply side-mount a pylon-type antenna on the leg of the tower that faces the greatest concentration of viewers. It is understood that viewers located behind the tower will receive a poorer signal; however, given the location of most TV transmitting towers — usually on the outskirts of their licensed city or on a mountaintop — this practice is often acceptable.

Broadband Antennas

Radiation of multiple channels from a single antenna requires the antenna to be broadband in both pattern and impedance (VSWR) characteristics. As a result, a broadband TV antenna represents a significant departure from the narrowband, single-channel pole antennas commonly used for VHF and UHF. The typical single-channel UHF antenna uses a series feed to the individual radiating elements, while a broadband antenna has a branch feed arrangement. The two feed configurations are shown in Fig. 17.23.

- At the design frequency, the series feed provides co-phased currents to its radiating elements. As the frequency varies, the electrical length of the series line feed changes such that the radiating elements are no longer in-phase outside of the designed channel. This electrical length change causes significant beam tilt out of band, and an input VSWR that varies widely with frequency.
- In contrast, the branch feed configuration employs feed lines that are nominally of equal length. Therefore the phase relationships of the radiating elements are maintained over a wide span of frequencies. This provides vertical patterns with stable beam tilt, a requirement for multi-channel applications.
- The basic building block of the multi-channel antenna is the broadband panel radiator. The individual radiating elements within a panel are fed by a branch feeder system that provides the panel with a single input cable connection. These panels are then stacked vertically and arranged around a supporting spine or existing tower to produce the desired vertical and horizontal radiation patterns.

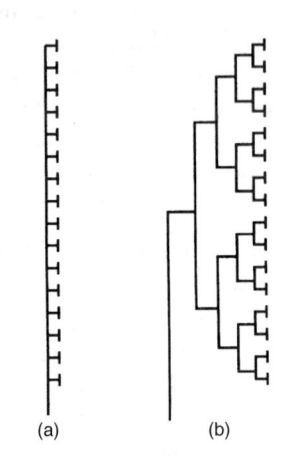

FIGURE 17.23 Antenna feed configurations: (a) series feed, and (b) branch feed.

Bandwidth.

The ability to combine multiple channels in a single transmission system depends on the bandwidth capabilities of the antenna and waveguide or coax. The antenna must have the necessary bandwidth in both pattern and impedance (VSWR). It is possible to design an antenna system for low-power applications using coaxial transmission line that provides whole-band capability. For high-power systems, waveguide bandwidth sets the limits of channel separation.

Antenna pattern performance is not a significant limiting factor. As frequency increases, the horizontal pattern circularity deteriorates, but this effect is generally acceptable. Also, the electrical aperture increases with frequency, which narrows the vertical pattern beamwidth. If a high-gain antenna were used over a wide bandwidth, the increase in electrical aperture might make the vertical pattern beamwidth unacceptably narrow. This, however, is usually not a problem because of the channel limits set by the waveguide.

Horizontal Pattern.

Because of the physical design of a broadband panel antenna, the cross-section is larger than the typical narrowband pole antenna. Therefore, as the operating frequencies approach the high end of the UHF band, the *circularity* (average circle to minimum or maximum ratio) of an omnidirectional broadband antenna generally deteriorates.

Improved circularity is possible by arranging additional panels around the supporting structure. Typical installations have used five, six, and eight panels per bay. These are illustrated in Fig. 17.24 along with measured patterns at different operating channels. These approaches are often required for power handling considerations, especially when three or four transmitting channels are involved.

The flexibility of the panel antenna allows directional patterns of unlimited variety. Two of the more common applications are shown in Fig. 17.25. The peanut and cardioid types are often constructed on square support spines (as indicated). A cardioid pattern may also be produced by side-mounting on a

Practical Antenna Systems

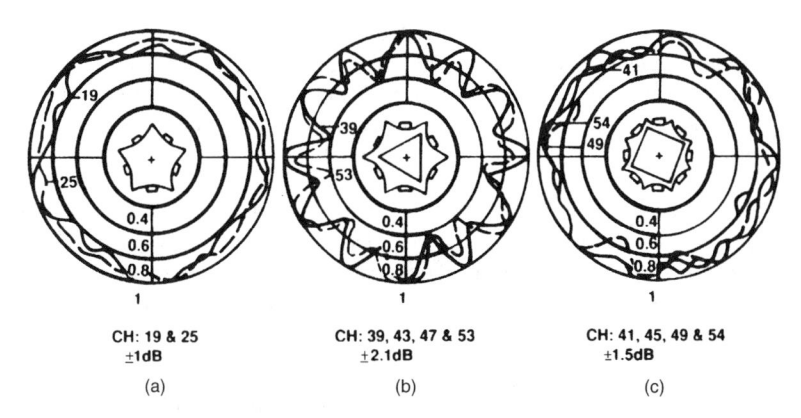

CH: 19 & 25
±1dB
(a)

CH: 39, 43, 47 & 53
±2.1dB
(b)

CH: 41, 45, 49 & 54
±1.5dB
(c)

FIGURE 17.24 Measured antenna patterns for three types of panel configurations at various operating frequencies: (a) five panels per bay, (b) six panels per bay, and (c) eight panels per bay.

(a)

(b)

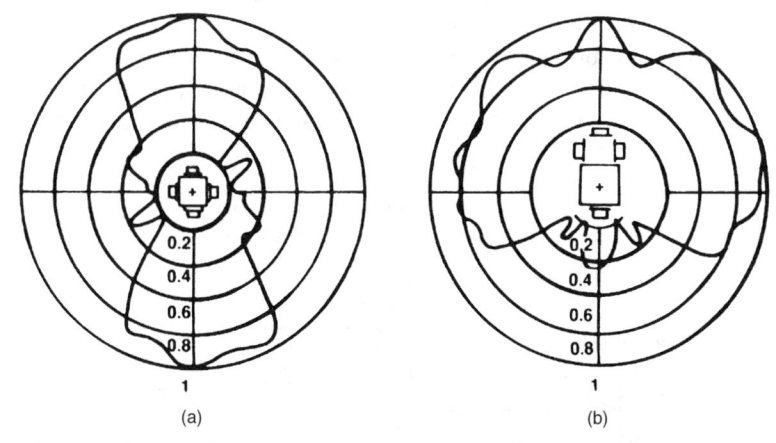

FIGURE 17.25 Common directional antenna patterns: (a) peanut, and (b) cardioid.

triangular tower. Different horizontal radiation patterns for each channel may also be provided, as indicated in Fig. 17.26. This is accomplished by changing the power and/or phase to some of the panels in the antenna with frequency.

Most of these antenna configurations are also possible using a circularly polarized panel. If desired, the panel can be adjusted for elliptical polarization with the vertical elements receiving less than 50% of the power. Using a circularly polarized panel will reduce the horizontally polarized ERP by half (assuming the same transmitter power).

17.4 Phased-Array Antenna Systems

Phased-array antennas are steered by tilting the phase front independently in two orthogonal directions called the *array coordinates*. Scanning in either array coordinate causes the beam to move along a cone whose center is at the center of the array. As the beam is steered away from the array normal, the projected aperture in the beam's direction varies, causing the beamwidth to vary proportionately.

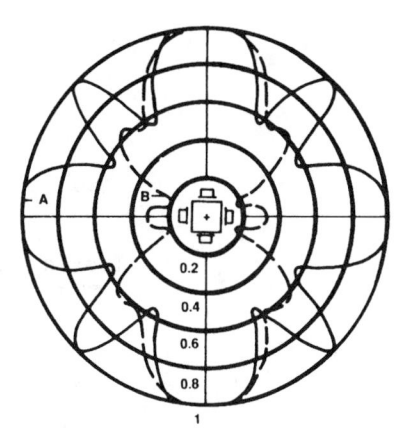

FIGURE 17.26 Use of a single antenna to produce two different radiation patterns, omnidirectional (trace "A") and peanut (trace "B").

Arrays can be classified as either *active* or *passive*. Active arrays contain duplexers and amplifiers behind every element or group of elements of the array. Passive arrays are driven from a single feed point. Active arrays are capable of high-power operation. Both passive and active arrays must divide the signal from a single transmission line among all the elements of the system. This can be accomplished through one of the following methods:

- *Optical feed*: a single feed, usually a monopulse horn, is used to illuminate the array with a spherical phase front (illustrated in Fig. 17.27). Power collected by the rear elements of the array is transmitted through the phase shifters to produce a planar front and steer the array. The energy can then be radiated from the other side of the array, or reflected and reradiated through the collecting elements. In the latter case, the array acts as a steerable reflector.
- *Corporate feed*: a system utilizing a series-feed network (Fig. 17.28) or parallel-feed network (Fig. 17.29). Both designs use transmission-line components to divide the signal among the elements. Phase shifters can be located at the elements or within the dividing network. Both the series- and parallel-feed systems have several variations, as shown in the figures.
- *Multiple-beam network*: a system capable of forming simultaneous beams with a given array. The Butler matrix, shown in Fig. 17.30, is one such technique. It connects the N elements of a linear array to N feed points corresponding to N beam outputs. The phase shifter is one of the most critical components of the system. It produces controllable phase shift over the operating band of the array. Digital and analog phase shifters have been developed using both ferrites and pin diodes.

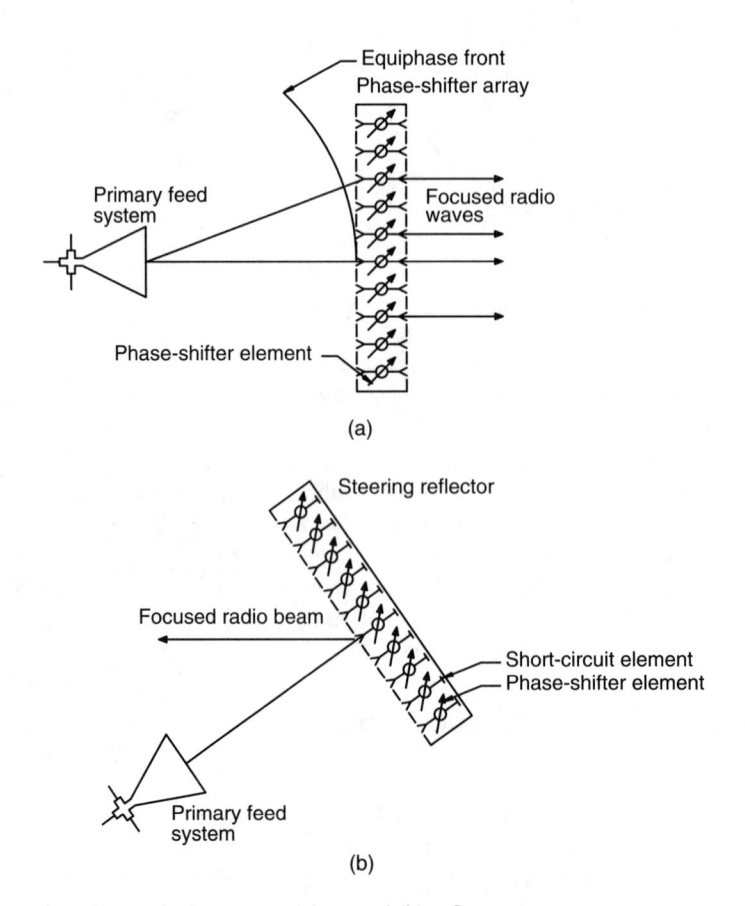

FIGURE 17.27 Optical antenna feed systems: (a) lens, and (b) reflector.

Practical Antenna Systems

Frequency scan is another type of multiple-beam network, but one that does not require phase shifters, dividers, or beam-steering computers. Element signals are coupled from points along a transmission line. The electrical path length between elements is longer than the physical separation, so a small frequency change will cause a phase change between elements that is large enough to steer the beam. This technique can be applied only to one array coordinate. If a two-dimensional array is required, phase shifters are normally used to scan the other coordinate.

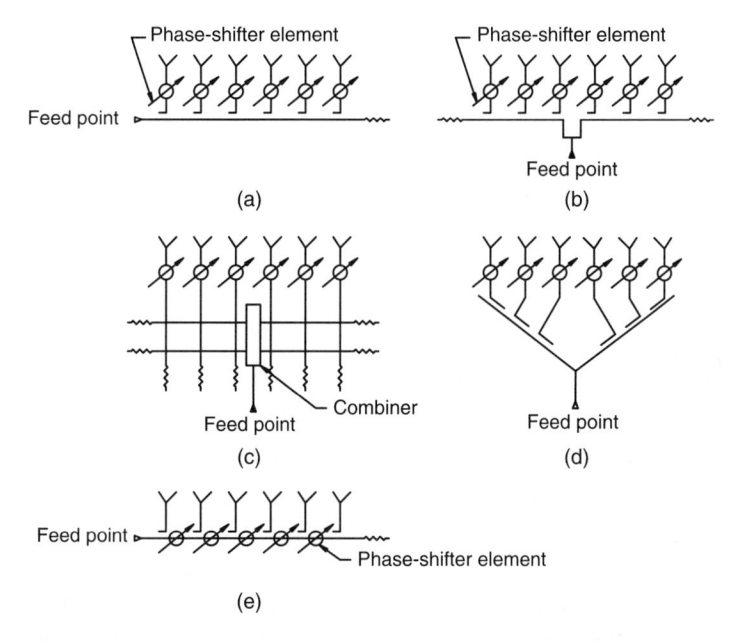

FIGURE 17.28 Series-feed networks: (a) end feed, (b) center feed, (c) separate optimization, (d) equal path length, and (e) series phase shifters.

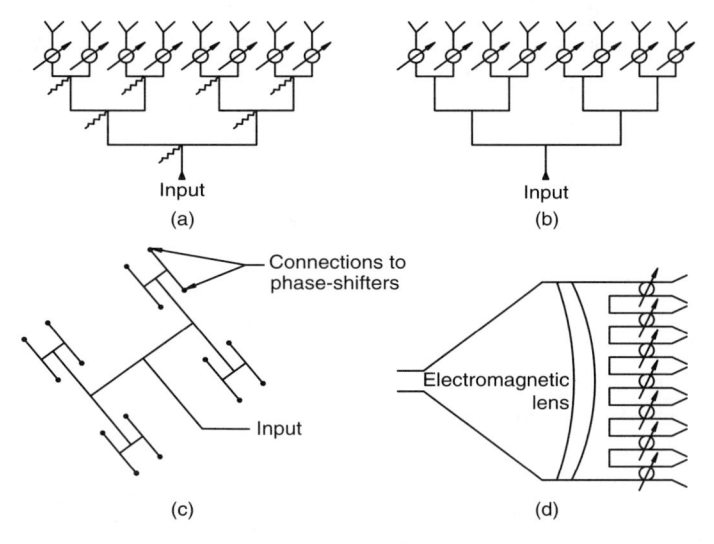

FIGURE 17.29 Types of parallel-feed networks: (a) matched corporate feed, (b) reactive corporate feed, (c) reactive stripline, and (d) multiple reactive divider.

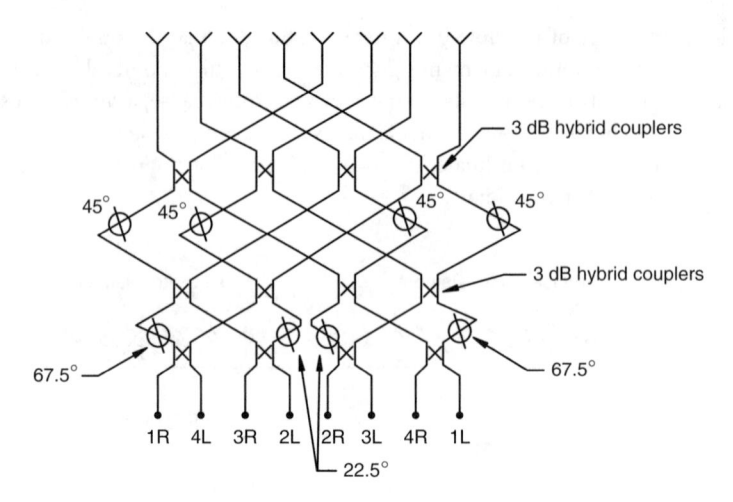

FIGURE 17.30 The Butler beam-forming network.

Phase-Shift Devices

The design of a phase shifter must meet two primary criteria:

- Low transmission loss
- High power-handling capability

The Reggia-Spencer phase shifter meets both requirements. The device, illustrated in Fig. 17.31, consists of a ferrite rod mounted inside a waveguide that delays the RF signal passing through the waveguide, permitting the array to be steered. The amount of phase shift can be controlled by the current in the solenoid, because of the effect a magnetic field has on the permeability of the ferrite. This design is a *reciprocal phase shifter*, meaning that the device exhibits the same phase shift for signals passing in either direction (forward or reverse). Nonreciprocal phase shifters are also available, where phase-shift polarity reverses with the direction of propagation.

Phase shifters may also be developed using pin diodes in transmission line networks. One configuration, shown in Fig. 17.32, uses diodes as switches to change the signal path length of the network. A second type uses pin diodes as switches to connect reactive loads across a transmission line. When equal loads are connected with 1/4-wave separation, a pure phase shift results.

Radar System Duplexer

The duplexer is an essential component of any radar system. The switching elements used in a duplexer include gas tubes, ferrite circulators, and pin diodes. Gas tubes are the simplest. A typical gas-filled TR tube is shown in Fig. 17.33. Low-power RF signals pass through the tube with little attenuation. Higher-power signals, however, cause the gas to ionize and present a short-circuit to the RF energy.

Figure 17.34 illustrates a balanced duplexer using hybrid junctions and TR tubes. When the transmitter is on, the TR tubes fire and reflect the RF power to the antenna port of the input hybrid. During the receive portion of the radar function, signals picked up by the antenna are passed through the TR tubes and on to the receiver port of the output hybrid.

Newer radar systems often use a ferrite circulator as the duplexer. A TR tube is required in the receiver line to protect input circuits from transmitter power reflected by the antenna because of an imperfect match. A four-port circulator generally is used with a load between the transmitter and receiver ports so that power reflected by the TR tube is properly terminated.

Practical Antenna Systems

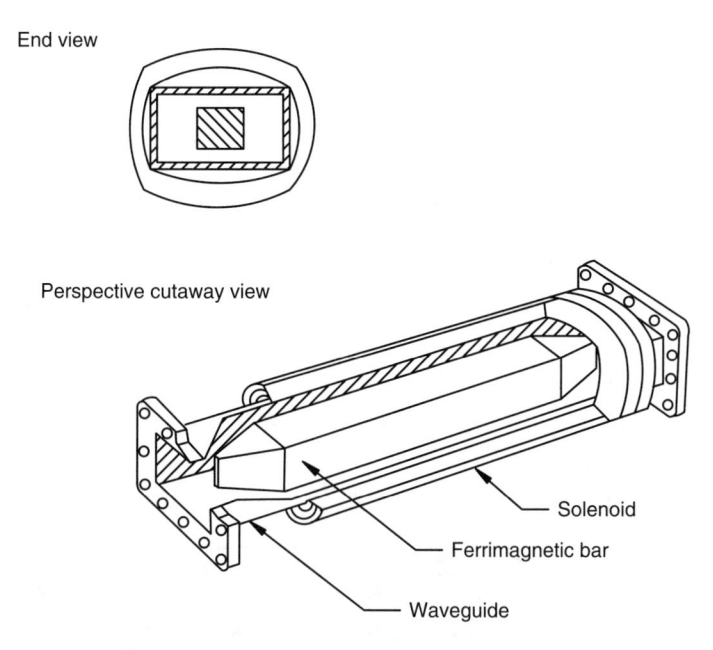

FIGURE 17.31 Basic concept of a Reggia-Spencer phase shifter.

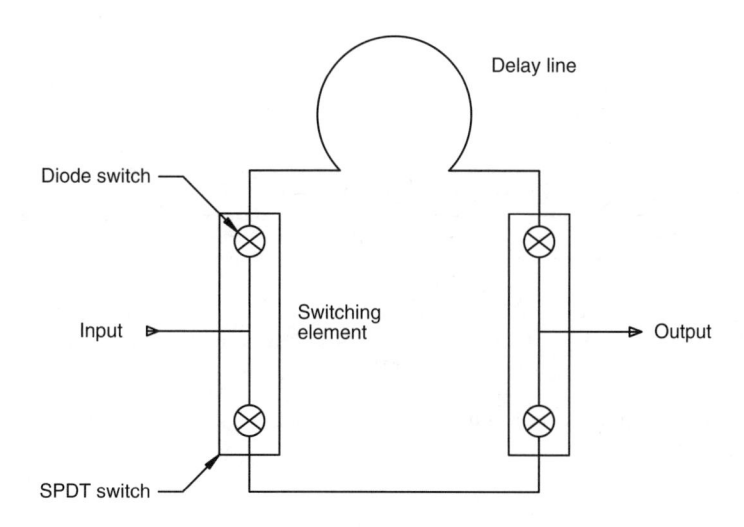

FIGURE 17.32 Switched-line phase shifter using pin diodes.

Pin diode switches have also been used in duplexers to perform the protective switching function of TR tubes. Pin diodes are more easily applied in coaxial circuitry, and at lower microwave frequencies. Multiple diodes are used when a single diode cannot withstand the expected power.

Microwave filters are sometimes used in the transmit path of a radar system to suppress spurious radiation, or in the receive signal path to suppress spurious interference.

Harmonic filters commonly are used in the transmission chain to absorb harmonic energy output by the system, preventing it from being radiated or reflected back from the antenna. Figure 17.35 shows a filter in which harmonic energy is coupled out through holes in the walls of the waveguide to matched loads.

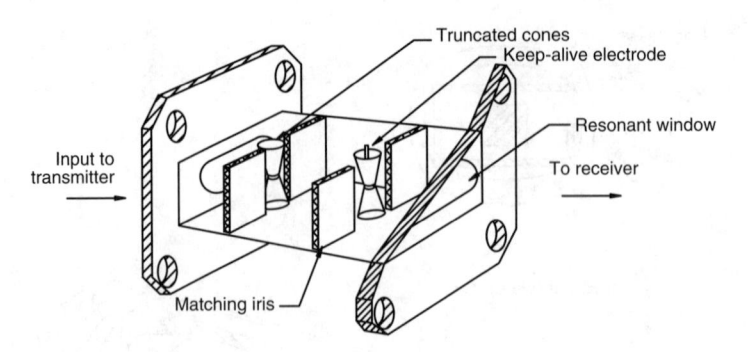

FIGURE 17.33 Typical construction of a TR tube.

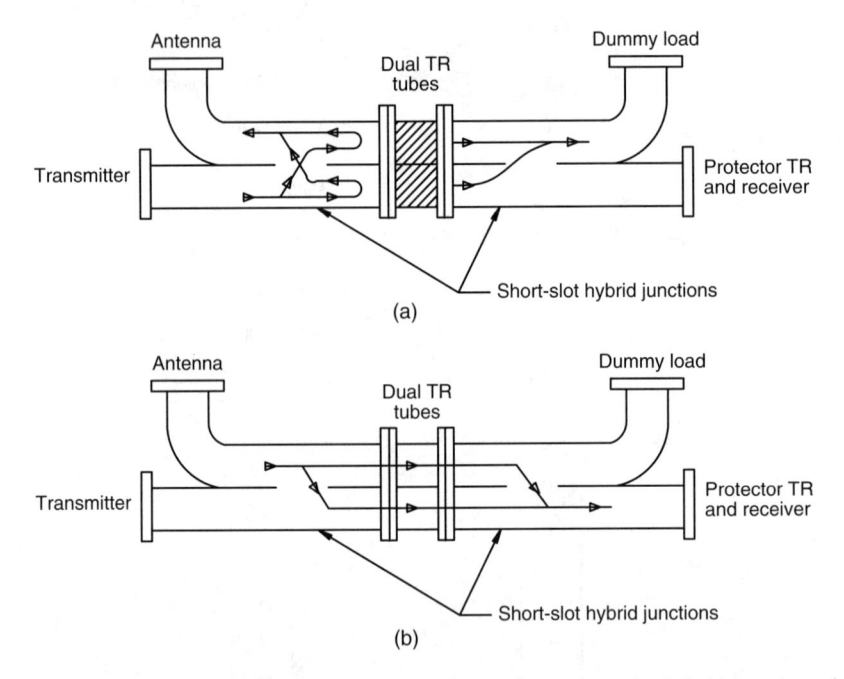

FIGURE 17.34 Balanced duplexer circuit using dual TR tubes and two short-slot hybrid junctions: (a) transmit mode, and (b) receive mode.

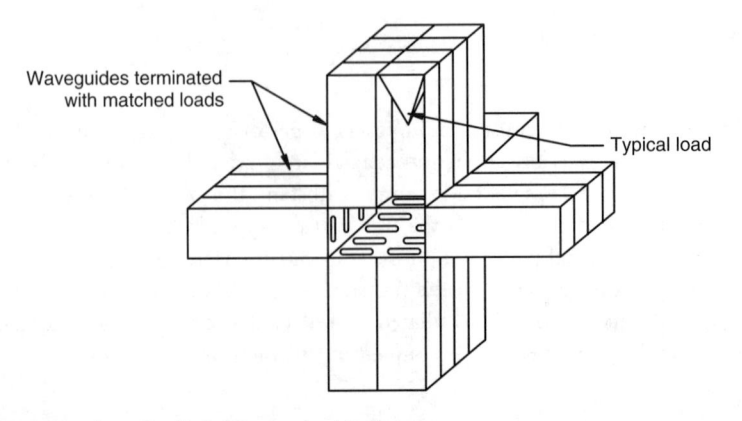

FIGURE 17.35 Construction of a dissipative waveguide filter.

Narrow-band filters in the receive path, often called *preselectors*, are built using mechanically tuned cavity resonators or electrically tuned TIG resonators. Preselectors can provide up to 80 dB suppression of signals from other radar transmitters in the same RF band, but at a different operating frequency.

Bibliography

Anders, M. B., A case for the use of multi-channel broadband antenna systems, *NAB Engineering Conference Proceedings*, National Association of Broadcasters, Washington, D.C., 1985.

Benson, Blair, Ed., *Television Engineering Handbook*, McGraw-Hill, New York, 1986.

Benson, B. and J. Whitaker, Ed., *Television and Audio Handbook for Technicians and Engineers*, McGraw-Hill, New York, 1989.

Bingeman, Grant, AM tower impedance matching, in *Broadcast Engineering*, Intertec Publishing, Overland Park, KS, July 1985.

Bixby, Jeffrey, AM DAs: doing it right, in *Broadcast Engineering*, Intertec Publishing, Overland Park, KS, February 1984.

Chick, Elton B., Monitoring directional antennas, in *Broadcast Engineering*, Intertec Publishing, Overland Park, KS, July 1985.

Crutchfield, E. B., Ed., *NAB Engineering Handbook*, 7th ed., Washington, D.C., 1985.

Dienes, Geza, Circularly and elliptically polarized UHF television transmitting antenna design, in *Proceedings of the NAB Engineering Conference*, National Association of Broadcasters, Washington, D.C., 1988.

Fink, D. and D. Christiansen, Eds., *Electronics Engineers' Handbook*, 3rd ed., McGraw-Hill, New York, 1989.

Jordan, Edward C., Ed., *Reference Data for Engineers: Radio, Electronics, Computer and Communications*, 7th ed., Howard W. Sams Company, Indianapolis, IN, 1985.

Howard, George P., The Howard AM Sideband Response Method, in *Radio World*, Falls Church, VA, August 1979.

Mullaney, John H., *The Consulting Radio Engineer's Notebook*, Mullaney Engineering, Gaithersburg, MD, 1985.

Mullaney, John H., The Folded Unipole Antenna, *Broadcast Engineering*, Intertec Publishing, Overland Park, KS, July 1986.

Mullaney, John H., The Folded Unipole Antenna for AM Broadcast, *Broadcast Engineering*, Intertec Publishing, Overland Park, KS, January 1960.

Mullaney, John H. and George P. Howard, SBNET: A Fortran Program for Analyzing Sideband Response and Design of Matching Networks, Mullaney Engineering, Gaithersburg, MD, 1970.

Mayberry, E. and J. Stenberg, UHF Multi-Channel Antenna Systems, *Broadcast Engineering*, Intertec Publishing, Overland Park, KS, March 1989.

Raines, J. K., Folded Unipole Studies, *Think Book Series*, Multronics, 1968–1969.

Raines, J. K., Unipol: A Fortran Program for Designing Folded Unipole Antennas, Mullaney Engineering, Gaithersburg, MD, 1970.

18
Preventing RF System Failures

18.1	Introduction	**18**-1
18.2	Routine Maintenance..	**18**-2
	The Maintenance Log · Preventive Maintenance Routine	
18.3	Klystron Devices...	**18**-8
18.4	Power Grid Tubes...	**18**-9
	Tube Dissipation · Air-Handling System · Tube Changing Procedure · Extending Vacuum Tube Life · PA Stage Tuning	
18.5	Preventing RF System Failures.....................................	**18**-18
	Common Mode Failures · Modifications and Updates · Spare Parts	
18.6	Transmission Line/Antenna Problems.........................	**18**-20
	Effects of Modulation · Maintenance Considerations	
18.7	High-Voltage Power Supply Problems.........................	**18**-24
	Power Supply Maintenance · Power Supply Metering · Transient Disturbances	
18.8	Temperature Control	**18**-28
	Cooling System Maintenance · Air Cooling System Design · Klystron Water Cooling Systems	

Jerry C. Whitaker
Editor

18.1 Introduction

Radio frequency (RF) equipment is unfamiliar to many persons entering the electronics industry. Colleges do not routinely teach RF principles, favoring instead digital technology. Unlike other types of products, however, RF equipment often must receive preventive maintenance to achieve its expected reliability. Maintaining RF gear is a predictable, necessary expense that facilities must include in their operating budgets. Tubes (if used in the system) will have to be replaced from time to time, no matter what the engineer does; components fail every now and then; and time must be allocated for cleaning and adjustments. By planning for these expenses each month, unnecessary downtime can be avoided.

Although the reason generally given for minimum RF maintenance is a lack of time and/or money, the cost of such a policy can be deceptively high. Problems that could be solved for a few dollars may, if left unattended, result in considerable damage to the system and a large repair bill. A standby system can often be a lifesaver; however, its usefulness is sometimes overrated. The best standby RF system is a *main system* in good working order.

18.2 Routine Maintenance

Most RF system failures can be prevented through regular cleaning and inspection, and close observation. The history of the unit is also important in a thorough maintenance program so that trends can be identified and analyzed.

The Maintenance Log

The control system front panel can tell a great deal about what is going on inside an RF generator. Develop a maintenance log and record all front-panel meter readings, as well as the positions of critical tuning controls, on a regular basis (as illustrated in Fig. 18.1). This information provides a history of the system and can be a valuable tool in noting problems at an early stage. The most obvious application of such logging is to spot failing power tubes, but any changes occurring in components can be identified.

Creating a history of the line and tank pressure for a pressurized transmission line can identify developing line or antenna problems. After the regulator is set for the desired line pressure, record the tank and line readings each week and chart the data. If possible, make the observations at the same time of day each week. Ambient temperature can have a significant effect on line pressure; note any temperature extremes in the transmission line log when the pressure is recorded. The transmission line pressure will usually change slightly between carrier-on and carrier-off conditions (depending on the power level). The presence of RF can heat the inner conductor of the line, causing the pressure to increase. After a few months of charting the gradual loss of tank pressure, a pattern should become obvious. Investigate any deviation from the normal amount of tank pressure loss over a given period.

Whenever a problem occurs with the RF system, make a complete entry describing the failure in the maintenance log. Include a description of all maintenance activities required to return the system to operational condition. Make all entries complete and clear. Include the following data:

- Description of the nature of the malfunction, including all observable symptoms and performance characteristics
- Description of the actions taken to return the system to a serviceable condition
- Complete list of the components replaced or repaired, including the device schematic number and part number
- Total system downtime as a result of the failure
- Name of the engineer who made the repairs

The importance of regular, accurate logging can best be emphasized through the following examples:

Case Study 1

Improper neutralization is detected on an AM broadcast transmitter IPA (intermediate power amplifier), as shown in Fig. 18.2. The neutralization adjustment is made by moving taps on a coil, and none have been changed. The history of the transmitter (as recorded in the maintenance record) reveals, however, that the PA grid tuning adjustment has, over the past 2 years, been moving slowly into the higher readings. An examination of the schematic diagram leads to the conclusion that C-601 is the problem.

The tuning change of the stage was so gradual that it was not thought significant until an examination of the transmitter history revealed that continual retuning in one direction only was necessary to achieve maximum PA grid drive. Without a record of the history of the unit, time could have been wasted in substituting expensive capacitors in the circuit, one at a time. Worse yet, the engineer might have changed the tap on coil L-601 to achieve neutralization, further hiding the real cause of the problem.

Case Study 2

A UHF broadcast transmitter is found to exhibit decreasing klystron body-current. The typical reading with average picture content is 50 mA, but over a 4-week period, the reading dropped to 30 mA. No other parameters show deviation from normal, yet the decrease in the reading indicates an alternate path

Preventing RF System Failures   18-3

Parameter	Typical value	Measured value
RF power output	18.3 kW	_____
Plate current	2.8 A	_____
Plate voltage	7.55 kV	_____
Screen current	380 mA	_____
Screen voltage	650 V	_____
PA grid current	110 mA	_____
PA bias voltage	490 V	_____
PA filament voltage	6 V	_____
Left driver cathode current	142 mA	_____
Right driver cathode current	142 mA	_____
Driver screen voltage	275 V	_____
Driver screen current	35 mA	_____
Driver grid current	1 mA	_____
Driver plate voltage	1.85 kV	_____
28-V power supply	27 V	_____
Reflected power	15 W	_____
Transmission-line pressure	3.9 psi	_____
Tank pressure	1500 psi	_____
Transmitter hours	5412	_____
Exciter AFC	Center scale	_____

FIGURE 18.1 Example of a transmitter operating log that should be completed regularly by maintenance personnel.

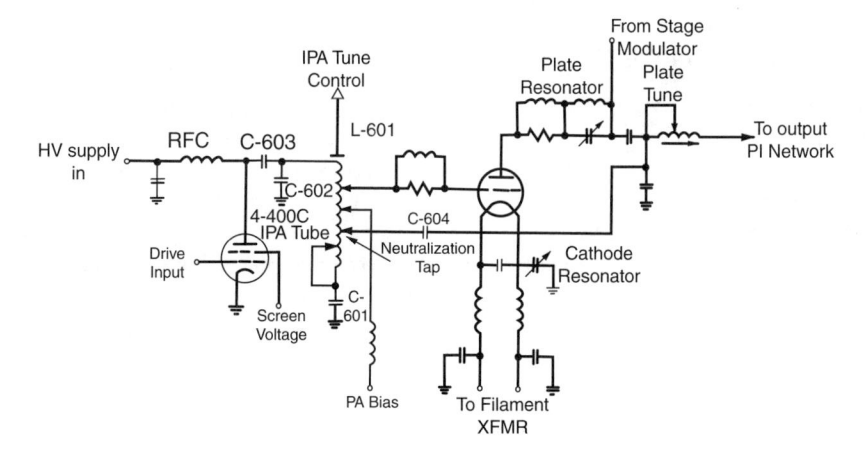

FIGURE 18.2 AM transmitter IPA/PA stage exhibiting neutralization problems. A history of IPA returning (through adjustment of L-601) helped determine that loss of neutralization was the result of C-601 changing in value.

(besides the normal body-current circuitry) by which electrons return to the beam power supply. A schematic diagram of the system is shown in Fig. 18.3. Several factors could cause the body-current variation, including water leakage into the body-to-collector insulation of the klystron. In time, this water can corrode the klystron envelope, possibly leading to a loss of vacuum and klystron failure.

Water leakage can also cause partial bypassing of the body-current circuitry, an important protection system in the transmitter. It is essential that the circuit functions normally at all times and at full sensitivity in order to detect change when a fault condition occurs. Regular logging of transmitter parameters ensures that developing problems are caught early.

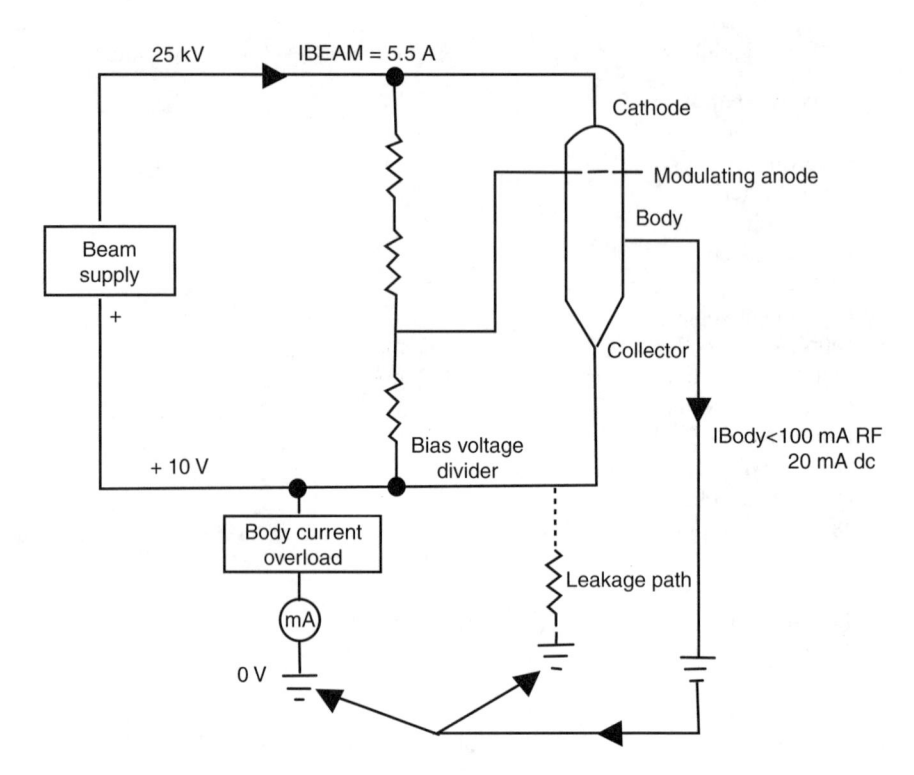

FIGURE 18.3 Simplified high-voltage schematic of a klystron amplifier showing the parallel leakage path that can cause a reduction in protection sensitivity of the body-current circuit.

Preventive Maintenance Routine

A thorough inspection of the RF system on a regular basis is the key to minimizing equipment downtime. Component problems can often be spotted at an early stage by regular inspection of the system. Remember to discharge all capacitors in the circuit with a grounding stick before touching any component in the high-voltage sections of the system. Confirm that all primary power has been removed before any maintenance work begins.

Special precautions must be taken with systems that receive ac power from two independent feeds. Typically, one ac line provides 208-V three-phase service for the high-voltage section of the system, and a separate ac line provides 120-V power for low-voltage circuits. Older transmitters or high-power transmitters often utilize this arrangement. Check to see that all ac is removed before any maintenance work begins.

Consider the following preventive maintenance procedures.

Resistors and Capacitors

- Inspect resistors and RF capacitors for signs of overheating (see Fig. 18.4).
- Inspect electrolytic or oil-filled capacitors for signs of leakage.
- Inspect feedthrough capacitors and other high-voltage components for signs of arcing (see Fig. 18.5).

Transmitting capacitors — mica vacuum and doorknob types — should never run hot. They may run warm, but usually as the result of thermal radiation from nearby components (such as power tubes) in the circuit. An overheated transmitting capacitor is often a sign of incorrect tuning. Vacuum capacitors present special requirements for the maintenance technician. Care in handling is a prime requisite for maximum service life. Because the vacuum capacitor is evacuated to a higher degree than most vacuum

Preventing RF System Failures

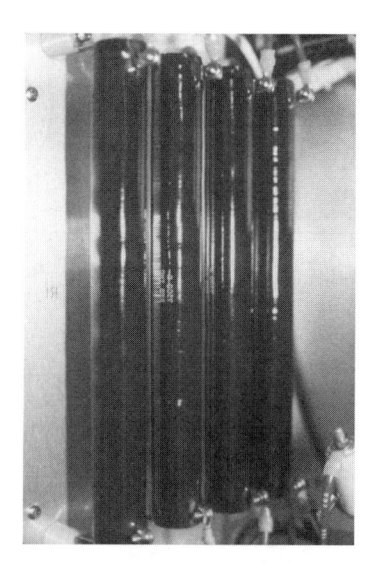

FIGURE 18.4 Check resistors, particularly high-power units, regularly for signs of premature wear caused by excessive heating.

FIGURE 18.5 Inspect high-voltage capacitors for signs of leakage around the case feedthrough terminals.

tubes, it is particularly susceptible to shock and rough handling. Provide adequate protection to vacuum capacitors whenever maintenance is performed. The most vulnerable parts of the capacitor are the glass-to-metal seals on each end of the unit. Exercise particular care during removal or installation.

The current ratings of vacuum capacitors are limited by the glass-to-metal seal temperature and the temperature of the solder used to secure the capacitor plates. Seal temperature is increased by poor connecting clip pressure, excessive ambient temperatures, corrosion of the end caps and/or connecting clip, excessive dust and dirt accumulation, or excessive currents. Dust accumulation on sharp points in high-voltage circuitry near the vacuum capacitor can cause arcs or corona that may actually burn a hole through the glass envelope.

Power Supply Components

- Inspect the mechanical operation of circuit breakers. Confirm that they provide a definite *snap* to the off position (remove all ac power for this test) and that they firmly reseat when restored. Replace any circuit breaker that is difficult to reset.
- Inspect power transformers and reactors for signs of overheating or arcing (see Fig. 18.6).
- Inspect oil-filled transformers for signs of leakage.
- Inspect transformers for dirt build-up, loose mounting brackets and rivets, and loose terminal connections.
- Inspect high-voltage rectifiers and transient suppression devices for overheating and mechanical problems (see Fig. 18.7).

Power transformers and reactors normally run hot to the touch. Check both the transformer frame and the individual windings. On a three-phase transformer, each winding should produce about the same amount of heat. If one winding is found to run hotter than the other two, further investigation is warranted. Dust, dirt, or moisture between terminals of a high-voltage transformer may cause flashover failures. Insulating compound or oil around the base of a transformer indicates overheating or leakage.

Coils and RF Transformers

- Inspect coils and RF transformers for indications of overheating (see Fig. 18.8).
- Inspect connection points for arcing or loose terminals.

FIGURE 18.6 Inspect power transformers just after shutdown for indications of overheating, or leakage in the case of oil-filled transformers.

FIGURE 18.7 Check the heating of individual rectifiers in a stack assembly. All devices should generate approximately the same amount of heat.

FIGURE 18.8 Clean RF coils and inductors as often as needed to keep contaminants from building up on the device loops.

Coils and RF transformers operating in a well-tuned system will rarely heat appreciably. If discoloration is noticed on several loops of a coil, consult the factory service department to see if the condition is normal. Pay particular attention to variable tap inductors, often found in AM broadcast transmitters and phasers. Closely inspect the roller element and coil loops for overheating or signs of arcing.

Relay Mechanisms

- Inspect relay contacts, including high-voltage or high-power RF relays, for signs of pitting or discoloration.
- Inspect the mechanical linkage to confirm proper operation. The contactor arm (if used) should move freely, without undue mechanical resistance.
- Inspect vacuum contactors for free operation of the mechanical linkage (if appropriate) and for indications of excessive dissipation at the contact points and metal-to-glass (or metal-to-ceramic) seals.

Preventing RF System Failures 18-7

FIGURE 18.9 Check the tightness of connectors on an occasional basis, but do not stress the connection points.

Unless problems are experienced with an enclosed relay, do not attempt to clean it. More harm than good can be done by disassembling properly working components for detailed inspection.

Connection Points

- Inspect connections and terminals that are subject to vibration. Tightness of connections is critical to the proper operation of high-voltage and RF circuits (see Fig. 18.9).
- Inspect barrier strip and printed circuit board contacts for proper termination.

Although it is important that all connections are tight, be careful not to over-tighten. The connection points on some components, such doorknob capacitors, can be damaged by excessive force. There is no section of an RF system where it is more important to keep connections tight than in the power amplifier stage. Loose connections can result in arcing between components and conductors that can lead to system failure. The cavity access door is a part of the outer conductor of the coaxial transmission line circuit in FM and TV transmitters, and in many RF generators operating at VHF and above. High-potential RF circulating currents flow along the inner surface of the door, which must be fastened securely to prevent arcing.

Cleaning the System

Cleaning is a large part of a proper maintenance routine. A shop vacuum and clean brush are generally all that are required. Use isopropyl alcohol and a soft cloth for cleaning insulators on high-voltage components (see Fig. 18.10). Cleaning affords the opportunity to inspect each component in the system and observe any changes.

Regular maintenance of insulators is important to the proper operation of RF final amplifier stages because of the high voltages usually present. Pay particular attention to the insulators used in the PA tube socket (see Fig. 18.11). Because the supply of cooling air is passed through the socket, airborne contaminants can be deposited on various sections of the assembly. These can create a high-voltage arc path across the socket insulators. Perform any cleaning work around the PA socket with extreme care. Do not use compressed air to clean out a power tube socket. Blowing compressed air into the PA or IPA stage of a transmitter will merely move the dirt from places where you *can* see it to places where you *cannot* see it. Use a vacuum instead. When cleaning the socket assembly, do not disturb any components in the PA circuit (see Fig. 18.12). Visually check the tube anode to see if dirt is clogging any of the heat-radiating fins.

Cleaning is also important to proper cooling of solid-state components in the transmitter. A layer of dust and dirt can create a thermal insulator effect and prevent proper heat exchange from a device into the cabinet (see Fig. 18.13).

FIGURE 18.10 Keep all high-voltage components, such as this rectifier bank, free of dust and contamination that might cause short-circuit paths.

FIGURE 18.11 Carefully inspect the PA tube socket assembly. Do not remove the PA tube unless necessary.

18.3 Klystron Devices

Klystrons are expensive to buy and expensive to operate. Compared to tetrodes, they require larger auxiliary components (such as power supplies and heat exchangers) and are physically larger, yet they are stable, provide high gain, and can be easily driven by solid-state circuitry. Klystrons are relatively simple to cool and are capable of long life with a minimum of maintenance. Two different types of klystrons are in use today:

- *Integral cavity* klystron, in which the resonant cavities are built into the body.
- *External cavity* klystron, in which the cavities are mechanically clamped onto the body and are outside the vacuum envelope of the device.

This difference in construction requires different maintenance procedures. The klystron body (the RF *interaction region* of the integral cavity device) is cooled by the same liquid that is fed to the collector. Required maintenance involves checking for leaks and adequate coolant flow. Although the cavities of the external cavity unit are air-cooled, the body can be water- or air-cooled. Uncorrected leaks in a water-cooled body can lead to cavity and tuning mechanism damage. Look inside the magnet frame with a flashlight once a week. Correct leaks immediately and clean away coolant residues.

FIGURE 18.12 Clean the PA cavity assembly to prevent an accumulation of dust and dirt. Check hardware for tightness.

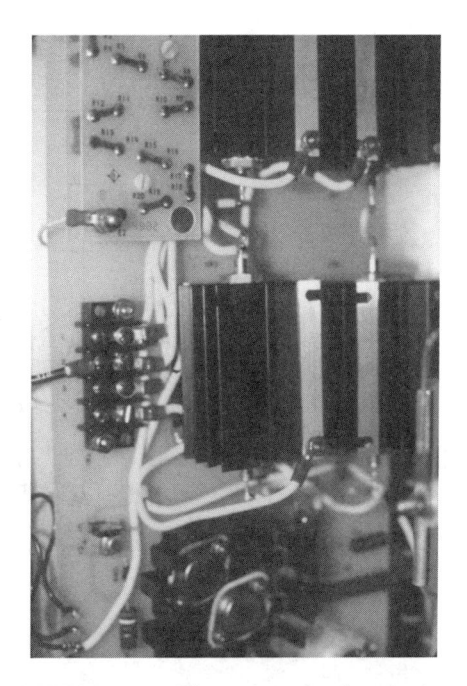

FIGURE 18.13 Clean power semiconductor assemblies to ensure efficient dissipation of heat from the devices.

The air-cooled body requires only sufficient airflow. The proper supply of air can be monitored with one or two adhesive temperature labels and close visual inspection. Look for discoloration of metallic surfaces. The external cavities need a clean supply of cooling air. Dust accumulation inside the cavities will cause RF arcing. Check air supply filters regularly. Some cavities have a mesh filter at the inlet flange. Inspect this point as required.

It is possible to make a visual inspection of the cavities of an external cavity device by removing the loading loops and/or air loops. This procedure is recommended only when unusual behavior is experienced, and not as part of routine maintenance. Generally, there is no need to remove a klystron from its magnet frame and cavities during routine maintenance.

18.4 Power Grid Tubes

The power tubes used in RF generators and transmitter are perhaps the most important and least understood components in the system. The best way to gain an understanding of the capabilities of a PA tube is to secure a copy of the tube manufacturer's data sheet for each type of device. They are available either from the tube or transmitter manufacturer. The primary value of the data sheets to the end user is the listing of maximum permissible values. These give the maintenance engineer a clear rundown of the maximum voltages and currents that the tube can withstand under normal operation. Note these values and avoid them.

An examination of the data sheet will show that a number of operating conditions are possible, depending on the class of service required by the application. As long as the maximum ratings of the tube are not exceeded, a wide choice of operating parameters, including plate voltage and current, screen voltage, and RF grid drive, are possible. When studying the characteristic curves of each tube, remember that they represent the performance of a *typical* device. All electronic products have some tolerance among devices of a single type. Operation of a given device in a particular system may be different than that specified on the data sheet or in the transmitter instruction manual. This effect is more pronounced at VHF and above.

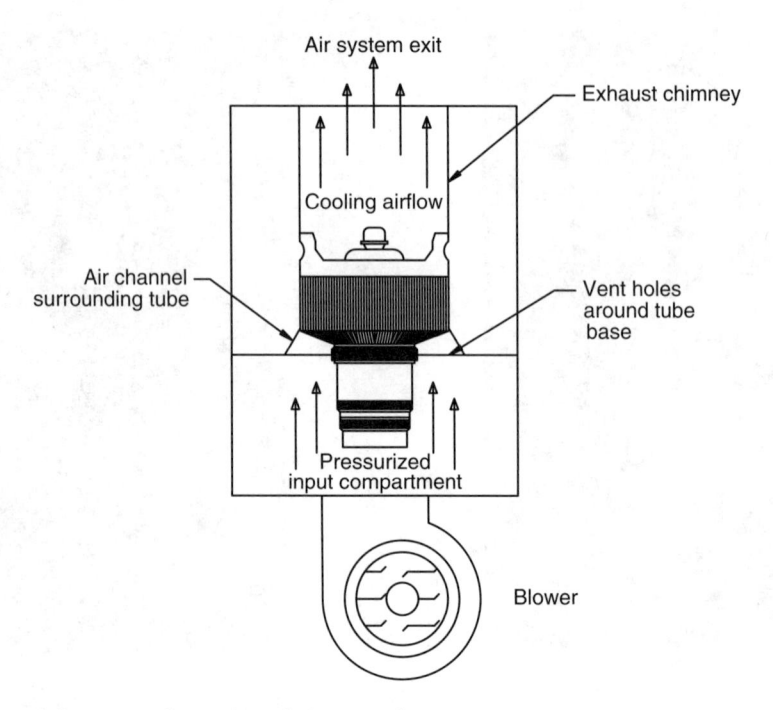

FIGURE 18.14 Airflow system for an air-cooled power tube.

Tube Dissipation

Proper cooling of the tube envelope and seals is a critical parameter for long tube life. Deteriorating effects that result in shortened tube life and reduced performance increase as the temperature increases. Excessive dissipation is perhaps the single greatest cause of catastrophic failure in a power tube. PA tubes used in broadcast, industrial, and research applications can be cooled using one of three methods: forced-air, liquid, and vapor-phase cooling. In radio and VHF-TV transmitters, forced-air cooling is by far the most common method used. Forced-air systems are simple to construct and easy to maintain.

The critical points of almost every PA tube type are the metal-to-ceramic junctions or seals. At temperatures below 250°C, these seals remain secure; but above this temperature, the bonding in the seal may begin to disintegrate. Warping of grid structures may also occur at temperatures above the maximum operating level of the tube. The result of prolonged overheating is shortened tube life or catastrophic failure. Several precautions are usually taken to prevent damage to tube seals under normal operating conditions. Air directors or sections of tubing may be used to provide spot-cooling to critical surface areas of the device. Airflow sensors prevent operation of the system in the event of a cooling system failure.

Tubes that operate in the VHF and UHF bands are inherently subject to greater heating action than devices operated at lower frequencies (such as AM service). This effect is the result of larger RF charging currents into the tube capacitances, dielectric losses, and the tendency of electrons to bombard parts of the tube structure other than the grid and plate in high-frequency applications. Greater cooling is required at higher frequencies.

The technical data sheet for a given power tube will specify cooling requirements. The end user is not normally concerned with this information; it is the domain of the transmitter manufacturer. The end user, however, is responsible for proper maintenance of the cooling system.

Air-Handling System

All modern air-cooled PA tubes use an air-system socket and matching chimney for cooling. Never operate a PA stage unless the air-handling system provided by the manufacturer is complete and in place. For example, the chimney for a PA tube often can be removed for inspection of other components in the circuit. Operation without the chimney, however, may significantly reduce airflow through the tube and result in over-dissipation of the device. It also is possible that operation without the proper chimney could damage other components in the circuit because of excessive radiated heat. Normally, the tube socket is mounted in a pressurized compartment so that cooling air passes through the socket and then is guided to the anode cooling fins, as illustrated in Fig. 18.14. Do not defeat any portion of the air-handling system provided by the manufacturer.

Cooling of the socket assembly is important for proper cooling of the tube base and for cooling of the contact rings of the tube itself. The contact fingers used in the *collet* assembly of a socket are typically made of beryllium copper. If subjected to temperatures above 150°C for an extended period of time, the beryllium copper will lose its temper (springy characteristic) and will no longer make good contact with the base rings of the device. In extreme cases, this type of socket problem can lead to arcing, which can burn through the metal portion of the tube base ring. Such an occurrence can ultimately lead to catastrophic failure of the device because of a loss of the vacuum envelope. Other failure modes for a tube socket include arcing between the collet and tube ring, which can weld a part of the socket and tube together. The end result is failure of both the tube and the socket.

Ambient Temperature

The temperature of the intake air supply is a parameter that is usually under the control of the maintenance engineer. The preferred cooling air temperature is no higher than 75°F and no lower than the room dew point. The air temperature should not vary because of an oversized air conditioning system or because of the operation of other pieces of equipment at the transmission facility. Monitoring the PA exhaust stack temperature is an effective method of evaluating overall RF system performance. This can be easily accomplished. It also provides valuable data on the cooling system and final-stage tuning.

Another convenient method for checking the efficiency of the transmitter cooling system over a period of time involves documenting the back-pressure that exists within the PA cavity. This measurement is made with a *manometer*, a simple device that is available from most heating, ventilation, and air-conditioning (HVAC) suppliers. The connection of a simplified manometer to a transmitter PA input compartment is illustrated in Fig. 18.15.

When using the manometer, be careful that the water in the device is not allowed to backflow into the PA compartment. Do not leave the manometer connected to the PA compartment when the transmitter is on the air. Make the necessary measurement of PA compartment back-pressure and then disconnect the device. Seal the connection point with a subminiature plumbing cap or other appropriate hardware.

By charting the manometer readings, it is possible to accurately measure the performance of the transmitter cooling system over time. Changes resulting from the buildup of small dust particles (microdust) may be too gradual to be detected except through back-pressure charting. Be certain to take the manometer readings during periods of calm weather. Strong winds can result in erroneous readings because of pressure or vacuum conditions at the transmitter air intake or exhaust ports.

Deviations from the typical back-pressure value, either higher or lower, could signal a problem with the air-handling system. Decreased PA input compartment back-pressure could indicate a problem with the blower motor or a buildup of dust and dirt on the blades of the blower assembly. Increased back-pressure, on the other hand, could indicate dirty PA tube anode cooling fins or a buildup of dirt on the PA exhaust ducting. Either condition is cause for concern. A system suffering from reduced air pressure into the PA compartment must be serviced as soon as possible. Failure to restore the cooling system to proper operation may lead to premature failure of the PA tube or other components in the input or output compartments. Cooling problems do not improve. They always get worse.

Failure of the PA compartment air-interlock switch to close reliably may be an early indication of impending cooling system trouble. This situation could be caused by normal mechanical wear or vibration of the switch assembly, or it may signal that the PA compartment air pressure has dropped. In such a case, documentation of manometer readings will show whether the trouble is caused by a failure of the air pressure switch or a decrease in the output of the air-handling system.

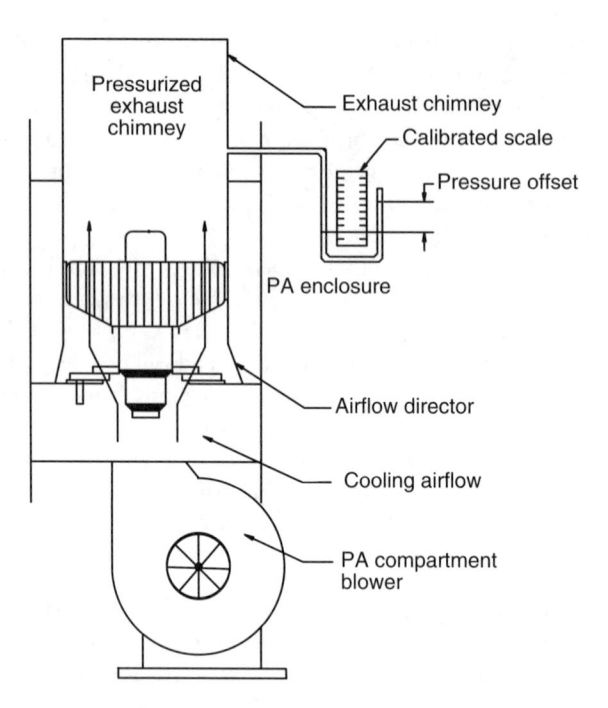

FIGURE 18.15 A manometer device used for measuring back-pressure in the PA compartment of a transmitter.

Thermal Cycling

Most power grid tube manufacturers recommend a warm-up period between application of *filament-on* and *plate-on* commands. Most RF equipment manufacturers specify a warm-up period of about 5 minutes. The minimum warm-up time is 2 minutes. Some RF generators include a time delay relay to prevent the application of a plate-on command until predetermined warm-up cycle is completed. Do not defeat these protective circuits. They are designed to extend PA tube life. Most manufacturers also specify a recommended cool-down period between the application of *plate-off* and *filament-off* commands. This cool-down, generally about 10 minutes, is designed to prevent excessive temperatures on the PA tube surfaces when the cooling air is shut off. Large vacuum tubes contain a significant mass of metal, which stores heat effectively. Unless cooling air is maintained at the base of the tube and through the anode cooling fins, excessive temperature rise can occur. Again, the result can be shortened tube life, or even catastrophic failure because of seal cracks caused by thermal stress.

Most tube manufacturers suggest that cooling air continue to be directed toward the tube base and anode cooling fins after filament voltage has been removed to further cool the device. Unfortunately, however, not all control circuits are configured to permit this mode of operation.

Tube Changing Procedure

Plug-in power tubes must be seated firmly in their sockets, and the connections to the anodes of the tubes must be tight. Once in place, do not remove a tube assembly for routine inspection unless it is malfunctioning. Whenever a tube is removed from its socket, carefully inspect the fingerstock for signs of overheating or arcing. Keep the socket assembly clean and all connections tight. If any part of a PA tube socket is found to be damaged, replace the defective portion immediately. In many cases, the damaged fingerstock ring can be ordered and replaced. In other cases, however, the entire socket must be replaced. This type of work is a major undertaking, requiring an experienced engineer.

Extending Vacuum Tube Life

RF power tubes are probably the most expensive replacement part that a transmitter or RF generator will need on a regular basis. With the cost of new and rebuilt tubes continually rising, maintenance engineers should do everything possible to extend tube life.

Conditioning a Power Tube

Whenever a new tube is installed in a transmitter, inspect the device for cracks or loose connections (in the case of tubes that do not socket-mount). Also check for interelectrode short circuits with an ohmmeter. Tubes must be seated firmly in their sockets to allow a good, low-resistance contact between the fingerstock and contact rings. After a new tube, or one that has been on the shelf for some time, is installed in the transmitter, run it with *filaments only* for at least 30 minutes, after which plate voltage may be applied. Next, slowly bring up the drive (modulation), in the case of an AM or TV visual transmitter. Residual gas inside the tube may cause an interelectrode arc (usually indicated by the transmitter as a plate overload) unless it is burned off in such a warm-up procedure.

Keep an accurate record of performance for each tube. Shorter than normal tube life could point to a problem in the RF amplifier itself. The average life that may be expected from a power grid tube is a function of many parameters, including:

- Filament voltage
- Ambient operating temperature
- RF power output
- Operating frequency
- Operating efficiency

The best estimate of life expectancy for a given system at a particular location comes from on-site experience. As a general rule of thumb, however, at least 12 months of service can be expected from most power tubes. Possible causes of short tube life include:

- Improper transmitter tuning
- Inaccurate panel meters or external wattmeter, resulting in more demand from the tube than is actually required
- Poor filament voltage regulation
- Insufficient cooling system airflow
- Improper stage neutralization

Filament Voltage

A *true-reading* RMS voltmeter is required to accurately measure filament voltage. Make the measurement directly from the tube socket connections. Secure the voltmeter test leads to the socket terminals and carefully route the cables outside the cabinet. Switch off the plate power supply circuit breaker. Close all interlocks and apply a *filament-on* command. Do not apply the high voltage during filament voltage tests. Serious equipment damage and/or injury to the maintenance engineer might result.

A true-reading RMS meter, instead of the more common *average-responding* RMS meter, is suggested because the true-reading meter can accurately measure a voltage despite an input waveform that is not a pure sine wave. Some filament voltage regulators use silicon-controlled rectifiers (SCRs) to regulate the output voltage. Do not put too much faith in the front-panel filament voltage meter. It is seldom a true-reading RMS device; most are average-responding meters.

Long tube life requires filament voltage regulation. Many RF systems have regulators built into the filament supply. Older units without such circuits can often be modified to provide a well-regulated supply by adding a ferroresonant transformer or motor-driven autotransformer to the ac supply input. A tube whose filament voltage is allowed to vary along with the primary line voltage will not achieve the life expectancy possible with a tightly regulated supply. This problem is particularly acute at mountain-top installations, where utility regulation is generally poor.

To extend tube life, some broadcast engineers leave the filaments on at all times, not shutting down at sign-off. If the sign-off period is 3 hours or less, this practice can be beneficial. Filament voltage regulation is a must in such situations because the primary line voltages may vary substantially from the carrier-on to carrier-off value. Do not leave voltage on the filaments of a klystron for a period of more than 2 hours if no beam voltage is applied. The net rate of evaporation of emissive material from the cathode surface of a klystron is greater without beam voltage. Subsequent condensation of the material on gun components may lead to voltage holdoff problems and an increase in body current.

Filament Voltage Management

By accurately managing the filament voltage of a thoriated tungsten power tube, the useful life of the device can be extended considerably, sometimes to twice the normal life expectancy. The following procedure is recommended:

- Operate the filament at its full-rated voltage for the first 200 hours following installation.
- Following the burn-in period, reduce the filament voltage by 0.1 V per step until power output begins to fall (for frequency modulated systems), or until modulating waveform distortion begins to increase (for amplitude modulated systems).
- When the *emissions floor* has been reached, raise the filament voltage 0.2 V.

Long-term operation at this voltage can result in a substantial extension in the usable life of the tube, as illustrated in Fig. 18.16.

Do not operate the tube with a filament voltage that is at or below 90% of its rated value. At regular intervals, about every 3 months, check the filament voltage and increase it if power output begins to fall or distortion begins to rise. Filament voltage should never be increased to more than 105% of rated voltage. Some tube manufacturers place the minimum operating point at 94%. Others recommend that the tube be set for 100% filament voltage and left there. The choice of which approach to follow is left to the user.

When it becomes necessary to boost filament voltage to more than 103%, order a new tube. If the old device is replaced while it still has some life remaining, the facility will have a standby tube that will perform well as a spare.

Check the filament current when the tube is first installed, and at annual intervals thereafter, to ensure that the filament draws the desired current. Tubes can fail early in life because of an open filament bar that would have been discovered during the warranty period if a current check had been made upon installation.

For 1 week of each year of tube operation, run the filament at full-rated voltage. This will operate the *getter* and clean the tube of gas.

Filament voltage is an equally important factor in achieving long life in a klystron. The voltages recommended by the manufacturer must be set and checked on a regular basis. Measure the voltage at the filament terminals and calibrate the front-panel meter as needed.

PA Stage Tuning

There are probably as many ways to tune the PA stage of an RF generator or transmitter as there are types of systems. Experience is the best teacher when it comes to adjusting for peak efficiency and performance. Compromises must often be made among various operating parameters. Some engineers follow the tuning procedures contained in the transmitter instruction manual to the letter. Others never open the manual, preferring to tune according to their own methods. Whatever procedure is used, document the operating parameters and steps for future reference. Do not rely on memory for a listing of the typical operating limits and tuning procedures for the system. Write down the information and post it at the facility. The manufacturer's service department can be an excellent source of information

Preventing RF System Failures   **18**-15

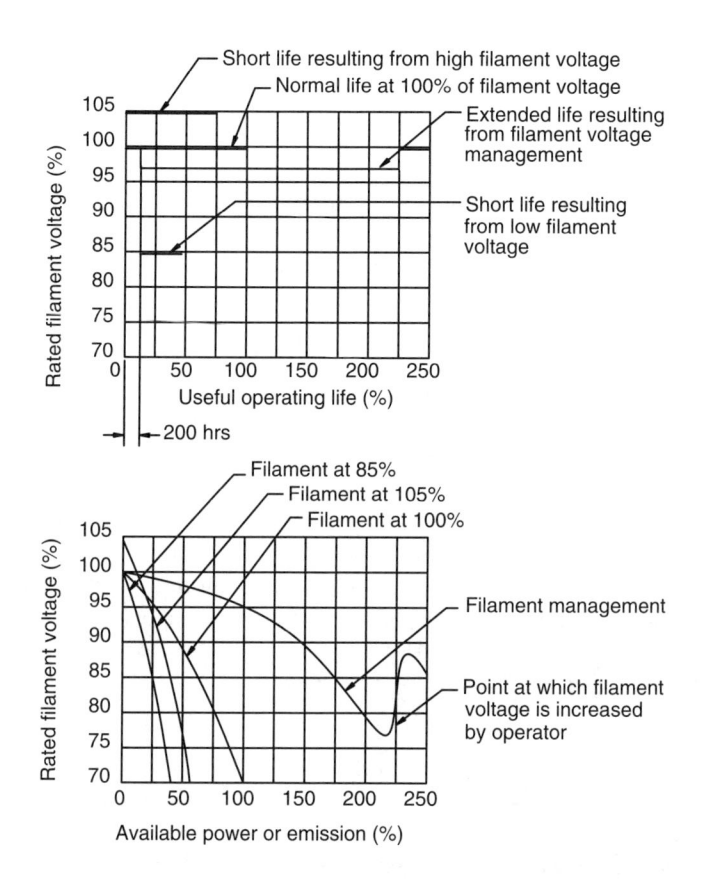

FIGURE 18.16 The effects of filament voltage management on the useful life of a thoriated tungsten filament power tube. Note the dramatic increase in emission hours when filament voltage management is practiced.

about tuning a particular unit. Many times, the factory can provide pointers on how to simplify the tuning process or what interaction of adjustments may be expected. Whatever information is learned from such conversations, write it down.

Table 18.1 shows a typical tuning procedure for an FM transmitter. The actual steps vary, of course, from transmitter to transmitter. However, when the tuning characteristics of a given unit are documented in a detailed manner, future repair work is simplified. This record can be of great value to an engineer who is fortunate enough to have a reliable system that does not require regular service. Many of the

TABLE 18.1 Sample Documented Transmitter Tuning Procedure

PA Tuning Adjustment

Unload the transmitter (switch the loading control to *lower*) to produce a PA screen current of 400 to 600 mA.
Peak the PA screen current with the plate-tuning control.
Maintain screen current at or below 600 mA by adjusting the loading control (switch it to *raise*).
Position the plate-tuning control in the center of travel by moving the coarse-tune shorting plane up or down as needed.
  If the screen current peak is reached near the raise end of plate-tune travel, raise the shorting plane slightly.
  If the peak is reached near the lower end of travel, lower the plane slightly.
After the screen current has been peaked, adjust the loading control for maximum power output and minimum synchronous AM.
Peak the driver screen current with C-37.
The driver screen peak should coincide with PA screen peak and PA grid peak, and with a dip in the left and right driver cathode currents.

tuning tips learned during the last service session may be forgotten by the time maintenance work must be performed again.

When to Tune

Tuning can be affected by any number of changes in the PA stage. Replacing the final tube in an AM transmitter or low- to-medium frequency RF generator usually does not significantly alter stage tuning. It is advisable, however, to run through a touch-up tuning procedure just to be sure. Replacing a tube in an FM or TV transmitter or high-frequency RF generator, on the other hand, can significantly alter stage tuning. At high frequencies, normal tolerances and variations in tube construction result in changes in element capacitance and inductance. Likewise, replacing a component in the PA stage may cause tuning changes because of normal device tolerances.

Stability is one of the primary objectives of transmitter tuning. Avoid tuning positions that do not provide stable operation. Adjust for broad peaks or dips, as required. Tune so the system is stable from a cold start-up to normal operating temperature. Readings should not vary measurably after the first minute of operation.

Adjust tuning not only for peak efficiency, but also for peak performance. Unfortunately, these two elements of transmitter operations do not always coincide. Trade-offs must sometimes be made to ensure proper operation of the system. For example, FM or TV aural transmitter loading can be critical to wide system bandwidth and low synchronous AM. Loading beyond the point required for peak efficiency must often be used to broaden cavity bandwidth. Heavy loading lowers the PA plate impedance and cavity Q. A low Q also reduces RF circulating currents in the cavity.

Vacuum Tube Life

Failures in semiconductor components result primarily from deterioration of the device caused by exposure to environmental fluctuations and voltage extremes. The vacuum tube, on the other hand, suffers wear-out because of a predictable chemical reaction. Life expectancy is one of the most important factors to be considered in the use of vacuum tubes. In general, manufacturers specify maximum operating parameters for power grid tubes so that operation within the ratings will provide for a minimum useful life of 1000 hours.

The cathode is the heart of any power tube. The device is said to *wear out* when filament emissions are inadequate for full power output or acceptable distortion levels. In the case of a thoriated tungsten filament tube, three primary factors determine the number of hours a device will operate before reaching this condition:

- The rate of evaporation of thorium from the cathode
- The quality of the tube vacuum
- The operating temperature of the filament

In the preparation of thoriated tungsten, 1 to 2% of thorium oxide (thoria) is added to the tungsten powder before it is sintered and drawn into wire form. After being mounted in the tube, the filament is usually *carburized* by being heated to a temperature of about 2000 Kelvin in a low-pressure atmosphere of hydrocarbon gas or vapor until its resistance increases by 10 to 25%. This process allows the reduction of the thoria to metallic thorium. The life of the filament as an emitter is increased because the rate of evaporation of thorium from the carburized surface is several times smaller than from a surface of pure tungsten.

Despite the improved performance obtained by carburization of a thoriated-tungsten filament, they are susceptible to deactivation by the action of positive ions. Although the deactivation process is negligible for anode voltages below a critical value, a trace of residual gas pressure too small to affect the emission from a pure tungsten filament can cause rapid deactivation of a thoriated-tungsten filament. This restriction places stringent requirements on vacuum processing the tube.

These factors taken together determine the wear-out rate of the tube. Catastrophic failures because of interelectrode shorts or failure of the vacuum envelope are considered abnormal and are usually the result of some external influence. Catastrophic failures not the result of the operating environment are usually caused by a defect in the manufacturing process. Such failures generally occur early in the life of the component.

The design of the equipment can have a substantial impact on the life expectancy of a vacuum tube. Protection circuitry must remove applied voltages rapidly to prevent damage to the tube in the event of a failure external to the device. The filament turn-on circuit can also have an effect on PA tube life expectancy. The surge current of the filament circuit must be maintained at a low level to prevent thermal cycling problems. This consideration is particularly important in medium- and high-power PA tubes. When the heater voltage is applied to a cage-type cathode, the tungsten wires expand immediately because of their low thermal inertia. However, the cathode support, which is made of massive parts (relative to the tungsten wires) expands more slowly. The resulting differential expansion can cause permanent damage to the cathode wires. It can also cause a modification of the tube operating characteristics, and occasionally arcs between the cathode and the control grid.

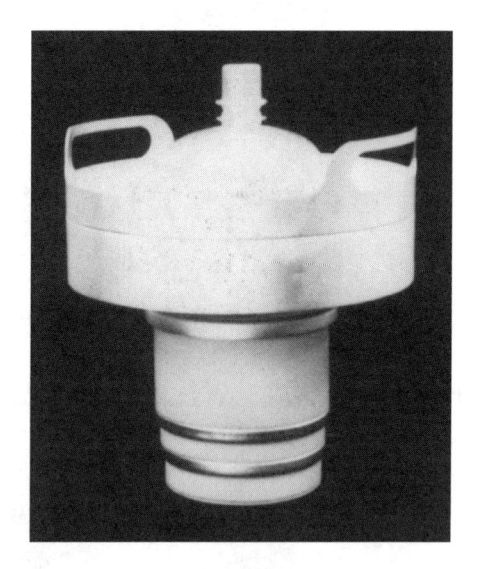

FIGURE 18.17 A new, unused 4CX15,000A tube. Contrast the appearance of this device with the example tubes that follow. (Courtesy of Varian/Eimac.)

Examining Tube Performance

Examining a power tube after it has been removed from a transmitter or other type of RF generator can tell a great deal about how well the transmitter–tube combination is working. Contrast the appearance of a new power tube, shown in Fig. 18.17, with a component at the end of its useful life. If a power tube fails prematurely, examine the device to determine if an abnormal operating condition exists within the transmitter. Consider the following examples:

- *Figure 18.18.* Two 4CX15,000A power tubes with differing anode heat dissipation patterns. Tube *a* experienced excessive heating because of a lack of PA compartment cooling air or excessive dissipation because of poor tuning. Tube *b* shows a normal thermal pattern for a silver-plated 4CX15,000A. Nickel-plated tubes do not show signs of heating because of the high heat resistance of nickel.
- *Figure 18.19.* Base heating patterns on two 4CX15,000A tubes. Tube *a* shows evidence of excessive heating because of high filament voltage or lack of cooling air directed toward the base of the device. Tube *b* shows a typical heating pattern with normal filament voltage.
- *Figure 18.20.* A 4CX5,000A with burning on the screen-to-anode ceramic. Exterior arcing of this type generally indicates a socketing problem, or another condition external to the tube.
- *Figure 18.21.* The stem portion of a 4CX15,000A that had gone down to air while the filament was on. Note the deposits of tungsten oxide formed when the filament burned up. The grids are burned and melted because of the ionization arcs that subsequently occurred. A failure of this type will trip overload breakers in the transmitter. It is indistinguishable from a shorted tube in operation.
- *Figure 18.22.* A 4CX10,000D that experienced arcing typical of a bent fingerstock, or exterior arcing caused by components other than the tube.

FIGURE 18.18 Anode dissipation patterns on two 4CX15,000A tubes: tube *a*, on the left, shows excessive heating, and tube *b*, on the right, shows normal wear. (Courtesy of Econco Broadcast Service, Woodland, CA.)

FIGURE 18.19 Base heating patterns on two 4CX15,000A tubes: tube *a*, on the left, shows excessive heating, and tube *b*, on the right, shows normal wear. (Courtesy of Econco Broadcast Service, Woodland, CA.)

18.5 Preventing RF System Failures

The reliability and operating costs over the lifetime of an RF system can be significantly impacted by the effectiveness of the preventive maintenance program designed and implemented by the engineering staff. When dealing with a *critical-system unit* such as a broadcast transmitter or other RF generator that must operate on a daily basis, maintenance can have a major impact — either positive or negative — on downtime and bottom-line profitability of the facility. The sections of a transmitter most vulnerable to failure are those exposed to the outside world: the ac-to-dc power supplies and RF output stage. These circuits are subject to high energy surges from lightning and other sources.

The reliability of a communications system may be compromised by an *enabling event phenomenon*. An enabling event phenomenon is an event that, while not causing a failure by itself, sets up (or enables)

Preventing RF System Failures

FIGURE 18.20 A 4CX5,000A tube that appears to have suffered socketing problems. (Courtesy of Econco Broadcast Service, Woodland, CA.)

FIGURE 18.21 The interior elements of a 4CX15,000A tube that had gone to air while the filament was lit. (Courtesy of Econco Broadcast Service, Woodland, CA.)

FIGURE 18.22 A 4CX15,000A tube showing signs of external arcing. (Courtesy of Econco Broadcast Service, Woodland, CA.)

a second event that can lead to failure of the communications system. This phenomenon is insidious because the enabling event is often not self-revealing. Examples include:

- A warning system that has failed or been disabled for maintenance
- One or more controls set incorrectly so that false readouts are provided for operations personnel
- Redundant hardware that is out of service for maintenance
- Remote metering that is out of calibration

Common Mode Failures

A common mode failure is one that can lead to the failure of all paths in a redundant configuration. In the design of redundant systems, therefore, it is important to identify and eliminate sources of common mode failures, or to increase their reliability to at least an order of magnitude above the reliability of the redundant system. Common mode failure points in a transmission system include the following:

- Switching circuits that activate standby or redundant hardware
- Sensors that detect a hardware failure
- Indicators that alert personnel to a hardware failure
- Software that is common to all paths in a redundant system

The concept of software reliability in control and monitoring has limited meaning in that a good program will always run, and copies of the program will always run. On the other hand, a program with one or more errors will always fail, and so will copies of the program, given the same input data. The reliability of software, unlike hardware, cannot be improved through redundancy if the software in the parallel path is identical to the primary path.

Modifications and Updates

If problems are experienced with a system, examine what can be done to prevent the failure from occurring again. A repeat of the problem can often be avoided by installing various protection devices or consulting the factory for updates to the hardware. If the transmitter is several years old, the factory service department can detail any changes that may have been made in the unit to provide more reliable operation. Many of these modifications are minor and can be incorporated into older models with little cost or effort. Modifications could include the following items:

- Changing a variable capacitor in a critical tuning stage to a vacuum variable for more stability
- Installing additional filtering in the high-voltage power supply to improve AM noise performance
- Replacing older-technology transistorized circuit boards with newer IC and power semiconductor PWBs to improve reliability and performance
- Improving the overload protection circuitry through the addition of solid-state logic
- Adding transient protection devices at critical stages of the transmitter

Spare Parts

The spare parts inventory is a key aspect of any successful equipment maintenance program. Having adequate replacement components on hand is important not only to correct equipment failures, but in identifying those failures as well. Many parts — particularly in the high-voltage power supply and RF chain — are difficult to test under static conditions. The only reliable way to test the component may be to substitute one of known quality. If the system returns to normal operation, then the substituted component is defective. Substitution is also a valuable tool in troubleshooting intermittent failures caused by component breakdown under peak power conditions.

18.6 Transmission Line/Antenna Problems

The *voltage standing wave ratio* (VSWR) of an antenna and its transmission line is a vital parameter that has a considerable effect on the performance and reliability of a transmission system. VSWR is a measure of the amount of power reflected back to the transmitter because of an antenna and/or transmission line mismatch. Figure 18.23 provides a chart for VSWR calculation. A mismatched or defective transmission system will result in a high degree of reflected power, or a higher VSWR.

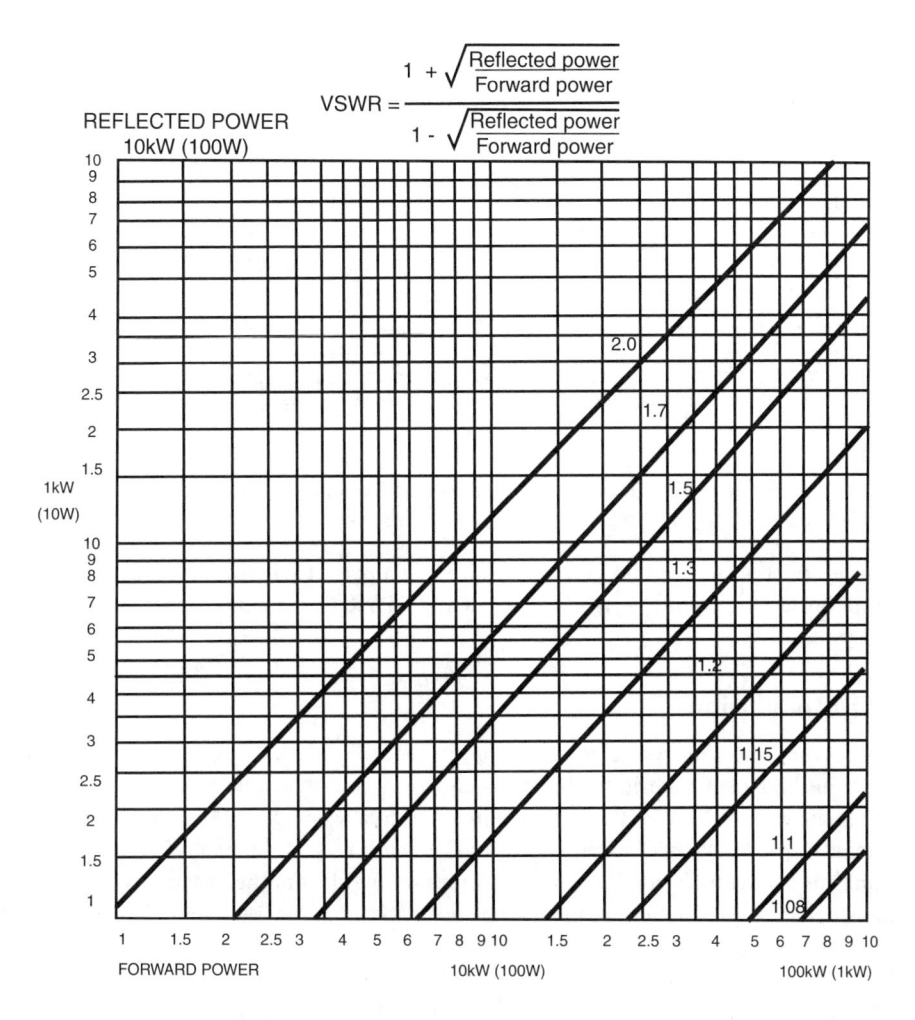

FIGURE 18.23 Transmission line VSWR graph. For low-power operation, use the values in parentheses.

The amount of reflected power that a given system can accept is a function of the application. For example, it is common practice in FM broadcasting to maintain a VSWR of 1.1:1 as the maximum level within the transmission channel that can be tolerated without degrading the quality of the on-air signal. For conventional TV broadcasting, a VSWR into the antenna feeder of more than 1.04:1 will start to degrade picture quality, particularly on systems that use a long transmission line. Reflections down the line from a mismatch at the antenna disrupt the performance of the transmitter output stage. The reflections also cause multipath distortion *within* the transmission line itself. When power is reflected back to the transmitter, it causes the RF output stage to look into a mismatched load with unpredictable phase and impedance characteristics. Because of the reflective nature of VSWR on a transmission system, the longer the transmission line (assuming the reflection is originating at the antenna), the more severe the problem may be for a given VSWR. A longer line means that reflected power seen at the RF output stage has greater time (phase) delays, increasing the reactive nature of the load.

VSWR is affected not only by the rating of the antenna and transmission line as individual units, but also by the combination of the two as a system. The worst-case system VSWR is equal to the antenna VSWR multiplied by the transmission line VSWR. For example, if an antenna with a VSWR of 1.05:1 is connected to a line with a VSWR of 1.05:1, the resulting worst-case system VSWR would be 1.1025:1. Given the right set of conditions, an interesting phenomenon can occur in which the VSWR of the

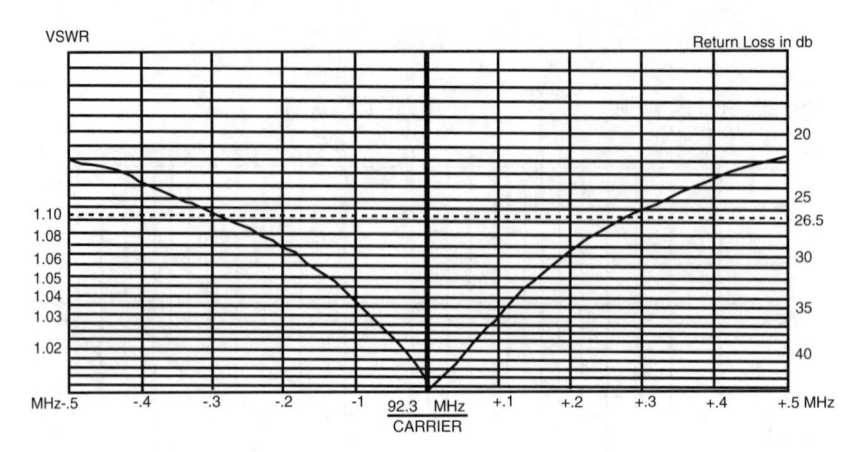

FIGURE 18.24 The measured performance of a single-channel FM antenna (tuned to 92.3 MHz). The antenna provides a VSWR of below 1.1:1 over a frequency range of nearly ±300 kHz. (Courtesy of Jampro Antenna Company.)

antenna cancels the transmission line VSWR, resulting in a perfect 1:1 match. The determining factors for this condition are the point of origin of the antenna VSWR, the length of transmission line, and the observation point.

Effects of Modulation

The VSWR of a transmission system is a function of frequency and changes with carrier modulation. This change can be large or small, but it will occur to some extent. The cause can be traced to the frequency dependence of the VSWR of the antenna (and to a lesser extent, the transmission line). The effects of frequency on VSWR can be observed in Fig. 18.24. Although the plot of VSWR-versus-frequency for a common FM antenna is good, notice that with no modulation, the system VSWR is one figure. VSWR measurements are different with *positive modulation* (carrier plus modulation) and *negative modulation* (carrier minus modulation).

VSWR is further complicated because power reflected back to the transmitter from the antenna may not come from a single point, but from a number of different points. One reflection might be caused by the antenna-matching unit, another by various flanges in the line, and a third by a damaged part of the antenna system. Because these reflection points are different lengths from the transmitter PA plate, a variety of standing waves can be generated along the line, varying with the modulating frequency.

Energy reflected back to the transmitter from the antenna is not lost. A small percentage of the energy is turned into heat, but the majority of it is radiated by the antenna, delayed in time by the length of the transmission line.

Maintenance Considerations

To maintain low VSWR, the transmission line and antenna system should be serviced regularly. Use the following guidelines:

- Inspect the antenna elements, interconnecting cables, impedance transformers, and support braces at least once each year. Falling ice can damage FM, TV, and communications antenna elements if proper precautions are not taken. Icing on the elements of an antenna will degrade the antenna VSWR because ice lowers the frequency of the electrical resonance of the antenna. Two methods are commonly used to prevent a buildup of ice on high-power transmitting antennas: electrical de-icers and *radomes*.

Preventing RF System Failures   **18**-23

- Check AM antennas regularly for structural integrity. Because the tower itself is the radiator, bond together each section of the structure for good electrical contact.
- Clean base insulators and guy insulators (if used) as often as required.
- Keep lightning ball gaps or other protective devices clean and properly adjusted.
- Inspect the transmission line for signs of damage. Check supporting hardware and investigate any indication of abnormal heating of the line.
- Keep a detailed record of VSWR in the facility maintenance log and investigate any increase above the norm.
- Regularly check the RF system test load. Inspect the coolant filters and flow rate, as well as the resistance of the load element.

UHF Transmission Systems

The RF transmission system of most high-power UHF stations is externally diplexed after the final RF amplifiers. Either coaxial or waveguide-type diplexers can be used. Maintenance of the combining sections is largely a matter of careful observation and record-keeping:

- Monitor the reject loads on diplexers and power combiners regularly to ensure adequate cooling.
- Check the temperature of the transmission line and components, particularly coaxial elements. Keep in mind that coax of the same size and carrying the same RF power runs warmer in UHF systems than in VHF systems. This phenomenon is caused by the reduced penetration depth (the *skin effect*) of UHF signal currents. Hot spots in the transmission line can be caused by poor contact areas or by high VSWR. If they are the result of a VSWR condition, the hot spots will be repeated every 1/2-wavelength toward the transmitter.
- Monitor the reverse power/VSWR meters closely. Some daily variation is not unusual, in small amounts. Greater variations that are cyclical in nature are an indication of a long-line problem — most likely at the antenna. Because transmission lines are usually long at UHF (a taller tower allows greater coverage) and the wavelength is small leads to large phase changes of a mismatch at the antenna. Mismatches inside the building do not cause the same cyclical variation. If reverse power is observed to vary significantly, run the system with the test load to see whether the problem disappears. If the variations are not present with a test load, arrange for an RF sweep of the line.

A change in klystron output power is another effect of VSWR variation on a UHF transmitter. The output coupler transforms the line characteristic impedance upward to approximately match the beam impedance. This provides maximum power transfer from the cavity. Large VSWR phase variations associated with long lines change the impedance that the output coupler sees. This causes the output power to vary, sometimes more significantly than the reverse power metering indicates. *Ghosting* on the output waveform is a common indication of antenna VSWR problems in a long-line conventional (analog) TV system. If the input signal is clean and the output has a ghost, arrange for an RF sweep.

18.7 High-Voltage Power Supply Problems

The high-voltage plate supply is the first line of defense between external ac line disturbances and the power amplifier stage(s). Next to the output circuit itself, the plate supply is the second section of a transmitter most vulnerable to damage because of outside influences.

Power Supply Maintenance

Figure 18.25 shows a high-reliability power supply of the type common in transmission equipment. Many transmitters use simpler designs, without some of the protection devices shown, but the principles of preventive maintenance are the same:

- Thoroughly examine every component in the high-voltage power supply. Look for signs of leakage on the main filter capacitor (C2).
- Check all current-carrying meter/overload shunt resistors (R1–R3) for signs of overheating.
- Carefully examine the wiring throughout the power supply for loose connections.
- Examine the condition of the filter capacitor series resistors (R4 and R5), if used, for indications of overheating. Excessive current through these resistors could point to a pending failure in the associated filter capacitor.
- Examine the condition of the bleeder resistors (R6–R8). A failure in one of the bleeder resistors could result in a potentially dangerous situation for maintenance personnel by leaving the main power supply filter capacitor (C2) charged after the removal of ac input power.
- Examine the plate voltage meter multiplier assembly (A1) for signs of resistor overheating. Replace any resistors that are discolored with the factory-specified type.

When changing components in the transmitter high-voltage power supply, be certain to use parts that meet with the approval of the manufacturer. Do not settle for a close match of a replacement part. Use the exact replacement part. This ensures that the component will work as intended and will fit in the space provided in the cabinet.

Power Supply Metering

Proper metering is one of the best ways to prevent failures in transmission equipment. Accurate readings of plate voltage and current are fundamental to RF system maintenance. Check each meter for proper mechanical and electrical operation. Replace any meter that sticks or will not zero.

With most transmitter plate current meters, accuracy of the reading can be verified by measuring the voltage drop across the shunt element (R2 in Fig. 18.25) and using Ohm's law to determine the actual current in the circuit. Be certain to take into consideration the effects of the meter coil itself. Contact the transmitter manufacturer for suggestions on how best to confirm the accuracy of the plate current meter.

The plate voltage meter can be checked for accuracy using a high-voltage probe and a high-accuracy external voltmeter. Be extremely careful when making such a measurement. Follow instructions to the letter regarding the use of a high-voltage probe. Do not defeat transmitter interlocks to make this measurement. Instead, fashion a secure connection to the point of measurement and carefully route the meter cables out of the transmitter. Never use common test leads to measure a voltage of more than 600 V. Standard test lead insulation for most meters is not rated for use above 600 V.

Overload Sensor

The plate supply overload sensor in most transmitters is arranged as shown in Fig. 18.25. An adjustable resistor — either a fixed resistor with a movable tap or a potentiometer — is used to set the sensitivity of the plate overload relay. Check potentiometer-type adjustments periodically. Fixed-resistor-type adjustments rarely require additional attention. Most manufacturers have a chart or mathematical formula that can be used to determine the proper setting of the adjustment resistor (R9) by measuring the voltage across the overload relay coil (K1) and observing the operating plate current value. Clean the overload relay contacts periodically to ensure proper operation. If mechanical problems are encountered with a relay, replace it.

Transmitter control logic for a high-power UHF system is usually configured for two states of operation:

- An *operational level*, which requires all the "life-support" systems to be present before the HV command is enabled
- An *overload level*, which removes HV when one or more fault conditions occur

At least one a month, inspect the logic ladder for correct operation. At longer intervals, perhaps annually, check the speed of the trip circuits. (A storage oscilloscope is useful for this measurement.)

Preventing RF System Failures

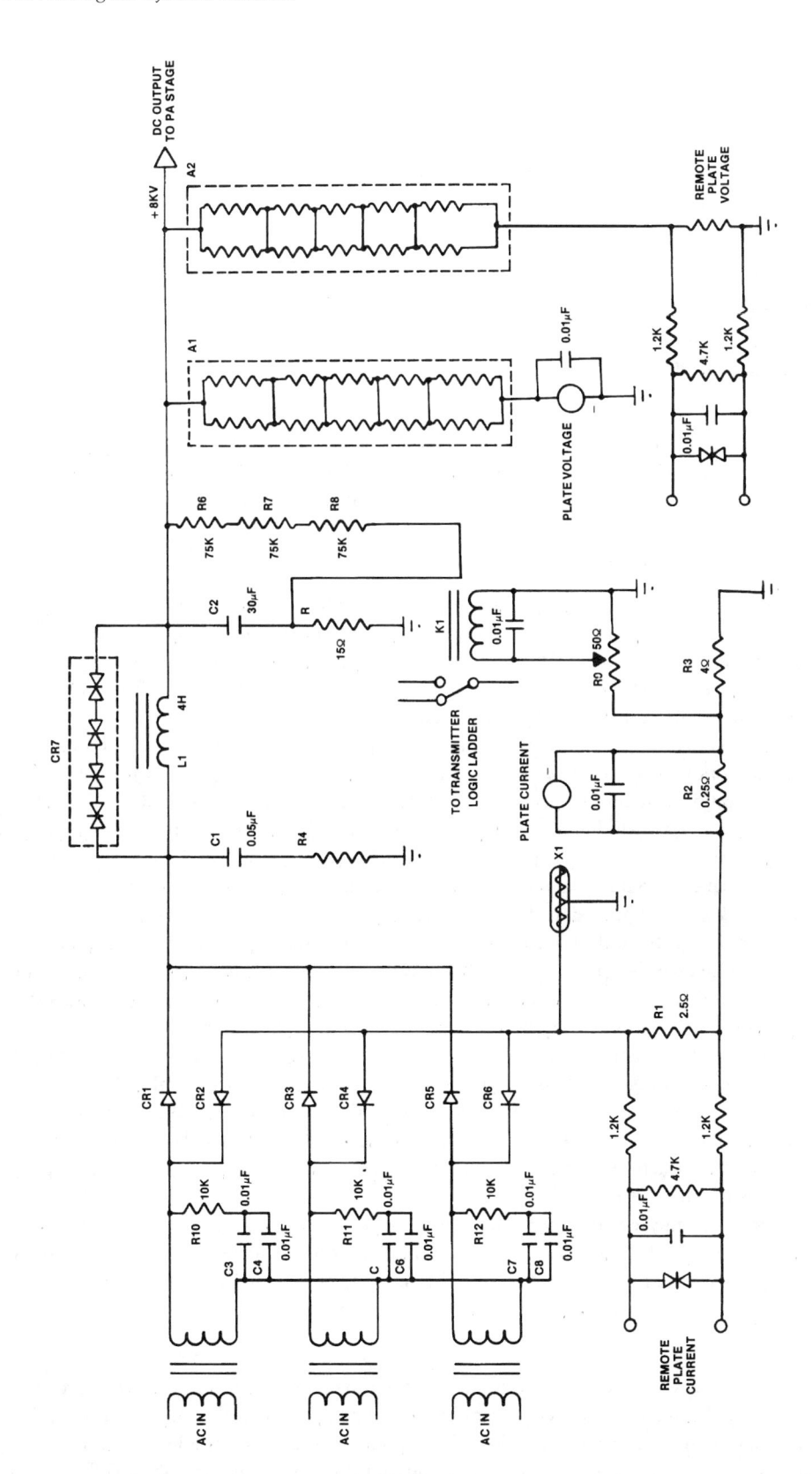

FIGURE 18.25 A common high-voltage transmitter power supply circuit design.

Most klystrons require an HV removal time of less than 100 msec from the occurrence of an overload. If the trip time is longer, damage may result to the klystron. Pay particular attention to the body-current overload circuits. Occasionally check the body-current without applied drive to ensure that the dc value is stable. A relatively small increase in dc body-current can lead to overheating problems.

The RF arc detectors in a UHF transmitter also require periodic monitoring. External cavity klystrons generally have one detector in each of the third and fourth cavities. Integral devices use one detector at the output window. A number of factors can cause RF arcing, including:

- Overdriving the klystron
- Mistuning the cavities
- Poor cavity fit (external type only)
- Under coupling of the output
- High VSWR

Regardless of the cause, arcing can destroy the vacuum seal, if drive and/or HV are not removed quickly. A lamp is included with each arc detector photocell for test purposes. If the lamp fails, a flashlight can provide sufficient light to trigger the cell until a replacement can be obtained.

Transient Disturbances

Every electronic installation requires a steady supply of clean power to function properly. Recent advances in technology have made the question of ac power quality even more important, as microcomputers are integrated into transmission equipment. Different types and makes of transmitters have varying degrees of transient overvoltage protection. Given the experience of the computer industry, it is difficult to overprotect electronic equipment from ac line disturbances.

Figure 18.25 shows surge suppression at two points in the power supply circuit. C1 and R4 make up an R/C snubber network that is effective in shunting high-energy, fast-rise time spikes that may appear at the output of the rectifier assembly (CR1–CR6). Similar R/C snubber networks (R10–R12 and C3–C8) are placed across the secondary windings of each section of the three-phase power transformer. Any signs of resistor overheating or capacitor failure are an indication of excessive transient activity on the ac power line. Transient disturbances should be suppressed before the ac input point of the transmitter.

Assembly CR7 is a surge suppression device that should be given careful attention during each maintenance session. CR7 is typically a selenium thyrector assembly that is essentially inactive until the voltage across the device exceeds a predetermined level. At the *trip point*, the device will break over into a conducting state, shunting the transient overvoltage. CR7 is placed in parallel with L1 to prevent damage to other components in the transmitter in the event of a loss of RF excitation to the final stage. A sudden drop in excitation will cause the stored energy of L1 to be discharged into the power supply and PA circuits in the form of a high-potential pulse. Such a transient can damage or destroy filter, feedthrough, or bypass capacitors; damage wiring; or cause PA tube arcing. CR7 prevents these problems by dissipating the stored energy in L1 as heat. Investigate discoloration or other outward signs of damage to CR7. Such an occurrence could indicate a problem in the exciter or IPA stage of the transmitter. Immediately replace CR7 if it appears to have been stressed.

Check spark gap surge suppressor X1 periodically for signs of overheating. X1 is designed to prevent damage to circuit wiring in the event that one of the meter/overload shunt resistors (R1–R3) opens. Because the spark gap device is nearly impossible to accurately test in the field and is relatively inexpensive, it is an advisable precautionary measure to replace the component every few years.

Single Phasing

Any transmitter using a three-phase ac power supply is subject to the problem of *single phasing*, the loss of one of the three legs from the primary ac power distribution source. Single phasing is usually a utility company problem, caused by a downed line or a blown pole-mounted fuse. The loss of one leg of a three-phase line results in a particularly dangerous situation for three-phase motors, which will overheat

Preventing RF System Failures 18-27

and sometimes fail. AM transmitters utilizing *pulse-width modulation* (PWM) systems are also vulnerable to single-phasing faults. PWM AM transmitters can suffer catastrophic failure of the plate power supply transformer as a result of the voltage regulation characteristics of the modulation system. The PWM circuit will attempt to maintain carrier and sideband power through the remaining legs of the three-phase supply. This forces the active transformer section and its associated rectifier stack to carry as much as three times the normal load.

Figure 18.26 shows a simple protection scheme that has been used to protect transmission equipment from damage caused by single phasing. Although at first glance the system looks as if it would easily handle the job, operational problems can result. The loss of one leg of a three-phase line rarely results in zero (or near-zero) voltages in the legs associated with the problem line. Instead, a combination of leakage currents caused by *regeneration* of the missing legs in inductive loads and the system load distribution usually results in voltages of some sort on the fault legs of the three-phase supply. It is possible, for example, to have phase-to-phase voltages of 220 V, 185 V, and 95 V on the legs of a three-phase, 208-V ac line experiencing a single-phasing problem. These voltages often change, depending on the equipment turned on at the transmitter site.

Integrated circuit technology has provided a cost-effective solution to this common design problem in medium- and high-power RF equipment. Phase-loss protection modules are available from several

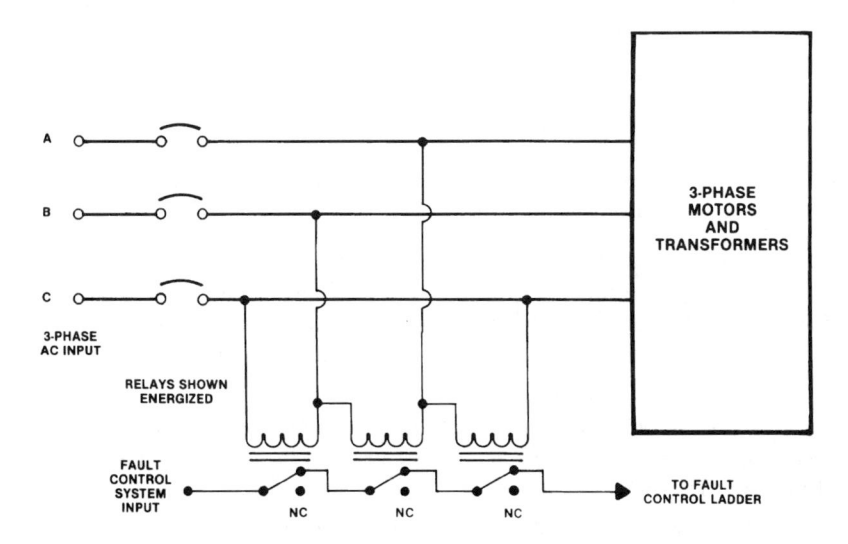

FIGURE 18.26 A protection circuit using relays for utility company ac phase loss protection.

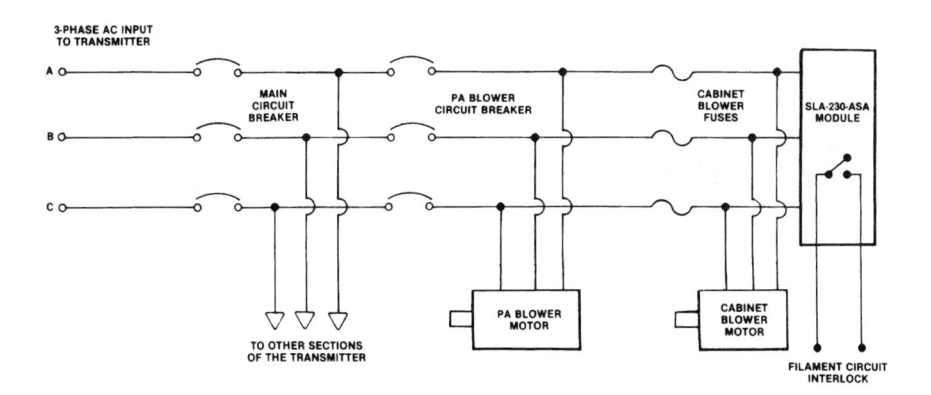

FIGURE 18.27 A high-performance, single-phasing protection circuit using a phase-loss module as the sensor.

manufacturers that provide a contact closure when voltages of proper magnitude and phase are present on the monitored line. The relay contacts can be wired into the logic control ladder of the transmitter to prevent the application of primary ac power during a single-phasing condition. Figure 18.27 shows the recommended connection method. Note that the input to the phase monitor module is taken from the final set of three-phase blower motor fuses. In this way, any failure inside the transmitter that might result in a single-phasing condition is taken into account. Because three-phase motors are particularly sensitive to single-phasing faults, the relay interlock is tied into the filament circuit logic ladder. For AM transmitters utilizing PWM schemes, the input of the phase-loss protector is connected to the load side of the plate circuit breaker. The phase-loss protector shown in Fig. 18.27 includes a sensitivity adjustment for various nominal line voltages. The unit is small and relatively inexpensive. If your transmitter does not have such a protection device, consider installing one. Contact the factory service department for recommendations on the connection methods that should be used.

18.8 Temperature Control

The environment in which the transmitter is operated is a key factor in determining system reliability. Proper temperature control must be provided for the transmitter to prevent thermal fatigue in semiconductor components and shortened life in vacuum tubes. Problems can be avoided if preventive maintenance work is performed on a regular basis.

Cooling System Maintenance

Each RF transmission system is unique and requires an individual assessment of cooling system needs. Still, a number of common preventive maintenance tasks apply to nearly all systems, including:

- Keep all fans and blowers clear of dirt, dust, and other foreign material that might restrict airflow. Check the fan blades and blower impellers for any imbalance conditions that could result in undue bearing wear or damage. Inspect belts for proper tension, wear, and alignment.
- Regularly clean the blower motor. Motors are usually cooled by the passage of air over the component. If the ambient air temperature is excessive or the air flow is restricted, the lubricant will gradually be vaporized from the motor bearings and bearing failure will occur. If dirty air passes over the motor, the accumulation of dust and dirt must be blown out of the device before the debris impairs cooling.
- Follow the manufacturer's recommendations for suggested frequency and type of lubrication. Bearings and other moving parts may require periodic lubrication. Carefully follow any special instructions on operation or maintenance of the cooling equipment.
- Inspect motor-mounting bolts periodically. Even well-balanced equipment experiences some vibration, which can cause bolts to loosen over time.
- Inspect air filters weekly and replace or clean them as necessary. Replacement filters should meet original specifications.
- Clean dampers and all ducting to avoid airflow restrictions. Lubricate movable and mechanical linkages in dampers and other devices as recommended. Check actuating solenoids and electromechanical components for proper operation. Movement of air throughout the transmitter causes static electrical charges to develop. Static charges can result in a buildup of dust and dirt in ductwork, dampers, and other components of the system. Filters should remove the dust before it gets into the system, but no filter traps every dust particle.
- Check thermal sensors and temperature system control devices for proper operation.

Air Cooling System Design

Transmitter cooling system performance is not necessarily related to airflow volume. The cooling capability of air is a function of its mass, not its volume. The designer must determine an appropriate airflow rate within the equipment and establish the resulting resistance to air movement. A specified static pressure that should be present within the ducting of the transmitter can be a measure of airflow. For any given combination of ducting, filters, heat sinks, RFI honeycomb shielding, tubes, tube sockets, and other elements in the transmitter, a specified system resistance to airflow can be determined. It is important to realize that any changes in the position or number of restricting elements within the system will change the system resistance, and therefore the effectiveness of the cooling. The altitude of operation is also a consideration in cooling system design. As altitude increases, the density (and cooling capability) of air decreases. A calculated increase in airflow is required to maintain the cooling effectiveness that the system was designed to achieve.

Transmitter room cooling requirements vary considerably from one location to another, but some general statements on cooling apply to all installations:

- A transmitter with a power output greater than 1 kW must have its exhaust ducted to the outside whenever the outside temperature is greater than 50°F.
- Transmitter buildings must be equipped with refrigerated air-conditioning units when the outside temperature is greater than 80°F. The exact amount of cooling capacity needed is subject to a variety of factors, such as actual transmitter efficiency, thermal insulation of the building itself, and size of the transmitter room.
- Radio transmitters up to and including 5 kW usually can be cooled (if the exhaust is efficiently ducted outside) by a 10,000 BTU air conditioner; 10-kW installations will require a minimum of 17,500 BTU of air conditioning; and 20-kW plants need at least 25,000 BTU of air conditioning. For larger radio installations or TV systems, consult an air-conditioning expert.
- Figure 18.28 shows a typical 20-kW FM transmitter plant. The building is oriented so that the cooling activity of the blowers is aided by normal wind currents during the summer months. Air brought in from the outside for cooling is well-filtered in a hooded air intake assembly that holds several filter panels. The building includes two air conditioners, one 15,000 BTU and the other 10,000 BTU. The thermostat for the smaller unit is set for slightly greater sensitivity than the larger air conditioner, allowing small temperature increases to be handled more economically.
- It is important to keep the transmitter room warm during the winter, as well as cool during the summer. Install heaters and PA exhaust recycling blowers as needed. A transmitter that runs 24 hours a day normally will not need additional heating equipment, but stations that sign off for several hours during the night should be equipped with electric room heaters (baseboard types, for example) to keep the room temperature above 50°F. PA exhaust recycling can be accomplished using a thermostat, relay logic circuit, and solenoid-operated register or electric blower. By controlling the room temperature to between 60°F and 70°F, tube and component life will be improved substantially.

Layout Considerations

The layout of a transmitter room HVAC (heating, ventilation, and air-conditioning) system can have a significant impact on the life of the PA tube(s) and the ultimate reliability of the transmitter. Air intake and output ports must be designed with care to avoid airflow restrictions and back-pressure problems. This process, however, is not as easy as it may seem. The science of airflow is complex and generally requires the advice of a qualified HVAC consultant.

To help illustrate the importance of proper cooling system design and the real-world problems that some facilities have experienced, consider the following examples taken from actual case histories:

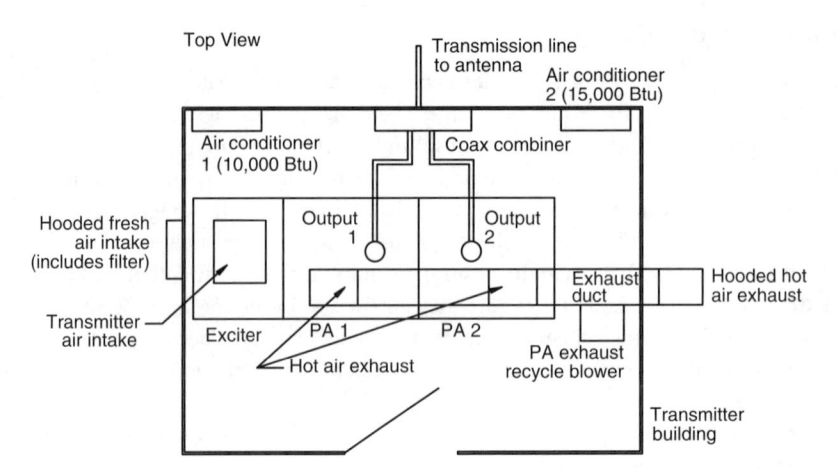

FIGURE 18.28 A typical heating and cooling arrangement for a 20-kW FM transmitter installation. Ducting of PA exhaust air should be arranged so that it offers minimum resistance to airflow.

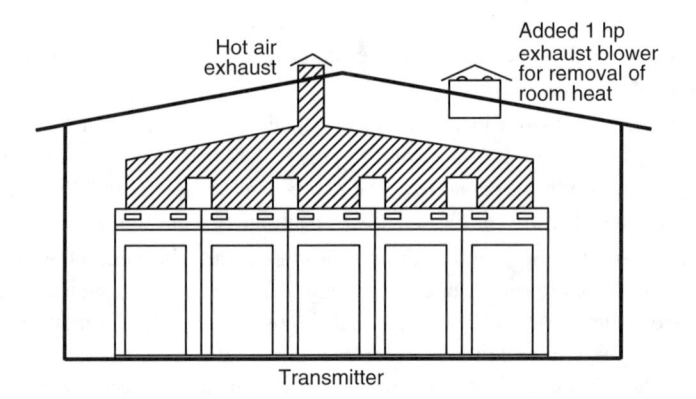

FIGURE 18.29 Case study in which excessive summertime heating was eliminated through the addition of a 1-hp exhaust blower to the building.

Case 1.
A fully automatic building ventilation system (Fig. 18.29) was installed to maintain room temperature at 20°C during the fall, winter, and spring. During the summer, however, ambient room temperature would increase to as much as 60°C. A field survey showed that the only building exhaust route was through the transmitter. Therefore, air entering the room was heated by test equipment, people, solar radiation on the building, and radiation from the transmitter itself before entering the transmitter. The problem was solved through the addition of an exhaust fan (3000 CFM). The 1-hp fan lowered room temperature by 20°C.

Case 2.
A simple remote installation was constructed with a heat recirculating feature for the winter (Fig. 18.30). Outside supply air was drawn by the transmitter cooling system blowers through a bank of air filters and hot air was exhausted through the roof. A small blower and damper were installed near the roof exit point. The damper allowed hot exhaust air to blow back into the room through a tee duct during winter months. For summer operation, the roof damper was switched open and the room damper closed. For winter operation, the arrangement was reversed. The facility, however, experienced short tube life during winter operation, although the ambient room temperature during winter was not excessive.

The solution involved moving the roof damper 12 ft down to just above the tee. This eliminated the stagnant "air cushion" above the bottom heating duct damper and significantly improved air flow in the region. Cavity back-pressure was, therefore, reduced. With this relatively simple modification, the problem of short tube life disappeared.

Case 3.

An inconsistency regarding test data was discovered within a transmitter manufacturer's plant. Units tested in the engineering lab typically ran cooler than those at the manufacturing test facility. Fig. 18.31 shows the test station difference, a 4-ft exhaust stack that was used in the engineering lab. The addition of the stack increased airflow by up to 20% because of reduced air turbulence at the output port, resulting in a 20°C decrease in tube temperature.

These examples point out how easily a cooling problem can be caused during HVAC system design. All power delivered to the transmitter is either converted to RF energy and sent to the antenna, or becomes heated air. Proper design of a cooling system, therefore, is a part of transmitter installation that should not be taken lightly.

Air Filters

Once the transmitter is clean, keeping it that way for long periods of time may require improving the air filtering system. Most filters are inadequate to keep out very small dirt particles (microdust), which can become a serious problem in an unusually dirty environment. Microdust can also become a problem in a relatively clean environment after a number of years of operation. In addition to providing a well-filtered air intake port for the transmitter building, an additional air filter can be placed in front of the normal transmitter filter assembly. A computer system filter panel can be secured to the air intake port to provide additional protection. With the extra filter in place, it is generally necessary only to replace or clean the outer filter panel. The transmitter's integral filter assembly will stay clean, eliminating the work and problems associated with pulling the filter out while the transmitter is operating. Be certain that the addition of supplemental filtering does not restrict airflow into the transmitter.

Klystron Water Cooling Systems

The cooling system is vital to any transmitter. In a UHF unit, the cooling system may dissipate as much as 70% of the input ac power in the form of waste heat in the klystron collector. For vapor phase-cooled klystrons, pure (distilled or demineralized) water must be used. Because the collector is only several volts above ground potential, it is not necessary to use deionized water. The collector and its water jacket act like a distillery. Any impurities in the water will eventually find their way into the water jacket and cause corrosion of the collector. It is essential to use high-purity water with low conductivity, less than 10 mS/cm (millisiemens per centimeter), and to replace the water in the cooling jacket as needed. Efficient heat transfer from the collector surface into the water is essential for long klystron life. Oil, grease, soldering flux residue, and pipe sealant containing silicone compounds must be excluded from the cooling system. This applies to both vapor- and liquid-conduction cooling systems, although it is usually more critical in the vapor-phase type. The sight glass in a vapor-phase water jacket provides a convenient checkpoint for coolant condition. Look for unusual residues, oil on the surface, foaming, and discoloration. If any of these appear, contact the manufacturer for advice on how to flush the system.

Water quality is essential to proper operation of a liquid-cooled klystron. In general, greater flows and greater pressures are inherent in liquid-cooled-versus-vapor-phase systems; and when a leak occurs, large quantities of coolant can be lost before the problem is discovered. Inspect the condition of gaskets, seals, and fittings regularly. Most liquid-cooled klystrons use a distilled water/ethylene glycol mixture. Do not exceed a 50/50 mix by volume. The heat transfer of the mixture is lower than that of pure water, requiring the flow to be increased, typically by 20 to 25%. Greater coolant flow means higher pressure and suggests close observation of the cooling system after adding the glycol. Allow the system to heat and cool several times. Then check all plumbing fittings for tightness.

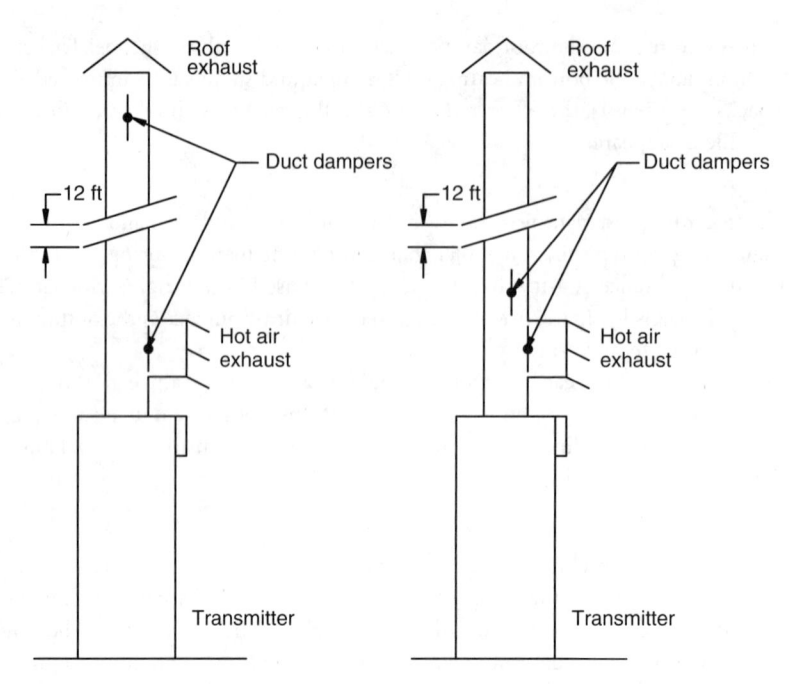

FIGURE 18.30 Case study in which there was excessive back-pressure to the PA cavity during winter periods, when the rooftop damper was closed. The problem was eliminated by repositioning the damper as shown.

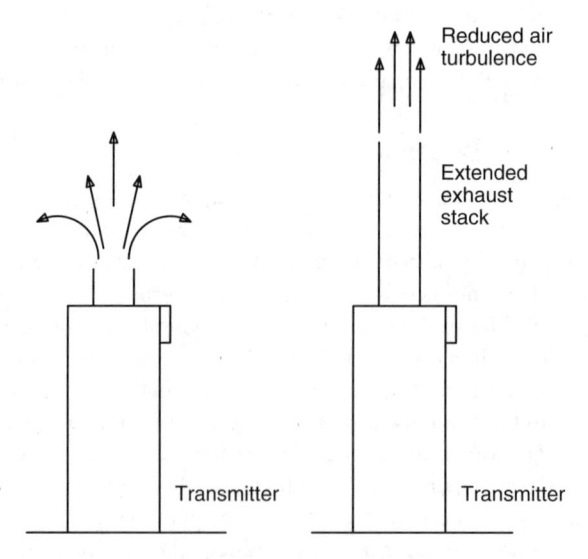

FIGURE 18.31 Case study in which air turbulence at the exhaust duct resulted in reduced airflow through the PA compartment. The problem was eliminated by adding a 4-ft extension to the output duct.

The action of heat and air on ethylene glycol causes the formation of acidic products. The acidity of the coolant can be checked with litmus paper. Buffers can and should be added with the glycol mixture. Buffers are alkaline salts that neutralize acid forms and prevent corrosion. Because they are ionizable chemical salts, the buffers cause the conductivity of the coolant to increase. Measure the collector-to-ground resistance periodically. Coolant conductivity is acceptable if the resistance caused by the coolant is greater than 20 times the resistance of the body-metering circuitry.

Experience has shown that the only practical way to ensure good coolant condition is to drain, flush, and recharge the system every spring. The equipment manufacturer can provide advice on how this procedure should be carried out and can recommend types of glycol to use. Maintain unrestricted airflow over the heat exchanger coils and follow the manufacturers' instructions on pump and motor maintenance.

Bibliography

Power Grid Tubes for Radio Broadcasting, Thomson-CSF publication #DTE-115, Thomson-CSF, Dover, NJ, 1986.

The Care and Feeding of Power Grid Tubes, Varian EIMAC, San Carlos, CA, 1984.

High Power Transmitting Tubes for Broadcasting and Research, Phillips technical publication, Eindhoven, the Netherlands, 1988.

Gray, T. S., *Applied Electronics*, Massachusetts Institute of Technology, 1954.

Svet, Frank A., Factors affecting on-air reliability of solid state transmitters, *Proceedings of the SBE Broadcast Engineering Conference*, Society of Broadcast Engineers, Indianapolis, IN, October 1989.

19
Troubleshooting RF Equipment

Jerry C. Whitaker
Editor

19.1 Introduction **19**-1
   Troubleshooting Procedure
19.2 Plate Overload Fault **19**-2
   Troubleshooting Procedure · Process of Elimination
19.3 RF System Faults **19**-5
   Troubleshooting Procedure · Component Substitution · Inside the PA Cavity
19.4 Power Control Faults **19**-11
   Thyristor Control System · Interlock Failures · Step-Start Faults · Protection Circuits

19.1 Introduction

Problems will occur from time to time with any piece of equipment. The best way to prepare for a transmitter failure is to know the equipment well. Study the transmitter design and layout. Know the schematic diagram and what each component does. Examine the history of the transmitter by reviewing old maintenance logs to see what components have failed in the past.

Troubleshooting Procedure

When a problem occurs, the first task is to keep the transmitter on the air. If a standby transmitter is available, the solution is obvious. If the facility does not have a standby, quick thinking will be needed to minimize downtime and keep the unit running until repairs can be made. Most transmitters have sufficient protective devices so that it is impossible to operate them with serious problems. If the transmitter will not stay on the air on normal power, try lowering the power output and see if the trip-offs are eliminated. Failing this, many transmitters have driver outputs that can be connected to the antenna on a temporary basis, thereby bypassing the final amplifiers — provided, of course, the failure is in one of the PA stages. Do not allow the transmitter to operate at any power level if the meter readings are out of tolerance. Serious damage can result to the system.

When presented with a problem, proceed in an orderly manner to trace it. Many failures are simple to repair if you stop and think about what is happening. Examine the last set of transmitter readings and make a complete list of meter readings in the failure mode. Note which overload lamps are lit and what other indicators are in an alarm state. With this information assembled, the cause of the failure can often be identified. Looking over the available data and the schematic diagram for 10 to 15 minutes can save hours of trial-and-error troubleshooting.

When checking inside the unit, look for changes in the physical appearance of components in the problem area. An overheated resistor or leaky capacitor may be the cause of the problem, or may point to the cause. Devices never fail without a reason. Try and piece together the sequence of events that led to the problem. Then, the cause of the failure — not just the more obvious symptoms — can be corrected. When working with direct-coupled transistors, a failure in one device will often cause a failure in another, so check all semiconductors associated with one found to be defective.

In high-power transmitters, look for signs of arcing in the RF compartments. Loose connections and clamps can cause failures that are difficult to locate. Never rush through a troubleshooting job. A thorough knowledge of the theory of operation and history of the transmitter is a great aid in locating problems in the RF sections. Do not overlook the possibility of tube failure when troubleshooting a transmitter. Tubes can fail in unusual ways; substitution may be the only practical test for power tubes used in modern transmitters.

Study the control ladder of the transmitter to identify interlock or fail-safe system problems. Most newer transmitters have troubleshooting aids built in to help locate problems in the control ladder. Older transmitters, however, often require a moderate amount of investigation before repairs can be accomplished.

The "Quick Fix"

There is no such thing as a "quick fix" when it comes to transmission equipment. Think out any problem and allow ample time for repair. It makes little sense to rush through a repair job just to get the system back on the air if another failure occurs immediately after the technician walks out the door. Careful analysis of the cause and effects of the failure will ensure that the original problem is solved, not just its obvious symptoms. If temporary repairs must be made to return the transmitter to a serviceable condition, make them and then finish the job as soon as the needed replacement parts are available.

Factory Service Assistance

Factory service engineers are available to aid in troubleshooting transmission equipment, but such services have their limits. No factory engineer can fix a transmitter over the phone. The factory can suggest areas of the system to investigate and relate the solutions to similar failure modes, but the facility engineer is the person who does the repair work. If the engineer knows the equipment, and has done a good job of analyzing the problem, the factory can help. When calling the factory service department, have the following basic items on hand:

- The type of transmitter and the exact failure mode. The service department will need to know the meter readings before and after the problem, and whether any unusual circumstances preceded the failure. For example, it would be important for the factory to know that the failure occurred after a brief power outage, or during an ice storm.
- A list of what has already been done in an effort to correct the problem. All too often the factory is called *before* any repair efforts are made. The service engineer will need to know what happens when the high voltage is applied, and what overloads may occur.
- A copy of the transmitter diagram and component layout drawings on hand. A thorough knowledge of the transmitter design and construction allows the maintenance engineer to intelligently converse with the factory service representative.

19.2 Plate Overload Fault

Of all the problems that can occur in a transmitter, probably the best known — and most feared — is the plate supply overload. Occasional plate trip-offs (one or two a month) are not generally cause for concern. Most of these occurrences can be attributed to power line transients. More frequent trip-offs require closer inspection of the transmission system. For the purposes of this discussion, we assume that the plate supply overload occurs frequently enough to make continued operation of the transmitter difficult.

Troubleshooting Procedure

The first step in any transmitter troubleshooting procedure is to switch the system to *local control* so that the maintenance technician, not an off-site operator, has control over the unit. This is important for safety reasons. Next, switch off the transmitter automatic recycle circuit. While troubleshooting, the transmitter should not be allowed to cycle through an overload any more times than absolutely necessary. Such action only increases the possibility of additional component damage. Use a logical, methodical approach to finding the problem. The following procedure is recommended:

- Determine the fault condition. When the maintenance engineer arrives at the transmitter site, the unit will probably be down. The carrier will be off but the filaments will still be lit. Check all multimeter readings on the transmitter and exciter. If they indicate a problem in a low-voltage stage, troubleshoot that failure before bringing the high voltage up.
- Assuming that all low-voltage systems are operating normally, switch off the filaments and make a quick visual check inside the transmitter cabinet. Determine whether there is any obvious problem. Pay particular attention to the condition of power transformers and high-voltage capacitors. Check for signs of arcing in the PA compartment. Look on the floor of the transmitter and in the RF compartments to see if there are any pieces of components laying around. Sniff inside the cabinet for hints of smoke. Check the circuit breakers and fuses to see what failures might be indicated.
- After running through these preliminary steps, restart the filaments. Then bring up the high voltage. Watch the front panel meters to see how they react. Observe what happens and listen for any sound of arcing. If the transmitter will come up, quickly run through the PA and IPA meter readings. Check the VSWR meter for excessive reflected power.

Assuming that problems persist, determine whether the plate supply overload is RF or dc based. With the plate off, switch off the exciter. Bring up the high voltage (plate supply). If the overload problem remains, the failure is based in the dc high-voltage power supply. If the problem disappears, the failure is centered in the transmitter RF chain. Proper bias must be present on all vacuum tube stages of the RF system when this test is performed. The PA tube bias supply is usually switched on with the filaments, and can generally be read from the front panel of the transmitter. Confirm proper bias before applying high voltage with no excitation. It is also important that the exciter is switched off while the high voltage is off. Removing excitation from a transmitter while it is on the air can result in a large transient overvoltage, which can cause arcing or component damage.

If the overload is based in the high-voltage dc power supply, shut down the transmitter and check the schematic diagram for the location in the circuit of the plate overload sensor relay (or comparator circuit). This will indicate within what limits component checking will be required. The plate overload sensor is usually found in one of two locations: the PA cathode dc return or the high-voltage power supply negative connection to ground. Transmitters using a cathode overload sensor generally have a separate high-voltage dc overload sensor in the plate power supply.

A sensor in the cathode circuit will substantially reduce the area of component checking required. A plate overload with no excitation in such an arrangement would almost certainly indicate a PA tube failure, because of either an interelectrode short-circuit or a loss of vacuum. Do not operate the transmitter when the PA tube is out of its socket. This is not an acceptable method of determining whether a problem exits with the PA tube. Instead, substitute a spare tube. Operating a transmitter with the PA tube removed can result in damage to other tubes in the transmitter when the filaments are on, and damage to the driver tubes and driver output/PA input circuit components when the high voltage is on.

If circuit analysis indicates a problem in the high-voltage power supply itself, use an ohmmeter to check for short circuits. Remove all power from the transmitter and discharge all filter capacitors before beginning any troubleshooting work inside the unit. When checking for short circuits with an ohmmeter, take into account the effects that bleeder resistors and high-voltage meter multiplier assemblies can have on resistance readings. Most access panels on transmitters use an interlock system that will remove the high voltage and ground the high-voltage supplies when a panel is removed. For the purposes of ohm-meter tests, these interlocks may have to be temporarily defeated. Never defeat the interlocks unless all ac power has been removed from the transmitter and all filter capacitors have been discharged using the grounding stick supplied with the transmitter.

Following the preliminary ohmmeter tests, check the following components in the dc plate supply:

- Oil-filled capacitors for signs of overheating or leakage
- Feed-through capacitors for signs of arcing or other damage
- The dc plate blocking capacitor for indications of insulation breakdown or arcing
- All transformers and chokes for signs of overheating or winding failure
- Transient suppression devices for indications of overheating or failure
- Bleeder resistors for signs of overheating
- Any surge-limiting resistors placed in series with filter capacitors in the power supply for indications of overheating or failure (a series resistor that shows signs of overheating can be an indication that the associated filter capacitor has failed)

If the plate overload trip-off occurs only at elevated voltage levels, ohmmeter checks will not reveal the cause of the problem. It may therefore be necessary to troubleshoot the problem using the process of elimination.

Process of Elimination

Troubleshooting through the process of elimination involves isolating various portions of the circuit — one section at a time — until the defective component is found. Special precautions are required before performing such work, including:

- Never touch anything inside the transmitter without first removing all ac power and then discharging all filter capacitors with the grounding stick.
- Never perform troubleshooting work alone; another person should be present.
- Whenever a wire is disconnected, temporarily wrap it with electrical tape and secure the connector so it will not arc over to ground or another component when power is applied.
- Analyze each planned test before it is conducted. Every test in the troubleshooting process requires time, and so steps should be arranged to provide the greatest amount of information about the problem.
- Check with the transmitter manufacturer to find out what testing procedures the company recommends. Ask what special precautions should be taken.

Troubleshooting the high-voltage plate supply is usually done under the following conditions:

- Exciter off
- Plate and screen IPA voltages off
- PA screen voltage off

Individual transmitters may require different procedures. Check with the manufacturer first.

Figure 19.1 shows a typical transmitter high-voltage power supply. Begin the troubleshooting process by breaking the circuit at point *A*. If the overload condition persists, the failure is caused by a problem in the power supply itself, not in the PA compartment. If, on the other hand, the overload condition disappears, a failure in the feedthrough capacitor (C1), decoupling capacitors (C2, C3), or blocking capacitor (C4) is indicated.

If a problem is indicated in the PA compartment, reconnect the high-voltage supply line at point *A* and break the circuit at point *B*. A return of the overload problem would indicate a failure in one of the decoupling capacitors or the feedthrough capacitor.

Troubleshooting RF Equipment

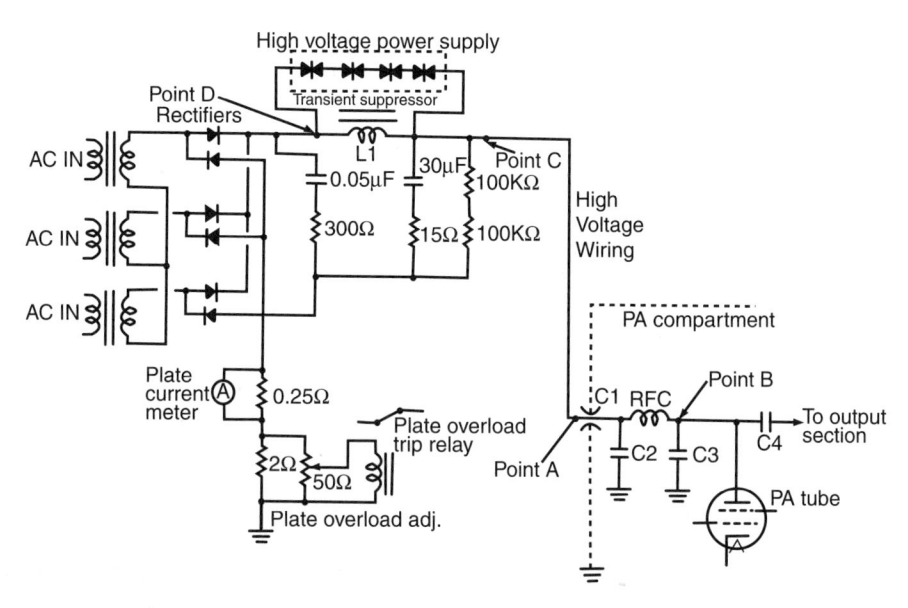

FIGURE 19.1 A typical transmitter high-voltage, three-phase power supply circuit.

To avoid unnecessary effort and time in troubleshooting, use the process of elimination to identify sections of the circuit to be examined. If, for example, the test at point A had indicated the problem was not in the load but in the power supply, a logical spot to perform the next test would be at point C (for long high-voltage cable runs). This test would identify or eliminate the interconnecting cable as a cause of the fault condition. If the cable run from the high-voltage supply to the PA compartment is short, point D might be the next logical point to check. Breaking the connection at the input to the power supply filter allows the rectifiers and interconnecting cables to be checked. Components protected by transient suppression devices (L1 as shown above) should be considered a part of the component they are designed to protect. If a choke is removed from the circuit for testing, its protective device must also be removed. Failure to remove both connections will usually result in damage to the protective device.

To avoid creating a new problem while trying to correct the original failure, break the circuit in only one point at a time. Also study the possible adverse effects of each planned step in the process. Disconnecting certain components from the circuit may cause overvoltages or power supply ripple that can damage other components in the transmitter. Consult the manufacturer to be sure.

Perform any troubleshooting work on a transmitter with extreme care. Transmitter high voltages can be lethal. Work inside the transmitter only after all ac power has been removed and after all capacitors have been discharged using the grounding stick provided with the transmitter. Remove primary power from the unit by tripping the appropriate power distribution circuit breakers in the transmitter building. Do not rely on internal contractors or SCRs to remove all dangerous ac. Do not defeat protective interlock circuits. Although defeating an access panel interlock switch may save work time, the consequences can be tragic.

19.3 RF System Faults

Although RF troubleshooting might seem intimidating, there is no secret to it. Patient examination of the circuit and careful study of the schematic diagram will go a long way toward locating the problem. The first step in troubleshooting an RF problem is to determine whether the fault is RF based or dc based.

Troubleshooting Procedure

Check the load by examining the transmitter overload indicators. Most transmitters monitor reflected power from the antenna and will trip-off if excessive VSWR is detected. If the VSWR fault indicator is

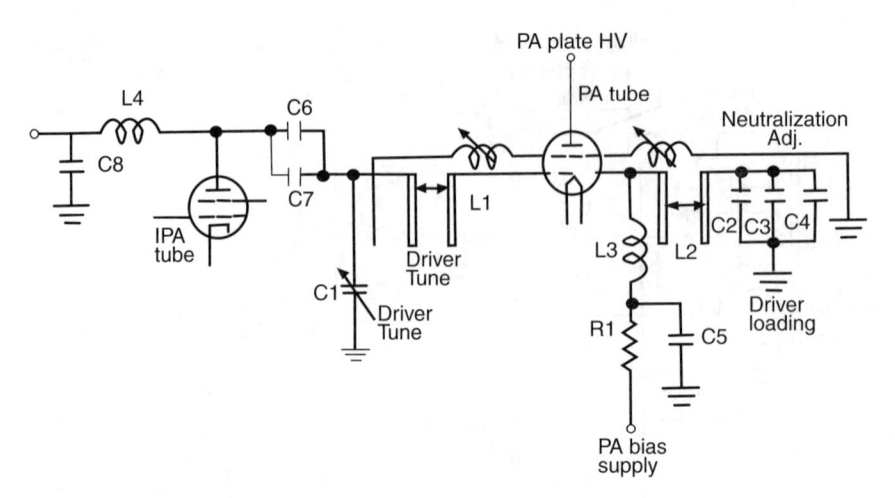

FIGURE 19.2 The PA grid input circuit of a grounded screen transmitter.

not lit, the load is not likely the cause of the problem. A definitive check of the load can be made by switching the transmitter output to a dummy load and bringing up the high voltage. The PA tube can be checked by substituting one of known quality. When the tube is changed, carefully inspect the contact fingerstock for signs of overheating or arcing. Be careful to protect the socket from damage when removing and inserting the PA tube. Do not change the tube unless there is good reason to believe that it may be defective.

If problems with the PA stage persist, examine the grid circuit of the tube. Figure 19.2 shows the input stage of a grounded screen FM transmitter. A short circuit in any of the capacitors in the grid circuit (C1–C5) will effectively ground the PA grid. This will cause a dramatic increase in plate current because the PA bias supply will be shorted to ground along with the RF signal from the IPA stage.

The process of finding a defective capacitor in the grid circuit begins with a visual inspection of the suspected components. Look for signs of discoloration because of overheating, loose connections, and evidence of package rupture. The voltage and current levels found in a transmitter PA stage are often sufficient to rupture a capacitor if an internal short circuit occurs. Check for component overheating right after shutting down the transmitter. (As mentioned previously, remove all ac power and discharge all capacitors first.) A defective capacitor will often overheat. Such heating can also occur, however, because of improper tuning of the PA or IPA stage, or a defective component elsewhere in the circuit.

Before replacing any components, study the transmitter schematic diagram to determine which parts in the circuit could cause the failure condition that exists. By knowing how the transmitter works, many hours can be saved in checking components that an examination of the fault condition and the transmitter design would show to be an unlikely cause of the problem.

Check blocking capacitors C6 and C7. A breakdown in either component would have serious consequences. The PA tube would be driven into full conduction, and could arc internally. The working voltages of capacitors C1–C5 could also be exceeded, damaging one or more of the components. Because most of the wiring in the grid circuit of a PA stage consists of wide metal straps (required because of the skin effect), it is not possible to view stress points in the circuit to narrow the scope of the troubleshooting work. Areas of the system that are interconnected using components that have low power dissipation capabilities, however, should be closely examined. For example, the grid bias decoupling components shown in Fig. 19.2 (R1, L3, and C5) include a low-wattage (2 W) resistor and a small RF choke. Because of the limited power dissipation ability of these two devices, a failure in decoupling capacitor C5 would likely cause R1 (and possibly L3) to burn out. The failure of C5 in a short circuit would pull the PA grid to near ground potential, causing the plate current to increase and trip off the transmitter high voltage. Depending on the sensitivity and speed of the plate overload sensor, L3 could be damaged or destroyed by the increased current it would carry to C5, and therefore to ground.

Troubleshooting RF Equipment

If L3 were able to survive the surge currents that resulted in PA plate overload, the choke would continue to keep the plate supply off until C5 was replaced. Bias supply resistor R1, however, would likely burn out because the bias power supply is generally switched on with the transmitter filament supply. Therefore, unless the PA bias power supply line fuse opened, R1 would overheat and probably fail.

Because of the close spacing of components in the input circuit of a PA stage, carefully check for signs of arcing between components or sections of the tube socket. Keep all components and the socket itself clean at all times. Inspect all interconnecting wiring for signs of damage, arcing to ground, or loose connections.

Component Substitution

Substituting a new component for a suspected part can save valuable time when troubleshooting. With some components, it is cost-effective to replace a group of parts that may include one defective component because of the time involved in gaining access to the damaged device. For example, the grid circuit of the PA stage shown in Fig. 19.2 includes three *doorknob* capacitors (C2–C4) formed into a single assembly. If one device was found to be defective, it might be advantageous to simply replace all three capacitors. These types of components are often integrated into a single unit that may be difficult to reach. Because doorknob capacitors are relatively inexpensive, it would probably be best to replace the three as a group. This way, the entire assembly is eliminated as a potential cause of the fault condition.

A good supply of spare parts is a valuable troubleshooting tool. In high-power transmitting equipment, substitution is sometimes the only practical means of finding the cause of a problem. The manufacturer's factory service department can usually recommend a minimum spare parts inventory. Obvious candidates for inventory include components that are not available locally, such as high-voltage fixed-value capacitors, vacuum variable capacitors, and specialized semiconductors.

Inside the PA Cavity

One of the things that makes troubleshooting a cavity-type power amplifier difficult is the nature of the major component elements. The capacitors do not necessarily look like capacitors, and the inductors do not necessarily look like inductors. It is often difficult to relate the electrical schematic diagram to the mechanical assembly that exists within the transmitter output stage. At VHF and UHF frequencies — the domain of cavity PA designs — inductors and capacitors can take on unusual mechanical forms.

Consider the PA cavity schematic diagram shown in Fig. 19.3. The grounded-screen stage is of conventional design. Decoupling of the high-voltage power supply is accomplished by C1, C2, C3, and L1. Capacitor C3 is located inside the PA chimney (cavity inner conductor). The RF sample lines provide two low-power RF outputs for a modulation monitor or other test instruments. Neutralization inductors L3 and L4 consist of adjustable grounding bars on the screen grid ring assembly. The combination of L2 and C6 prevents spurious oscillations within the cavity. Figure 19.4 shows the electrical equivalent of the PA cavity schematic diagram. The $^1/_4$-wavelength cavity acts as the resonant tank for the PA. Coarse tuning of the cavity is accomplished by adjustment of the shorting plane. Fine tuning is performed by the PA tuning control, which acts as a variable capacitor to bring the cavity into resonance. The PA loading control consists of a variable capacitor that matches the cavity to the load. There is one value of plate loading that will yield optimum output power, efficiency, and PA tube dissipation. This value is dictated by the cavity design and values of the various dc and RF voltages and currents supplied to the stage.

The logic of a PA stage often disappears when the maintenance engineer is confronted with the actual physical design of the system. As illustrated in Fig. 19.5, many of the components take on an unfamiliar form. Blocking capacitor C4 is constructed of a roll of Kapton insulating material sandwiched between two circular sections of aluminum. (Kapton is a registered trademark of DuPont.) PA plate tuning control C5 consists of an aluminum plate of large surface area that can be moved in or out of the cavity to reach resonance. PA loading control C7 is constructed much the same as the PA tuning assembly, with a large-area paddle feeding the harmonic filter, located external to the cavity. The loading paddle can be moved

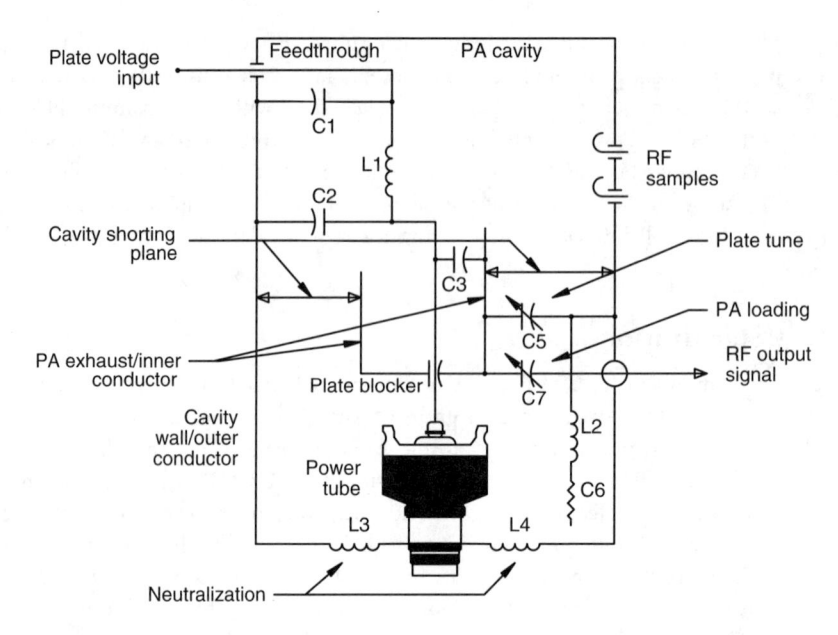

FIGURE 19.3 An FM transmitter PA output stage built around a 1/4-wavelength cavity with capacitive coupling to the load.

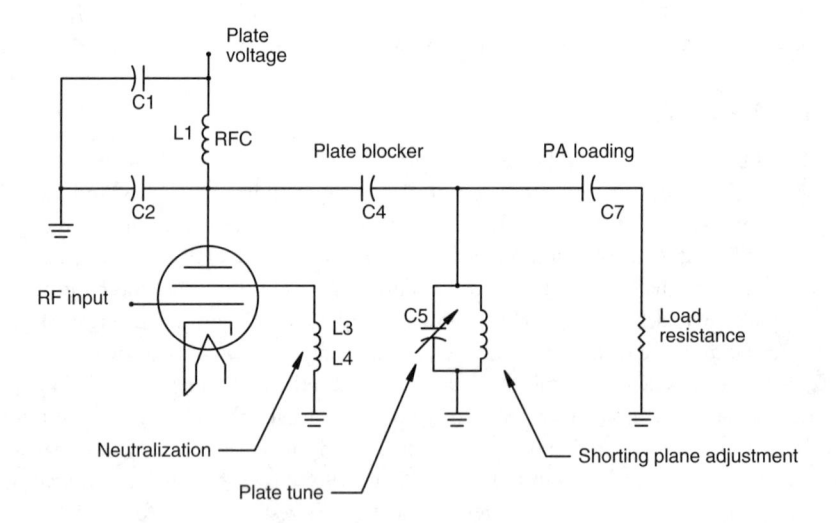

FIGURE 19.4 The equivalent electrical circuit of the PA stage shown in Fig. 19.3.

toward the PA tube or away from it to achieve the required loading. The L2–C6 damper assembly actually consists of a 50-Ω noninductive resistor mounted on the side of the cavity wall. Component L2 is formed by the inductance of the connecting strap between the plate tuning paddle and the resistor. Component C6 is the equivalent stray capacitance between the resistor and the surrounding cavity box.

From this example it can be seen that many of the troubleshooting techniques that work well with low-frequency RF and dc do not necessarily apply in cavity stages. It is therefore critically important to understand how the system operates and what each component does. Because many of the cavity components (primarily inductors and capacitors) are mechanical elements more than electrical ones, troubleshooting a cavity stage generally focuses on checking the mechanical integrity of the box.

Most failures resulting from problems within a cavity are the result of poor mechanical connections. All screws and connections must be kept tight. Every nut and bolt in a PA cavity was included for a

Troubleshooting RF Equipment 19-9

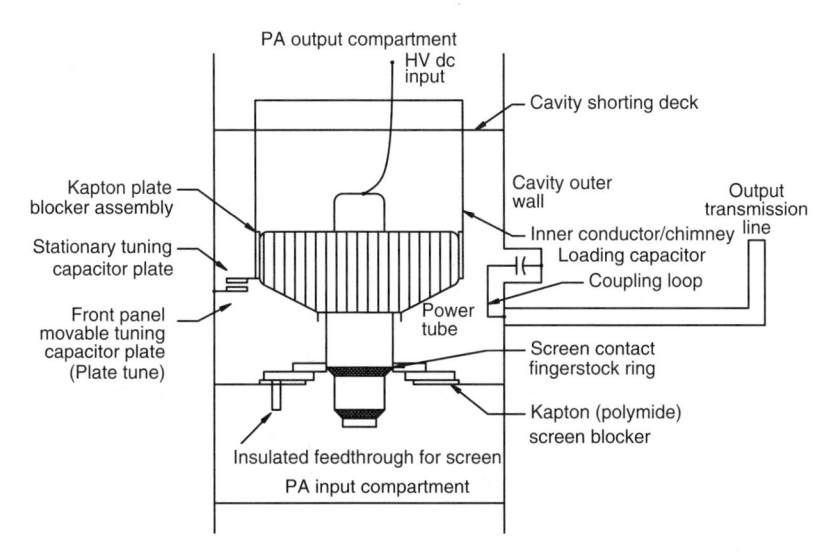

FIGURE 19.5 The mechanical equivalent of the PA stage shown in Fig. 19.3.

reason. There are no insignificant screws that do not need to be tight. However, do not over tighten either. Stripped threads and broken component connection lugs will only cause additional grief.

When a problem occurs in a PA cavity, it is usually difficult to determine which individual element (neutralization inductor, plate tuning capacitor, loading capacitor, etc.) is defective from the symptoms the failure will display. A fault within the cavity is usually a catastrophic event that will take the transmitter off the air. It is often impossible to bring the transmitter up for even a few seconds to assess the fault situation. The only way to get at the problem is to shut the transmitter down and take a look inside.

Closely inspect every connection, using a trouble light and magnifying glass. Look for signs of arcing or discoloration of components or metal connections. Check the mechanical integrity of each element in the circuit. Be certain the tuning and loading adjustments are solid, and without excessive mechanical play. Look for signs of change in the cavity. Check areas of the cavity that may not seem like vital parts of the output stage, such as the maintenance access door fingerstock and screws. Any failure in the integrity of the cavity, whether at the base of the PA tube or on part of the access door, will cause high circulating currents to develop and may prevent proper operation of the stage. If a problem is found that involves damaged fingerstock, replace the affected sections. Failure to do so will likely result in future problems because of the currents that can develop at any discontinuity in the cavity inner or outer conductor.

VSWR Overload

A VSWR overload in transmission equipment can result from a number of different problems. The following checklist presents some common problems and solutions:

1. VSWR overloads are usually caused by an improper impedance match external to the transmitter. The first step in the troubleshooting procedure is to substitute a dummy load for the entire antenna and transmission line system. Connect the dummy load at the transmitter output port, thereby eliminating all coax, external filters, and other RF hardware that might be present in the system.
2. If the VSWR trip fault is eliminated in step 1, the problem lies somewhere in the transmission line or antenna. The dummy load can next be moved to the point at which the transmission line leaves the building and heads for the tower (if different than the point checked in step 1). This test will check any RF plumbing, switches, or filter assemblies. If the VSWR overload condition is still absent, the problem is centered in the transmission line or the antenna.
3. If a standby antenna is not available, operation may still be possible at reduced power on a temporary basis. For example, if arcing occurs in the antenna or line at full power, emergency

operation may be possible at half power. Inspect the antenna and line for signs of trouble. Repair work beyond this point normally requires specialized equipment and a tower crew. This discussion assumes that the problem is not caused by ice buildup on the antenna, which can be alleviated by reducing the transmitter power output until VSWR trips do not occur.

4. If step 1 shows the VSWR overload to be internal to the transmitter, determine whether the problem is caused by an actual VSWR overload or by a failure in the VSWR control circuitry. Check for this condition by disabling the transmitter exciter and bringing up the high voltage. Under these conditions, RF energy will not be generated. (It is assumed that the transmitter has proper bias on all stages and is properly neutralized.) If a VSWR overload is indicated, the problem is centered in the VSWR control circuitry and not the RF chain. Possible explanations for control circuitry failure include loose connections, dirty switch contacts, dirty calibration potentiometers, poor PC board edge connector contacts, defective IC amplifiers or logic gates, and intermittent electrolytic capacitors.

5. If step 4 shows that the VSWR overload is real, and not the result of faulty control circuitry, check all connections in the output and coupling sections of the final stage. Look for signs of arcing or loose hardware, particularly on any movable tuning components. Inspect high-voltage capacitors for signs of overheating, which might indicate failure; and check coils for signs of dust buildup, which might cause a flash-over. In some transmitters, VSWR overloads can be caused by improper final-stage tuning or loading. Consult the equipment instruction book for this possibility. Also, certain transmitters include glass-enclosed spark-gap lightning protection devices (*gas-gaps*) that can be disconnected for testing.

6. If VSWR overload conditions resulting from problems external to the transmitter are experienced at an AM radio station, check the following items:

    a. *Component dielectric breakdown*. If a normal (near zero) reflected power reading is indicated at the transmitter under carrier-only conditions, but VSWR overloads occur during modulation, component dielectric breakdown may be the problem. A voltage breakdown could be occurring within one of the capacitors or inductors at the antenna tuning unit (ATU) or phasor. Check all components for signs of damage. Clean insulators as required. Carefully check any open-air coils or transformers for dust buildup or loose connections.

    b. *Narrow-band antenna*. If the overload occurs with any modulating frequency, the probable cause of the fault is dielectric breakdown. If, on the other hand, the overload seems particularly sensitive to high-frequency modulation, then narrow antenna bandwidth is indicated. Note the action of the transmitter forward/reflected power meter. An upward deflection of reflected power with modulation is a symptom of limited antenna bandwidth. The greater the upward deflection, the more limited the bandwidth. If these symptoms are observed, conduct an antenna impedance sweep of the system.

    c. *Static buildup*. Tower static buildup is characterized by a gradual increase in reflected power as shown on the transmitter front panel. The static buildup, which usually occurs prior to or during thunderstorms and other bad weather conditions, continues until the tower base ball gaps arc-over and neutralize the charge. The reflected power reading then falls to zero. A static drain choke at the tower base to ground will generally prevent this problem.

    d. *Guy wire arc-over*. Static buildup on guy wires is similar to a nearby lightning strike in that no charge is registered during the buildup of potential on the reflected power meter. Instead, the static charge builds on the guys until it is of sufficient potential to arc across the insulators to the tower. The charge is then removed by the static drain choke and/or ball gaps at the base of the tower. Static buildup on guy wires can be prevented by placing RF chokes across the insulators, or by using non-metallic guys. Arcing across the insulators may also be reduced or eliminated by regular cleaning.

19.4 Power Control Faults

A failure in the transmitter ac power control circuitry can result in problems ranging from zero RF output to a fire inside the unit. Careful, logical troubleshooting of the control system is mandatory. Two basic types of primary ac control are used in transmitters today:

- Silicon controlled rectifier or SCR (thyristor) system
- Relay logic

Thyristor Control System

A failure in the thyristor power control system of a transmitter is not easy to overlook. In the worst case, no high voltage whatsoever will be produced by the transmitter. In the best case, power control may be erratic or uneven when using the continuously variable power adjustment mode. Understanding how the servo circuit works and how it is interconnected is the first step in correcting such a problem.

Figure 19.6 shows a block diagram of a typical thyristor control circuit. Three gating cards are used to drive back-to-back SCR pairs, which feed the high-voltage power transformer primary windings. Although the applied voltage is three-phase, the thyristor power control configuration simulates a single-phase design for each phase-to-phase leg. This allows implementation of a control circuit that basically consists of a single-phase gating card duplicated three times (one for each load phase). This approach has advantages from the standpoint of design simplicity, and also from the standpoint of field troubleshooting. In essence, each power control circuit is identical, allowing test voltages and waveforms from one gating card to be directly compared with a gating card experiencing problems.

If the high-voltage supply will not come up at all, the problem involves more than a failure in just one of the three gating cards. The failure of any one gating board would result in reduced power output (and other side effects), but not in zero output. Begin the search with the interlock system.

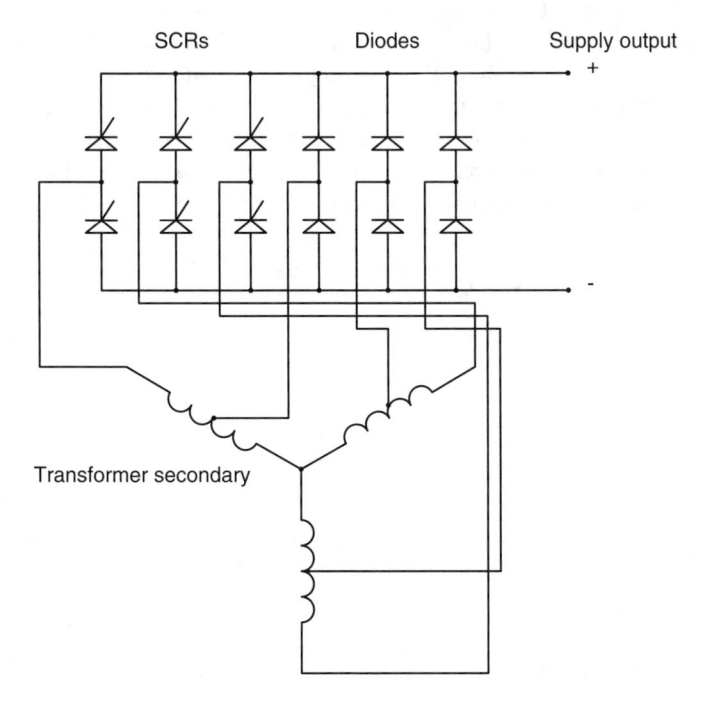

FIGURE 19.6 Block diagram of a three-phase thyristor power control system.

Interlock Failures

Newer transmitters provide the engineer with built-in diagnostic readouts on the status of the transmitter interlock circuits. These may involve discrete LEDs or a microcomputer-driven visual display of some type. If the transmitter has an advanced status display, the process of locating an interlock fault is relatively simple. For an older transmitter that is not so equipped, substantially more investigation will be needed.

Make a close observation of the status of all fuses, circuit breakers, transmitter cabinet doors, and access panels. Confirm that all doors are fully closed and secured. Switch the transmitter from *remote* to *local control* (if operated remotely) to eliminate the remote control system as a possible cause of the problem. Observe the status of all control-panel indicator lamps. Some transmitters include an *interlocks open* lamp; other units provide an indication of an open interlock through the *filament on* or *plate off* push-button lamps. These indicators can save valuable minutes or even hours of troubleshooting, so pay attention to them. Replace any burned-out indicator lamps as soon as they are found to be defective.

If the front-panel indicators point to an interlock problem, pull out the schematic diagram of the transmitter, get out the digital multimeter (DMM), and shut down the transmitter. If the transmitter interlock circuit operates from a low-voltage power supply, such as 24 V dc, use a voltmeter to check for the loss of continuity. Remove ac power from all sections of the transmitter except the low-voltage supply by tripping the appropriate front-panel circuit breakers. Be extremely careful when working on the transmitter to avoid any line voltage ac. If the layout of the transmitter does not permit safe troubleshooting with only the low-voltage power supply active, remove all ac from the unit by tripping the wall-mounted main breaker. Then use an ohmmeter to check for the loss of continuity.

If the transmitter interlock circuit operates from 120 V ac or 220 V ac, remove all power from the transmitter by tripping the wall-mounted main breaker. Use an ohmmeter to locate the problem. Many older transmitters use line voltages in the interlock system. Do not try to troubleshoot such transmitters with ac power applied.

Finding a problem such as an open control circuit interlock is basically a simple procedure, despite the time involved. Do not rush through such work. When searching for a break in the interlock system, use a methodical approach to solving the problem. Consider the circuit configuration shown in Fig. 19.7. The most logical approach to finding a break in the control ladder is to begin at the source of the 24 V dc input and, step by step, work toward the input of the power controller. Although this approach may be logical, it can also be time-consuming. Instead, eliminate stages of the interlock circuit. For example, make the first test at terminal *A*. A correct voltage reading at this point in the circuit will confirm that all of the interlock door switches are operating properly.

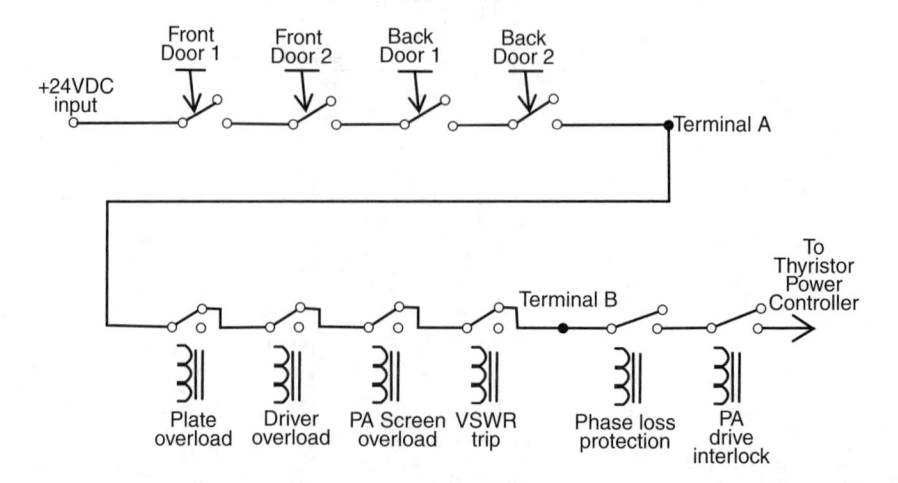

FIGURE 19.7 A typical transmitter interlock circuit. Terminals *A* and *B* are test points used for troubleshooting the system in the event of an interlock system failure.

Troubleshooting RF Equipment

With the knowledge that the problem is after terminal *A*, move on to terminal *B*. If the 24-V supply voltage disappears, check the fault circuit overload relays to find where the control signal is lost. Often, such interlock problems can be attributed to dirty contacts in one of the overload relays. If a problem is found with one set of relay contacts, clean all of the other contacts in the overload interlock string for good measure. Be sure to use the proper relay-contact cleaning tools. If sealed relays are used, do not attempt to clean them. Instead, replace the defective unit.

Step-Start Faults

The high-voltage power supply of any medium- or high-power transmitter must include provisions for in-rush current-limiting upon application of a *plate-on* command. The filter capacitor(s) in the power supply will appear as a virtual short circuit during a sudden increase in voltage from the rectifier stacks. To avoid excessive current surges through the rectifiers, capacitor(s), choke, and power transformer, nearly all transmitters use some form of *step-start* arrangement. Such circuits are designed to limit the in-rush current to a predictable level. This can be accomplished in a variety of ways.

For transmitters using a thyristor power control system, the step-start function can be easily designed into the gate firing control circuit. An R-C network at the input point of the gating cards is used to ramp the thyristor pairs from a zero conduction angle to full conduction (or a conduction angle preset by the user). This system provides an elegant solution to the step-start requirement, allowing plate voltage to be increased from zero to full value within a period of approximately 5 seconds.

Transmitters employing a conventional ac power control system usually incorporate a step-start circuit consisting of two sets of contactors: the *start contactor* and the *run contactor*, as illustrated in Fig. 19.8. Surge-limiting resistors provide sufficient voltage drop upon application of a *plate-on* command to limit the surge current to a safe level. Auxiliary contacts on the start contactor cause the run contactor to close as soon as the start contacts have seated.

A fault in the step-start circuit of a transmitter is often evidenced — initially at least — by random tripping of the plate supply circuit breaker at high-voltage turn-on. If left uncorrected, this condition can lead to problems such as failed power rectifiers or filter capacitors.

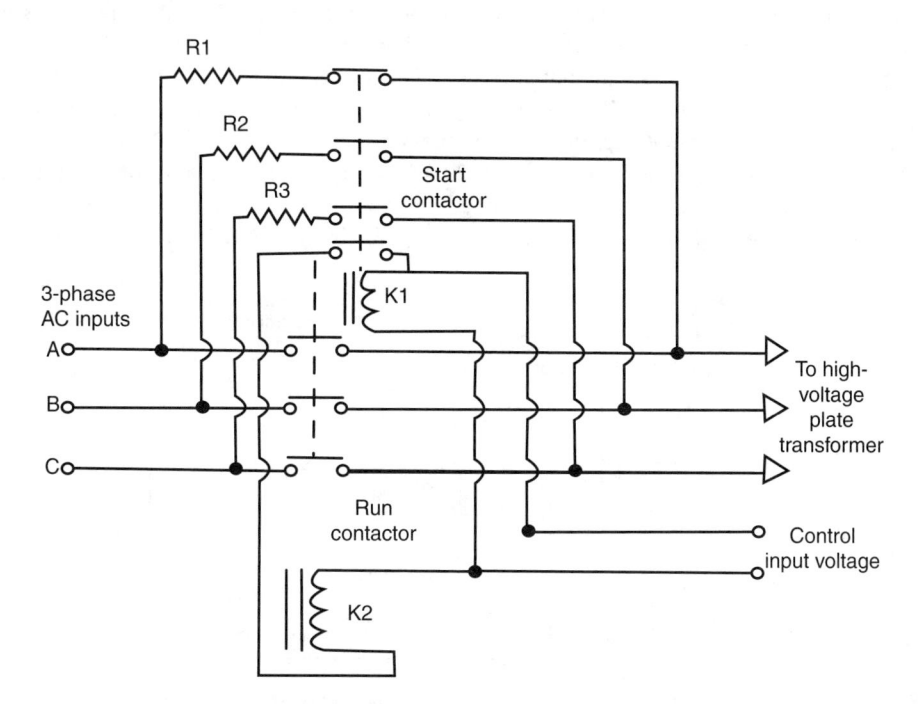

FIGURE 19.8 A typical three-phase step-start power control system.

Troubleshooting a step-start fault in a system employing thyristor power control should begin at the R-C ramp network. Check the capacitor to see if it has opened. Monitor the control voltage to the thyristor gating cards to confirm that the output voltage of the controller slowly increases to full value. If it does and the turn-on problem persists, the failure involves one or more of the gating cards.

When troubleshooting a step-start fault in a transmitter employing the dual contactor arrangement, begin with a close inspection of all contact points on both contactors. Pay careful attention to the auxiliary relay contacts of the start contactor. If the contacts fail to properly close, the full load of the high-voltage power supply will be carried through the resistors and start contactor. These devices are normally sized only for intermittent duty. They are not intended to carry the full load current for any length of time. Look for signs of arcing or overheating of the contact pairs and current-carrying connector bars. Check the current-limiting resistors for excessive dissipation and continuity.

Protection Circuits

Many engineers enjoy a false sense of security with transmission equipment because of the protection devices included in most designs. Although conventional overload circuits provide protection against most common failure modes, they are not foolproof. The first line of defense in the transmitter — the ac power system circuit breakers — can allow for potentially disastrous currents to flow under certain fault conditions.

Consider the thyristor ac power serve system shown in Fig. 19.9. This common type of voltage-regulator circuit adjusts the condition angle of the SCR pairs to achieve the desired dc output from the high-voltage power supply. An alternative configuration could have the output voltage sample derived from a transmission line RF pickup and amplifier/detector. In this way, the primary power control is adjusted to match the desired RF output from the transmitter. If one of the high-voltage rectifier stacks of this system failed in a short-circuit condition, the output voltage (and RF output) would fall, causing the thyristor circuit to increase the conduction period of the SCR pairs. Depending on the series resistance of the failed rectifier stack and the rating of the primary side circuit breaker, the breaker may or may not trip. Remember that the circuit breaker was chosen to allow operation at full transmitter power with the necessary headroom to prevent random tripping. The primary power system can therefore dissipate a significant amount of heat under reduced power conditions, such as those that would be experienced with a drop in the high-voltage supply output. The difference between the maximum designed power output of the supply (and therefore the transmitter) and the failure-induced power demand of the system can be dissipated as heat without tripping the main breaker.

Operation under such fault conditions, even for 20 seconds or less, can cause considerable damage to power-supply components, such as the power transformer, rectifier stack, thyristors, or system wiring. Damage can range from additional component failures to a fire in the affected section of the transmitter.

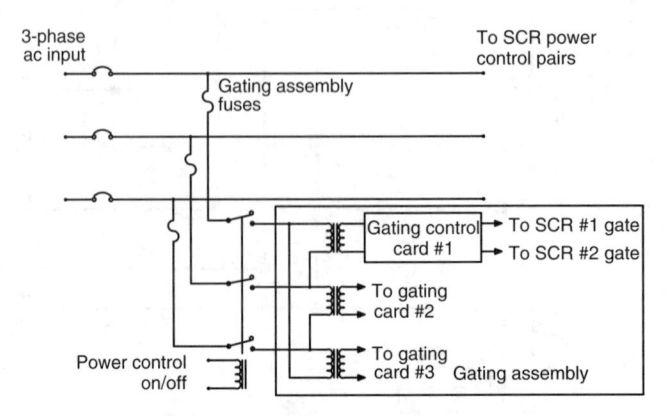

FIGURE 19.9 A common three-phase thyristor servo ac power control system.

Case in Point

The failure mode just outlined represents a real threat to high-power systems. This author is aware of a case in which just such a failure resulted in a fire inside a 20-kW FM transmitter. The following sequence of events led to the destruction of the unit:

- One or more transient overvoltages hit the transmitter site, causing an arc to occur within the driver stage plate transformer. The arcing continued until particles from the secondary winding broke free from the transformer.
- The failure of the driver transformer caused the driver output voltage to drop significantly, which decreased the RF output of the transmitter to about 25% of normal. Because the failure occurred between windings of the secondary of the driver plate transformer (and before the driver stage over current sensor), plate voltage remained on. Also, because of the point where the secondary winding short circuit occurred, the transformer primary did not draw sufficient current to initially trip the driver circuit breaker. As a result, ac power continued to flow to the damaged transformer.
- Small pieces of molten metal continued to drop from the driver transformer, landing on the PA plate transformer. These particles dropped into the windings, causing the plate transformer to short and starting a localized fire.
- When the smoke finally cleared, the entire high-voltage power supply had been damaged. In addition to the two ruined transformers, logic relays were melted away, rectifier stacks were fried, and most of the wiring harness was destroyed. The transmitter was determined to be damaged beyond repair.

Disasters such as this are rare, but they do occur. Be prepared to respond to any emergency condition by thoroughly understanding how the transmitter works and by identifying potential weak points in the system. Troubleshooting is far too important to be left to chance, or to the inexperienced.

20
RF Voltage and Power Measurement

Jerry C. Whitaker
Editor

20.1 Introduction 20-1
Root Mean Square · Average-Response Measurement
· Peak-Response Measurement · Measurement Bandwidth
· Meter Accuracy

20.2 RF Power Measurement.. 20-5
Decibel Measurement · Noise Measurement · Phase
Measurement · Nonlinear Distortion · Intermodulation
Distortion · Measurement Techniques · Addition and
Cancellation of Distortion Components

20.1 Introduction

The simplest definition of a level measurement is "the alternating current amplitude at a particular place in the system under test." However, in contrast to direct current measurements, there are many ways of specifying ac voltage in a circuit. The most common methods include:

- Average response
- Root mean square (rms)
- Peak response

Strictly speaking, the term *level* refers to a logarithmic, or decibel, measurement. However, common parlance employs the term for an ac amplitude measurement, and that convention will be followed in this chapter.

Root Mean Square

The root-mean-square (rms) technique measures the effective power of the ac signal. It specifies the value of the dc equivalent that would dissipate the same power if either were applied to a load resistor. This process is illustrated in Fig. 20.1 for voltage measurements. The input signal is squared, and the average value is found. This is equivalent to finding the average power. The square root of this value is taken to transfer the signal from a power value back to a voltage. For the case of a sine wave, the rms value is 0.707 of its maximum value.

Assume that the signal is no longer a sine wave but rather a sine wave and several harmonics. If the rms amplitude of each harmonic is measured individually and added, the resulting value will be the same as an rms measurement of the signals together. Because rms voltages cannot be added directly, it is necessary to perform an rms addition. Each voltage is squared, and the squared values are added as follows:

$$V_{rms} = \sqrt{V_{rms1}^2 + V_{rms2}^2 + V_{rms3}^2 + V_{rmsn}^2} \qquad (20.1)$$

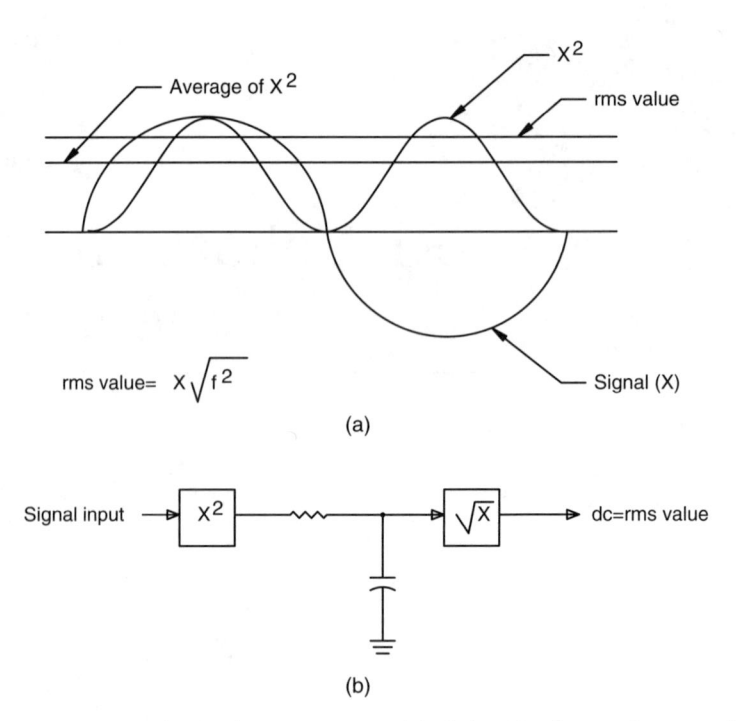

FIGURE 20.1 Root-mean-square (rms) voltage mesurement: (a) relationship of rms and average values, and (b) rms measurement circuit.

Note that the result is not dependent on the phase relationship of the signal and its harmonics. The rms value is determined completely by the amplitude of the components. This mathematical predictability is useful in practical applications of level measurement, enabling correlation of measurements made at different places in a system. It is also important in correlating measurements with theoretical calculations.

Average-Response Measurement

The average-responding voltmeter measures ac voltage by rectifying it and filtering the resulting waveform to its average value, as shown in Fig. 20.2. This results in a dc voltage that can be read on a standard dc voltmeter. As shown in the figure, the average value of a sine wave is 0.637 of its maximum amplitude. Average-responding meters are usually calibrated to read the same as an rms meter for the case of a single sine wave signal. This results in the measurement being scaled by a constant K of 0.707/0.637, or 1.11. Meters of this type are called *average-responding, rms calibrated*. For signals other than sine waves, the response will be different and difficult to predict.

If multiple sine waves are applied, the reading will depend on the phase shift between the components and will no longer match the rms measurement. A comparison of rms and average-response measurements is made in Fig. 20.3 for various waveforms. If the average readings are adjusted as previously described to make the average and rms values equal for a sine wave, all the numbers in the average column would be increased by 11.1%, and the rms-average numbers would be reduced by 11.1%.

Peak-Response Measurement

Peak-responding meters measure the maximum value that the ac signal reaches as a function of time. This approach is illustrated in Fig. 20.4. The signal is full-wave-rectified to find its absolute value, then passed through a diode to a storage capacitor. When the absolute value of the voltage rises above the value stored on the capacitor, the diode will conduct and increase the stored voltage. When the voltage decreases, the capacitor will maintain the old value. Some method of discharging the capacitor is required so that a new peak value can be measured. In a true peak detector, this is accomplished by a solid-state

RF Voltage and Power Measurement

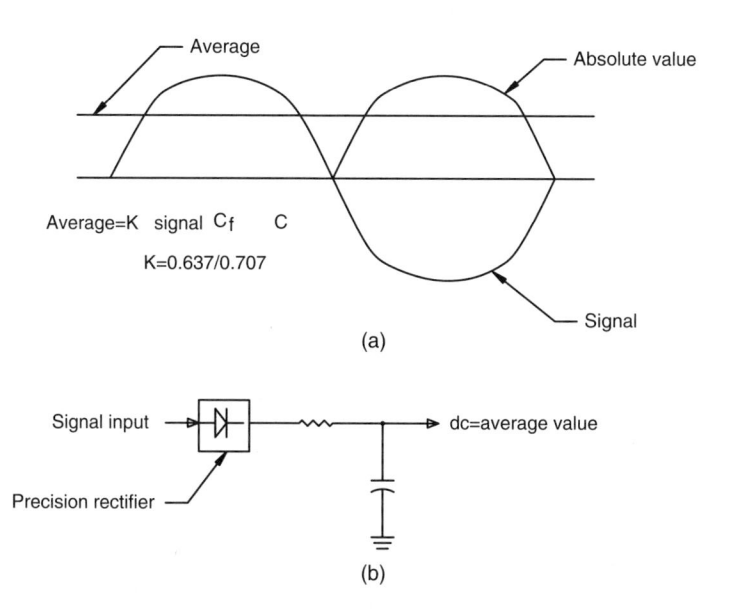

FIGURE 20.2 Average voltage measurement: (a) average detection, and (b) average measurement circuit.

Waveform		rms	Average	rms average	Crest factor
	Sine wave	$\frac{V_m}{\sqrt{2}}$ = 0.707 V_m	$\frac{2}{\pi} V_m$ = 0.637 V_m	$\frac{\pi}{2\sqrt{2}}$ = 1.111	$\sqrt{2}$ = 1.414
	Square wave	V_m	V_m	1	1
	Triangular wave or sawtooth wave	$\frac{V_m}{\sqrt{3}}$	$\frac{V_m}{2}$	$\frac{2}{\sqrt{3}}$ = 1.155	$\sqrt{3}$ = 1.732

FIGURE 20.3 Comparison of rms and average voltage characteristics.

switch. Practical peak detectors usually include a large-value resistor to discharge the capacitor gradually after the user has had a chance to read the meter.

The ratio of the true peak to the rms value is called the *crest factor*. For any signal but an ideal square wave, the crest factor will be greater than 1, as demonstrated in Fig. 20.5. As the measured signal becomes more peaked, the crest factor increases.

By introducing a controlled charge and discharge time, a quasi-peak detector is achieved. The charge and discharge times can be selected, for example, to simulate the transmission pattern for a digital carrier. The gain of a quasi-peak detector is normally calibrated so that it reads the same as an rms detector for sine waves.

The *peak-equivalent sine* is another method of specifying signal amplitude. This value is the rms level of a sine wave having the same peak-to-peak amplitude as the signal under consideration. This is the peak value of the waveform scaled by the correction factor 1.414, corresponding to the peak-to-rms ratio

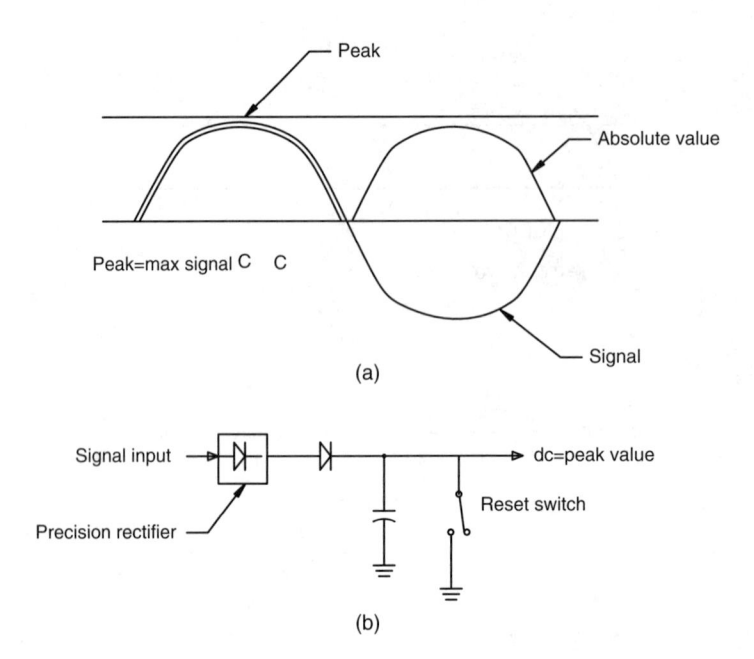

FIGURE 20.4 Peak voltage measurement: (a) peak detection, and (b) peak measurement circuit.

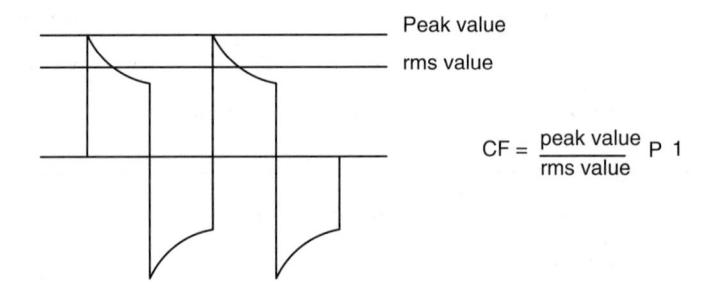

FIGURE 20.5 Crest factor in voltage measurements.

of a sine wave. Peak-equivalent sine is useful when specifying test levels of waveforms in distortion measurements. If the distortion of a device is measured as a function of amplitude, a point will be reached at which the output level cannot increase any further. At this point the peaks of the waveform will be clipped, and the distortion will rise rapidly with further increases in level. If another signal is used for distortion testing on the same device, it is desirable that the levels at which clipping is reached correspond. Signal generators are normally calibrated in this way to allow changing between waveforms without clipping or readjusting levels.

Measurement Bandwidth

The bandwidth of the level-measuring instrument can have a significant effect on the accuracy of the reading. For a meter with a single-pole rolloff (one bandwidth-limiting component in the signal path), significant measurement errors can occur. Such a meter with a specified bandwidth of 1 MHz, for example, will register a 10% error in the measurement of signals at 500 kHz. To obtain 1% accurate measurements (disregarding other error sources in the meter), the signal frequency must be less than 100 kHz.

Figure 20.6 illustrates another problem associated with limited-bandwidth measuring devices. In the figure, a distorted sine wave is measured by two meters with different bandwidths. The meter with the narrower bandwidth does not respond to all the harmonics and gives a lower reading. The severity of

RF Voltage and Power Measurement

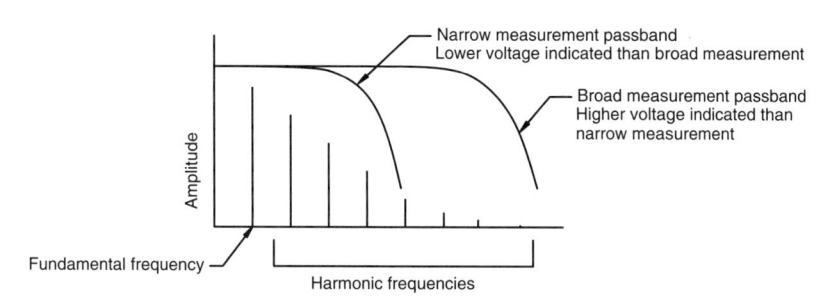

FIGURE 20.6 The effects of instrument bandwidth on voltage measurement.

this effect varies with the frequency being measured and the bandwidth of the meter; it can be especially pronounced in the measurement of wideband noise. Peak measurements are particularly sensitive to bandwidth effects. Systems with restricted low-frequency bandwidth will produce tilt in a square wave, and bumps in the high-frequency response will produce an overshoot. The effect of either will be an increase in the peak reading.

Meter Accuracy

Accuracy is a measure of how well an instrument quantifies a signal at a midband frequency. This sets a basic limit on the performance of the meter in establishing the absolute amplitude of a signal. It is also important to look at the *flatness* specification to see how well this performance is maintained with changes in frequency. Flatness describes how well the measurements at any other frequency track those at the reference. If a meter has an accuracy of 2% at 1 MHz and a flatness of 1 dB (10%) from 20 kHz to 20 MHz, the inaccuracy can be as great as 12% at 20 MHz.

Meters often have a specification of accuracy that changes with voltage range, being most accurate only in the range in which the instrument was calibrated. A meter with 1% accuracy on the 2-V range and 1% accuracy per step would be 3% accurate on the 200-V scale. Using the flatness specification given previously, the overall accuracy for a 100-V, 20-MHz sine wave is 14%. In many instruments, an additional accuracy derating is given for readings as a percentage of full scale, making readings at less than full scale less accurate.

However, the accuracy specification is not normally as important as the flatness. When performing frequency response or gain measurements, the results are relative and are not affected by the absolute voltage used. When measuring gain, however, the attenuator accuracy of the instrument is a direct error source. Similar comments apply to the accuracy and flatness specifications for signal generators. Most are specified in the same manner as voltmeters, with the inaccuracies adding in much the same manner.

20.2 RF Power Measurement

Measurement of RF power output is typically performed with a directional wattmeter, calibrated for use over a specified range of frequencies and power levels. Various grades of accuracy are available, depending on the requirements of the application. The most accurate measurements are usually made by determining the temperature rise of the cooling through a calibrated dummy load. The power absorbed by the coolant, usually water, can be calculated from the following equation:

$$P = K \times Q \times \Delta T \tag{20.2}$$

where P is the power dissipated in kilowatts (kW), K is a constant, determined by the coolant (for pure water at 30°C, $K = 0.264$), Q is coolant flow in gallons per minute (gpm), and ΔT is the difference between inlet and outlet water temperature in degrees Celsius (°C). This procedure is often used to verify the accuracy of in-line directional wattmeters, which are more convenient to use but typically offer less accuracy.

The main quality characteristics of a voltmeter or power meter are high measurement accuracy and short measurement time. Both can be achieved through care in the design of the probe or sensor and through the use of microprocessors for computed correction of frequency response, temperature effect, and linearity errors.

Even a top-quality measuring instrument can fail, either due to obvious functional faults or — with severe consequences — out-of-tolerance conditions that remain unnoticed. An increase in the measurement uncertainty is very difficult to detect, in particular with power meters, because there are usually no reference instruments that are considerably more accurate.

Decibel Measurement

Measurements in RF work are often expressed in decibels. Radio frequency signals span a wide range of levels, too wide to be accommodated on a linear scale. The decibel (dB) is a logarithmic unit that compresses this wide range down to one that is easier to handle. Order-of-magnitude (factor of 10) changes result in equal increments on a decibel scale. A decibel can be defined as the logarithmic ratio of two power measurements, or as the logarithmic ratio of two voltages:

$$db = 20\log\left\{\frac{E_1}{E_2}\right\} \qquad (20.3)$$

$$db = 10\log\left\{\frac{P_1}{P_2}\right\} \qquad (20.4)$$

Decibel values from power measurements and decibel values from voltage measurements are equal if the impedances are equal. In both equations, the denominator variable is usually a stated reference, as illustrated by the example in Fig. 20.7. Whether the decibel value is computed from the power-based equation or from the voltage-based equation, the same result is obtained.

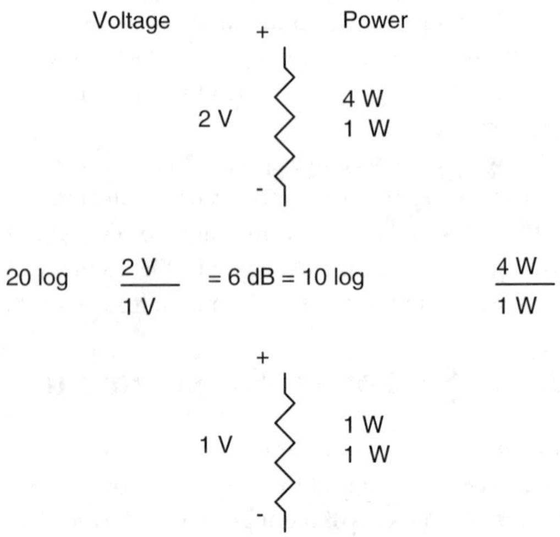

A doubling of voltage will yield a value of 6.02 dB, and a doubling of power will yield 3.01 dB. This is true because the doubling of voltage results in an increase in power by a factor of 4. Table 20.1 lists the decibel values for some common voltage and power ratios.

RF engineers often express the decibel value of a signal relative to some standard reference instead of another signal. The reference for decibel measurements can be predefined as a power level, as in dBk (decibels above 1 kW), or it can be a voltage reference. It is often desirable to specify levels in terms of a reference transmission level somewhere in the system under test. These measurements are designated dBr, where the reference point or level must be separately conveyed.

FIGURE 20.7 Example of the equivalence of voltage and power decibels.

RF Voltage and Power Measurement

TABLE 20.1 Common Decibel Values and Conversion Ratios

dB Value	Voltage Ratio	Power Ratio
0	1	1
+1	1.122	1.259
+2	1.259	1.586
+3	1.412	1.995
+6	1.995	3.981
+10	3.162	10
+20	10	100
+40	100	10,000
−1	0.891	0.794
−2	0.794	0.631
−3	0.707	0.501
−6	0.501	0.251
−10	0.3163	0.1
−20	0.1	0.01
−40	0.01	0.0001

Noise Measurement

Noise measurements are specialized level measurements. Noise can be expressed as an absolute level by simply measuring the voltage at the desired point in the system. This approach, however, is often not very meaningful. Specifying the noise performance as the signal-to-noise ratio (S/N) is a better approach. S/N is a decibel measurement of the noise level using the signal level measured at the same point as a reference. This makes measurements at different points in a system or in different systems directly comparable. A signal with a given S/N can be amplified with a perfect amplifier or attenuated with no change in the S/N. Any degradation in S/N at a later point in the system is the result of limitations of the equipment that follows.

Noise performance is an important parameter in the operation of any RF amplifier or oscillator. All electric conductors contain free electrons that are in continuous random motion. It can be expected that, by pure chance, more electrons will be moving in one direction than in another at any instant. The result is that a voltage will be developed across the terminals of the conductor if it is an open circuit, or a current will be delivered to any connected circuit. Because this voltage (or current) varies in a random manner, it represents noise energy distributed throughout the frequency spectrum, from the lowest frequencies well into the microwave range. This effect is commonly referred to as *thermal-agitation noise* because the motion of electrons results from thermal action. It is also referred to as *resistance noise*. The magnitude of the noise depends on the following:

- Resistance across which the noise is developed
- Absolute temperature of the resistance
- Bandwidth of the system involved

Random noise, similar in character to that produced in a resistance, is generated in active devices as a result of irregularities in electron flow. Vacuum tube noise can be divided into the following general classes:

- *Shot effect*, representing random variations in the rate of electron emission from the cathode
- *Partition noise*, arising from chance variations in the division of current between two or more positive electrodes
- *Induced grid noise*, produced as a result of variations in the electron stream passing adjacent to a grid
- *Gas noise*, generated by random variations in the rate of ion production by collision

- *Secondary emission noise*, arising from random variations in the rate of production of secondary electrons
- *Flicker effect*, a low-frequency variation in emission that occurs with oxide-coated cathodes

Shot effect in the presence of space charge, partition noise, and induced grid noise is the principal source of tube noise that must be considered in RF work.

Phase Measurement

When a signal is applied to the input of a device, the output will appear at some later point in time. For sine wave excitation, this delay between input and output can be expressed as a proportion of the sine wave cycle, usually in degrees. One cycle is 360°, one half-cycle is 180°, etc. This measurement is illustrated in Fig. 20.8. The phasemeter input signal number 2 is delayed from, or is said to be lagging, input number 1 by 45°.

Most RF test equipment checks phase directly by measuring the proportion of one signal cycle between zero crossings of the signals. Phase typically is measured and recorded as a function of frequency over a specified range. For most vacuum tube devices, phase and amplitude responses are closely coupled. Any change in amplitude that varies with frequency will produce a corresponding phase shift.

Relation to Frequency

When dealing with complex signals, the meaning of "phase" can become unclear. Viewing the signal as the sum of its components according to Fourier theory, a different value of phase shift is found at each frequency. With a different phase value on each component, the one to be used as the reference is unclear. If the signal is periodic and the waveshape is unchanged passing through the device under test, a phase value can still be defined. This can be done by using the shift of the zero crossings as a fraction of the waveform period. Indeed, most commercial phase-measuring instruments will display this value. However, in the case of *differential phase shift* with frequency, the waveshape will be changed. It is then impossible to define any phase-shift value, and phase must be expressed as a function of frequency.

Group delay is another useful expression of the phase characteristics of an RF device. Group delay is the slope of the phase response. It expresses the relative delay of the spectral components of a complex waveform. If the group delay is flat, all components will arrive together at a given point. A peak or rise in the group delay indicates that those components will arrive later by the amount of the peak or rise. Group delay is computed by taking the derivative of the phase response vs. frequency. Mathematically:

$$\text{Group delay} = \frac{\theta_2 - \theta_1}{f_2 - f_1} \tag{20.5}$$

where θ_1 is the phase at f_1 and θ_2 is the phase at f_2. This definition requires that the phase be measured over a range of frequencies to give a curve that can be differentiated. It also requires that the phase measurements be performed at frequencies sufficiently close to provide a smooth and accurate derivative.

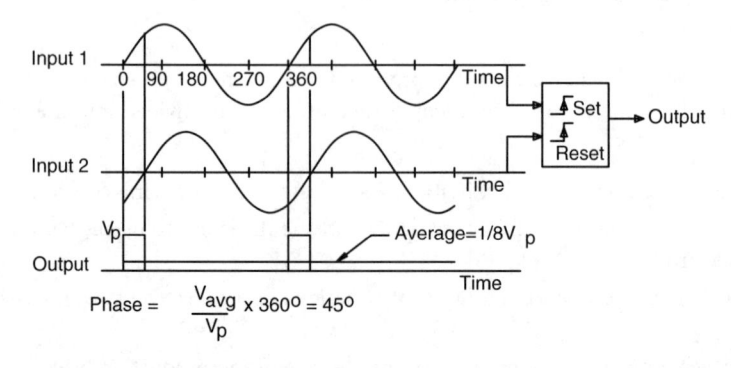

FIGURE 20.8 Measurement of a phase shift between two signals.

Nonlinear Distortion

Distortion is a measure of signal impurity, a deviation from ideal performance of a device, stage, or system. Distortion is usually expressed as a percentage or decibel ratio of the undesired components to the desired components of a signal. Distortion of an amplifying device or stage is measured by inputting one or more sine waves of various amplitudes and frequencies. In simplistic terms, any frequencies at the output that were not present at the input are distortion. However, strictly speaking, components caused by power supply ripple or another spurious signal are not distortion but, rather, noise. Many methods of measuring distortion are in common use, including harmonic distortion and several types of intermodulation distortion.

Harmonic Distortion

The transfer characteristic of a typical RF amplifier is shown in Fig. 20.9. The transfer characteristic represents the output voltage at any point in the signal waveform for a given input voltage; ideally, this is a straight line. The output waveform is the projection of the input sine wave on the device transfer characteristic. A change in the input produces a proportional change in the output. Because the actual transfer characteristic is nonlinear, a distorted version of the input waveshape appears at the output.

Harmonic distortion measurements excite the device under test with a sine wave and measure the spectrum of the output. Because of the nonlinearity of the transfer characteristic, the output is not sinusoidal. Using Fourier series, it can be shown that the output waveform consists of the original input sine wave plus sine waves at integer multiples (harmonics) of the input frequency. The spectrum of a distorted signal is shown in Fig. 20.10. The harmonic amplitudes are proportional to the amount of

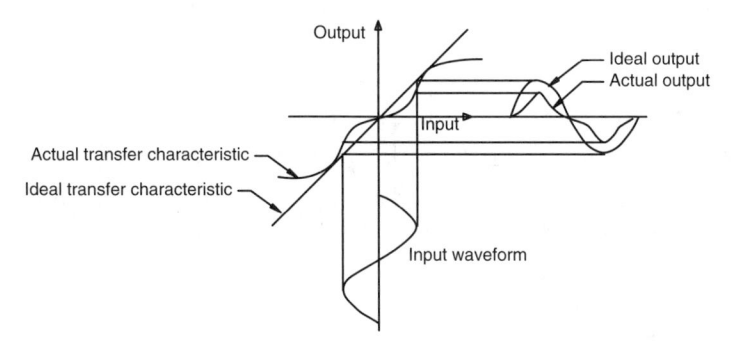

FIGURE 20.9 Illustration of total harmonic distortion (THD) measurement of an amplifier transfer characteristic.

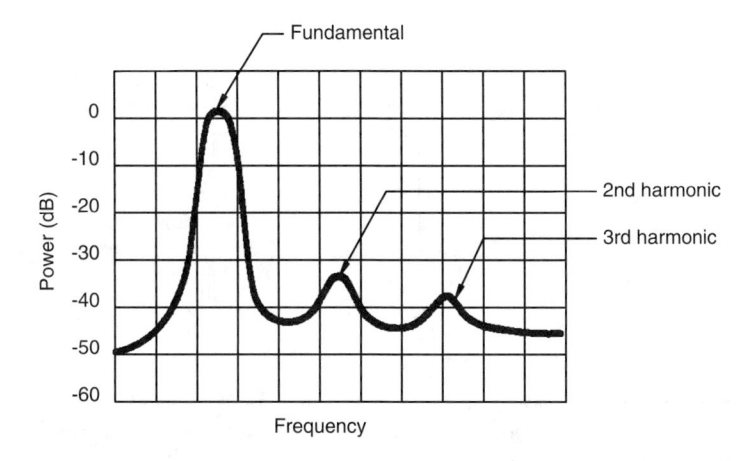

FIGURE 20.10 Example of reading THD from a spectrum analyzer.

distortion in the device under test. The percentage of harmonic distortion is the rms sum of the harmonic amplitudes divided by the rms amplitude of the fundamental.

Harmonic distortion can be measured with a spectrum analyzer or a distortion test set. Figure 20.11 shows the setup for a spectrum analyzer. As shown in Fig. 20.10, the fundamental amplitude is adjusted to the 0-dB mark on the display. The amplitudes of the harmonics are then read and converted to linear scale. The rms sum of these values is taken, which represents the THD.

A simpler approach to the measurement of harmonic distortion can be found in the notch-filter distortion analyzer illustrated in Fig. 20.12. This device, commonly referred to as simply a *distortion analyzer*, removes the fundamental of the signal to be investigated and measures the remainder. Fig. 20.13 shows the notch-filter approach applied to a spectrum analyzer for distortion measurements.

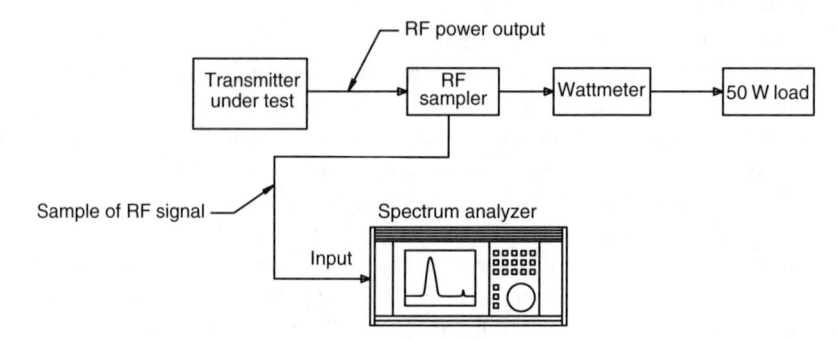

FIGURE 20.11 Common test setup to measure harmonic distortion with a spectrum analyzer.

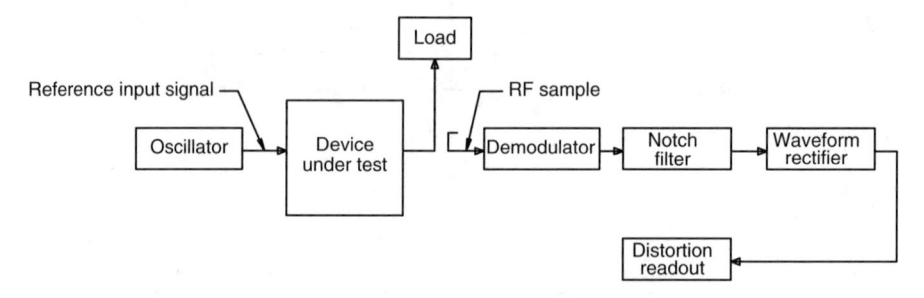

FIGURE 20.12 Simplified block diagram of a harmonic distortion analyzer.

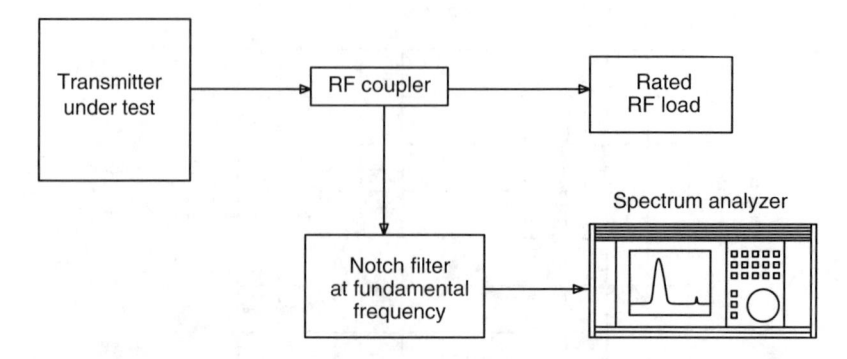

FIGURE 20.13 Typical test setup for measuring the harmonic and spurious output of a transmitter. The notch filter is used to remove the fundamental frequency to prevent overdriving the spectrum analyzer input and to aid in evaluation of distortion components.

The correct method of representing percentage distortion is to express the level of the harmonics as a fraction of the fundamental level. However, many commercial distortion analyzers use the total signal level as the reference voltage. For small amounts of distortion, these two quantities are essentially equivalent. At large values of distortion, however, the total signal level will be greater than the fundamental level. This makes distortion measurements on such units lower than the actual value. The relationship between the measured distortion and true distortion is given in Fig. 20.14. The errors are not significant until about 20% measured distortion.

Because of the notch-filter response, any signal other than the fundamental will influence the results, not just harmonics. Some of these interfering signals are illustrated in Fig. 20.15. Any practical signal contains some noise, and the distortion analyzer will include this noise in the reading. Because of these added components, the correct term for this measurement is *total harmonic distortion and noise* (THD+N). Additional filters are included on most distortion analyzers to reduce unwanted noise and permit a more accurate reading.

The use of a sine wave test signal and a notch-type distortion analyzer provides the distinct advantage of simplicity in both design and use. This simplicity has the additional benefit of ease of interpretation. The shape of the output waveform from a notch-type analyzer indicates the slope of the nonlinearity. Displaying the residual components on the vertical axis of an oscilloscope and the input signal on the horizontal axis provides a plot of the deviation of the transfer characteristic from a best-fit straight line. This technique is diagrammed in Fig. 20.16. The trace will be a horizontal line for a perfectly linear device. If the transfer characteristic curves upward on positive input voltages, the trace will bend upward at the right-hand side.

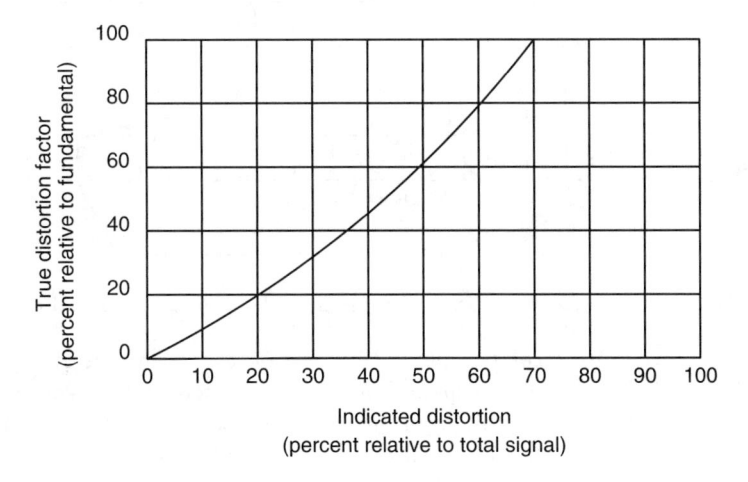

FIGURE 20.14 Conversion graph for indicated distortion and true distortion.

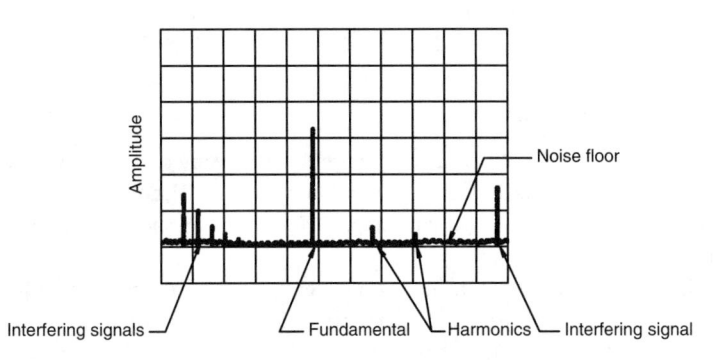

FIGURE 20.15 Example of interference sources in distortion and noise measurements.

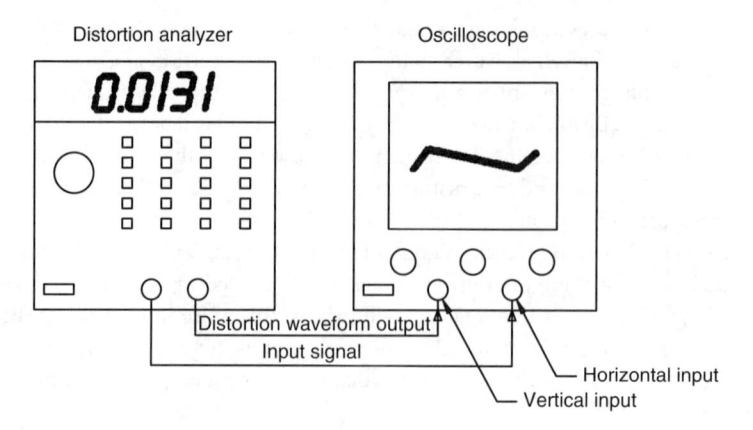

FIGURE 20.16 Transfer-function monitoring configuration using an oscilloscopre and distortion analyzer.

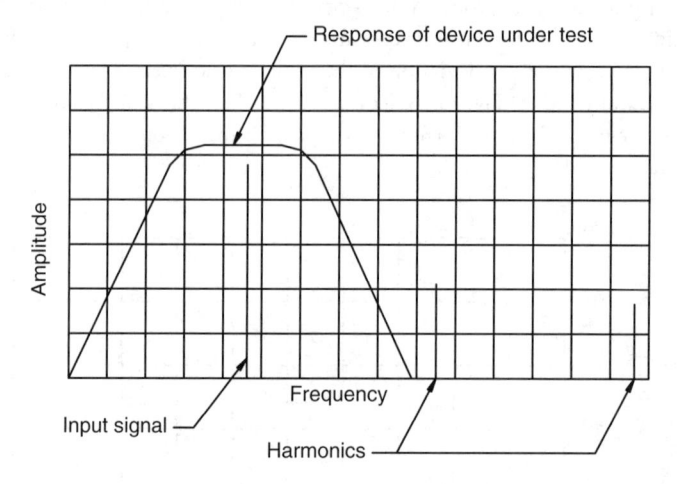

FIGURE 20.17 Illustration of problems that occur when measuring harmonic distortion in band-limited systems.

Examination of the distortion components in real time on an oscilloscope allow observation of oscillation on the peaks of the signal and clipping. This is a valuable tool in the design and development of RF circuits, and one that no other distortion measurement method can fully match. Viewing the residual components in the frequency domain using a spectrum analyzer also reveals significant information about the distortion mechanism inside the device or stage under test.

When measuring distortion at high frequencies, bandwidth limitations are an important consideration, as illustrated in Fig. 20.17. Because the components being measured are harmonics of the input frequency, they may fall outside the passband of the device under test.

Intermodulation Distortion

Intermodulation distortion (IMD) describes the presence of unwanted signals that are caused by interactions between two or more desired signals. IMD, as it applies to an RF amplifier, refers to the generation of spurious products in a nonlinear amplifying stage. IMD can be measured using a distortion monitor or a spectrum analyzer.

Nonlinearities in a circuit can cause it to act like a mixer, generating the sum and difference frequencies of two signals that are present. These same imperfections generate harmonics of the signals, which then can be mixed with other fundamental or harmonic frequencies. Figure 20.18 shows the relationship of these signals in the frequency domain. The order of a particular product of IMD is defined as the number

RF Voltage and Power Measurement

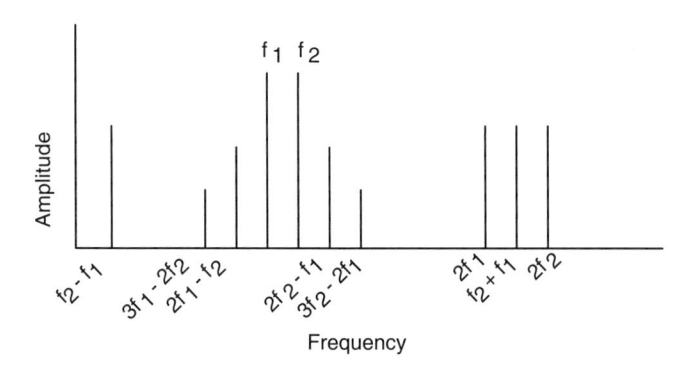

[Figure showing spectrum with peaks labeled from left to right: f2−f1, 3f1−2f2, 2f1−f2, 2f2−f1, 3f2−2f1, f1, f2, 2f1, f2+f1, 2f2; axes: Amplitude vs Frequency]

FIGURE 20.18 Frequency-domain relationships of the desired signals and low-order IMD products.

of "steps" between it and a single fundamental signal. Harmonics add steps. For example, the second harmonic = $2f_1 = f_1 + f_1$, which is two steps, making it a second-order product. The product resulting from $2f_1 - f_2$ is, then, a third-order product. Some of the IMD products resulting from two fundamental frequencies include:

$$\text{Second-order} = 2f_1 \tag{20.6}$$

$$= 2f_2 \tag{20.7}$$

$$= f_1 + f_2 \tag{20.8}$$

$$= f_1 - f_2 \tag{20.9}$$

$$\text{Third-order} = 2f_1 + f_2 \tag{20.10}$$

$$= 2f_1 - f2 \tag{20.11}$$

$$= 2f_2 + f_1 \tag{20.12}$$

$$= 2f_2 - f_1 \tag{20.13}$$

and so on.

The order is important because, in general, the amplitudes of IMD products fall off as the order increases. Therefore, low-order IMD products have the greatest potential to cause problems if they fall within the band of interest, or on frequencies used by other services or equipment. Odd-order IMD products are particularly troublesome because they fall closest to the signals that cause them, usually within the operating frequency band.

Measurement Techniques

A number of techniques have been devised to measure the intermodulation (IM) of two or more signals passing through a device simultaneously. The most common of these involves the application of a test waveform consisting of a low-frequency signal and a high-frequency signal mixed in a specified amplitude ratio. The signal is applied to the device under test, and the output waveform is examined for modulation of the upper frequency by the low-frequency signal. The amount by which the low-frequency signal modulates the high-frequency signal indicates the degree of nonlinearity. As with harmonic distortion measurement, this test can be done with a spectrum analyzer or a dedicated distortion analyzer. The test setup for a spectrum analyzer is shown in Fig. 20.19. Two independent signal sources are connected using a power combiner to drive the device under test. The sources are set at the same output level, but at different frequencies. A typical spectrum analyzer display of the two-tone distortion test is shown in

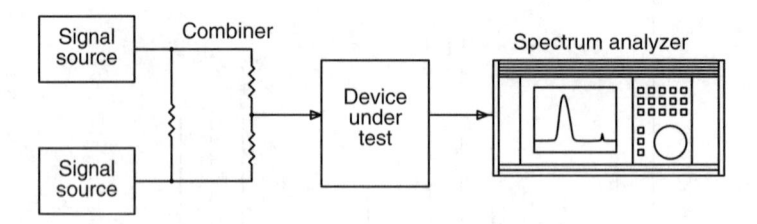

FIGURE 20.19 Test setup for measuring the intermodulation distortion of an RF amplifier using a spectrum analyzer.

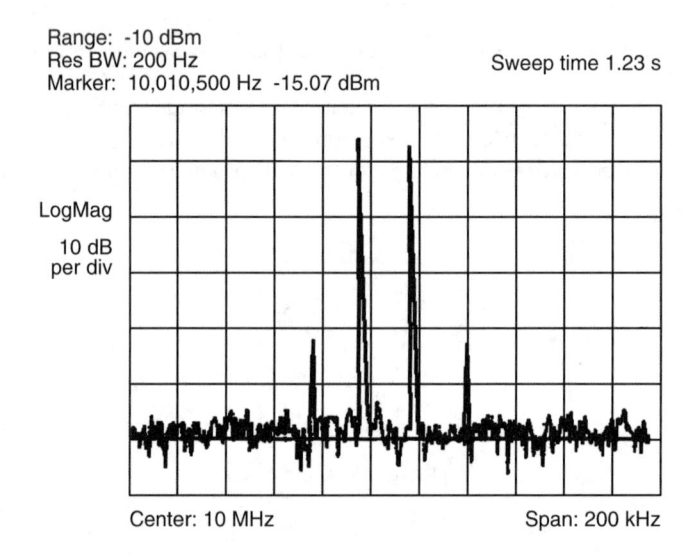

FIGURE 20.20 Typical two-tone IMD measurement, which evaluates the third-order products within the passband of the original tones.

Fig. 20.20. As shown, the third-order products that fall close to the original two tones are measured. This measurement is a common, and important, one because such products are difficult to remove by filtering.

Addition and Cancellation of Distortion Components

The addition and cancellation of distortion components in test equipment or the device under test is an often-overlooked problem in measurement. Consider the examples in Figs. 20.21 and 20.22. Assume that one device under test has a transfer characteristic similar to that diagrammed at the top left of Fig. 20.21a, and another has the characteristic diagrammed at the center. If the devices are cascaded, the resulting transfer-characteristic nonlinearity will be magnified as shown. The effect on sine waves in the time domain is illustrated in Fig. 20.21b. The distortion components generated by each nonlinearity are in phase and will sum to a component of twice the original magnitude. However, if the second device under test has a complementary transfer characteristic, as shown in Fig. 20.22, quite a different result is obtained. When the devices are cascaded, the effects of the two curves cancel, yielding a straight line for the transfer characteristic (Fig. 20.22a). The corresponding distortion products are out of phase, resulting in no measured distortion components in the final output (Fig. 20.22b).

This problem is common at low levels of distortion, especially between the test equipment and the device under test. For example, if the test equipment has a residual of 0.002% when connected to itself, and readings of 0.001% are obtained from the circuit under test, cancellations are occurring. It is also possible for cancellations to occur in the test equipment itself, with the combined analyzer and signal generator giving readings lower than the sum of their individual residuals. If the distortion is the result

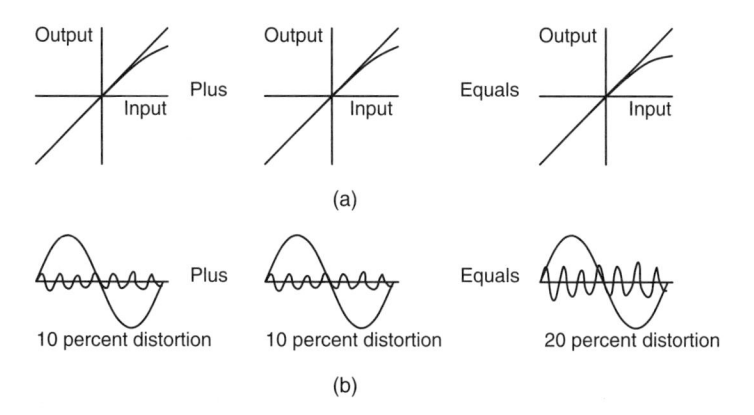

FIGURE 20.21 The addition of distortion components: (a) addition of transfer-function nonlinearities, and (b) addition of distortion components.

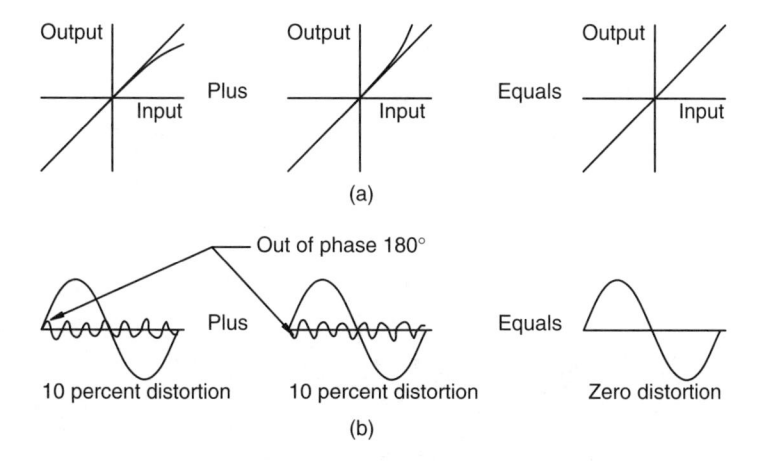

FIGURE 20.22 Cancellation of distortion components: (a) cancellation of transfer-characteristic nonlinearities, and (b) cancellation of the distortion waveform.

of even-order (asymmetrical) nonlinearity, reversing the phase of the signal between the offending devices will change a cancellation to an addition. If the distortion is the result of an odd-order (symmetrical) nonlinearity, phase inversions will not affect the cancellation.

Bibliography

Benson, K. B. and Jerry C. Whitaker, *Television and Audio Handbook for Engineers and Technicians*, McGraw-Hill, New York, 1989.
Crutchfield, E. B., Ed., *NAB Engineering Handbook*, 8th Ed., National Association of Broadcasters, Washington, D.C., 1992.
Terman, F. E., *Radio Engineering*, 3rd ed., McGraw-Hill, New York, 1947.
Whitaker, Jerry C., *Maintaining Electronic Systems*, CRC Press, Boca Raton, FL, 1992.
Whitaker, Jerry C., *Radio Frequency Transmission Systems: Design and Operation*, McGraw-Hill, New York, 1990.
Witte, Robert A., Distortion measurements using a spectrum analyzer, in *RF Design*, Cardiff Publishing, Denver, CO, September 1992, 75–84.

21
Spectrum Analysis

Jerry C. Whitaker
Editor

21.1 Introduction 21-1
   Principles of Operation · Applications · Spurious Harmonic
   Distortion

21.1 Introduction

An oscilloscope-type instrument displays voltage levels referenced to time and a spectrum analyzer indicates signal levels referenced to frequency. The frequency components of the signal applied to the input of the analyzer are detected and separated for display against a frequency-related time base. Spectrum analyzers are available in a variety of ranges, with some models designed for use with audio or video frequencies and others intended for use with RF frequencies.

The primary application of a spectrum analyzer is the measurement and identification of RF signals. When connected to a small receiving antenna, the analyzer can measure carrier and sideband power levels. By expanding the sweep width of the display, offset or multiple carriers can be observed. By increasing the vertical sensitivity of the analyzer and adjusting the center frequency and sweep width, it is possible to observe the occupied bandwidth of the RF signal. Convention dictates that the vertical axis displays amplitude and the horizontal axis displays frequency. This frequency-domain presentation allows the user to glean more information about the characteristics of an input signal than is possible from an oscilloscope. Figure 21.1 compares the oscilloscope and spectrum analyzer display formats.

Principles of Operation

A spectrum analyzer intended for use at RF frequencies is shown in block diagram form in Fig. 21.2. The instrument includes a superheterodyne receiver with a swept-tuned local oscillator (LO) that feeds a CRT display. The tuning control determines the center frequency (F_c) of the spectrum analyzer, and the *scan-width* selector determines how much of the frequency spectrum around the center frequency will be covered. Full-feature spectrum analyzers also provide front-panel controls for scan-rate selection and bandpass filter selection. Key specifications for a spectrum analyzer include:

- *Resolution*: the frequency separation required between two signals so that they may be resolved into two distinct and separate displays on the screen. Resolution is usually specified for equal-level signals. When two signals differ significantly in amplitude and are close together in frequency, greater resolution is required to separate them on the display.
- *Scan width*: the amount of frequency spectrum that can be scanned and shown on the display. Scan width is usually stated in kilohertz or megahertz per division. The minimum scan width available is usually equal to the resolution of the instrument.
- *Dynamic range*: the maximum amplitude difference that two signals can have and still be viewed on the display. Dynamic range is usually stated in decibels.

- *Sensitivity*: the minimum signal level required to produce a usable display on the screen. If low-level signal tracing is planned, as in receiver or off-air monitoring, the sensitivity of the spectrum analyzer is important.

When using the spectrum analyzer, care must be taken not to overload the front-end with a strong input signal. Overloading can cause "false" signals to appear on the display. These false signals are the result of nonlinear mixing in the front end of the instrument. False signals can be identified by changing the RF attenuator setting to a higher level. The amplitude of false signals (caused by overloading) will drop much more than the amount of increased attenuation.

The spectrum analyzer is useful in troubleshooting receivers as well as transmitters. As a tuned signal tracer, it is well adapted to stage-gain measurements and other tests. There is one serious drawback,

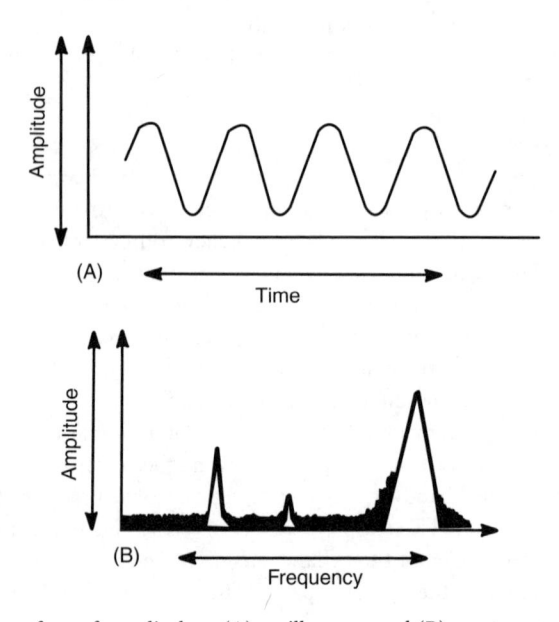

FIGURE 21.1 Comparison of waveform displays: (A) oscilloscope, and (B) spectrum analyzer.

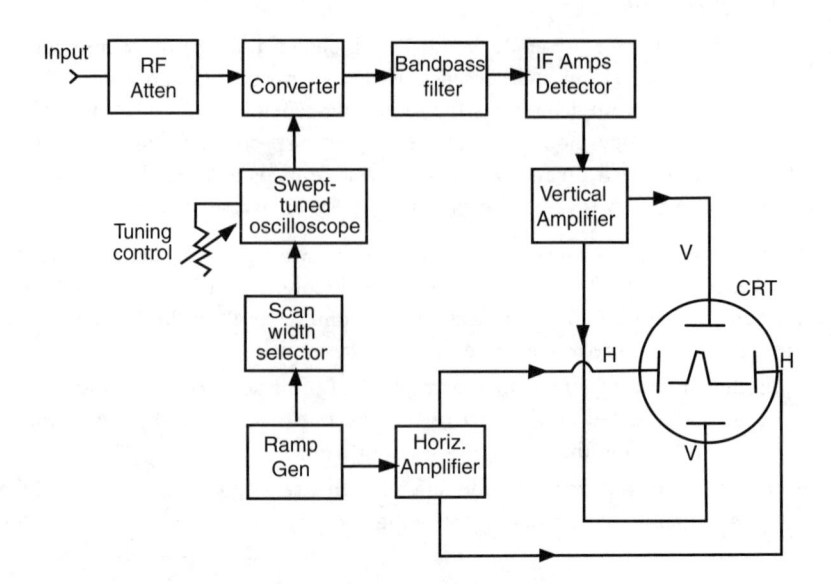

FIGURE 21.2 Block diagram of a spectrum analyzer.

Spectrum Analysis

however. The 50-Ω spectrum analyzer input can load many receiver circuits too heavily, especially high impedance circuits such as FET amplifiers. Isolation probes are available to overcome loading problems. Such probes, however, also attenuate the input signal, and unless the spectrum analyzer has enough reserve gain to overcome the loss caused by the isolation probe, the instrument will fail to provide useful readings. Isolation probes with 20 to 40 dB attenuation are typical. As a rule of thumb, probe impedance should be at least ten times the impedance of the circuit to which it is connected.

Applications

The primary application for a spectrum analyzer centers around measuring the occupied bandwidth of an input signal. Harmonics and spurious signals can be checked and potential causes investigated. Figure 21.3 shows a typical test setup for making transmitter measurements.

The spectrum analyzer is also well-suited for making accurate transmitter FM deviation measurements. This is accomplished using the *Bessel null method*. The Bessel null is a mathematical function that describes the relationship between spectral lines in frequency modulation. The Bessel null technique is highly accurate; it forms the basis for modulation monitor calibration. The concept behind the Bessel null method is to drive the carrier spectral line to zero by changing the modulating frequency. When the carrier amplitude is zero, the modulation index is given by a Bessel function. Deviation can be calculated from:

$$\Delta f_c = MI \times f$$

where Δf_c is the deviation frequency, MI is the modulation index, and f_m is the modulating frequency. The carrier frequency "disappears" at the Bessel null point, with all power remaining in the FM sidebands.

A *tracking generator* can be used in conjunction with the spectrum analyzer to check the dynamic response of frequency-sensitive devices, such as transmitter isolators, cavities, ring combiners, duplexers,

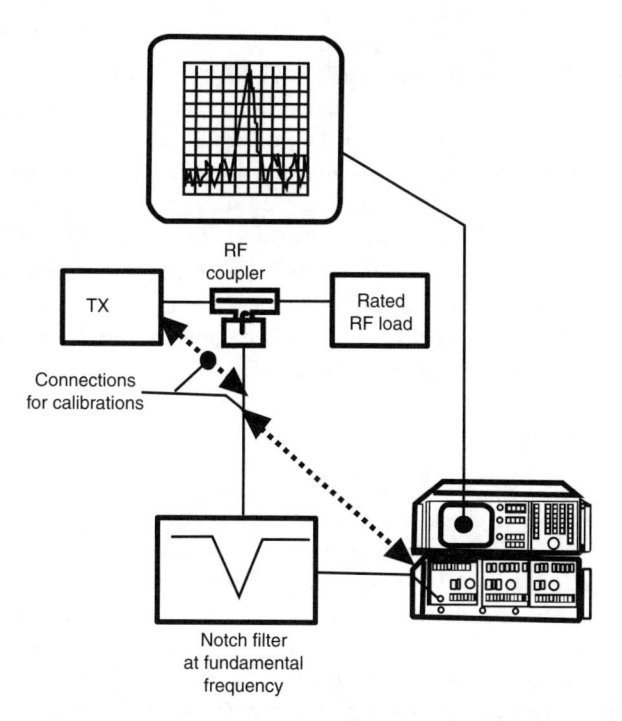

FIGURE 21.3 Typical test setup for measuring the harmonic and spurious output of a transmitter. The notch filter is used to remove the fundamental frequency to prevent overdriving the spectrum analyzer input.

and antenna systems. A tracking generator is a frequency source that is locked in step with the spectrum analyzer horizontal trace rate. The resulting display shows the relationship of the amplitude-vs.-frequency response of the device under test. The spectrum analyzer can also be used to perform gain-stage measurements. The combination of a spectrum analyzer and a tracking generator makes filter passband measurements possible. As measurements are made along the IF (intermediate frequency) chain of a receiver, the filter passbands become increasingly narrow, as illustrated in Fig. 21.4.

On-Air Measurements

The spectrum analyzer is used for three primary on-air tests:

- *Measuring unknown signal frequencies.* A spectrum analyzer can be coupled to the output of a transmitter to determine its exact operating frequency.
- *Intermod and interference signal tracking.* A directional Yagi antenna can be used to identify the source of an interfering signal. If the interference is on-frequency but carries little intelligence, chances are good that it is an intermod being produced by another transmitter. Use the wide dispersion display mode of the analyzer and note signal spikes that appear simultaneously with the interference. A troubleshooter, armed with a spectrum analyzer, a directional antenna, and the knowledge of how intermod signals are generated, can usually locate a suspected transmitter rapidly. Each suspected unit can then be tested individually by inserting an isolator between the transmitter PA and duplexer (or PA and antenna). When the intermod signal amplitude drops the equivalent of the reverse insert loss of the isolator, the offending transmitter has been located.
- *Field strength measurements.* With an external antenna, a spectrum analyzer can be used for remote field strength measurements. Omnidirectional, broadband antennas are the most versatile because they allow several frequency bands to be checked simultaneously. Obviously, the greater the external antenna height, the greater the testing range.

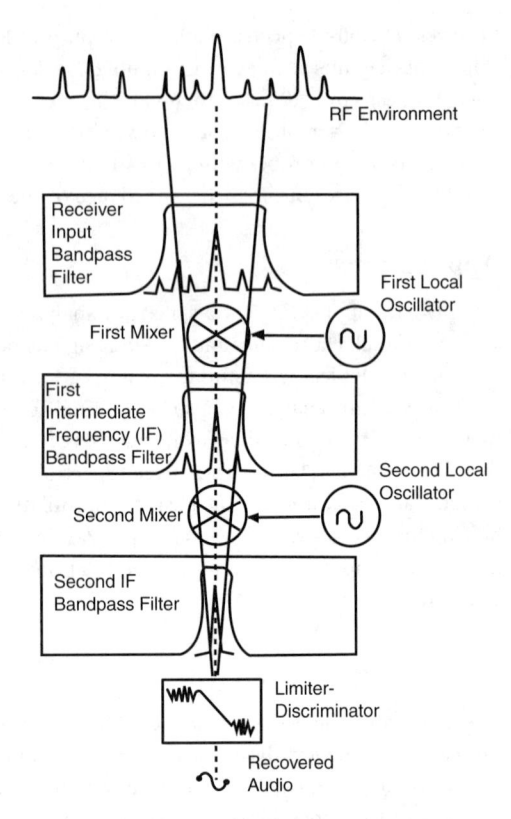

FIGURE 21.4 Using a spectrum analyzer to measure the tunnel effect of bandpass filters in a receiver.

Spurious Harmonic Distortion

An incorrectly tuned or malfunctioning transmitter can produce spurious harmonics. The spectrum analyzer provides the best way to check for spectral purity. Couple a sample of the transmitter output signal to the analyzer input, either by loop coupling or RF sampling, as illustrated in Fig. 21.5. For low power levels from portable and mobile units, transmit into a dummy load and use a flexible rubber antenna on the analyzer input. For maximum accuracy when measuring larger amounts of power, use an RF sampler to control the input level to the analyzer. Use maximum RF attenuation initially on the analyzer front end to prevent overload damage and internal intermod. False signals on the display can also be observed if covers or shields are not in place on the radio under test. The oscillator, doubler, or tripler levels may radiate sufficient signals to register on the analyzer display.

Spectrum Analysis

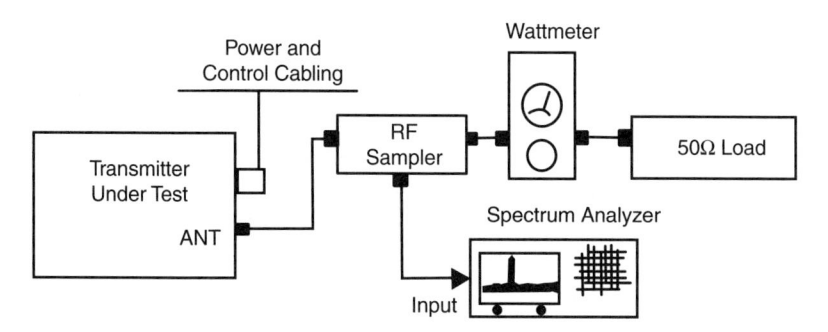

FIGURE 21.5 Common test setup to measure transmitter harmonic distortion with a spectrum analyzer.

Energize the transmitter and observe any spurious harmonics on the analyzer display. After centering the main signal of interest and adjusting the input attenuation for maximum display amplitude, calculate the spurious radiation. Spurious signal attenuation is measured in decibels, referenced to the amplitude of the transmitter fundamental. When a radio transmitter has a harmonic distortion problem, the defective stage can usually be isolated by tuning each stage and observing the spurious harmonics on the analyzer display. When the defective stage is tuned, the harmonics either shift frequency or change in amplitude. Signal tracing, with a probe and heavy input attenuation, also helps to determine where the distortion first occurs. Overdriven stages are prime culprits.

Selective-Tuned Filter Alignment

The high input sensitivity and visual frequency-selective display of the spectrum analyzer provide the technician with an efficient analog tuning instrument. Injection circuits can be peaked quickly by loop coupling the analyzer input to the mixer section of a receiver. This is most helpful in tuning older radios when a service manual is not available. Radios without test points, or that require elaborate, specialized test sets, can be tuned in a similar manner. Portable radio transmitters can be tuned directly into their antennas for maximum signal strength. This reveals problems such as improper signal transfer to the antenna or a defective antenna.

Cavities, combiners, and duplexers are all selective-tuned filters. They are composed of three types of passive filters:

- Bandpass
- Band-reject
- Combination pass/reject

The spectrum analyzer is well-suited to tuning these types of filters, particularly duplexers. Couple the cavity between the tracking generator and the spectrum analyzer. Calibrate the tuning frequency in the center of the CRT display and adjust the generator output and analyzer input level for an optimal display trace. Then tune the cavity onto frequency. Measure the various filter characteristics and compare them to the manufacturer's specifications. Figure 21.6 shows a typical spectrum analyzer display of a single-cavity device. Important filter characteristics include frequency, bandwidth, insertion loss, and selectivity. Measure the bandwidth at the points 3 dB down from maximum amplitude. Insertion loss is equal to the cavity's attenuation of the center-tune frequency. Calculate filter Q (quality factor), which is directly proportional to selectivity, by dividing the filter center frequency (f_{CT}) by the bandwidth (BW) of the filter.

Duplexer tuning is similar to cavity tuning but is complicated by interaction between the multiple cavities. Using the test setup shown in Fig. 21.7, couple the tracking generator output to the duplexer antenna input. Alternately couple the receive and transmit ports to the spectrum analyzer while tuning the pass or reject performance of each cavity. When adjusting a combination cavity, tune the reject last because it will tend to follow the pass tuning. Alternate tuning between the receive and transmit sides

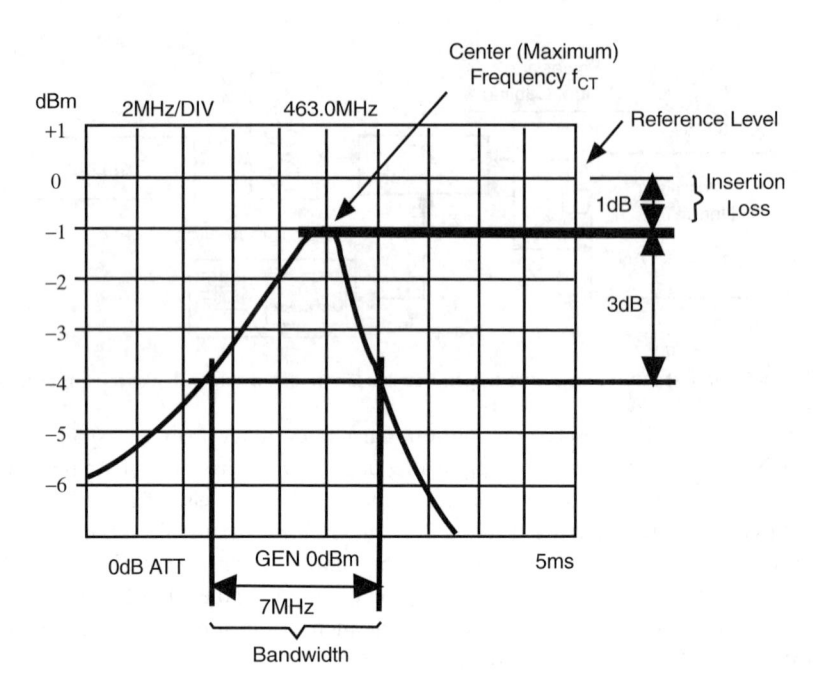

FIGURE 21.6 Typical spectrum analyzer display of a single-cavity filter. Bandwidth (BW) = 458 MHz to 465 MHz = 7 MHz. Filter $Q = f_{CT}/BW$ = 463 MHz/7 MHz = 66. The trace shows 1-dB insertion loss.

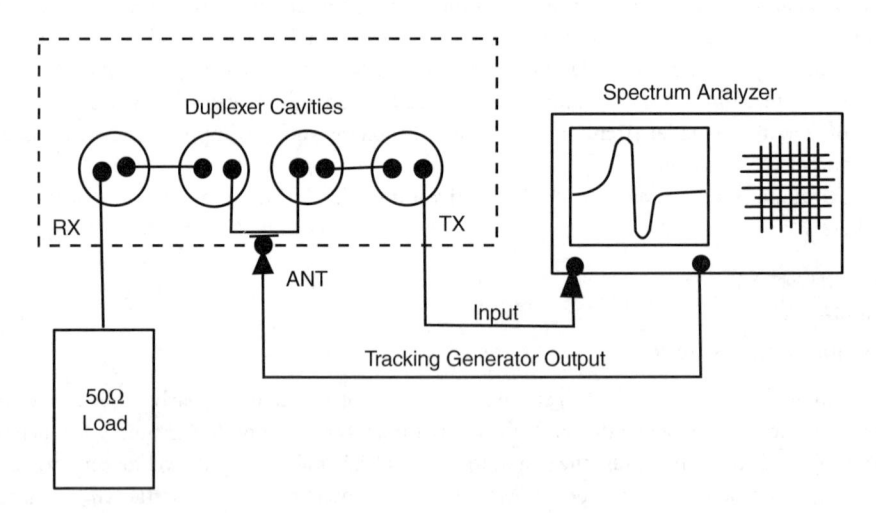

FIGURE 21.7 Test setup for duplexer tuning using a spectrum analyzer.

of the duplexer until no further improvement can be gained. To ensure proper alignment, terminate all open ports into a 50-Ω load while tuning.

The duplexer insertion loss of each port can be measured as the difference between a reference amplitude and the pass frequency amplitude. The reference signal is measured by shorting the generator output cable to the analyzer input cable. This reference level nulls any cable losses to prevent them from being included in duplexer insertion loss measurements.

Small-Signal Troubleshooting

The spectrum analyzer is well-suited for use in small-signal RF troubleshooting. Stage gain, injection level, and signal loss can be easily checked.

Receiver sensitivity loss of less than 6 dB is usually a difficult problem to trace with most test instruments. Such losses are usually associated with the front end of the radio, either as a stage gain or injection amplifier deficiency. Before the spectrum analyzer was commonly available, there was little to aid a technician in tracing such a problem. An RF millivolt meter, with typical usable sensitivity of 1 mV, was of little help in troubleshooting such microvolt-level signals. Oscilloscopes are restricted in the same way, as well as having inadequate bandwidth for the high frequencies used in the front-end circuits of most receivers. A spectrum analyzer, with a typical 1 µV sensitivity and broad RF bandwidth, is well-adapted for RF troubleshooting. Microvolt-level signals can be accurately traced, allowing signal losses to be examined at each individual stage.

To conduct measurements, inject a test signal generated by a communications monitor into the antenna input of the radio under test. Use an appropriate probe to trace the signal from stage to stage. Stage gain, mixer output, injection level, and filter loss can be checked from the antenna to the discriminator. Compare measurements taken with those from a correctly functioning unit to isolate the problem. A standard test signal level of 100 µV usually works well. It is large enough to allow testing of the RF amplifier without causing overload. If the radio includes an automatic gain control (AGC) circuit, disable it temporarily during testing to permit measurement of the true stage gain.

When tracing a loss of sensitivity, start with the RF mixer. The mixer provides a junction point for narrowing the direction of troubleshooting. If the output of the mixer is correct, the defective stage lies beyond the mixer, probably in the IF amplifier. If the mixer output is low, check the input signal to the mixer. Check the RF amplifier, input cavities, and receive/transmit (RX/TX) switching circuit (if used). If the mixer IF carrier injection is low, check the multipliers and oscillators along that path.

The bandwidth of IF filters can be measured in-circuit by varying the receiver input frequency and measuring the bandwidth of the 3-dB roll-off points. A crystal bandpass filter can be tested for flaws by injecting an over-deviated signal into the receiver. A cracked filter crystal will produce a sharp, deep notch on the analyzer display. The key to comparative troubleshooting is accurate recording of test signal levels of a radio known to be working correctly prior to troubleshooting.

Remember to protect the front end of the analyzer from dc voltages and overload when tracing signals with a probe. When testing unknown signal levels, use maximum input attenuation and external attenuators as needed to avoid analyzer damage. Isolate the analyzer input from dc voltages using a capacitively coupled probe.

Defining Terms[1]

1 dB compression point: The point approaching saturation at which the output is 1 dB less than it should be if the output linearly followed input.
Antenna sweep: A technique for measuring the return loss of an antenna to determine the antenna tuning.
Average detection: A detection scheme wherein the average (mean) amplitude of a signal is measured and displayed.
B-SAVE A (or B, C MINUS A): Waveform subtraction mode wherein a waveform in memory is subtracted from a second, active waveform and the result displayed on screen.
Band switching: Technique for changing the total range of frequencies (band) to which a spectrum analyzer can be tuned.
Baseband: The lowest frequency band in which a signal normally exists (often the frequency range of a receiver's output or a modulator's input); the band from dc to a designated frequency.
Baseline clipper: A means of blanking the bright baseline portion of the analyzer display.
Bessel functions: Solutions to a particular type of differential equation; predicts the amplitudes of FM signal components.

---

[1]Definitions courtesy of Tektronix, Beauerton, Oregon.

Bessel null method: A technique most often used to calibrate FM deviation meters. A modulating frequency is chosen such that some frequency component of the FM signal nulls at a specified peak deviation.

Calibrator: A signal generator producing a specified output used for calibration purposes.

Carrier-to-noise ratio (C/N): The ratio of carrier signal power to average noise power in a given bandwidth surrounding the carrier; usually expressed in decibels.

Center frequency: The frequency at the center of a given spectrum analyzer display.

Coax bands: The range of frequencies that can be satisfactorily passed via coaxial cable.

Comb generator: A source producing a fundamental frequency component and multiple components at harmonics of the fundamental.

Component: In spectrum analysis, usually denotes one of the constituent sine waves making up electrical signals.

Decibel (dB): Ten times the logarithm of the ratio of one electrical power to another.

Delta F (ΔF): A mode of operation on a spectrum analyzer wherein a difference in frequency may be read out directly.

Distortion: Degradation of a signal, often a result of nonlinear operations, resulting in unwanted signal components. Harmonic and intermodulation distortion are common types.

Dynamic range: The maximum ratio of two simultaneously present signals that can be measured to a specified accuracy.

Emphasis: Deliberate shaping of a signal spectrum or some portion thereof, often used as a means of overcoming system noise. Pre-emphasis is often used before signal transmission and de-emphasis after reception.

Envelope: The limits of an electrical signal or its parameters. For example, the modulation envelope limits the amplitude of an AM carrier.

Equivalent noise bandwidth: The width of a rectangular filter that produces the same noise power at its output as an actual filter when subjected to a spectrally flat input noise signal. Real filters pass different noise power than implied by their nominal bandwidths because their skirts are not infinitely sharp.

External mixers: A mixer, often in a waveguide format, that is used external to a spectrum analyzer.

Filter: A circuit that separates electrical signals or signal components based on their frequencies.

Filter loss: The insertion loss of a filter is the minimum difference, in decibels (dB), between the input signal level and the output level.

First mixer input level: Signal amplitude at the input to the first mixer stage of a spectrum analyzer. An optimum value is usually specified by the manufacturer.

Flatness: Unwanted variations in signal amplitude over a specified bandwidth, usually expressed in decibels (dB).

Fourier analysis: A mathematical technique for transforming a signal from the time domain to the frequency domain, and vice versa.

Frequency: The rate at which a signal oscillates, or changes polarity, expressed as hertz or number of cycles per second.

Frequency band: A range of frequencies that can be covered without switching.

Frequency deviation: The maximum difference between the instantaneous frequency and the carrier frequency of an FM signal.

Frequency domain representation: The portrayal of a signal in the frequency domain; representing a signal by displaying its sine wave components; the signal spectrum.

Frequency marker: An intensified or otherwise distinguished spot on a spectrum analyzer display that indicates a specified frequency point.

Frequency range: That range of frequencies over which the performance of the instrument is specified.

Fundamental frequency: The basic rate at which a signal repeats itself.

Grass: Noise or a noise-like signal giving the ragged, hashy appearance of grass seen close-up at eye level.

Graticule: The calibrated grid overlaying the display screen of spectrum analyzers, oscilloscopes, and other test instruments.

Harmonic distortion: The distortion that results when a signal interacts with itself, often because of nonlinearities in the equipment, to produce sidebands at multiples, or harmonics, of the frequency components of the original signal.

Harmonic mixing: A technique wherein harmonics of the local oscillator signal are deliberately mixed with the input signal to achieve a large total input bandwidth. Enables a spectrum analyzer to function at higher frequencies than would otherwise be possible.

Harmonics: Frequency components of a signal occurring at multiples of the signal's fundamental frequency.

Heterodyne spectrum analyzer: A type of spectrum analyzer that scans the input signal by sweeping the incoming frequency band past one of a set of fixed RBW filters and measuring the signal level at the output of the filter.

Intermediate frequency (IF): In a heterodyne process, the sum or difference frequency at the output of a mixer stage which will be used for further signal processing.

IF gain: The gain of an amplifier stage operating at IF.

Instantaneous frequency: The rate of change of the phase of a sinusoidal signal at a particular instant.

Intermodulation distortion: The distortion that results when two or more signals interact, usually because of nonlinearities in the equipment, to produce new signals.

Linear scale: A scale wherein each increment represents a fixed difference between signal levels.

LO output: A port on a spectrum analyzer where a signal from the local oscillator (LO) is made available; used for tracking generators and external mixing.

Local oscillator (LO): An oscillator that produces the internal signal that is mixed with an incoming signal in a mixer to produce the IF signal.

Logarithmic scale: A scale wherein each scale increment represents a fixed ratio between signal levels.

Magnitude-only measurement: A measurement that responds only to the magnitude of a signal and is insensitive to its phase.

MAX HOLD: A spectrum analyzer feature that captures the maximum signal amplitude at all displayed frequencies over a series of sweeps.

Max span: The maximum frequency span that can be swept and displayed by a spectrum analyzer.

MAX/MIN: A display mode on some spectrum analyzers that shows the maximum and minimum signal levels at alternate frequency points; its advantage is its resemblance to an analog display.

Maximum input level: The maximum input signal amplitude that can be safely handled by a particular instrument.

MIN HOLD: A spectrum analyzer feature that captures the minimum signal amplitude at all displayed frequencies over a series of sweeps.

Mixing: The process whereby two or more signals are combined to produce sum and difference frequencies of the signals and their harmonics.

Noise: Unwanted random disturbances superimposed on a signal that tends to obscure it.

Noise bandwidth: The frequency range of a noise-like signal. For white noise, the noise power is directly proportional to the bandwidth of the noise.

Noise floor: The self-noise of an instrument or system that represents the minimum limit at which input signals can be observed. The spectrum analyzer noise floor appears as a "grassy" baseline in the display, even when no signal is present.

Noise sideband: Undesired response caused by noise internal to the spectrum analyzer appearing on the display immediately around a desired response, often having a pedestal-like appearance.

Peak/average cursor: A manually controllable function that enables the user to set the threshold at which the type of signal processing changes prior to display in a digital storage system.

Peak detection: A detection scheme wherein the peak amplitude of a signal is measured and displayed. In spectrum analysis, 20 log(peak) is often displayed.

Peaking: The process of adjusting a circuit for maximum amplitude of a signal by aligning internal filters. In spectrum analysis, peaking is used to align preselectors.

Period: The time interval at which a process recurs; the inverse of the fundamental frequency.

Phase lock: The control of an oscillator or signal generator so as to operate at a constant phase angle relative to a stable reference signal source. Used to ensure frequency stability in spectrum analyzers.

Preselector: A tracking filter located ahead of the first mixer that allows only a narrow band of frequencies to pass into the mixer.

Products: Signal components resulting from mixing or from passing signals through other nonlinear operations such as modulation.

Pulse stretcher: Pulse shaper that produces an output pulse whose duration is greater than that of the input pulse and whose amplitude is proportional to that of the peak amplitude of the input pulse.

Pulse repetition frequency (PRF): The frequency at which a pulsing signal recurs; equal to the fundamental frequency of the pulse train.

Reference level: The signal level required to deflect the CRT display to the top graticule line.

Reference-level control: The control used to vary the reference level on a spectrum analyzer.

Resolution bandwidth (RBW): The width of the narrowest filter in the IF stages of a spectrum analyzer. The RBW determines how well the analyzer can resolve or separate two or more closely spaced signal components.

Return loss: The ratio of power sent to a system to that returned by the system. In the case of antennas, the return loss can be used to find the SWR.

Ring: A transient response wherein a signal initially performs a damped oscillation about its steady-state value.

SAVE function: A feature of spectrum analyzers incorporating display storage that enables them to store displayed spectra.

Sensitivity: Measure of a spectrum analyzer's ability to display minimum level signals at a given IF bandwidth, display mode, and any other influencing factors.

Shape factor: In spectrum analysis, the ratio of an RBW filter's 60-dB bandwidth to its 3-dB or 6-dB width (depending on manufacturer).

Sideband: Signal components observable on either or both sides of a carrier as a result of modulation or distortion processes.

Sideband suppression: An amplitude modulation technique wherein one of the AM sidebands is deliberately suppressed, usually to conserve bandwidth.

Single sweep: Operating mode in which the sweep generator must be reset for each sweep. Especially useful for obtaining single examples of a signal spectrum.

Span per division, Span/Div: Frequency difference represented by each major horizontal division of the graticule.

Spectrum: The frequency domain representation of a signal wherein it is represented by displaying its frequency distribution.

Spurious response: An undesired extraneous signal produced by mixing, amplification, or other signal-processing technique.

Stability: The property of retaining defined electrical characteristics for a prescribed time and in specified environments.

Starting frequency: The frequency at the left-hand edge of the spectrum analyzer display.

Sweep speed, sweep rate: The speed or rate, expressed in time per horizontal divisions, at which the electron beam of the CRT sweeps the screen.

Time-domain representation: Representation of signals by displaying the signal amplitude as a function of time. Typical of oscilloscope and waveform monitor displays.

Time-varying signal: A signal whose amplitude changes with time.

Total span: The total width of the displayed spectrum. The Span/Div times the number of divisions.

Tracking generator: A signal generator whose output frequency is synchronized to the frequency being analyzed by the spectrum analyzer.

Ultimate rejection: The ratio, in dB, between a filter's passband response and its response beyond the slopes of its skirts.
Vertical scale factor, vertical display factor: The number of dB, volts, etc. represented by one vertical division of a spectrum analyzer display screen.
Waveform memory: Memory dedicated to storing a digital replica of a spectrum.
Waveform subtraction: A process wherein a saved waveform can be subtracted from a second, active waveform.
Zero hertz peak: A fictitious signal peak occurring at zero hertz that conveniently marks zero frequency. The peak is present regardless of whether or not there is an input signal.
Zero span: A spectrum analyzer mode of operation in which the RBW filter is stationary at the center frequency; the display is essentially a time-domain representation of the signal propagated through the RBW filter.

Bibliography

Kinley, Harold, Using service monitor/spectrum analyzer combos, *Mobile Radio Technology*, Intertec Publishing, Overland Park, KS, July 1987.

Pepple, Carl, How to use a spectrum analyzer at the cell site, *Cellular Business*, Intertec Publishing, Overland Park, KS, March 1989.

Whitaker, Jerry C., *Maintaining Electronic Systems*, CRC Press, Boca Raton, FL, 1991.

Wolf, Richard J., Spectrum analyzer uses for two-way technicians, *Mobile Radio Technology*, Intertec Publishing, Overland Park, KS, July 1987.

22
Testing Coaxial Transmission Line

Jerry C. Whitaker
Editor

22.1 Introduction 22-1
   Measuring VSWR · Antenna Measurements
22.2 Testing Coaxial Lines 22-3
   Crimps and Mismatches

22.1 Introduction

Antenna and transmission line performance measurements are among the most neglected and least understood parameters at most transmission facilities. Many facilities do not have the equipment to perform useful measurements. Experience is essential because much of the knowledge obtained from such tests is derived by interpreting the raw data. In general, transmission systems measurements should be made:

- *Before and during installation of the antenna and transmission line.* Barring unforeseen operational problems, this will be the only time that the antenna is at ground level. Ready access to the antenna allows a variety of key measurements to be performed without climbing the tower.
- *During system troubleshooting when attempting to locate a problem.* Following installation, these measurements usually concern the transmission line itself.

To ensure that the transmission line is operating normally, many facilities check the transmission line and antenna system on a regular basis. A quick sweep of the line with a network analyzer and a time-domain reflectometer (TDR) may disclose developing problems before they can cause a transmission line failure.

Ideally, the measurements should be used to confirm a good impedance match, which can be interpreted as minimum VSWR or maximum *return loss*. Return loss is related to the level of signal that is returned to the input connector after the signal has been applied to the transmission line and reflected from the load. A line perfectly matched to the load would transfer all energy to the load. No energy would be returned, resulting in an infinite return loss, or an ideal VSWR of 1:1. The benefits of matching the transmission line system for minimum VSWR include:

- Most efficient power transfer from the transmitter to the antenna system
- Best performance with regard to overall bandwidth
- Improved transmitter stability with tuning following accepted procedures more closely
- Minimum transmitted signal distortions

The network analyzer allows the maintenance engineer to perform a number of critical measurements in a short period of time. The result is an antenna system that is tuned as close as practical for uniform

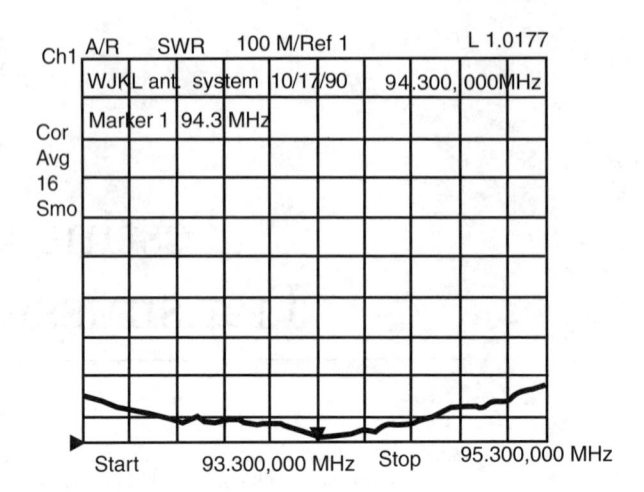

FIGURE 22.1 Network analyzer plot of an FM broadcast antenna.

impedance across the operating bandwidth. A well-matched system increases operating efficiency by properly coupling the signal from the transmitter to the antenna. Figure 22.1 shows a network analyzer plot of an FM broadcast antenna.

Measuring VSWR

Historically, a *slotted line* device has commonly been used to measure VSWR on a transmission line. A slotted line includes a probe that penetrates the outer conductor of the line through a slot. The probe, in close proximity with the inner conductor, measures the voltage or samples the field along the center conductor. The sample is detected, which results in a voltage proportional to the actual signal on the center conductor. It is an accurate, reliable instrument. However, the slotted line procedure takes a considerable amount of time to accurately sweep a transmission line over a wide bandwidth, and then to plot the resulting data.

A network analyzer incorporating a return loss bridge performs antenna measurements more quickly. The analog network analyzer consists of a sweep generator coupled to a tracking receiver. A sample of the signal applied to the transmission line is compared to the return signal through a return loss bridge or directional coupler. By adding a storage or *normalizer* device to store the signal digitally, the instrument can provide a stable display while sweeping the line at low speed to find all irregularities that may exist at discrete frequencies.

Digital designs do not use a sweep generator, but instead an integral synthesizer. In this way, the return loss is measured at discrete frequencies. Software-calibration procedures correct each measurement for system frequency and phase response errors, delay irregularities, and directivity errors in the return loss bridge or directional coupler. By calibrating a software-controlled unit at the top of the transmission line, measurements will accurately show antenna characteristics without effects of the transmission line. Results are plotted on an X-Y plotter or defined and stored for later printout.

One particularly desirable feature of a network analyzer is its capability to display either a Smith chart or a more simple Cartesian X-Y presentation of return loss-vs.-frequency. (Some units may provide both displays simultaneously.) The Smith chart is useful, but interpretation can be confusing. The Cartesian presentation is usually easier to interpret, but technically is not better.

Calibration

Calibration methods vary for different instruments. For one method, a short circuit is placed across the network analyzer terminals, producing a return loss of zero (the short reflects all signals applied to it). The instrument is then checked with a known termination. This step often causes the inexperienced technician to go astray. The termination should have known characteristics and full documentation. It is acceptable procedure to check the equipment by examining more than one termination, where the operator knows the characteristics of the devices used. Significant changes from the known characteristics suggest that additional tests should be performed. After the test unit is operating correctly, check to ensure that the adapters and connectors to be used in the measurement do not introduce errors of their own. An accepted practice for this involves the use of a piece of transmission line of known quality. A 20-ft section of line should sufficiently separate the input and output connectors. The results of any adjustment at either end will be noticeable on the analyzer. Also, the length allows adjustments to be made quite easily. The section of line used should include tuning stubs or tuners to permit the connectors to be matched to the line across the operating channel.

The facility's dummy load must next be matched to the transmission line. Do not assume that the dummy load is an appropriate termination by itself, or a station reference. The primary function of a dummy load is to dissipate power in a manner that allows easy measurement. It is neither a calibration standard nor a reference. Experience proves it is necessary to match dummy and transmission line sections to maintain a good reference. The load is matched by looking into the transmission line at the patch panel (or other appropriate point). Measurements are then taken at locations progressively closer to the transmitter, until the last measurement is made at the output connection of the transmitter. After the dummy load is checked, it serves as a termination.

Antenna Measurements

An antenna should be properly tuned before placing into service. Any minor tuning adjustments to the antenna should be made at its base, not by compensation at the input to the transmission line. Impedance adjustments are typically made with tuning rings on the center conductor or with an impedance-matching section. Adjustments are performed while observing the return loss on the network analyzer. Transmission line rings are less convenient, but less expensive than an impedance-matching section. The rings can be used for matching short runs or the overall line between the transmitting equipment and the antenna. Either tuning method can be used at the antenna.

Both tuning systems operate by introducing a discontinuity into the transmission line. The ring effectively changes the diameter of the center conductor, causing an impedance change at that point on the line. This introduces a reflection into the line, the magnitude of which is a function of the size of the ring. The phase of the reflection is a function of the location of the ring along the length of the center conductor.

Installing the ring is usually a "cut-and-try" process. It may be necessary to open, adjust, close, and test the line several times. However, after a few cuts, the effect of the ring will become apparent. It is not uncommon to need more than one ring on a given piece of transmission line for a good match over the required bandwidth. When a match is obtained, the ring is normally soldered into place.

Impedance-matching hardware is also available for use with waveguide. A piece of material is placed into the waveguide and its location is adjusted to create the desired mismatch. For any type of line, the goal is to create a mismatch equal in magnitude, but opposite in phase, to the existing undesirable mismatch. The overall result is a minimum mismatch and minimum VSWR.

A tuner alters the line characteristic impedance at a given point by changing the distance between the center and outer conductors by effectively moving the outer conductor. In reality, it increases the capacity between the center and outer conductors to produce a change in the impedance and introduce a reflection at that point.

22.2 Testing Coaxial Lines

When dealing with transmission lines, a few electrical parameters have great influence on the ability of the cable to transfer energy from one place to another. Among these are surge impedance (Z_O), the dielectric constants (k) of the insulating materials, and the velocity of propagation (V_p) of electromagnetic waves, both in free space and within the transmission lines.

In a coaxial cable with a single center conductor and an outer conductor, the surge impedance is determined by the relationship

$$Z_O = \frac{138}{\sqrt{k}} \times \log_{10} \frac{D}{d} \tag{22.1}$$

where k is the dielectric constant of the insulating material, D is the inside diameter of the outer conductor, and d is the outside diameter of the center conductor.

In two-conductor transmission lines, the surge impedance is determined by

$$Z_O = \frac{276}{\sqrt{k}} \times \log_{10} \frac{2S}{s} \tag{22.2}$$

where k is the dielectric constant of the insulating material, S is the spacing between the centers of the two conductors, and s is the diameter of the conductors. (Diameters and spacings can be measured in either centimeters or inches.)

If the inductance and capacitance per foot (or meter) of the transmission lines are given by the manufacturer, the surge impedance can be calculated from

$$Z_O = \sqrt{L/C} \tag{22.3}$$

where L is the quoted inductance per unit length, and C is the quoted capacitance per unit length. Note that the actual *length* of the line is not a factor in any of these formulas. The surge impedance of the line is independent of cable length and wholly dependent on cable type.

One common variety of 1/2-inch foam dielectric coaxial cable has an inductance per foot of 0.058 microhenries (mH) and a capacitance per foot of 23.1 picofarads (pf). Its surge impedance is therefore

$$Z_O = \sqrt{(0.058 \times 10^{-6}/23.1 \times 10^{-12})} = 50 \ \Omega \tag{22.4}$$

Most transmitters use 50-Ω or 52-Ω coaxial transmission lines to feed energy to an antenna. Receivers traditionally employ 75-Ω or 300-Ω impedances.

Crimps and Mismatches

Because the surge impedance of a transmission line is based on its cross-sectional characteristics, its impedance can be changed if the cable is compressed or bent past its recommended bend radius. The inner and outer conductors would no longer have the same spacing between them at that point, causing a change in impedance there (Eq. 22.1.) This change of impedance causes some of the energy in the line to be reflected in both directions. Some of the energy never reaches the load. Standing waves are created, which in turn cause energy to be radiated by the transmission line. (A perfectly matched transmitter, line, and load system has no standing waves, no lost energy except for dielectric losses, and no radiation from the transmission line itself.)

A crimp in a line will cause an increase in capacitance at that point. This will cause the impedance to be reduced. Mismatches can also occur between connectors and sections of transmission line. Manufacturers will often specify what type of cable a connector is to be used with instead of listing its surge impedance. Although this may be adequate for ordering connectors for a given installation, consider the spare parts stock issue.

Multiple Mismatches

Unfortunately, if there is more than one discontinuity in the cable system, it becomes more difficult to determine the impedance of the second and subsequent mismatches.

Some TDR manufacturers have attempted to provide correction factors to account for multiple mismatch errors, but they may not work for all situations. In general, the larger the first mismatch, the greater the error in calculating the second mismatch. When the values of the mismatched cable sections and/or loads are known, the technician can work backward and calculate what the values should have been. When the cable values and/or loads are unknown, only the first mismatch can be calculated accurately.

Bibliography

Cable Testing with Time Domain Reflectometry, Application Note 67, Hewlett Packard, Palo Alto, CA, 1988.

Kennedy, George, *Electronic Communication Systems*, 3rd ed., McGraw-Hill, New York, 1985.

Kolbert, Don, Testing Coaxial Lines, *Broadcast Engineering*, Intertec Publishing, Overland Park, KS, November 1991.

Improving Time Domain Network Analysis Measurements, Application Note 62, Hewlett Packard, Palo Alto, CA, 1988.

Strickland, James A., Time Domain Reflectometry Measurements, Measurement Concepts Series, Tektronix, Beaverton, OR, 1970.

TDR Fundamentals, Application Note 62, Hewlett Packard, Palo Alto, CA, 1988.

23
Safety Issues for RF Systems

Jerry C. Whitaker
Editor

23.1	Introduction 23-1
	Facility Safety Equipment
23.2	Electric Shock 23-4
	Effects on the Human Body · Circuit-Protection Hardware · Working with High Voltage · First-Aid Procedures
23.3	Polychlorinated Biphenyls 23-11
	Health Risk · Governmental Action · PCB Components · Identifying PCB Components · Labeling PCB Components · Record-Keeping · Disposal · Proper Management
23.4	OSHA Safety Requirements............................. 23-16
	Protective Covers · Identification and Marking · Extension Cords · Grounding
23.5	Beryllium Oxide Ceramics 23-19
23.6	Corrosive and Poisonous Compounds........... 23-19
	FC-75 Toxic Vapor
23.7	Nonionizing Radiation 23-20
	NEPA Mandate · Revised Guidelines · Multiple-User Sites · Operator Safety Considerations
23.8	X-Ray Radiation Hazard.................................... 23-22
	Implosion Hazard
23.9	Hot Coolant and Surfaces 23-22
23.10	Management Responsibility 23-22

23.1 Introduction

Safety is critically important to engineering personnel who work around powered hardware, especially if they work under considerable time pressures. Safety is not something to be taken lightly. *Life safety* systems are those designed to protect life and property. Such systems include emergency lighting, fire alarms, smoke exhaust and ventilating fans, and site security.

Facility Safety Equipment

Personnel safety is the responsibility of the facility manager. Proper life safety procedures and equipment must be installed. Safety-related hardware includes the following:

- *Emergency power off* (EPO) *button*. EPO push buttons are required by safety code for data processing centers. One must be located at each principal exit from the data processing (DP) room. Other EPO buttons may be located near operator workstations. The EPO system, intended only for emergencies, disconnects all power to the room, except for lighting.

- *Smoke detector.* Two basic types of smoke detectors commonly are available. The first compares the transmission of light through air in the room with light through a sealed optical path into which smoke cannot penetrate. Smoke causes a differential or *backscattering* effect that, when detected, triggers an alarm after a preset threshold has been exceeded. The second type of smoke detector senses the ionization of combustion products rather than visible smoke. A mildly radioactive source, usually nickel, ionizes the air passing through a screened chamber. A charged probe captures ions and detects the small current that is proportional to the rate of capture. When combustion products or material other than air molecules enter the probe area, the rate of ion production changes abruptly, generating a signal that triggers the alarm.
- *Flame detector.* The flame sensor responds not to heated surfaces or objects, but to infrared when it flickers with the unique characteristics of a fire. Such detectors, for example, will respond to a lighted match, but not to a cigarette. The ultraviolet light from a flame also is used to distinguish between hot, glowing objects and open flame.
- *Halon.* The Halon fire-extinguishing agent is a low-toxicity, compressed gas that is contained in pressurized vessels. Discharge nozzles in data processing (DP) rooms and other types of equipment rooms are arranged to dispense the entire contents of a central container or of multiple smaller containers of Halon when actuated by a command from the fire control system. The discharge is sufficient to extinguish flame and stop combustion of most flammable substances. Halon is one of the more common fire-extinguishing agents used for DP applications. Halon systems usually are not practical, however, in large, open-space computer centers.
- *Water sprinkler.* Although water is an effective agent against a fire, activation of a sprinkler system will cause damage to the equipment it is meant to protect. Interlock systems must drop all power (except for emergency lighting) before the water system is discharged. Most water systems use a two-stage alarm. Two or more fire sensors, often of different design, must signal an alarm condition before water is discharged into the protected area. Where sprinklers are used, floor drains and EPO controls must be provided.
- *Fire damper.* Dampers are used to block ventilating passages in strategic parts of the system when a fire is detected. This prevents fire from spreading through the passages and keeps fresh air from fanning the flames. A fire damper system, combined with the shutdown of cooling and ventilating air, enables Halon to be retained in the protected space until the fire is extinguished.

Many life safety system functions can be automated. The decision of what to automate and what to operate manually requires considerable thought. If the life safety control panels are accessible to a large number of site employees, most functions should be automatic. Alarm-silencing controls should be maintained under lock and key. A mimic board can be used to readily identify problem areas. Figure 23.1 illustrates a well-organized life safety control system. Note that fire, HVAC (heating, ventilation, and air-conditioning), security, and EPO controls all are readily accessible. Note also that operating instructions are posted for life safety equipment, and an evacuation route is shown. Important telephone numbers are posted, and a direct-line telephone (not via the building switchboard) is provided. All equipment is located adjacent to a lighted emergency exit door.

Life safety equipment must be maintained just as diligently as the computer system that it protects. Conduct regular tests and drills. It is, obviously, not necessary or advisable to discharge Halon or water during a drill.

Configure the life safety control system to monitor not only the premises for dangerous conditions, but also the equipment designed to protect the facility. Important monitoring points include HVAC machine parameters, water and/or Halon pressure, emergency battery-supply status, and other elements of the system that could compromise the ability of life safety equipment to carry out its functions. Basic guidelines for life safety systems include the following:

- Carefully analyze the primary threats to life and property within the facility. Develop contingency plans to meet each threat.

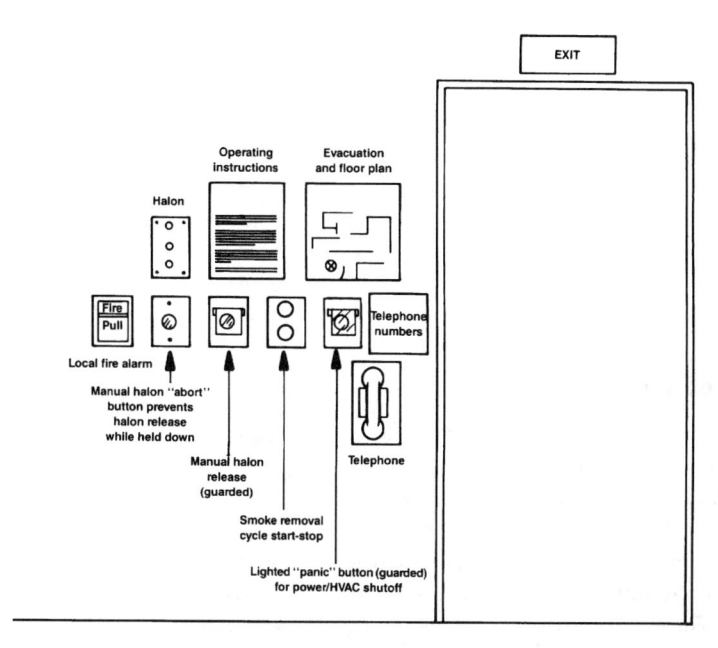

FIGURE 23.1 A well-organized life safety control station. (Adapted from Federal Information Processing Standards Publication No. 94, *Guideline on Electrical Power for ADP Installations*, U.S. Department of Commerce, National Bureau of Standards, Washington, D.C., 1983.)

- Prepare a life safety manual and distribute it to all employees at the facility. Require them to read it.
- Conduct drills for employees at random times without notice. Require acceptable performance from employees.
- Prepare simple, step-by-step instructions on what to do in an emergency. Post the instructions in a conspicuous place.
- Assign after-hours responsibility for emergency situations. Prepare a list of supervisors who operators should contact if problems arise. Post the list with phone numbers. Keep the list accurate and up-to-date. Always provide the names of three individuals who can be contacted in an emergency.
- Work with a life safety consultant to develop a coordinated control and monitoring system for the facility. Such hardware will be expensive, but it must be provided. The facility may be able to secure a reduction in insurance rates if comprehensive safety efforts can be demonstrated.
- Interface the life safety system with automatic data-logging equipment so that documentation can be assembled on any event.
- Insist upon complete, up-to-date schematic diagrams for all hardware at the facility. Insist that the diagrams include any changes made during installation or subsequent modification.
- Provide sufficient emergency lighting.
- Provide easy-access emergency exits.

The importance of providing standby power for sensitive loads at commercial and industrial facilities has been outlined previously. It is equally important to provide standby power for life safety systems. A lack of ac power must not render the life safety system inoperative. Sensors and alarm control units should include their own backup battery supplies. In a properly designed system, all life safety equipment will be fully operational despite the loss of all ac power to the facility, including backup power for sensitive loads.

Place cables linking the life safety control system with remote sensors and actuators in separate conduit containing only life safety conductors. Study the National Electrical Code and all applicable local and federal codes relating to safety. Follow them to the letter.

23.2 Electric Shock

It takes surprisingly little current to injure a person. Studies at Underwriters' Laboratories (UL) show that the electrical resistance of the human body varies with the amount of moisture on the skin, the muscular structure of the body, and the applied voltage. The typical hand-to-hand resistance ranges from 500 Ω to 600 kΩ, depending on the conditions. Higher voltages have the capability to break down the outer layers of the skin, which can reduce the overall resistance value. UL uses the lower value, 500 Ω, as the standard resistance between major extremities, such as from the hand to the foot. This value generally is considered the minimum that would be encountered. In fact, it may not be unusual because wet conditions or a cut or other break in the skin significantly reduce human body resistance.

Effects on the Human Body

Table 23.1 lists some effects that typically result when a person is connected across a current source with a hand-to-hand resistance of 2.4 kΩ. The table shows that a current of 50 mA will flow between the hands if one hand is in contact with a 120-V ac source and the other hand is grounded. The table also indicates that even the relatively small current of 50 mA can produce *ventricular fibrillation* of the heart, and maybe even cause death. Medical literature describes ventricular fibrillation as very rapid, uncoordinated contractions of the ventricles of the heart resulting in loss of synchronization between heartbeat and pulse beat. The electrocardiograms shown in Fig. 23.2 compare a healthy heart rhythm with one in ventricular fibrillation. Unfortunately, once ventricular fibrillation occurs, it will continue. Barring resuscitation techniques, death will ensue within a few minutes.

TABLE 23.1 The Effects of Current on the Human Body

1 mA or less	No sensation, not felt
More than 3 mA	Painful shock
More than 10 mA	Local muscle contractions, sufficient to cause "freezing" to the circuit for 2.5% of the population
More than 15 mA	Local muscle contractions, sufficient to cause "freezing" to the circuit for 50% of the population
More than 30 mA	Breathing is difficult, can cause unconsciousness
50 mA to 100 mA	Possible ventricular fibrillation of the heart
100 mA to 200 mA	Certain ventricular fibrillation of the heart
More than 200 mA	Severe burns and muscular contractions; heart more apt to stop than to go into fibrillation
More than a few amperes	Irreparable damage to body tissues

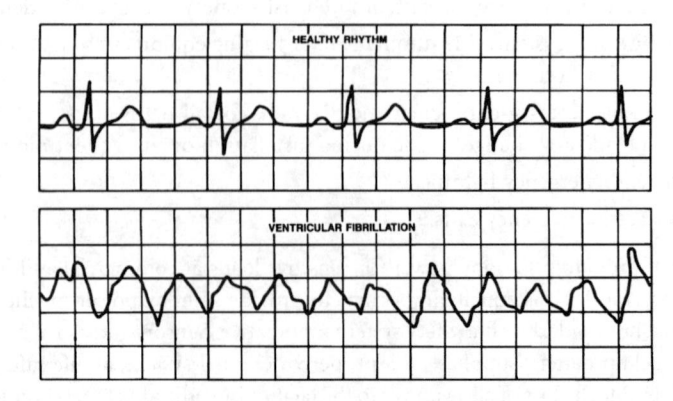

FIGURE 23.2 Electrocardiogram traces: (upper panel) healthy heart rhythm, (lower panel) ventricular fibrillation of the heart.

Safety Issues for RF Systems

The route taken by the current through the body greatly affects the degree of injury. Even a small current, passing from one extremity through the heart to another extremity, is dangerous and capable of causing severe injury or electrocution. There are cases in which a person has contacted extremely high current levels and lived to tell about it. However, when this happens, it is usually because the current passes only through a single limb and not through the entire body. In these instances, the limb is often lost but the person survives.

Current is not the only factor in electrocution. Figure 23.3 summarizes the relationship between current and time on the human body. The graph shows that 100 mA flowing through an adult human body for 2 sec will cause death by electrocution. An important factor in electrocution, the *let-go range*, is also shown on the graph. This point marks the amount of current that causes *freezing*, or the inability to let go of a conductor. At 10 mA, 2.5% of the population would be unable to let go of a live conductor; at 15 mA, 50% of the population would be unable to let go of an energized conductor. It is apparent from the graph that even a small amount of current can freeze someone to a conductor. The objective for those who must work around electric equipment is to protect themselves from electric shock. Table 23.2 lists required precautions for maintenance personnel working near high voltages.

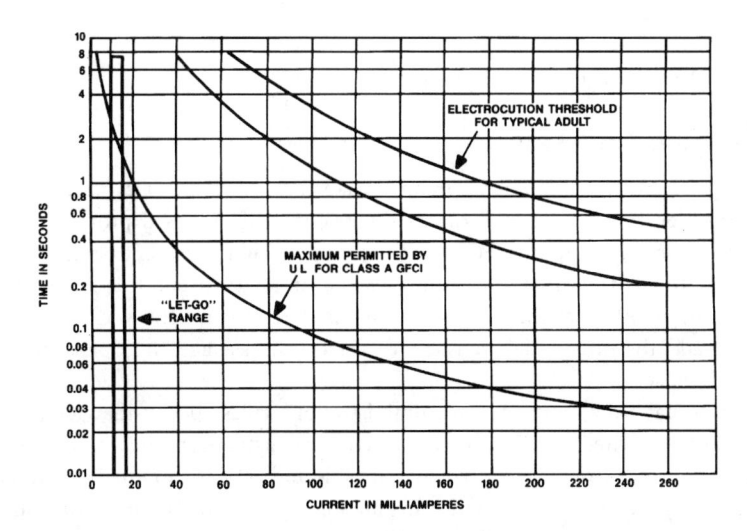

FIGURE 23.3 Effects of electric current and time on the human body. Note the "let-go" range.

TABLE 23.2 Required Safety Practices for Engineers Working around High-Voltage Equipment

- ✓ Remove all ac power from the equipment. Do not rely on internal contactors or SCRs to remove dangerous ac.
- ✓ Trip the appropriate power-distribution circuit breakers at the main breaker panel.
- ✓ Place signs as needed to indicate that the circuit is being serviced.
- ✓ Switch the equipment being serviced to the *local control* mode as provided.
- ✓ Discharge all capacitors using the discharge stick provided by the manufacturer.
- ✓ Do not remove, short-circuit, or tamper with interlock switches on access covers, doors, enclosures, gates, panels, or shields.
- ✓ Keep away from live circuits.
- ✓ Allow any component to cool completely before attempting to replace it.
- ✓ If a leak or bulge is found on the case of an oil-filled or electrolytic capacitor, do not attempt to service the part until it has cooled completely.
- ✓ Know which parts in the system contain PCBs. Handle them appropriately.
- ✓ Minimize exposure to RF radiation.
- ✓ Avoid contact with hot surfaces within the system.
- ✓ Do not take chances.

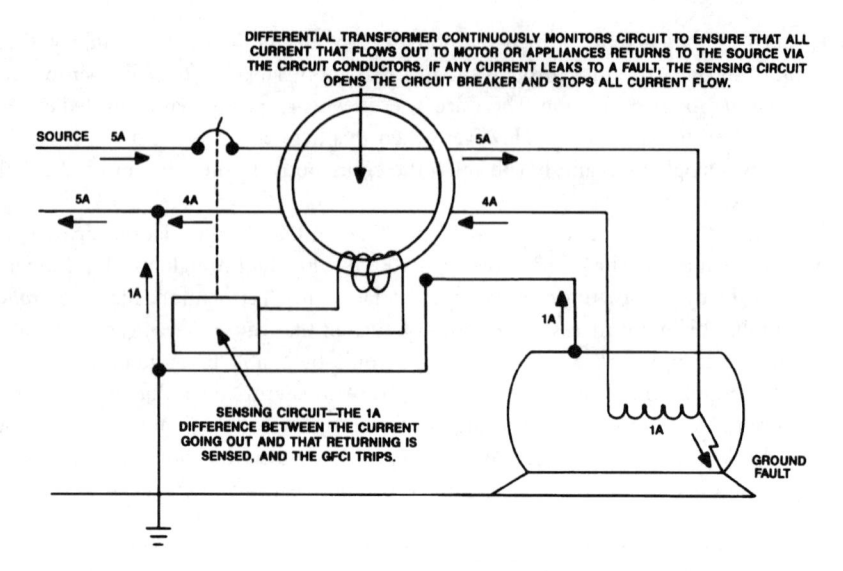

FIGURE 23.4 Basic design of a ground-fault current interrupter (GFCI).

Circuit-Protection Hardware

A common primary panel or equipment circuit breaker or fuse will not protect an individual from electrocution. However, the *ground-fault current interrupter* (GFCI), used properly, can help prevent electrocution. Shown in Fig. 23.4, the GFCI works by monitoring the current being applied to the load. It uses a differential transformer that senses an imbalance in load current. If a current (typically 5 mA, ±1 mA on a low-current 120 V ac line) begins flowing between the neutral and ground or between the hot and ground leads, the differential transformer detects the leakage and opens the primary circuit (typically within 2.5 msec).

OSHA (Occupational Safety and Health Administration) rules specify that temporary receptacles (those not permanently wired) and receptacles used on construction sites be equipped with GFCI protection. Receptacles on two-wire, single-phase portable and vehicle-mounted generators of not more than 5 kW, where the generator circuit conductors are insulated from the generator frame and all other grounded surfaces, need not be equipped with GFCI outlets.

GFCIs will not protect a person from every type of electrocution. If you become connected to both the neutral and the hot wire, the GFCI will treat you as if you are merely a part of the load and will not open the primary circuit.

For large, three-phase loads, detecting ground currents and interrupting the circuit before injury or damage can occur is a more complicated proposition. The classic method of protection involves the use of a zero-sequence current transformer (CT). Such devices are basically an extension of the single-phase GFCI circuit shown in Fig. 23.4. Three-phase CTs have been developed to fit over bus ducts, switchboard buses, and circuit-breaker studs. Rectangular core-balanced CTs are able to detect leakage currents as small as several milliamperes when the system carries as much as 4 kA. "Doughnut-type" toroidal zero-sequence CTs are also available in varying diameters.

The zero-sequence current transformer is designed to detect the magnetic field surrounding a group of conductors. As shown in Fig. 23.5, in a properly operating three-phase system, the current flowing through the conductors of the system, including the neutral, goes out and returns along those same conductors. The net magnetic flux detected by the CT is zero. No signal is generated in the transformer winding, regardless of current magnitudes — symmetrical or asymmetrical. If one phase conductor is faulted to ground, however, the current balance will be upset. The ground-fault-detection circuit then will trip the breaker and open the line.

Safety Issues for RF Systems

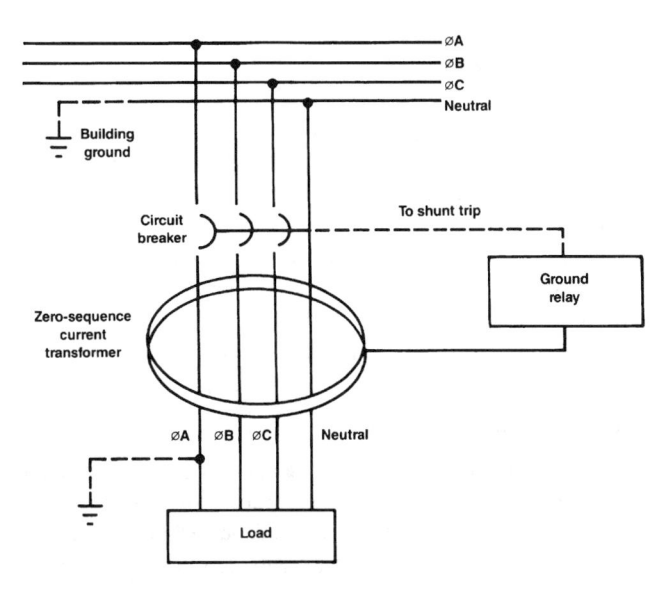

FIGURE 23.5 Ground-fault detection in a three-phase ac system.

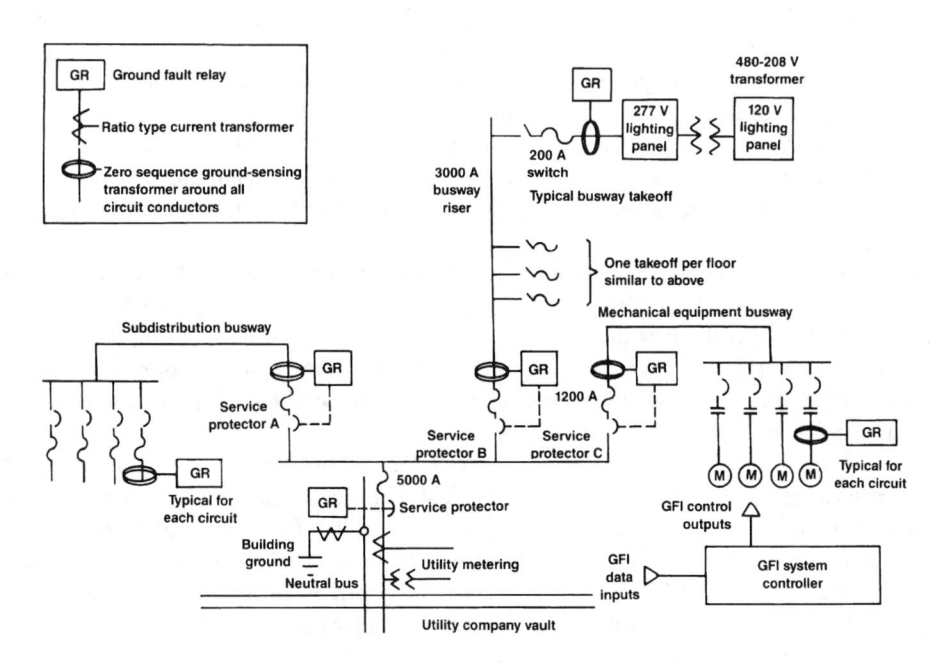

FIGURE 23.6 Ground-fault protection system for a large, multistory building.

For optimum protection in a large facility, GFCI units are placed at natural branch points of the ac power system. It is obviously preferable to lose only a small portion of a facility in the event of a ground fault than it is to have the entire plant dropped. Figure 23.6 illustrates such a distributed system. Sensors are placed at major branch points to isolate any ground fault from the remainder of the distribution network. In this way, the individual GFCI units can be set for higher sensitivity and shorter time delays than would be practical with a large, distributed load. The technology of GFCI devices has improved significantly within the past few years. New integrated circuit devices and improved CT designs have provided improved protection components at a lower cost.

Sophisticated GFCI monitoring systems are available that analyze ground-fault currents and isolate the faulty branch circuit. This feature prevents needless tripping of GFCI units up the line toward the utility service entrance. For example, if a ground fault is sensed in a fourth-level branch circuit, the GFCI system controller automatically locks out first-, second-, and third-level devices from operating to clear the fault. The problem, therefore, is safely confined to the fourth-level branch. The GFCI control system is designed to operate in a fail-safe mode. In the event of a control-system shutdown, the individual GFCI trip relays would operate independently to clear whatever fault currents may exist.

Any facility manager would be well-advised to hire an experienced electrical contractor to conduct a full ground-fault protection study. Direct the contractor to identify possible failure points and to recommend corrective actions.

An extensive discussion of GFCI principles and practices can be found in Ref. 2.

Working with High Voltage

Rubber gloves are a common safety measure used by engineers working on high-voltage equipment. These gloves are designed to provide protection from hazardous voltages when the wearer is working on "hot" circuits. Although the gloves may provide some protection from these hazards, placing too much reliance on them poses the potential for disastrous consequences. There are several reasons why gloves should be used only with a great deal of caution and respect. A common mistake made by engineers is to assume that the gloves always provide complete protection. The gloves found in some facilities may be old and untested. Some may even have been "repaired" by users, perhaps with electrical tape. Few tools could be more hazardous than such a pair of gloves.

Know the voltage rating of the gloves. Gloves are rated differently for ac and dc voltages. For instance, a *class 0* glove has a minimum dc breakdown voltage of 35 kV; the minimum ac breakdown voltage, however, is only 6 kV. Furthermore, high-voltage rubber gloves are not tested at RF frequencies, and RF can burn a hole in the best of them. Working on live circuits involves much more than simply wearing a pair of gloves. It involves a frame of mind — an awareness of everything in the area, especially ground points.

Gloves alone may not be enough to protect an individual in certain situations. Recall the axiom of keeping one hand in your pocket while working on a device with current flowing? The axiom actually is based on simple electricity. It is not the hot connection that causes the problem; it is the ground connection that permits current flow. Studies have showed that more than 90% of electric equipment fatalities occurred when the grounded person contacted a live conductor. Line-to-line electrocution accounted for less than 10% of the deaths.

When working around high voltages, always look for grounded surfaces — and keep away from them. Even concrete can act as a ground if the voltage is high enough. If work must be conducted in live cabinets, consider using — in addition to rubber gloves — a rubber floor mat, rubber vest, and rubber sleeves. Although this may seem to be a lot of trouble, consider the consequences of making a mistake. Of course, the best troubleshooting methodology is never to work on any circuit unless you are sure no hazardous voltages are present. In addition, any circuits or contactors that normally contain hazardous voltages should be grounded firmly before work begins.

Another important safety rule is to never work alone. Even if a trained assistant is not available when maintenance is performed, someone should accompany you and be available to help in an emergency.

First-Aid Procedures

Be familiar with first-aid treatment for electric shock and burns. Always keep a first-aid kit on hand at the facility. Figure 23.7 illustrates the basic treatment for electric shock victims. Copy the information, and post it in a prominent location. Better yet, obtain more detailed information from your local heart association or Red Cross chapter. Personalized instruction on first aid is usually available locally. Table 23.3 lists basic first-aid procedures for burns.

Safety Issues for RF Systems

TREATMENT OF ELECTRICAL SHOCK

1. IF VICTIM IS NOT RESPONSIVE, FOLLOW THE A-B-Cs OF BASIC LIFE SUPPORT.

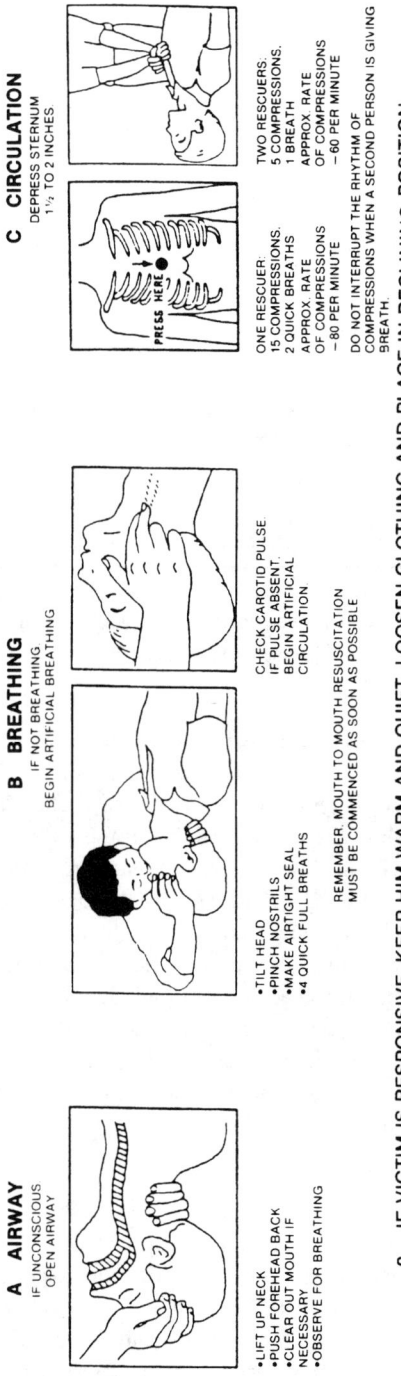

A AIRWAY
IF UNCONSCIOUS
OPEN AIRWAY

- LIFT UP NECK
- PUSH FOREHEAD BACK
- CLEAR OUT MOUTH IF NECESSARY
- OBSERVE FOR BREATHING

B BREATHING
IF NOT BREATHING
BEGIN ARTIFICIAL BREATHING

- TILT HEAD
- PINCH NOSTRILS
- MAKE AIRTIGHT SEAL
- 4 QUICK FULL BREATHS

REMEMBER, MOUTH TO MOUTH RESUSCITATION MUST BE COMMENCED AS SOON AS POSSIBLE

CHECK CAROTID PULSE
IF PULSE ABSENT,
BEGIN ARTIFICIAL CIRCULATION

C CIRCULATION
DEPRESS STERNUM
1½ TO 2 INCHES

ONE RESCUER:
15 COMPRESSIONS.
2 QUICK BREATHS
APPROX. RATE
OF COMPRESSIONS
~80 PER MINUTE

DO NOT INTERRUPT THE RHYTHM OF COMPRESSIONS WHEN A SECOND PERSON IS GIVING BREATH

TWO RESCUERS:
5 COMPRESSIONS.
1 BREATH
APPROX. RATE
OF COMPRESSIONS
~60 PER MINUTE

2. IF VICTIM IS RESPONSIVE, KEEP HIM WARM AND QUIET, LOOSEN CLOTHING AND PLACE IN RECLINING POSITION.
PLACE VICTIM FLAT ON HIS BACK ON A HARD SURFACE.
CALL FOR MEDICAL ASSISTANCE AS SOON AS POSSIBLE

FIGURE 23.7 Basic first-aid treatment for electric shock.

TABLE 23.3 Basic First-Aid Procedures

For extensively burned and broken skin:

- ✓ Cover affected area with a clean sheet or cloth.
- ✓ Do not break blisters, remove tissue, remove adhered particles of clothing, or apply any salve or ointment.
- ✓ Treat victim for shock as required.
- ✓ Arrange for transportation to a hospital as quickly as possible.
- ✓ If victim's arms or legs are affected, keep them elevated.
- ✓ If medical help will not be available within an hour and the victim is conscious and not vomiting, prepare a weak solution of salt and soda. Mix 1 teaspoon of salt and 1/2-teaspoon of baking soda to each quart of tepid water. Allow the victim to sip slowly about 4 oz (half a glass) over a period of 15 min. Discontinue fluid intake if vomiting occurs. (Do not allow alcohol consumption.)

For less severe burns (first- and second-degree):

- ✓ Apply cool (not ice-cold) compresses using the cleanest available cloth article.
- ✓ Do not break blisters, remove tissue, remove adhered particles of clothing, or apply salve or ointment.
- ✓ Apply clean, dry dressing if necessary.
- ✓ Treat victim for shock as required.
- ✓ Arrange for transportation to a hospital as quickly as possible.
- ✓ If victim's arms or legs are affected, keep them elevated.

For electric shock, the best first aid is prevention. In the event that an individual has sustained or is sustaining an electric shock at the workplace, several guidelines are suggested, as detailed next.

Shock in Progress

For the case when a co-worker is receiving an electric shock and cannot let go of the electrical source, the safest action is to trip the circuit breaker that energizes the circuit involved, or to pull the power-line plug on the equipment involved if the latter can be accomplished safely.[2] Under no circumstances should the rescuer touch the individual who is being shocked, because the rescuer's body may then also be in the dangerous current path. If the circuit breaker or equipment plug cannot be located, then an attempt can be made to separate the victim from the electrical source through the use of a nonconducting object such as a wooden stool or a wooden broom handle. Use only an *insulating* object and nothing that contains metal or other electrically conductive material. The rescuer must be very careful not to touch the victim or the electrical source and thus become a second victim.

If such equipment is available, hot sticks *used in conjunction with lineman's gloves* may be applied to push or pull the victim away from the electrical source. Pulling the hot stick normally provides the greatest control over the victim's motion and is the safest action for the rescuer. After the electrical source has been turned off, or the victim can be reached safely, immediate first-aid procedures should be implemented.

Shock No Longer in Progress

If the victim is conscious and moving about, have the victim sit down or lie down. Sometimes there is a delayed reaction to an electrical shock that causes the victim to collapse. Call 911 or the appropriate plant-site paramedic team immediately. If there is a delay in the arrival of medical personnel, check for electrical burns. In the case of severe shock, there will normally be burns at a minimum of two sites: the entry point for the current and the exit point(s). Cover the burns with dry (and sterile, preferably) dressings. Check for possible bone fractures if the victim was violently thrown away from the electrical source and possibly impacted objects in the vicinity. Apply splints as required if suitable materials are available and you have appropriate training. Cover the victim with a coat or blanket if the environmental temperature is below room temperature or the victim complains of feeling cold.

If the victim is unconscious, call 911 or the appropriate plant-site paramedic team immediately. In the interim, check to see if the victim is breathing and if a pulse can be felt at either the inside of a wrist above the thumb joint (radial pulse) or in the neck above and to either side of the Adam's apple (carotid pulse). It is usually easier to feel the pulse in the neck as opposed to the wrist pulse, which may be weak. The index and middle fingers should be used to sense the pulse, and not the thumb. Many individuals have an apparent thumb pulse that can be mistaken for the victim's pulse. If a pulse can be detected but the victim is not breathing, begin mouth-to-mouth respiration if you know how to do so. If no pulse can be detected (presumably the victim will not be breathing), carefully move the victim to a firm surface and begin cardiopulmonary resuscitation if you have been trained in the use of CPR. Respiratory arrest and cardiac arrest are crisis situations. Because of loss of the oxygen supply to the brain, permanent brain damage can occur after several minutes even if the victim is successfully resuscitated.

Ironically, the treatment for cardiac arrest induced by an electric shock is a massive counter-shock, which causes the entire heart muscle to contract. The random and uncoordinated ventricular fibrillation contractions (if present) are thus stilled. Under ideal conditions, normal heart rhythm is restored once the shock current ceases. The counter shock is generated by a cardiac defibrillator, various portable models of which are available for use by emergency medical technicians and other *trained* personnel. Although portable defibrillators may be available at industrial sites where there is a high risk of electrical shock to plant personnel, they should be used only by *trained* personnel. Application of a defibrillator to an unconscious subject whose heart is beating can induce cardiac standstill or ventricular fibrillation — just the conditions that the defibrillator was designed to correct.

23.3 Polychlorinated Biphenyls

Polychlorinated biphenyls (PCBs) belong to a family of organic compounds known as *chlorinated hydrocarbons*. Virtually all PCBs in existence today have been synthetically manufactured. PCBs are of a heavy, oil-like consistency and have a high boiling point, a high degree of chemical stability, low flammability, and low electrical conductivity. These characteristics led to the past widespread use of PCBs in high-voltage capacitors and transformers. Commercial products containing PCBs were distributed widely from 1957 to 1977 under several trade names, including:

- Aroclor
- Pyroclor
- Sanotherm
- Pyranol
- Askarel

Askarel is also a generic name used for nonflammable dielectric fluids containing PCBs. Table 23.4 lists some common trade names for Askarel. These trade names are typically listed on the nameplate of a PCB transformer or capacitor.

TABLE 23.4 Commonly Used Names for PCB Insulating Material

Apirolio	Abestol	Askarel	Aroclor B	Chlorextol	Chlophen
Chlorinol	Clorphon	Diaclor	DK	Dykanol	EEC-18
Elemex	Eucarel	Fenclor	Hyvol	Inclor	Inerteen
Kanechlor	No-Flamol	Phenodlor	Pydraul	Pyralene	Pyranol
Pyroclor	Sal-T-Kuhl	Santothern FR	Santovac	Solvol	Therminal

Health Risk

PCBs are harmful because, once they are released into the environment, they tend not to break apart into other substances. Instead, PCBs persist, taking several decades to slowly decompose. By remaining in the environment, they can be taken up and stored in the fatty tissues of all organisms, from which they are released slowly into the bloodstream. Therefore, because of the storage in fat, the concentration of PCBs in body tissues can increase with time, although PCB exposure levels may be quite low. This process is called *bioaccumulation*. Furthermore, as PCBs accumulate in the tissues of simple organisms, which are consumed by progressively higher organisms, the concentration increases. This process is referred to as *biomagnification*. These two factors are especially significant because PCBs are harmful even at low levels. Specifically, PCBs have been shown to cause chronic (long-term) toxic effects in some species of animals and aquatic life. Well-documented tests on laboratory animals show that various levels of PCBs can cause reproductive effects, gastric disorders, skin lesions, and cancerous tumors.

PCBs can enter the body through the lungs, the gastrointestinal tract, and the skin. After absorption, PCBs are circulated in the blood throughout the body and stored in fatty tissues and skin, as well as in a variety of organs, including the liver, kidneys, lungs, adrenal glands, brain, and heart.

The health risk lies not only in the PCB itself, but also in the chemicals developed when PCBs are heated. Laboratory studies have confirmed that PCB by-products, including *polychlorinated dibenzofurans* (PCDFs) and *polychlorinated dibenzo-p-dioxins* (PCDDs), are formed when PCBs or chlorobenzenes are heated to temperatures ranging from approximately 900°F to 1300°F. Unfortunately, these products are more toxic than PCBs themselves.

The problem for the owner of PCB equipment is that the liability from a PCB spill or fire contamination can be tremendous. A fire involving a PCB large transformer in Binghamton, NY, resulted in $20 million in cleanup expenses. The consequences of being responsible for a fire-related incident with a PCB transformer can be monumental.

Governmental Action

The U.S. Congress took action to control PCBs in October 1975 by passing the Toxic Substances Control Act (TSCA). A section of this law specifically directed the EPA to regulate PCBs. Three years later, the EPA issued regulations to implement a congressional ban on the manufacture, processing, distribution, and disposal of PCBs. Since that time, several revisions and updates have been issued by the EPA. One of these revisions, issued in 1982, specifically addressed the type of equipment used in industrial plants. Failure to properly follow the rules regarding the use and disposal of PCBs has resulted in high fines and some jail sentences.

Although PCBs no longer are being produced for electric products in the United States, significant numbers still exist. The threat of widespread contamination from PCB fire-related incidents is one reason behind the EPA's efforts to reduce the number of PCB products in the environment. The users of high-power equipment are affected by the regulations, primarily because of the widespread use of PCB transformers and capacitors. These components are usually located in older (pre-1979) systems, so this is the first place to look for them. However, some facilities also maintain their own primary power transformers. Unless these transformers are of recent vintage, it is quite likely that they too contain a PCB dielectric. Table 23.5 lists the primary classifications of PCB devices.

PCB Components

The two most common PCB components are transformers and capacitors. A PCB transformer is one containing at least 500 ppm (parts per million) PCBs in the dielectric fluid. An Askarel transformer generally has 600,000 ppm or more. A PCB transformer can be converted to a *PCB-contaminated device* (50 to 500 ppm) or a *non-PCB device* (less than 50 ppm) by being drained, refilled, and tested. The testing must not take place until the transformer has been in service for a minimum of 90 days. Note that this is *not* something that a maintenance technician can do. It is the exclusive domain of specialized remanufacturing companies.

PCB transformers must be inspected quarterly for leaks. However, if an impervious dike (sufficient to contain all the liquid material) is built around the transformer, the inspections can be conducted yearly. Similarly, if the transformer is tested and found to contain less than 60,000 ppm, a yearly inspection is sufficient. Failed PCB transformers cannot be repaired; they must be disposed of properly.

Safety Issues for RF Systems

TABLE 23.5 Definition of PCB Terms as Identified by the EPA

Term	Definition	Examples
PCB	Any chemical substance that is limited to the biphenyl molecule that has been chlorinated to varying degrees, or any combination of substances that contain such substances.	PCB dielectric fluids, PCB heat-transfer fluids, PCB hydraulic fluids, 2,2′,4-trichlorobiphenyl
PCB article	Any manufactured article, other than a PCB container, that contains PCBs and whose surface has been in direct contact with PCBs.	Capacitors, transformers, electric motors, pumps, pipes
PCB container	A device used to contain PCBs or PCB articles, and whose surface has been in direct contact with PCBs.	Packages, cans, bottles, bags, barrels, drums, tanks
PCB article container	A device used to contain PCB articles or equipment, and whose surface has not been in direct contact with PCBs.	Packages, cans, bottles, bags, barrels, drums, tanks
PCB equipment	Any manufactured item, other than a PCB container or PCB article container, which contains a PCB article or other PCB equipment.	Microwave ovens, fluorescent light ballasts, electronic equipment
PCB item	Any PCB article, PCB article container, PCB container, or PCB equipment that deliberately or unintentionally contains, or has as a part of it, any PCBs.	See PCB article, PCB article container, PCB container, and PCB equipment
PCB transformer	Any transformer that contains PCBs in concentrations of 500 ppm or greater.	High-power transformers
PCB contaminated	Any electric equipment that contains more than 50 ppm, but less than 500 ppm, of PCBs. (Oil-filled electric equipment other than circuit breakers, reclosers, and cable whose PCB concentration is unknown must be assumed to be PCB-contaminated electric equipment.)	Transformers, capacitors, circuit breakers, reclosers, voltage regulators, switches, cable, electromagnets

If a leak develops, it must be contained and daily inspections must begin. A cleanup must be initiated as soon as possible, but no later than 48 hours after the leak is discovered. Adequate records must be kept of all inspections, leaks, and actions taken for 3 years after disposal of the component. Combustible materials must be kept a minimum of 5 m from a PCB transformer and its enclosure.

As of October 1, 1990, the use of PCB transformers (500 ppm or greater) was prohibited in or near commercial buildings with secondary voltages of 480 V ac or higher. The use of radial PCB transformers was allowed if certain electrical protection was provided.

The EPA regulations also require that the operator notify others of the possible dangers. All PCB transformers (including those in storage for reuse) must be registered with the local fire department. Supply the following information:

- The location of the PCB transformer(s)
- Address(es) of the building(s) (for outdoor PCB transformers, provide the outdoor location)
- Principal constituent of the dielectric fluid in the transformer(s)
- Name and telephone number of the contact person in the event of a fire involving the equipment

Any PCB transformers used in a commercial building must be registered with the building owner. All owners of buildings within 30 m of such PCB transformers must also be notified. In the event of a fire-related incident involving the release of PCBs, immediately notify the Coast Guard National Spill Response Center at 1-800-424-8802. Also take appropriate measures to contain and control any possible PCB release into water.

Capacitors are divided into two size classes: *large* and *small*. The following are guidelines for classification:

- A PCB small capacitor contains less than 1.36 kg (3 lb) dielectric fluid. A capacitor having less than 100 in.³ is also considered to contain less than 3 lb dielectric fluid.
- A PCB large capacitor has a volume of more than 200 in.³ and is considered to contain more than 3 lb dielectric fluid. Any capacitor having a volume from 100 to 200 in.³ is considered to contain 3 lb dielectric, provided the total weight is less than 9 lb.

- A PCB large low-voltage capacitor contains 3 lb or more dielectric fluid and operates below 2 kV.
- A PCB large high-voltage capacitor contains 3 lb or more dielectric fluid and operates at 2 kV or greater voltages.

The use, servicing, and disposal of PCB small capacitors is not restricted by the EPA unless there is a leak. In that event, the leak must be repaired or the capacitor disposed of. Disposal can be performed by an approved incineration facility, or the component can be placed in a specified container and buried in an approved chemical waste landfill. Currently, chemical waste landfills are only for disposal of liquids containing 50 to 500 ppm PCBs and for solid PCB debris. Items such as capacitors that are leaking oil containing greater than 500 ppm PCBs should be taken to an EPA-approved PCB disposal facility.

Identifying PCB Components

The first task for the facility manager is to identify any PCB items on the premises. Equipment built after 1979 probably does not contain any PCB-filled devices. Even so, inspect all capacitors, transformers, and power switches to be sure. A call to the manufacturer may also help. Older equipment (pre-1979) is more likely to contain PCB transformers and capacitors. A liquid-filled transformer usually has cooling fins, and the nameplate may provide useful information about its contents. If the transformer is unlabeled or the fluid is not identified, it must be treated as a PCB transformer. Untested (not analyzed) mineral-oil-filled transformers are assumed to contain at least 50 ppm, but less than 500 ppm PCBs. This places them in the category of PCB-contaminated electric equipment, which has different requirements than PCB transformers. Older high-voltage systems are likely to include both large and small PCB capacitors. Equipment rectifier panels, exciter/modulators, and power-amplifier cabinets may contain a significant number of small capacitors. In older equipment, these capacitors often are Askarel-filled. Unless leaking, these devices pose no particular hazard. If a leak does develop, follow proper disposal techniques. Also, liquid-cooled rectifiers may contain Askarel. Although their use is not regulated, treat them as a PCB article, as if they contain at least 50 ppm PCBs. Never make assumptions about PCB contamination; check with the manufacturer to be sure.

Any PCB article or container being stored for disposal must be date-tagged when removed, and inspected for leaks every 30 days. It must be removed from storage and disposed of within 1 year from the date it was placed in storage. Items being stored for disposal must be kept in a storage facility meeting the requirements of 40 CFR (*Code of Federal Regulations*), Part 761.65(b)(1), unless they fall under alternative regulation provisions. There is a difference between PCB items stored for disposal and those stored for reuse. Once an item has been removed from service and tagged for disposal, it cannot be returned to service.

Labeling PCB Components

After identifying PCB devices, proper labeling is the second step that must be taken by the facility manager. PCB article containers, PCB transformers, and large high-voltage capacitors must be marked with a standard 6-in. × 6-in. large marking label (ML) as shown in Fig. 23.8. Equipment containing these transformers or capacitors also should be marked. PCB large low-voltage (less than 2 kV) capacitors need not be labeled until removed from service. If the capacitor or transformer is too small to hold the large label, a smaller 1-in. × 2-in. label is approved for use. Labeling each PCB small capacitor is not required. However, any equipment containing PCB small capacitors should be labeled on the outside of the cabinet or on access panels. Properly label any spare capacitors and transformers that fall under the regulations. Identify with

FIGURE 23.8 Marking label (ML) used to identify PCB transformers and PCB large capacitors.

Safety Issues for RF Systems

TABLE 23.6 The Inspection Schedule Required for PCB Transformers and Other Contaminated Devices

PCB Transformers	Standard PCB transformer	Quarterly
	If full-capacity impervious dike is added	Annually
	If retrofitted to <60,000 ppm PCB	Annually
	If leak is discovered, clean up ASAP (retain these records for 3 years)	Daily
PCB article or container stored for disposal (remove and dispose of within 1 year)		Monthly
Retain all records for 3 years after disposing of transformers		

TABLE 23.7 Inspection Checklist for PCB Components

Transformer location:

Date of visual inspection:
Leak discovered? (Yes/No):
If yes, date discovered (if different from inspection date):

Location of leak:

Person performing inspection:
Estimate of the amount of dielectric fluid released from leak:
Date of cleanup, containment, repair, or replacement:
Description of cleanup, containment, or repair performed:

Results of any containment and daily inspection required for uncorrected active leaks:

---

the large label any doors, cabinet panels, or other means of access to PCB transformers. The label must be placed so that it can be read easily by firefighters. All areas used to store PCBs and PCB items for disposal must be marked with the large (6-in. × 6-in.) PCB label.

Record-Keeping

Inspections are a critical component in the management of PCBs. EPA regulations specify a number of steps that must be taken and the information that must recorded. Table 23.6 summarizes the schedule requirement, and Table 23.7 can be used as a checklist for each transformer inspection. This record must be retained for 3 years. In addition to the inspection records, some facilities may need to maintain an annual report. This report details the number of PCB capacitors, transformers, and other PCB items on the premises. The report must contain the dates when the items were removed from service, their disposition, and detailed information regarding their characteristics. Such a report must be prepared if the facility uses or stores at least one PCB transformer containing greater than 500 ppm PCBs, 50 or more PCB large capacitors, or at least 45 kg of PCBs in PCB containers. Retain the report for 5 years after the facility ceases using or storing PCBs and PCB items in the prescribed quantities. Table 23.8 lists the information required in the annual PCB report.

Disposal

Disposing of PCBs is not a minor consideration. Before contracting with a company for PCB disposal, verify its license with the area EPA office. That office can also supply background information on the company's compliance and enforcement history.

The fines levied for improper disposal are not mandated by federal regulations. Rather, the local EPA administrator, usually in consultation with local authorities, determines the cleanup procedures and costs. Civil penalties for administrative complaints issued for violations of the PCB regulations are determined according to a matrix provided in the PCB penalty policy. This policy, published in the *Federal Register*, considers the amount of PCBs involved and the potential for harm posed by the violation.

Proper Management

Properly managing the PCB risk is not difficult. The keys are to understand the regulations and to follow them carefully. A PCB management program should include the following steps:

- Locate and identify all PCB devices. Check all stored or spare devices.
- Properly label PCB transformers and capacitors according to EPA requirements.
- Perform the required inspections, and maintain an accurate log of PCB items, their location, inspection results, and actions taken. These records must be maintained for 3 years after disposal of the PCB component.
- Complete the annual report of PCBs and PCB items by July 1 of each year. This report must be retained for 5 years.
- Arrange for any necessary disposal through a company licensed to handle PCBs. If there are any doubts about the company's license, contact the EPA.
- Report the location of all PCB transformers to the local fire department and owners of any nearby buildings.

The importance of following the EPA regulations cannot be overstated.

23.4 OSHA Safety Requirements

The federal government has taken a number of steps to help improve safety within the workplace. OSHA, for example, helps industries to monitor and correct safety practices. The agency's records show that electrical standards are among the most frequently violated of all safety standards. Table 23.9 lists 16 of the most common electrical violations, which include these areas:

- Protective covers
- Identification and marking
- Extension cords
- Grounding

TABLE 23.8 Required Information for PCB Annual Report

I. PCB device background information:
   a. Dates when PCBs and PCB items are removed from service.
   b. Dates when PCBs and PCB items are placed into storage for disposal, and are placed into transport for disposal.
   c. The quantities of the items removed from service, stored, and placed into transport are to be indicated using the following breakdown:
      (1) Total weight, in kilograms, of any PCB and PCB items in PCB containers, including identification of container contents (such as liquids and capacitors).
      (2) Total number of PCB transformers and total weight, in kilograms, of any PCBs contained in the transformers.
      (3) Total number of PCB large high- or low-voltage capacitors.
II. The location of the initial disposal or storage facility for PCBs and PCB items removed from service, and the name of the facility owner or operator.
III. Total quantities of PCBs and PCB items remaining in service at the end of calendar year per the following breakdown:
   a. Total weight, in kilograms, of any PCB and PCB items in PCB containers, including the identification of container contents (such as liquids and capacitors).
   b. Total number of PCB transformers and total weight, in kilograms, of any PCBs contained in the transformers.
   c. Total number of PCB large high- or low-voltage capacitors.

Safety Issues for RF Systems

TABLE 23.9 Sixteen Common OSHA Violations

Fact Sheet No.	Subject	NEC Ref.
1	Guarding of live parts	110-17
2	Identification	110-22
3	Uses allowed for flexible cord	400-7
4	Prohibited uses of flexible cord	400-8
5	Pull at joints and terminals must be prevented	400-10
6-1	Effective grounding, Part 1	250-51
6-2	Effective grounding, Part 2	250-51
7	Grounding of fixed equipment, general	250-42
8	Grounding of fixed equipment, specific	250-43
9	Grounding of equipment connected by cord and plug	250-45
10	Methods of grounding, cord and plug-connected equipment	250-59
11	AC circuits and systems to be grounded	250-5
12	Location of overcurrent devices	240-24
13	Splices in flexible cords	400-9
14	Electrical connections	110-14
15	Marking equipment	110-21
16	Working clearances about electric equipment	110-16

Source: Adapted from the *National Electrical Code*, NFPA No. 70.

Protective Covers

Exposure of live conductors is a common safety violation. All potentially dangerous electric conductors should be covered with protective panels. The danger is that someone can come into contact with the exposed, current-carrying conductors. It also is possible for metallic objects such as ladders, cable, or tools to contact a hazardous voltage, creating a life-threatening condition. Open panels also present a fire hazard.

Identification and Marking

Properly identify and label all circuit breakers and switch panels. The labels for breakers and equipment switches may be years old, and may no longer describe the equipment that is actually in use. This confusion poses a safety hazard. Improper labeling of the circuit panel can lead to unnecessary damage — or worse, casualties — if the only person who understands the system is unavailable in an emergency. If there are a number of devices connected to a single disconnect switch or breaker, provide a diagram or drawing for clarification. Label with brief phrases, and use clear, permanent, and legible markings.

Equipment marking is a closely related area of concern. This is not the same thing as equipment identification. Marking equipment means labeling the equipment breaker panels and ac disconnect switches according to device rating. Breaker boxes should contain a nameplate showing the manufacturer name, rating, and other pertinent electrical factors. The intent of this rule is to prevent devices from being subjected to excessive loads or voltages.

Extension Cords

Extension (flexible) cords often are misused. Although it may be easy to connect a new piece of equipment with a flexible cord, be careful. The National Electrical Code lists only eight approved uses for flexible cords.

The use of a flexible cord where the cable passes through a hole in the wall, ceiling, or floor is an often-violated rule. Running the cord through doorways, windows, or similar openings is also prohibited. A flexible cord should not be attached to building surfaces or concealed behind building walls or ceilings. These common violations are illustrated in Fig. 23.9

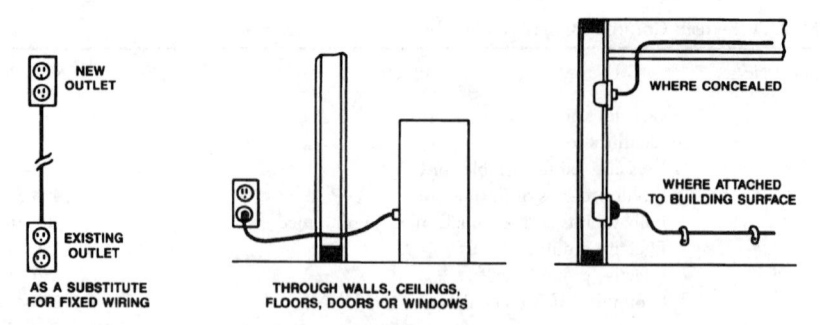

FIGURE 23.9 Flexible cord uses prohibited under NEC rules.

Along with improper use of flexible cords, failure to provide adequate strain relief on connectors is a common problem. Whenever possible, use manufactured cable connections.

Grounding

OSHA regulations describe two types of grounding: *system grounding* and *equipment grounding*. System grounding actually connects one of the current-carrying conductors (such as the terminals of a supply transformer) to ground (see Fig. 23.10.) Equipment grounding connects all the noncurrent-carrying metal surfaces together and to ground. From a grounding standpoint, the only difference between a grounded electrical system and an ungrounded electrical system is that the *main-bonding jumper* from the service equipment ground to a current-carrying conductor is omitted in the ungrounded system.

The system ground performs two tasks:

- It provides the final connection from equipment-grounding conductors to the grounded circuit conductor, thus completing the ground-fault loop.
- It solidly ties the electrical system and its enclosures to their surroundings (usually Earth, structural steel, and plumbing). This prevents voltages at any source from rising to harmfully high voltage-to-ground levels.

It should be noted that equipment grounding — bonding all electric equipment to ground — is required whether or not the system is grounded. System grounding should be handled by the electrical contractor installing the power feeds.

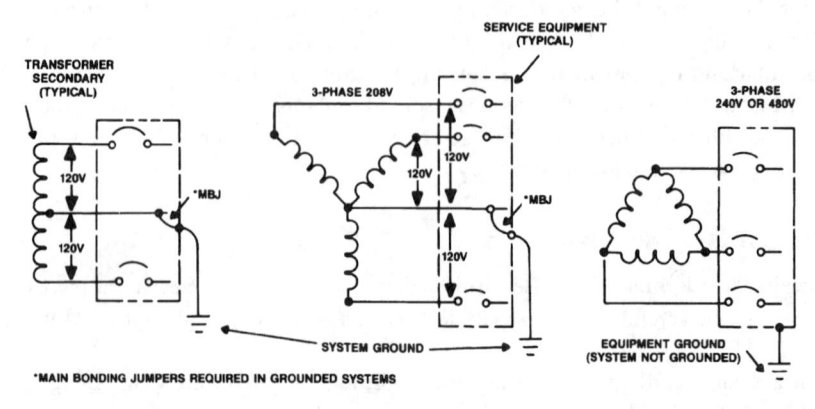

FIGURE 23.10 Although regulations have been in place for many years, OSHA inspections still uncover violations in the grounding of primary electrical service systems.

Equipment grounding serves two important functions:

- It bonds all surfaces together so that there can be no voltage differences among them.
- It provides a ground-fault current path from a fault location back to the electrical source so that if a fault current develops, it will operate the breaker, fuse, or GFCI.

The National Electrical Code is complex and it contains numerous requirements concerning electrical safety. If the facility electric wiring system has gone through many changes over the years, have the entire system inspected by a qualified consultant. The fact sheets listed in Table 23.9 provide a good starting point for a self-evaluation. The fact sheets are available from any local OSHA office.

23.5 Beryllium Oxide Ceramics

Some tubes, both power grid and microwave, contain beryllium oxide (BeO) ceramics, typically at the output waveguide window or around the cathode. Never perform any operations on BeO ceramics that produce dust or fumes, such as grinding, grit blasting, or acid cleaning. Beryllium oxide dust and fumes are highly toxic, and breathing them can result in serious personal injury or death.

If a broken window is suspected on a microwave tube, carefully remove the device from its waveguide, and seal the output flange of the tube with tape. Because BeO warning labels may be obliterated or missing, maintenance personnel should contact the tube manufacturer before performing any work on the device. Some tubes have BeO internal to the vacuum envelope.

Take precautions to protect personnel working in the disposal or salvage of tubes containing BeO. All such personnel should be made aware of the deadly hazards involved and the necessity for great care and attention to safety precautions. Some tube manufacturers will dispose of tubes without charge, provided they are returned to the manufacturer prepaid, with a written request for disposal.

23.6 Corrosive and Poisonous Compounds

The external output waveguides and cathode high-voltage bushings of microwave tubes are sometimes operated in systems that use a dielectric gas to impede microwave or high-voltage breakdown. If breakdown does occur, the gas may decompose and combine with impurities, such as air or water vapor, to form highly toxic and corrosive compounds. Examples include Freon gas, which may form lethal *phosgene*, and sulfur hexafluoride (SF_6) gas, which may form highly toxic and corrosive sulfur or fluorine compounds such as *beryllium fluoride*. When breakdown does occur in the presence of these gases, proceed as follows:

- Ventilate the area to outside air.
- Avoid breathing any fumes or touching any liquids that develop.
- Take precautions appropriate for beryllium compounds and for other highly toxic and corrosive substances.

If a coolant other than pure water is used, follow the precautions supplied by the coolant manufacturer.

FC-75 Toxic Vapor

The decomposition products of FC-75 are highly toxic. Decomposition may occur as a result of any of the following:

- Exposure to temperatures above 200°C
- Exposure to liquid fluorine or alkali metals (lithium, potassium, or sodium)
- Exposure to ionizing radiation

Known thermal decomposition products include *perfluoroisobutylene* [PFIB; $(CF_3)_2 C = CF_2$], which is highly toxic in small concentrations.

If FC-75 has been exposed to temperatures above 200°C through fire, electric heating, or prolonged electric arcs, or has been exposed to alkali metals or strong ionizing radiation, take the following steps:

- Strictly avoid breathing any fumes or vapors.
- Thoroughly ventilate the area.
- Strictly avoid any contact with the FC-75.

Under such conditions, promptly replace the FC-75 and handle and dispose of the contaminated FC-75 as a toxic waste.

23.7 Nonionizing Radiation

Nonionizing radio frequency radiation (RFR) resulting from high-intensity RF fields is a growing concern to engineers who must work around high-power transmission equipment. The principal medical concern regarding nonionizing radiation involves heating of various body tissues, which can have serious effects, particularly if there is no mechanism for heat removal. Recent research has also noted, in some cases, subtle psychological and physiological changes at radiation levels below the threshold for heat-induced biological effects. However, the consensus is that most effects are thermal in nature.

High levels of RFR can affect one or more body systems or organs. Areas identified as potentially sensitive include the ocular (eye) system, reproductive system, and the immune system. Nonionizing radiation is also thought to be responsible for metabolic effects on the central nervous system and the cardiac system.

Despite these studies, many of which are ongoing, there is still no clear evidence in Western literature that exposure to medium-level nonionizing radiation results in detrimental effects. Russian findings, on the other hand, suggest that occupational exposure to RFR at power densities above 1.0 mW/cm^2 does result in symptoms, particularly in the central nervous system.

Clearly, the jury is still out as to the ultimate biological effects of RFR. Until the situation is better defined, however, the assumption must be made that potentially serious effects can result from excessive exposure. Compliance with existing standards should be the minimum goal to protect members of the public as well as facility employees.

NEPA Mandate

The National Environmental Policy Act of 1969 required the Federal Communications Commission to place controls on nonionizing radiation. The purpose was to prevent possible harm to the public at large and to those who must work near sources of the radiation. Action was delayed because no hard and fast evidence existed that low- and medium-level RF energy is harmful to human life. Also, there was no evidence showing that radio waves from radio and TV stations did not constitute a health hazard.

During the delay, many studies were carried out in an attempt to identify those levels of radiation that might be harmful. From the research, suggested limits were developed by the American National Standards Institute (ANSI) and stated in the document known as ANSI C95.1-1982. The protection criteria outlined in the standard are shown in Fig. 23.11.

The energy-level criteria were developed by representatives from a number of industries and educational institutions after performing research on the possible effects of nonionizing radiation. The projects focused on absorption of RF energy by the human body, based upon simulated human body models. In preparing the document, ANSI attempted to determine those levels of incident radiation that would cause the body to absorb less than 0.4 W/kg of mass (averaged over the whole body) or peak absorption values of 8 W/kg over any 1 gram of body tissue.

From the data, the researchers found that energy would be absorbed more readily at some frequencies than at others. The absorption rates were found to be functions of the size of a specific individual and

Safety Issues for RF Systems 23-21

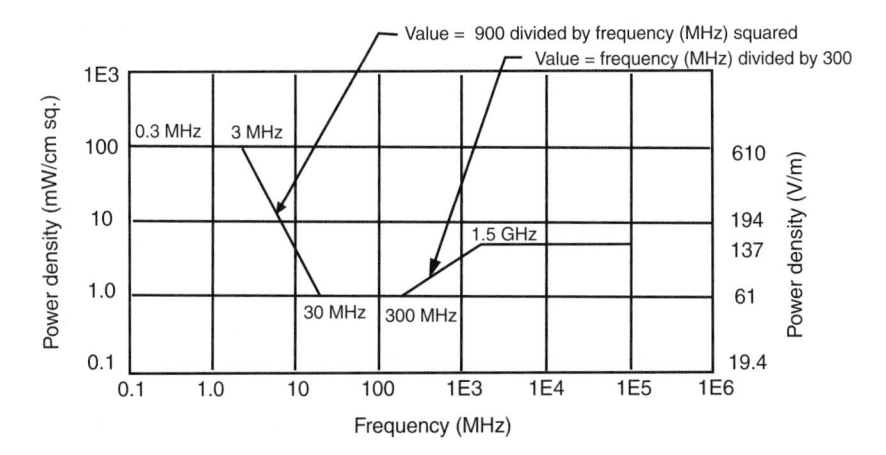

FIGURE 23.11 The power density limits for nonionizing radiation exposure for humans.

the frequency of the signal being evaluated. It was the result of these absorption rates that culminated in the shape of the *safe curve* shown in Fig. 23.11. ANSI concluded that no harm would come to individuals exposed to radio energy fields, as long as specific values were not exceeded when averaged over a period of 0.1 hour. It was also concluded that higher values for a brief period would not pose difficulties if the levels shown in the standard document were not exceeded when averaged over the 0.1-hour time period.

The FCC adopted ANSI C95.1-1982 as a standard that would ensure adequate protection to the public and to industry personnel involved in working around RF equipment and antenna structures.

Revised Guidelines

The ANSI C95.1-1982 standard was intended to be reviewed at 5-year intervals. Accordingly, the 1982 standard was due for reaffirmation or revision in 1987. The process was indeed begun by ANSI, but was handed off to the Institute of Electrical and Electronics Engineers (IEEE) for completion. In 1991, the revised document was completed and submitted to ANSI for acceptance as ANSI/IEEE C95.1-1992.

The IEEE standard incorporated changes from the 1982 ANSI document in four major areas:

- An additional safety factor was provided in certain situations. The most significant change was the introduction of new *uncontrolled* (public) exposure guidelines, generally established at one-fifth of the *controlled* (occupational) exposure guidelines. Figure 23.11 illustrates the concept for the microwave frequency band.
- For the first time, guidelines were included for body currents; examination of the electric and magnetic fields were determined to be insufficient to determine compliance.
- Minor adjustments were made to occupational guidelines, including relaxation of the guidelines at certain frequencies and the introduction of *breakpoints* at new frequencies.
- Measurement procedures were changed in several aspects, most notably with respect to spatial averaging and to minimum separation from reradiating objects and structures at the site.

The revised guidelines are complex and beyond the scope of this handbook. Refer to the ANSI/IEEE document for details.

Multiple-User Sites

At a multiple-user site, the responsibility for assessing the RFR situation — although officially triggered by either a new user or the license renewal of all site tenants — is, in reality, the joint responsibility of all the site tenants. In a multiple-user environment involving various frequencies, and various protection criteria, compliance is indicated when the fraction of the RFR limit within each pertinent frequency band is established and added to the sum of all the other fractional contributions. The sum must not be greater than 1.0. Evaluating the multiple-user environment is not a simple matter, and corrective actions, if indicated, can be quite complex.

Operator Safety Considerations

RF energy must be contained properly by shielding and transmission lines. All input and output RF connections, cables, flanges, and gaskets must be RF leakproof. The following guidelines should be followed at all times:

- Never operate a power tube without a properly matched RF energy absorbing load attached.
- Never look into or expose any part of the body to an antenna or open RF generating tube, circuit, or RF transmission system that is energized.
- Monitor the RF system for radiation leakage at regular intervals and after servicing.

23.8 X-Ray Radiation Hazard

The voltages typically used in microwave tubes are capable of producing dangerous x-rays. As voltages increase beyond 15 kV, metal-body tubes are capable of producing progressively more dangerous radiation. Adequate x-ray shielding must be provided on all sides of such tubes, particularly at the cathode and collector ends, as well as at the modulator and pulse transformer tanks (as appropriate). High-voltage tubes should never be operated without adequate x-ray shielding in place. The x-ray radiation of the device should be checked at regular intervals and after servicing.

Implosion Hazard

Because of the high internal vacuum in power grid and microwave tubes, the glass or ceramic output window or envelope can shatter inward (implode) if struck with sufficient force or exposed to sufficient mechanical shock. Flying debris could result in bodily injury, including cuts and puncture wounds. If the device is made of beryllium oxide ceramic, implosion may produce highly toxic dust or fumes.

In the event of such an implosion, assume that toxic BeO ceramic is involved unless confirmed otherwise.

23.9 Hot Coolant and Surfaces

Extreme heat occurs in the electron collector of a microwave tube and the anode of a power grid tube during operation. Coolant channels used for water or vapor cooling also can reach high temperatures (boiling — 100°C — and above), and the coolant is typically under pressure (as high as 100 psi). Some devices are cooled by boiling the coolant to form steam.

Contact with hot portions of the tube or its cooling system can scald or burn. Carefully check that all fittings and connections are secure, and monitor back-pressure for changes in cooling system performance. If back-pressure is increased above normal operating values, shut down the system and clear the restriction.

For a device whose anode or collector is air-cooled, the external surface normally operates at a temperature of 200 to 300°C. Other parts of the tube may also reach high temperatures, particularly the cathode insulator and the cathode/heater surfaces. All hot surfaces remain hot for an extended time after the tube is shut off. To prevent serious burns, take care to avoid bodily contact with these surfaces during operation and for a reasonable cool-down period afterward.

23.10 Management Responsibility

The key to operating a safe facility is diligent management. A carefully thought-out plan ensures a coordinated approach to protecting staff members from injury and the facility from potential litigation.

TABLE 23.10 Major Points to Consider when Developing a Facility Safety Program

- ✓ Management assumes the leadership role regarding safety policies.
- ✓ Responsibility for safety- and health-related activities is clearly assigned.
- ✓ Hazards are identified, and steps are taken to eliminate them.
- ✓ Employees at all levels are trained in proper safety procedures.
- ✓ Thorough accident/injury records are maintained.
- ✓ Medical attention and first aid is readily available.
- ✓ Employee awareness and participation is fostered through incentives and an ongoing, high-profile approach to workplace safety.

TABLE 23.11 Sample Checklist of Important Safety Items

Refer regularly to this checklist to maintain a safe facility. For each category shown, be sure that:

Electrical Safety

- ✓ Fuses of the proper size have been installed.
- ✓ All ac switches are mounted in clean, tightly closed metal boxes.
- ✓ Each electrical switch is marked to show its purpose.
- ✓ Motors are clean and free of excessive grease and oil.
- ✓ Motors are properly maintained and provided with adequate overcurrent protection.
- ✓ Bearings are in good condition.
- ✓ Portable lights are equipped with proper guards.
- ✓ All portable equipment is double-insulated or properly grounded.
- ✓ The facility electrical system is checked periodically by a contractor competent in the NEC.
- ✓ The equipment-grounding conductor or separate ground wire has been carried all the way back to the supply ground connection.
- ✓ All extension cords are in good condition, and the grounding pin is not missing or bent.
- ✓ Ground-fault interrupters are installed as required.

Exits and Access

- ✓ All exits are visible and unobstructed.
- ✓ All exits are marked with a readily visible, properly illuminated sign.
- ✓ There are sufficient exits to ensure prompt escape in the event of an emergency.

Fire Protection

- ✓ Portable fire extinguishers of the appropriate type are provided in adequate numbers.
- ✓ All remote vehicles have proper fire extinguishers.
- ✓ Fire extinguishers are inspected monthly for general condition and operability, which is noted on the inspection tag.
- ✓ Fire extinguishers are mounted in readily accessible locations.
- ✓ The fire alarm system is tested annually.

Facilities that have effective accident-prevention programs follow seven basic guidelines. Although the details and overall organization can vary from workplace to workplace, these practices — summarized in Table 23.10 — still apply.

If managers are concerned about safety, it is likely that employees also will be. Display safety pamphlets and recruit employee help in identifying hazards. Reward workers for good safety performance. Often, an incentive program will help to encourage safe work practices. Eliminate any hazards identified, and obtain OSHA forms and any first-aid supplies that would be needed in an emergency. The OSHA *Handbook for Small Business* outlines the legal requirements imposed by the Occupational Safety and Health Act of 1970. The handbook, which is available from OSHA, also suggests ways in which a company can develop an effective safety program.

Free on-site consultations are also available from OSHA. A consultant will tour the facility and offer practical advice about safety. These consultants do not issue citations, propose penalties, or routinely provide information about workplace conditions to the federal inspection staff. Contact the nearest OSHA office for additional information. Table 23.11 provides a basic checklist of safety points for consideration.

Maintaining safety standards is difficult in any size organization. A written safety manual that has specific practices and procedures for normal workplace hazards as well as the emergency-related hazards you identify is a good idea, and may lower your insurance rates.[4] If outside workers set foot in your facility, prepare a special Safety Manual for Contractors. Include in it installation standards, compliance with *Lock-Out/Tag-Out*, and emergency contact names, and phone numbers. *Lock-Out/Tag-Out* is a set of standard safety policies that assure that energy is removed from equipment during installation and maintenance. It ensures that every member of a work detail is clear before power is reapplied. Make sure outside contractors carry proper insurance, and are qualified, licensed, or certified to do the work for which you contract.

References

1. Federal Information Processing Standards Publication No. 94, *Guideline on Electrical Power for ADP Installations*, U.S. Department of Commerce, National Bureau of Standards, Washington, D.C., 1983.
2. *Practical Guide to Ground Fault Protection*, PRIMEDIA Intertec, Overland Park, KS, 1995.
3. *National Electrical Code*, NFPA No. 70.
4. Rudman, Richard, Disaster Planning and Recovery, in *The Electronics Handbook*, Jerry C. Whitaker, Ed., CRC Press, Boca Raton, FL, 1996, 2266–2267.

Bibliography

Code of Federal Regulations, 40, Part 761.
Current Intelligence Bulletin #45, National Institute for Occupational Safety and Health Division of Standards Development and Technology Transfer, February 24, 1986.
Electrical Standards Reference Manual, U.S. Department of Labor, Washington, D.C.
Hammar, Willie, *Occupational Safety Management and Engineering*, Prentice-Hall, New York.
Lawrie, Robert, *Electrical Systems for Computer Installations*, McGraw-Hill, New York, 1988.
Pfrimmer, Jack, Identifying and managing PCBs in broadcast facilities, *NAB Engineering Conference Proceedings*, National Association of Broadcasters, Washington, D.C., 1987.
Occupational Injuries and Illnesses in the United States by Industry, OSHA Bulletin 2278, U.S. Department of Labor, Washington, D.C., 1985.
OSHA, *Handbook for Small Business*, U.S. Department of Labor, Washington, D.C.
OSHA, *Electrical Hazard Fact Sheets*, U.S. Department of Labor, Washington, D.C., January 1987.

Index

A

ACE pulsing. See Annular control electrode (ACE) pulsing
Active arrays, **17**-24
Active region, **9**-7, **9**-14
Additive white Gaussian noise (AWGN), **6**-2, **6**-23
AGC circuits. See Automatic gain control circuits
Air cooling, **18**-10
    air filters, **18**-31
    power tubes, **7**-15, **7**-16–**7**-17
    system design, **18**-29–**18**-30
Air-dielectric cable, **12**-4, **12**-6
Air filters, **18**-31
Aliasing, **5**-2, **5**-11
Alignment, of filters, **14**-3–**14**-4
All-pass filters, **14**-2
All-pole filters, **14**-3
AM. See AM broadcasting; Amplitude modulation
AM broadcasting, **1**-8–**1**-11
    antennas, **17**-11–**17**-17
    bearing, **17**-13
Amplifiers
    broadband amplifier design, **1**-6
    cavity-type power amplifiers, **19**-7
    Class A amplifiers, **1**-4, **11**-10–**11**-12
    Class B amplifiers, **1**-5, **11**-12–**11**-17
    Class C amplifiers, **1**-5, **11**-17–**11**-31
    Class D amplifiers, **1**-5, **11**-31–**11**-36
    configuration, **9**-3–**9**-5
    crossed-field amplifiers (CFA), **8**-4–**8**-5
    gyrotron traveling wave tube amplifiers, **8**-29
    gyrotwystron amplifiers, **8**-29-**8**-30
    junction field-effect transistors (JFETs) as, **9**-8–**9**-10
    linear amplifiers, **11**-2–**11**-8, **11**-10–**11**-17, **11**-37
    nonlinear amplifiers, **11**-2, **11**-8–**11**-10, **11**-17–**11**-36
    nonsaturated class C amplifiers, **11**-17–**11**-23
    push-pull amplifiers, **11**-13
    quiescent point (q-point), **9**-8, **11**-15
    saturated class C amplifiers, **11**-23–**11**-31
    solid-state, **11**-1–**11**-36
    stagger tuning, **1**-6
    switch-mode amplifiers, **11**-31–**11**-36
    two-cavity klystron amplifiers, **8**-13
Amplitron, **8**-4, **8**-26
Amplitude modulation (AM), **3**-1–**3**-15
    bandwidth, **3**-2
    defined, **3**-15
    double sideband-suppressed carrier (DSB-SC) modulation, **3**-4, **3**-14, **3**-15
    linear filters, **3**-3
    noise effects, **3**-12
    quadrature amplitude modulation (QAM), **3**-11–**3**-12, **3**-14, **3**-15
    single sideband, **3**-7–**3**-10, **3**-14, **3**-15
    superheterodyne receivers, **3**-12–**3**-14
    vestigial sideband (VSB), **3**-10–**3**-11, **3**-14, **3**-15
Amplitude-shift keying (ASK), **6**-4, **6**-23
Analog pulse modulation, **5**-10–**5**-11
Analog-to-digital conversion, **5**-4–**5**-5, **5**-11
Angle modulation, **3**-4, **3**-15
Annular cathode, **8**-17
Annular control electrode (ACE) pulsing, **1**-19
Annular slot, **17**-10
Anode pulsing, **8**-17
Antennas, **16**-1–**16**-33
    AM broadcasting, **17**-11–**17**-17
    antenna sweep, **21**-7
    aperture antennas, **16**-26–**16**-27
    array antenna. See Array antennas
    bandwidth of, **16**-2, **17**-2
    beamwidth of, **17**-3
    biconical antennas, **16**-2
    broadband antennas, **16**-2, **17**-22–**17**-23
    butterfly antennas, **17**-21
    circular loop antennas, **16**-20, **16**-22–**16**-24
    circularly polarized antennas, **17**-20–**17**-21
    coax slot antennas, **17**-19, **17**-20
    conical log spiral antennas, **17**-8
    corner-reflector antennas, **17**-7, **17**-9
    current distribution, **16**-3–**16**-4
    dipole antennas, **16**-2, **16**-33, **17**-6
    dipole characteristics of, **16**-16–**16**-18, **16**-19
    directive gain of, **16**-11
    directivity of, **16**-3, **16**-11, **16**-12, **16**-33, **17**-3
    earth's effect on, **16**-13–**16**-16
    efficiency of, **1**-2, **16**-11, **16**-12–**16**-13
    equivalence theorem, **16**-26
    fields of, **16**-5–**16**-6
    FM broadcasting, **17**-17
    folded dipole antennas, **17**-6, **17**-9
    folded unipole antennas, **17**-16–**17**-17
    frequency-independent antennas, **16**-2, **16**-28–**16**-29, **16**-33
    frequency independent phased arrays (FIPA), **16**-30–**16**-31
    gain of, **17**-3
    height above average terrain (HAAT), **1**-15–**1**-16
    helix antennas, **16**-2, **17**-20
    horn antennas, **17**-10
    Huygens' sources, **16**-26–**16**-27
    impedance, **16**-8–**16**-9
    impedance matching, **17**-4–**17**-5
    interlaced traveling wave arrays, **17**-20, **17**-21

I-1

isotropic antennas, **15**-2, **15**-16
linear array antennas, **17**-11
log-periodic antennas, **16**-2, **16**-33, **17**-8, **17**-10
log-periodic dipole array (LPDA), **16**-29–**16**-30, **16**-31, **17**-8
log-periodic V antennas, **17**-8
loop antennas, **16**-1, **16**-2, **16**-18, **16**-20–**16**-24, **16**-33
multi-arm short helix antennas, **17**-17
multi-slot traveling-wave antennas, **17**-20
mutual impedance, **16**-9–**16**-10
narrow-band antennas, **16**-3
narrowband antennas, **16**-3, **19**-10
operating characteristics of, **17**-1–**17**-2
panel antennas, **17**-17, **17**-20, **17**-21
parabolic reflector antennas, **17**-10–**17**-11
pattern multiplication in, **16**-7–**16**-8
phased-array antennas, **17**-23–**17**-29
phase-shift devices, **17**-26
planar array antennas, **17**-11
polarization of, **16**-4–**16**-5, **17**-2
power gain of, **16**-11
preventive maintenance for, **18**-20–**18**-23
pylon antennas, **17**-19
pyramidal horn antennas, **16**-2, **17**-10
quarter-wave monopole antennas, **17**-7
radar, **1**-30, **1**-32
radar system duplexer, **17**-26–**17**-29
radiated fields, **16**-18, **16**-20–**16**-22
radiation patterns of, **16**-6–**16**-7
radiation resistance, **17**-2
reflector antennas, **17**-10–**17**-11
ring stub antennas, **17**-17
satellite transmission, **1**-25–**1**-29
self-similar fractal antennas, **16**-31–**16**-33
series-fed slanted dipole antennas, **17**-17
shunt-fed slanted dipole antennas, **17**-17
side-mounted antennas, **17**-21
slot antennas, **17**-10
Sommerfield-Norton solution, **16**-13, **16**-15, **16**-33
space regions, **17**-4
television broadcasting, **1**-15–**1**-16, **17**-17–**17**-23
three-dimensional frequency-independent phased array (3D-FIPA), **16**-32
top-mounted antennas, **17**-18–**17**-20
tower-top, pole-type antennas, **17**-18
tuning of, **22**-3
turnstile antennas, **17**-19, **17**-20
twisted ring antennas, **17**-17
types of, **16**-1–**16**-3, **17**-6–**17**-11
UHF side-mounted antennas, **17**-21
uplink antennas for satellite systems, **1**-25–**1**-26
V-dipole antennas, **17**-6
VHF multi-slot antennas, **17**-20
waveguide antennas, **17**-10
waveguide slot antennas, **17**-19, **17**-20
wideband antennas, **16**-27–**16**-33
wideband panel antennas, **17**-17
Yagi-Uda arrays, **16**-24–**16**-26, **16**-34, **17**-9, **17**-10
zigzag antennas, **17**-20
zigzag panel antennas, **17**-21
Antenna sweep, **21**-7

Antipodal signaling, **6**-2, **6**-4, **6**-23
Aperture, **16**-1
Aperture antennas, **16**-26–**16**-27
Array antennas, **16**-1, **16**-2, **17**-11
  critical arrays, **17**-15
  end-fire arrays, **17**-10
  frequency independent phased arrays (FIPA), **16**-30–**16**-31
  interlaced traveling wave arrays, **17**-20, **17**-21
  linear array antennas, **17**-11
  log-periodic dipole array (LPDA), **16**-29–**16**-30, **16**-31
  phased-array antennas, **17**-23–**17**-29
  planar array antennas, **17**-11
  Weierstrass arrays, **16**-32–**16**-33
  Yagi-Uda arrays, **16**-24–**16**-26, **16**-34, **17**-9, **17**-10
Array coordinates, **17**-23
Arrays, classification of, **17**-24
ASK. See Amplitude-shift keying
Attenuation, **11**-36
  coaxial transmission lines, **12**-6, **12**-9–**12**-12
  rain, **15**-15
Attenuators, **11**-1
Aural subcarrier, **1**-16, **1**-18, **11**-5, **11**-36
Automatic gain control circuits (AGC circuits), **9**-13
Average detection, **21**-7
Average-response measurement, power measurement, **20**-2
AWGN. See Additive white Gaussian noise

B

Backward wave oscillator (BWO), **8**-24–**8**-25
Balun transformer, **1**-7
Bandpass constant-impedance diplexer, **14**-10–**14**-11
Bandpass filters (BPF), **3**-3
Bandpass multiplexer module, group delay, **14**-12
Band-reject filter, **14**-2
Band-stop diplexer, **14**-8–**14**-10
Band switching, **21**-7
Bandwidth
  amplitude modulation, **3**-2
  of antennas, **16**-2, **17**-2
  of filters, **14**-3
  frequency modulation, **4**-5
  modulation and, **1**-2
Bandwidth efficiency
  binary modulation, **6**-6–**6**-8
  defined, **6**-23
  M-ary modulation, **6**-15–**6**-17
Bandwidth occupancy, **5**-6, **5**-11
Baseband, **21**-7
Baseband digital pulse modulation, **5**-5–**5**-10
Baseband section, satellite system, **1**-25
Baseband signal, **3**-15
Baseband systems, **3**-1
Baseline clipper, **21**-7
Basis set, **6**-9
Beam focusing, traveling wave tube (TWT), **8**-17–**8**-18

Index I-3

Beam pulsing, **1**-19
Beamwidth, of antennas, **17**-3
Bearing, AM broadcasting, **17**-13
Beryllium fluoride, **23**-19
Beryllium oxide ceramics, safety, **23**-19
Bessel alignment, **14**-3
Bessel functions, **4**-5, **21**-7
Bessel null method, **21**-3, **21**-8
Biconical antennas, **16**-2
Binary modulation, **6**-1–**6**-8
 bandwidth efficiency, **6**-6–**6**-8
 coherent, **6**-1–**6**-5
 noncoherent, **6**-5–**6**-6
Binary representation, **5**-4, **5**-11
Bioaccumulation, **23**-12
Biomagnification, **23**-12
Biphase baseband data modulation, **5**-6
Biphase-shift keying (BPSK), **6**-4
Bipolar junction transistors (BJTs), **9**-1–**9**-3, **9**-5
Bit period, **6**-2, **6**-23
BJTs. See Bipolar junction transistors
Block code, **6**-21, **6**-23
Body, **10**-14
BPF. See Bandpass filters
BPSK. See Biphase-shift keying
Branch diplexer, **14**-6
Brillouin flow, **8**-18
Broadband amplifiers, design, **1**-6
Broadband antennas, **16**-2, **17**-22–**17**-23
Broadcasting
 AM radio, **1**-8–**1**-11
 FM radio, **1**-12–**1**-15
 shortwave, **1**-12
 television, **1**-15–**1**-23
B-SAVE A, **21**-7
Buried-channel MOSFET, **10**-1
Butler beam-forming network, **17**-24, **17**-26
Butterfly antennas, **17**-21
Butterworth alignment, **14**-3
BWO. See Backward wave oscillator

C

Cable jacketing, **12**-13
Calibrator, **21**-8
Capacitors
 doorknob capacitors, **19**-7
 polychlorinated biphenyls (PCBs) and, **23**-12, **23**-13–**23**-14
 preventive maintenance for, **18**-4
 troubleshooting, **19**-6
Carcinotron, **8**-25
Carrier, **1**-2, **3**-15
Carrier frequency, **3**-1, **3**-15
Carrier signal, **3**-4
Carrier swing, **4**-3
Carrier-to-noise ratio (C-N), **21**-8
Carson's rule, **11**-9
Cascade circuit, **9**-3

Cathode pulsing, **8**-16
Cathodes
 back heating of, **7**-15
 triode, **7**-6
Cauer filter, **14**-3
Cavity magnetron oscillator, **8**-22
Cavity resonators, waveguides, **13**-8–**13**-10
Cavity-type power amplifier, troubleshooting, **19**-7
C-band, **2**-2, **2**-5
C-band systems, **1**-24
Center frequency, **14**-3, **21**-8
CFA. See Crossed-field amplifiers
Channel, MOSFET, **10**-13
Channel encoding, **5**-5, **5**-11
Chebyshev alignment, **14**-3
Chirp spread spectrum, **1**-3
Circuit breakers, OSHA safety requirements, **23**-17
Circularity, **17**-22
Circular loop antennas, **16**-20, **16**-22–**16**-24
Circularly polarized antennas, **17**-20–**17**-21
Circular polarization, **16**-5, **17**-2
Circular waveguide, **13**-4, **13**-5, **13**-8
Circulating currents, power tubes, **8**-5
Circulators, **1**-26, **14**-18
Class A amplifiers, **1**-4, **11**-10–**11**-12
Class B amplifiers, **1**-5, **11**-12–**11**-17
Class C amplifiers, **1**-5
 nonsaturated, **11**-17–**11**-23
 saturated, **11**-23–**11**-31
Class D amplifiers, **1**-5, **11**-31–**11**-36
Class E enhancement mode, **11**-35
Class of operation, RF amplifiers, **1**-4–**1**-5, **11**-10–**11**-36
Cloverleaf, **8**-16
CMFSK. See Coherent M-ary frequency-shift keying
CMOS (complementary MOS), **10**-3
C-N. See Carrier-to-noise ratio
Coax bands, **21**-8
Coaxial magnetron, **8**-23–**8**-24
Coaxial transmission lines, **12**-1–**12**-13
 attenuation, **12**-9–**12**-12
 connector effects, **12**-8–**12**-9
 electrical parameters, **12**-3–**12**-7
 fire-retardant jacket, **12**-13
 mechanical parameters, **12**-12–**12**-13
 phase stability, **12**-12
 power rating, **12**-7–**12**-8, **12**-9
 skin effect, **12**-1–**12**-2
 testing, **22**-1–**22**-5
 voltage standing wave ratio (VSWR), **12**-6
Coax slot antenna, **17**-19, **17**-20
Cochannel interference, **3**-12
Code rate, **6**-23
Coding, for modulation, **6**-21–**6**-22
Coding gain, **6**-21, **6**-23
Coherent binary modulation, **6**-1–**6**-5, **6**-23
Coherent detection, **3**-5, **3**-15
Coherent M-ary frequency-shift keying (CMFSK), **6**-12
Coils, preventive maintenance for, **18**-5–**18**-6
Collector, traveling wave tube (TWT), **8**-19
Collector depression, **8**-19–**8**-20

Color notch filter, television broadcasting, 1-22
Color subcarrier, **11**-5, **11**-36
Color television, transmission standards, 1-17
Comb generator, **21**-8
Combiners, **14**-1
    four-port hybrid combiner, **14**-4–**14**-6
    hot-switching combiners, **14**-13–**14**-17
    microwave combiners, **14**-12–**14**-13
Common amplification, 1-19
Common-collector circuit, **9**-4
Common point, **17**-16
Companders, **5**-5
Companding, **5**-5, **5**-11, **5**-12
Compensation, RF amplifiers, 1-6
Complementary MOS. See CMOS
Component, **21**-8
Compression, amplifiers, **11**-2–**11**-3, **11**-36
Conduction angle, **11**-10, **11**-36
Conical horn, **17**-10
Conical log spiral antenna, **17**-8
Connection points, preventive maintenance of, **18**-7
Connector effects, coaxial transmission lines, **12**-8–**12**-9
Constant-impedance diplexer, **14**-8–**14**-12
Constraint length, **6**-22, **6**-23
Contact potential, **9**-13, **9**-14
Continuous-phase modulation (CPM), **6**-19–**6**-20, **6**-23
Convolutional code, **6**-21, **6**-23
Cooling systems
    air cooling, **18**-29–**18**-30
    of klystrons, **18**-31–**18**-33
    of power tubes, **7**-15–**7**-18
    preventive maintenance of, **18**-28
    vapor-phase cooling of, **7**-15, **7**-18
    water cooling systems, **18**-31–**18**-33
Cooperative electronic navigation systems, 1-32
Corner-reflector antenna, **17**-7, **17**-9
Corporate feed, **17**-24
Correlation coefficient, **6**-23
Correlation implementation, **6**-2–**6**-3
Correlation receiver, **5**-10, **5**-12
Correlator implementation, **6**-9, **6**-23
Corrosive compounds, safety, **23**-19–**23**-20
Coupled-cavity circuit, **8**-16
Coupling-filtering system, television broadcasting, 1-22
Critical arrays, **17**-15
Critical frequency, **12**-6
Crossed-field amplifiers (CFA), **8**-4–**8**-5
Crossed-field microwave tubes, **8**-3–**8**-5, **8**-20–**8**-30
Crossover distortion, **11**-15
Crosstalk, 3-12
Crush strength, of coaxial transmission lines, **12**-13
Crystal resonators, 1-3
Crystals, 1-3, 1-4
Current distribution, antennas, **16**-3–**16**-4
Current regulator diode, **9**-11, **9**-12
Cutoff region, **9**-5
CW magnetron, **8**-20

D

Decibel (dB), **20**-6, **21**-8
Decibel measurement, RF power measurement, **20**-6
Degree of modulation, **4**-4
Delta F, **21**-8
Delta-matched, **17**-17
Demodulation, 3-15
Demodulation equations, 3-5
Demodulator, 3-5
Demultiplexing, 1-2
Depletion layer, **10**-5
Depletion-mode MOSFET, **10**-1
Detection, 3-15, **5**-10
Detector, 3-5
DF. See Direction finding
Diacrode, **8**-7–**8**-9
Dicode, **5**-12
Dielectric construction, coaxial transmission lines, **12**-4
Dielectric heating, 1-36
Dielectric posts, **14**-17
Dielectric slugs, impedance matching, **13**-6
Differential distance ranging, 1-34
Differential encoding, **5**-6, **5**-12, **6**-5, **6**-23
Differential gain distortion, linear amplifiers, **11**-7
Differentially coherent phase-shift keying (DPSK), **6**-5, **6**-14, **6**-24
Differential phase distortion, linear amplifiers, **11**-7–**11**-8
Differential phase shift, **20**-8
Diffraction
    defined, **15**-16
    radio wave propagation, **15**-6
Digital modulation, 1-11, **6**-1–**6**-23
    analog-to-digital conversions, **5**-4–**5**-5
    binary modulation, **6**-1–**6**-8
    error correction coding, **6**-20–**6**-22
    higher-order schemes, **6**-20–**6**-23
    M-ary modulation, **6**-8–**6**-17
    pulse modulation, **5**-5–**5**-10
    trellis-coded modulation (TCM), **6**-22–**6**-23
Digital television (DTV), 1-8
Diode modulators, **4**-7
Diodes, **7**-5–**7**-6, **9**-11, **9**-12
Diplexer-combiner, television broadcasting, 1-23
Diplexers, **14**-1
    bandpass constant-impedance diplexer, **14**-10–**14**-11
    band-stop diplexer, **14**-8–**14**-10
    constant-impedance diplexer, **14**-8–**14**-12
    non-constant-impedance diplexer, **14**-6–**14**-8
Dipole antennas, **16**-2, **16**-33, **17**-6
Dipole characteristics, of antennas, **16**-16–**16**-18, **16**-19
Dipoles, **16**-1
Direct FM, **4**-2, **4**-7, **4**-14
Direct FM modulators, **4**-7–**4**-9
Directional array, **17**-13
Direction finding (DF), 1-33
Directive gain, antennas, **16**-11

Index I-5

Directivity, of antennas, **16**-3, **16**-11, **16**-12, **16**-33, **17**-3
Direct modulation, **1**-13
Directors, **16**-24
Direct paralleling, **1**-7
Direct-sequence spread spectrum (DSSS), **6**-20, **6**-24
Discriminator, **4**-11, **4**-14
Dissipative waveguide filter, **17**-28
Distance-measuring equipment (DME) systems, **1**-34
Distortion
    crossover distortion, **11**-15
    defined, **21**-8
    differential gain distortion, **11**-7
    differential phase distortion, **11**-7–**11**-8
    group delay distortion, **11**-9
    harmonic distortion, **20**-9–**20**-12
    intermodulation distortion (IMD), **11**-4, **11**-37, **20**-12–**20**-14
    linear amplifiers, **11**-3–**11**-7, **11**-8–**11**-10
    nonlinear amplifiers, **11**-9–**11**-10
    total harmonic distortion and noise (THD+N), **20**-11
Distortion analyzer, **20**-10
Distributed amplification, **1**-6
DME systems. See Distance-measuring equipment (DME) systems
Dominant mode, **13**-1
Doorknob capacitors, **19**-7
Dopant-ion control, MOSFETs, **10**-13
Double-coil helical beam gyrotron, **8**-27
Double reflector, satellite systems, **1**-26
Double sideband plus carrier (DSB+C), **3**-5
Double sideband-suppressed carrier (DSB-SC) modulation, **3**-4, **3**-14, **3**-15
Doubly truncated waveguide (DTW), **13**-4–**13**-5
Downlink antennas, satellite transmission, **1**-28–**1**-29
Downlink transmitting station, **1**-24
DPSK. See Differentially coherent phase-shift keying
Drain, **1**-01
Drain characteristics, **9**-8
Drain conductance, MOSFETs, **10**-10
Drain engineering, **10**-12
Drain-induced barrier lowering, **10**-11
Drain response time, **10**-9
Drain source, **10**-13
Driving ability, MOSFETs, **10**-7–**10**-9, **10**-11
DSB+C. See Double sideband plus carrier
DSB-SC modulation. See Double sideband-suppressed carrier (DSB-SC) modulation
DSSS. See Direct-sequence spread spectrum
DTV. See Digital television
DTW. See Doubly truncated waveguide
Dual-polarity waveguide, **13**-2
Ducting
    defined, **15**-16
    tropospheric, **15**-13
Duobinary encoding, **5**-12
Duplexers
    radar system, **17**-26–**17**-29
    tuning, **21**-5–**21**-6

Dynamic range, **21**-8
Dynamic transfer conductance, **9**-9

E

Early voltage, **9**-10
Earth, effect on antennas, **16**-13–**16**-16
Effective Earth radius, **15**-6, **15**-16
Effective power, rms measurement, **20**-1
Effective radiated power (ERP), **1**-12, **17**-3
Efficiency
    of antennas, **1**-2, **16**-11, **16**-12–**16**-13
    of waveguides, **13**-3
EHF band (extremely high-frequency band), **2**-2, **2**-3
Electrical beam tilt, **17**-3
Electrical length, **17**-2
Electric current, effect on human body, **23**-4–**23**-5
Electric shock, **23**-4–**23**-11, **23**-4–**23**-5
    circuit-protection hardware, **23**-6–**23**-8
    first aid, **23**-8–**23**-11
    working with high voltage, **23**-8
Electrocution, **23**-5, **23**-6
    See also Electric shock
Electromagnetic spectrum, **1**-2, **2**-1–**2**-5
Electron gun, traveling wave tube (TWT), **8**-17
Electronic navigation, **1**-32–**1**-35
Electron optics, **7**-2–**7**-5
Electrons, **7**-1–**7**-2
Electron transit time, power tubes, **7**-13–**7**-14, **7**-15, **8**-5
Elliptic alignment, **14**-3
Elliptical polarization, **17**-20
ELNEC program, **16**-4, **16**-6
Emergency power off (EPO) button, **23**-1
Emitter-follower circuit, **9**-4
Emitting-sole tubes, **8**-20
Emphasis, **21**-8
Encoder, **1**-2, **6**-21
End-fire array, **17**-10
End-to-end load, **11**-14
Enhancement-mode device, **10**-1
Envelope, **21**-8
Envelope detection, **3**-15
E-plane plots, **16**-6
E-plane sectoral horn, **17**-10
EPO button. See Emergency power off (EPO) button
Equipment marking, OSHA safety requirements, **23**-17
Equivalence theorem, antennas, **16**-26
Equivalent current, **16**-3
Equivalent noise bandwidth, **21**-8
ERP. See Effective radiated power
Error correction encoding, **5**-5, **6**-20–**6**-22
Euclidian distance, **6**-11, **6**-24
Extension cords, OSHA safety requirements, **23**-17–**23**-18
External cavity klystron, **18**-8
External mixers, **21**-8
Extremely high-frequency band. See EHF band

F

Fast frequency hopping, 1-3
FC-75 toxic vapor, **23**-19–**23**-20
FDM. See Frequency-domain multiplexing
Feed-point, **16**-1
Ferrite isolators, **14**-17
FETs. See Field-effect transistors
FHSS. See Frequency-hop spread spectrum
Field-effect transistors (FETs), **10**-1–**10**-2, **11**-31
Fields, of antennas, **16**-5–**16**-6
Filament-on command, **18**-13
Filament voltage, power tubes, **18**-13–**18**-14
Filter loss, **21**-8
Filters
    air filters, **18**-31
    alignment of, **14**-3–**14**-4
    all-pass filters, **14**-2
    all-pole filters, **14**-3
    bandpass filters (BPF), **3**-3
    Cauer filter, **14**-3
    color notch filter, **1**-22
    defined, **14**-2, **21**-8
    dissipative waveguide filter, **17**-28
    harmonic filters, **1**-22, **17**-27
    linear filters, **3**-3
    low-pass filters (LPF), **3**-3
    matched filters, **5**-10, **6**-3, **6**-9, **6**-24
    narrowband filters, **17**-29
    notch filter, **14**-2
    order of, **14**-4
    passive filters, **14**-2
    preselectors, **17**-29
    radar system, **17**-27
    types, **14**-2–**14**-3
    waveguide filters, **13**-7–**13**-8
FIPA. See Frequency independent phased arrays
Fire damper, **23**-2
Fire-retardant jacket, coaxial transmission lines, **12**-13
First aid, electric shock, **23**-8–**23**-11
First mixer input level, **21**-8
Flame detector, **23**-2
Flatband voltage, **10**-7
Flatness, **21**-8
Flat top sampling, **5**-2, **5**-3, **5**-12
Flicker effect, **20**-8
Fly-wheel effect, **1**-21
FM. See FM broadcasting; FM receiver; Frequency modulation
FM broadcasting, **1**-12–**1**-15
    antennas, **17**-17
    auxiliary services, **1**-14
    modulation circuits, **1**-13–**1**-14
    power amplifiers, **1**-14–**1**-15
FM receiver, **4**-11–**4**-14
Foam-dielectric cable, **12**-4, **12**-8, **12**-13
Focus electrode pulsing, **8**-17
Folded dipole antenna, **17**-6, **17**-9
Folded unipole antenna, **17**-16–**17**-17
Forward-biased base-emitter junction, **9**-10
Forward fundamental circuit, **8**-16

Forward transform, **3**-1
Foster-Seeley discriminator, **4**-13–**4**-14
FOT. See Frequence optimum de travail
Fourier analysis, **21**-8
Fourier transform, **3**-1, **5**-2
Four-port hybrid combiner, **14**-4–**14**-6
Fractals, **16**-31
Fraunhofer region, **16**-5, **16**-6, **16**-33
Free space path loss, **15**-3, **15**-16
Frequence optimum de travail (FOT), **15**-11, **15**-16
Frequency, **21**-8
Frequency-agile magnetron, **8**-23
Frequency band, **21**-8
Frequency deviation, **4**-3–**4**-4, **4**-14, **21**-8
Frequency-domain multiplexing (FDM), **1**-2, **3**-1
Frequency domain representation, **21**-8
Frequency hopping, **1**-3
Frequency-hop spread spectrum (FHSS), **6**-20, **6**-24
Frequency-independent antenna, **16**-2, **16**-28–**16**-29, **16**-33
Frequency independent phased arrays (FIPA), **16**-30–**16**-31
Frequency marker, **21**-8
Frequency modulation (FM), **3**-4, **4**-1–**4**-15
    bandwidth, **4**-5
    broadcasting, **1**-12–**1**-15
    defined, **4**-15
    detector circuits, **4**-13
    direct FM, **4**-2, **4**-7, **4**-14
    direct FM modulators, **4**-7–**4**-9
    frequency deviation, **4**-3–**4**-4
    indirect FM, **4**-7
    indirect-FM modulators, **4**-10–**4**-11
    limiters, **4**-11–**4**-13, **4**-15
    modulated FM carrier, **4**-1–**4**-3
    modulation index, **4**-4
    narrow-band FM (NBFM), **4**-7
    percent of modulation, **4**-4
    reception, **4**-11–**4**-14
    sideband frequency, **4**-5–**4**-7
    voltage controlled oscillator (VCO) direct-FM modulators, **4**-10
    wide-band FM (WBFM), **4**-7
Frequency range, **21**-8
Frequency scan, **17**-25
Frequency-shift keying (FSK), **6**-4, **6**-24
Frequency swing, **4**-3
Frequency translation, **1**-2
Fresnel region, **15**-5, **15**-16, **16**-6, **16**-33
FSK. See Frequency-shift keying
Full-wave depletion, **11**-33
Full-wave H-bridge, **11**-33–**11**-35
Fundamental frequency, **21**-8

G

Gain, **11**-1, **11**-36, **17**-3
Gas noise, **20**-7
Gate, MOSFET, **10**-1, **10**-13

Index

Gate insulator, **10**-1
Gate voltage, **10**-7
Gaussian-filtered MSK (GMSK), **6**-18–**6**-19, **6**-24
Generic flow, **8**-18
Geostationary satellite, **1**-23
GFCI. See Ground-fault current interrupter
"Ghosting," **15**-13
Global positioning system (GPS), **6**-20, **6**-24
GMSK. See Gaussian-filtered MSK
GPS. See Global positioning system
Gram-Schmidt procedure, **6**-8–**6**-9, **6**-24
Grass, **21**-8
Graticule, **21**-9
Gray code, **5**-4, **5**-12
Grid, triode, **7**-6
Grid pulsing, **8**-17
Grid vacuum tubes, **8**-5–**8**-9
 diacrode, **8**-7–**8**-9
 high-power UHF tetrodes, **8**-6–**8**-7, **8**-8
 planar triodes, **7**-9, **7**-10, **8**-5–**8**-6
Ground-fault current interrupter (GFCI), **23**-6–**23**-8
Grounding, OSHA safety requirements, **23**-18–**23**-19
Groundwave, **15**-4, **15**-16
Group delay, **14**-12, **20**-8
Group delay distortion, linear amplifiers, **11**-9
Guy wire arc-over, **19**-10
Gyroklystron amplifier, **8**-28
Gyrotron, **8**-26–**8**-30
Gyrotron backward oscillator, **8**-29
Gyrotron traveling wave tube amplifier, **8**-29
Gyrotwystron amplifier, **8**-29–**8**-30

H

Half-power-beamwidth (HPBW), **16**-7
Half-power frequency, **14**-3
Halon, **23**-2
Hard decision, **6**-22, **6**-24
Harmonic distortion
 defined, **21**-9
 RF power measurement, **20**-9–**20**-12
 spurious harmonic distortion, **21**-4–**21**-7
Harmonic filters
 radar system, **17**-27
 television broadcasting, **1**-22
Harmonic mixing, **21**-9
Harmonics, **21**-9
H-bridge, **11**-33
Heat dissipation
 power tubes, **8**-5
 See also Cooling
Heating, dielectric heating, **1**-36
Height above average terrain (HAAT), **1**-15–**1**-16
Helix antenna, **16**-2, **17**-20
Helix circuit, TWT, **8**-16
Heterodyne spectrum analyzer, **21**-9
Heterostructure field-effect transistors (HFETs), **10**-2
HFETs. See Heterostructure field-effect transistors
HF propagation, **15**-9–**15**-12

High-band VHF, **1**-15
High-frequency delay, linear amplifiers, **11**-8
High-frequency radio band, **2**-2, **2**-3
High-level AM modulation, **1**-8–**1**-9
High-level modulation, **11**-29
High-power amplifier (HPA), satellite systems, **1**-25
High-power isolators, **14**-17–**14**-20
High-power pulsed magnetrons, **8**-21
High-power UHF tetrodes, **8**-6–**8**-7, **8**-8
High-voltage equipment, safety practices for, **23**-5, **23**-8
High-voltage plate supply, troubleshooting, **19**-2–**19**-5
High-voltage power supply
 metering, **18**-24
 preventive maintenance, **18**-23–**18**-24
 single phasing, **18**-26–**18**-28
 transient disturbances, **18**-26–**18**-28
Hilbert transform, **3**-9
Horizontal polarization, **16**-15, **17**-2
Horn antenna, **17**-10
Hot coolant and surfaces, safety, **23**-22
Hot switch, high-power isolator, **14**-19–**14**-20
Hot switching, **14**-13
Hot-switching combiners, **14**-13–**14**-17
HPA. See High-power amplifier
HPBW. See Half-power-beamwidth
H-plane plots, **16**-6
H-plane sectoral horn, **17**-10
Huffman code, **5**-5, **5**-12
Huygens' sources, antennas, **16**-26–**16**-27
Hybrid combiners, **14**-4–**14**-5, **14**-7, **14**-8
 four-port hybrid combiner, **14**-4–**14**-6
 microwave combiners, **14**-12–**14**-13
Hybrid coupler, **1**-7
Hybrid splitting-combining, **1**-7
Hyperbolic positioning, **1**-34
Hyperbolic systems, **1**-34

I

ICPM. See Incidental carrier phase modulation
Ideal sampling waveform, Fourier transform of, **5**-2
IF. See Intermediate frequency
IF gain, **21**-9
IGBJTs. See Insulated gate bipolar junction transistors
Image antenna, **17**-7
Image stations, **3**-15
IMD. See Intermodulation distortion
Immersed flow, **8**-18
Impedance, **4**-8
 antennas, **16**-8–**16**-9
 coaxial cable, **12**-4
Impedance matching
 antennas, **17**-4–**17**-5
 waveguides, **13**-6–**13**-10
Incidental carrier phase modulation (ICPM), linear amplifiers, **11**-9
Indirect FM, **4**-7
Indirect-FM modulators, **4**-10–**4**-11

Induced grid noise, 20-7
Induction heating, 1-36, 7-9
Inductive output tube (IOT), 1-19
Infrared light, 2-4
Injected-beam crossed-field tubes, 8-20
In-line array, 17-13
Input signal-to-noise ratio, 3-12
Instantaneous frequency, 21-9
Insulated gate bipolar junction transistors (IGBJTs), 11-31
Insulators, preventive maintenance of, 18-7–18-8
Integral cavity klystron, 18-8
Integrate-and-dump detector, 5-10, 5-12
INTELSAT system, 1-23
Interaction circuit, TWT, 8-15
Interlaced traveling wave array, 17-20, 17-21
Interlock failures, troubleshooting, 19-12–19-13
Intermediate frequency (IF), 1-25, 3-15, 21-9
Intermediate power amplifier (IPA), 1-21
Intermodulation distortion (IMD), 11-4, 11-37, 20-12–20-14, 21-9
Intermodulation products, 14-11–14-12
Intersymbol-interference property, 5-8, 5-12
Inverse Chebyshev alignment, 14-3
Ionosphere, HF propagation in, 15-11, 15-12
IOT. See Inductive output tube
IPA. See Intermediate power amplifier
Isolators, 14-17–14-20
Isotropic antenna, 15-2, 15-16
Isotropic elements, 16-7, 16-33

J

JFETs. See Junction field-effect transistors
Junction field-effect transistors (JFETs), 9-5–9-14
    as amplifiers, 9-8–9-10
    as constant current source, 9-10–9-12
    current regulation in, 9-10–9-11
    early voltage, 9-10
    as voltage-variable resistor, 9-12–9-14

K

Ka-band, 2-2, 2-5
K-band, 2-2
Klystrons, 1-19, 8-9–8-14
    external cavity klystron, 18-8
    integral cavity klystron, 18-8
    multicavity klystron, 8-13–8-14
    multi-stage depressed collector (MSDC) klystron, 1-19
    preventive maintenance of, 18-8–18-9
    reflex klystron, 8-9–8-11
    two-cavity klystron, 8-11–8-12
    two-cavity klystron amplifier, 8-13
    two-cavity klystron oscillator, 8-12–8-13
    water cooling systems, 18-31–18-33

Ku-band, 2-2, 2-5
Ku-band systems, 1-24

L

Laws of reciprocity, 16-1
L-band, 2-2, 2-5
Let-go range, 23-5
Level measurement, 20-1
LF. See Low frequency
Life safety equipment, 23-2
Lightly doped drain, 10-12
Limiters, 4-11–4-13, 4-15
Linear amplifiers, 11-37
    Class A amplifier, 1-4, 11-10–11-12
    Class B amplifiers, 1-5, 11-12–11-17
    compression, 11-2–11-3
    odd order intermodulation distortion, 11-3–11-7
    other distortions, 11-7–11-8
Linear array antenna, 17-11
Linear-beam microwave tubes, 8-2–8-3
Linear filters, 3-3
Linear magnetron, 8-23–8-24
Linear scale, 21-9
Line code, 5-7, 5-12
Line stretcher, 17-5, 17-8
LNA. See Low noise amplifier
L network, 17-4, 17-5
LO. See Local oscillator
Load resistance, 11-11
Local oscillator (LO), 3-5, 21-9
Lock-Out-Tag-Out, 23-24
Logarithmic scale, 21-9
Log-periodic antennas, 16-2, 16-33, 17-8, 17-10
Log-periodic dipole array (LPDA), 16-29–16-30, 16-31, 17-8
Log-periodic V antenna, 17-8
Loop antenna, 16-1, 16-2, 16-18, 16-20–16-24, 16-33
LO output, 21-9
Loran C navigation systems, 1-34
Loss, in connectors, 12-10
Loss factor, 1-36
Low-band VHF, 1-15, 1-19
Low-end radio frequency band, 2-1–2-3
Lower sideband (LSB), 3-4, 3-15
Lowest usable frequency (LUF), 15-11
Low-frequency delay, linear amplifiers, 11-8
Low frequency (LF), radio wave propagation, 15-7–15-9
Low noise amplifier (LNA), 1-29
Low noise conversion unit, 1-29
Low-pass filters (LPF), 3-3
Lowpass uniform sampling theorem, 5-1
Low-power television (LPTV), 1-16
LPDA. See Log-periodic dipole array
LPF. See Low-pass filters
LPTV. See Low-power television
LSB. See Lower sideband
LUF. See Lowest usable frequency

M

Magnetic declination, **17**-13
Magnetic field effects, electron optics, **7**-3
Magnetrons, **8**-20–**8**-24
  cavity magnetron oscillator, **8**-22
  coaxial magnetron, **8**-23–**8**-24
  frequency-agile magnetron, **8**-23
  linear magnetron, **8**-23–**8**-24
Magnitude-only measurement, **21**-9
Maintenance, **18**-2–**18**-8
Maintenance log, **18**-2–**18**-3
Major lobe, **17**-13
Manchester baseband data modulation, **5**-6
Manometer, **18**-11
M-ary modulation, **6**-1, **6**-8–**6**-17, **6**-24
M-ary phase-shift keying (MPSK), **6**-10–**6**-12, **6**-15
M-ary quadrature-amplitude-shift keying (MQASK), **6**-12–**6**-13
Matched filters, **5**-10, **6**-2, **6**-9, **6**-24
Matching circuits, **1**-6–**1**-7, **5**-12
MAX HOLD, **21**-9
Maximum input level, **21**-9
Maximum-likelihood receiver, **6**-2, **6**-24
Maximum usable frequency (MUF), **15**-11
MAX-MIN, **21**-9
Max span, **21**-9
Medium-end radio frequency band, **2**-1–**2**-3
Medium frequency (MF), radio wave propagation, **15**-7–**15**-9
MESFETs. See Metal-semiconductor field-effect transistors
Message corruption, **1**-3
Message signal, **3**-15
Metal-oxide-semiconductor field-effect transistors (MOSFETs), **9**-10, **10**-1–**10**-14
  current-voltage characteristics, **10**-3–**10**-4
  dopant-ion control, **10**-13
  drain conductance, **10**-10
  driving ability, **10**-7–**10**-9, **10**-11
  hot-electron effects, **10**-12
  miniaturization, limits on, **10**-11–**10**-13
  output resistance, **10**-10
  strong-inversion characteristics, **10**-3–**10**-4
  subthreshold characteristics, **10**-4
  subthreshold control, **10**-11–**10**-12
  thin oxides, **10**-12
  threshold voltage, **10**-4–**10**-7
  transconductance, **10**-10
Metals, thermal emission from, **7**-3
Metal-semiconductor field-effect transistors (MESFETs), **10**-1
Meteor burst propagation, **15**-13–**15**-14
Metering, power supply, **18**-24–**18**-26
Method of moments (MoM), **16**-3, **16**-18, **16**-33
MF. See Medium frequency
Microwave band
  applications of, **2**-4–**2**-5
  frequencies, **2**-3
Microwave combiners, **14**-12–**14**-13
Microwave filters, radar system, **17**-27

Microwave power tubes, **8**-1–**8**-30
  applications, **8**-2
  backward wave oscillator (BWO), **8**-24–**8**-25
  crossed-field tubes, **8**-3–**8**-5, **8**-20–**8**-30
  diacrode, **8**-7–**8**-9
  gyrotron, **8**-26–**8**-30
  high-power UHF tetrodes, **8**-6–**8**-7, **8**-8
  implosion hazard, **23**-22
  klystrons, **8**-9–**8**-14
  linear-beam tubes, **8**-2–**8**-3
  magnetrons, **8**-20–**8**-24
  multicavity klystron, **8**-13–**8**-14
  planar triodes, **7**-9, **7**-10, **8**-5–**8**-6
  reflex klystron, **8**-9–**8**-11
  strap-fed devices, **8**-25–**8**-26
  traveling wave tube (TWT), **8**-14–**8**-20
  two-cavity klystron, **8**-11–**8**-12
  two-cavity klystron amplifier, **8**-13
  two-cavity klystron oscillator, **8**-12–**8**-13
  X-ray shielding, **23**-22
Microwave propagation, **15**-14–**15**-15
Microwave radio relay systems, **1**-35–**1**-36
Miller effect, **4**-8
MIN HOLD, **21**-9
Miniaturization, MOSFETs, **10**-11–**10**-13
Minimum shift keying (MSK), **6**-18, **6**-24
MININEC, **16**-4
Minority-carrier current, **10**-1
Mixer, **3**-4
Mixing, **3**-15, **21**-9
Mod-anode pulsing, **1**-19
MODFETs. See Modulation doped FETs
Modulation, **1**-1–**1**-2
  advantages of, **3**-1
  amplitude modulation, **3**-1–**3**-15
  analog pulse modulation, **5**-10–**5**-11
  angle modulation, **3**-4, **3**-15
  baseband digital pulse modulation, **5**-5–**5**-10
  continuous-phase modulation (CPM), **6**-19–**6**-20, **6**-23
  defined, **1**-2, **3**-15
  digital modulation, **1**-11, **6**-1–**6**-23
  direct modulation, **1**-13
  frequency modulation, **3**-4, **4**-1–**4**-15
  incidental carrier phase modulation (ICPM), **11**-9
  negative modulation, **18**-22
  phase modulation, **4**-7, **4**-15
  positive modulation, **18**-22
  pulse modulation, **5**-1–**5**-13
  pulse-width modulation (PWM), **1**-9–**1**-11, **18**-27
  television transmission, **1**-19
  transmission system, **18**-22
  trellis-coded modulation (TCM), **6**-22–**6**-23, **6**-25
  velocity modulation, **8**-10
  vestigial sideband (VSB) modulation, **3**-10–**3**-11, **3**-14, **3**-15
Modulation circuits, FM broadcasting, **1**-13–**1**-14
Modulation doped FETs (MODFETs), **10**-2
Modulation index, **4**-4, **4**-15
Modulation theorem, **3**-2
Modulator tube, **1**-11

Molnya satellites, 1-23
MoM. See Method of moments
Monitor points, 17-16
MOSFETs. See Metal-oxide-semiconductor field-effect transistors
MPSK. See M-ary phase-shift keying
MQASK. See M-ary quadrature-amplitude-shift keying
MSDC klystron. See Multi-stage depressed collector klystron
MSK. See Minimum shift keying
M-tubes, 8-20
M-type radial BWO, 8-25
MUF. See Maximum usable frequency
m-law compressor, 5-5, 5-12
Multi-arm short helix antennas, 17-17
Multicavity klystron, 8-13–8-14
Multifunction array, radar, 1-32
Multipath fading, 15-15
Multipath interference, 15-13
Multiple-beam network, 17-24
Multiple-user sites, radio frequency radiation (RFR), 23-21–23-22
Multiplexer, high-power isolator, 14-20
Multiplexing, 1-2
Multiplication theorem, 5-2
Multi-slot traveling-wave antennas, 17-20
Multi-stage depressed collector (MSDC) klystron, 1-19
Mutual impedance, antennas, 16-9–16-10

N

Narrowband antennas, 16-3, 19-10
Narrowband filters, 17-29
Narrow-band FM (NBFM), 4-7
National Electrical Code (NEC), 23-17, 23-19
National Environmental Policy Act (NEPA), nonionizing radiation, 23-20–23-21
National Television Systems Committee signal. See NTSC signal
Natural sampling, 5-2, 5-4, 5-12
Navigation systems, based on radio transmission, 1-32–1-35
NBFM. See Narrow-band FM
Near vertical incidence skywave propagation (NVIS propagation), 15-10–15-11
NEC-2. See Numerical Electromagnetic Code
Negative modulation, 18-22
NEPA. See National Environmental Policy Act
Noise, 21-9
Noise bandwidth, 21-9
Noise effects, amplitude modulation, 3-12
Noise floor, 21-9
Noise measurement, RF power measurement, 20-7–20-8
Noise sideband, 21-9
Noise temperature, 1-29
Noncoherent binary modulation, 6-5–6-6, 6-24

Noncoherent M-ary FSK, 6-14–6-15
Non-constant-impedance diplexer, 14-6–14-8
Nonionizing radiation, 23-20–23-22
Nonlinear amplifiers, 11-2, 11-8–11-9
 Class D amplifiers, 1-5, 11-31–11-36
 distortion, 11-9–11-10
 nonsaturated class C amplifiers, 11-17–11-23
 saturated class C amplifiers, 11-23–11-31
 switch-mode amplifier, 11-31–11-36
Nonlinear distortion, RF power measurement, 20-9–20-12
Nonreturn-to-zero (NRZ) format, 5-6, 5-12
Nonsaturated class C amplifiers, 11-17–11-23
Nonsaturated region, 9-7, 9-14
Norm, 6-24
Normally off device, 10-1
Normally on MOSFET, 10-1
Normal mode helix, 17-20
Notch filter, 14-2
NRZ format. See Nonreturn-to-zero (NRZ) format
NTSC signal, 1-17, 1-18
Numerical Electromagnetic Code (NEC-2), 16-4, 16-33
NVIS propagation. See Near vertical incidence skywave propagation
Nyquist's pulse-shaping criterion, 5-8, 5-12

O

Odd order intermodulation distortion, linear amplifiers, 11-3–11-7
Offset quadriphase-shift keying (OQPSK), 6-17, 6-24
Offset reflector, 1-26
Ohmic region, 9-7, 9-14
Omega navigation, 1-35
On-resistance, 11-24, 11-37
Operating class. See Class of operation
Operating efficiency
 RF amplifiers, 1-5–1-6
 traveling wave tube (TWT), 8-19–8-20
OQPSK. See Offset quadriphase-shift keying
Order, of filters, 14-4
Orthogonal signaling, 6-2, 6-24
Oscillators
 backward wave oscillator (BWO), 8-24–8-25
 cavity magnetron oscillator, 8-22
 gyrotron backward oscillator, 8-29
 local oscillator (LO), 3-5, 21-9
 oven-controlled crystal oscillator, 1-4
 ring oscillator, 10-9
 temperature-compensated crystal oscillator (TCXO), 1-4
 two-cavity klystron oscillator, 8-12–8-13
 voltage controlled oscillators, 4-7
OSHA safety requirements, 23-16–23-19
Output characteristics, 9-8, 9-14
Output devices, 1-8
Output resistance, MOSFETs, 10-10
Output signal-to-noise ratio, 3-12

Index

Oven-controlled crystal oscillator, **1**-4
Overload sensor, **18**-24–**18**-26

P

PAL signal, **1**-17, **1**-18
Panel antennas, **17**-17, **17**-20, **17**-21
Parabolic reflector antenna, **17**-10–**17**-11
Parallel-feed networks, **17**-24, **17**-25
Parasitic director elements, **17**-10
Parasitic energy, circular waveguide, **13**-4
Partition noise, **20**-7
Passband signal, **3**-15
Passband systems, **3**-1
Passive arrays, **17**-24
Passive filters, **14**-2
Patch antenna, **16**-2
Pattern multiplication, in antennas, **16**-7–**16**-8
Pattern nulls (minima), **17**-13
Pattern optimization, **17**-3
PCBs. See Polychlorinated biphenyls
PCDDs. See Polychlorinated dibenzo-p-dioxins
PCDFs. See Polychlorinated dibenzofurans
PDM. See Pulse-duration modulation
Peak-average cursor, **21**-9
Peak detection, **21**-9
Peak envelope power (PEP), **11**-4, **11**-37
Peak-equivalent sine, **20**-3
Peaking, **21**-10
Peak of sync, **1**-18
Peak-response measurement, power measurement, **20**-2–**20**-4
Pentodes, **7**-12–**7**-13
PEP. See Peak envelope power
Percentage modulation, **4**-4, **4**-15
Perfluoroisobutylene (PFIB), **23**-20
Period, **21**-10
PFIB. See Perfluoroisobutylene
Phantom tee network, **17**-4
Phase Alternation each Line signal. See PAL signal
Phased-array antennas, **17**-23–**17**-29
Phase lock, **21**-10
Phase-locked loops, **3**-5
Phase measurement, RF power measurement, **20**-8
Phase modulation, **4**-7, **4**-15
Phase shifter, **17**-26
Phase-shifting method, **3**-9
Phase-shift keying (PSK), **6**-4, **6**-15, **6**-24
Phase stability, coaxial transmission lines, **12**-12
Phosgene, **23**-19
p-4-differential QPSK (p4-DQPSK), **6**-19, **6**-24
Pierce gun, **8**-17
Pinch-off voltage, **9**-5, **9**-14
Pin diode switches, radar system, **17**-27
Pi network, **17**-5, **17**-7
Planar array antenna, **17**-11
Planar triodes, **7**-9, **7**-10, **8**-5–**8**-6
Plate, triode, **7**-6
Plate overload fault, troubleshooting, **19**-2–**19**-5

Platinotron, **8**-25–**8**-26
Pocklington's integral equation, **16**-16
Poisonous compounds, safety, **23**-19–**23**-20
Polarization, of antennas, **16**-4–**16**-5, **17**-2
Polychlorinated biphenyls (PCBs), **23**-11–**23**-16
    components containing, **23**-12–**23**-15
    disposal of, **23**-15–**23**-16
    health risk, **23**-11–**23**-12
    identifying PCB components, **23**-14
    labeling PCB components, **23**-14–**23**-15
    management program, **23**-16
    record-keeping, **23**-15
Polychlorinated dibenzofurans (PCDFs), **23**-12
Polychlorinated dibenzo-p-dioxins (PCDDs), **23**-12
Positive modulation, **18**-22
Power amplifiers
    FM broadcasting, **1**-14–**1**-15
    television broadcasting, **1**-21–**1**-22
Power combining, **1**-7–**1**-8
Power control faults, troubleshooting, **19**-11–**19**-15
Power efficiency, **6**-6, **6**-16–**6**-17, **6**-24
Power factor, **12**-5
Power gain, antennas, **16**-11
Power measurement
    average-response measurement, **20**-2
    peak-response measurement, **20**-2–**20**-4
    root mean square, **20**-1–**20**-2
Power rating, coaxial transmission lines, **12**-7–**12**-8, **12**-9
Power supply
    metering, **18**-24–**18**-26
    preventive maintenance for, **18**-5–**18**-7, **18**-23–**18**-24
    single phasing, **18**-26–**18**-28
    transient disturbances, **18**-26–**18**-28
Power tubes, **7**-1–**7**-19
    air cooling, **7**-15, **7**-16–**7**-17, **18**-10
    air-handling system, **18**-10–**18**-12
    conditioning of, **18**-13
    crossed-field microwave tubes, **8**-3–**8**-5, **8**-20–**8**-30
    device cooling, **7**-15–**7**-18
    diacrode, **8**-7–**8**-9
    diodes, **7**-5–**7**-6
    electron optics, **7**-2–**7**-5
    electron transit time, **7**-13–**7**-14, **7**-15, **8**-5
    extending vacuum tube life, **18**-13–**18**-14
    filament voltage, **18**-13–**18**-14
    grid vacuum tubes, **8**-5–**8**-9
    gyrotron, **8**-26–**8**-30
    high-frequency operating limits, **7**-13–**7**-15
    high-power UHF tetrodes, **8**-6–**8**-7, **8**-8
    implosion hazard, **23**-22
    linear-beam microwave tubes, **8**-2–**8**-3
    magnetrons, **8**-20–**8**-24
    multicavity klystron, **8**-13–**8**-14
    PA stage tuning, **18**-14–**18**-17
    pentads, **7**-12–**7**-13
    planar triodes, **7**-9, **7**-10, **8**-5–**8**-6
    preventive maintenance of, **18**-9–**18**-18
    reflex klystron, **8**-9–**8**-11
    strap-fed devices, **8**-25–**8**-26

tetrodes, 7-9–7-12
thermal cycling, 18-12
traveling wave tube (TWT), 8-14–8-20
triodes, 7-6–7-9
tube changing procedure, 18-12
tube dissipation, 18-10
two-cavity klystron, 8-11–8-12
two-cavity klystron amplifier, 8-13
two-cavity klystron oscillator, 8-12–8-13
vacuum tube life, 18-16
water cooling, 7-15, 7-17–7-18
Preselectors, 17-29, 21-10
Preventive maintenance, 18-2–18-8
    antennas, 18-20–18-23
    capacitors, 18-4
    cleaning the system, 18-7–18-8
    coils, 18-5–18-6
    connection points, 18-7
    cooling system, 18-28
    high-voltage power supply, 18-23–18-24
    klystrons, 18-8–18-9
    power grid tubes, 18-9–18-18
    power supply, 18-5–18-7, 18-23–18-24
    relay mechanisms, 18-6
    resistors, 18-4
    RF transformers, 18-5–18-6
    temperature control, 18-28–18-33
    transmission lines, 18-20–18-23
    UHF transmission systems, 18-23
PRF. See Pulse repetition frequency
Primary reflector, satellite systems, 1-25–1-26
Probability of error, 5-10, 5-12, 6-2
Probability of symbol error, 6-11, 6-24
Probe, impedance matching, 13-6
Product detection, 3-5, 3-15
Products, 21-10
Propagation, radio waves, 15-7–15-15
Protection circuits, troubleshooting, 19-14
Protective covers, for electric conductors, 23-17
Pseudorandom number (PN) sequence, 1-3
PSK. See Phase-shift keying
Pulse-duration modulation (PDM), 1-9
Pulse modulation, 5-1–5-13
    analog pulse modulation, 5-10–5-11
    analog-to-digital conversion, 5-4–5-5
    baseband digital pulse modulation, 5-5–5-10
    defined, 5-12
    detection scheme, 5-10
    sampling theorem, 5-1–5-4
    traveling wave tube (TWT), 8-16–8-17
Pulse repetition frequency (PRF), 21-10
Pulse stretcher, 21-10
Pulse-width modulation (PWM), 1-9–1-11, 18-27
Pulsing, 1-19
Punch through, 10-11
Puncturing, 6-21, 6-25
Push-pull amplifier, 11-13
PWM. See Pulse-width modulation
Pylon antenna, 17-19
Pyramidal horn antenna, 16-2, 17-10

Q

QAM. See Quadrature amplitude modulation
QASK. See Quadrature-amplitude-shift keying
Q-band, 2-2, 2-5
Q-point, of amplifier, 9-8, 11-15
QPSK. See Quadriphase-shift keying
Quadrature, 14-4
Quadrature amplitude modulation (QAM), 3-11–3-12, 3-14, 3-15
Quadrature-amplitude-shift keying (QASK), 6-25
Quadriphase-shift keying (QPSK), 6-10, 6-25
Quantization, 5-5, 5-12
Quarter-wave monopole antenna, 17-7
Quarter-wave transformer, 1-6–1-7
Quartz, 1-3
Quiescent point, of amplifier, 9-8, 11-15

R

Radar, 1-30–1-32
    antennas, 1-30, 1-32
    applications of, 1-31
    duplexer, 17-26–17-29
    frequency bands, 1-31
    transmission equipment, 1-31
Radial beam power tetrad, 7-11–7-12
Radiated fields, antennas, 16-18, 16-20–16-22
Radiating near-field region, 17-4
Radiation resistance, 17-2
Radio frequency bands, 2-1–2-3
    See also under RF
Radio frequency radiation (RFR), safety, 23-20–23-22
Radio wave propagation, 15-1–15-15
    diffraction, 15-6
    free space path loss, 15-3
    HF propagation, 15-9–15-12
    low frequency (LF), 15-7–15-9
    medium frequency (MF), 15-7–15-9
    microwave propagation, 15-14–15-15
    reflection, 15-4
    refraction, 15-4–15-6
    very low frequency (VLF), 15-7–15-9
    VHF and UHF propagation, 15-12–15-14
    See also under RF
Radio waves
    basics of, 15-1–15-3
    See also under RF
Raised-cosine spectra, 5-9, 5-12
Rapid thermal processing, 10-13
Ratio detector, 4-11, 4-13, 4-14, 4-15
RBW. See Resolution bandwidth
Reactance modulator, 1-13, 4-7–4-9
Reactance transistor, 4-7
Reactance tube, 4-7
Reactive near-field region, 17-4
Real antenna, 17-7
Receiver oscillator, 3-5
Reciprocity, laws of, 16-1

Reed-Solomon codes, **6**-21
Reference level, **21**-10
Reference-level control, **21**-10
Reference tower, **17**-12
Reflection, radio wave propagation, **15**-4
Reflection coefficient, **16**-14
Reflector antenna, **17**-10–**17**-11
Reflectors, **16**-24
Reflex klystron, **8**-9–**8**-11
Refraction, **15**-4–**15**-6, **15**-16
Reggie-Spencer phase shifter, **17**-26, **17**-27
Reject port, **1**-7
Relative velocity of propagation, **12**-3
Relay mechanisms, preventive maintenance for, **18**-6
Resistors, preventive maintenance for, **18**-4
Resolution bandwidth (RBW), **21**-10
Resonant characteristics, coaxial transmission line, **12**-5–**12**-6
Return loss, **21**-10
Reverse-biased gate-to-channel PN junction, **9**-10
RF amplifiers, **1**-3–**1**-6
RF combiner, **14**-1
RF equipment
  preventing system failures, **18**-1–**18**-33
  routine maintenance, **18**-2–**18**-8
  safety issues, **23**-1–**23**-24
  troubleshooting, **19**-1–**19**-15
RF interaction region, **18**-8
RF power measurement, **20**-2–**20**-15
  addition and cancellation of distortion components, **20**-14–**20**-15
  decibel measurement, **20**-6
  harmonic distortion, **20**-9–**20**-12
  intermodulation distortion (IMD), **20**-12–**20**-14
  noise measurement, **20**-7–**20**-8
  nonlinear distortion, **20**-9–**20**-12
  phase measurement, **20**-8
RFR. See Radio frequency radiation
RF system failures, preventing, **18**-1–**18**-33
  antennas, **18**-20–**18**-23
  high-voltage power supplies, **18**-23–**18**-28
  klystrons, **18**-8–**18**-9
  power grid tubes, **18**-9–**18**-18
  routine maintenance, **18**-2–**18**-8
  temperature control, **18**-28–**18**-33
  transmission lines, **18**-20–**18**-23
RF technology
  AM radio, **1**-8–**1**-11
  coaxial transmission lines, **12**-1–**12**-13
  electronic navigation, **1**-32–**1**-35
  FM broadcasting, **1**-12–**1**-15
  induction heating, **1**-36
  microwave radio relay systems, **1**-35–**1**-36
  preventing system failures. See RF system failures, preventing
  radar, **1**-30–**1**-32
  routine maintenance, **18**-1–**18**-33
  safety, **23**-1–**23**-24
  satellite transmission, **1**-23–**1**-29
  shortwave broadcasting, **1**-12
  television broadcasting, **1**-15–**1**-23

RF transformers, preventive maintenance for, **18**-5–**18**-6
Ridged waveguide, **13**-3–**13**-4
Rigid coaxial cable, **12**-2
Ring, **21**-10
Ring oscillator, **10**-9
Ring stub antennas, **17**-17
Rolloff, **14**-3
Root mean square, power measurement, **20**-1–**20**-2
Routine maintenance, **18**-2–**18**-8
Rubber gloves, working with high voltage, **23**-8
RZ format. See Unipolar return-to-zero (RZ) format

S

Safety, **23**-1–**23**-24
  beryllium oxide ceramics, **23**-19
  corrosive and poisonous compounds, **23**-19–**23**-20
  electric shock, **23**-4–**23**-11
  hot coolant and surfaces, **23**-22
  management responsibility, **23**-22–**23**-24
  nonionizing radiation, **23**-20–**23**-22
  OSHA safety requirements, **23**-16–**23**-19
  polychlorinated biphenyls (PCBs), **23**-11–**23**-16
  safety-related hardware, **23**-1
  X-ray radiation hazard, **23**-22
Safety-related hardware, **23**-1
Sampling, **5**-2
Sampling theorem, **5**-1–**5**-4, **5**-12
Satellite systems, **1**-23–**1**-29
  downlink antennas, **1**-28–**1**-29
  frequency allocations, **2**-5
  high-power amplifier (HPA), **1**-25
  satellite antennas, **1**-27–**1**-28
  satellite downlink, **1**-28
  satellite link, **1**-27
  signal formats, **1**-26–**1**-27
  uplink antennas, **1**-25–**1**-26
Saturated class C amplifiers, **11**-23–**11**-31
Saturated region, **9**-7
Saturation, **10**-3, **11**-23
Saturation angle, **11**-23, **11**-37
Saturation current, **11**-37
Saturation region, **9**-5
Saturation voltage, **11**-23
SAVE function, **21**-10
S-band, **2**-2, **2**-5
Scalar product, **6**-25
SCA services. See Subsidiary Communications Authorization (SCA) services
Search antenna, radar, **1**-32
SECAM signal, **1**-17, **1**-18
Secondary emission, of electrons, **7**-3–**7**-5
Secondary emission noise, **20**-8
Selective-tuned filter alignment, **21**-5
Self-contained electronic navigation systems, **1**-32
Self-similar fractal antennas, **16**-31–**16**-33
Semiflexible coaxial cable, **12**-3, **12**-13
Sensitivity, **21**-10

Sequential Color with Memory signal. See SECAM signal
Series-fed slanted dipole antennas, **17**-17
Series-feed networks, **17**-24, **17**-25
Set partitioning, **6**-22, **6**-25
Shape factor, **21**-10
SHF band (super high-frequency) band, **2**-2, **2**-3
Shock. See Electric shock
Short-channel effects, **10**-9
Shortwave broadcasting, **1**-12
Shortwave fade (SWF), **15**-12
Shot effect, **19**-7
Shunt-fed slanted dipole antennas, **17**-17
Sideband, **3**-15, **21**-10
Sideband frequency, frequency modulation, **4**-5–**4**-7
Sideband regeneration, **11**-7
Sideband suppression, **21**-10
Side-mounted antennas, **17**-21
Signal detection
    binary modulation, **6**-1–**6**-8
    geometric view of, **6**-9–**6**-10
    M-ary modulation, **6**-8–**6**-17
Signaling interval, **6**-8
Signal multiplexing, **1**-2
Signal-to-noise ratio (SNR), **3**-12
Silicon MOSFETs, **10**-2, **10**-3
Single sideband, **3**-7–**3**-10, **3**-14, **3**-15
Single-slot space harmonic circuit, **8**-16
Single sweep, **21**-10
Skin effect
    coaxial transmission lines, **12**-1–**12**-2
    UHF transmission systems, **18**-23
Skywave, **15**-4, **15**-9, **15**-16
Skywave propagation, **15**-8
Slot antenna, **17**-10
Slotted line device, **22**-2
Slow frequency hopping, **1**-3
Slugs, impedance matching, **13**-6
Small signal behavior, **11**-10
SNR. See Signal-to-noise ratio
Soft decision, **6**-22, **6**-25
Sokut phase baseband data modulation, **5**-6
Solid-state amplifiers, **11**-1–**11**-36
    linear, **11**-2–**11**-8, **11**-10–**11**-17
    nonlinear, **11**-2, **11**-8–**11**-10, **11**-17–**11**-36
    satellite systems, **1**-25
Sommerfield-Norton solution, **16**-13, **16**-15, **16**-33
Source code, **5**-12
Source encoding, **5**-5
Span per division, **21**-10
Spectral diagrams, **4**-5
Spectrum, **21**-10
Spectrum analysis, **21**-1–**21**-7
    spectrum analyzer, **21**-1–**21**-5
    spurious harmonic distortion, **21**-4–**21**-7
Spectrum analyzer, **21**-1–**21**-3
    applications, **21**-3–**21**-5
    troubleshooting, **21**-6–**21**-7
Split phase, **5**-12
Spread-spectrum systems, **1**-3
Spurious response, **21**-10

Square law transfer equation, **9**-8, **9**-14
Stability, **21**-10
Stabilotron, **8**-25
Stagger tuning, **1**-6
Standard load, **11**-37
Standing wave ratio, **12**-4
Starting frequency, **21**-10
Stationary array, **1**-30
Step-start faults, troubleshooting, **19**-13
Strap-fed devices, **8**-25–**8**-26
Straps, **8**-25
Strong inversion, **10**-3–**10**-4, **10**-6, **10**-13
Subsidiary Communications Authorization (SCA) services, **1**-14
Substrate, **10**-14
Subthreshold, **10**-4, **10**-14
Subthreshold control, MOSFETs, **10**-11–**10**-12
Sulfur hexafluoride, **23**-19
Superheterodyne, **3**-12–**3**-14, **3**-15
Super high-frequency band. See SHF band
Suppressed carrier systems, **3**-15
Surface impedance, **16**-12, **16**-34
Surface potential, **10**-4
Sweep rate, **21**-10
Sweep speed, **21**-10
SWF. See Shortwave fade
Switch-mode Class D amplifiers, **11**-31–**11**-36
Switch panels, OSHA safety requirements, **23**-17
Switch tube, **1**-11
Symbol interval, **6**-8
Symbol period, **6**-2, **6**-25

T

TCM. See Trellis-coded modulation
TCT. See Torroidal current transformer
TCXO. See Temperature-compensated crystal oscillator
TDM. See Time-division multiplexing; Time-domain multiplexing
Tee network, **17**-4, **17**-6
Television broadcasting, **1**-15–**1**-23
    analog transmitter, **1**-20
    antennas, **1**-15–**1**-16, **17**-17–**17**-23
    color notch filter, **1**-22
    coupling-filtering system, **1**-22
    diplexer-combiner, **1**-23
    harmonic filters, **1**-22
    intermediate power amplifier (IPA), **1**-21
    linear amplifier for, **11**-5–**11**-6
    low-power television (LPTV), **1**-16
    power amplifier (PA), **1**-21–**1**-22
    translators, **1**-16
    transmission standards, **1**-17–**1**-18
    visual carrier, **11**-5
TEM. See Transverse electromagnetic mode
Temperature-compensated crystal oscillator (TCXO), **1**-4

Index

Temperature control, preventive maintenance, **18**-28–**18**-33
Tensile strength, of coaxial transmission lines, **12**-13
Testing, coaxial transmission lines, **22**-1–**22**-5
Tetrodes, **1**-19, **7**-9–**7**-12
   high-power UHF tetrodes, **8**-6–**8**-7, **8**-8
   radial beam power tetrode, **7**-11–**7**-12
   television broadcasting, **1**-21
THD+N. See Total harmonic distortion and noise
Thermal cycling, power tubes, **18**-12
Thermal emission, from metals, **7**-3
Thermal voltage, **9**-10, **9**-14
Thin oxides, MOSFETs, **10**-12
Thin-wire antennas, modeling of, **16**-16
Three-dimensional frequency-independent phased array (3D-FIPA), **16**-32
Three-tone test, **11**-6
Threshold, **10**-1–**10**-2, **10**-14
Threshold effect, **3**-12
Threshold voltage, MOSFETs, **10**-4–**10**-7
Thyristor control system, troubleshooting, **19**-11
Time-division multiplexing (TDM), **5**-1, **5**-12
Time-domain multiplexing (TDM), **1**-2
Time-domain representation, **21**-10
Time-varying signal, **21**-10
Top-mounted antennas, **17**-18–**17**-20
Tornadotron, **8**-27
Torroidal current transformer (TCT), **17**-14
Total harmonic distortion and noise (THD+N), **20**-11
Total span, **21**-10
Tower-top, pole-type antennas, **17**-18
Toxic Substances Control Act (TSCA), **23**-12
Tracking antenna, radar, **1**-32
Tracking generator, **21**-3, **21**-10
Transconductance, MOSFETs, **10**-10
Transfer characteristics, **9**-8
Transfer conductance, **9**-14
Transfer equation, **9**-8, **9**-14
Transfer function, **3**-3
Transformers, polychlorinated biphenyls (PCBs) and, **23**-12–**23**-13
Transient disturbances, high-voltage power supply, **18**-26–**18**-28
Transistors
   bipolar junction transistors (BJTs), **9**-1–**9**-3, **9**-5
   field-effect transistors (FETs), **10**-1–**10**-2, **11**-31
   heterostructure field-effect transistors (HFETs), **10**-2
   insulated gate bipolar junction transistors (IGBJTs), **11**-31
   junction field-effect transistors (JFETs), **9**-5–**9**-14
   metal-oxide-semiconductor field-effect transistors (MOSFETs), **9**-10, **10**-1–**10**-14
   metal-semiconductor field-effect transistors (MESFETs), **10**-1
   modulation doped FETs (MODFETs), **10**-2
Transit-time effects, power tubes, **7**-13–**7**-14, **7**-15, **8**-5
Translators, **1**-16
Transmission line impedance, **12**-4
Transmission lines, preventive maintenance, **18**-20–**18**-23

Transmission standards, television broadcasting, **1**-17–**1**-18
Transponders, **1**-24
Transverse-electric (TE) waves, **13**-1
Transverse electromagnetic mode (TEM), coaxial transmission lines, **12**-4
Transverse-magnetic (TM) waves, **13**-1–**13**-2
Traveling wave tube (TWT), **1**-25, **8**-14–**8**-20
Trellis-coded modulation (TCM), **6**-22–**6**-23, **6**-25
Triode region, **9**-7, **9**-14
Triodes, **7**-6–**7**-9
   defined, **9**-14
   planar triodes, **7**-9, **7**-10, **8**-5–**8**-6
Troposcatter, **15**-13, **15**-14
Tropospheric ducting, **15**-13
Tropospheric scatter, **15**-13, **15**-14
Troubleshooting, **19**-1–**19**-15
   capacitors, **19**-6
   cavity-type power amplifier, **19**-7
   factory service assistance, **19**-2
   high-voltage plate supply, **19**-2–**19**-5
   interlock failures, **19**-12–**19**-13
   plate overload fault, **19**-2–**19**-5
   power control faults, **19**-11–**19**-15
   protection circuits, **19**-14
   RF system faults, **19**-5–**19**-10
   spectrum analyzer, **21**-6–**21**-7
   step-start faults, **19**-13
   thyristor control system, **19**-11
TSCA. See Toxic Substances Control Act
Tuning
   of antennas, **22**-3
   PA stage, **18**-14–**18**-17
   of waveguides, **13**-8
Turnstile antennas, **17**-19, **17**-20
Twisted ring antennas, **17**-17
Two-cavity klystron, **8**-11–**8**-12
Two-cavity klystron amplifier, **8**-13
Two-cavity klystron oscillator, **8**-12–**8**-13
Two-tone test, **11**-5
Two-way distance ranging, **1**-33
TWT. See Traveling wave tube

U

U-band, frequency, **2**-2
Ultimate rejection, **21**-11
Ultra high-frequency (UHF), **1**-15, **1**-19, **2**-2, **2**-3
   high-power UHF tetrodes, **8**-6–**8**-7, **8**-8
   radio wave propagation, **15**-12–**15**-14
   transmission systems, preventive maintenance, **18**-23
   UHF side-mounted antennas, **17**-21
Ultraviolet light, **2**-4
Uniform theory of diffraction (UTD), **16**-16, **16**-34
Unipolar return-to-zero (RZ) format, **5**-6, **5**-13
Uplink antennas, satellite systems, **1**-25–**1**-26
Uplink transmitting station, **1**-24
Upper sideband (USB), **3**-4, **3**-15

USB. See Upper sideband
UTD. See Uniform theory of diffraction

V

Vacuum devices
    microwave devices, 8-1–8-30
    power tubes, 7-1–7-19
Vacuum tubes. See Power tubes
Vapor-phase cooling, 7-15, 7-18
Variable-dielectric vane, 14-16
Variable-phase hybrid, 14-17
V-band, 2-2, 2-5
VCO. See Voltage controlled oscillator
V-dipole antenna, 17-6
Vector potential integrals, 16-20–16-21
Velocity modulation, 8-10
Velocity saturation, 10-8
Vertical display factor, 21-11
Vertical polarization, 16-15, 17-2
Vertical scale factor, 21-11
Very high frequency (VHF)
    radio wave propagation, 15-12–15-14
    satellite frequency allocations, 2-5
    VHF multi-slot antenna, 17-20
Very low frequency (VLF), radio wave propagation, 15-7–15-9
Vestigial sideband (VSB) modulation, 3-10–3-11, 3-14, 3-15
Vestigial symmetry, 3-11
VHF. See Very high frequency
Visible light, 2-4
Visual carrier, 11-5
Visual section, television transmission, 1-16, 1-18
Viterbi algorithm, 6-22, 6-25
VLF. See Very low frequency
Voltage compliance range, 9-11
Voltage controlled oscillator (VCO), 4-7, 4-10
Voltage controlled oscillator (VCO) direct-FM modulators, 4-10
Voltage gain, of filter, 14-2
Voltage standing wave ratio (VSWR)
    of antenna and transmission lines, 18-20–18-21
    coaxial transmission lines, 12-6
    measuring, 22-2–22-3
    troubleshooting excessive, 19-5–19-6, 19-9–19-10
Voltage standoff, power tubes, 8-5
Voltage-variable resistor, junction field-effect transistors (JFETs) as, 9-12–9-14
VSB. See Vestigial sideband modulation
VSWR. See Voltage standing wave ratio

W

Water cooling, power tubes, 7-15, 7-17–7-18
Water sprinkler, 23-2
Waveform memory, 21-11
Waveform subtraction, 21-11
Waveguide antenna, 17-10
Waveguide filters, 13-7–13-8
Waveguides, 13-1–13-10
    cavity resonators, 13-8–13-10
    circular waveguide, 13-4, 13-5, 13-8
    doubly truncated waveguide (DTW), 13-4–13-5
    dual-polarity waveguide, 13-2
    efficiency, 13-3
    hardware, 13-8
    impedance matching, 13-6–13-10
    installation of, 13-8
    propagation modes, 13-1–13-2
    ridged waveguide, 13-3–13-4
    tuning of, 13-8
    waveguide antenna, 17-10
    waveguide filters, 13-7–13-8
    waveguide slot antenna, 17-19, 17-20
Waveguide slot antenna, 17-19, 17-20
Wavelength, defined, 15-1–15-2, 15-16, 17-1
Wave tilt, 15-8
W-band, 2-2
WBFM. See Wide-band FM
Weierstrass arrays, 16-32–16-33
Wideband antennas, 16-27–16-33
Wide-band FM (WBFM), 4-7
Wideband panel antennas, 17-17
Wordlength, 5-4, 5-13

X

X-band, 2-2, 2-5
X-rays, 2-4, 23-22
X-ray shielding, 23-22

Y

Yagi-Uda arrays, 16-24–16-26, 16-34, 17-9, 17-10

Z

Zero hertz peak, 21-11
Zero-sequence current transformer, 23-6
Zero span, 21-11
Zigzag antennas, 17-20
Zigzag panel antennas, 17-21